Lecture Notes in Computer Science   13185

Matthias Hagen · Suzan Verberne ·
Craig Macdonald · Christin Seifert ·
Krisztian Balog · Kjetil Nørvåg ·
Vinay Setty (Eds.)

Advances in Information Retrieval

44th European Conference on IR Research, ECIR 2022
Stavanger, Norway, April 10–14, 2022
Proceedings, Part I

Springer

Editors
Matthias Hagen ⓘD
Martin Luther University Halle-Wittenberg
Halle, Germany

Craig Macdonald ⓘD
University of Glasgow
Glasgow, UK

Krisztian Balog ⓘD
University of Stavanger
Stavanger, Norway

Vinay Setty ⓘD
University of Stavanger
Stavanger, Norway

Suzan Verberne ⓘD
Leiden University
Leiden, The Netherlands

Christin Seifert ⓘD
University of Duisburg-Essen
Essen, Germany

Kjetil Nørvåg ⓘD
Norwegian University of Science
and Technology
Trondheim, Norway

ISSN 0302-9743   ISSN 1611-3349 (electronic)
Lecture Notes in Computer Science
ISBN 978-3-030-99735-9   ISBN 978-3-030-99736-6 (eBook)
https://doi.org/10.1007/978-3-030-99736-6

This Springer imprint is published by the registered company Springer Nature Switzerland AG
The registered company address is: Gewerbestrasse 11, 6330 Cham, Switzerland

Preface

The 44th European Conference on Information Retrieval (ECIR 2022) was held in Stavanger, Norway, during April 10–14, 2022, and brought together hundreds of researchers from Europe and abroad. The conference was organized by the University of Stavanger, in cooperation with the British Computer Society's Information Retrieval Specialist Group (BCS IRSG).

These proceedings contain the papers related to the presentations, workshops, and tutorials given during the conference. This year's ECIR program boasted a variety of novel work from contributors from all around the world. In total, 395 papers from authors in 53 countries were submitted to the different tracks.

The final program included 35 full papers (20% acceptance rate), 29 short papers (22% acceptance rate), 12 demonstration papers (55% acceptance rate), 11 reproducibility papers (61% acceptance rate), 12 doctoral consortium papers (71% acceptance rate), and 13 invited CLEF papers. All submissions were peer-reviewed by at least three international Program Committee members to ensure that only submissions of the highest relevance and quality were included in the final program. The acceptance decisions were further informed by discussions among the reviewers for each submitted paper, led by a senior Program Committee member.

The accepted papers cover the state of the art in information retrieval: advances in ranking models, applications of entities and knowledge graphs, evaluation, multimodal retrieval, recommender systems, query understanding, user simulation studies, etc. As in previous years, the ECIR 2022 program contained a high proportion of papers with students as first authors, as well as papers from a variety of universities, research institutes, and commercial organizations.

In addition to the papers, the program also included three keynotes, four tutorials, five workshops, a doctoral consortium, the presentation of selected papers from the 2021 issues of the Information Retrieval Journal, and an industry day. Keynote talks were given by Isabelle Augenstein (University of Copenhagen), Peter Flach (University of Bristol), and this year's BCS IRSG Karen Spärck Jones Award winner, Ivan Vulić (University of Cambridge & PolyAI). The tutorials covered a range of topics including high recall retrieval, incrementally testing for online advertising, information extraction from social media, and keyphrase identification, while the workshops brought together participants to discuss algorithmic bias in search and recommendation (BIAS), bibliometrics (BIR), online misinformation (ROMCIR), narrative extraction (Text2Story), and technology-assisted review systems (ALTARS).

The success of ECIR 2022 would not have been possible without all the help from the team of volunteers and reviewers. We wish to thank all the reviewers and meta-reviewers who helped to ensure the high quality of the program. We also wish to thank the reproducibility chairs: Faegheh Hasibi and Carsten Eickhoff; the demo chairs: Theodora Tsikrika and Udo Kruschwitz; the workshop chairs: Lucie Flek and Javier Parapar; the tutorial chairs: Nazli Goharian and Shuo Zhang; the industry chairs: Jiyin

He and Marcel Worring; the doctoral consortium chairs: Asia Biega and Alistair Moffat; and the awards chair: Maarten de Rijke. We would like to thank our local administration chair, Russel Wolff, along with all the student volunteers who helped to create an excellent online and offline experience for participants and attendees.

ECIR 2022 was sponsored by Amazon, Bloomberg, Cobrainer, Elsevier, Google, the L3S Research Center, MediaFutures, the Norwegian University of Science and Technology, NorwAI, Schibsted, SIGIR, Signal AI, Spotify, Springer, Textkernel, Thomson Reuters, the University of Stavanger, Vespa AI, and Wayfair. We thank them all for their support.

Finally, we wish to thank all the authors and contributors to the conference.

April 2022

Matthias Hagen
Suzan Verberne
Craig Macdonald
Christin Seifert
Krisztian Balog
Kjetil Nørvåg

Organization

General Chairs

Krisztian Balog   University of Stavanger, Norway
Kjetil Nørvåg   NTNU, Norway

Program Chairs – Full Papers

Suzan Verberne   Leiden University, The Netherlands
Matthias Hagen   Martin-Luther-Universität Halle-Wittenberg, Germany

Program Chairs – Short Papers

Christin Seifert   University of Duisburg-Essen, Germany
Craig Macdonald   University of Glasgow, UK

Reproducibility Track Chairs

Faegheh Hasibi   Radboud University, The Netherlands
Carsten Eickhoff   Brown University, USA

Demo Chairs

Theodora Tsikrika   Information Technologies Institute, CERTH, Greece
Udo Kruschwitz   University of Regensburg, Germany

Workshop Chairs

Lucie Flek   University of Marburg, Germany
Javier Parapar   Universidade da Coruña, Spain

Tutorials Chairs

Nazli Goharian   Georgetown University, USA
Shuo Zhang   Bloomberg, UK

Industry Day Chairs

Jiyin He   Signal AI, UK
Marcel Worring   University of Amsterdam, The Netherlands

Doctoral Consortium Chairs

Asia Biega	Max Planck Institute for Security and Privacy, Germany
Alistair Moffat	University of Melbourne, Australia

Awards Chair

Maarten de Rijke	University of Amsterdam, The Netherlands

Local Organization Chairs

Vinay Setty	University of Stavanger, Norway
Russel Wolff	University of Stavanger, Norway

Program Committee

Qingyao Ai	University of Utah, USA
Dyaa Albakour	Signal AI, UK
Mohammad Aliannejadi	University of Amsterdam, The Netherlands
Satya Almasian	Heidelberg University, Germany
Omar Alonso	Instacart, USA
Sophia Althammer	TU Vienna, Austria
Ismail Sengor Altingovde	Bilkent University, Turkey
Giuseppe Amato	ISTI-CNR, Italy
Enrique Amigó	UNED, Spain
Avishek Anand	L3S Research Center, Germany
Linda Andersson	Artificial Researcher IT GmbH and TU Wien, Austria
Negar Arabzadeh	University of Waterloo, Canada
Ioannis Arapakis	Telefonica Research, Spain
Jaime Arguello	University of North Carolina at Chapel Hill, USA
Arian Askari	Shahid Beheshti University, Iran
Maurizio Atzori	University of Cagliari, Italy
Sandeep Avula	Amazon, USA
Leif Azzopardi	University of Strathclyde, UK
Mossaab Bagdouri	Walmart Labs, USA
Ebrahim Bagheri	Ryerson University, Canada
Seyed Ali Bahreinian	IDSIA, Swiss AI Lab, Switzerland
Georgios Balikas	Salesforce Inc, France
Valeriia Baranova	RMIT University, Australia
Alvaro Barreiro	University of A Coruña, Spain
Alberto Barrón-Cedeño	University of Bologna, Italy
Alejandro Bellogin	Universidad Autonoma de Madrid, Spain
Patrice Bellot	CNRS, LSIS, Aix-Marseille Université, France
Michael Bendersky	Google, USA
Alessandro Benedetti	Sease, UK

Klaus Berberich	Saarbruecken University of Applied Sciences, Germany
Sumit Bhatia	Adobe Inc., India
Paheli Bhattacharya	Indian Institute of Technology Kharagpur, India
Roi Blanco	Amazon, Spain
Alexander Bondarenko	Martin-Luther-Universität Halle-Wittenberg, Germany
Ludovico Boratto	University of Cagliari, Italy
Gloria Bordogna	CNR, Italy
Mohand Boughanem	IRIT, Université Toulouse III - Paul Sabatier, France
Leonid Boytsov	BCAI, USA
Alex Brandsen	Leiden University, The Netherlands
Pavel Braslavski	Ural Federal University, Russia
Timo Breuer	TH Köln, Germany
Fidel Cacheda	Universidade da Coruña, Spain
Jamie Callan	Carnegie Mellon University, USA
Rodrigo Calumby	State University of Feira de Santana, Brazil
Ricardo Campos	Polytechnic Institute of Tomar and INESC TEC, Portugal
Zeljko Carevic	GESIS Leibniz Institute for the Social Sciences, Germany
Ben Carterette	Spotify, USA
Shubham Chatterjee	University of New Hampshire, USA
Tao Chen	Google Research, USA
Xuanang Chen	University of Chinese Academy of Sciences, China
Adrian-Gabriel Chifu	CNRS, LIS, Aix-Marseille Université and Université de Toulon, France
Charles Clarke	University of Waterloo, Canada
Maarten Clements	TomTom, The Netherlands
Stephane Clinchant	Xerox Research Centre Europe, France
Paul Clough	University of Sheffield, UK
Juan Soler Company	Pompeu Fabra University, Spain
Alessio Conte	University of Pisa, Italy
Gordon Cormack	University of Waterloo, Canada
Anita Crescenzi	University of North Carolina at Chapel Hill, USA
Fabio Crestani	University of Lugano, Switzerland
Bruce Croft	University of Massachusetts Amherst, USA
Arthur Câmara	Delft University of Technology, The Netherlands
Arjen de Vries	Radboud University, The Netherlands
Yashar Deldjoo	Polytechnic University of Bari, Italy
Elena Demidova	University of Bonn, Germany
Emanuele Di Buccio	University of Padua, Italy
Giorgio Maria Di Nunzio	University of Padua, Italy
Gaël Dias	Normandy University, France
Laura Dietz	University of New Hampshire, USA
Anne Dirkson	Leiden University, The Netherlands
Vlastislav Dohnal	Masaryk University, Czech Republic

Shiri Dori-Hacohen	University of Connecticut, USA
Dennis Dosso	University of Padua, Italy
Pan Du	Thomson Reuters Labs, Canada
Tamer Elsayed	Qatar University, Qatar
Liana Ermakova	Université de Bretagne Occidentale, France
Ralph Ewerth	L3S Research Center, Germany
Anjie Fang	Amazon, USA
Hui Fang	University of Delaware, USA
Hossein Fani	University of Windsor, Canada
Juan M. Fernández-Luna	University of Granada, Spain
Nicola Ferro	University of Padua, Italy
John Foley	Smith College, USA
Thibault Formal	Naver Labs Europe, France
Ophir Frieder	Georgetown University, USA
Ingo Frommholz	University of Wolverhampton, UK
Maik Fröbe	Martin-Luther-Universität Halle-Wittenberg, Germany
Norbert Fuhr	University of Duisburg-Essen, Germany
Michael Färber	Karlsruhe Institute of Technology, Germany
Ujwal Gadiraju	Delft University of Technology, The Netherlands
Debasis Ganguly	University of Glasgow, UK
Dario Garigliotti	Aalborg University, Denmark
Eric Gaussier	LIG-UJF, France
Kripabandhu Ghosh	IISER Kolkata, India
Anastasia Giachanou	Universitat Politècnica de València, Spain
Lukas Gienapp	Leipzig University, Germany
Lorraine Goeuriot	Université Grenoble Alpes, France
Simon Gog	Karlsruhe Institute of Technology, Germany
Marcos Goncalves	Federal University of Minas Gerais, Brazil
Julio Gonzalo	UNED, Spain
Michael Granitzer	University of Passau, Germany
Adrien Guille	Université de Lyon, France
Jiafeng Guo	Institute of Computing Technology, China
Cathal Gurrin	Dublin City University, Ireland
Martin Halvey	University of Strathclyde, UK
Lei Han	University of Queensland, Australia
Allan Hanbury	Vienna University of Technology, Austria
Preben Hansen	Stockholm University, Sweden
Donna Harman	NIST, USA
Morgan Harvey	University of Sheffield, UK
Maram Hasanain	Qatar University, Qatar
Faegheh Hasibi	Radboud University, The Netherlands
Claudia Hauff	Delft University of Technology, The Netherlands
Ben He	University of Chinese Academy of Sciences, China
Daniel Hienert	GESIS - Leibniz Institute for the Social Sciences, Germany
Michiel Hildebrand	Spinque, The Netherlands

Gilles Hubert   IRIT, France
Bogdan Ionescu   Politehnica University of Bucharest, Romania
Radu Tudor Ionescu   University of Bucharest, Romania
Adam Jatowt   University of Innsbruck, Austria
Faizan Javed   Kaiser Permanente, USA
Shiyu Ji   University of California, Santa Barbara, USA
Jiepu Jiang   University of Wisconsin-Madison, USA
Hideo Joho   University of Tsukuba, Japan
Gareth Jones   Dublin City University, Ireland
Joemon Jose   University of Glasgow, UK
Chris Kamphuis   Radboud University, The Netherlands
Jaap Kamps   University of Amsterdam, The Netherlands
Nattiya Kanhabua   SCG CBM, Thailand
Sumanta Kashyapi   NIT Hamirpur, India
Liadh Kelly   Maynooth University, Ireland
Roman Kern   Graz University of Technology, Austria
Oren Kurland   Technion - Israel Institute of Technology, Israel
Mucahid Kutlu   TOBB University of Economics and Technology,
                           Turkey
Saar Kuzi   Amazon, USA
Jochen L. Leidner   Refinitiv Labs and University of Sheffield, UK
Mark Levene   Birkbeck, University of London, UK
Elisabeth Lex   Graz University of Technology, Austria
Xiangsheng Li   Tsinghua University, China
Shangsong Liang   Sun Yat-sen University, China
Jimmy Lin   University of Waterloo, Canada
Matteo Lissandrini   Aalborg University, Denmark
Haiming Liu   University of Bedfordshire, UK
Yiqun Liu   Tsinghua University, China
Sean MacAvaney   University of Glasgow, UK
Andrew Macfarlane   City, University of London, UK
Joel Mackenzie   University of Melbourne, Australia
Eddy Maddalena   King's College London, UK
Joao Magalhaes   Universidade NOVA de Lisboa, Portugal
Maria Maistro   University of Copenhagen, Denmark
Antonio Mallia   New York University, USA
Behrooz Mansouri   University of Tehran, Iran
Jiaxin Mao   Renmin University of China, China
Stefano Marchesin   University of Padua, Italy
Mirko Marras   University of Cagliari, Italy
Monica Marrero   Europeana Foundation, The Netherlands
Bruno Martins   University of Lisbon, Portugal
Yosi Mass   IBM Haifa Research Lab, Israel
Jeanna Matthews   Clarkson University, USA
David Maxwell   TU Delft, The Netherlands

Philipp Mayr	GESIS - Leibniz-Institute for the Social Sciences, Germany
Richard McCreadie	University of Glasgow, UK
Graham McDonald	University of Glasgow, UK
Edgar Meij	Bloomberg L.P., UK
Ida Mele	IASI-CNR, Italy
Massimo Melucci	University of Padua, Italy
Zaiqiao Meng	University of Glasgow, UK
Donald Metzler	Google, USA
Stefano Mizzaro	University of Udine, Italy
Ali Montazeralghaem	University of Massachusetts Amherst, USA
Jose Moreno	IRIT, Université Toulouse III - Paul Sabatier, France
Yashar Moshfeghi	University of Strathclyde, UK
Josiane Mothe	IRIT, France
Philippe Mulhem	LIG-CNRS, France
Cristina Ioana Muntean	ISTI-CNR, Italy
Vanessa Murdock	Amazon, USA
Henning Müller	HES-SO, Switzerland
Franco Maria Nardini	ISTI-CNR, Italy
Wolfgang Nejdl	L3S Research Center, Germany
Jian-Yun Nie	University de Montreal, Canada
Michael Oakes	University of Wolverhampton, UK
Doug Oard	University of Maryland, USA
Harrie Oosterhuis	Radboud University, The Netherlands
Salvatore Orlando	Università Ca' Foscari Venezia, Italy
Iadh Ounis	University of Glasgow, UK
Pooja Oza	University of New Hampshire, USA
Deepak Padmanabhan	Queen's University Belfast, Ireland
Panagiotis Papadakos	FORTH-ICS, Greece
Javier Parapar	Universidade da Coruña, Spain
Pavel Pecina	Charles University in Prague, Czech Republic
Gustavo Penha	Delft University of Technology, The Netherlands
Giulio Ermanno Pibiri	ISTI-CNR, Italy
Karen Pinel-Sauvagnat	IRIT, France
Florina Piroi	TU Wien, Austria
Benjamin Piwowarski	CNRS, Sorbonne Université, France
Martin Potthast	Leipzig University, Germany
Chen Qu	Google and University of Massachusetts Amherst, USA
Pernilla Qvarfordt	FX Palo Alto Laboratory, USA
Filip Radlinski	Google, UK
Gábor Recski	TU Wien, Austria
David Reiley	Google, USA
Zhaochun Ren	Shandong University, China
Jean-Michel Renders	Naver Labs Europe, France

Yannis Tzitzikas   University of Crete and FORTH-ICS, Greece
Md Zia Ullah   CNRS, France
Manisha Verma   Amazon, UK
Vishwa Vinay   Adobe Research, India
Marco Viviani   Università degli Studi di Milano-Bicocca, Italy
Michael Völske   Bauhaus-Universität Weimar, Germany
Xi Wang   University of Glasgow, UK
Zhihong Wang   Tsinghua University, China
Zhijing Wu   Tsinghua University, China
Xiaohui Xie   Tsinghua University, China
Eugene Yang   Johns Hopkins University, USA
Andrew Yates   University of Amsterdam, The Netherlands
Ran Yu   GESIS - Leibniz Institute for the Social Sciences,
                           Germany
Eva Zangerle   University of Innsbruck, Austria
Richard Zanibbi   Rochester Institute of Technology, USA
Fattane Zarrinkalam   University of Guelph, Canada
Sergej Zerr   L3S Research Center, Germany
Junqi Zhang   Tsinghua University, China
Min Zhang   Tsinghua University, China
Rongting Zhang   Amazon, USA
Ruqing Zhang   Chinese Academy of Sciences, China
Yongfeng Zhang   Rutgers, The State University of New Jersey, USA
Liting Zhou   Dublin City University, Ireland
Steven Zimmerman   University of Essex, UK
Justin Zobel   University of Melbourne, Australia
Guido Zuccon   University of Queensland, Australia

Additional Reviewers

Aumiller, Dennis   Gerritse, Emma
Bartscherer, Frederic   Ghauri, Junaid
Belém, Fabiano   Gottschalk, Simon
Bigdeli, Amin   Gémes, Kinga
Biswas, Debanjali   Gérald, Thomas
Cheema, Gullal   Haak, Fabian
Chen, Fumian   Hamidi Rad, Radin
Cunha, Washington   Hoppe, Anett
Dashti, Arman   Huang, Jin
Ebrahimzadeh, Ehsan   Kamateri, Eleni
Engelmann, Björn   Kanase, Sameer
Feher, Gloria   Khawar, Farhan
Ferreira, Thiago   Knyazev, Norman
Fortes, Reinaldo   Leonhardt, Jurek
França, Celso   Liu, Siwei

Lodhia, Zeeshan Ahmed
Mahdavimoghaddam, Jalehsadat
Mamedov, Murad
Mangaravite, Vítor
Marcia, Diego
Mayerl, Maximilian
Mountantonakis, Michalis
Müller-Budack, Eric
Navarrete, Evelyn
Nguyen, Hoang
Paiva, Bruno
Pérez Vila, Miguel Anxo
Ramos, Rita
Renders, Jean-Michel

Saier, Tarek
Sanguinetti, Manuela
Santana, Brenda
Seyedsalehi, Shirin
Shliselberg, Michael
Soprano, Michael
Springstein, Matthias
Stamatis, Vasileios
Su, Ting
Tempelmeier, Nicolas
Viegas, Felipe
Vo, Duc-Thuan
Zhang, Yue
Ziaeinejad, Soroush

Sponsors

Platinum Sponsors

amazon | science

Bloomberg
Engineering

Google

SIGIR
Special Interest Group
on Information Retrieval

Gold Sponsor

✖wayfair

Silver Sponsors

University
of Stavanger

vespa

Bronze Sponsors

NTNU
Department of Computer Science

NorwAI

Schibsted

**THOMSON
REUTERS®**

Industry Impact Award Sponsor

With Generous Support From

Contents – Part I

Reproducibility Papers

Contents – Part II

Demonstration Papers

CLEF 2022 Lab Descriptions

Doctoral Consortium

Workshops

Tutorials

Full Papers

Supercalifragilisticexpialidocious: Why Using the "Right" Readability Formula in Children's Web Search Matters

Garrett Allen[1]([✉])[iD], Ashlee Milton[3][iD], Katherine Landau Wright[2][iD],
Jerry Alan Fails[1][iD], Casey Kennington[1][iD], and Maria Soledad Pera[1][iD]

[1] Department of Computer Science, Boise State University, Boise, ID, USA
GarrettAllen@u.boisestate.edu, solepera@boisestate.edu
[2] Department of Literacy, Language and Culture,
Boise State University, Boise, ID, USA
cast-group@boisestate.edu
[3] University of Minnesota, Minneapolis, MN 55455, USA
milto064@umn.edu

Abstract. Readability is a core component of information retrieval (IR) tools as the complexity of a resource directly affects its relevance: a resource is only of use if the user can comprehend it. Even so, the link between readability and IR is often overlooked. As a step towards advancing knowledge on the influence of readability on IR, we focus on *Web search* for *children*. We explore how traditional formulas–which are simple, efficient, and portable–fare when applied to estimating the readability of Web resources for children written in English. We then present a formula well-suited for readability estimation of child-friendly Web resources. Lastly, we empirically show that readability can sway children's information access. Outcomes from this work reveal that: (i) for Web resources targeting children, a simple formula suffices as long as it considers contemporary terminology and audience requirements, and (ii) instead of turning to Flesch-Kincaid– a popular formula–the use of the "right" formula can shape Web search tools to best serve children. The work we present herein builds on three pillars: Audience, Application, and Expertise. It serves as a blueprint to place readability estimation methods that best apply to and inform IR applications serving varied audiences.

Keywords: readability · web search · information retrieval · relevance

1 Introduction

Readability, or "the overall effect of language usage and composition on readers' ability to easily and quickly comprehend the document" [58], has a rich history of research surrounding its methods of estimation. These methods range from traditional formulas to advanced lexical and semantic models [16, 31, 58]. Traditional formulas, based on shallow features and developed using highly-curated printed materials like novels and journal articles [31], are routinely applied in real-world environments [15, 26]. They target varied audience groups [43, 71],

M. Hagen et al. (Eds.): ECIR 2022, LNCS 13185, pp. 3–18, 2022.
https://doi.org/10.1007/978-3-030-99736-6_1

languages [35,39,72], and content domains [75]. State-of-the-art counterparts leverage complex models [16,37,62] based on feature engineering and/or neural-network architectures. They can also adopt a featureless design approach [55,62]. Yet, how non-traditional models estimate readability is not intuitively understood, nor are these models as easy to deploy as the traditional formulas.

Readability plays a prominent role in *Information Retrieval* (**IR**) for children. In the literature focused on studying and facilitating information access for children, readability is strongly intertwined with the concept of *relevance*. Children must be able to read and understand resource content for it to be deemed relevant, i.e., children must comprehend the text presented to them to extract information that satisfies their needs [10,56,68]. The relationship between relevance and readability is discernible in the design of search and recommendation tools that explicitly target children, such as EmSe [32], Read-X [61], and Rabbit [66]. This association is not limited to informing algorithm design but also serves as a perspective for exploratory studies. For instance, a recent study uses readability as a performance measure when inspecting how Web search engines respond to children's queries in the classroom [14]. Bilal et al. [17,18] rely on readability to examine search result snippets generated by commercial search engines, i.e., Google or Bing, for children's queries. These are meaningful explorations in view of works showing that materials retrieved in response to Web search tasks are inaccessible to many users [28,83]. In general, top-ranked Web pages retrieved by Google are easier to read than those ranked lower [13]. Still, the average readability of top pages is around the 12th grade [13,14], which exceeds children's reading skills. This is a concern, as children often browse Search Engine Result Pages (**SERP**) from top to bottom [41]. Despite how interconnected readability and IR for children are, there is no consensus as to what formula to use for readability estimation, nor is there careful consideration about the link between IR applications and the formulas they use.

In this paper, we examine the connection between readability and IR to deepen understanding among researchers and practitioners. We anchor our exploration on three pillars that enable us to study the natural interactions of users with differing skill-sets and the IR applications they use to access information: (i) Target Audience, (ii) Application, and (iii) Expertise. Among other traits, resource relevance depends on the requirements of a user. The diversity in reading ability among *children* in Kindergarten–12th grade allows them to serve as an opportune demographic for our Target Audience.[1] Due to the ubiquitous presence of search engines like Google and the fact that children commonly turn to these tools to access online information [14], we designate *Web search tools* as our Application. For Expertise we use *readability*. We favor traditional formulas for estimation of *English texts*, as opposed to neural methods, due to their simplicity of calculation, portability, prevalence among IR tools [32,36,50,70], and use in real-world general settings [15,26,74]. With the analysis presented in this paper, we seek to answer two research questions.

---

[1] Grade levels according to the United States' educational system.

RQ1: Do traditional formulas effectively estimate the readability of resources targeting children? To answer this question, we undertake an empirical exploration to gauge the applicability of ten traditional formulas on resources written in English targeting children. We first compare and contrast the performance of these formulas across grade levels when applied to books, the medium they were intended to assess. Given our `Application` we further analyze the performance of these formulas when applied to digital resources, not print. We find that the effectiveness of these formulas greatly varies across grades and that lexicon-based formulas fare better than the most popular ones, e.g., Flesch-Kincaid [43], when predicting the readability of Web resources for children. This leads us to another question.

RQ2: Does the choice of readability formula impact the performance of Web search? We investigate if and how readability influences different scenarios related to Web search. We quantify the differences in performance observed by solely exchanging formulas when (i) estimating the readability of children's queries and snippets generated by search engines in response to children's inquiries, (ii) providing query suggestions for children, and (iii) re-ranking resources retrieved in response to children's queries to prioritize those suitable to them. Results from this analysis showcase that the choice of readability formula has the potential to affect children's online information discovery.

The findings emerging from our study highlight the importance of choosing the "right" formula for readability estimation when dealing with children's Web resources, and how that decision exerts influence on Web search for children. The study also results in Spache-Allen, a new formula that extends Spache [71] by explicitly considering terminology familiar to children.[2] With our three pillars, we create a foundation for the investigation of the interaction between readability and IR; particularly the need to appraise the readability formulas used when designing information access tools and how to do so. These tools should be architected to provide user-friendly versions of resources, particularly for domains that use advanced technical jargon. This work has implications for the future development of fair and equitable resource access tools serving all users [38] and reinforces research on IR applications that leverage different readability approaches. Burgeoning research features (multi-modal) conversational applications that interact with users to clarify their information needs [6]. We envision readability playing a role in equipping these applications to formulate response utterances fitting disparate users' skills. In the spirit of accessibility [7,59], these applications could support users beyond children who may have issues comprehending text, e.g., users with dyslexia or English language learners.

2 Background and Related Work

Readability has been a heavily-investigated area within the last century. Earlier works focused on traditional formulas that take a statistical approach considering

---

[2] The script used for analysis purposes, along with the Spache-Allen itself can be found at https://github.com/BSU-CAST/ecir22-readability.

shallow features like the number of complex words, the number of syllables, or the length of sentences [31]. Among the many formulas in this group, the more well-known include the Flesch-Kincaid Reading Ease [43], the Coleman-Liau Index [24], the Dale-Chall Readability Formula [22,27], the Gunning Fog Index [40], and the Spache Readability Formula [71]. With the advent of machine learning and neural networks, readability formulas transitioned to readability models, incorporating lexical, semantic, and even multilingual features alongside traditional shallow features to produce estimations [16,37,51,62]. At the same time, we would be remiss not to mention existing commercial efforts, such as Wizenoze Readability Index [80,81] and Lexile [1]. Unfortunately, there is a lack of standardization of reading levels used for estimations, with differing "scales" in readability prediction. For instance, some use grade levels, others binary labels (simple vs. complex), or varied categorical labels [43,55,84]. Consequently, it is increasingly difficult to explore which formula works best and why. Even with recent advancements, traditional formulas tend to be the ones most used in real-world scenarios [15,26]. Still, traditional formulas are not without flaws. They can produce results that are inaccurate when assessing text that contains many simple, short terms that are highly technical in nature or build a complex, or subtle, story [21,51,72,79]. Further, a critical evaluation of the predicted reading levels of passages used in academic readiness exams revealed that estimations yielded by traditional formulas were 1–3 grades higher than the intended grade levels [73,74].

Works related to **readability and IR** that also align with our `Target Audience` and `Application` of interest include that of Bilal et al. [18] and Anuyah et al. [14], who study the complexity of resources retrieved by search engines in response to children's queries. Both agree that the reading levels of snippets and resources are too high for children to comprehend. Still, both explorations base their findings on traditional formulas, which can offer misleading estimations and might not be suitable for analyzing Web resources. The impact of readability is not constrained to IR for children. Literature shows that readability is far-reaching within IR. Lately, we see readability support a broad range of IR-related applications, from easing information access [36] and helping teachers locate news articles aligning with the readability levels of their students, to supporting classroom instruction [34] and fake news detection [64]. Through a Firefox plugin, Yu and Miller [85] provide readability support for Asian users who are not fluent in English by enhancing the readability of Web pages. Focusing on recommendation systems, researchers have considered readability as a trait for determining helpful reviews [70] as well as influencing algorithms that recommend books [5,66,82] and learning resources [50]. Readability also benefits question answering (QA). For example, researchers have used readability estimated via traditional formulas to identify high-quality developer chats [23] and educational answers in community QA [48], as well as aid detection of the "best" answers to questions in health QA communities [49], and the ranking of answers in community QA sites [29]. Concerning Web search, readability is a trait that has been considered to predict knowledge gain during Web search [65]. It has also been used as a means to personalize retrieved resources [25, 61] and assess learning as a result of engaging with Web search tasks [69].

This brief overview exhibits the pervasive nature of readability within IR, making the pursuit of understanding its impact a must. With the analysis we discuss in this manuscript, we take initial steps towards that goal.

3 The Fit of Readability Formulas on Web Text

The lack of consensus around which readability formula to use on IR tools makes it uncertain which formula best suits complexity estimation of general Web texts, let alone those intended for young searchers (Target Audience). To address this concern, we examine the efficacy of readability formulas for their originally intended purpose: estimating the reading levels of published texts. We then probe their performance when applied to Web resources (Application). We study popular traditional formulas (Expertise): (i) **DC** - New Dale-Chall [22]; (ii) **SMOG** [57]; (iii) **GF** - Gunning-FOG Index [4]; (iv) **LIX** [20]; (v) **RIX** [12]; (vi) **CL** - Coleman-Liau Index [24], designed for digital texts; (vii) **FK** - Flesch-Kincaid [43], due to its widespread adoption; (viii) **Spache** - Spache Readability Formula [71], meant for texts targeting grades 1st–3rd; and (ix) **SS** - Spache-Sven [52], an enhanced version of Spache that augments its vocabulary with terms that frequently occur on children's websites. For formula details, see [16, 31]. It is apparent in traditional formulas which and how shallow features impact estimation. Instead, neural solutions often lack interpretability on how estimations are produced. Thus, traditional formulas, which are broadly adopted for research and mainstream applications alike, are the focus of this exploration.

For this empirical exploration we use two datasets built using existing corpora. We explicitly examine printed and digital mediums. DSBOOK is comprised of 235 book excerpts extracted from the appendices of the Common Core State Standards[3] [42], each associated with a range of grade levels. We use the minimum grade level from these ranges as the label, as children reading below their level experience less difficulty with comprehension versus when reading above their level [11]. DSBOOK also includes 2,084 books from Reading A-Z (RAZ) labeled with their corresponding reading level[4]. DSWEB is made up of 22,689 resources. It includes resources from the WeeBit corpus [77], which consists of samples extracted from WeeklyReader (an educational newspaper), each labeled with their corresponding grade level, and the NewsELA corpus [63], a set of curated news articles with their corresponding grade labels. Given the few resources targeting Kindergarten and 1st graders, DSWEB also incorporates Web resources expertly curated from sites offering content for younger children.

In our experiment, we use Python's Textstat library [2] to estimate the readability of resources in DSBOOK and DSWEB. We quantify performance via Mean Error Rate (**MER**) and Root Mean Squared Error (**RMSE**). RMSE and MER

---

[3] A set of learning outcomes to inform curriculum for schools in the United States.

[4] RAZ uses a 26-letter scale assigned by experts for readability [47]. To enable fair comparison, we map letter labels to grade labels, using RAZ's conversion table [46].

exhibited similar trends, thus we omit detailed discussions on the former for brevity. To enable fair assessment for those formulas that provide a score rather than a grade, i.e., LIX, RIX, and DC, we map their outputs to a grade according to conversion tables from their original publications [12,20,27]. Through comparison of the results in each medium, we can discern disparities in performance and identify the formulas that better suit estimation of text difficulty of online resources for children. Significance of results are determined using the Kruskal-Wallis H-test [44] with a $p < 0.05$. Unless otherwise stated, results reported in this section are significant.

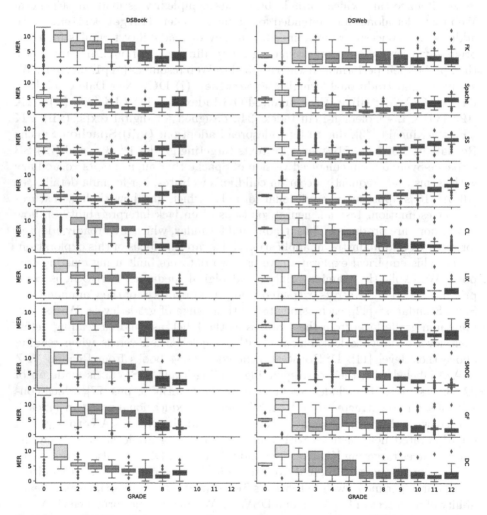

Fig. 1. MER across grades for readability formulas applied to DSBOOK and DSWEB. Resources in DSBOOK are labeled with a grade range indicating the corresponding target audience, so we take the lowest grade as ground truth.

We first investigate the capabilities of readability formulas using DSBOOK. As shown in Fig. 1, Spache and SS tend to produce lower MER towards the middle grades, whereas the other formulas yield lower MER for the later grades. CL exhibits the lowest MER at the 9th grade and above. Interestingly, even though it is commonly used [17], FK is one of the formulas that produces the highest MER, as compared to Spache and DC. Overall, Spache and SS are the least error-prone for resources intended for grades K–6th.

To see if this performance translates to Web resources, we repeat the experiment using DSWEB. As illustrated in Fig. 1, the results for Web resources are similar to those obtained for books in that Spache and SS are the least error-prone formula for resources targeting younger readers. With the exception of Spache and SS, traditional formulas are inconsistent when estimating the complexity of texts for earlier grades (K–6th). Outcomes from the presented analysis serve as an indication of Spache and SS being formulas particularly well-suited for estimating the readability of Web resources for young readers.

Regardless of its effectiveness for our audience and resource type, Spache's static vocabulary–consisting of 1,064 words that are considered "easy" for children to comprehend [3]–is limited and includes terminology from the 1970s. As language changes over time [67], an outdated vocabulary may not capture easy terms for children in today's world, potentially pushing the formula to misleading text complexity estimations. The benefit of changing the 1940s vocabulary used in the original Dale-Chall formula [27] to the one used by the DC formula, more aligned to the 1990s, is apparent [22]. Similar boosts are seen with SS [52], which augments Spache's original vocabulary list through the inclusion of a dictionary of 48,000 non-stop lemmatized terms the authors extracted from children-related websites. Nevertheless, this enhancement relies on word frequency analysis and assumes that terms added to the vocabulary are understood by children, which may not always be the case.

To include vocabulary that children learn through instruction, we take advantage of the Age of Acquisition (**AoA**) dataset. This dataset contains acquisition ratings in the form of ages, ranging from 1–17 years, for ∼30,000 English words [45]. We posit that there is a benefit to simultaneously accounting for terminology that children have been exposed to through websites as well as terminology that has been taught. Thus, we merge the original Spache vocabulary with the terms from AoA and the dictionary from [52]; we call this updated formula Spache-Allen (**SA**), which is computed as in Eq. 1.

$$Spache\text{-}Allen(R) = (0.141 \times w_R/s_R) + (0.086 * dif(R)) + 0.839 \quad (1)$$

where R is a resource, w_R and s_R are the number of words and sentences in R, respectively. The function $dif(R)$ determines the percentage of difficult words in R, where a word is deemed difficult if it does not appear in the "easy" vocabulary–in this case it includes 65,669 unique terms that children learn through instruction and/or are exposed to online, in addition to the original Spache's term list.

Regardless of the dataset considered, augmenting Spache's original vocabulary has a positive effect on readability estimation as it leads to decreases in MER

(Fig. 1). SA consistently outperforms all other investigated traditional formulas through grade 5; it's performance is comparable to that of Spache and SS on higher grades. For grades 9 and above, formulas like CL yield the lowest MER, which is anticipated, given that Spache, SS, and SA have the express purpose of determining the difficulty of texts targeting younger readers.

With RQ1, we aimed to answer: *Do traditional formulas effectively estimate the readability of resources targeting children?* From trends in MER and RMSE, it is evident that the reliability of some formulas differs upon the source material they are applied to (e.g., DC averages a MER of 6.88 for books vs. 4.12 for Web resources). We see that, on average, book resources result in larger errors than Web resources; this is also prevalent among material targeting early readers, i.e., grades K–4th. Interestingly, the MER and RMSE per formula varies depending on the grade of the text being assessed. This is particularly salient among early readers, both numerically and visibly in Fig. 1. Even more so in Fig. 2, when contrasting performance on the DSBOOK and DSWEB versus respective subsets of the datasets consisting of materials till the 4th grade. The RMSE reported in Fig. 2(b) is particularly telling as it doubles for CL and more than triples for GF, and FK, when contrasting overall performance for K–4th grade resources. In the end, the Spache, SS, and SA formulas are the least error-prone when applied to Web resources targeting younger audiences. Though these three formulas perform similarly, the differences across them are significant (Kruskal-Wallis H-test, $p < 0.05$). Therefore, the formula we see as most suitable to support tasks related to Web search for children is SA.

4   The Effect of Readability on Web Search for Children

It emerges from Sect. 3 that readability formulas do falter. With readability playing a prominent role in Web search for children, we investigate the cascading effect that the choice of readability formula can have on Web search. To do so, we consider four scenarios that spotlight different stages of the search process. In each scenario, we quantify the fluctuations in performance that result from using traditional readability formulas. As in Sect. 3, we use Python's Textstat library for readability estimation. A cursory search (on ACM Digital Library and Google Scholar) for recent literature focused on readability and IR applications reveal a plethora of recommender systems, QA, search, and text simplification strategies, to name a few, that depend upon readability as one of their components. Many of these applications default to FK as the readability formula of choice. For this reason, in each scenario, we treat performance based on FK as a baseline. For significance, we use a two-tailed student t-test with a Bonferroni correction (with $\alpha = 0.05$ and the number of tests $N = 10$, which is the number of formulas) with $p < 0.05$; all results are significant unless reported otherwise.

Scenario 1. In this scenario, we consider readability as a means to facilitate personalization, e.g., filtering and/or prioritizing retrieved resources. We posit that the readability of a query could serve as a proxy for the reading skills of the user initiating the search. In turn, this information can be used as a signal

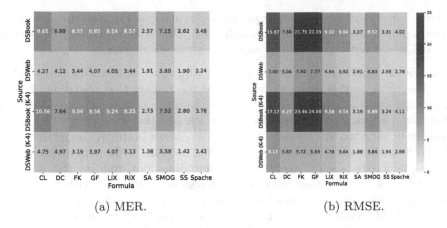

(a) MER.   (b) RMSE.

Fig. 2. Error rate analysis by source and formula. We pay special attention to errors yielded by traditional formulas when applied to material targeting early readers.

Table 1. Fluctuations in performance observed in scenarios related to Web search for children when applying traditional readability formulas. Bold denotes best performing formula for the corresponding scenario, and '*' indicates significant w.r.t. Flesch-Kincaid (FK), a formula that is often used as a component of IR applications.

Scenario	Metric	Formulas									
		DC	SMOG	GF	LIX	RIX	CL	FK	Spache	SS	SA
1	MER	5.94*	4.68*	4.37*	4.49*	**2.1***	3.84	3.72	2.74*	**2.3***	2.34*
2	MRR	**0.39**	0.38	**0.39**	0.36	**0.39**	0.38	0.36	0.4	n/a	0.37
3	MER	4.08*	5.23*	3.49	3.09	3.02*	3.61	3.31	**2.85***	3.0*	3.02*
4	MRR	0.36*	0.2*	0.27*	0.3*	0.27*	0.27*	0.42	**0.43**	0.3*	0.29*
4 (K–4)	MRR	0.26*	0.14*	0.26*	0.22*	0.22*	0.14*	0.49	0.48	**0.50**	0.48

to filter and/or re-rank retrieved resources to match the users' inferred skills [25,53,76]. We use the 168 queries made available by the authors in [8,52], each labeled with the grade of the child formulating the corresponding query. We estimate the readability of each query using the formulas in Sect. 3 and then compare their estimations with respect to the ground truth. For evaluation, we use MER (excluding RMSE for brevity, given similar trends). We are aware that the nature of traditional formulas makes them less suitable for short texts such as queries. Nevertheless, this is a limitation that affects all formulas (possibly with the exception of SS, which has been proven successful in identifying if a query was child-like [52]), therefore reported observations are not affected.

As reported in Table 1, it is clear that the choice of the formula used to estimate the readability of queries has the potential to skew the inference of the users' reading skills [76]. For example, formulas like RIX, Spache, SS, and SA lead to a MER of approximately ±2 grades, whereas DC or SMOG can predict up to 4 grades above or below the grade of the child who initiated the query.

While we expected discrepancy w.r.t. ground truth given that queries are short, if readability is used to enable personalization, the latter would be a concern as the tool would not be adequately supporting the target user.

Scenario 2. Children struggle with formulating queries when searching online [54]. Therefore, in this scenario, we examine the impact that readability has on query suggestion as a means to alleviate children's query formulation issues. We study the performance of ReQuIK [52], a state-of-the-art strategy that offers suggestions targeted at children. As ReQuIK utilizes both FK (popularity) and SS (Web applicability), we first observe fluctuations in its performance when exchanging FK for each of the remaining formulas. Motivated by the outcomes reported in Sect. 3, we also retain FK but instead replace SS with SA. In this experiment, we rely on ReQuik's implementation provided by the authors, and also use the 95 queries used in the original experiments [52]. To generate candidate query suggestions for each of the aforementioned queries, we use an $N - 1$ approach: in each case, we use the prefix of each query (consisting of $N - 1$ terms) to trigger Google's query suggestions via its API. Treating the original query as the ground truth, i.e., what should be ranked first, we calculate the Mean Reciprocal Rank (**MRR**) of the top-10 query suggestions ranked by ReQuIK.

Based on the results reported for Scenario 1, we expected the wide range of estimation errors to impact query suggestion generation. However, from the analysis of results reported in row 2 of Table 1, as well as the experiment using SA and FK to power ReQuIK (MRR of 0.37), it emerges that variations on ReQuIK's performance caused by swapping traditional formulas are not significant. Upon in-depth inspection, we attribute this to ReQuIK's design that incorporates neural architectures. Even though readability is an important trait considered in the wide model component of ReQuIK, it is the deep neural model component that most contributes to ReQuIK's overall success (cf. [52]).

Scenario 3. Snippets are meant to offer children a glimpse into the resources retrieved as they navigate a SERP. For the snippets to facilitate relevant resource selection, they must offer content that children can comprehend. We conduct a new experiment following the procedure outlined in Scenario 1, but on snippets instead of queries. We consider the snippets generated using Google's Custom Search API for a sample of 395 NewsELA resource titles acting as queries. We estimate snippet readability using the formulas in Sect. 3. Treating the original grade label for the corresponding NewsELA resource as ground truth, we compute the respective MER for assessment purposes.

As reported in row 3 of Table 1, there are significant performance fluctuations. As anticipated, Spache, SS, and SA lead to the lowest errors in estimation. On the other hand, SMOG and DC lead to more erroneous estimations. If SERP were to be personalized to ensure children could comprehend presented snippets, then the misleading readability estimations caused by some formulas could result in a SERP that excludes relevant resources. Additionally, as snippets act as proxies for resource content, they could be used in lieu of Web page content for the purpose of re-ranking SERP for children and thus the use of misleading formulas could cause unhelpful changes to the SERP.

Scenario 4. Readability is a key relevance trait informing ranking, particularly given that children have different expectations and needs when it comes to retrieved resources [19]. In this scenario, we examine the effect that readability has on the performance of KORSCE [60], a re-ranking strategy that prioritizes resources for children in the classroom setting. Following the experimental protocol of Scenario 2, we exchange FK, the formula originally used by KORSCE, with each of the formulas under study enabling us to gauge potential performance implications. In this experiment, we sample 193 NewsELA resources and use their titles as queries. Using Google's Custom Search API, we collect the top-10 corresponding resources per query. We re-rank the resources associated with each query using KORSCE, treating the original resource as ground truth. To quantify performance, we use MRR.[5]

The results reported in row 4 of Table 1 show that just by interchanging the readability formula embedded in KORSCE's architecture, relevant resources move from position 5 in the ranking (i.e., SMOG's MRR is 0.2) to position ~2 (based on MRR for Spache and FK, which are 0.43 and 0.42, respectively).[6] This is even more evident among rankings of resources for early readers, who would need the most help from tools when pursuing online information discovery tasks (row K–4 in Table 1). In their case, the relevant resources could move from position 7 in the ranking to 2, simply by exchanging SMOG or CL with Spache, SS, SA, or FK. As children tend to linearly examine SERP [41], the choice of readability formula could prompt the ranking algorithm to inadvertently position higher on the SERP resources children are unable to read or understand, thus negatively affecting their search experience.

With RQ2, we sought to answer: *Does the choice of readability formula impact the performance of Web search?* From the findings discussed in this section, we can surmise that yes, the choice of readability formula affects in a meaningful manner Web search for children. Altering the formula used leads to variations in performance across most of the scenarios examined for Web search for children. Variations were not significant for Scenario 2. We attribute this to ReQuIK's deep model dominating its wide counterpart. Overall, the shift in performance caused by the choice of formula matters, as retrieving resources at appropriate reading levels positively impacts user satisfaction [25,33]. A further concern related to this shift is that searchers could be deterred from engaging with query suggestions or resources that are assumed to be above searchers' skills when the queries and resources could very well be comprehensible and hence relevant. Formulas underestimating difficulty could mistakenly direct searchers to query suggestions or prioritize resources that are far beyond what searchers can comprehend, thus unintentionally setting them up for a failed search.

---

[5] We use KORSCE's implementation made available by the authors.

[6] In Scenario 4, FK's performance is not unexpected as KORSCE is optimized for FK.

5   Conclusion and Future Work

In this paper, we aimed to highlight the natural connection between readability and IR tools. In particular, we focused our analysis on the impact readability has on Web search for children. We gauged the performance of traditional formulas when applied to estimate the readability of printed and digital material targeting children. Moreover, through different scenarios intended to draw attention to different stages of the search process, we studied performance fluctuations that are a direct consequence of exchanging readability formulas.

Analysis of the experimental results suggests that even though Flesch-Kincaid is commonly used to determine the readability of Web resources, it is not the one that best captures their level of difficulty, especially when these Web resources target younger audiences. We have shown that variations of the well-known Spache formula, which explicitly considers terminology children are exposed to online and/or learn as they grow, are better suited to estimate the readability of Web resources for young searchers (RQ1). Of note, we introduced Spache-Allen, which emerged as a result of the explorations conducted in pursuit of RQ1. The effect of readability on algorithms empowering information discovery for young searchers also became apparent during our explorations; making it imperative for developers and researchers to consider using the "right" formula, one best serving the target audience and application, as it directly translates to performance improvements (RQ2). From reported findings we surmise that (i) the performance of IR applications can indeed change based on the readability formula used and (ii) by carefully considering which readability formula supports the target audience of interest, IR applications can be optimized for performance or personalization with respect to an audience (echoing the reports in [78] on general Web resources, not just those targeting children).

Lessons learned from this work inform ongoing efforts related to better enabling children's information discovery through Web search. These include algorithmic solutions that rely on readability as one of their components to suggest queries [52], determine search intent [30], identify resources that are relevant to children [25,60], aid teachers seeking texts for their classrooms [34], or offer teachers insights on students' abilities via search [9]. As decisions related to readability impact all areas of IR, the applicability of this work is far reaching. Further, the pillars introduced can serve as a blueprint that researchers can turn to as a guide for their own explorations towards finding a well-suited readability estimation solution for their intended tasks and audiences.

We limited our examination to traditional formulas applied to Web resources written in English. In the future, we plan to extend our analysis to state-of-the-art counterparts to identify the benefits and constraints inherent to dealing with these more complex models. As a step towards making information accessible worldwide, and given the rise of multilingual strategies for readability estimation, we will extend our exploration to written languages beyond English [56].

Acknowledgments. Work partially funded by NSF Award #1763649. The authors would like to thank Dr. Ion Madrazo Azpiazu and Dr. Michael D. Ekstrand for their valuable feedback.

References

1. https://www.lexile.com/
2. https://github.com/shivam5992/textstat
3. https://github.com/cdimascio/py-readability-metrics/blob/master/readability/data/spache_easy.txt
4. Albright, J., de Guzman, C., Acebo, P., Paiva, D., Faulkner, M., Swanson, J.: Readability of patient education materials: implications for clinical practice. Appl. Nurs. Res. **9**(3), 139–143 (1996)
5. Alharthi, H., Inkpen, D.: Study of linguistic features incorporated in a literary book recommender system. In: ACM/SIGAPP SAC, pp. 1027–1034 (2019)
6. Aliannejadi, M., Zamani, H., Crestani, F., Croft, W.B.: Asking clarifying questions in open-domain information-seeking conversations. In: ACM SIGIR, pp. 475–484 (2019)
7. Allan, J., Croft, B., Moffat, A., Sanderson, M.: Frontiers, challenges, and opportunities for information retrieval: report from SWIRL 2012. In: ACM SIGIR Forum, vol. 46, pp. 2–32 (2012)
8. Allen, G., et al.: Engage!: co-designing search engine result pages to foster interactions. In: ACM IDC, pp. 583–587 (2021)
9. Allen, G., Wright, K.L., Fails, J.A., Kennington, C., Pera, M.S.: Casting a net: supporting teachers with search technology. arXiv preprint arXiv:2105.03456 (2021)
10. Amendum, S.J., Conradi, K., Hiebert, E.: Does text complexity matter in the elementary grades? A research synthesis of text difficulty and elementary students' reading fluency and comprehension. Educ. Psychol. Rev. **30**(1), 121–151 (2018)
11. Amendum, S.J., Conradi, K., Liebfreund, M.D.: The push for more challenging texts: an analysis of early readers' rate, accuracy, and comprehension. Read. Psychol. **37**(4), 570–600 (2016)
12. Anderson, J.: Lix and Rix: variations on a little-known readability index. J. Read. **26**(6), 490–496 (1983)
13. Antunes, H., Lopes, C.T.: Readability of web content. In: CISTI, pp. 1–4 (2019)
14. Anuyah, O., Milton, A., Green, M., Pera, M.S.: An empirical analysis of search engines' response to web search queries associated with the classroom setting. Aslib J. Inf. Manage. **72**(1), 88–111 (2020)
15. Begeny, J.C., Greene, D.J.: Can readability formulas be used to successfully gauge difficulty of reading materials? Psychol. Sch. **51**(2), 198–215 (2014)
16. Benjamin, R.G.: Reconstructing readability: recent developments and recommendations in the analysis of text difficulty. Educ. Psychol. Rev. **24**(1), 63–88 (2012)
17. Bilal, D.: Comparing Google's readability of search results to the Flesch readability formulae: a preliminary analysis on children's search queries. Am. Soc. Inf. Sci. Technol. **50**(1), 1–9 (2013)
18. Bilal, D., Huang, L.-M.: Readability and word complexity of SERPs snippets and web pages on children's search queries: Google vs Bing. Aslib J. Inf. Manage. **71**(2), 241–259 (2019)
19. Bilal, D., Kirby, J.: Differences and similarities in information seeking: children and adults as web users. IPM **38**(5), 649–670 (2002)
20. Björnsson, C.H.: Läsbarhet: hur skall man som författare nå fram till läsarna? Bokförlaget Liber (1968)
21. Bruce, B., Rubin, A., Starr, K.: Why readability formulas fail. IEEE Trans. Prof. Commun. **1**, 50–52 (1981)

22. Chall, J.S., Dale, E.: Readability Revisited: The New Dale-Chall Readability Formula. Brookline Books (1995)
23. Chatterjee, P., Damevski, K., Kraft, N.A., Pollock, L.: Automatically identifying the quality of developer chats for post hoc use. ACM TOSEM **30**(4), 1–28 (2021)
24. Coleman, M., Liau, T.L.: A computer readability formula designed for machine scoring. J. Appl. Psychol. **60**(2), 283 (1975)
25. Collins-Thompson, K., Bennett, P.N., White, R.W., De La Chica, S., Sontag, D.: Personalizing web search results by reading level. In: ACM CIKM, pp. 403–412 (2011)
26. Crossley, S.A., Skalicky, S., Dascalu, M.: Moving beyond classic readability formulas: new methods and new models. J. Res. Read. **42**(3–4), 541–561 (2019)
27. Dale, E., Chall, J.S.: A formula for predicting readability: instructions. Educ. Res. Bull. **27**, 37–54 (1948)
28. D'Alessandro, D.M., Kingsley, P., Johnson-West, J.: The readability of pediatric patient education materials on the world wide web. Arch. Pediatr. Adolesc. Med. **155**(7), 807–812 (2001)
29. Dalip, D.H., Gonçalves, M.A., Cristo, M., Calado, P.: Exploiting user feedback to learn to rank answers in q&a forums: a case study with stack overflow. In: ACM SIGIR, pp. 543–552 (2013)
30. Dragovic, N., Madrazo Azpiazu, I., Pera, M.S.: "Is Sven Seven?" A search intent module for children. In: ACM SIGIR, pp. 885–888 (2016)
31. DuBay, W.H.: Smart Language: Readers, Readability, and the Grading of Text (2007)
32. Eickhoff, C., et al.: EmSe: initial evaluation of a child-friendly medical search system. In: IIiX, pp. 282–285 (2012)
33. Eickhoff, C., de Vries, A.P., Collins-Thompson, K.: Copulas for information retrieval. In: ACM SIGIR, pp. 663–672 (2013)
34. Ekstrand, M.D., Wright, K.L., Pera, M.S.: Enhancing classroom instruction with online news. Aslib J. Inf. Manage. **72**(5), 725–744 (2020)
35. El-Haj, M., Rayson, P.: Osman–a novel Arabic readability metric. In: LREC, pp. 250–255 (2016)
36. Ermakova, L., et al.: Text simplification for scientific information access. In: ECIR (2021)
37. François, T., Miltsakaki, E.: Do NLP and machine learning improve traditional readability formulas? In: 1st Workshop on Predicting and Improving Text Readability for Target Reader Populations, pp. 49–57 (2012)
38. Garcia-Febo, L., Hustad, A., Rösch, H., Sturges, P., Vallotton, A.: IFLA code of ethics for librarians and other information workers. https://www.ifla.org/publications/ifla-code-of-ethics-for-librarians-and-other-information-workers-short-version-/
39. Gonzalez-Dios, I., Aranzabe, M.J., de Ilarraza, A.D., Salaberri, H.: Simple or complex? Assessing the readability of Basque Texts. In: COLING, pp. 334–344 (2014)
40. Gunning, R.: The fog index after twenty years. J. Bus. Commun. **6**(2), 3–13 (1969)
41. Gwizdka, J., Bilal, D.: Analysis of children's queries and click behavior on ranked results and their thought processes in Google search. In: CHIIR, pp. 377–380 (2017)
42. Common Core Stat Standards Initiative: Appendix B: text exemplars and sample performance tasks (2020). http://www.corestandards.org/assets/Appendix_B.pdf
43. Kincaid, J.P., Fishburne, R.P., Jr., Rogers, R.L., Chissom, B.S.: Derivation of new readability formulas (automated readability index, fog count and Flesch reading ease formula) for navy enlisted personnel. Technical report, Naval Technical Training Command Millington TN Research Branch (1975)

44. Kruskal, W.H., Wallis, W.A.: Use of ranks in one-criterion variance analysis. J. Am. Stat. Assoc. **47**(260), 583–621 (1952)
45. Kuperman, V., Stadthagen-Gonzalez, H., Brysbaert, M.: Age-of-acquisition ratings for 30,000 English words. Behav. Res. Meth. **44**(4), 978–990 (2012)
46. Lazel, I.: Level correlation chart (2021). https://www.readinga-z.com/learninga-z-levels/level-correlation-chart/. Accessed 18 Jan 2021
47. Lazel, I.: Reading A-Z: the online reading program with downloadable books to print and assemble (2021). https://www.readinga-z.com/. Accessed 18 Jan 2021
48. Le, L.T., Shah, C., Choi, E.: Evaluating the quality of educational answers in community question-answering. In: IEEE/ACM JCDL, pp. 129–138 (2016)
49. Lin, C.Y., Wu, Y.-H., Chen, A.L.P.: Selecting the most helpful answers in online health question answering communities. J. Intell. Inf. Syst. **57**(2), 271–293 (2021)
50. Liu, L., Koutrika, G., Wu, S.: LearningAssistant: a novel learning resource recommendation system. In: IEEE ICDE, pp. 1424–1427 (2015)
51. Madrazo Azpiazu, I.: Towards multipurpose readability assessment. Master's thesis, Boise State University (2016). https://scholarworks.boisestate.edu/td/1210/
52. Madrazo Azpiazu, I., Dragovic, N., Anuyah, O., Pera, M.S.: Looking for the movie Seven or Sven from the movie frozen? A multi-perspective strategy for recommending queries for children. In: ACM CHIIR, pp. 92–101 (2018)
53. Madrazo Azpiazu, I., Dragovic, N., Pera, M.S.: Finding, understanding and learning: making information discovery tasks useful for children and teachers. In: SAL Workshop co-located with ACM SIGIR (2016)
54. Madrazo Azpiazu, I., Dragovic, N., Pera, M.S., Fails, J.A.: Online searching and learning: YUM and other search tools for children and teachers. Inf. Retr. J. **20**(5), 524–545 (2017)
55. Madrazo Azpiazu, I., Pera, M.S.: Multiattentive recurrent neural network architecture for multilingual readability assessment. TACL **7**, 421–436 (2019)
56. Madrazo Azpiazu, I., Pera, M.S.: An analysis of transfer learning methods for multilingual readability assessment. In: Adjunct Publication of the 28th ACM UMAP, pp. 95–100 (2020)
57. Mc Laughlin, G.H.: Smog grading-a new readability formula. J. Read. **12**(8), 639–646 (1969)
58. Meng, C., Chen, M., Mao, J., Neville, J.: ReadNet: a hierarchical transformer framework for web article readability analysis. In: Jose, J.M., et al. (eds.) Advances in Information Retrieval: 42nd European Conference on IR Research, ECIR 2020, Lisbon, Portugal, April 14–17, 2020, Proceedings, Part I, pp. 33–49. Springer, Cham (2020). https://doi.org/10.1007/978-3-030-45439-5_3
59. Milton, A., Allen, G., Pera, M.S.: To infinity and beyond! Accessibility is the future for kids' search engines. arXiv preprint arXiv:2106.07813 (2021)
60. Milton, A., Anuya, O., Spear, L., Wright, K.L., Pera, M.S.: A ranking strategy to promote resources supporting the classroom environment. In: IEEE/WIC/ACM WI-IAT, pp. 121–128 (2020)
61. Miltsakaki, E., Troutt, A.: Read-X: automatic evaluation of reading difficulty of web text. In: E-Learn, pp. 7280–7286. AACE (2007)
62. Mohammadi, H., Khasteh, S.H.: Text as environment: a deep reinforcement learning text readability assessment model. arXiv preprint arXiv:1912.05957 (2019)
63. Newsela: Newsela article corpos (2016). https://newsela.com/data
64. Ngada, O., Haskins, B.: Fake news detection using content-based features and machine learning. In: IEEE CSDE, pp. 1–6 (2020)
65. Otto, C., et al.: Predicting knowledge gain during web search based on multimedia resource consumption. In: AIED, pp. 318–330 (2021)

66. Pera, M.S., Ng, Y.K.: Automating readers' advisory to make book recommendations for k-12 readers. In: ACM RecSys, pp. 9–16 (2014)
67. Ramiro, C., Srinivasan, M., Malt, B.C., Xu, Y.: Algorithms in the historical emergence of word senses. Nat. Acad. Sci. **115**(10), 2323–2328 (2018)
68. Reed, D.K., Kershaw-Herrera, S.: An examination of text complexity as characterized by readability and cohesion. J. Exp. Educ. **84**(1), 75–97 (2016)
69. Roy, N., Torre, M.V., Gadiraju, U., Maxwell, D., Hauff, C.: Note the highlight: incorporating active reading tools in a search as learning environment. In: ACM CHIIR, pp. 229–238 (2021)
70. Saptono, R., Mine, T.: Time-based sampling methods for detecting helpful reviews. In: IEEE/WIC/ACM WI-IAT, pp. 508–513 (2020)
71. Spache, G.D.: The Spache readability formula. In: Good Reading for Poor Readers, pp. 195–207 (1974)
72. Spaulding, S.: A Spanish readability formula. Mod. Lang. J. **40**(8), 433–441 (1956)
73. Szabo, S., Sinclair, B.: STAAR reading passages: the readability is too high. Schooling **3**(1), 1–14 (2012)
74. Szabo, S., Sinclair, B.B.: Readability of the STAAR test is still misaligned. Schooling **10**(1), 1–12 (2019)
75. Tahir, M., et al.: Evaluation of quality and readability of online health information on high blood pressure using DISCERN and Flesch-Kincaid tools. Appl. Sci. **10**(9), 3214 (2020)
76. Taranova, A., Braschler, M.: Textual complexity as an indicator of document relevance. In: Hiemstra, D., Moens, M.-F., Mothe, J., Perego, R., Potthast, M., Sebastiani, F. (eds.) ECIR 2021. LNCS, vol. 12657, pp. 410–417. Springer, Cham (2021). https://doi.org/10.1007/978-3-030-72240-1_42
77. Vajjala, S., Meurers, D.: On improving the accuracy of readability classification using insights from second language acquisition. In: 7th Workshop on Building Educational Applications using NLP, pp. 163–173 (2012)
78. Vajjala, S., Meurers, D.: On the applicability of readability models to web texts. In: 2nd Workshop on Predicting and Improving Text Readability for Target Reader Populations, pp. 59–68 (2013)
79. Wang, H.X.: Developing and testing readability measurements for second language learners. Ph.D. thesis, Queensland University of Technology (2016)
80. Westervelf, T.: Wizenoze search white paper (2021). https://cdn.theewf.org/uploads/pdf/Wizenoze-white-paper.pdf
81. Wizenoze: Wizenoze readability index (2021). http://www.wizenoze.com
82. Wojciechowski, A., Gorzynski, K.: A method for measuring similarity of books: a step towards an objective recommender system for readers. In: Vetulani, Z., Uszkoreit, H., Kubis, M. (eds.) LTC 2013. LNCS (LNAI), vol. 9561, pp. 161–174. Springer, Cham (2016). https://doi.org/10.1007/978-3-319-43808-5_13
83. Wong, K., Levi, J.R.: Readability of pediatric otolaryngology information by children's hospitals and academic institutions. Laryngoscope **127**(4), E138–E144 (2017)
84. Xia, M., Kochmar, E., Briscoe, T.: Text readability assessment for second language learners. arXiv preprint arXiv:1906.07580 (2019)
85. Yu, C.H., Miller, R.C.: Enhancing web page readability for non-native readers. In: CHI 2010, pp. 2523–2532 (2010)

PARM: A Paragraph Aggregation Retrieval Model for Dense Document-to-Document Retrieval

Sophia Althammer[1(✉)], Sebastian Hofstätter[1], Mete Sertkan[1], Suzan Verberne[2], and Allan Hanbury[1]

[1] Institute for Information Systems Engineering, TU Wien, Vienna, Austria
{sophia.althammer,sebastian.hofstatter,mete.sertkan,
allan.hanbury}@tuwien.ac.at
[2] Leiden University, Leiden, The Netherlands
s.verberne@liacs.leidenuniv.nl

Abstract. Dense passage retrieval (DPR) models show great effectiveness gains in first stage retrieval for the web domain. However in the web domain we are in a setting with large amounts of training data and a query-to-passage or a query-to-document retrieval task. We investigate in this paper dense document-to-document retrieval with limited labelled target data for training, in particular legal case retrieval. In order to use DPR models for document-to-document retrieval, we propose a Paragraph Aggregation Retrieval Model (PARM) which liberates DPR models from their limited input length. PARM retrieves documents on the paragraph-level: for each query paragraph, relevant documents are retrieved based on their paragraphs. Then the relevant results per query paragraph are aggregated into one ranked list for the whole query document. For the aggregation we propose vector-based aggregation with reciprocal rank fusion (VRRF) weighting, which combines the advantages of rank-based aggregation and topical aggregation based on the dense embeddings. Experimental results show that VRRF outperforms rank-based aggregation strategies for dense document-to-document retrieval with PARM. We compare PARM to document-level retrieval and demonstrate higher retrieval effectiveness of PARM for lexical and dense first-stage retrieval on two different legal case retrieval collections. We investigate how to train the dense retrieval model for PARM on limited target data with labels on the paragraph or the document-level. In addition, we analyze the differences of the retrieved results of lexical and dense retrieval with PARM.

1 Introduction

Dense passage retrieval (DPR) models brought substantial effectiveness gains to information retrieval (IR) tasks in the web domain [14,19,39]. The promise of DPR models is to boost the recall of first stage retrieval by leveraging the semantic information for retrieval as opposed to traditional retrieval models [31], which rely on lexical matching. The web domain is a setting with query-to-passage or query-to-document retrieval tasks and a large amount of training data, while training data is much more limited in other domains. Furthermore we see recent advances in neural retrieval remain neglected for document-to-document retrieval despite the task's importance in several, mainly professional, domains [24,28–30].

© The Author(s), under exclusive license to Springer Nature Switzerland AG 2022
M. Hagen et al. (Eds.): ECIR 2022, LNCS 13185, pp. 19–34, 2022.
https://doi.org/10.1007/978-3-030-99736-6_2

In this paper we investigate the effectiveness of dense retrieval models for document-to-document tasks, in particular legal case retrieval. We focus on first stage retrieval with dense models and therefore aim for a high recall. The first challenge for DPR models in document-to-document retrieval tasks is the input length of the query documents and of the documents in the corpus. In legal case retrieval the cases tend to be long documents [35] with an average length of 1269 words in the COLIEE case law corpus [29]. However the input length of DPR models is limited to 512 tokens [19] and theoretically bound of how much information of a long text can be compressed into a single vector [25]. Furthermore we reason in accordance with the literature [7,33,37,38] that relevance between two documents is not only determined by the complete text of the documents, but that a candidate document can be relevant to a query document based on one paragraph that is relevant to one paragraph of the query document. In the web domain DPR models are trained on up to $500k$ training samples [6], whereas in most domain-specific collections only a limited amount of hundreds of labelled samples is available [13,15,29].

In this paper we address these challenges by proposing a **paragraph aggregation retrieval model (PARM)** for dense document-to-document retrieval. PARM liberates dense passage retrieval models from their limited input length without increasing the computational cost. Furthermore PARM gives insight on which paragraphs the document-level relevance is based, which is beneficial for understanding and explaining the retrieved results. With PARM the documents are retrieved on the paragraph-level: the query document and the documents in the corpus are split up into their paragraphs and for each query paragraph a ranked list of relevant documents based on their paragraphs is retrieved. The ranked lists of documents per query paragraph need to be aggregated into one ranked list for the whole query document. As PARM provides the dense vectors of each paragraph, we propose **vector-based aggregation with reciprocal rank fusion weighting (VRRF)** for PARM. VRRF combines the merits of rank-based aggregation [10,16] with semantic aggregation with dense embeddings. We investigate:

RQ1 *How does VRRF compare to other aggregation strategies within PARM?*

We find that our proposed aggregation strategy of VRRF for PARM leads to the highest retrieval effectiveness in terms of recall compared to rank-based [10,34] and vector-based aggregation baselines [21]. Furthermore we investigate:

RQ2 *How effective is PARM with VRRF for document-to-document retrieval?*

We compare PARM with VRRF to document-level retrieval for lexical and dense retrieval methods on two different test collections for the document-to-document task of legal case retrieval. We demonstrate that PARM consistently improves the first stage retrieval recall for dense document-to-document retrieval. Furthermore, dense document-to-document retrieval with PARM and VRRF aggregation outperforms lexical retrieval methods in terms of recall at higher cut-off values.

The success of DPR relies on the size of labelled training data. As we have a limited amount of labelled data as well as paragraph and document-level labels we investigate:

RQ3 *How can we train dense passage retrieval models for PARM for document-to-document retrieval most effectively?*

For training DPR for PARM we compare training with relevance labels on the paragraph or document-level. We find that despite the larger size of document-level labelled datasets, the additional training data is not always beneficial compared to training DPR on smaller, but more accurate paragraph-level samples. Our contributions are:

- We propose a **paragraph aggregation retrieval model (PARM)** for dense document-to-document retrieval and demonstrate higher retrieval effectiveness for dense retrieval with PARM compared to retrieval without PARM and to lexical retrieval with PARM.
- We propose **vector-based aggregation with reciprocal rank fusion weighting (VRRF)** for dense retrieval with PARM and find that VRRF leads to the highest recall for PARM compared to other aggregation strategies.
- We investigate training DPR for PARM and compare the impact of fewer, more accurate paragraph-level labels to more, potentially noisy document-level labels.
- We publish the code at https://github.com/sophiaalthammer/parm

2   Related Work

Dense Passage Retrieval. Improving the first stage retrieval with DPR models is a rapidly growing area in neural IR, mostly focusing on the web domain. Karpukhin et al. [19] propose dense passage retrieval for open-domain QA using BERT models as bi-encoder for the query and the passage. With ANCE, Xiong et al. [39] train a DPR model for open-domain QA with sampling negatives from the continuously updated index. Efficiently training DPR models with distillation [17] and balanced topic aware sampling [18] has demonstrated to improve the retrieval effectiveness. As opposed to this prior work, we move from dense passage to dense document-to-document retrieval and propose PARM to use dense retrieval for document-to-document tasks.

Document Retrieval. The passage level influence for retrieval of documents has been analyzed in multiple works [7,22,37,38] and shown to be beneficial, but in these works the focus lies on passage-to-document retrieval. Cohan et al. [9] present document-level representation learning strategies for ranking, however the input length remains bounded by 512 tokens and only title and abstract of the document are considered. Abolghasemi et al. [1] present multi-task learning for document-to-document retrieval. Liu et al. [40] propose similar document matching for documents up to a length of 2048 however here the input length is still bounded and the computational cost of training and using the model is increased. Different to this prior work, the input length of PARM is not bounded without increasing the computational complexity of the retrieval.

Aggregation Strategies. Aggregating results from different ranked lists has a long history in IR. Shaw et al. [20,34] investigate the combination of multiple result lists by summing the scores. Different rank aggregation strategies like Condorcet [26] or Borda count [36] are proposed, however it is demonstrated [10,41] that reciprocal rank fusion outperforms them. Ai et al. [2] propose a neural passage model for scoring passages for a passage-to-document retrieval task. Multiple works [3,4,11,42] propose score aggregation for re-ranking with BERT on a passage-to-document task ranging from taking the first passage of a document to the passage of the document with the highest score. Different to rank/score-based aggregation approaches, Li et al. [21] propose vector-based aggregation for re-ranking for a passage-to-document task. Different to our approach they concatenate query and passage and learn a representation for binary classification of the relevance score. The focus of score/rank aggregation is mainly on federated search or passage-to-document tasks, however we focus on document-to-document retrieval. We have not seen a generalization of aggregation strategies for the query and candidate paragraphs for document-to-document retrieval yet. Different to previous work, we propose to combine rank and vector-based aggregation methods for aggregating the representation of query and candidate documents independently.

3 Paragraph Aggregation Retrieval Model (PARM)

In this section we propose PARM as well as the aggregation strategy VRRF for PARM for dense document-to-document retrieval and training strategies.

3.1 Workflow

We use the DPR model [19] based on BERT [12] bi-encoders, of which one encodes the query passage q, the other one the candidate passage p. After storing the encoded candidate passages $\hat{p}$ in the index, the relevance score between a query q and a candidate passage p is computed by the dot-product between the encoded query passage $\hat{q}$ and $\hat{p}$.

As the input length of BERT [12] is limited to 512 tokens, the input length for the query and the candidate passage for DPR [19] is also limited by that. The length of query and candidate documents for document-to-document tasks exceeds this input length. For example the average length of a document is 1296 words for the legal case retrieval collection COLIEE [29]. We reason that for document-to-document tasks a single paragraph or multiple paragraphs can be decisive for the relevance of a document to another one [7,22,37,38] and that different paragraphs contain different topics of a document. Therefore we propose a **paragraph aggregation retrieval model (PARM)**, in order to use DPR models for dense document-to-document retrieval. PARM retrieves relevant documents based on the paragraph-level relevance.

The workflow of PARM is visualized in Fig. 1. For the documents in the corpus we split each document d into paragraphs $p_1, ..., p_{m_d}$ with m_d the number of paragraphs of document d. We take the paragraphs of the document as passages for DPR. We index each paragraph $p_j, j \in 1, ..., m_d$ of each document d in the corpus and attain a paragraph-level index containing the encoded paragraphs $\hat{p}_j$ for all documents d in the corpus. At query time, the query document q is also split up into paragraphs $q_1, ..., q_{n_q}$,

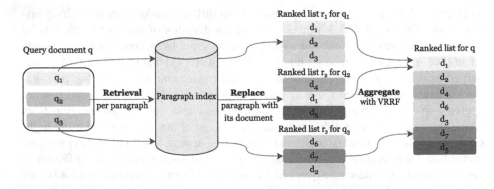

Fig. 1. PARM workflow for query document q and retrieved documents $d_1, .., d_7$

where n_q is the number of paragraphs of q. For each query paragraph q_i with $i \in 1, ..., n_q$ the top N most relevant paragraphs are retrieved from the paragraph-level corpus. The result is a ranked list r_i with $i \in 1, ..., n_q$ per query paragraph q_i with N relevant paragraphs. The paragraphs in the ranked lists r_i with $i \in 1, ..., n_q$ are replaced by the documents that contain the paragraphs. Therefore it is possible that one document occurs multiple times in the list. In order to attain one ranked list for the whole query document q, the ranked paragraph lists of retrieved documents $r_1, ..., r_{n_q}$ of each query paragraph q_i with $i \in 1, ..., n_q$ need to be aggregated to one ranked list.

3.2 Vector-Based Aggregation with Reciprocal Rank Fusion Weighting (VRRF)

Multiple works have demonstrated the benefit of reciprocal rank fusion [10, 16, 27] for rank-based aggregation of multiple ranked retrieved lists. Using dense retrieval with PARM we have more information than the ranks and scores of the retrieved paragraphs: we have dense embeddings, which encode the semantic meaning of the paragraphs, for each query paragraph and the retrieved paragraphs. In order to make use of this additional information for aggregation, we propose **vector-based aggregation with reciprocal rank fusion weighting (VRRF)**, which extends reciprocal rank fusion for neural retrieval. VRRF combines the advantages of reciprocal rank fusion with relevance signals of semantic aggregation using the dense vector embeddings.

In **VRRF** we aggregate documents using the dense embeddings $\hat{p}_i$ of the passages p_i, which are from the same document d and which are in the retrieved list r_i with $i \in 1, ..., n_q$, with a weighted sum, taking the reciprocal rank fusion score [10] as weight. The dense embeddings $\hat{q}_i$ of each query paragraph q_i with $i \in 1, ..., n_q$ are aggregated by adding the embeddings without a weighting:

$$\hat{q} = \sum_{i=1}^{n_q} \hat{q}_i \qquad \hat{d} = \sum_{i=1}^{n_q} \sum_{p \in d, d \in r_i} rrf(q_i, p_i)\, \hat{p}_i$$

We compute the relevance score between query and candidate document with the dot-product between the aggregated embedding of query $\hat{q}$ and candidate document $\hat{d}$.

To confirm the viability of VRFF aggregation, we propose simple baselines: **VRanks** and **VScores**, where the paragraph embeddings $\hat{p}_i$ of d are aggregated with the rank or the score of the passage p_i as weight.

3.3 Training Strategies

As we have a limited amount of labelled target data, we examine how to effectively train a DPR model for PARM with the training collections at hand. We assume that we have test collections consisting of documents with clearly identifiable paragraphs, with relevance assessments on either the paragraph or the document-level.

Paragraph-Level Training. For the paragraph-level labelled training we take the relevant paragraphs in the training set as positives and sample random negatives from the paragraphs in the corpus. Here we sample as many negatives as we have positive samples for each query paragraph, thereby balancing the training data.

Document-Level Training. For the document-level labelled training the collection contains query documents and a corpus of documents with relevance assessments for each query document. We sample negative documents randomly from the corpus. In order to use the document-level labelled collection for training the DPR model, we split up the query document as well as the positive documents into its paragraphs and consider each paragraph of the query document relevant to each paragraph of each positive document. Equivalently we consider each paragraph of a negative document irrelevant to each query paragraph. As on average each document in the COLIEE dataset [29] contains 42.44 paragraphs, one relevant document leads to $42 \cdot 20 = 840$ paragraph-level labels containing one positive and one negative sample to a query paragraph. Therefore this method greatly increases the number of paragraph-level annotations, however this comes with the risk of potentially noisy labels [5].

4 Experiment Design

4.1 Training and Test Collections

We focus on the document-to-document task of legal case retrieval because of the importance for the legal domain [23,24,32,33] which facilitates the availability of training collections with relevance annotations on the paragraph and the document-level [29]. For training the DPR models, we introduce paragraph and document-level labelled collections. For the evaluation we use the document-level collections.

Paragraph-Level Labelled Collections. COLIEE [29] is a competition for legal information extraction and retrieval which provides datasets for legal case retrieval and case entailment. Task 2 of COLIEE 2020 [29] provides a training and test collection for legal case entailment. It contains relevance labels on the legal case paragraph level, given a query claim, a set of candidate claims to the query claim as well as relevance labels for the candidate claims. We denote these sets with *COLIEEPara train/test*.

Document-Level Labelled Collections. In Task 1 of COLIEE 2021 [29], the legal case retrieval task, query cases with their relevance judgements on the document-level are provided together with a corpus of candidate documents. We divide the training set of COLIEEDoc into a training and validation set. The validation set contains the last 100 queries of the training set from query case 550 to 650. We will denote the training, validation and test collection with *COLIEEDoc train/val/test*. For a broader evaluation, we evaluate our models additionally on the CaseLaw collection [24]. It contains a corpus of legal cases, query cases and their relevance judgements for legal case retrieval.

Data Pre-processing. For COLIEEDoc, we remove the French versions of the cases, we divide the cases into an introductory part, a summary, if it contains one, and its claims, which are indicated by their numbering. As indicated in Table 1, the paragraphs have an average length of 84 words and 96.2% of the paragraphs are not longer than 512 words. The CaseLaw dataset is split along the line breaks of the text and merged to paragraphs by concatenating sentences until the paragraphs exceed the length of 200 words.

4.2 Baselines

As baseline we use the lexical retrieval model BM25 [31]. For BM25 we use Elas-ticSearch[1] with parameters $k = 1.3$ and $b = 0.8$, which we optimized on COLIEE-Docval. *VRRF aggregation for PARM (RQ1).* In order to investigate the retrieval effec-tiveness of our proposed aggregation strategy VRFF for PARM, we compare VRRF to the commonly used score-based aggregation strategy CombSum [34] and rank-based aggregation strategy of reciprocal rank fusion (RRF) [10] for PARM. As baselines for vector-based aggregation, we investigate VSum, VMin, VMax, VAvg, which are origi-nally proposed by Li et al. [21] for re-ranking on a passage-to-document retrieval task. In order to use VSum, VMin, VMax, VAvg in the context of PARM, we aggregate independently the embeddings of both, the query and the candidate document. In con-trast to Li et al. [21] we aggregate the query and paragraph embeddings independently and score the relevance between aggregated query and aggregated candidate embedding after aggregation. The learned aggregation methods of CNN and Transformer proposed by Liu et al. [21] are therefore not applicable to PARM, as they learn a classification on the embedding of the concatenated query and paragraph.

Table 1. Statistics of paragraph- and document-level labelled collections.

Labels	Dataset	Train/ Test	Statistics					
			# queries	$\varnothing$ # docs	$\varnothing$ # rel docs	$\varnothing$ para length	% para < 512 words	$\varnothing$ # para
Para	COLIEEPara	Train	325	32.12	1.12	102	95.5%	-
	COLIEEPara	Test	100	32.19	1.02	117	95.2%	-
Doc	COLIEEDoc	Train	650	4415	5.17	84	96.2%	44.6
	COLIEEDoc	Test	250	4415	3.60	92	97.8%	47.5
	CaseLaw	Test	100	63431	7.2	219	91.3%	7.5

PARM VRRF for Dense Document-to-Document Retrieval (RQ2). In order to investi-gate the retrieval effectiveness of PARM with VRRF for dense document-to-document retrieval, we compare PARM to document-level retrieval on two document-level collec-tions (COLIEEDoc and CaseLaw). Because of the limited input length, the document-level retrieval either reduces to retrieval based on the First Passage (FirstP) or the pas-sage of the document with the maximum score (MaxP) [3,42]. In order to separate the impact of PARM for lexical and dense retrieval methods, we also use PARM with BM25 as baseline. For PARM with BM25 we also investigate which aggregation strategy leads to the highest retrieval effectiveness in order to have a strong baseline. As BM25 does not provide dense embeddings only rank-based aggregation strategies are applicable.

---

[1] https://github.com/elastic/elasticsearch.

Paragraph and Document-Level Labelled Training (RQ3). We train a DPR model on a paragraph- and another document-level labelled collection and compare the retrieval performance of PARM for document-to-document retrieval. As bi-encoders for DPR we choose BERT [12] and LegalBERT [8]. We train DPR on the paragraph-level labelled collection COLIEEPara train and additionally on the document-level labelled collection COLIEEDoc train as described in Sect. 3.3. We use the public code[2] and train DPR according to Karpukhin et al. [19]. We sample the negative paragraphs randomly from randomly sampled negative documents and take the 20 paragraphs of a positive document as positive samples, which have the highest BM25 score to the query paragraph. This training procedure lead to the highest recall compared to training with all positive paragraphs or with BM25 sampled negative paragraphs. We also experimented with the DPR model pre-trained on open-domain QA as well as TAS-balanced DPR model [18], but initial experiments did not show a performance improvement. We train each DPR model for 40 epochs and take the best checkpoint according to COLIEEPara test/COL-IEEDoc val. We use batch size of 22 and a learning rate of $2 * 10^{-5}$, after comparing three commonly used learning rates $(2 * 10^{-5}, 1 * 10^{-5}, 5 * 10^{-6})$ for [19].

5   Results and Analysis

We evaluate the first stage retrieval performance with nDCG@10, recall@100, recall@500 and recall@1k using pytrec_eval. We focus our evaluation on recall because the recall performance of the first stage retrieval bounds the ranking performance after re-ranking the results in the second stage for a higher precision. We do not compare our results to the reported state-of-the-art results as they rely on re-ranked results and do not report evaluation results after the first stage retrieval.

5.1   RQ1: VRRF Aggregation for PARM

As we propose vector-based aggregation with reciprocal rank fusion weighting (VRRF) for PARM, we first investigate:

(RQ1) *How does VRRF compare to other aggregation strategies within PARM?*
We compare VRRF, which combines dense-vector-based aggregation with rank-based weighting, to score/rank-based and vector-based aggregation methods for PARM. The results in Table 2 show that VRRF outperforms all rank and vector-based aggregation approaches for the dense retrieval results of DPR PARM with BERT and LegalBERT. For the lexical retrieval BM25 with PARM, only rank-based aggregation approaches are feasible, here RRF shows the best performance, which will be our baseline for RQ2.

---

[2] https://github.com/facebookresearch/DPR.

Table 2. Aggregation comparison for PARM on COLIEEval, VRRF shows best results for dense retrieval, stat. sig. difference to RRF w/ paired t-test ($p < 0.05$) denoted with †, Bonferroni correction with $n = 7$. For BM25 only rank-based methods applicable.

Aggregation	BM25			DPR BERT			DPR LegalBERT		
	R@100	R@500	R@1K	R@100	R@500	R@1K	R@100	R@500	R@1K
Rank-based									
CombSum [34]	.5236	.7854	.8695	.4460	.7642	.8594	.5176	.7975	.8882
RRF [10]	**.5796**	**.8234**	**.8963**	.5011	.8029	.8804	**.5830**	.8373	.9049
Vector-based									
VAvg [21]	-	-	-	.1908†	.4668†	.6419†	.2864†	.4009†	.7466†
VMax [21]	-	-	-	.3675†	.6992†	.8273†	.4071†	.6587†	.8418†
VMin [21]	-	-	-	.3868†	.6869†	.8295†	.4154†	.6423†	.8465†
VSum [21]	-	-	-	.4807	.7496†	.8742	.5182†	.8069	.8882
Vector-based with rank-based weights (Ours)									
VScores	-	-	-	.4841	.7616†	.8709	.5195†	.8075†	.8882†
VRanks	-	-	-	.4826	.7700†	.8804	.5691†	.8212	.8980
VRRF	-	-	-	**.5035**	**.8062†**	**.8806**	**.5830†**	**.8386†**	**.9091†**

Table 3. Document-to-document retrieval results for PARM and Document-level retrieval. No comparison to results reported in prior work as those rely on re-ranking, while we evaluate only first stage retrieval evaluation. *nDCG cutoff at 10, stat. sig. difference to BM25 Doc w/ paired t-test ($p < 0.05$) denoted with † and Bonferroni correction with $n = 12$, effect size > 0.2 denoted with ‡.*

Model	Retrieval	COLIEEDoc test				CaseLaw			
		nDCG	R@100	R@500	R@1K	nDCG	R@100	R@500	R@1K
BM25									
BM25	Doc	**.2435**	.6231	.7815	.8426	**.2653**	.4218	.5058	.5438
	PARM RRF	.1641†‡	**.6497†‡**	.8409†‡	.8944†‡	.0588†‡	.3362†‡	.5716†‡	.6378†‡
DPR									
BERT para	Doc FirstP	.0427†‡	.3000†‡	.5371†‡	.6598†‡	.0287†‡	.0871†‡	.1658†‡	.2300†‡
	Doc MaxP	.0134†‡	.1246†‡	.5134†‡	.6201†‡	.0000†‡	.0050†‡	.4813†‡	.4832†‡
	PARM RRF	.0934†‡	.5765†‡	.8153†‡	.8897†‡	.0046†‡	.1720†‡	.5019†‡	.5563†
	PARM VRRF	.0952†‡	.5786†‡	.8132†‡	.8909†‡	.1754†‡	.3855†‡	.5328†‡	.5742†‡
LegalBERT para	Doc FirstP	.0553†‡	.2447†‡	.4598†‡	.5657†‡	.0397†‡	.0870†‡	.1844†‡	.2248†‡
	Doc MaxP	.0073†‡	.0737†‡	.3970†‡	.5670†‡	.0000†‡	.0050†‡	.4846†‡	.4858†‡
	PARM RRF	.1280†‡	.6370	.8308†‡	.8997†‡	.0177†‡	.2595†‡	.5446†‡	.6040†‡
	PARM VRRF	.1280†‡	.6396	.8310†‡	.9023†‡	.0113†‡	**.4986†‡**	.5736†‡	.6340†‡
LegalBERT doc	Doc FirstP	.0682†‡	.3881†‡	.6187†‡	.7361†‡	.0061†‡	.0050†‡	.4833†‡	.4866†‡
	Doc MaxP	.0008†‡	.0302†‡	.2069†‡	.2534†‡	.0022†‡	.0050†‡	.4800†‡	.4833†‡
	PARM RRF	.1248†‡	.6086†	.8394†‡	.9114†‡	.0117†‡	.2277†‡	.5637†‡	.6265†‡
	PARM VRRF	.1256†‡	.6127†	**.8426†‡**	**.9128†‡**	.2284†‡	.4620†‡	**.5847†‡**	**.6402†‡**

5.2   RQ2: PARM VRRF vs Document-Level Retrieval

As we propose PARM VRRF for document-to-document retrieval, we investigate:
(**RQ2**) *How effective is PARM with VRRF for document-to-document retrieval?*
We evaluate and compare PARM and document-level retrieval for lexical and dense
retrieval methods on the two test collections (COLIEEDoc and CaseLaw) for document-
to-document retrieval in Table 3. For BM25 we find that PARM-based retrieval outperforms document-level retrieval at each recall stage, except for R@100 on CaseLaw.

For dense retrieval we evaluate DPR models with BERT trained solely on the
paragraph-level labels and with LegalBERT trained on the paragraph-level labels
(denoted with LegalBERT para) and with additional training on the document-level
labels (denoted with LegalBERT doc). For dense document-to-document retrieval
PARM consistently outperforms document-level retrieval for all performance metrics
for both test collections. Furthermore PARM aggregation with VRRF outperforms
PARM RRF in nearly all cases. Overall we find that LegalBERTdoc-based dense
retrieval with PARM VRRF achieves the highest recall at high ranks. When comparing
the nDCG@10 evaluation we find that PARM lowers the nDCG@10 score for BM25 as
well as for dense retrieval. Therefore we suggest that PARM is beneficial for first stage
retrieval, so that in the re-ranking stage the overall ranking can be improved.

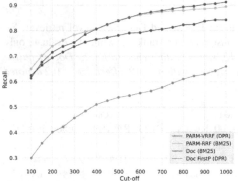

	COLIEEDoc		CaseLaw	
	BM25	DPR	BM25	DPR
Total				
relevant	900	900	720	720
PARM	892	896	578	545
Doc	751	662	419	199
Sets				
PARM ∩ Doc	750	661	417	196
PARM\Doc	142	235	161	349
Doc\PARM	1	1	2	3

Fig. 2. Recall at different cut-off values for PARM-VRRF (DPR) and PARM-RRF (BM25) and Document-level retrieval with BM25 and DPR for COLIEEDoc test.

Fig. 3. Number of relevant documents retrieved in comparison between PARM and Doc-level retrieval for COLIEEDoc and CaseLaw with BM25 or LegalBERT_doc-based DPR.

In Fig. 2 we show the recall at different cut-off values for PARM-VRRF with DPR
(based on LegalBERTdoc) and PARM-RRRF with BM25 compared to document-level
retrieval (Doc FirstP) of BM25/DPR. When comparing PARM to document-retrieval,
we can see a clear gap between the performance of document-level retrieval and PARM
for BM25 and for DPR. Furthermore we see that dense retrieval (PARM-VRRF DPR)
outperforms lexical retrieval (PARM-RRF BM25) at cut-off values above 500.

In order to analyze the differences between PARM and document-level retrieval further, we analyze in Fig. 3, how many relevant documents are retrieved with PARM or with document-level retrieval with lexical (BM25) or dense methods (DPR). Furthermore we investigate how many relevant documents are retrieved by both PARM and document-level retrieval (PARM ∩ Doc), and how many relevant documents are retrieved only with PARM and not with document-level retrieval (PARM\Doc) and vice versa (Doc\PARM). When comparing the performance of PARM and document-level retrieval, we find that PARM retrieves more relevant documents in total for both test collections. PARM retrieves 142–380 of the relevant documents that did not get retrieved with document-level retrieval (PARM\Doc), which are 15–52% of the total number of relevant documents. This analysis demonstrates that PARM largely retrieves many of relevant documents that are not retrieved with document-level retrieval. We conclude that PARM is not only beneficial for dense but also for lexical retrieval.

5.3 RQ3: Paragraph-Level vs Document-level Labelled Training

As labelled in-domain data for document-to-document retrieval tasks is limited, we ask: (**RQ3**) *How can we train dense passage retrieval models for PARM for document-to-document retrieval most effectively?* We compare the retrieval performance for BERT-based and LegalBERT-based dense retrieval models in Table 4, which are either trained solely on the paragraph-level labelled collection or additionally trained on the document-level labelled collection. The upper part of the table shows that for BERT the additional training data on document-level improves the retrieval performance for document-level retrieval, but harms the performance for PARM RRF and PARM VRRF. For LegalBERT the additional document-level training data highly improves the performance of document-retrieval. For PARM the recall is improved at higher cut-off values (@500, @1000) for a cut-off. Therefore we consider the training on document-level labelled data beneficial for dense retrieval based on LegalBERT. This reveals that it is not always better to have more, potentially noisy data, for BERT-based dense retrieval the training with fewer, but accurate paragraph-level labels is more beneficial for overall document-to-document retrieval with PARM.

5.4 Analysis of Paragraph Relations

With our proposed paragraph aggregation retrieval model for dense document-to-document retrieval we can analyze on which paragraphs the document-level relevance is based. To gain more insight in what the dense retrieval model learned to retrieve on the paragraph-level with PARM, we analyze which query paragraph retrieves which paragraphs from relevant documents with dense retrieval with PARM and compare it to lexical retrieval with PARM. In Fig. 4, a heatmap visualizes which query paragraph how often retrieves which paragraph from a relevant document with PARM BM25 or PARM DPR on the COLIEEDoc test set. As introduced in Sect. 4.1, the legal cases in COLIEEDoc contain an introduction, a summary and claims as paragraphs. For the introduction (I) and the summary (S) we see the paragraph relation for lexical and dense retrieval that both methods retrieve also more introductions and summaries from the relevant documents. We reason this is due to the special structure of the introduction and

Table 4. Paragraph- and document-level labelled training of DPR. Document-level labelled training improves performance at high ranks for LegalBERT.

Model	Retrieval	Train Labels	COLIEEDoc val				
			R@100	R@200	R@300	R@500	R@1K
DPR Retrieval							
BERT	Doc FirstP	para	.3000	.4018	.4566	.5371	.6598
	Doc FirstP	+ doc	.3800	.4641	.5160	.6054	.7211
	PARM RRF	para	.5765	.6879	.7455	.8153	.8897
	PARM RRF	+ doc	.5208	.6502	.7100	.7726	.8660
	PARM VRRF	para	.5786	.6868	.7505	.8132	.8909
	PARM VRRF	+ doc	.5581	.6696	.7298	.7970	.8768
LegalBERT	Doc FirstP	para	.2447	.3286	.3853	.4598	.5657
	Doc FirstP	+ doc	.3881	.4665	.5373	.6187	.7361
	PARM RRF	para	.6350	.7323	.7834	.8308	.8997
	PARM RRF	+ doc	.6086	.7164	.7561	.8394	.9114
	PARM VRRF	para	**.6396**	**.7325**	**.7864**	.8310	.9023
	PARM VRRF	+ doc	.6098	.7152	.7520	**.8396**	**.9128**

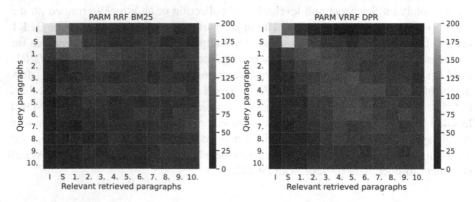

Fig. 4. Heatmap for PARM retrieval with BM25 or DPR visualizing which query paragraph how often retrieves which paragraph from a relevant document. I denotes the introduction, S the summary, 1.–10. denote the claims 1.–10. of COLIEEDoc test.

the summary which is distinct to the claims. For the query paragraphs 1.–10. we see that PARM DPR seems to focus on to the diagonal different to PARM BM25. This means for example that the first paragraph retrieves more first paragraphs from relevant documents than they retrieve other paragraphs. As the claim numbers are removed in the data pre-processing, this focus relies on the textual content of the claims. This paragraph relation suggests that there is a topical or hierarchical structure in the claims of legal cases, which is learned by DPR and exhibited with PARM. This structural component can not be exhibited with document-level retrieval.

6   Conclusion

In this paper we address the challenges of using dense passage retrieval (DPR) in first stage retrieval for document-to-document tasks with limited labelled data. We propose the paragraph aggregation retrieval model (PARM), which liberates dense passage retrieval models from their limited input length and which takes the paragraph-level relevance for document retrieval into account. We demonstrate on two test collections higher first stage recall for dense document-to-document retrieval with PARM than with document-level retrieval. We also show that dense retrieval with PARM outperforms lexical retrieval with BM25 in terms of recall at higher cut-off values. As part of PARM we propose the novel vector-based aggregation with reciprocal rank fusion weighting (VRFF), which combines the advantages of rank-based aggregation with RRF [10] and topical aggregation with dense embeddings. We demonstrate the highest retrieval effectiveness for PARM with VRRF aggregation compared to rank and vector-based aggregation baselines. Furthermore we investigate how to train dense retrieval models for dense document-to-document retrieval with PARM. We find the interesting result that training DPR models on more, but noisy document-level data does not always lead to overall higher retrieval performance compared to training on less, but more accurate paragraph-level labelled data. Finally, we analyze how PARM retrieves relevant paragraphs and find that the dense retrieval model learns a structural paragraph relation which it exhibits with PARM and therefore benefits the retrieval effectiveness.

References

1. Abolghasemi, A., Verberne, S., Azzopardi, L.: Improving BERT-based query-by-document retrieval with multi-task optimization. In: Hagen, M. et al. (Eds.) ECIR 2022. LNCS, vol. 13185, pp. xx–yy. Springer, Heidelberg (2022). https://doi.org/10.1007/978-3-030-99736-6_2

2. Ai, Q., O'Connor, B., Croft, W.B.: A neural passage model for ad-hoc document retrieval. In: Pasi, G., Piwowarski, B., Azzopardi, L., Hanbury, A. (eds.) ECIR 2018. LNCS, vol. 10772, pp. 537–543. Springer, Cham (2018). https://doi.org/10.1007/978-3-319-76941-7_41

3. Akkalyoncu Yilmaz, Z., Wang, S., Yang, W., Zhang, H., Lin, J.: Applying BERT to document retrieval with birch. In: Proceedings of the 2019 Conference on Empirical Methods in Natural Language Processing and the 9th International Joint Conference on Natural Language Processing (EMNLP-IJCNLP): System Demonstrations, Hong Kong, China, November 2019, pp. 19–24. Association for Computational Linguistics (2019). https://doi.org/10.18653/v1/D19-3004. https://aclanthology.org/D19-3004

4. Akkalyoncu Yilmaz, Z., Yang, W., Zhang, H., Lin, J.: Cross-domain modeling of sentence-level evidence for document retrieval. In: Proceedings of the 2019 Conference on Empirical Methods in Natural Language Processing and the 9th International Joint Conference on Natural Language Processing (EMNLP-IJCNLP), Hong Kong, China, November 2019, pp. 3490–3496. Association for Computational Linguistics (2019). https://doi.org/10.18653/v1/D19-1352. https://aclanthology.org/D19-1352

5. Akkalyoncu Yilmaz, Z., Yang, W., Zhang, H., Lin, J.: Cross-domain modeling of sentence-level evidence for document retrieval. In: Proceedings of the 2019 Conference on Empirical Methods in Natural Language Processing and the 9th International Joint Conference on Natural Language Processing (EMNLP-IJCNLP), Hong Kong, China, November 2019, pp. 3490–3496. Association for Computational Linguistics (2019). https://doi.org/10.18653/v1/D19-1352. https://www.aclweb.org/anthology/D19-1352

6. Bajaj, P., et al.: MS MARCO: a human generated MAchine Reading COmprehension dataset. In: Proceedings of the NIPS (2016)

7. Bendersky, M., Kurland, O.: Utilizing passage-based language models for document retrieval. In: Macdonald, C., Ounis, I., Plachouras, V., Ruthven, I., White, R.W. (eds.) ECIR 2008. LNCS, vol. 4956, pp. 162–174. Springer, Heidelberg (2008). https://doi.org/10.1007/978-3-540-78646-7_17

8. Chalkidis, I., Fergadiotis, M., Malakasiotis, P., Aletras, N., Androutsopoulos, I.: LEGAL-BERT: the Muppets straight out of law school. In: Findings of the Association for Computational Linguistics, EMNLP 2020, Online, November 2020, pp. 2898–2904. Association for Computational Linguistics (2020). https://doi.org/10.18653/v1/2020.findings-emnlp.261. https://www.aclweb.org/anthology/2020.findings-emnlp.261

9. Cohan, A., Feldman, S., Beltagy, I., Downey, D., Weld, D.: SPECTER: document-level representation learning using citation-informed transformers. In: Proceedings of the 58th Annual Meeting of the Association for Computational Linguistics, Online, July 2020, pp. 2270–2282. Association for Computational Linguistics (2020). https://doi.org/10.18653/v1/2020.acl-main.207. https://aclanthology.org/2020.acl-main.207

10. Cormack, G.V., Clarke, C.L.A., Buettcher, S.: Reciprocal rank fusion outperforms condorcet and individual rank learning methods. In: Proceedings of the 32nd International ACM SIGIR Conference on Research and Development in Information Retrieval, SIGIR 2009, pp. 758–759. Association for Computing Machinery, New York (2009). https://doi.org/10.1145/1571941.1572114

11. Dai, Z., Callan, J.: Deeper text understanding for IR with contextual neural language modeling. In: Proceedings of the 42nd International ACM SIGIR Conference on Research and Development in Information Retrieval, SIGIR 2019, pp. 985–988. Association for Computing Machinery, New York (2019). https://doi.org/10.1145/3331184.3331303

12. Devlin, J., Chang, M.W., Lee, K., Toutanova, K.: BERT: pre-training of deep bidirectional transformers for language understanding. In: Proceedings of the 2019 Conference of the North American Chapter of the Association for Computational Linguistics: Human Language Technologies, Volume 1 (Long and Short Papers), Minneapolis, Minnesota, June 2019, pp. 4171–4186. Association for Computational Linguistics (2019). https://doi.org/10.18653/v1/N19-1423. https://www.aclweb.org/anthology/N19-1423

13. Gao, J., et al.: FIRE 2019@AILA: legal retrieval based on information retrieval model. In: Proceedings of the Forum for Information Retrieval Evaluation, FIRE 2019 (2019)

14. Gao, L., Dai, Z., Chen, T., Fan, Z., Durme, B.V., Callan, J.: Complementing lexical retrieval with semantic residual embedding. arXiv arXiv:2004.13969 (April 2020)

15. Hedin, B., Zaresefat, S., Baron, J., Oard, D.: Overview of the TREC 2009 legal track. In: Proceedings of the 18th Text REtrieval Conference, TREC 2009 (January 2009)

16. García Seco de Herrera, A., Schaer, R., Markonis, D., Müller, H.: Comparing fusion techniques for the ImageCLEF 2013 medical case retrieval task. Comput. Med. Imaging Graph. **39**, 46–54 (2014). http://publications.hevs.ch/index.php/attachments/single/676

17. Hofstätter, S., Althammer, S., Schröder, M., Sertkan, M., Hanbury, A.: Improving efficient neural ranking models with cross-architecture knowledge distillation (2021)

18. Hofstätter, S., Lin, S.C., Yang, J.H., Lin, J., Hanbury, A.: Efficiently teaching an effective dense retriever with balanced topic aware sampling (2021)

19. Karpukhin, V., et al.: Dense passage retrieval for open-domain question answering. In: Proceedings of the 2020 Conference on Empirical Methods in Natural Language Processing (EMNLP), pp. 6769–6781. Association for Computational Linguistics, Online, November 2020 (2020). https://doi.org/10.18653/v1/2020.emnlp-main.550. https://www.aclweb.org/anthology/2020.emnlp-main.550

20. Lee, J.H.: Analyses of multiple evidence combination. SIGIR Forum **31**(SI), 267–276 (1997). https://doi.org/10.1145/278459.258587

21. Li, C., Yates, A., MacAvaney, S., He, B., Sun, Y.: Parade: passage representation aggregation for document reranking. arXiv preprint arXiv:2008.09093 (2020)

22. Liu, X., Croft, W.B.: Passage retrieval based on language models. In: Proceedings of the 11th International Conference on Information and Knowledge Management, CIKM 2002, pp. 375–382. Association for Computing Machinery, New York (2002). https://doi.org/10.1145/584792.584854

23. Locke, D., Zuccon, G.: A test collection for evaluating legal case law search. In: 41st International ACM SIGIR Conference on Research and Development in Information Retrieval, SIGIR 2018, pp. 1261–1264. Association for Computing Machinery, Inc. (June 2018). https://doi.org/10.1145/3209978.3210161

24. Locke, D., Zuccon, G., Scells, H.: Automatic query generation from legal texts for case law retrieval. In: Sung, W.-K., et al. (eds.) Information Retrieval Technology, pp. 181–193. Springer, Cham (2017). https://doi.org/10.1007/978-3-319-70145-5_14

25. Luan, Y., Eisenstein, J., Toutanova, K., Collins, M.: Sparse, dense, and attentional representations for text retrieval. arXiv preprint arXiv:2005.00181 (2020)

26. Montague, M., Aslam, J.A.: Condorcet fusion for improved retrieval. In: Proceedings of the 11th International Conference on Information and Knowledge Management, CIKM 2002, pp. 538–548. Association for Computing Machinery, New York (2002). https://doi.org/10.1145/584792.584881

27. Mourão, A., Martins, F., Magalhães, J.: Multimodal medical information retrieval with unsupervised rank fusion. Comput. Med. Imaging Graph. **39**, 35–45 (2015). Medical visual information analysis and retrieval. https://doi.org/10.1016/j.compmedimag.2014.05.006. https://www.sciencedirect.com/science/article/pii/S0895611114000664

28. Piroi, F., Tait, J.: CLEF-IP 2010: retrieval experiments in the intellectual property domain. In: Proceedings of CLEF 2010 (2010)

29. Rabelo, J., Kim, M.-Y., Goebel, R., Yoshioka, M., Kano, Y., Satoh, K.: A summary of the COLIEE 2019 competition. In: Sakamoto, M., Okazaki, N., Mineshima, K., Satoh, K. (eds.) JSAI-isAI 2019. LNCS (LNAI), vol. 12331, pp. 34–49. Springer, Cham (2020). https://doi.org/10.1007/978-3-030-58790-1_3

30. Risch, J., Alder, N., Hewel, C., Krestel, R.: PatentMatch: a dataset for matching patent claims & prior art (2020)

31. Robertson, S., Zaragoza, H.: The probabilistic relevance framework: BM25 and beyond. Found. Trends Inf. Retr. **3**(4), 333–389 (2009). https://doi.org/10.1561/1500000019

32. Shao, Y., Liu, B., Mao, J., Liu, Y., Zhang, M., Ma, S.: THUIR@COLIEE-2020: leveraging semantic understanding and exact matching for legal case retrieval and entailment. CoRR abs/2012.13102 (2020). https://arxiv.org/abs/2012.13102

33. Shao, Y., et al.: BERT-PLI: modeling paragraph-level interactions for legal case retrieval. In: Bessiere, C. (ed.) Proceedings of the 29th International Joint Conference on Artificial Intelligence, IJCAI-20, pp. 3501–3507. International Joint Conferences on Artificial Intelligence Organization (July 2020). Main track. https://doi.org/10.24963/ijcai.2020/484

34. Shaw, J.A., Fox, E.A.: Combination of multiple searches. In: The 2nd Text Retrieval Conference, TREC-2, pp. 243–252 (1994)

35. Van Opijnen, M., Santos, C.: On the concept of relevance in legal information retrieval. Artif. Intell. Law **25**(1), 65–87 (2017). https://doi.org/10.1007/s10506-017-9195-8
36. Wu, S.: Ranking-based fusion. In: Data Fusion in Information Retrieval. Adaptation, Learning, and Optimization, vol. 13, pp 135–147. Springer, Heidelberg (2012). https://doi.org/10.1007/978-3-642-28866-1_7
37. Wu, Z., et al.: Leveraging passage-level cumulative gain for document ranking. In: Proceedings of the Web Conference 2020, WWW 2020, pp. 2421–2431. Association for Computing Machinery, New York (2020). https://doi.org/10.1145/3366423.3380305
38. Wu, Z., Mao, J., Liu, Y., Zhang, M., Ma, S.: Investigating passage-level relevance and its role in document-level relevance judgment. In: Proceedings of the 42nd International ACM SIGIR Conference on Research and Development in Information Retrieval, SIGIR 2019, pp. 605–614. Association for Computing Machinery, New York (2019). https://doi.org/10.1145/3331184.3331233
39. Xiong, L., et al.: Approximate nearest neighbor negative contrastive learning for dense text retrieval. In: International Conference on Learning Representations (2021). https://openreview.net/forum?id=zeFrfgyZln
40. Yang, L., Zhang, M., Li, C., Bendersky, M., Najork, M.: Beyond 512 tokens: Siamese multi-depth transformer-based hierarchical encoder for long-form document matching. In: Proceedings of the 29th ACM International Conference on Information & Knowledge Management, CIKM 2020, pp. 1725–1734. Association for Computing Machinery, New York (2020). https://doi.org/10.1145/3340531.3411908
41. Zhang, X., Yates, A., Lin, J.: Comparing score aggregation approaches for document retrieval with pretrained transformers. In: Hiemstra, D., Moens, M.-F., Mothe, J., Perego, R., Potthast, M., Sebastiani, F. (eds.) ECIR 2021. LNCS, vol. 12657, pp. 150–163. Springer, Cham (2021). https://doi.org/10.1007/978-3-030-72240-1_11
42. Zhang, X., Yates, A., Lin, J.J.: Comparing score aggregation approaches for document retrieval with pretrained transformers. In: ECIR (2021)

Recommendation of Compatible Outfits Conditioned on Style

Debopriyo Banerjee[1,2](✉) (iD), Lucky Dhakad[2] (iD), Harsh Maheshwari[2] (iD),
Muthusamy Chelliah[2], Niloy Ganguly[1,3] (iD), and Arnab Bhattacharya[2] (iD)

[1] IIT Kharagpur, Kharagpur, India
debopriyo@iitkgp.ac.in, niloy@cse.iitkgp.ac.in
[2] Flipkart Internet Pvt. Ltd., Bangalore, India
[3] Leibniz University Hannover, Hannover, Germany

Abstract. Recommendation in the fashion domain has seen a recent surge in research in various areas, for example, shop-the-look, context-aware outfit creation, personalizing outfit creation, etc. The majority of state of the art approaches in the domain of outfit recommendation pursue to improve compatibility among items so as to produce high quality outfits. Some recent works have realized that *style* is an important factor in fashion and have incorporated it in compatibility learning and outfit generation. These methods often depend on the availability of fine-grained product categories or the presence of rich item attributes (e.g., long-skirt, mini-skirt, etc.). In this work, we aim to generate outfits conditional on styles or themes as one would dress in real life, operating under the practical assumption that each item is mapped to a high level category as driven by the taxonomy of an online portal, like outdoor, formal etc. and an image. We use a novel style encoder network that renders outfit styles in a smooth latent space. We present an extensive analysis of different aspects of our method and demonstrate its superiority over existing state of the art baselines through rigorous experiments.

Keywords: complete the look · neural networks · outfit compatibility · style

1 Introduction

Recommendation of outfits having compatible fashion items is a well studied research topic in the fashion domain [1,3,19,22,23,32,38]. Recent research in this regard explores graph neural networks (GNN) to connect users, items and outfits [6,21,26,37,39,42] based on historical purchases as well as personalization [4,5,11,16,25,27,28] and explainability [7,8,24,40]. An apparent shortcoming of the current research on compatibility learning is the complete disregard for the

Supported by Flipkart Internet Pvt. Ltd.
D. Banerjee, L. Dhakad and H. Maheshwari—These authors contributed equally to this work.

explicit style associated with an outfit. However in real life, a person, say a user on an e-commerce platform, would typically have an explicit style in mind while choosing items for an outfit. The main objective of this paper is to learn compatibility between items given a specific style which in turn helps to generate style-specific outfits.

We illustrate the importance of style-guided outfit generation through an example figure. Three sets of outfits are shown in Fig. 1 with a white top-wear, an item that a user likes but is doubtful about making the final purchase (the reader is requested to ignore the values at the bottom of the figure for the time being). The platform may have the capability to showcase or to provide the user an option of generating outfits specific to various styles (this example showcases *Athleisure*, *Formal* and *Casual*). Given this setup, a style-guided algorithm has two advantages: (a) it can generate compatible outfits from different styles, hence, providing the choice to the user, and (b) it will not generate an outfit which may be otherwise compatible but not in accordance with the desired style. The concept of jointly modelling for explicit style and compatibility is lacking in the area of fashion recommendation and current research have mostly treated them in separate silos. Having said this, one should be mindful of the fact that a style-independent compatibility algorithm followed by a style classification method, say Style2Vec [17], can allocate outfits to their relevant styles post the generation step. Thus in principle it is possible to combine existing work to generate the outfits in Fig. 1. It is however easy to see that such a technique is not efficient, since a large set of outfits need to be generated of which only a subset will be relevant to a particular style.

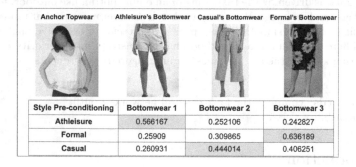

Fig. 1. Given a top-wear liked by a user, a style-guided method is able to create outfits conditional on various styles (*athleisure*, *formal* and *casual*) while a style-independent compatibility model will typically generate outfits from dominant style. The values indicate the style-conditional compatibility scores for each item. Note that for a given style, the bottom-wear corresponding to that style gets the highest score.

In recent times there have been some attempts at connecting style and outfit recommendation. Kuhn et al. [14] does not consider the presence of explicit styles and rather learn compatibility while inferring the presence of latent style associated with each item. Jeon et al. [12] use extracted fashion attributes of full-body outfit images for modelling style classification, ignoring compatibility learning in the process. Learning outfit level theme or style from item descriptions, done by Li et al. [20] is a weak approach and fails when the descriptions do not exhaustively cover different styles. Singhal et al. [33] models style between item pairs using an autoencoder, thus treating style as an implicit idea. A common deficiency in all of these works is the ability to generate style guided outfits. Theme Matters [15], authored by Lai et al. is an archived work which comes closest to our model. It proposes a supervised approach that applies theme-aware attention to item pairs having fine-grained category tags (e.g., long-skirt, mini-skirt, etc.). The main handicap of their approach is that the size of the model increases exponentially with the number of fine-grained categories which was validated by our experiments.

We propose a **Style-Attention-based Compatible Outfit Recommendation** (SATCORec) framework that uses high-level categories like top-wear, bottom-wear etc. (general e-commerce taxonomy) and explicit outfit-level style information (*formal, casual, sporty* etc.) to learn compatibility among items in an outfit. It consists of two components, namely a Style-Compatibility-Attention Network (SCA Net) [23] and a novel Style Encoder Network (SE-Net). SE-Net considers an outfit to be a *set* of items and makes use of the Set Transformer [18] architecture to model a style specific distribution for each outfit. We believe that we are the first to adopt the set transformer, which is state-of-the-art technique to model data points that have the properties of a set, in a framework to project an outfit into a latent style space. Several variations of extracting a style representation from the learnt distribution have been investigated. We make use of this representation to estimate style-specific subspace attention within SCA Net which helps to learn compatibility conditional on style. Finally, we use the beam search approach [2] to generate outfits based on a parent item, a template and a style.

We have created an in-house dataset of size approx. 100k corresponding to women's western wear outfits, taking items from an e-commerce portal. Various experiments have been performed on this data, comparing compatibility and style-specific metrics between baseline methods and SATCORec. Our method has been found to excel in compatibility learning, even when outfits are generated conditional on style. Most importantly, SATCORec is seen to outperform all the baselines in style metrics by a large margin.

2 Methodology

SATCORec is a deep learning model, developed to learn the compatibility between lifestyle items present within an outfit, contingent on the style to which the outfit belongs. The model first infers the style of the outfit which is subsequently used to learn compatibility between items within it.

We start with proposing a novel *Style Encoder Network* (SE-Net) which learns a parametric probability distribution representing outfit style using the set transformer [18], followed by a style classification task further downstream. We extend the compatibility framework of Lin et al. [23] to allocate differential importance to features extracted from the image of an item not just based on category information but also on the outfit style, thus complementing SE-Net. We have further modified the compatibility loss in [23] to incorporate style. The

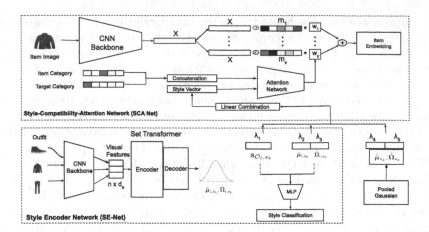

Fig. 2. Architecture of SATCORec. The lower module combines the SE-Net and Style classifier and is trained separately. Item images of an outfit are fed to a CNN to extract visual features which are subsequently passed onto a Set transformer to output a Gaussian distribution. The style classifier is trained using either a random sample or the parameters of the Gaussian. A linear combination of these two along with the parameters of style-specific pooled Gaussian is passed as a feature in the SCA Net module which learns compatibility via attention.

entire architecture is shown in Fig. 2. Details of SE-Net and the Style Classifier are provided in Sects. 2.1 and 2.2 respectively. SCA Net and the modified compatibility loss are explained in Sect. 2.3. We explain the generation of outfits based on individual or mixture of style in Sect. 2.4.

To introduce the notations, let us assume that m explicit styles, say $\mathcal{S} \equiv \{s_1, s_2, \ldots, s_m\}$, are defined in an online portal recommending complete outfits for a user. For an outfit $\mathcal{O}_i$ belonging to style s_k (say $\mathcal{O}_i | s_k$), we assume one of the items within the outfit to be the *anchor item* and the rest is defined as *query set*. We call this <anchor item, query set> as a *positive example* of compatibility. A *negative instance* is one where the anchor item is changed so that it no longer stays compatible with the query set.

2.1 Style Encoder Network

The process of encoding style of an outfit starts with acknowledging the fact that denoting an outfit as an ordered sequence of items, as is done in some recent work [9,30], can be seen to be unrealistic. In this paper, we portray an outfit as a set of items which serves two important properties, (i) items within an outfit can be termed as permutation invariant, and (ii) an outfit is allowed to be of varying length. This characterization makes the *set transformer* approach [18] an appropriate candidate for our style encoder job. This approach consists of an encoder and a decoder, both of which rely on attention mechanisms to produce a representative output vector.

The idea of representing an individual outfit style by a specific embedding is apt for compatibility training but found to be lacking in the generation context. Since outfit generation is hinged on a single parent item, a pre-defined template and a style, we may not be able to pass any reference outfit to the style encoder. To circumvent this problem, we make the assumption that each $\mathcal{O}_i|s_k$, is generated from some parametric continuous probability distribution thus representing a latent style space. In this paper, we assume that this distribution is Gaussian, although we acknowledge that it can be any other continuous distribution. The parameters of this Gaussian distribution is estimated by the set transformer. In this framework, as can be seen in Fig. 2, the images of an outfit are passed through a pre-trained ResNet18 [10] and the corresponding visual feature vectors ($\in \mathbb{R}^{d_s}$) are fed into the set transformer to provide estimates for the mean vector and co-variance matrix (we assume this to be diagonal). To summarise, the set transformer produces an unique Gaussian distribution for each outfit $\mathcal{O}_i|s_k$,

$$\mathcal{O}_i|s_k \sim \mathcal{N}(\mu_{i,s_k}, \Omega_{i,s_k}), \text{ where } \Omega_{i,s_k} = \text{diag}(\sigma^2_{il,s_k}), \quad l = 1,\ldots,d_s \text{ and } \mu \in \mathbb{R}^{d_s}.$$

Here, we additionally impose the restriction that the inferred Gaussian distributions are close to the unit Normal $\mathcal{N}(0, \mathbb{1})$, so that the learnt style space is smooth across the various styles. We achieve this via the KL divergence loss defined in Eq. 1.

$$\mathcal{L}_{Style} = \text{KL}(\mathcal{N}(\hat{\mu}_{i,s_k}, \hat{\Omega}_{i,s_k}) \,\|\, \mathcal{N}(0, \mathbb{1})) \tag{1}$$

Figure 3 demonstrates a t-SNE visualisation of random samples drawn from outfit specific Gaussians for 4 different styles. A common and smooth representation space is formed after introducing the KL-loss even though clusters are maintained. A smooth space is necessary particularly in the generation of outfits with style mixing, as we will see later.

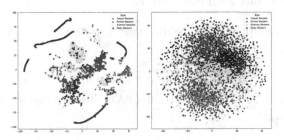

Fig. 3. t-SNE plots of the sample vectors ($s_{\mathcal{O}_i,s_k}$) for 4 styles (Casual, Formal, Summer, Party). The plot on the left is when these vectors are generated without the KL-divergence loss. Existence of a smooth yet identifiable style latent space is evident in the plot on the right when we introduce the loss. Best viewed in colour. (Color figure online)

The output emanating from the set transformer is passed on for a style classifier job. Depending on the specific variation, an outfit $\mathcal{O}_i|s_k$, we pass either the parameters of the Gaussian ($\theta_{i,s_k} \equiv [\hat{\mu}_{i,s_k}, \hat{\Omega}_{i,s_k}]$) or a random sample from the Gaussian $s_{\mathcal{O}_i,s_k} \sim \mathcal{N}(\hat{\mu}_{i,s_k}, \hat{\Omega}_{i,s_k})$ to the style classifier. We have elaborated on the exact process in Sect. 2.2.

2.2   Style classifier

The SE-Net output vector is passed as a feature to an MLP used to classify the style of the outfit. This supervision ensures that SE-Net captures specific and correct information about the outfit style. The style classification module solves an m-class classification problem using an MLP with N layers. The classification loss is thus,

$$\mathcal{L}_{\text{classif}} = - \sum_{i=1}^{m} y_{s_k} log(\hat{p}(O_i \mid s_k)) \tag{2}$$

where $y_{s_k} = 1$, if outfit O_i has style s_k and $\hat{p}(O_i \mid s_k) = \text{MLP}(\mathbf{s}_{\mathcal{O}_i,s_k}$ or $\theta_{i,s_k})$. The SE-Net and style classifier are trained jointly as a separate module. Post training, we extract a vector $(\mathbf{r}_{\mathcal{O}_i,s_k})$ from this module as style representation of the outfit $\mathcal{O}_i|s_k$ to be passed as a feature to SCA Net. Further, a *global style* representation for a style s_k is given by a pooled Gaussian distribution, aggregating over the parameters of all outfits belonging to that style: $\hat{\boldsymbol{\mu}}_{s_k} = \frac{1}{n_{s_k}} \sum_{i=1}^{n_{s_k}} \hat{\mu}_{i,s_k}$ and $\hat{\boldsymbol{\Omega}}_{s_k} = \text{diag}(\hat{\sigma}_l^2)$ where $\hat{\sigma}_l^2 = \frac{1}{n_{s_k}^2} \sum_{i=1}^{n_{s_k}} \hat{\sigma}_{il}^2$. These global distribution parameters will be used again in the outfit generation step. Equation (3) shows a generic form of style representation vector,

$$\mathbf{r}_{\mathcal{O}_i,s_k} \equiv \left[\lambda_1\,\mathbf{s}_{\mathcal{O}_i,s_k} + \lambda_2\,\hat{\mu}_{i,s_k} + \lambda_4\,\hat{\boldsymbol{\mu}}_{s_k}, \lambda_3\hat{\boldsymbol{\Omega}}_{i,s_k}, + \lambda_5\hat{\boldsymbol{\Omega}}_{s_k} \right]. \tag{3}$$

SATCORec variations, defined in Table 1, are created by setting values for each λ_j. Also note that, we pass $\mathbf{s}_{\mathcal{O}_i,s_k}$ to the style classifier for *SATCORec-r*, *SATCORec-(p_m+g_m)* and *SATCORec-(r+g_m)* and θ_{i,s_k} for the rest. It is possible to set λ as unknown and learn it.

Table 1. Variations of SATCORec that have been experimented with.

	λ_1	λ_2	λ_3	λ_4	λ_5		λ_1	λ_2	λ_3	λ_4	λ_5
SATCORec-r	1	0	0	0	0	SATCORec-p	0	1	1	0	0
SATCORec-(p_m+g_m)	0	λ	0	1	0	SATCORec-(p+g)	0	λ	λ	1	1
SATCORec-(r+g_m)	λ	0	0	1	0						

2.3   SCA Net

We have extended the CSA-Net framework developed by Lin et al. in [23] to incorporate the concept of style while learning item-item compatibility. In [23], the image of an anchor item (I^a) within an outfit is passed through a ResNet18, which acts as the CNN backbone. The embedding output vector $(\mathbf{x})$ of size 64 is multiplied by k learnt masks $(\mathbf{m}_1, \ldots, \mathbf{m}_\tau)$ that help to learn the subspaces. The anchor item category (c^a) and a query set item (referred to as *target*) category (c^t) information are consumed as 1-hot encoded vectors to estimate a set of subspace attention weights $(\omega_1, \ldots, \omega_\tau)$. A weighted average of the masked embeddings results in the final embedding of the anchor item.

We simply extend the CSA-Net algorithm by providing the style representation $(\mathbf{r}_{\mathcal{O}_i, s_k})$ from SE-Net as an additional input in the estimation of attention weights. Thus, we define the final embedding as,

$$f^{s_k}_{\mathcal{O}_i, a} = \psi(I^a, c^a, c^t, \mathbf{r}_{\mathcal{O}_i, s_k}) = \sum_{j=1}^{\tau} (\mathbf{x} \odot \mathbf{m}_j) \times \omega_{j, \mathbf{r}_{\mathcal{O}_i, s_k}}.$$

Here, $\psi(\cdot)$ represents the SCA network.

The SCA net uses the triplet loss for learning compatibility, similar to some current methods [35,36]. We represent the average distance between a positive item and remaining items in the outfit as $D^{s_k}_p$, same as CSA-Net. The multiple distances corresponding to the negatives are aggregated as $D^{s_k}_N$. The overall compatibility loss conditional on style is thus defined as,

$$\mathcal{L}_{compat} = \max(0, D^{s_k}_p - D^{s_k}_N + m), \tag{4}$$

We introduce one more loss function to account for penalisation when the wrong style is specified for an outfit. Given $\mathcal{O}_i | s_k$, we pass the style representation vector corresponding to a different style s_q, and compute the same distance metrics as above, and use them in the following loss function:

$$\mathcal{L}_{stylecompat} = \max(0, D^{s_k}_p - D^{s_q}_p + m). \tag{5}$$

The overall loss is defined as the weighted sum of these four individual losses:

$$\mathcal{L}_{overall} = \sum_p \alpha_p \mathcal{L}_p, \quad p \in \{\text{KL, classification, compatibility, style-compatibiliy}\}$$

2.4 Outfit Generation

A globally optimal outfit generation task is non-trivial since it is infeasible to look into all possible combinations. An approximate solution based on the well known *beam search method* [43] is provided in this case. Note that to create an outfit for a user based on a chosen parent item, a given template and a specific style, we need a style representation vector to rank compatible items. If there is a reference outfit present, then this job is trivial. In the alternative case, we assume the pooled parameters to be representative of style for all the variations within SE-Net. To generate an outfit based on mixing of styles, we simply pass a linear combination of style representation vectors $(\alpha \mathbf{r}_{\mathcal{O}_i, s_k} + \beta \mathbf{r}_{\mathcal{O}_i, s_l})$ and rank compatible items.

3 Experimental Evaluation

In this section, we elaborate the dataset, metrics, baselines, implementation details and the different results testing the compatibility as well as style preservation power of the algorithms.

Dataset Creation and Metrics: We have annotated ∼100K outfits in two stages. At **first**, we worked with fashion experts to get approximately 5000 outfits curated with 8 style annotations, namely [*party, outdoor, summer, formal, athleisure, winter, causal, celeb*]. Each annotated outfit consists of the images of individual items and and its style. There 6 high level item categories, [top-wear, bottom-wear, foot-wear,accessory, clothing-accessory, wholebody]. In the **second** stage, we augmented the outfit set using a simple attribute based similarity algorithm, where we used attributes like brand, colour, pattern, sleeve, etc. to get top-k similar products for an item in an outfit. Given an outfit, we removed one item from the original outfit and gave approx. top-10 similar candidates as options for replacement to human taggers for verification of compatibility and style of the new outfit. We repeated this for all item in an outfit and for all outfits in the outfit set. This operation expanded the data to ∼100K outfits, which are then divided into train, test and validation splits in 70:20:10 ratio. The overall frequency for each style type is given in Table 2.

Fill-in-the-blank (FITB) [36] and Compatibility AU-ROC are well known metrics used to evaluate an outfit compatibility model [9,29]. Both these approaches involve creating negative items corresponding to each item of an outfit. To test performance at various levels of difficulty, we generate two types of negative items, *soft negatives* where negative sampling is done from existing categories; and *hard negatives* where we sample negatives from more fine-grained categories such as tops, t-shirts, heels etc. For each outfit, 5 replications for negative sampling are done and the mean metric values are reported. Note that the fine-grained category information is not used for training.

Table 2. Distribution of curated outfits across different styles

	Party	Outdoor	Summer	Formal	Athleisure	Winter	Casual	Celebrity	Total
# of Train Outfits	8183	6280	7061	5136	16232	16028	5194	5424	69538
# of Valid Outfits	1174	1001	1204	840	1981	2135	791	808	9934
# of Test Outfits	3018	1937	2551	1648	2506	4695	2034	1480	19869

Implementation Details: We used ResNet18 as the CNN backbone to extract visual features in both the modules of SATCORec. We do not train the entire ResNet18 but instead only the last convolutional block and an additional fully connected layer. Embeddings are of size 64 as is conventional in other state-of-the-art compatibility learning methods [23,36].

Inside SE-Net, we use the SAB Set Transformer [18] with hidden dimension $d_z = 32$ and 2 heads. We use 2 fully connected MLP layers for classification. An Adam optimizer [13] with mini batches of 128 outfits and a learning rate of 5×10^{-5} is used. Note that, we have trained and frozen the SE-Net module separately. We used the Adam optimizer again to train SCA Net with a mini-batch size of 32 triplets, learning rate of 1×10^{-5} and 5 subspaces. The *Attention network* first transforms the concatenated one-hot-encoded category and the style representations to 32 dimensions each using a single fully connected layer and then concats the two to pass it to 2 fully connected layers which output the

5 subspace attention weights. The margin within the triplet loss was set to 0.3 and the weights for $\mathcal{L}_{compat}$, $\mathcal{L}_{stylecompat}$ and $\mathcal{L}_{Style}$ were set to 1, 0.5 and 0.05 respectively.

Baselines: We compare the performance of SATCORec against that of state-of-the-art techniques on the basis of the multiple metrics to demonstrate its efficacy in style conditional outfit generation and compatibility learning. Note that we use the same CNN backbone and embedding size for all the baselines. Additionally, the same 6 categories have been used for all the methods, even for those requiring fine-grained category information. The following are used as baselines (a). **CSA-Net** [23], (b). **Type Aware** [36], (c). **TransNFCM** [41], (d). **Theme Matters** [15], (e). **BPR-DAE** [34]. For each of the methods we follow the same architecture parameters which the paper specifies. Except **Type aware**, whose code was available, we have implemented all of the baselines from scratch. For **Theme Matters** we have first taken the type aware code and built upon it as is defined in the paper. In **BPR-DAE**, the method is specified for only for 2 categories, and we extend it for outfits with multiple items.

Table 3. Comparison of compatibility learning for the baselines and SATCORec variations. We compute FITB and compatibility AU-ROC with hard and soft negatives separately. The style entropy for each methods are also tabulated. Using parameters or random sample from outfit style specific Gaussian is clearly the leader with respect to compatibility measures.

Method	FITB		Compat. AU-ROC		Entropy
	HN	**SN**	**HN**	**SN**	
TypeAware	30.7 ± 0.17	34.85 ± 0.25	52.62 ± 0.06	55.51 ± 0.21	0.49
BPR-DAE	31.16 ± 0.15	31.21 ± 0.12	55.83 ± 0.09	55.76 ± 0.08	0.43
TransNFCM	31.53 ± 0.17	36.47 ± 0.33	51.84 ± 0.07	57.78 ± 0.08	0.50
Theme Matters	38.53 ± 0.17	63.2 ± 0.21	85.4 ± 0.15	93.85 ± 0.1	0.61
CSA-Net	53.14 ± 0.17	67.05 ± 0.25	94.42 ± 0.03	96.3 ± 0.03	0.48
SATCORec-r	$\mathbf{53.32 \pm 0.18}$	66.63 ± 0.15	94.47 ± 0.02	95.99 ± 0.04	**1.09**
SATCORec-p	52.06 ± 0.10	$\mathbf{67.31 \pm 0.14}$	$\mathbf{94.78 \pm 0.02}$	$\mathbf{96.47 \pm 0.02}$	0.97
SATCORec-(p+g)	46.56 ± 0.05	61.03 ± 0.17	88.41 ± 0.02	90.10 ± 0.02	0.78
SATCORec-(r+g_m)	47.61 ± 0.12	60.70 ± 0.06	88.88 ± 0.06	91.34 ± 0.02	0.12
SATCORec-(p_m+g_m)	49.73 ± 0.05	63.02 ± 0.11	90.96 ± 0.05	92.25 ± 0.02	0.63

3.1   Compatibility Experiment

FITB and compatibility AU-ROC are computed separately on the hard and soft negative datasets for variations of SATCORec and the baselines and presented in Table 3. A preliminary sweep of the results clearly differentiates the performance of Theme Matters, CSA-Net and SATCORec variations from the rest. CSA-Net is based on subspace based attention mechanism, which is the state-of-the-art in learning outfit item compatibility, and SATCORec makes use of the same framework. It is surprising that Theme Matters performs better than TypeAware since both have the same compatibility learning framework. This performance bump is caused due to these methods incorporating complete outfit loss in their learning [23].

SATCORec-p is the best performing model in the group, winning in 3 out of 4 cohorts. We think that the outfit-level Gaussian parameters capture sufficient information about the parent style of the outfit as well as variations within. The random sampling of the space can also capture the basic information of a style category, resulting in the healthy performance of SATCORec-r. The other variations do not perform well, probably because of ignoring individual or overall uncertainty.

3.2 Style Experiments

Given that our methods show better performance than others in compatibility learning, we now compare their performance vis-a-vis style. We look at two specific style comparison metrics and discuss a characteristic that our method has, but is absent in style-independent methods. Statistical comparisons for our metrics and further qualitative results will be added over time in this link: https:// harshm121.github.io/project_pages/satco_rec.html.

Style Entropy: A user would get maximum utility if her top-wear can be part of outfits belonging to a large number of style categories, i.e. the portal is able to recommend from a wide range of styles. Say given an anchor item, we want to recommend a total of n outfits from k styles. SATCORec, using the style-handle, can produce a ranked lists of outfits conditioned on each of the k styles. We choose the top $\lfloor n/k \rfloor$ or $\lceil n/k \rceil$ outfits from each style specific list. Style independent methods will get its top-n outfits as per the general compatibility rank, thus oblivious to their reference styles. We use the entropy measure on style to compare the final lists. A higher entropy would mean that the compatibility framework is not restrictive to a single or small number of styles. For this, we select the list of all those outfits which have the same anchor item, but belong to different styles. From this list, we pick those instances where SATCORec is able to correctly predict the items of an outfit given a style. We then choose the top outfit from each each style, thus forcing $n = k = 6$, and present the result in Table 3, column *Entropy*. Again, SATCORec-r (slightly better) and SATCORec-p outperform all other methods, implying that they are able to recommend outfits corresponding to most of the styles feasible for the anchor item. On manual inspection we also find that style-independent methods are biased towards the most prevalent style in the training data set. Henceforth, we will consider only the top performing variations, SATCORec-r and SATCORec-p.

Style-Specific Selection Accuracy and Ranking: SACTORec-r is also seen to be superior in some other metrics we compute like MRR, Avg rank etc. Table 4 - *Metric* captures this for the three style-dependent methods. Given a method, we have taken each outfit and calculated the compatibility scores conditional on all the available styles. We record the outfit rank corresponding to the style it actually belongs compute the metrics based on them. Figure 1 presents an example for an anchor top-wear and the style conditional compatibility scores for each of the outfits comprising only of bottom-wear. We see that the scores are highest (top ranked) for the style in which the outfit actually belongs.

Other Metrics: We make use of the list of outfits used in the calculation of style entropy again to understand the efficacy of the algorithms. Note that this list has outfits from different styles but common anchor item. For each such anchor item, conditional on the style, we check the top-1 accuracy of selecting the right child item in the outfit. To understand *accuracy*, we refer again to Fig. 1, where accuracy for Bottomwear1 equals 1 since the inferred rank corresponding to actual style is lowest. Table 4 - *Parent-Child* shows the results for various parent-child category combinations. Here SATCORec-p performs much better, although when we were checking column-wise ranking, it was behind SATCORec-r.

Table 4. The upper section of the table contains metrics on outfit ranks conditional on style while the lower section provides the percentage of correct selection of compatible item for anchor items with outfits across various styles.

		SATCORec-r	SATCORec-p	Theme Matters
Metric	MRR of correct style	**0.8844**	0.7676	0.6213
	Correct style on 1st rank	**80.94**	59.36	42.37
	Correct style in top 3 ranks	95.00	**95.10**	76.51
	Avg rank of the correct style	**1.4**	1.7	2.5
Parent-child	Topwear - Bottomwear	66.74	**77.33**	50.32
	Bottomwear - Topwear	72.02	**86.65**	57.92
	Topwear - Footwear	65.79	**75.97**	59.73
	Bottomwear - Footwear	69.81	**80.13**	62.79

Style-Specific Fine-Grained Category Selection in Outfit Generation: For each style, there can be multiple child-items which may match an anchor item, however, a good recommendation system would mostly output the items which differentiate the outfit from other styles. To check this phenomenon, for each style, we determine the most discriminating child-items [31], in terms of fine-grained categories e.g. *skirt* is a fine-grained category in bottomwears which most prominently shapes a casual style. Note that this is different from the most popular item across styles, say for example *jeans*. We posit that a superior algorithm would more frequently output such discriminative categories as a likely-match for a style. Style specific and overall results are shown in Table 5, we see in almost all the cases, SATCORec's output chooses discriminative fine-grained categories significantly higher number of times than the other baselines.

Table 5. Comparison of style-specific fine-grained categories chosen by different methods.

Method	Party	Outdoor	Summer	Formal	Athleisure	Winter	Casual	Celeb	Overall
TypeAware	28.33	29.22	10.24	33.54	19.52	18.10	2.67	15.92	19.30
BPR-DAE	28.19	17.07	17.74	36.26	31.64	29.05	23.42	19.05	25.64
TransNFCM	12.78	25.72	3.09	23.84	30.01	21.21	0.00	27.86	18.36
CSA-Net	34.63	26.79	13.98	35.44	28.69	26.94	11.00	27.11	25.38
Theme Matters	34.26	24.20	7.48	24.68	14.21	30.05	18.00	9.95	21.39
SATCORec-r	**50.56**	**32.12**	19.84	45.78	**38.65**	39.31	18.17	25.62	**34.27**
SATCORec-p	38.59	21.89	**23.06**	**47.26**	37.18	**40.92**	**24.09**	**28.09**	32.96

Blending of Styles: We have also checked the ability of SATCORec to generate outfits that are a linear combinations of different styles. We observe a smooth blending of the styles, also a higher (lesser) weight of a particular style (in the linear combination) results in the presence of more (less) items resembling that style in the generated outfits (Fig. 4). We will provide an web-based app along with the final version of the paper if accepted where a user would be able to explore different such combinations.

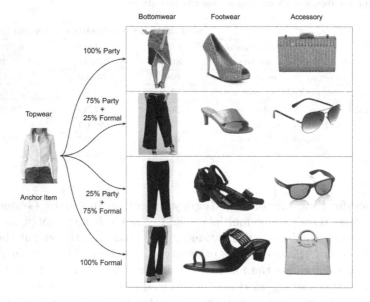

Fig. 4. Here we demonstrate the ability of our method to mix styles in outfit generation. Given the anchor item from top-wear, the top and bottom rows correspond to outfits generated from the two very separate styles: *Party* and *Formal*. The outfits in between them are generated by passing a weighted style vector for each of those two styles, thereby creating a nice blend.

4   Conclusion

The novelty of the paper lies in developing a Style-Attention-based Compatible Outfit recommendation and generation framework, SATCORec, utilizing high-level categories. SATCORec employs a Style-Compatibility-Attention Network - SCA Net and a Style Encoder Network - SE-Net. The SE-Net uses the Set Transformer to extract outfit style features, which is used to provide style-specific sub-space attention to individual items. The extensive style experiments establish the power of SATCORec in recommending with high accuracy a broader collection of compatible outfits across different styles to users. More interestingly, SATCORec chooses items which can make a pronounced style statement. Since in this paper we have focused on compatibility and employed a traditional beam search for outfit generation, an immediate future work would be to explore more sophisticated generation algorithms.

References

1. Bettaney, E.M., Hardwick, S.R., Zisimopoulos, O., Chamberlain, B.P.: Fashion outfit generation for E-commerce. In: Dong, Y., Ifrim, G., Mladenić, D., Saunders, C., Van Hoecke, S. (eds.) ECML PKDD 2020. LNCS (LNAI), vol. 12461, pp. 339–354. Springer, Cham (2021). https://doi.org/10.1007/978-3-030-67670-4_21
2. Bettaney, E.M., Hardwick, S.R., Zisimopoulos, O., Chamberlain, B.P.: Fashion outfit generation for e-commerce. In: Proceedings of the 2020 European Conference on Machine Learning and Principles and Practice of Knowledge Discovery in Databases (ECML PKDD 2020), pp. 339–354 (2021)
3. Chen, H., Lin, Y., Wang, F., Yang, H.: Tops, bottoms, and shoes: building capsule wardrobes via cross-attention tensor network. In: Fifteenth ACM Conference on Recommender Systems, pp. 453–462 (2021)
4. Chen, W., et al.: POG: personalized outfit generation for fashion recommendation at Alibaba i fashion. In: Proceedings of the 25th ACM SIGKDD International Conference on Knowledge Discovery & Data Mining, pp. 2662–2670 (2019)
5. Chen, X., et al.: Personalized fashion recommendation with visual explanations based on multimodal attention network: towards visually explainable recommendation. In: Proceedings of the 42nd International ACM SIGIR Conference on Research and Development in Information Retrieval, pp. 765–774 (2019)
6. Cui, Z., Li, Z., Wu, S., Zhang, X.Y., Wang, L.: Dressing as a whole: outfit compatibility learning based on node-wise graph neural networks. In: The World Wide Web Conference, pp. 307–317 (2019)
7. Dong, X., Wu, J., Song, X., Dai, H., Nie, L.: Fashion compatibility modeling through a multi-modal try-on-guided scheme. In: Proceedings of the 43rd International ACM SIGIR Conference on Research and Development in Information Retrieval, pp. 771–780 (2020)
8. Han, X., Song, X., Yin, J., Wang, Y., Nie, L.: Prototype-guided attribute-wise interpretable scheme for clothing matching. In: Proceedings of the 42nd International ACM SIGIR Conference on Research and Development in Information Retrieval, pp. 785–794 (2019)

9. Han, X., Wu, Z., Jiang, Y.G., Davis, L.S.: Learning fashion compatibility with bidirectional LSTMs. In: Proceedings of the 25th ACM International Conference on Multimedia (MM 2017), New York, pp. 1078–1086 (2017)

10. He, K., Zhang, X., Ren, S., Sun, J.: Deep residual learning for image recognition. In: Proceedings of the 2016 IEEE Conference on Computer Vision and Pattern Recognition (CVPR 2016), pp. 770–778 (2016)

11. Jaradat, S., Dokoohaki, N., Matskin, M.: Outfit2Vec: incorporating clothing hierarchical MetaData into outfits' recommendation. In: Dokoohaki, N. (ed.) Fashion Recommender Systems. LNSN, pp. 87–107. Springer, Cham (2020). https://doi.org/10.1007/978-3-030-55218-3_5

12. Jeon, Y., Jin, S., Han, K.: FANCY: human-centered, deep learning-based framework for fashion style analysis. In: Proceedings of the 2021 World Wide Web Conference (WWW 2021), pp. 2367–2378 (2021)

13. Kingma, D.P., Ba, J.: Adam: a method for stochastic optimization. In: Proceedings of the 3rd International Conference on Learning Representations (ICLR 2015), pp. 1–15 (2015)

14. Kuhn, T., et al.: Supporting stylists by recommending fashion style. In: recSysX-Fashion, pp. 1–6. CoRR arxiv:1908.09493 (2019)

15. Lai, J.H., Wu, B., Wang, X., Zeng, D., Mei, T., Liu, J.: Theme-matters: fashion compatibility learning via theme attention, pp. 1–15. CoRR arxiv:1912.06227 (2020)

16. Landia, N.: Personalised outfit recommendations: use cases, challenges and opportunities. In: Fifteenth ACM Conference on Recommender Systems, pp. 572–574 (2021)

17. Lee, H., Seol, J., Goo Lee, S.: Style2Vec: representation learning for fashion items from style sets, pp. 1–6. CoRR arxiv:1708.04014 (2017)

18. Lee, J., Lee, Y., Kim, J., Kosiorek, A., Choi, S., Teh, Y.W.: Set transformer: a framework for attention-based permutation-invariant neural networks. In: Proceedings of the 36th International Conference on Machine Learning (PMLR 2019), vol. 97, pp. 3744–3753 (2019)

19. Li, E., Kim, E., Zhai, A., Beal, J., Gu, K.: Bootstrapping complete the look at pinterest. In: Proceedings of the 26th ACM SIGKDD International Conference on Knowledge Discovery & Data Mining, pp. 3299–3307 (2020)

20. Li, K., Liu, C., Forsyth, D.: Coherent and controllable outfit generation, pp. 1–9. CoRR arxiv:1906.07273 (2019)

21. Li, X., Wang, X., He, X., Chen, L., Xiao, J., Chua, T.S.: Hierarchical fashion graph network for personalized outfit recommendation. In: Proceedings of the 43rd International ACM SIGIR Conference on Research and Development in Information Retrieval, pp. 159–168 (2020)

22. Li, Z., Wu, B., Liu, Q., Wu, L., Zhao, H., Mei, T.: Learning the compositional visual coherence for complementary recommendations. In: Proceedings of the 29th International Joint Conference on Artificial Intelligence (IJCAI 2020), pp. 3536–3543 (2020)

23. Lin, Y.L., Tran, S., Davis, L.S.: Fashion outfit complementary item retrieval. In: Proceedings of the 2020 IEEE Conference on Computer Vision and Pattern Recognition (CVPR 2020), pp. 3308–3316 (2020)

24. Lin, Y., Ren, P., Chen, Z., Ren, Z., Ma, J., de Rijke, M.: Explainable outfit recommendation with joint outfit matching and comment generation. IEEE Trans. Knowl. Data Eng. **32**(8), 1502–1516 (2019)

25. Lin, Y., Moosaei, M., Yang, H.: OutfitNet: fashion outfit recommendation with attention-based multiple instance learning. In: Proceedings of the 2020 World Wide Web Conference (WWW 2020), pp. 77–87 (2020)
26. Liu, X., Sun, Y., Liu, Z., Lin, D.: Learning diverse fashion collocations via neural graph filtering. IEEE Trans. Multim. (2020)
27. Lu, Z., Hu, Y., Chen, Y., Zeng, B.: Personalized outfit recommendation with learnable anchors. In: Proceedings of the 2021 IEEE Conference on Computer Vision and Pattern Recognition (CVPR 2021), pp. 12722–12731 (2021)
28. Lu, Z., Hu, Y., Jiang, Y., Chen, Y., Zeng, B.: Learning binary code for personalized fashion recommendation. In: Proceedings of the IEEE/CVF Conference on Computer Vision and Pattern Recognition, pp. 10562–10570 (2019)
29. McAuley, J., Targett, C., Shi, Q., van den Hengel, A.: Image-based recommendations on styles and substitutes. In: Proceedings of the 38th International ACM SIGIR Conference on Research and Development in Information Retrieval (SIGIR 2015), pp. 43–52 (2015)
30. Nakamura, T., Goto, R.: Outfit generation and style extraction via bidirectional LSTM and autoencoder, pp. 1–9. CoRR arxiv:1807.03133 (2018)
31. Ramos, J.: Using TF-IDF to determine word relevance in document queries. In: Proceedings of the 1st Instructional Conference on Machine Learning, pp. 29–48 (2003)
32. Revanur, A., Kumar, V., Sharma, D.: Semi-supervised visual representation learning for fashion compatibility. In: Fifteenth ACM Conference on Recommender Systems, pp. 463–472 (2021)
33. Singhal, A., Chopra, A., Ayush, K., Patel, U., Krishnamurthy, B.: Towards a unified framework for visual compatibility prediction. In: Proceedings of the 2020 IEEE Winter Conference on Applications of Computer Vision (WACV 2020), pp. 3596–3605 (2020)
34. Song, X., Feng, F., Liu, J., Li, Z., Nie, L., Ma, J.: NeuroStylist: neural compatibility modeling for clothing matching. In: Proceedings of the 25th ACM International Conference on Multimedia (MM 2017), pp. 753–761 (2017)
35. Tan, R., Vasileva, M.I., Saenko, K., Plummer, B.A.: Learning similarity conditions without explicit supervision. In: Proceedings of the 2019 IEEE International Conference on Computer Vision (ICCV 2019), pp. 10373–10382 (2019)
36. Vasileva, M.I., Plummer, B.A., Dusad, K., Rajpal, S., Kumar, R., Forsyth, D.: Learning type-aware embeddings for fashion compatibility. In: Ferrari, V., Hebert, M., Sminchisescu, C., Weiss, Y. (eds.) ECCV 2018. LNCS, vol. 11220, pp. 405–421. Springer, Cham (2018). https://doi.org/10.1007/978-3-030-01270-0_24
37. Wang, J., Cheng, X., Wang, R., Liu, S.: Learning outfit compatibility with graph attention network and visual-semantic embedding. In: Proceedings of the 2021 IEEE International Conference on Multimedia and Expo (ICME 2021), pp. 1–6 (2021)
38. Yang, X., Xie, D., Wang, X., Yuan, J., Ding, W., Yan, P.: Learning tuple compatibility for conditional outfit recommendation. In: Proceedings of the 28th ACM International Conference on Multimedia, pp. 2636–2644 (2020)
39. Yang, X., Du, X., Wang, M.: Learning to match on graph for fashion compatibility modeling. In: Proceedings of the AAAI Conference on Artificial Intelligence, pp. 287–294 (2020)
40. Yang, X., et al.: Interpretable fashion matching with rich attributes. In: Proceedings of the 42nd International ACM SIGIR Conference on Research and Development in Information Retrieval, pp. 775–784 (2019)

41. Yang, X., Ma, Y., Liao, L., Wang, M., Chua, T.S.: Transnfcm: translation-based neural fashion compatibility modeling. In: Proceedings of the AAAI Conference on Artificial Intelligence, pp. 403–410 (2019)
42. Zhan, H., Lin, J., Ak, K.E., Shi, B., Duan, L.Y., Kot, A.C.: A3-FKG: attentive attribute-aware fashion knowledge graph for outfit preference prediction. IEEE Trans. Multim. 1–13 (2021)
43. Zhang, A., Lipton, Z.C., Li, M., Smola, A.J.: Dive into deep learning. CoRR arxiv:2106.11342 (2021)

Why Did You Not Compare with That? Identifying Papers for Use as Baselines

Manjot Bedi[1], Tanisha Pandey[1], Sumit Bhatia[2(✉)], and Tanmoy Chakraborty[1]

[1] IIIT Delhi, New Delhi, India
{manjotb,tanisha17116,tanmoy}@iiitd.ac.in
[2] Media and Data Science Research Lab, Adobe Systems, Noida, India
sumit.bhatia@adobe.com

Abstract. We propose the task of automatically identifying papers used as baselines in a scientific article. We frame the problem as a binary classification task where all the references in a paper are to be classified as either baselines or non-baselines. This is a challenging problem due to the numerous ways in which a baseline reference can appear in a paper. We develop a dataset of $2,075$ papers from ACL anthology corpus with all their references manually annotated as one of the two classes. We develop a multi-module attention-based neural classifier for the baseline classification task that outperforms four state-of-the-art citation role classification methods when applied to the baseline classification task. We also present an analysis of the errors made by the proposed classifier, eliciting the challenges that make baseline identification a challenging problem.

Keywords: Baseline recommendation · Dataset search · Scientific documents · Faceted search

1 Introduction

One of the common criticisms received by the authors of a scientific article during the paper review is that the method proposed in the submitted paper has not been compared with appropriate baselines. The reviewers often suggest a list of existing papers which, according to them, should have been used as baselines by the submitted work. Oftentimes, the authors find the suggestions unexpected and surprising as they have never encountered these papers before. The reasons behind the lack of awareness of the state-of-the-art of a specific research area are two-fold – (i) the authors have not done due diligence to explore the field completely, and/or (ii) due to the exponential growth of the number of papers published per year, many relevant papers get unnoticed. Both these problems can be addressed if we have a recommendation system that collects all the papers published in a certain field, analyzes them, and recommends a set of selected papers for a given topic/task that needs to be considered for the purpose of comparison. The current work is the first step towards the goal of building such an *intelligent baseline recommendation system* that can assist the authors to find and select suitable baselines for their work.

© The Author(s), under exclusive license to Springer Nature Switzerland AG 2022
M. Hagen et al. (Eds.): ECIR 2022, LNCS 13185, pp. 51–64, 2022.
https://doi.org/10.1007/978-3-030-99736-6_4

With the availability of online tools such as CiteSeerX [33], Google Scholar [16], and Semantic Scholar [15], it has become convenient for researchers to search for related articles. However, these search engines provide flat recommendations and do not distinguish between the recommended papers based on how and why the recommendations are relevant to the query. For example, if the query is 'citation classification models', how do we know, among the set of recommendations returned by the search engines, which one would be used to understand the *background* of the area, which one to explore to know the *datasets* used in the past to address the problem, which one to use for the purpose of *comparison*, etc. In short, the existing systems do not provide *faceted recommendations* where a facet can determine the role of a recommendation with respect to the query.

In order to build an intelligent baseline recommendation system, the first requirement is the capability to automatically identify the references in a given paper used by the paper as baselines. This capability allows creating the training corpus as well as automatically process the ever-growing stream of new papers. One may think that this problem of automatic baseline identification is trivial as a baseline reference is likely to be cited in the experiment and/or the result sections of the paper; therefore, the position information of a reference may give a precise cue about its usage in the paper. Surprisingly, we observe that this assumption does not work satisfactorily – out of $2,075$ papers we analyze in this work, the probability of a baseline citation to appear in the experiment section is 0.73. It indicates that around 30% baseline references lie in some other sections of the paper. More importantly, only 23% of the references placed in the experiment section are actually used as baselines in the paper. We further observe that only 7.13% papers have keywords such as 'baseline', 'state-of-the-art', 'gold standard' present in the headings of different sections or subsections (see discussion on error analysis in Sect. 5 for the other challenges). These obstacles make the problem of accurately classifying references of a given paper into baselines or non-baselines non-trivial.

The problem of *baseline classification* is closely related to the task of *citation role classification* studied extensively in the literature. Notable contributions include the works by Chakraborty et al. [5] who proposed a faceted scientific paper recommendation system by categorizing the references into four major facets; Dong and Schfer [12] who proposed an ensemble model to figure out different roles of references in a paper; ; Jurgens et al. [18] who unfolded the evolution of research in a scientific field by understanding why a paper is being cited; Cohan et al. [7] who outperformed the methods developed by Jugens et al. [18] in the task of citation role classification. (See Sect. 2 for more details of the related literature.) However, none of these methods are explicitly developed to address the problem of baseline recommendation. Our experiments (Sect. 5) reveal that these methods do not work well to distinguish the baseline references from other references in a given paper.

In this paper, we consider the ACL Anthology dataset, select a subset of papers and employ human annotators to identify the references corresponding to the baselines used in the papers (Sect. 3). We present a series of issues encountered during the annotation phase that illustrate the non-trivial nature of the problem. We then develop a multi-module attention (MMA) based neural architecture to classify references into baselines and non-baselines (Sect. 4). We also adopt state-of-the-art approaches for citation role classification for a fair comparison with our methods. A detailed comparative analysis shows that the neural attention based approach outperforms others with 0.80 F1-score. We present a thorough error analysis to understand the reasons behind the failures of the proposed models and identify challenges that need to be addressed to build better baseline identification systems (Sect. 5). The dataset developed and code for our proposed model is available at https://github.com/sumit-research/baseline-search.

2 Related Work

Understanding the Role of Citations. Stevens et al. [25] first proposed that papers are cited due to 15 different reasons. Singh et al. [24] presented the role of citation context in predicting the long term impact of researchers. Pride and Knoth [22] and Teufel et al. [28] attempted to classify the roles of citations. Chakraborty and Narayanam [6] and Wan and Liu [32] argued that all citations are not equally important for a citing paper, and proposed models to measure the intensity of a citation. Doslu and Bingol [13] analysed the context around a citation to rank papers from similar topics. Cohen et al. [8] showed that the automatic classification of citations could be a useful tool in systematic reviews. Chakraborty et al. [5] presented four reasons/tags associated with citations of a given paper – 'background' (those which are important to understand the background literature of the paper), 'alternative approaches' (those which deal with the similar problem as that of the paper), 'methods' (those which helped in designing the model in the paper) and 'comparison' (those with which the paper is compared). Therefore, one can simply assume that the citations with 'comparison' tag are the baselines used in the paper. Dong and Schfer [12] classified citations into four categories i.e., 'background', 'fundamental idea', 'technical basis' and 'comparison'. They employed ensemble learning model for the classification. We also consider this as a relevant method for our task assuming that the citations tagged as 'comparison' are the baselines of the paper. Chakraborty and Narayanam [6] measured how relevant a citation is w.r.t the citing paper and assigned five granular levels to the citations. Citations with level-5 are those which are extremely relevant and occur multiple times within the citing paper. We treat this work as another competing method for the current paper by considering citations tagged with level-5 as the baselines of the citing paper. Jurgens et al. [18] built a classifier to categorize citations based on their functions in the text. The 'comparison or contrast' category expresses the similarity/differences to the cited paper. This category might include some citations which are not considered for direct comparison, but they are the closest category to be considered as baseline. However, we have not compared with this method as as the proposed approach by Cohan et al. [7], which is a baseline for the current work, already claimed to achieve better performance than this classifier. Su et al. [26] used a single-layer convolutional neural network to classify citations and showed that it outperforms state-of-the-art methods. We also consider this as a baseline for our work. Cohan et al. [7] used a multi-task learning framework (using BiLSTM and Attention) and outperformed the approach of Jurgens et al. [18] on the citation classification task. Their 'results comparison' category can be thought of as equivalent to the baseline class. This model achieved state-of-the-art performance on citation classification and we consider it as another baseline for our work.

Recommending Citations for Scholarly Articles. A survey presented by Beel et al. [1] showed that among 200 research articles dealing with citation recommendation, more than half used content-based filtering on authors, citations and topics of the paper. Few such models include topic-based citation recommendation [27] and content-based recommendation [3, 11] that work even when the metadata information about the paper being queried is missing. Yang et al. [34] used the LSTM model to develop a context-aware citation recommendation system. Recently, Jeong et al. [17] developed a context-aware neural citation recommendation model. While there are a lot of new methods coming in the domain of citation recommendation systems, the problem of identifying and recommending baselines of a paper has been untouched. Citation recommendation can help researchers to efficiently write a scientific article, while baseline recommendation can further enable to get a glance at the work done in a particular domain.

3   Dataset for Baseline Classification

We used ACL Anthology Reference Corpus (ARC) [4] as the base data source for preparing the annotated dataset for our study. The ARC corpus consists of scholarly papers published at various Computational Linguistics up to December 2015. The corpus consists of 22, 875 articles and provides the original PDFs, extracted text and logical document structure (section information) of the papers, and parsed citations using the ParsCit tool [9].

The complete ARC corpus contains all types of papers presented at various conferences under the ACL banner such as long and short research papers, system and demonstration papers, workshop and symposium papers. We noted that a significant fraction of short and workshop papers, and system and demonstration papers are not useful for our purpose as these papers often do not contain rigorous comparative evaluation. They generally are position papers, describe tools/systems, or work in progress. Therefore, we discarded such articles from the dataset by removing papers having keywords such as *short papers, workshops, demo, tutorial, poster, project notes, shared task, doctoral consortium, companion volume*, and *interactive presentation* in the title/venue fields of the papers. This filtering resulted in a final set of 8, 068 papers.

We recruited two annotators, A_1 and A_2, for annotating the references of papers as baseline references. A_1 was a senior year undergraduate student, and $A2$ was a graduate student. Both the annotators were from the Computer Science discipline and had a good command of the English language (English being the primary medium of education). A_1 provided annotations for a total of 1, 200 documents selected randomly from the filtered list of 8, 068 papers. A_2 worked independently of A_1 and provided annotations for a total of 1, 000 papers. The set of documents annotated by A_2 had 875 randomly selected new documents from the filtered ARC corpus and 125 documents chosen randomly from the documents annotated by A_1. We used this set of 125 papers annotated by both A_1

Table 1. Summary of the annotated dataset. Annotators $A1$ and $A2$ provided annotations for a total of 1, 200 and 1, 000 papers, respectively.

	# Papers	# Baseline references	# Non-baseline references
Annotator 1 (A1)	1,200	3,048	29,474
Annotator 2 (A2)	1,000	2,246	24,831
Common Papers	125	305	3,252
Unique Papers	2,075	4,989	51,053

Table 2. Distribution of papers in the dataset across different time periods.

	1980–2000	2001–2005	2006–2010	2011–2015
# Papers	125	179	589	1,182
# References	2,339	3,534	13,976	36,193
# Baselines	192	406	1,295	3.096
Mean references per paper	18.71	19.74	23.73	30.62
Mean baselines per paper	1.53	2.27	2.20	2.62

and A_2 to measure the inter-annotator agreement between them. The value of Cohen's Kappa was found to be 0.913 indicating near-perfect agreement between the two annotators.

We now discuss some of the challenges faced and observations made by the annotators while examining the assigned papers. The annotators noted that there were no associated citations for the baseline methods in the paper in many cases. This often happens when a well-established technique (such as tf-idf for document retrieval) or a simple method (such as a majority class baseline, a random classifier, a heuristic as a baseline) is used as a baseline. Second, there were cases where the authors reported that it was difficult for them to compare their methods with other published techniques due to the novelty of the problem making published techniques unsuitable for their task. Finally, there were many cases where ideas from multiple papers were combined to create a suitable baseline for the task considered, making it hard and challenging to identify the baseline reference.

Table 1 summarizes the statistics of the annotated dataset. The final dataset consists of $2,075$ unique papers. These papers have a total of $56,052$ references, out of which $4,989$ references were marked as baselines, and the remaining $51,053$ references were non-baseline references.

3.1   Observations and Characteristics of the Dataset

Year-wise Distribution of Annotated Papers: Table 2 presents the year-wise distribution of the $2,075$ papers in the final dataset. The oldest paper in the dataset is from 1980, and the latest paper is from 2015. Table 2 shows that papers published in the period $2011 - 2015$ cite more papers and have more baselines on an average compared to the papers published in the earlier years. This observation is consistent with the trend of an increased number of citations in papers [30] and the increased focus on empirical rigor and reproducibility.

Section-wise Distribution of Baseline Citations: We now present the distribution of baseline references in different sections of papers in the dataset. Due to the diversity of writing styles and author preferences, there are no standardized section headers that are used in literature, and it is common to use simple rules, regular expressions [10], or simple feature-based classification methods [29] to identify section headers from document text. We use a simple keyword-based approach to group all the sections into five categories – Introduction, Related Work, Methods and Results, Conclusions, and Others. A section of a paper containing a keyword as specified in Table 3 would be mapped to its corresponding section category.

Table 3. List of keywords used to identify the five section categories.

Section Heading	Keywords
Introduction	introduction
Related Work	related work; background; previous work; study
Methods and Results	method; approach; architect; experiment; empiric; evaluat; result; analys; compar; perform; discussion
Conclusion	conclusion; future work
Other sections	everything else

Table 4 reports the distribution of baseline citations in different sections of the papers in our dataset. Note that a paper can be cited multiple times in the citing paper. Thus, a given citation can occur in multiple sections in a paper. We provide both the statistics, i.e., the total number of baseline citations in a section and the number of baseline citations that appear exclusively in the section in parenthesis.

Interestingly, we note that there are a few cases where the baseline citations appear exclusively in the Introduction (117) and Conclusion (3) sections. One would expect the baseline citations not to appear exclusively in these sections. However, it turned out that the citations occurring exclusively in the conclusion section were part of a comparison table placed at the end of the paper.

Table 4. Distribution of baselines and non-baselines in different sections. Numbers in parentheses are the count of baselines appearing exclusively in the section.

Section	# baselines	#non-baselines
Introduction	2,138 (117)	13,930 (7,360)
Related	1,755 (105)	14,917 (9,217)
Experiment	3,664 (534)	11,939 (6,173)
Conclusion	203 (3)	873 (360)
Other Sections	1,769 (181)	13,283 (7,646)

Therefore, they were counted under the conclusion section. Further, the citations in the Introduction and Related Work section were given an alias name when they were first mentioned in the paper (e.g. LocLDA for location based LDA, see Table 8 for example) and were referred to by the aliases in other sections. Therefore, their presence in other sections of the paper could not be easily counted.

From Table 4, we observe that most of the baseline citations appear in the experiment section. Therefore, classifying a reference as a baseline if it occurs in the experiment section may be considered as a naive solution and a very simple baseline. In Table 5, we present the results obtained by hypothetical classifiers that classify all the citations in a given section as a baseline. Note that we

Table 5. Precision and recall values obtained by a naïve classifier that considers all citations in a specific section or table as baseline citations.

Section Heading	Precision	Recall
Introduction	0.13	0.42
Related	0.10	0.35
Experiment	0.234	0.734
Conclusion	0.18	0.040
Other Sections	0.11	0.35
Table	0.72	0.18

also report numbers for a classifier that considers all citations appearing in a Table as baselines.

We note that while such a simple classifier will be able to recover a large number of baselines from the Experiment section (high recall value of 0.734), it will miss out on

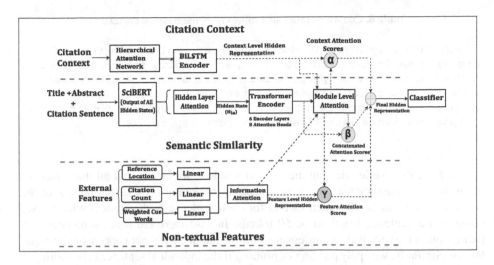

Fig. 1. Our proposed multi-module attention based neural classification model for the baseline classification task.

about 30% baselines and will suffer from a very high number of false positives (very low precision of 0.234). An opposite trend can be observed in the case of Tables – most citations in Tables are baseline references (high precision of 0.72); however, due to a very low recall (0.18), most of the baselines are missed by this simple classifier.

4 Multi-module Attention Based Baseline Classifier

We now describe our approach for classifying the citations of a paper as baselines. Our model utilizes contextual and textual signals present in the text around a citation to classify it as a baseline. We use Transformer encodings [31] to capture the nuances of the language and uses neural attention mechanisms [31,35] to learn to identify key sentences and words in the citation context of a citation. Further, given the vagaries of the natural language and varied writing styles of different authors, we also utilize non-textual signals such as popularity of a paper (in terms of its overall citations) to have a more robust classifier.

Figure 1 describes our proposed neural architecture for the baseline classification task. The proposed architecture is designed to capture different context signals in which a paper is cited to learn to differentiate between baselines and non-baseline citations. The proposed model utilizes a Transformer-based architecture consisting of three modules to handle different signals and uses the representations obtained from these modules together to classify a citation into a baseline.

The first module (top row in Fig. 1) tries to capture the intuition that the context around a citation in the paper can help in determining if the cited paper is being used as a baseline or not. Therefore, we take a fixed size context window and pass it through a hierarchical attention network [35] that learns to identify and focus on sentences in the context window that can provide contextual clues about the cited paper being a baseline

Table 6. Cue words (after stemming) from the baseline contexts.

```
among base origin precis modifi highest implement extend
signific maximum metric higher experi baselin fscore strategi
accord compar overal perform best previou model evalu correl
recal result calcul standard stateoftheart achiev figur
accuraci gold comparison method top yield procedur obtain
outperform score significantli increas report
```

or not. Note that while selecting the context window, we ensure that all the sentences lie in the same paragraph as the citation under consideration. We select the size of the context window to be 10 sentences and for each sentence in the context window, we consider the sentence length to be 50 tokens. In case there are fewer sentences in a paragraph, we apply padding to ensure that the input to the network is of the same length. Similarly, we apply padding or pruning if the individual sentences are shorter or longer, respectively than 50 tokens. The citation context window thus obtained is then converted to a vector representation using SciBERT embeddings [2] that provide word embeddings trained specifically for NLP applications using scholarly data.

The input vector representations thus obtained are fed to the hierarchical attention based encoder that outputs the hidden model representation of the context window after applying a series of localized attentions to learn the significance of constituent sentences and words in the input context vector. We show an example of the sentence level attention in Fig. 2. The sentence containing the baseline citation (the middle sentence of the document) obtains the highest attention scores with rest of the attention distributed towards the other important sentences in the paragraph. This finally produces a better semantic understanding for the model in order to correctly classify it as a baseline. The output of the hierarchical attention encoder model is then passed through a bidirectional LSTM encoder in order to capture any sequential relationships present in the citation context. This yields the final learned representation of the context surrounding the citation under consideration.

The second module (middle row in Fig. 1) is designed to capture the semantic similarity and relations between a given citation and the overall content of the citing paper. We consider the title and abstract of the citing paper as a concise summary of the citing paper. For a given citation, we take the title and abstract of the citing paper and the citation sentence and pass them through the pre-trained SciBERT language model that outputs a fine-tuned representation for the concatenated text. Further, we consider all the output hidden states for all the thirteen hidden layers in SciBERT. Different layers learn different feature representations of the input text. These representations from all the hidden layers, thus obtained are then passed through an attention module that learns attention weights for different hidden states. The resulting attention-weighted representation is then passed through a Transformer encoder layer[1] to capture any sequential dependencies between input tokens yielding the final representation capturing relations between the cited paper and the title and abstract of the citing paper.

---

[1] We use a six layer Transformer encoder with eight attention heads. This was found to be the best performing configuration.

Note that the two modules discussed so far can capture the linguistic variations in the citation context and semantic relations between the cited and citing papers. In the third module (bottom row in Fig. 1), we utilize the following three additional non-textual signals that might indicate whether a paper is being cited as a baseline.

1. **Reference location:** Intuitively, if a paper is used as a baseline, it is more likely to be discussed (and cited) in the experiment section of the paper. Hence, we define five features that record the number of times a given reference is cited in each of the five sections defined in Table 3. In addition, we also define a feature to capture if a reference is cited in one of the tables as many times, baseline papers are also (and often exclusively) mentioned in the result-related tables.

2. **Cue words:** There are certain cue words and phrases that authors frequently use while discussing the baseline methods. Thus, their presence (or absence) in citation contexts can help differentiate between baseline and non-baseline references. We create a list of such cue words (as shown in Table 6) by manually inspecting the citation contexts of baseline references in 50 papers (separate from the papers in the dataset). Thus, the cue word features capture the presence (or absence) of each cue word in the citation context of a reference. Further, each cue word w present in the citation context is assigned a weight $w = 1/d_w$, where d_w is the number of words between w and the citation mention. Thus, cue words that appear near the citation mention are given a higher weight. If a cue word appear multiple times in the citation context, we consider its nearest occurrence to the citation mention (maximum weight).

3. **Citation count:** We use the total number of citations received by a paper as a feature to capture the intuition that highly-cited (and hence, more popular and impactful) papers have a higher chance of being used as a baseline than papers with low citations.

Each of these features is then passed through a linear layer followed by a feature level attention module that yields the final attention weighted representation of all the features.

The output of the three modules described above provides three different representations capturing different information signals that can help the network classify the given citation as baseline. The three representations thus obtained are passed through a module-level attention unit that learns attention weights to be given to the output of the three representations and outputs a 128 dimensional attention-weighted representation which is then passed through a linear classifier that outputs if the input citation is a baseline citation or not.

5   Empirical Results and Discussions

Baselines for Citation Classification: We select following methods for citation classification and adopt them for the task of baseline classification. We use author provided source-code where available; otherwise, we implement the methods using details and parameter settings as provided in the respective papers.

Table 7. Performance on baseline classification task for the different methods. We report overall precision, recall, and F-1 values as well as the numbers for each class.

Models	Baselines			Non-baselines			Overall		
	Precision	Recall	F-1	precision	Recall	F-1	Precision	Recall	F-1
Dong and Schäfer [12]	0.33	0.67	0.44	0.96	0.87	0.91	0.65	0.77	0.68
Chakraborty and Narayanam [6]	0.26	**0.74**	0.39	0.96	0.78	0.86	0.61	0.76	0.62
Su et al. [26]	0.69	0.16	0.26	0.63	0.95	0.76	0.66	0.55	0.51
Cohan et al. [7]	0.47	0.48	0.47	0.96	0.95	0.95	0.71	0.71	0.71
Proposed MMA classifier	**0.69**	0.57	**0.63**	**0.96**	**0.98**	**0.97**	**0.82**	**0.78**	**0.80**

1. Dong and Schäfer [12] proposed an ensemble-style self-training classifier to classify the citations of a paper into four categories – *background, fundamental idea, technical basis* and *comparison*. We implemented their classifier (using their feature set) and used it for baseline classification task.
2. Chakraborty and Narayanam [6] proposed a method for measuring relevance of a citation to the citing paper on a five point scale with level-5 citations being the most relevant. We consider the citations identified as level-5 as the baselines of the citing paper.
3. Su et al. [26] proposed a CNN based architecture for citation function classification that we use for our binary classification task.
4. Cohan et al. [7] proposed a multi-task learning framework for the citation classification task. We implement the model using the settings as recommended in the paper and use it for baseline classification.

Experimental Settings: For evaluating different classification methods, we split the developed dataset (Sect. 3) into training, development, and test sets in 70 : 10 : 20 ratio. Different hyper-parameters involved are fine-tuned using the development set. Consequently, the size of the input citation context vectors is set to 768, the size of the hidden layer for the BiLSTM layer is 64 and the dropout rate is set to 0.2. The Transformer encoder has 6 layers and 8 attention heads. The batch size and learning rate are set to 32 and 0.001, respectively. The model was trained for 20 epochs. For our proposed model, we used cross-entropy loss and Adam Optimizer [19] to minimize the overall loss of the model. As our dataset is unbalanced, we incorporated class weights in our loss function fine-tuned the class weights.

Results and Discussions: Table 7 summarizes the results as achieved by different methods on the test set. We note that four state-of-the-art methods for citation classification achieve only moderate performance on the baseline classification task indicating their inadequacy at this task, and hence, the need for developing specialized methods for baseline classification. Our proposed model, outperforms the state-of-the-art citation role classifiers in terms of F-1 measure. Further, note that the performance of the proposed Multi-module Attention based model is more balanced with relatively high recall (0.57) and the highest precision(0.69) among all the methods studied.

The 5 sentences are manually selected in a way to cover as many nuggets as possible with higher priority for the nuggets with higher frequencies
We also created random summaries using Mead (Radev et al., 2004).
These summaries 900 are basically a random selection of 5 sentences from the pool of sentences in the citation summary
Generally we expect the summaries created by the greedy method to be significantly better than random ones.
In addition to the gold and random summaries, we also used 4 baseline state of the art summaries: LexRank, the clustering C-RR and C-LexRank, and MMR
LexRank (Erkan and Radev, 2004) works based on a random walk on the cosine similarity of sentences and prints out the most frequently visited sentences.
Said differently, LexRank first builds a network in which nodes are sentences and edges are cosine similarity values
It then uses the eigenvalue centralities to find the most central sentences
For each set, the top 5 sentences on the list are chosen for the summary
The clustering methods, C-RR and C-LexRank, work by clustering the cosine similarity network of sentences.
In such a network, nodes are sentences and edges are cosine similarity of node pairs

Fig. 2. Example of a sentence-level attention distribution (Red) obtained from the Attention Encoder. (Color figure online)

Integrating Phrase-based Reordering Features into a Chart-based Decoder for Machine Translation Hiero translation models have two limitations compared to phrase-based models:
1) Limited hypothesis space; 2) No lexicalised reordering model. We propose an extension of Hiero called PhrasalHiero to address Hiero's second problem. Phrasal-Hiero still
has the same hypothesis space as the original Hiero but incorporates a phrase-based distance cost feature and lexicalized reordering features into the chart decoder. The work
consists of two parts: 1) for each Hiero translation derivation, find its corresponding discontinuous phrase-based path. 2) Extend the chart decoder to incorporate features from
the phrase-based path. We achieve significant improvement over both Hiero and phrase-based baselines for ArabicEnglish, Chinese-English and GermanEnglish translation. To
implement Phrasal-Hiero, we extented Moses chart decoder (Koehn et al., 2007) to include distance-based reordering as well as the lexicalized phrase orientation reordering model

Fig. 3. Illustrative example of an attention weight distribution (red) from the Attention Encoder in the semantic similarity module of the proposed network. (Color figure online)

Table 8. Example of false positives treated as baselines by the classifier. Paper IDs are the IDs used in the dataset.

Paper Id	Citation text
N12-1051	We evaluated our taxonomy induction algorithm using McRae et al.'s (2005) dataset which consists of for 541 basic level nouns.
P08-1027	For each parameter we have estimated its desired range using the (Nastase and Szpakowicz 2003) set as a development set.
D13-1083	In the future work, we will compare structural SVM and c-MIRA under decomposable metrics like WER or SSER (Och and Ney 2002).
E09-1027	For comparison purposes, we plan to implement other features that have been used in earlier readability assessment systems. For example, Petersen and Ostendorf (2009) created lists of the most common words from the Weekly Reader articles,
P10-1116	This is in line with results obtained by previous systems (Griffiths et al. 2005; Boyd-Graber and Blei 2008; Cai et al. 2007). While the performance on verbs can be increased to outperform the most frequent sense baseline.
D10-1006	This is the model used in (Brody and Elhadad 2010) to identify aspects, and we refer to this model as LocLDA.
D11-1115	we compare Chart Inference to the two baseline methods: Brute Force (BF), derived from Watkinson and Manandhar, and Rule-Based (RB), derived from Yao et al.

Figure 2 shows an illustrative example of the hierarchical attention module in the proposed network. The figure shows the citation context as extracted from the paper by Qazvinian et al. [23] where the LexRank method by Erkan and Radev [14] is being used as a baseline. The attention given to different sentences in the context window is illustrated by shades of red where a sentence in darker shade is given a higher weight. We note that the sentence which the LexRank paper is cited, is given the highest weight and other sentences that talk about the task of summarization are also given some weights whereas the fourth sentence ("Generally we expect...") is being given no weight as the network did not find it to be useful for the classification task. Likewise, Fig. 3 presents an example of the role of the attention encoder in the semantic similarity module in the proposed network. The figure shows the concatenated title and abstract of the paper by Nguyen and Vogel [21] that uses the MOSES decoder [20] for machine translation (last sentence in the figure is the citation sentence). Note that the network is able to identify keywords like *reordering, distance-based, translation,* and *lexicalized* that indicate the similarity between the content of the citing paper with the citation context.

Error Analysis: We now present representative examples of hard cases and the types of errors made by the classifiers.

Confusion with Datasets: We observed that often the citation for datasets used in the experiments were classified as baselines by the classifier. Such citations are often made in the experiment section, and the language patterns in their citation contexts are often very similar to contexts of baseline citations (rows 1 and 2 in Table 8).

Citations for Future Work: Often, authors discuss the results of papers that are not explicitly used as baselines in the current work but are discussed for the sake of completeness and could be used as baselines as part of the future work. One could argue that such citations should be easy to classify as they must be part of the *Conclusions and Future Work* sections. However, as we observed, this does not always hold true. Such citations could be found in the *Experiment* or *Other* custom section headers (e.g. rows 3, 4 in Table 8).

Context Overlap of Multiple Citations: The key assumption that the methods studied in this work make is that the baseline and non-baseline citations differ in the language patterns in their respective citation contexts. However, we noted that multiple papers are often cited together, and thus, share the same citation contexts (and other properties represented by different features). For instance, row 5 in Table 8 presents an example of non-baseline citations sharing the context with baseline *(Cai et al. 2007)*.

Citation Aliases and Table Citations: Often, authors give an alias to a particular method (as shown in rows 6, 7 in Table 8) and then use the alias to refer to that method in the rest of the paper. As a result, it becomes challenging to capture the context around the alias mentions in the text. Further, many errors were made in cases where the baseline references are not cited and discussed extensively in the running text but are mentioned directly in the results table. Hence, we lose out on the context for such baseline citations.

6   Conclusions

We introduced the task of identifying the papers that have been used as baselines in a given scientific article. We framed the task as a reference classification problem and developed a dataset out of ACL anthology corpus for the baseline classification task. We empirically evaluated four state-of-the-art methods for citation classification and found that they do not perform well for the current task. We then developed custom classifiers for the baseline classification task. While the proposed methods outperformed the state-of-the-art citation classification methods, there is still a significant performance gap that needs to be filled. We further presented error analysis illustrating the challenges and examples that the proposed systems found difficult to classify.

Acknowledgement. T. Chakraborty would like to acknowledge the support of the Ramanujan Fellowship, and ihub-Anubhuti-iiitd Foundation set up under the NM-ICPS scheme of the Department of Science and Technology, and the Infosys Centre for AI at IIIT-Delhi.

References

1. Beel, J., Gipp, B., Langer, S., Breitinger, C.: Paper recommender systems: a literature survey. Int. J. Digit. Lib. **17**(4), 305–338 (2016)
2. Beltagy, I., Lo, K., Cohan, A.: SciBERT: pretrained language model for scientific text. In: EMNLP (2019)
3. Bhagavatula, C., Feldman, S., Power, R., Ammar, W.: Content-based citation recommendation. arXiv preprint arXiv:1802.08301 (2018)
4. Bird, S., et al.: The ACL anthology reference corpus: a reference dataset for bibliographic research in computational linguistics. In: LREC. European Language Resources Association (2008)
5. Chakraborty, T., Krishna, A., Singh, M., Ganguly, N., Goyal, P., Mukherjee, A.: FeRoSA: a faceted recommendation system for scientific articles. In: Bailey, J., Khan, L., Washio, T., Dobbie, G., Huang, J.Z., Wang, R. (eds.) PAKDD 2016. LNCS (LNAI), vol. 9652, pp. 528–541. Springer, Cham (2016). https://doi.org/10.1007/978-3-319-31750-2_42
6. Chakraborty, T., Narayanam, R.: All fingers are not equal: intensity of references in scientific articles. In: EMNLP, pp. 1348–1358 (2016)
7. Cohan, A., Ammar, W., van Zuylen, M., Cady, F.: Structural scaffolds for citation intent classification in scientific publications. arXiv preprint arXiv:1904.01608 (2019)
8. Cohen, A.M., Hersh, W.R., Peterson, K., Yen, P.Y.: Reducing workload in systematic review preparation using automated citation classification. J. Am. Med. Inf. Assoc. **13**(2), 206–219 (2006)
9. Councill, I.G., Giles, C.L., Kan, M.Y.: Parscit: an open-source CRF reference string parsing package. In: LREC. European Language Resources Association (2008). http://www.lrec-conf.org/proceedings/lrec2008/
10. Ding, Y., Liu, X., Guo, C., Cronin, B.: The distribution of references across texts: some implications for citation analysis. J. Informetrics **7**(3), 583–592 (2013)
11. Ding, Y., Zhang, G., Chambers, T., Song, M., Wang, X., Zhai, C.: Content-based citation analysis: the next generation of citation analysis. J. Assoc. Inf. Sci. Technol. **65**(9), 1820–1833 (2014)
12. Dong, C., Schäfer, U.: Ensemble-style self-training on citation classification. In: IJCNLP, pp. 623–631 (2011)
13. Doslu, M., Bingol, H.O.: Context sensitive article ranking with citation context analysis. Scientometrics **108**(2), 653–671 (2016). https://doi.org/10.1007/s11192-016-1982-6
14. Erkan, G., Radev, D.R.: Lexrank: graph-based lexical centrality as salience in text summarization. J. Artif. Intell. Res. **22**, 457–479 (2004). https://doi.org/10.1613/jair.1523
15. Fricke, S.: Semantic scholar. J. Med. Lib. Assoc. **106**(1), 145 (2018)
16. Jacsó, P.: Google scholar: the pros and the cons. Online information review (2005)
17. Jeong, C., Jang, S., Shin, H., Park, E., Choi, S.: A context-aware citation recommendation model with bert and graph convolutional networks. arXiv preprint arXiv:1903.06464 (2019)
18. Jurgens, D., Kumar, S., Hoover, R., McFarland, D., Jurafsky, D.: Measuring the evolution of a scientific field through citation frames. TACL **6**, 391–406 (2018)

19. Kingma, D.P., Ba, J.: Adam: a method for stochastic optimization. arXiv preprint arXiv:1412.6980 (2014)
20. Koehn, P., et al.: Moses: open source toolkit for statistical machine translation. In: Proceedings of the 45th Annual Meeting of the Association for Computational Linguistics Companion Volume Proceedings of the Demo and Poster Sessions, pp. 177–180 (2007)
21. Nguyen, T., Vogel, S.: Integrating phrase-based reordering features into a chart-based decoder for machine translation. In: Proceedings of the 51st Annual Meeting of the Association for Computational Linguistics, Vol. 1, Long Papers, pp. 1587–1596 (2013)
22. Pride, D., Knoth, P.: An authoritative approach to citation classification. In: JCDL (2020)
23. Qazvinian, V., Radev, D.R., Özgür, A.: Citation summarization through keyphrase extraction. In: Coling, pp. 895–903 (2010)
24. Singh, M., Patidar, V., Kumar, S., Chakraborty, T., Mukherjee, A., Goyal, P.: The role of citation context in predicting long-term citation profiles: an experimental study based on a massive bibliographic text dataset. In: CIKM, pp. 1271–1280 (2015)
25. Stevens, M.E., Giuliano, V.E., Garfield, E.: Can citation indexing be automated? (1964)
26. Su, X., Prasad, A., Kan, M.Y., Sugiyama, K.: Neural multi-task learning for citation function and provenance. In: 2019 ACM/IEEE Joint Conference on Digital Libraries (JCDL), pp. 394–395. IEEE (2019)
27. Tang, J., Zhang, J.: A discriminative approach to topic-based citation recommendation. In: Theeramunkong, T., Kijsirikul, B., Cercone, N., Ho, T.-B. (eds.) PAKDD 2009. LNCS (LNAI), vol. 5476, pp. 572–579. Springer, Heidelberg (2009). https://doi.org/10.1007/978-3-642-01307-2_55
28. Teufel, S., Siddharthan, A., Tidhar, D.: Automatic classification of citation function. In: EMNLP, pp. 103–110 (2006)
29. Tuarob, S., Bhatia, S., Mitra, P., Giles, C.L.: Algorithmseer: a system for extracting and searching for algorithms in scholarly big data. IEEE Trans. Big Data 2(1), 3–17 (2016)
30. Ucar, I., López-Fernandino, F., Rodriguez-Ulibarri, P., Sesma-Sanchez, L., Urrea-Micó, V., Sevilla, J.: Growth in the number of references in engineering journal papers during the 1972–2013 period. Scientometrics 98(3), 1855–1864 (2013). https://doi.org/10.1007/s11192-013-1113-6
31. Vaswani, A., et al.: Attention is all you need. arXiv preprint arXiv:1706.03762 (2017)
32. Wan, X., Liu, F.: Are all literature citations equally important? automatic citation strength estimation and its applications. JASIST 65(9), 1929–1938 (2014)
33. Wu, J., et al.: Citeseerx: AI in a digital library search engine. AI Magazine 36(3), 35–48 (2015)
34. Yang, L., et al.: A ISTM based model for personalized context-aware citation recommendation. IEEE Access 6, 59618–59627 (2018)
35. Yang, Z., Yang, D., Dyer, C., He, X., Smola, A., Hovy, E.: Hierarchical attention networks for document classification. In: NAACL, pp. 1480–1489 (2016)

Exploring Entities in Event Detection
as Question Answering

Emanuela Boros[1]($\boxtimes$) (iD), Jose G. Moreno[1,2] (iD), and Antoine Doucet[1] (iD)

[1] University of La Rochelle, L3i, 17000 La Rochelle, France
{emanuela.boros,antoine.doucet}@univ-lr.fr
[2] University of Toulouse, IRIT, UMR 5505 CNRS, 31000 Toulouse, France
jose.moreno@irit.fr
https://www.univ-larochelle.fr

Abstract. In this paper, we approach a recent and under-researched paradigm for the task of event detection (ED) by casting it as a question-answering (QA) problem with the possibility of multiple answers and the support of entities. The extraction of event triggers is, thus, transformed into the task of identifying answer spans from a context, while also focusing on the surrounding entities. The architecture is based on a pre-trained and fine-tuned language model, where the input context is augmented with entities marked at different levels, their positions, their types, and, finally, their argument roles. Experiments on the ACE 2005 corpus demonstrate that the proposed model properly leverages entity information in detecting events and that it is a viable solution for the ED task. Moreover, we demonstrate that our method with different entity markers is particularly able to extract unseen event types in few-shot learning settings.

Keywords: Event detection · Question answering · Few-shot learning

1 Introduction

Event extraction (EE) is a crucial and challenging task of information extraction (IE) that aims at identifying the instances of specified types of events in a text, generally referred to as event detection (ED), and the detection and classification of the corresponding arguments (participants). For instance, according to the ACE 2005 annotation guidelines[1], an event is described as having the following characteristics:

- the *event mention* is an occurrence of an event with a particular type. This is usually a sentence or a phrase that describes an event; the *event trigger* is the word that most clearly expresses the event mention, e.g. *Attack*;

---

[1] https://www.ldc.upenn.edu/sites/www.ldc.upenn.edu/files/english-events-guidelines-v5.4.3.pdf.

This work has been supported by the European Union's Horizon 2020 research and innovation program under grants 770299 (NewsEye) and 825153 (Embeddia), and by the ANNA and Termitrad projects funded by the Nouvelle-Aquitaine Region.

M. Hagen et al. (Eds.): ECIR 2022, LNCS 13185, pp. 65–79, 2022.
https://doi.org/10.1007/978-3-030-99736-6_5

- the *event argument* is an entity mention or temporal expression (e.g., *Crime*, *Job-Title*) that serves as a participant with a specific role in an event mention. Event arguments have an *entity type*, e.g. persons (PER), locations (LOC), organizations (ORG), etc.; and the *argument role* that is the relationship between an argument and the event in which it participates.

Following this description, from the sentence "*Police have arrested four people in connection with the killings.*", an event extraction system should be able to recognize the word killings as a trigger for an event of type *Die*, with the person (PER) entity *Police* as an argument with the role of an Agent and the person *four people* as an argument of type Person, and the word arrested as a trigger for an *Arrest-Jail* event type with no arguments.

In this paper, we approach the task of event detection (ED) by studying the usage of entities in a recent and under-researched paradigm for the task of event detection (ED) by casting it as a question-answering (QA) problem.

There have been several deep learning-based major techniques applied for approaching the ED task while taking advantage of entity or argument information in the literature. First, systems extensively utilized linguistic analysis, entity information, entity coreference, and other knowledge resources to capture the discrete structures for ED, focusing on the combination of these discriminative features to build statistical models [7,11,13]. Next, neural-based approaches were based on convolutional and recurrent neural networks (CNNs and RNNs) that utilized effective feature representations from entity type embeddings [4,16,21–25].

Recent approaches adopt the usage of pre-trained language models [28]. Since BERT [5] broke records for several natural language processing (NLP) tasks (part-of-speech tagging, named entity recognition, etc.) and received a lot of attention, recent advances in ED imply architectures based on fine-tuning this type of models [2,8,29,31], these methods holding the state of the art for ED.

Differently from these Transformer-based methods, where event and argument detection were considered as classification tasks, a new paradigm was introduced [6,15] formulating EE as a question answering (QA)/machine reading comprehension (MRC[2]) task, where events can be extracted by responding to the 5W1H questions (who did what, when, where, why, and how). While these recent advances claim to cast the EE task as an MRC task [6,15], they mostly focus on argument extraction as QA, while for ED, the models remain formulated as a sequential classification problem that aims at detecting event triggers of specific types.

Thus, in this paper, we focus on the event detection task, and we first cast it as a QA task with the possibility of multiple answers, in the case where more than one event is present in the text. By approaching it as a QA model, not only are we able to leverage the recent advances in MRC, we also avoid the classification based-methods that can either require lots of training data and are challenged by the annotation cost or data scarcity.

---

[2] In one view, the recent tasks titled MRC can also be seen as the extended tasks of question answering (QA).

Second, we take advantage of the presence of entities for extracting the events[3], considering that informative features can be brought by additional entity markers for better distinguishing the event triggers. We agree that "Entities of the consistent type normally participate in similar events as the same role." [7].

In addition, modeling the task as QA can improve the ED task in regard to this challenge due to the fact that the answers are only considered in relation to the context and the question, which could reduce trigger ambiguity. Furthermore, compared to classification based-methods [4,6,12,15,15] that generally lack this ability, we demonstrate that our proposed QA models are more effective in few-shot scenarios by showing that they are able to extract unseen event types. The work of [9] is distinguished in the literature, where the authors prove that zero-shot learning for detecting events can be possible and efficient. They proposed to leverage existing human-constructed event schemas and manual annotations for a small set of seen types, and transfer the knowledge from the existing event types to the extraction of unseen event types. We consider this paper as our reference method for the few-shot learning setting, and we prove that modeling the ED task as QA with entity information can obtain higher performance results.

Our proposed method with entity information obtains state-of-the-art results when compared with previous models that utilize entity or argument information. Moreover, these methods could foster further research and help to study transfer learning from QA models to boost the performance of existing information extraction systems. Furthermore, compared to classification based-methods that lack this ability, we demonstrate that our proposed QA models are more effective in few-shot scenarios by showing that they are able to extract unseen event types.

Next, we continue with the related work in Sect. 2, and we detail the QA model with entity markers in Sect. 3. The experimental setup and the results are presented in Sect. 4. We provide a discussion of the results by analyzing the output in Sect. 5 and we draw conclusions in Sect. 6.

2   Related Work

Event Detection with Entity Information. In the context of event detection, some works made use of gold-standard entities in different manners. Higher results can be obtained with gold-standard entity types [23], by concatenating randomly initialized embeddings for the entity types. A graph neural network (GNN) based on dependency trees [25] has also been proposed to perform event detection with a pooling method that relies on entity mentions aggregation. Arguments provided significant clues to this task in the supervised attention

---

[3] We note here that event extraction generally depends on previous phases as, for example, named entity recognition, entity mention coreference, and classification. Thereinto, the named entity recognition is another hard task in the ACE evaluation and not the focus of this paper. Therefore, we will temporarily skip the phase and instead directly use the entities provided by ACE, following previous work [4,7,10, 12,14,15].

mechanism proposed to exploit argument information explicitly for ED proposed by [17]. Other methods that took advantage of argument information were joint-based approaches.

The architecture adopted by [18] was jointly extracting multiple event triggers and event arguments by introducing syntactic shortcut arcs derived from the dependency parsing trees. [7]'s cross-entity feature-based method extracted events by using gold standard cross-entity inference in order to take advantage of the consistency of entity mentions while achieving efficient sentence-level trigger and argument (role) classification. [13] utilized the contextual entities in a joint framework based on a structured prediction that extracted triggers and arguments together so that the local predictions can be mutually improved.

Approaches presented by [23] and [4] experimented with the integration of entities in ED models based on CNNs. These models utilized effective feature representations from pre-trained word embeddings, position embeddings as well as entity type embeddings. [24] improve the previous model proposed by [23] by taking into account the possibility to have non-consecutive n-grams as basic features instead of continuous n-grams.

A different technique was explored by [1] and it consisted in marking the entities in the relation extraction task and by studying the ability of the Transformer-based neural networks to encode relations between entity pairs. They identified a method of representation based on marking the present entities that outperform previous work in supervised relation extraction. [20] also explored the use of pre-trained neural models into the relation validation problem by explicitly using a triplet-sentence representation with marked entities, proving that the relation extraction performance could be further improved by using this additional information. Furthermore, [2] also proposed the use of pre-trained neural models in a BERT-based classification-based architecture for detecting events.

Event Detection as Question Answering. While QA for event detection is roughly under-researched, Transformer-based models have led to striking gains in performance on MRC tasks recently, as measured on the SQuAD v1.1[4] [27] and SQuAD v2.0[5] [26] leaderboards.

A recent work proposed by [6] introduced this new paradigm for event extraction by formulating it as a QA task, which extracts the event triggers and arguments in an end-to-end manner. For detecting the event, they considered an approach based on BERT that is usually applied to sequential data. The task of ED is a classification-based method where the authors designed simple fixed templates as in *what is the trigger, trigger, action, verb*, without specifying the event type. For example, if they chose *verb* template, the input sequence would be: [CLS] *verb* [SEP] sentence [SEP]. Next, they use a sequential fine-tuned BERT for detecting event trigger candidates.

---

[4] SQuAD v1.1 consists of reference passages from Wikipedia with answers and questions constructed by annotators after viewing the passage.

[5] SQuADv2.0 augmented the SQuAD v1.1 collection with additional questions that did not have answers in the referenced passage.

Another recent paper [15] also approaches the event extraction task as a question answering task, similar to the [6] method. The task remains classification-based (instead of the span-based QA method) for trigger extraction, jointly encode [EVENT] with the sentence to compute an encoded representation, as in the approach proposed by [6] where the special token was *verb* or *trigger*.

3 Event Question Answering Model with Entity Positions, Types, and Argument Roles

We formulate the ED task as a QA task, where, for every sentence, we ask if a particular event type is present, and we expect a response with an event trigger, multiple event triggers, or none. Our model extends the BERT [5] pre-trained model which is a stack of Transformer layers [28] that takes as input a sequence of subtokens, obtained by the WordPiece tokenization [30] and produces a sequence of context-based embeddings of these subtokens.

To feed a QA task into BERT, we pack both the question and the reference text into the input, as illustrated in Fig. 1. The input embeddings are the sum of the token embeddings and the segment embeddings. The input is processed in the following manner: token embeddings (a [CLS] token is added to the input word tokens at the beginning of the question and a [SEP] token is inserted at the end of both the question and the reference text) and segment embeddings (a marker indicating the question or the reference text is added to each token). This allows the model to distinguish between the question and the text.

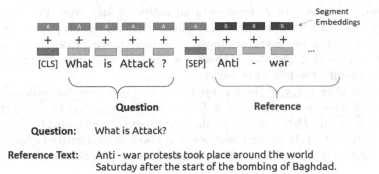

Question: What is Attack?

Reference Text: Anti - war protests took place around the world Saturday after the start of the bombing of Baghdad.

Fig. 1. Example of input modification to fit the QA paradigm for a sentence that contains an event of type *Attack*. The question is separated by [SEP] token from the reference text that contains the event trigger <u>war</u>.

To fine-tune BERT for a QA system, a start vector and an end vector are introduced. A linear layer is added at the top of BERT layers with two outputs for the start and end vectors of the answer. The probability of each word being the start or end word is calculated by taking a dot product between the final embedding of the word and the start or end vector, followed by a Softmax over all the words. The word with the highest probability value is considered. This method differs from the event detection approaches presented by [6] and [15] where the models are classification-based, instead of the span-based QA.

Next, for every type of event [**Event Type**] (*Demonstrate, Die, Attack*, etc.), we formulate the question by automatically generating them using the following template:

What is the [Event Type] ?

An example for a sentence containing an *Attack* event is illustrated in Fig. 1. We also consider questions that do not have an answer in the case where an event of a specific type is not present in the sentence. When there is more than one event of the same type in a sentence, we consider that the question has multiple answers. From the n best-predicted answers, we consider all those that obtained a probability higher than a selected threshold (established on the development set). When the predicted chunks are self-contained, we consider only the first predicted event trigger. For example, if the noun chunks *assault* and *air assault* are predicted, only *assault* is considered.

Next, for adding entity information, we augment the input data with a series of special tokens. Thus, if we consider a sentence $x = [x_0, x_1, \ldots, x_n]$ with n tokens, we augment x with two reserved word pieces to mark the beginning and the end of each event entity or argument mention in the sentence.

Next, we propose three types of markers: (1) *Entity Position Markers*, e.g. <E> and </E> where E represents an entity of any type, (2) *Entity Type Markers*, e.g. <PER> and </PER> where PER represents an entity of type Person, and (3) if the event argument roles are known beforehand, *Argument Role Markers*, e.g. <Agent>, </Agent> where Agent is an event argument role. Thus, we modify the following sentence:

"**Police** have <u>arrested</u> **four people** in connection with the <u>killings</u>."

where <u>killings</u> is a trigger for a *Die* event, and <u>arrested</u> is a trigger for an *Arrest-Jail* event, *Police* is one of the participants, a person (*PER*) with the argument role of an *Agent*, and *four people* is also a person entity (*PER*) with the *Person* argument role. The modified sentences with the three types of markers are:

(1) "< E > **Police** < /E > have <u>arrested</u> < E > **four people** < /E > in connection with the <u>killings</u>."

(2) "< PER > **Police** < /PER > have <u>arrested</u> < /PER > **four people** < /PER > in connection with the <u>killings</u>."

(3) "< Agent > **Police** < /Agent > have <u>arrested</u> < Person > **four people** < /Person > in connection with the <u>killings</u>."

Further, an ED system should detect in the presented sentence, the trigger word <u>killings</u> for an event of type Die (this event has two arguments *Police* and *four people*) and <u>arrested</u> for an event of type Arrest-Jail (this event has no arguments). For the *Argument Role Markers*, if an entity has different roles in different events that are present in the same sentence, we mark the entity with all the argument roles that it has.

4 Experiments

Table 1. Evaluation of our models and comparison with state-of-the-art systems for event detection on the blind test data. The models with ♣ utilized gold standard entity mentions. The models with ♡ utilized gold standard arguments. Statistical significance is measured with McNemar's test. * denotes a significant improvement at p ≤ 0.01.

Approaches	P	R	F1
MaxEnt with local features♣ [12]	74.5	59.1	65.9
Cross-entity♣ [7]	72.9	64.3	68.3
DMCNN♣ [4]	75.6	63.6	69.1
Word CNN♣ [23]	71.8	66.4	69.0
Joint RNN♣ [21]	66.0	73.0	69.3
BERT-QA-base-uncased	68.4	70.5	69.5
BERT-base [6]	67.1	73.2	70.0
Non-Consecutive CNN♣ [22]	–	–	71.3
Attention-based♣♡ [16]	78.0	66.3	71.7
BERT_QA_Trigger [6]	71.1	73.7	72.3
Graph CNN♣ [25]	77.9	68.8	73.1
BERT-QA-base-uncased + Entity Position Markers♣	78.0	70.7	74.2*
RCEE_ER♣ [15]	75.6	74.2	74.9
BERT-QA-base-uncased + Entity Type Markers♣	78.5	77.2	77.8*
BERT-QA-base-uncased + Argument Role Markers♡	83.2	80.5	81.8*

The evaluation is conducted on the ACE 2005 corpus provided by ACE program[6]. For comparison purposes, we use the same test set with 40 news articles (672 sentences), the same development set with 30 other documents (863 sentences) and the same training set with the remaining 529 documents (14,849 sentences) as in previous studies of this dataset [10,14]. The ACE 2005 corpus has 8 types of events, with 33 subtypes (e.g. the event type *Conflict* has two subtypes *Attack, Demonstrate*). In this paper, we refer only to the subtypes of the events, without diminishing the meaning of main event types.

Evaluation Metrics. Following the same line of previous works, we consider that a trigger is correct if its event type, subtype, and offsets match those of a reference trigger. We use Precision (P), Recall (R), and F-measure (F1) to evaluate the overall performance.

---

[6] https://catalog.ldc.upenn.edu/LDC2006T06.

Hyperparameters. We used the Stanford CoreNLP toolkit[7] to pre-process the data, including tokenization and sentence splitting[8]. For fine-tuning the BERT-based models, we followed the selection of hyperparameters presented by [5]. We found that 3×10^{-5} learning rate and a mini-batch of dimension 12 for the *base* models provided stable and consistent convergence across all experiments as evaluated on the development set. The maximum sequence length is set to 384 and the document stride of 128. For selecting the event triggers, we generate $n = 10$ candidates, and we use the same threshold for all the experiments, with a value of 0.2 that was decided on the development set.

General Evaluation. In Table 1, we present the comparison between our model and state-of-the-art approaches that utilised entity or argument information.

We compare with the MaxEnt-based model with local features in [12], the cross-entity feature-based method extracted events by using gold standard cross-entity inference [7] and the models proposed by [4,22,23], and the joint framework with bidirectional RNNs [21] that experimented with the integration of entities in ED models based on a CNN-based architectures.

We also compare with the method proposed by [16] that also exploited entity information explicitly for ED via supervised attention mechanisms, and the graph CNN by [25] that investigated a CNN based on dependency trees for ED with pooling method that relied on entity mentions to aggregate the convolution vectors.

We also compare with the models where the task has been approached as a QA task but still formulated as a sequential classification problem that aims at locating trigger candidates, the fine-tuned baseline BERT-base-uncased and the BERT_QA_Trigger [6], and the RCEE_ER (Reading Comprehension for Event Extraction, with *ER* that denotes that the model has golden entity refinement) [15].

When compared with the previous state-of-the-art models that included entity information, except for the RCEE_ER method, our models that use either the positions or the types of the entities bring a considerable improvement in the performance of trigger detection. It is clear that further marking the entities with their types can increase both precision and recall, balancing the final scores.

It is noteworthy that, while entities can be present in the entire document, arguments can only surround event triggers. Knowing the argument roles beforehand brings further improvements, we assume that an important reason for this is that, since the arguments are present only around event triggers, this could help the language model to be more aware of the existence of an event or multiple events in a sentence.

---

[7] http://stanfordnlp.github.io/CoreNLP/.
[8] The code is available at https://github.com/nlpcl-lab/ace2005-preprocessing as it consists of the same pre-processing as utilized in several other papers [21,23].

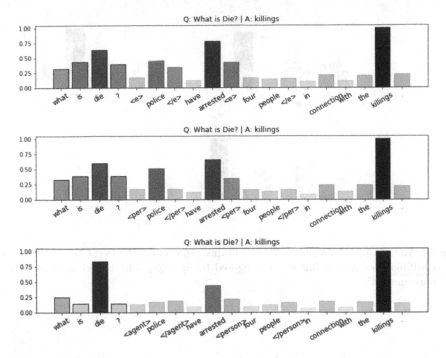

Fig. 2. An example for the *Die* event triggered by killings with three types of markers: *Entity Position, Entity Type,* and *Argument Role Markers.*

5 Discussion

5.1 Trigger Ambiguity Analysis

For a deeper analysis of the impact of entity information, we leverage the gradients in our proposed models to efficiently infer the relationship between the question, context, and the output response. [3] studied the identifiability of attention weights and token embeddings in Transformer-based models. They show that the self-attention distributions are not directly interpretable and suggest that simple gradient explanations are stable and faithful to the model and data generating process.

Thus, as applied by [19], to get a better idea of how well each model memorizes and uses memory for contextual understanding, we analyze the connectivity between the desired output and the input. This is calculated as:

$$\text{connectivity}(t, \tilde{t}) = \left\| \frac{\partial y_k^{\tilde{t}}}{\partial x^t} \right\|_2$$

where t is the time index, $\tilde{t}$ the output time index, and the result is the magnitude of the gradient between the logits for the desired output $y_k^{\tilde{t}}$ and the input x^t. The connectivity is computed with respect to both start position and end position

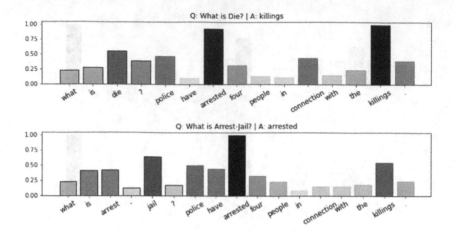

Fig. 3. An example of a sentence that contains two events: *Die* event triggered by the word killings and *Arrest-Jail* event triggered by arrested. The model used is BERT-QA-base-uncased.

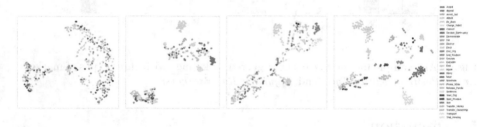

Fig. 4. [CLS] representation of each sentence in the test set that contains at least an event for BERT-QA-base-uncased, BERT-QA-base-uncased + *Entity Position Markers+*, BERT-QA-base-uncased + *Entity Type Markers+*, and BERT-QA-base-uncased + *Argument Role Markers*.

of the answer, then it is normalized, and it is visible as saliency maps for every word in Figs. 3 and 2[9].

By looking at the gradients in Fig. 3, where two events of different types are present, we can observe, in the upper part of the figure, that while the model sees the word killings and arrested as impactful, it also sees the words *police, connection* as impactful and selects an answer in that neighborhood. Even though both trigger candidates killings and arrested have a clear impact due to their gradient values, by looking at the probability values, killings is recognized with a 99.4% probability, while arrested obtained a probability of 2.3×10^{-7}, value that is lower than our selected threshold 0.2. In the lower part of the figure, for the question *What is Arrest-Jail?*, the words *die, police,* killings clearly influence the choice of the answer arrested.

---

[9] The sentence is lowercased for the *uncased* models.

In Fig. 2, we present the same sentence with the three types of input modifications: *Entity Position Markers, Entity Type Markers*, and *Argument Role Markers*, with the *What is Die?* question and the correct answer killings. In the upper part of the figure, where the sentence has been augmented with the entity position markers <E> and </E>, we notice that the words that impact the most in the result are killings along with *die*, arrested, and *police*. In this case, one can also see that the end marker </E> contributed too.

In the middle part of the figure, where the sentence has been augmented with the entity position markers <PER> and </PER> for the two entities *police* and *four people*, the influence of other words as in *die*, arrested, and *police* slightly decreased. In the bottom part of the image, the gradients of these words are visibly reduced.

When the sentence is augmented with argument roles, <Agent>, </Agent>, <Person> and </Person>, the noise around the correct answer has noticeably diminished, being reduced by the additional markers. The most impactful remaining words are the word *die* in the question and the correct answer killings.

In order to analyze the quality of the sentence representations, we extract the [CLS] representation of each sentence for BERT-QA-base-uncased and for BERT-QA-base-uncased + *Argument Role Markers*. Then, we plot these representations in two spaces where the labels (colors of the dots) are the event types, as illustrated in Fig. 4. On the right-hand side of the figure, where argument role markers are used, it is clear that the sentence representations clusters are more cohesive than when no entity information is considered (left-hand side), thus confirming our assumption regarding the importance of the entity informative features in a QA system.

5.2   Evaluation on Unseen Event Types

In the first scenario, we follow the same strategy as [6] where we keep 80% of event types (27) in the training set and 20% (6) unseen event types in the test set. More exactly, the unseen event types were chosen randomly, and they are: *Marry, Trial-Hearing, Arrest-Jail, Acquit, Attack*, and *Declare-Bankruptcy*. Table 2 presents the performance scores of our models for the unseen event types.

Table 2. Evaluation of our models on unseen event types. The models with ♣ utilized gold standard entity mentions. The models with ♡ utilized gold standard arguments.

Approaches	P	R	F1
BERT-QA-base-uncased (*not trained on ACE 2005*)	0.7	8.3	1.3
BERT-QA-base-uncased	47.7	26.7	31.1
BERT-QA-base-uncased + *Entity Position Markers*♣	44.0	47.5	37.3
BERT-QA-base-uncased + *Entity Type Markers*♣	53.6	54.4	50.4
BERT-QA-base-uncased + *Argument Role Markers*♡	83.3	47.4	53.6

We compare with BERT-QA-base-uncased which is our baseline that selects an event trigger in a sentence without being trained on ACE 2005 data. Since the majority of the models in Table 1 are classification-based in a sequential manner, they are not capable of handling unseen event types, and thus, we were not able to obtain performance values. From the results, without any event annotation, the BERT-QA-base-uncased obtains a low F1 value (1.38%). We observe that the performance values increase proportionally to the specificity of the markers. Thus, it is not surprising that the highest values are obtained when the argument roles are marked, also obtaining the highest precision.

In a second scenario, we consider larger amounts of unseen events, and we follow the strategy proposed by [9], where out of the total number of event types (33), we select the top-N most popular event types as seen, while the rest remain unseen. N is set as 1, 3, 5, 10 respectively. We perform experiments in four settings (A, B, C, and D). Table 3 shows the types that were selected for training in each experiment setting.

Table 3. Seen types in each experiment setting as proposed by [9].

Setting	N	Seen Event Types
A	1	Attack
B	3	Attack, Transport, Die
C	5	Attack, Transport, Die, Meet, Arrest-Jail
D	10	Attack, Transport, Die, Meet, Sentence, Arrest-Jail, Transfer-Money, Elect, Transfer-Ownership, End-Position

Table 4. Evaluation of our models on unseen event types (Hit@1 as in [9]). The models with ♣ utilized gold standard entity mentions. The models with ♡ utilized gold standard arguments.

Approaches	Settings			
	A	B	C	D
Huang et al. [9]	3.9	7.0	20.0	33.4
BERT-QA-base-uncased + *Entity Positions Markers*♣	2.3	4.9	18.8	21.7
BERT-QA-base-uncased + *Entity Type Markers*♣	2.3	8.8	21.8	25.8
BERT-QA-base-uncased + *Argument Role Markers*♡	2.4	10.0	26.2	32.0

Table 4 presents the performance scores of our models for the unseen event types. We focus on showing the effectiveness of our methods juxtaposed with the results of [9]. We first observe, for each model, that the performance values improve as the number of seen events types. Second, one can notice that the scores also increase proportionally to the specificity of the markers.

6   Conclusions and Perspectives

In this paper, we utilized a recent and under-researched paradigm for detecting events by modeling the ED as a QA task with the addition of entity and argument information. The questions were simplified to a pre-defined list with a question for every type of event present in the dataset, which allows the model to predict multiple events in a sentence. The additional informative features brought by the presence of entities and the argument roles in the same context of the events considerably increased the performance of the model, achieving state-of-the-art results. Moreover, this type of model that utilizes the entity information leveraged the ambiguity of the event triggers and demonstrate potential in detecting unseen event types.

In future work, we will focus on approaching the entity and argument detection tasks, in order to analyze the influence of the predicted event arguments and the error propagation from this task to the downstream event detection task. Furthermore, we will consider approaching both event extraction sub-tasks (ED and argument detection and classification) in a joint QA-based architecture for alleviating the aforementioned issue concerning the diffusion of detection errors.

References

1. Baldini Soares, L., FitzGerald, N., Ling, J., Kwiatkowski, T.: Matching the blanks: distributional similarity for relation learning. In: Proceedings of the 57th Annual Meeting of the Association for Computational Linguistics, pp. 2895–2905. Association for Computational Linguistics, Florence, Italy (2019). https://doi.org/10.18653/v1/P19-1279, https://www.aclweb.org/anthology/P19-1279
2. Boros, E., Moreno, J.G., Doucet, A.: Event detection with entity markers. In: Hiemstra, D., Moens, M., Mothe, J., Perego, R., Potthast, M., Sebastiani, F. (eds.) Advances in Information Retrieval - 43rd European Conference on IR Research, ECIR 2021, Virtual Event, 28 March 1 April 2021, Proceedings, Part II. Lecture Notes in Computer Science, vol. 12657, pp. 233–240. Springer, Heidelberg (2021). https://doi.org/10.1007/978-3-030-72240-1_20
3. Brunner, G., Liu, Y., Pascual, D., Richter, O., Ciaramita, M., Wattenhofer, R.: On identifiability in transformers. In: International Conference on Learning Representations (2019)
4. Chen, Y., Xu, L., Liu, K., Zeng, D., Zhao, J.: Event extraction via dynamic multi-pooling convolutional neural networks. In: Proceedings of the 53rd Annual Meeting of the Association for Computational Linguistics and the 7th International Joint Conference on Natural Language Processing (Volume 1: Long Papers), pp. 167–176 (2015)
5. Devlin, J., Chang, M.W., Lee, K., Toutanova, K.: BERT: pre-training of deep bidirectional transformers for language understanding. arXiv preprint arXiv:1810.04805 (2018)
6. Du, X., Cardie, C.: Event extraction by answering (almost) natural questions. In: Proceedings of the 2020 Conference on Empirical Methods in Natural Language Processing (EMNLP), pp. 671–683. Association for Computational Linguistics, Online (2020). https://www.aclweb.org/anthology/2020.emnlp-main.49

7. Hong, Y., Zhang, J., Ma, B., Yao, J., Zhou, G., Zhu, Q.: Using cross-entity inference to improve event extraction. In: Proceedings of the 49th Annual Meeting of the Association for Computational Linguistics: Human Language Technologies, vol. 1, pp. 1127–1136. Association for Computational Linguistics (2011)
8. Hong, Y., Zhou, W., Zhang, J., Zhou, G., Zhu, Q.: Self-regulation: employing a generative adversarial network to improve event detection. In: Proceedings of the 56th Annual Meeting of the Association for Computational Linguistics (Volume 1: Long Papers), pp. 515–526 (2018)
9. Huang, L., Ji, H., Cho, K., Dagan, I., Riedel, S., Voss, C.: Zero-shot transfer learning for event extraction. In: Proceedings of the 56th Annual Meeting of the Association for Computational Linguistics (Volume 1: Long Papers), Melbourne, Australia, pp. 2160–2170. Association for Computational Linguistics (2018). https://doi.org/10.18653/v1/P18-1201, https://www.aclweb.org/anthology/P18-1201
10. Ji, H., Grishman, R., et al.: Refining event extraction through cross-document inference. In: ACL, pp. 254–262 (2008)
11. Li, P., Zhu, Q., Zhou, G.: Argument inference from relevant event mentions in Chinese argument extraction. In: ACL, no. 1, pp. 1477–1487 (2013)
12. Li, Q., Ji, H., Huang, L.: Joint event extraction via structured prediction with global features. In: ACL, no. 1, pp. 73–82 (2013)
13. Li, W., Cheng, D., He, L., Wang, Y., Jin, X.: Joint event extraction based on hierarchical event schemas from FrameNet. IEEE Access **7**, 25001–25015 (2019)
14. Liao, S., Grishman, R.: Using document level cross-event inference to improve event extraction. In: Proceedings of the 48th Annual Meeting of the Association for Computational Linguistics, pp. 789–797. Association for Computational Linguistics (2010)
15. Liu, J., Chen, Y., Liu, K., Bi, W., Liu, X.: Event extraction as machine reading comprehension. In: Proceedings of the 2020 Conference on Empirical Methods in Natural Language Processing (EMNLP), pp. 1641–1651 (2020)
16. Liu, S., Chen, Y., Liu, K., Zhao, J.: Exploiting argument information to improve event detection via supervised attention mechanisms. In: Proceedings of the 55th Annual Meeting of the Association for Computational Linguistics (Volume 1: Long Papers), pp. 1789–1798 (2017)
17. Liu, S., Chen, Y., Liu, K., Zhao, J.: Exploiting argument information to improve event detection via supervised attention mechanisms. In: 55th Annual Meeting of the Association for Computational Linguistics (ACL 2017), Vancouver, Canada, pp. 1789–1798 (2017)
18. Liu, X., Luo, Z., Huang, H.: Jointly multiple events extraction via attention-based graph information aggregation. arXiv preprint arXiv:1809.09078 (2018)
19. Madsen, A.: Visualizing memorization in RNNs. Distill (2019). https://doi.org/10.23915/distill.00016, https://distill.pub/2019/memorization-in-rnns
20. Moreno, J.G., Doucet, A., Grau, B.: Relation classification via relation validation. In: Proceedings of the 6th Workshop on Semantic Deep Learning (SemDeep-6), pp. 20–27 (2021)
21. Nguyen, T.H., Cho, K., Grishman, R.: Joint event extraction via recurrent neural networks. In: Proceedings of NAACL-HLT, pp. 300–309 (2016)
22. Nguyen, T.H., Fu, L., Cho, K., Grishman, R.: A two-stage approach for extending event detection to new types via neural networks. ACL **2016**, 158 (2016)
23. Nguyen, T.H., Grishman, R.: Event detection and domain adaptation with convolutional neural networks. In: ACL, no. 2, pp. 365–371 (2015)
24. Nguyen, T.H., Grishman, R.: Modeling skip-grams for event detection with convolutional neural networks. In: Proceedings of EMNLP (2016)

25. Nguyen, T.H., Grishman, R.: Graph convolutional networks with argument-aware pooling for event detection. In: Thirty-Second AAAI Conference on Artificial Intelligence (AAAI 2018) (2018)
26. Rajpurkar, P., Jia, R., Liang, P.: Know what you don't know: unanswerable questions for squad. arXiv preprint arXiv:1806.03822 (2018)
27. Rajpurkar, P., Zhang, J., Lopyrev, K., Liang, P.: Squad: 100,000+ questions for machine comprehension of text. arXiv preprint arXiv:1606.05250 (2016)
28. Vaswani, A., et al.: Attention is all you need. In: Advances in Neural Information Processing Systems, pp. 5998–6008 (2017)
29. Wang, X., Han, X., Liu, Z., Sun, M., Li, P.: Adversarial training for weakly supervised event detection. In: Proceedings of the 2019 Conference of the North American Chapter of the Association for Computational Linguistics: Human Language Technologies, Volume 1 (Long and Short Papers), pp. 998–1008 (2019)
30. Wu, Y., et al.: Google's neural machine translation system: bridging the gap between human and machine translation. arXiv preprint arXiv:1609.08144 (2016)
31. Zhang, T., Ji, H., Sil, A.: Joint entity and event extraction with generative adversarial imitation learning. Data Intell. 1(2), 99–120 (2019)

Validating Simulations of User Query Variants

Timo Breuer[1]([✉]) [ID], Norbert Fuhr[2] [ID], and Philipp Schaer[1] [ID]

[1] TH Köln, Köln, Germany
{timo.breuer,philipp.schaer}@th-koeln.de
[2] Universität Duisburg-Essen, Duisburg, Germany
norbert.fuhr@uni-due.de

Abstract. System-oriented IR evaluations are limited to rather abstract understandings of real user behavior. As a solution, simulating user interactions provides a cost-efficient way to support system-oriented experiments with more realistic directives when no interaction logs are available. While there are several user models for simulated clicks or result list interactions, very few attempts have been made towards query simulations, and it has not been investigated if these can reproduce properties of real queries. In this work, we validate simulated user query variants with the help of TREC test collections in reference to real user queries that were made for the corresponding topics. Besides, we introduce a simple yet effective method that gives better reproductions of real queries than the established methods. Our evaluation framework validates the simulations regarding the retrieval performance, reproducibility of topic score distributions, shared task utility, effort and effect, and query term similarity when compared with real user query variants. While the retrieval effectiveness and statistical properties of the topic score distributions as well as economic aspects are close to that of real queries, it is still challenging to simulate exact term matches and later query reformulations.

Keywords: Query Simulation · Dynamic Test Collection · Reproducibility

1 Introduction

In accordance with the Cranfield paradigm, the underlying user model of system-oriented IR evaluations is an abstract representation of real search behavior. The simplified understanding of users is limited to a single query and the examination of the result list in its entirety. Real search behavior is more complex: searching is normally an iterative process with query reformulations, and not every search result is examined but rather picked out after judging its snippet text. To compensate for this shortcoming, it is common practice to include (logged) user interactions in the evaluation process. Industrial research is often supported by large datasets of user interactions that, unfortunately, cannot be shared publicly, e.g., due to privacy concerns [14]. Carterette et al. address the lack of user

M. Hagen et al. (Eds.): ECIR 2022, LNCS 13185, pp. 80–94, 2022.
https://doi.org/10.1007/978-3-030-99736-6_6

interaction data available to academic research by introducing the concept of Dynamic Test Collections [12]. Their framework expands test collections with simulated interactions comprising the entire sequence of interactions, including the simulation of queries, clicks, dwell times, and session abandonment.

Our work can be seen in the light of Dynamic Test Collections, but with a special focus on simulating user query variants (UQVs). While previous work on simulating interactions either focused on the completeness of interaction sequences [12,34,47], click interactions [13], or stopping rules [34,37], work on simulating queries is underrepresented [20]. To the best of our knowledge, the degree to which query simulators reproduce real user queries has not yet been analyzed with TREC test collections. As opposed to previous work in this regard, it is not our primary goal to generate the most effective queries but rather to validate simulated queries, since the query formulation is one of the first user interactions with the search system and as such it is a critical component for any subsequent simulated interactions like clicks and others. More specifically, our evaluations answer (**RQ1**) *How do real user queries relate to simulated queries made from topic texts and known-items in terms of retrieval effectiveness?* and (**RQ2**) *To which degree do simulated queries reproduce real queries provided that only resources of the test collection are considered for the query simulation?*

Our contributions are as follows. (**1**) We introduce an evaluation framework that is used to analyze to which extent simulations reproduce real queries and that reveals current limitations, (**2**) we compare and analyze conventional query simulation methods that do not rely on large-scale interaction logs, (**3**) we propose a new simulation method and hereby show that the parameterized query reformulation behavior results in a better approximation of real queries and resembles those of specific users, (**4**) we publish the code of the experiments and provide the simulated queries for follow-up studies.

2   Related Work

Carterette et al. introduced Dynamic Test Collections [12] by enriching test collections with simulated user interactions. Their outlined interaction sequences included the simulation of queries, clicks, dwell times, and session abandonment. Even though they implemented some specific simulators as part of their experiments, they intended to provide a general framework that covers all elements of user interactions which can also be implemented with various methods. More recently, similar frameworks were introduced by Pääkkönen et al. as Common Interaction Model [37] and Zhang et al. [47].

Most of the current methods for query simulations follow a two-stage approach including the *term candidate generation* and the *query modification strategy*. Usually, the term candidates are derived from a language model. Jordan et al. introduced Controlled Query Generation (CQG) [28] that exploited the relative entropy of a language model for query term generation. Azzopardi et al. applied CQG when generating queries for known-item search [3,4]. In a similar vein, Berendsen et al. used annotations to group documents [9] and Huurnik et al. simulated queries for purchased items [24]. When query term candidates are available, there exist some commonly used query modification strategies [2,5,28,29], which were also applied in follow-up studies [34,35,41] and followed a principled way resulting in controlled query reformulation patterns (cf. Sect. 3.1).

If large-scale user logs are available, different approaches propose, for instance, learning to rewrite queries [22], model syntactic and semantic changes between query reformulations [23], or replace old query terms with new phrases with the help of the point-wise mutual information [27]. In contrast to these examples, the query simulations analyzed in this study do not rely on large-scale user logs but make use of test collections, i.e., topics and relevance judgments.

As part of follow-up studies related to the TREC Session Track, Guan et al. improved session search results by introducing the Query Change Model (QCM) [19,44] according to which the session search is modeled as a Markov Decision Process that considers transitions between states, i.e., queries and other interactions, to improve search results for query reformulations. Van Gysel et al. found that QCM is especially effective for longer sessions while being on par with term-frequency based approaches for shorter sessions [21]. Our query simulation method draws inspiration from QCM, but generates queries instead of improving retrieval results throughout a session.

Simulated UQVs contribute to more diverse and more realistic user-oriented directives as part of system evaluations. Besides the actual simulation of session search, applications for simulated queries are manifold. For instance, UQVs enhance the pooling process [36], make rank fusion approaches possible [8], are used for query performance prediction [18], or assist users with query suggestions that improve the recall [42]. In this work, we compare simulated to real UQVs.

3   Approach

In this section, we introduce the analyzed approaches for query simulations (Sect. 3.1) featuring conventional methods of term candidate generation and query modification strategies and our new method. Furthermore, the evaluation framework (Sect. 3.2) and details about the datasets and implementations (Sect. 3.3) are described.

3.1   Query Simulation

Term Candidate Generation. Simulating queries based on topics of test collections most likely complies with exploitation search tasks [32], where users normally have a very concrete understanding of their information needs. Provided that real users have read the topic, they are very likely to include key terms of the topic texts when formulating queries. As a simplified implementation, the *TREC Topic Searcher* (TTS) considers only terms of the set $T_{\text{topic}} = \{t_1, ..., t_n\}$ composed of the topic's title, description and narrative with $t_1, ..., t_n$ being the term sequence in the concatenated text. For upper bound performance estimates, we simulate a *Known-item Searcher* (KIS). Here, we assume the simulated users to be familiar with the document collection. When reading the topics, they recall key terms of the relevant documents in the collection and use these as their query terms. In this case, the term candidates $T_{\text{rel}} = \{t_1, ..., t_n\}$ are derived from a language model based on CQG by Jordan et al. [28] according to $P(t|D_{\text{rel}}) = (1 - \lambda)P_{\text{topic}}(t|D_{\text{rel}}) + \lambda P_{\text{background}}(t)$, where the topic model $P_{\text{topic}}(t|D_{\text{rel}})$ is made from the relevant documents D_{rel} for a given topic, while the background model $P_{\text{background}}(t)$ is derived from the vocabulary of the entire corpus. λ is used to model the influence of the background model, and it is set to 0.4 to be consistent with previous work [16,28]. In this case, $t_1, ..., t_n$ are ordered by the decreasing term probabilities of the underlying language model.

Query Modification Strategy. We make use of the query generation techniques proposed by Baskaya et al. [5], that were also used in previous simulation studies [28,34,35,37,41]. More specifically, the following strategies are considered and used in combination with the term candidates of T_{topic} and T_{rel}: the strategy S1 outputs single term queries q_i following the ordering of term candidates ($q_1 = \{t_1\}; q_2 = \{t_2\}; q_3 = \{t_3\}; ...$); S2 keeps the first candidate term fixed and composes query strings by replacing the second term for reformulations ($q_1 = \{t_1, t_2\}; q_2 = \{t_1, t_3\}; q_3 = \{t_1, t_4\}; ...$); S2' is similar to S2, but keeps two candidate terms fixed ($q_1 = \{t_1, t_2, t_3\}; q_2 = \{t_1, t_2, t_4\}; q_3 = \{t_1, t_2, t_5\}; ...$); S3 starts with a single term query and incrementally adds query terms for reformulations ($q_1 = \{t_1\}; q_2 = \{t_1, t_2\}; q_3 = \{t_1, t_2, t_3\}; ...$); S3' is similar to S3, but starts with two candidate terms ($q_1 = \{t_1, t_2, t_3\}; q_2 = \{t_1, t_2, t_3, t_4\}; q_3 = \{t_1, t_2, t_3, t_4, t_5\}; ...$). In total, we analyze ten different query simulators that result from the two term candidate generators that are combined with five query modification strategies, denoted as $\text{TTS}_{\text{S1-S3}'}$ and $\text{KIS}_{\text{S1-S3}'}$, respectively. We hypothesize, that the system performance of real queries should range somewhere between those queries of the naive approach of TTS and those queries of KIS.

Controlled Query Generation Combined with Query Change Model. Compared to the previous query simulators, this approach adds an additional scoring stage for the generated query string candidates. These candidates are generated by considering every possible combination of n-grams made from a term set. The corresponding terms are either taken from T_{rel} or $T_{\text{topic+rel}} = (T_{\text{topic}} \cap T_{\text{rel}}) \cup (T_{\text{rel}} \setminus T_{\text{topic}})_k$, whereas $(T_{\text{topic}} \cap T_{\text{rel}})$ contains topic terms in T_{rel} and $(T_{\text{rel}} \setminus T_{\text{topic}})_k$ denotes the top k terms of T_{rel} that are not in the topic text. In this regard, k models the user's vocabulary and domain knowledge. Having a set of different query string candidates, we rank the queries by $\frac{\sum_{j=1}^{|q|} \Theta_j}{|q|}$, which is the sum over all query terms normalized by the query length $|q|$, whereas Θ_j is a term-dependent score inspired by QCM [19,44] and is implemented as follows.

$$
\Theta_j = \begin{cases}
\alpha(1 - P(t_j|D_{\text{rel}})), & t_j \in q_{\text{title}} \\
1 - \beta P(t_j|D_{\text{rel}}), & t_j \in +\Delta q \wedge t_j \in T_{\text{topic}} \\
\epsilon \, \text{idf}(t_j), & t_j \in +\Delta q \wedge t_j \notin T_{\text{topic}} \\
-\delta P(t_j|D_{\text{rel}}), & t_j \in -\Delta q
\end{cases} \tag{1}
$$

whereas q_{title} is the set of topic title terms and $+/-\Delta q$ denotes added or removed terms of a query reformulation that is made in reference to the previously simulated query, except for the first query formulation q_1 for which the topic title is used as a reference. In our experiments, we analyze 3-,4-,5-gram term candidates and analyze three different parametrizations of the simulators, which are defined as follows. First, we analyze the strategy S4 ($\alpha = 2.2, \beta = 0.2, \epsilon = 0.05, \delta = 0.6$), which tends to prefer topic terms and mostly keeps terms of previous queries. Second, we analyze the strategy S4' ($\alpha = 2.2, \beta = 0.2, \epsilon = 0.25, \delta = 0.1$), which mostly keeps terms of previous queries, but tends to include terms that are not in the topic text. Finally, we analyze the strategy S4'' ($\alpha = 0.2, \beta = 0.2, \epsilon = 0.025, \delta = 0.5$), which tends to stick to the topic terms, but does not necessarily keep terms of previous query formulations. In sum, we analyze six different instantiations of these simulators, which are either based on T_{rel} (denoted as $\text{KIS}_{\text{S4-S4''}}$), or based on $T_{\text{topic+rel}}$ with $k = 4$ (denoted as $\text{TTS}_{\text{S4-S4''}}$).

3.2 Evaluation Framework

In the following, we outline our evaluation framework used to validate the simulations in reference to real queries in different aspects. It includes evaluation of the average retrieval performance, shared task utility, effort and effect, and query term similarity between simulated and real queries.

Retrieval Performance. As shown by Tague and Nelson, simulated queries fall behind real queries in terms of retrieval performance [40]. For this reason, we evaluate the *Average Retrieval Performance* (ARP) as it is common practice in system-oriented IR experiments. The ARP is determined by the average of a measure over all topics in a test collection. Beyond comparing the averaged means of different queries, we propose a more in-depth analysis of the topic score distributions. Recently, the *Root Mean Square Error* (RMSE) and *paired t-tests* were introduced as reproducibility measures [10]. The RMSE measures the closeness between the topic score distributions, and low errors indicate a good reproduction. When using t-tests as a reproducibility measure, low p-values result from diverging score distributions and indicate a higher probability of failing the reproduction.

Shared Task Utility. According to Huurnik et al. [24], the ARP of the simulated queries alone is not an appropriate indicator of how well the simulations resemble the real queries since useful query simulators should identify the best system. As proposed by Huurnik et al., we analyze how the simulated queries reproduce *relative system orderings* by comparing them with the help of Kendall's τ as it is common practice as part of shared task evaluations [43]. We compare the simulated and real queries by determining how well the ordering of systems with different parametrizations (and different retrieval performance) can be reproduced by simulated queries.

Effort and Effect. In order to account for a more user-oriented evaluation, we simulate sessions and evaluate them with regards to the effort (number of queries) that has to be made and the resulting effects (cumulated gain). First, we simulate sessions using ten simulated queries and an increasing number of documents per query and evaluate the results by the *sDCG* measure [26], whereas the cumulated gain is discounted for each result and query. Second, we evaluate the simulation quality from another more economical point of view. Azzopardi applies economic theory to the retrieval process [2] and demonstrates that for a pre-defined level of cumulated gain, query reformulations can be compensated by browsing depth (or vice versa browsing depth by more query reformulations). Furthermore, he illustrates this relationship with isoquants - a visualization technique used in microeconomics. Thus, we evaluate the closeness between isoquants of simulated and real queries by the *Mean Squared Logarithmic Error* (MSLE).

Query Term Similarity. It is not the primary goal of this study to simulate query strings with exact term matches. Instead, simulated UQVs should result in diverse query strings for a fixed information need (topic). Nonetheless, it is worth analyzing the term overlap between the simulated and real queries. As Liu et al. [31] or Mackenzie and Moffat [33] propose, we determine the Jaccard similarity between the sets of unique terms made from the query reformulations. When compared with the other evaluations, the term similarities add more insights about the simulated UQVs. For instance, if it is possible to simulate query reformulations that adequately relate to the properties of real queries, but with other terms.

3.3 Datasets and Implementation Details

In our experimental setup, we use the user query variant (UQV) dataset provided by Benham and Culpepper [6][1]. Given the topic texts, eight users formulated up to ten query variants for each topic. Each user formulated at least one query for each topic, and the fifth user (denoted as UQV_5) formulated ten queries for each topic. More details about the query collection process are provided by Benham et al. [7]. Accordingly, we evaluate the system runs with The New York Times Annotated Corpus and the topics of TREC Common Core 2017 [1]. As part of our experiments, we exploit the interactive search possibilities of the Pyserini toolkit [30]. We index the Core17 test collection with the help of Anserini [45] and the default indexing options as provided in the regression guide[2]. Unless stated otherwise, all results were retrieved with the BM25 method and Anserini's default parameters ($b = 0.4$, $k = 0.9$). We evaluate the results with the `repro_eval` toolkit [11] that is a dedicated reproducibility framework featuring bindings to `trec_eval` measures. The source code of the experiments and the simulated queries are available in a public GitHub repository[3].

---

[1] https://culpepper.io/publications/robust-uqv.txt.gz.

[2] https://github.com/castorini/anserini/blob/master/docs/regressions-core17.md.

[3] ♉ https://github.com/irgroup/ecir2022-uqv-sim.

4   Experimental Results

Retrieval Performance. Regarding **RQ1**, we validate the retrieval performance of real (UQV) and simulated (TTS/KIS) queries. Table 1 shows the ARP including nDCG and AP scores that are determined by averaging results with 1000 documents per topic and P@10 scores over *all* queries, the *first*[4], or the *best* query of a topic. Our assumptions are confirmed. The retrieval performance of real queries ranges between that of the $TTS_{S1\text{-}S3'}$ and $KIS_{S1\text{-}S3'}$ simulators. Especially, the performance of the $TTS_{S1\text{-}S3'}$ queries stays below that of real queries. For instance, the average nDCG scores of the UQV queries range between 0.3787 and 0.4980, whereas the maximum score of the $TTS_{S1\text{-}S3'}$ queries is 0.3499 and the nDCG scores of $KIS_{S2'\text{-}S3'}$ lie above those of UQV. Similarly, the nDCG scores averaged over the first UQV queries reach 0.3979 at a minimum, whereas the maximum score of the $TTS_{S1\text{-}S3'}$ queries is 0.3895. When averaging over the best queries, most nDCG scores of TTS fall into the range of real queries, but there is also a higher probability of finding a good performing query since more TTS than UQV queries are available. Except for single term queries (S1), all KIS scores outperform the UQV queries when averaging over the best queries. With regard to the simulated queries based on the $TTS_{S4\text{-}S4''}$ approach, most of the nDCG, P@10, and AP scores fall into the range of the real queries, while $KIS_{S4\text{-}S4''}$ queries outperform UQV queries. Thus, we have a specific focus on $TTS_{S4\text{-}S4''}$.

Figure 4 shows the $RMSE_{nDCG}$ between queries with conventional query modification strategies ($TTS_{S1\text{-}S3'}$/$KIS_{S1\text{-}S3'}$) and the real queries (UQV). Especially for the TTS queries, the strategy S2' has the lowest RMSE scores and acceptable scores for the KIS queries. In the following experiments, we primarily use the strategy S2' for both the TTS and KIS queries since their term length complies with the typical length of real queries [25] and they serve as estimates of lower and upper bound retrieval performance. Additionally, we evaluate the $TTS_{S4\text{-}S4''}$ queries with the help of the RMSE and simulations in reference to the ten queries per topic of UQV_5. For each query reformulation, 100 documents are retrieved and contribute to the final ranking list of a topic if a previous query has not retrieved them. Figure 1 shows the RMSE instantiated with P@1000, nDCG, and AP along with an increasing number of documents retrieved with ten queries. For all measures, the error increases when more documents per query are retrieved. With regard to P@1000 and nDCG, the $TTS_{S2'}$ and $KIS_{S2'}$ queries have the largest error, while $KIS_{S2'}$ has a lower $RMSE_{AP}$ than $TTS_{S4'}$. For all measures, the $TTS_{S4\text{-}S4''}$ queries have the lowest error, which means they are the best approximation of UQV_5 among all analyzed query simulations.

Finally, we compare the topic score distributions of the simulated queries and all UQV queries by paired t-tests[5]. Since some users formulated no more than one query per topic, we limit our evaluations to the first query of each simulator.

---

[4]   S1 and S3, as well as S2 and S3', do not differ when averaging over the first queries.
[5]   Applying the Bonferroni correction adjusts the alpha level to $\alpha = \frac{0.05}{64} \approx 0.0008$ (considering eight users and eight query simulators for an alpha level of 0.05).

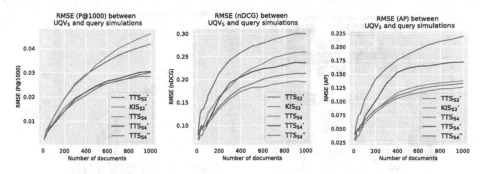

Fig. 1. RMSE instantiated with P@1000, nDCG, and AP

It means that each of the p-values shown in Fig. 5 is determined by t-tests with nDCG score distributions that result from 50 UQV and 50 simulated queries. The $TTS_{S2'}$ queries have the highest p-values when compared with $UQV_{\{2,3,8\}}$. These results align with the ARP scores reported in Table 1. The nDCG scores of UQV_2 (0.4096), UQV_3 (0.3979), and UQV_8 (0.4046) are the most similar to the nDCG score of TTS_{S3} (0.3895) in comparison to other simulators. In contrast, the p-values of $KIS_{S2'}$ queries are low for all UQV queries, which complies with the ARP scores in Table 1. The $KIS_{S2'}$ scores averaged over the first queries are substantially higher compared to the UQV scores (e.g., nDCG($KIS_{S2'}$)=0.5474 compared to the best UQV query with nDCG(UQV_7)=0.4980). The $UQV_{\{1,4,5,6,8\}}$ queries have comparably higher p-values with the $TTS_{\{S4,S4''\}}$ queries which align with similar ARP scores. Interestingly, the t-test with UQV_7 and $TTS_{S4'}$ results in the highest overall p-value of 0.9901 and similarly high p-values with $KIS_{S4\text{-}S4''}$. This lets us assume that the corresponding user of the UQV_7 queries diverged from the terms in the topic texts and had some prior knowledge about adequate queries for at least some of the topics. In sum, not only the ARP can be reproduced with the simulated $TTS_{S4\text{-}S4''}$ and $KIS_{S4\text{-}S4''}$ queries, but also statistical properties of the topic score distributions.

Shared Task Utility. Regarding **RQ2**, we validate to which degree the simulated queries reproduce properties of the real queries in several regards. First, we evaluate if the simulated queries can preserve the relative system orderings. To be consistent with Huurnik et al., we evaluate five systems and different parametrizations ($\mu = 50, 250, 500, 1250, 2500, 5000$) of the query likelihood model with Dirichlet smoothing (QLD) [46], but other retrieval methods and variations thereof can be reasonable as well. For each query formulation q_i, we determine the correlation by Kendall's τ averaged over all topics (cf. Fig. 2 (left)) in comparison to the UQV_5 queries. The $TTS_{S2'}$ queries do not preserve the relative system ordering. Especially for the first five query reformulations, there is a low correlation with the relative system orderings of the real queries. Interestingly, the $KIS_{S2'}$ queries result in acceptable Kendall's τ scores [43], while the scores beyond the sixth query formulation show low correlations. Similarly, the

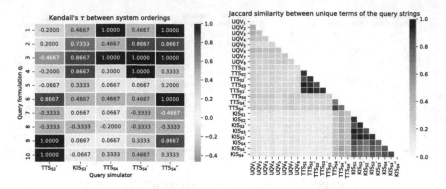

Fig. 2. Kendall's τ between system orderings of query reformulations in reference to UQV_5 (left). Jaccard similarity between unique terms of the queries (right).

$TTS_{S4\text{-}S4''}$ queries correlate with the system orderings of UQV_5 queries fairly well, even reaching the maximum score of 1.0. Beyond the sixth query reformulation, the correlation falls off. While it is out of this study's scope to reach any definitive conclusions, we assume that this is related to query drifts - an issue that is also known from term expansions as part of pseudo-relevance feedback [15,38].

Effort and Effect. Since most of the experiments validated single queries only, we simulate search sessions and evaluate these by sDCG (instantiated with b=2, bq=4). We compare sessions with 3, 5, or 10 queries and an increasing number of documents per query. Figure 3 (top) compares the queries of UQV_5 (made by a single user [7]) to ten simulated queries of $TTS_{S2'}$, $KIS_{S2'}$, and $TTS_{S4\text{-}S4''}$. As expected, the cumulative gain increases faster when more queries per session are used. Likewise, the $TTS_{S2'}$ and $KIS_{S2'}$ queries deliver lower and upper bound limits, respectively. In between, there are the cumulative gains by the UQV_5 and $TTS_{S4\text{-}S4''}$ queries. These results show that it is possible to fine-tune and to reproduce the cumulative gain close to that of real queries, in this particular case with $TTS_{S4''}$.

Figure 3 (bottom) shows the isoquants and illustrates how many documents have to be examined by a simulated user to reach pre-defined levels of nDCG (0.3, 0.4, 0.5). More queries compensate browsing depth, and as expected, the least documents have to be examined with $KIS_{S2'}$ queries and the most with $TTS_{S2'}$ queries. The $TTS_{S2'}$ isoquants lie above the others, which can be explained by the poorer retrieval performance as already shown in Table 1. As shown by the MSLE, the TTS_{S4} isoquant has the lowest error for all values of nDCG. Again, we see a better approximation of the UQV_5 isoquant with the $TTS_{S4\text{-}S4''}$ strategies and that it is possible to reproduce economic properties through parameterizing the query reformulation behavior.

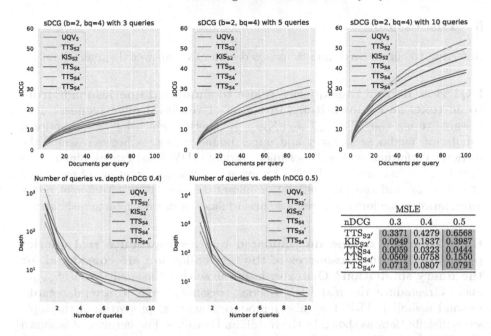

Fig. 3. Simulations with 3, 5, or 10 queries per session evaluated by sDCG (top). Isoquants and MSLE between simulations and UQV$_5$ with fixed nDCG (bottom).

Query Term Similarities. Figure 2 (right) shows the Jaccard similarities between the concatenated query strings. More specifically, only normalized unique terms are compared, and depending on the number of available queries for a specific topic, we include an equal number of simulated queries to avoid low Jaccard similarities when less than ten UQV queries are available. As the results show, the highest similarities are between the simulated queries. While the similarities between conventional strategies S1 to S3′ and the strategies S4 to S4″ are rather low for the TTS queries, there are higher similarities for the KIS queries. Compared to UQV and TTS queries, the KIS queries have the lowest similarities, which indicates that descriptive terms of relevant documents are very different from those used in real queries and the topic texts. Interestingly, the UQV$_{\{2,3,8\}}$ queries do not have a remarkably high Jaccard similarity with TTS$_{S2'}$ queries, despite the high p-values that are shown in Fig. 5. This shows that it is possible to simulate UQVs with different query terms than in the real queries, but with comparable statistical properties as indicated by the p-values even with the rather naive approach of TTS$_{S2'}$. There are slightly higher similarities between KIS queries and the TTS$_{S4\text{-}S4''}$ queries. In particular, there is a higher similarity between TTS$_{S4'}$ and the KIS queries since the simulator is parameterized to diverge from the topic terms. Overall, we conclude that the analyzed simulation methods do not result in query strings that exactly match the terms of real queries in this specific UQV dataset [6].

5 Discussion

Referring to our research questions posed earlier, we answer them as follows.

RQ1 How do real user queries relate to simulated queries made from topic texts and known-items in terms of retrieval effectiveness? It is possible to use the $TTS_{S1-S3'}$ and $KIS_{S1-S3'}$ queries, which follow conventional simulation methods, as lower and upper bound estimates between which the retrieval performance of real user query variants (UVQ_{1-8}) ranges. Simulations based on our new method ($TTS_{S4''}$) provide better approximations of real query effectiveness, and the parametrization allows the simulation of different query formulation behaviors and a retrieval performance better resembling real queries.

RQ2 To which degree do simulated queries reproduce real queries provided that only resources of the test collection are considered for the query simulation? Our experiments show that the simulated $TTS_{S4-S4''}$ queries reproduce the real UQV queries reasonably well in several regards. Beyond a similar ARP, they also reproduce statistical properties of the topic score distributions as shown by the RMSE and p-values. Furthermore, it is shown that the simulated queries also reproduce economic aspects of the real queries as evaluated with the sDCG experiments and the isoquants that compare trade-offs between the number of query reformulations and the browsing depth for a fixed level of gain. Furthermore, when evaluating the shared task utility, the queries of our new parameterized simulation approach preserve the relative system orderings up to the fifth reformulation, while the correlations fall off for later reformulations. We assume that this is related to topic drifts, and further analysis in this direction is required. Finally, even though it is not the primary goal to simulate exact term matches with UQVs, the analysis of the query term similarity showed that there is only a slight overlap between terms of simulated and real queries, and a more dedicated approach is required to reproduce exact term matches.

6 Conclusion

In this work, we present an evaluation framework and a new method for simulated user query variants. Our experiments showed that the retrieval performance of real queries ranges between that of simulated queries from conventional methods based on topic texts and known-items. As a better approximation of user queries, we introduce a simulation method that allows parameterizing the query reformulation behavior and thus better reproduces real queries from specific users. One limitation of our simulations is the exclusion of relevance feedback from previous search results. Users normally include terms of documents or snippets they consider as relevant [17,39] in their query reformulations. Likewise, the experiments neglect click simulations. We leave it for future work to complement and analyze simulations in this regard.

A   Appendix

Table 1. Average retrieval performance over q queries

	All queries				First queries				Best queries			
	q	nDCG	P@10	AP	q	nDCG	P@10	AP	q	nDCG	P@10	AP
UQV_1	150	.3787	.4507	.1581	50	.4293	.5040	.2003	50	.4969	.6320	.2429
UQV_2	52	.4221	.5058	.2020	50	.4096	.4880	.1894	50	.4103	.4900	.1896
UQV_3	68	.3922	.4353	.1780	50	.3979	.4560	.1813	50	.4117	.4800	.1878
UQV_4	123	.4126	.4894	.1888	50	.4469	.5220	.2099	50	.5146	.6300	.2644
UQV_5	500	.3922	.4330	.1649	50	.4447	.4920	.2043	50	.5353	.7240	.2807
UQV_6	136	.4030	.4713	.1843	50	.4488	.5080	.2197	50	.4980	.5980	.2515
UQV_7	50	.4980	.5720	.2418	50	.4980	.5720	.2418	50	.4980	.5720	.2418
UQV_8	156	.3814	.4545	.1645	50	.4046	.4500	.1799	50	.4556	.5620	.2193
TTS_{S1}	500	.0479	.0306	.0127	50	.1705	.1280	.0541	50	.3066	.2360	.0971
TTS_{S2}	500	.1964	.1716	.0688	50	.3592	.3900	.1604	50	.4391	.5100	.2097
$TTS_{S2'}$	500	.3387	.3426	.1413	50	.3895	.4020	.1821	50	.4639	.5940	.2283
TTS_{S3}	500	.3323	.3632	.1388	50	.1705	.1280	.0541	50	.4776	.6080	.2383
$TTS_{S3'}$	500	.3499	.3874	.1474	50	.3592	.3900	.1604	50	.4709	.6060	.2311
TTS_{S4}	500	.4493	.5168	.2088	50	.4409	.4920	.2072	50	.5945	.7620	.3282
$TTS_{S4'}$	500	.4788	.5626	.2288	50	.4976	.5940	.2429	50	.6207	.8040	.3554
$TTS_{S4''}$	500	.3780	.4224	.1644	50	.4393	.4860	.2065	50	.5812	.7680	.3222
KIS_{S1}	500	.1334	.1044	.0314	50	.2836	.2040	.0813	50	.4087	.4400	.1492
KIS_{S2}	500	.3969	.3972	.1615	50	.5096	.5400	.2535	50	.5988	.7460	.3429
$KIS_{S2'}$	500	.5114	.5666	.2507	50	.5474	.6220	.2870	50	.6336	.7980	.3762
KIS_{S3}	500	.5598	.6336	.3009	50	.2836	.2040	.0813	50	.6907	.8620	.4299
$KIS_{S3'}$	500	.5941	.6882	.3285	50	.5096	.5400	.2535	50	.6922	.8620	.4337
KIS_{S4}	500	.5216	.5976	.2604	50	.5146	.5960	.2630	50	.6461	.8200	.3902
$KIS_{S4'}$	500	.5008	.5888	.2416	50	.5033	.5980	.2400	50	.6269	.8080	.3703
$KIS_{S4''}$	500	.4859	.5584	.2293	50	.5191	.6020	.2644	50	.6401	.8360	.3781

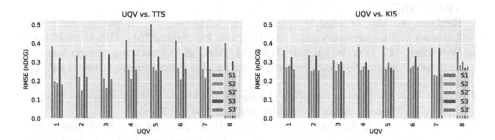

Fig. 4. RMSE between $TTS_{S1\text{-}S3'}$ and $KIS_{S1\text{-}S3'}$ queries and the UQV queries.

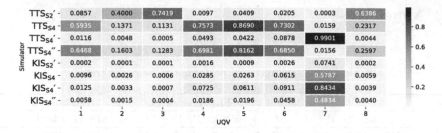

Fig. 5. p-values of paired t-tests between UQV and simulated queries.

References

1. Allan, J., Harman, D., Kanoulas, E., Li, D., Gysel, C.V., Voorhees, E.M.: TREC 2017 common core track overview. In: Proceedings of the TREC (2017)
2. Azzopardi, L.: The economics in interactive information retrieval. In: Proceedings of the SIGIR, pp. 15–24 (2011)
3. Azzopardi, L., de Rijke, M.: Automatic construction of known-item finding test beds. In: Efthimiadis, E.N., Dumais, S.T., Hawking, D., Järvelin, K. (eds.) Proceedings of the SIGIR, pp. 603–604 (2006)
4. Azzopardi, L., de Rijke, M., Balog, K.: Building simulated queries for known-item topics: an analysis using six European languages. In: Proceedings of the SIGIR, pp. 455–462 (2007)
5. Baskaya, F., Keskustalo, H., Järvelin, K.: Time drives interaction: simulating sessions in diverse searching environments. In: Proceedings of the SIGIR, pp. 105–114 (2012)
6. Benham, R., Culpepper, J.S.: Risk-reward trade-offs in rank fusion. In: Proceedings of the ADCS, pp. 1:1–1:8 (2017)
7. Benham, R., et al.: RMIT at the 2017 TREC CORE track. In: Proceedings of the TREC (2017)
8. Benham, R., Mackenzie, J.M., Moffat, A., Culpepper, J.S.: Boosting search performance using query variations. ACM Trans. Inf. Syst. **37**(4), 41:1-41:25 (2019)
9. Berendsen, R., Tsagkias, M., de Rijke, M., Meij, E.: Generating pseudo test collections for learning to rank scientific articles. In: Proceedings of the CLEF, pp. 42–53 (2012)
10. Breuer, T., et al.: How to measure the reproducibility of system-oriented IR experiments. In: Proceedings of the SIGIR, pp. 349–358 (2020)
11. Breuer, T., Ferro, N., Maistro, M., Schaer, P.: Repro_eval: a python interface to reproducibility measures of system-oriented IR experiments. In: Proceedings of the ECIR, pp. 481–486 (2021)
12. Carterette, B., Bah, A., Zengin, M.: Dynamic test collections for retrieval evaluation. In: Proceedings of the ICTIR, pp. 91–100. ACM (2015)
13. Chuklin, A., Markov, I., de Rijke, M.: Click models for web search. In: Retrieval, and Services, Morgan & Claypool Publishers, Synthesis Lectures on Information Concepts (2015)
14. Craswell, N., Campos, D., Mitra, B., Yilmaz, E., Billerbeck, B.: ORCAS: 20 million clicked query-document pairs for analyzing search. In: Proceedings of the CIKM, pp. 2983–2989 (2020)

15. Croft, W.B., Harper, D.J.: Using probabilistic models of document retrieval without relevance information. J. Document. **35**(4), 285–295 (1979)
16. Cronen-Townsend, S., Zhou, Y., Croft, W.B.: Predicting query performance. In: Proceedings of the SIGIR, pp. 299–306 (2002)
17. Eickhoff, C., Teevan, J., White, R., Dumais, S.T.: Lessons from the journey: a query log analysis of within-session learning. In: Proceedings of the WSDM, pp. 223–232 (2014)
18. Faggioli, G., Zendel, O., Culpepper, J.S., Ferro, N., Scholer, F.: An enhanced evaluation framework for query performance prediction. In: Proceedings of the ECIR, pp. 115–129 (2021)
19. Guan, D., Zhang, S., Yang, H.: Utilizing query change for session search. In: Proceedings of the SIGIR, pp. 453–462 (2013)
20. Günther, S., Hagen, M.: Assessing query suggestions for search session simulation. In: Proceedings of the Sim4IR (2021). http://ceur-ws.org/Vol-2911/paper6.pdf
21. Gysel, C.V., Kanoulas, E., de Rijke, M.: Lexical query modeling in session search. In: Proceedings of the ICTIR, pp. 69–72 (2016)
22. He, Y., Tang, J., Ouyang, H., Kang, C., Yin, D., Chang, Y.: Learning to rewrite queries. In: Proceedings of the CIKM, pp. 1443–1452 (2016)
23. Herdagdelen, A., et al.: Generalized syntactic and semantic models of query reformulation. In: Proceedings of the SIGIR, pp. 283–290 (2010)
24. Huurnink, B., Hofmann, K., de Rijke, M., Bron, M.: Validating query simulators: an experiment using commercial searches and purchases. In: Proceedings of the CLEF, pp. 40–51 (2010)
25. Jansen, B.J., Booth, D.L., Spink, A.: Patterns of query reformulation during web searching. J. Assoc. Inf. Sci. Technol. **60**(7), 1358–1371 (2009)
26. Järvelin, K., Price, S.L., Delcambre, L.M.L., Nielsen, M.L.: Discounted cumulated gain based evaluation of multiple-query IR sessions. In: Proceedings of the ECIR, pp. 4–15 (2008)
27. Jones, R., Rey, B., Madani, O., Greiner, W.: Generating query substitutions. In: Proceedings of the WWW, pp. 387–396 (2006)
28. Jordan, C., Watters, C.R., Gao, Q.: Using controlled query generation to evaluate blind relevance feedback algorithms. In: Proceedings of the JCDL, pp. 286–295 (2006)
29. Keskustalo, H., Järvelin, K., Pirkola, A., Sharma, T., Lykke, M.: Test collection-based IR evaluation needs extension toward sessions - a case of extremely short queries. In: Proceedings of the AIRS, pp. 63–74 (2009)
30. Lin, J., Ma, X., Lin, S., Yang, J., Pradeep, R., Nogueira, R.: Pyserini: a python toolkit for reproducible information retrieval research with sparse and dense representations. In: Proceedings of the SIGIR, pp. 2356–2362. ACM (2021)
31. Liu, B., Craswell, N., Lu, X., Kurland, O., Culpepper, J.S.: A comparative analysis of human and automatic query variants. In: Proceedings of the SIGIR, pp. 47–50 (2019)
32. Liu, J., Sarkar, S., Shah, C.: Identifying and predicting the states of complex search tasks. In: Proceedings of the CHIIR, pp. 193–202 (2020)
33. Mackenzie, J., Moffat, A.: Modality effects when simulating user querying tasks. In: Proceedings of the ICTIR, pp. 197–201 (2021)
34. Maxwell, D., Azzopardi, L.: Agents, simulated users and humans: an analysis of performance and behaviour. In: Proceedings of the CIKM, pp. 731–740. ACM (2016)

35. Maxwell, D., Azzopardi, L.: Simulating interactive information retrieval: simiir: a framework for the simulation of interaction. In: Proceedings of the SIGIR, pp. 1141–1144. ACM (2016)
36. Moffat, A., Scholer, F., Thomas, P., Bailey, P.: Pooled evaluation over query variations: users are as diverse as systems. In: Proceedings of the CIKM, pp. 1759–1762 (2015)
37. Pääkkönen, T., Kekäläinen, J., Keskustalo, H., Azzopardi, L., Maxwell, D., Järvelin, K.: Validating simulated interaction for retrieval evaluation. Inf. Ret. J. **20**(4), 338–362 (2017). https://doi.org/10.1007/s10791-017-9301-2
38. Ruthven, I., Lalmas, M.: A survey on the use of relevance feedback for information access systems. Knowl. Eng. Rev. **18**(2), 95–145 (2003)
39. Sloan, M., Yang, H., Wang, J.: A term-based methodology for query reformulation understanding. Inf. Retriev. J. **18**(2), 145–165 (2015). https://doi.org/10.1007/s10791-015-9251-5
40. Tague, J., Nelson, M.J.: Simulation of user judgments in bibliographic retrieval systems. In: Proceedings of the SIGIR, pp. 66–71 (1981)
41. Verberne, S., Sappelli, M., Järvelin, K., Kraaij, W.: User simulations for interactive search: Evaluating personalized query suggestion. In: Proceedings of the ECIR, pp. 678–690 (2015)
42. Verberne, S., Sappelli, M., Kraaij, W.: Query term suggestion in academic search. In: Proceedings of the ECIR, pp. 560–566 (2014)
43. Voorhees, E.M.: Variations in relevance judgments and the measurement of retrieval effectiveness. In: Proceedings of the SIGIR, pp. 315–323 (1998)
44. Yang, H., Guan, D., Zhang, S.: The query change model: modeling session search as a Markov decision process. ACM Trans. Inf. Syst. **33**(4), 20:1-20:33 (2015)
45. Yang, P., Fang, H., Lin, J.: Anserini: reproducible ranking baselines using Lucene. ACM J. Data Inf. Qual. **10**(4), 16:1-16:20 (2018)
46. Zhai, C., Lafferty, J.D.: A study of smoothing methods for language models applied to ad hoc information retrieval. In: Proceedings of the SIGIR, pp. 334–342 (2001)
47. Zhang, Y., Liu, X., Zhai, C.: Information retrieval evaluation as search simulation: a general formal framework for IR evaluation. In: Proceedings of the ICTIR, pp. 193–200 (2017)

Out-of-Domain Semantics to the Rescue! Zero-Shot Hybrid Retrieval Models

Tao Chen[✉], Mingyang Zhang, Jing Lu, Michael Bendersky, and Marc Najork

Google Research, Mountain View, CA 94043, USA
{taochen,mingyang,ljwinnie,bemike,najork}@google.com

Abstract. The pre-trained language model (*eg,* BERT) based deep retrieval models achieved superior performance over lexical retrieval models (*eg,* BM25) in many passage retrieval tasks. However, limited work has been done to generalize a deep retrieval model to other tasks and domains. In this work, we carefully select five datasets, including two in-domain datasets and three out-of-domain datasets with different levels of domain shift, and study the generalization of a deep model in a zero-shot setting. Our findings show that the performance of a deep retrieval model is significantly deteriorated when the target domain is very different from the source domain that the model was trained on. On the contrary, lexical models are more robust across domains. We thus propose a simple yet effective framework to integrate lexical and deep retrieval models. Our experiments demonstrate that these two models are complementary, even when the deep model is weaker in the out-of-domain setting. The hybrid model obtains an average of 20.4% relative gain over the deep retrieval model, and an average of 9.54% over the lexical model in three out-of-domain datasets.

Keywords: deep retrieval · lexical retrieval · zero-shot learning · hybrid model

1 Introduction

Traditionally, search engines have used lexical retrieval models (*eg,* BM25) to perform query-document matching. Such models are efficient and simple, but are vulnerable to vocabulary mismatch when queries use different terms to describe the same concept [4]. Recently, deep pre-trained language models (*eg,* BERT) have shown strong ability in modeling text semantics and have been widely adopted in retrieval tasks. Unlike lexical retrievers, deep/dense retrievers[1] capture the semantic relevance between queries and documents in a lower dimensional space, bridging the vocabulary mismatch gaps. Deep retrievers have been successful in many retrieval benchmarks. For instance, the most recent five winners in MS-MARCO passage [2] ranking leaderboard adopt deep retrievers as their first-stage retrieval model.

---

[1] While we recognize that in some cases the deep retrievers are not necessarily dense, and vice versa, we loosely use these two terms interchangeably throughout the paper.

© The Author(s), under exclusive license to Springer Nature Switzerland AG 2022
M. Hagen et al. (Eds.): ECIR 2022, LNCS 13185, pp. 95–110, 2022.
https://doi.org/10.1007/978-3-030-99736-6_7

However, training a deep retrieval model is computationally expensive and a sizable labeled dataset to guide model training is not always available. A natural question then arises, can we train a deep retrieval model in one domain, and then directly apply it to new datasets/domains in a zero-shot setting with no in-domain training data? To answer this question, we carefully select five datasets, including two in-domain, and three out-of-domain datasets with different levels of domain shift. Through comprehensive experiments, we find that a deep retriever model performs well on related domains, but deteriorates when the target domain is distinct from the model source domain. On the contrary, lexical models are rather robust across datasets and domains. Our further analysis shows that lexical and deep models can be complementary to each other, retrieving different sets of relevant documents.

Inspired by this, we propose a zero-shot hybrid retrieval model to combine lexical and deep retrieval models. For simplicity and flexibility, we train a deep retrieval model and a lexical model separately and integrate the two (or more) models via Reciprocal Rank Fusion. This non-parametric fusion framework can be easily applied to any new datasets or domains, without any fine-tuning. Our experiments demonstrate the effectiveness of the hybrid model in both in-domain and out-of-domain datasets. In particular, though the zero-shot deep model is weaker in out-of-domain datasets, the hybrid model brings an average of 20.4% of relative recall gain over the deep retrieval model, and an average of 9.54% gain over lexical model (BM25) in three out-of-domain datasets. It also outperforms a variety of stronger baselines including query and document expansion.

To summarize, in this paper we explore the following research questions:

- **RQ 1**: Can deep retrieval generalize to a new domain in a zero-shot setting?
- **RQ 2**: Is deep retrieval complementary to lexical matching and query and document expansion?
- **RQ 3**: Can lexical matching, expansion, and deep retrieval models be combined in a *non-parametric hybrid retrieval* model?

To the best of our knowledge, this paper is the first to propose a hybrid retrieval model that incorporates lexical matching, expansion and deep retrieval in a zero-shot setup. We demonstrate that the proposed hybrid model is simple yet effective in a variety of datasets and domains.

2   Related Work

Information retrieval systems usually contain of two main stages: (a) *candidate retrieval* (b) *candidate re-ranking*. The retrieval stage is aimed at optimizing the recall of relevant documents, while the re-ranking stage optimizes early precision metrics such as NDCG@k or MRR. Prior research (*eg,* [3]) found that the two stages are complementary – gains in retrieval recall often lead to better early precision. Therefore, in this paper, we focus on retrieval recall optimization, with the assumption that the findings can benefit the re-ranking stage as well.

Lexical Retriever. Traditionally, the first-stage retrieval has been a lexical based model such as BM25 [35], to capture the exact lexical match between queries and documents. Such simple and effective lexical models were the state-of-the-art for decades, and are still widely used in both academia and industry. One key issue with lexical models is the vulnerability to vocabulary mismatch, where queries and documents mention the same concept with different terms. One popular line to alleviate this is to expand terms in queries from pseudo relevance feedback (eg, [15] and [1]) or expand terms in documents from related documents (eg, [37]). As a result, queries and documents have a higher chance to match each other at the surface form.

Deep LM Augmented Lexical Retriever. More recently, pre-trained deep language models (LM) such as BERT [10] have been shown to be powerful in many natural language understanding tasks. The very first application of such models in IR is to augment lexical retrieval models. Dai et al. [7,9] proposed to learn context-aware term weights by BERT to replace the term frequencies used by lexical models. To remedy the vocabulary gap between queries and documents, Nogueira and Lin [28,29] employed seq2seq model transformer [39] and later T5 [33] to generate document expansions, which brings significant gains for BM25. In the same vein, Mao et al. [27] adopted seq2seq model BART [20] to generate query expansions, which outperforms RM3 [15], a highly performant lexical query expansion method.

Deep Retriever. In a separate line of research, deep neural retrieval models adopt LMs to build a new paradigm for first-stage retrieval: instead of performing exact lexical match, they aim at capturing the relevance of queries and documents in a lower dimensional semantic space. This paradigm can largely bridge the vocabulary gap between queries and documents. Since cross-attention models are cost-prohibitive for first-stage retrieval, most works adopt a dual-encoder architecture to learn two single vector representations for the query and the document separately, and then measure their relevance by a simple scoring function (eg, dot product or cosine similarity). In this way, finding most relevant documents can be formulated as a nearest neighbor search problem and can be accelerated with quantization techniques [14,16].

For model training, it is often the case that positive (query, document) pairs are available, while negative pairs need to be sampled from the dataset. Negative sampling strategy plays a crucial role for model performance. Earlier works adopt simple in-batch negative sampling [18,45], or mine negative pairs from top BM25 results [17]. Recent works propose more sophisticated sampling strategies to identify high-quality hard negatives, such as cross-batch negatives [32], demonised hard negatives [32] and semantic similarity based negatives [23].

Deep retriever model has shown superior performance over lexical models in several passage retrieval tasks (eg, MS-MARCO passage ranking [2]). However, training a deep model is expensive computationally but also in terms of labeled data creation. A simple remedy is to directly apply a trained deep retriever model to new domains in a zero-shot setting. However, little work has been conducted to uncover the generalization ability of deep retrievers. One exception is by

Thakur et al. [38] who introduce BEIR, an IR benchmark of 18 datasets with diverse domains and tasks, and evaluate several trained deep models in a zero-shot setup. They found that deep models exhibit a poor generalization ability, and are significantly worse than BM25 on datasets that have a large domain shift compared from what they have been trained on. In our work, we conduct similar studies, and observe the same performance deterioration for a deep model in zero-shot setting. We additionally propose a hybrid model to utilize a lexical model to alleviate the domain shift.

Hybrid Retriever. Deep retrievers are good at modeling semantic similarity, while could be weaker at capturing exact match or could have capacity issues when modeling long documents [24,42]. A few recent works attempt to build a hybrid model to take the strength of both deep and lexical retrievers. Most works train a deep model separately and then interpolate its score with a lexical model score [17,21,22,24,25,42], or use RM3 built on the top lexical results to select deep retriever results as the final list [18], or simply combine the results of the two models in an alternative way [45]. Gao et al. [13] is the only work that explicitly trains a deep model to encode semantics that lexical model fails to capture. In model inference, they interpolate the scores of deep and lexical models and generate the top retrieval results. While insightful, these prior works limit the model evaluation to a single task and a single domain. It is unclear how such hybrid model performs in a cross-domain setting, without any fine-tuning. Our work aims to fill this research gap, and demonstrates that a zero-shot hybrid retrieval model can be more effective than either of the two models alone.

3   Method

In this section, we describe our zero-shot hybrid retrieval model. For simplicity and flexibility, we train deep and lexical retrieval models separately, and propose a simple yet effective non-parametric framework to integrate the two.

3.1   Hybrid Retrieval Model

Both traditional lexical retrieval models [1,30,35], as well as deep neural retrieval models [13,18,41] represent queries and documents using vectors $\mathbf{q}, \mathbf{d} \in \mathbb{R}^N$, and score candidates based on the dot product $<\mathbf{q}, \mathbf{d}>$. Thus, the difference between deep and lexical models stems from how these vectors are constructed.

Lexical models represent queries and documents using sparse weight vectors $\mathbf{q}^{sparse}, \mathbf{d}^{sparse} \in \mathbb{R}^V$, respectively (where V denotes the vocabulary size). The vectors are sparse such that all the entries for vocabulary terms that do not appear in query and document are zeroed out. To combat issues of term mismatch, lexical models often include additional terms in queries and document through some form of expansion (*eg*, based on pseudo-relevance feedback [19]). However, the resulting vectors are still highly sparse, due to the high dimensionality of vocabulary size V.

In contrast, deep neural retrieval models represent queries and documents using dense embedding vectors $\mathbf{q}^{dense}, \mathbf{d}^{dense} \in \mathbb{R}^{E}$, where $E << V$. While theoretically dense embeddings overcome the term mismatch problem, they do have several shortcomings. First, they require large amounts of data and resources for training [32], and thus may not be directly trained over collections with fewer queries and relevance judgments. Second, they do not capture *exact* query-document matches as well as the sparse lexical scores. Therefore, a lexical and deep model combination is likely to yield the optimal relevance scores.

Most prior works [17,23,24,42] model this combination as a linear interpolation of the scores of deep and lexical retrieval models. This fusion method is sensitive to the score scales and the weights assigned to the different models [42], which needs careful score normalization and weight tuning, especially when multiple models are combined. We expect that the raw scores of the models can vary from one domain/dataset to another, and likewise the interpolation weights.

Since our goal is to build a hybrid model which can be easily applied to a new domain in a zero shot setting (with no in-domain training data), we would like to eliminate such domain-specific normalization and tuning. Therefore, we adopt Reciprocal Rank Fusion (RRF) [5] to generate the final ranking results by considering the *ranking positions* of each candidate generated by different models, instead of fusing their scores. RRF demonstrates robust and effective ensembles in prior works [3,5] and our experiments. Assuming a set of lexical and deep retrieval models M, we define $\pi^{m}(q, d)$ as the rank for document d, induced by its score for query q assigned by model $m \in M$. The RRF score is then defined as:

$$RRF(q, d, M) = \sum_{m \in M} \frac{1}{k + \pi^{m}(q, d)} \tag{1}$$

where $k = 60$, following the definition in the original paper [5].

In the remainder of this paper we demonstrate that this simple non-parametric approach generalizes well across domains, and can make an effective use of out-of-domain semantics of retrieval models trained on a different collection. In the remainder of this section, we describe the lexical and deep retrieval models used to instantiate Eq. 1.

3.2   Lexical Retrieval Model

We adopt **BM25** as the base lexical retrieval model, as it is widely used and shown to be robust [38]. To alleviate the vocabulary mismatch issue, we additionally apply popular query expansion and document expansion techniques to expand the query and the document, forming enhanced lexical models.

BM25+Query Expansion. Most conventional query expansion approaches follow the pseudo-relevance feedback (PRF) paradigm. It assumes the top K ranked documents for the original query to be relevant, and generates query expansions from these documents. In our work, we experiment with **RM3** [15] (a relevance-based language model) and **Bo1** [1] (a variant of Divergence From Randomness term weighting model), to obtain query expansions from PRF.

BM25+Document Expansion. Recently, generative models like T5 were shown to generate high-quality document expansions, and bring large gains to the BM25 model on retrieval tasks [28,29,31]. Following the **docT5query** approach [28,31], we fine-tune T5-base with identical setting as the prior works on (query, relevant passage) pairs from the MS-MARCO passage ranking training set, where the query is considered as pseudo document expansion. We adopt the top-k sampling decoder [11] to generate N (a tunable parameter) queries per passage. For each document, we append the expansions to each passage and aggregate them as the document expansion.

3.3 Deep Retrieval Model

We adopt **NPR** [23], a neural passage retrieval model with improved negative contrast as the deep retrieval model in our framework. Note that our framework is flexible, and NPR can be replaced with any other deep model. Aligned with many popular deep retrievers [17,32,43], NPR adopts a dual encoder architecture, learning dense embedding vectors representations, computing the relevance using the dot product $<\mathbf{q}^{dense}, \mathbf{d}^{dense}>$. The training of this model is enhanced with several negative sampling strategies, aiming at obtaining hard and high-quality negative (query, passage) pairs. This model is trained on MS-MARCO passage dataset (detailed in Sect. 4.1), and achieves a very competitive performance. To adapt NPR to document retrieval setting, we split documents into passages by applying sliding overlapping sentence windows. Following work by Dai and Callan [8], we use the max passage retrieval score as the document level score.

4   Experimental Setup

4.1   Datasets

As we are interested in exploring the performance of the deep retrieval model in a variety of out-of-domain settings, we choose to specifically focus on five datasets in our evaluation (summarized in Table 1).

1. **MS-MARCO passage** [2] dev set is the dataset we use for the *in-domain* model evaluation, as the NPR deep retrieval model, and the docT5query model are trained using the training portion of this dataset (see Sect. 4.2 for more details). The queries in this dataset are all questions.
2. **MS-MARCO doc** [2] is derived from MS-MARCO passage, but instead the retrieval is done using documents. We use the queries in dev set for evaluation (a subset of MS-MARCO passage dev set). This evaluates the generalization of the model to document retrieval.
3. **ORCAS** [6] is a click dataset based on an intersection of Bing search engine logs and the documents in MS-MARCO dataset. Compared to MS-MARCO, queries in ORCAS exhibit wider topics (not limited to questions) and shorter length (76% of queries have no more than 3 tokens after removing stopwords). Since it has a very large number of queries (10M), we evaluate our model using a stratified sample of 10k queries, based on query length.

Table 1. The five datasets used for model evaluation. "Avg. D/Q" denotes the average number of relevant docs per query.

Dataset	Domain	Task	#Query	#Corpus	Avg. D/Q
MS-MARCO passage [2]	Misc.	Passage retrieval	6980	8.8M	1.1
MS-MARCO doc [2]	Misc	Doc retrieval	5193	3.2M	1.1
ORCAS [6]	Misc.	Doc retrieval	9670	1.4M	1.8
Robust04 [40]	News	Doc retrieval	250	528K	69.9
TREC-COVID [34]	Bio-medical	Doc retrieval	50	191K	493.5

4. **Robust04** [40] is a dataset comprising 528K news stories and 250 queries. Each query consists of three fields, including title (keywords), description (a sentence-length statement of the information needs) and narrative (a paragraph-length text explaining what makes a document relevant). It evaluates how well the retrieval model generalizes to the news domain.

5. **TREC-COVID** [34] is based on the CORD19 [41] collection – PubMed articles and preprints about the COVID-19 pandemic. Each query contains a few keywords, along with a more specific natural language version of question, and a narrative which adds additional clarifications of user intent. As shown by Thakur et al. [38], it is quite distinct from the MS-MARCO dataset, and provides a good test case for whether an out-of-domain retrieval system can be useful in a bio-medical domain.

4.2 Data Processing and Benchmarking

In following, we detail our experimental setup to ensure the reproducibility of all the reported results.

Deep Retrieval Model. As described in Sect. 3, we train NPR on the training set of MS-MARCO passage dataset, and apply this model to the other four datasets without any fine-tuning. The documents in the other four datasets are long and may exceed the 512 token length limitation. Following prior work [31], we use a sliding window of ten sentences with a stride of five to split each document into passages. We run NPR on each passage, perform the nearest neighbor search via SCaNN [14] at passage-level and consider the best passage score as its document score. The query used for each dataset is the same as BM25 based lexical model (detailed in Table 2).

Lexical Retrieval Models. For implementing our lexical models, we use the Terrier search engine [26], and apply the default options for stemming and stop word removal provided by Terrier. We employ three fully lexical benchmarks. We carefully tune the parameters, and detail the settings in Table 2.

– BM25 is a commonly used bag-of-words retrieval method. We use the default parameters provided by Terrier, and verify that our results (in terms of MAP) are comparable to other previously reported BM25 benchmarks [44].

Table 2. The best setup for lexical retrieval models. "des./narr./ques." denotes description/narrative/question field and "#fk docs/terms" denotes the number of feedback documents/terms.

Model (→)	BM25		Bo1		docT5query
Dataset (↓)	index	query	#fk doc	#fk terms	#expansions
MS-MARCO passage	full text	query	5	10	40
MS-MARCO doc	full text	query	5	5	20/passage
ORCAS	full text	query	10	10	20/passage
Robust04	full text	query+des.+narr.	5	10	10/passage
TREC-COVID	abstract	query+ques.+narr.	20	40	10

We experiment with a few indexing options: 1) full text, 2) passage and 3) abstract for TREC-COVID only.

- Bo1 is a query expansion package implemented in Terrier. For each dataset, we carefully tune the number of feedback documents ([5, 10, ..., 50]) and the number of feedback terms (*ie*, expansions; [5, 10, ..., 60]). We also experiment with RM3 query expansion package by Terrier and carefully tune the two parameters. However, it yields lower performance than Bo1 in all the five datasets. We thus only report the results of Bo1 in the Sect. 5.
- docT5query is a T5 based document expansion model. As described in Sect. 3, we fine-tune T5 model on the MS-MARCO passage training set by strictly following the setup of prior works [28,31]. We feed each passage length text, namely, passage in the MS-MARCO passage collection, the abstract in TREC-COVID, or split passages of other three datasets, to T5 model and generate N (a tunable parameter; [10, 20, 40]) numbers of expansions. We append the expansions for all the passages to a document.

5   Evaluation

As our work focuses on the first stage retrieval, in this section we adopt Recall@1K as the primary evaluation metric and additionally report MAP score. In our evaluation, we aim to address the research questions posed in Sect. 1.

5.1   Generalization of the Deep Retrieval Model

We first focus on the results on two in-domain datasets (Table 3). As expected, the deep retrieval model NPR performs very well on MS-MARCO passage on which it is trained. It substantially beats BM25 by an absolute 10.77 (relative 12.35%) and 16.15 (relative 83.59%) in terms of Recall@1K and MAP, respectively. In MS-MARCO doc (the in-domain document retrieval task), NPR also performs well, and betters BM25 by 4.55 (5.0%) and 3.86 (14.57%) at Recall@1K and MAP, respectively. This indicates that a well-trained deep passage retrieval model generalizes well to an in-domain document retrieval task.

Table 3. Experimental results on two in-domain datasets. The improvements (R@1K) of all hybrid models (5-8) over baselines (1-4) are statistically significant via a paired two-tailed t-test ($p < 0.05$).

Dataset (→)	MS-MARCO passage		MS-MARCO doc	
Model (↓)	R@1K	MAP	R@1K	MAP
1. BM25	87.18	19.32	90.91	26.50
2. BM25+Bo1	88.27	17.95	91.64	22.69
3. BM25+docT5query	94.07	26.09	93.18	30.28
4. NPR	97.95	**35.47**	95.46	30.36
5. RRF(1, 4)	98.31	29.46	96.80	32.09
6. RRF(2, 4)	98.36	28.62	96.90	31.48
7. RRF(3, 4)	**98.65**	32.89	96.86	**33.50**
8. RRF(2, 3, 4)	98.48	29.58	**96.96**	32.48

Table 4. Experimental results on three out-of-domain datasets. The improvements (R@1K) of all hybrid models (5-8) over baselines (1-4) are statistically significant via a paired two-tailed t-test ($p < 0.05$), except 5/7 vs. 2 in Robust04 and TREC-COVID.

Dataset (→)	ORCAS		Robust04		TREC-COVID	
Model (↓)	R@1K	MAP	R@1K	MAP	R@1K	MAP
1. BM25	77.52	27.1	72.84	26.91	49.29	27.86
2. BM25+Bo1	78.85	23.53	79.02	30.83	52.58	30.98
3. BM25+docT5query	79.62	30.28	74.64	28.01	50.66	28.77
4. NPR	81.18	28.29	70.28	28.39	37.58	17.14
5. RRF(1, 4)	85.95	30.33	79.62	33.19	52.32	30.38
6. RRF(2, 4)	86.18	28.36	**82.82**	**34.60**	54.63	32.21
7. RRF(3, 4)	86.44	**31.39**	79.81	33.34	53.01	30.64
8. RRF(2, 3, 4)	**86.49**	29.74	82.65	34.51	**55.66**	**34.22**

In Table 4, we discuss the results of three out-of-domain document retrieval datasets. Compared to MS-MARCO doc, ORCAS dataset has the least domain shift (as the candidate documents stem from MS-MARCO doc albeit with different queries), followed by Robust04 (news domain). TREC-COVID contains COVID-19 specific topics and has the largest domain shift. We observe that NPR has a clear performance drop with the increased domain shift. NPR performs reasonably on ORCAS, and betters BM25 by relative 4.72% and 4.39% at Recall@1K and MAP, respectively. However, it still has an absolute drop of 14.28 and 2.07 in terms of Recall@1K and MAP, compared to its performance on MS-MARCO doc. In the news domain (Robust04 dataset), the performance of NPR is mixed: it outperforms BM25 by 5.5% of relative MAP improvement, but underperforms by relative 3.51% at Recall@1K. In TREC-COVID dataset, BM25 significantly beats NPR by 11.71 (23.76%) and 10.72 (38.48%) in terms of Recall@1K and MAP. This demonstrates that the generalization ability of deep retrieval models is poor, especially when the target domain is dramatically different from its training domain.

5.2   Utility of Query and Document Expansion

Lexical retrieval models are prone to vocabulary mismatch between queries and documents. We examine whether query and document expansion models could bridge this gap. From Table 3 and Table 4, we see that Bo1 query expansion model consistently brings recall gains, with 1% relative gain on MS-MARCO passage/doc and ORCAS, 8.48% on Robust04 and 6.67% on TREC-COVID.

Recall that docT5query document expansion model is trained on the training set of MS-MARCO passage dataset. In this dataset, it brings very large gains to BM25. In the other four datasets, docT5query shows a consistent, albeit smaller, improvement over BM25 (above 2.5% recall gain), similar to the analysis by Thakur et al. [38].

5.3   Complementarity of Lexical and Deep Retrieval Models

As with query/document expansion, deep retrieval model can narrow the vocabulary gap between queries and documents. One natural question is, are these models still complementary to each other? To answer this, we plot the unique relevant documents retrieved by BM25+Bo1, BM25+docT5query and NPR and their overlaps in Fig. 1 for Robust04 and TREC-COVID (other three datasets only have around one relevance document per query, ref Table 1). We see that each method is complementary to each other. In general, NPR retrieves the largest number of unique relevant results, though it retrieves less relevant results than the other two methods.

5.4   Effectiveness of the Proposed Hybrid Model

Our proposed hybrid framework provides a flexible mechanism for fusing multiple lexical or deep retrieval models. In Table 3 and Table 4 (row 5–8), we demonstrate the performance of our hybrid model which consistently outperforms either the lexical or deep retrieval model alone. In in-domain MS-MARCO passage dataset, the best performing hybrid model of BM25+docT5query and NPR (#7) obtains a Recall@1K of 98.65, betters BM25 and NPR by relative 12.94% and 0.52%, respectively. This hybrid model outperforms coCondenser (Recall@1K = 98.4) [12], the current MS-MARCO leaderboard winner (as of 2021/08/09) in the passage retrieval task.[2] In the in-domain document retrieval task, the best performing hybrid model is the one with all the three methods (Bo1, docT5query and NPR).

In three out-of-domain datasets, the advantage of hybrid model is more evident, given that NPR is weakened in datasets with a large domain shift (i.e.,

---

[2] Note that we focus solely on recall, since we do not apply a second re-ranking stage for optimizing early precision.

TREC-COVID). It consistently improves over BM25 by almost 10% relatively for the three datasets, and substantially outperforms NPR by 6.11%, 14.16% and 44.25% in ORCAS, Robust04, and TREC-COVID, respectively. This demonstrates that our proposed zero-shot hybrid retrieval model is effective and robust across different tasks and domains.

6   Discussion

Our zero-shot hybrid model has demonstrated its effectiveness in the experiments. For comparison, we implement the linear interpolation method that most prior works adopted [17,23,24,42], though such model is not zero-shot, and requires weight tuning. As weight tuning complexity increases with the number of models, we only interpolate BM25 and NPR as a case study: $s(d) = \alpha \times s_{BM25}(d) + (1 - \alpha) \times s_{NPR}(d)$. We perform min-max score normalization and carefully tune the weight $\alpha \in [0.1, \ldots, 0.9]$ via grid search for out-of-domain datasets Robust04 and TREC-COVID.

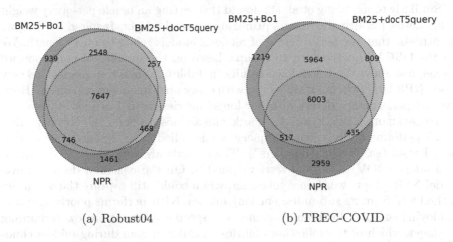

(a) Robust04   (b) TREC-COVID

Fig. 1. A Venn diagram of relevant results by the Bo1, docT5query, and NPR.

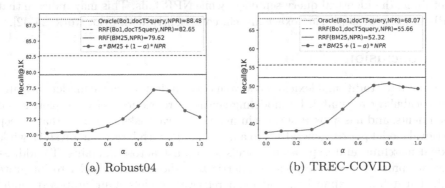

(a) Robust04   (b) TREC-COVID

Fig. 2. The comparisons of our hybrid model, oracle system and interpolation.

Table 5. Mean R@1K result by query length for ORCAS dataset (best result bolded).

Model \ Query Length	1	2	3	4	5	6	7	8	9	10
1. BM25	35.0	74.2	80.2	82.0	82.3	82.9	81.1	87.8	85.6	87.5
2. BM25+Bo1	39.5	76.6	81.3	83.1	83.3	83.0	82.6	88.1	**86.3**	**87.9**
3. BM25+docT5query	36.9	77.8	83.0	84.1	86.2	84.9	**83.8**	**88.6**	85.8	87.8
4. NPR	**59.8**	**82.5**	**85.8**	**86.6**	**87.6**	**85.6**	82.8	83.3	80.1	75.9

Figure 2 (bottom curve), shows that interpolation is weight-sensitive, and furthermore even the best setting underperforms our simple non-parametric hybrid model RRF(BM25, NPR) by a relative 3% in both datasets. The differences are even larger, when compared with the full RRF model (dashed line). We also explore the hybrid upper bound by fusing the retrieval results of BM25+Bo1, BM25+docT5query and NPR via an oracle, *ie,* merging all relevant results from each method regardless of their ranking positions. Figure 2 (dotted top line) illustrates the large potential headroom for designing an even better fusion model.

Similarly to us, Wang et al. [42] found that setting an oracle per-query weight yields better performance than optimizing a global weight. Inspired by this, we hypothesize that the performance of retrieval models relate to query length. We bin the ORCAS queries into 10 groups, based on the number of non-stopword tokens, and show the breakdown results in Table 5. When the queries are very short, NPR largely beats BM25, even with query and document expansion. However, its performance deteriorates for longer queries, with 7 or more tokens.

To gain more insights, we spot-check wins and losses. For single token queries, BM25 performs badly when the query is misspelled (*eg,* "ihpone6") or a compound word (*eg,* "tvbythenumbers"). These words are very likely to be out-of-vocabulary (OOV) in lexical retrieval models. On the contrary, deep retrieval model NPR adopts wordpiece tokenizer, which could still capture the semantics of the OOV from its sub-units. For long queries, NPR performs poorly for those employing complex logic and seeking very specific information, *eg,* "according to piaget, which of the following abilities do children gain during middle childhood?". In this example query, BM25 successfully retrieves relevant documents containing the identical query sentence, while NPR fails. This may indicate that NPR is worse at capturing exact match, consistently with prior work [24, 42].

7   Conclusion

Compared to traditional lexical retrieval models, a deep retrieval model mitigates the vocabulary mismatch by modeling semantic relevance between queries and documents, and has a great success in many retrieval tasks. We show that a deep retrieval model poorly generalizes to a new domain with large domain shift, while lexical matching and expansion models are robust across domains. To address this, we propose a simple non-parametric zero-shot hybrid model to integrate lexical matching, expansion, and deep retrieval models. Our proposed model demonstrates its effectiveness in both in-domain and out-of-domain datasets.

A recent work [36] found that deep retrieval models underperform lexical models for rare entities in an entity-centric QA task. As a future work, we plan to investigate the effectivenss of our hybrid model in this task. Additionally, we plan to parameterize the hybrid retrieval model using query structure, query length, the degree of domain shift, and other signals that may reflect the performance of each individual model. Finally, we plan to explore techniques that improve the utility of out-of-domain deep retrieval models via domain adaptation.

References

1. Amati, G., Rijsbergen, C.J.V.: Probabilistic models of information retrieval based on measuring the divergence from randomness. ACM Trans. Inf. Syst. **20**(4), 357–389 (2002)
2. Bajaj, P., et al.: MS MACRO: a human generated machine reading comprehension dataset (2018)
3. Bendersky, M., Zhuang, H., Ma, J., Han, S., Hall, K.B., McDonald, R.T.: RRF102: meeting the TREC-COVID challenge with a 100+ runs ensemble. CoRR, abs/2010.00200 (2020)
4. Berger, A., Caruana, R., Cohn, D., Freitag, D., Mittal, V.: Bridging the lexical chasm: statistical approaches to answer-finding. In: Proceedings of the 23rd Annual International ACM SIGIR Conference on Research and Development in Information Retrieval, SIGIR 2000, pp. 192–199. Association for Computing Machinery. New York (2000)
5. Cormack, G.V., Clarke, C.L.A., Buettcher, S.: Reciprocal rank fusion outperforms condorcet and individual rank learning methods. In: Proceedings of the 32nd International ACM SIGIR Conference on Research and Development in Information Retrieval, SIGIR 2009, pp, 758–759. Association for Computing Machinery, New York (2009)
6. Craswell, N., Campos, D., Mitra, B., Yilmaz, E., Billerbeck, B.: ORCAS: 20 million clicked query-document pairs for analyzing search. In: Proceedings of the 29th ACM International Conference on Information and Knowledge Management, pp. 2983–2989. Association for Computing Machinery, New York (2020)
7. Dai, Z., Callan, J.: Context-aware sentence/passage term importance estimation for first stage retrieval. CoRR, abs/1910.10687 (2019)
8. Dai, Z., Callan, J.: Deeper text understanding for IR with contextual neural language modeling. In: Proceedings of the 42nd International ACM SIGIR Conference on Research and Development in Information Retrieval, SIGIR 2019, pp. 985–988. Association for Computing Machinery, New York (2019)
9. Dai, Z., Callan, J.: Context-aware term weighting for first stage passage retrieval. In: Proceedings of the 43rd International ACM SIGIR Conference on Research and Development in Information Retrieval, SIGIR 2020, pp. 1533–1536. Association for Computing Machinery, New York (2020)
10. Devlin, J., Chang, M.-W., Lee, K., Toutanova, K.: BERT: pre-training of deep bidirectional transformers for language understanding. In: Proceedings of the 2019 Conference of the North American Chapter of the Association for Computational Linguistics: Human Language Technologies, Volume 1 (Long and Short Papers), pp. 4171–4186, Minneapolis, Minnesota, June 2019. Association for Computational Linguistics (2019)

11. Fan, A., Lewis, M., Dauphin, Y.N.: Hierarchical neural story generation. CoRR, abs/1805.04833 (2018)
12. Gao, L., Callan, J.: Unsupervised corpus aware language model pre-training for dense passage retrieval. CoRR, abs/2108.05540 (2021)
13. Gao, L., Dai, Z., Chen, T., Fan, Z., Van Durme, B., Callan, J.: Complement lexical retrieval model with semantic residual embeddings. In: Hiemstra, D., Moens, M.-F., Mothe, J., Perego, R., Potthast, M., Sebastiani, F. (eds.) ECIR 2021. LNCS, vol. 12656, pp. 146–160. Springer, Cham (2021). https://doi.org/10.1007/978-3-030-72113-8_10
14. Guo, R., Sun, P., Lindgren, E., Geng, Q., Simcha, D., Chern, F., Kumar, S.: Accelerating large-scale inference with anisotropic vector quantization. In: Proceedings of the 37th International Conference on Machine Learning, ICML 2020, Virtual Event, 13–18 July 2020, vol. 119, pp. 3887–3896. PMLR (2020)
15. Jaleel, N.A., et al.: UMass at TREC 2004: Novelty and HARD. In: Voorhees, E.M., Buckland, L.P. (eds.) Proceedings of the 13th Text REtrieval Conference, TREC 2004, Gaithersburg, Maryland, USA, 16–19 November 2004, vol. 500-261. National Institute of Standards and Technology (NIST) (2004)
16. Johnson, J., Douze, M., Jégou, H.: Billion-scale similarity search with GPUs. IEEE Trans. Big Data 7(3), 535–547 (2021)
17. Karpukhin, V., et al.: Dense passage retrieval for open-domain question answering. In: Proceedings of the 2020 Conference on Empirical Methods in Natural Language Processing (EMNLP), Online, November 2020, pp. 6769–6781. Association for Computational Linguistics (2020)
18. Kuzi, S., Zhang, M., Li, C., Bendersky, M., Najork, M.: Leveraging semantic and lexical matching to improve the recall of document retrieval systems: a hybrid approach. CoRR, abs/2010.01195 (2020)
19. Lavrenko, V., Croft, W.B.: Relevance based language models. In: Proceedings of the 24th Annual International ACM SIGIR Conference on Research and Development in Information Retrieval, SIGIR 2001, pp, 120–127. Association for Computing Machinery, New York (2001)
20. Lewis, M., et al.: BART: denoising sequence-to-sequence pre-training for natural language generation, translation, and comprehension. In: Proceedings of the 58th Annual Meeting of the Association for Computational Linguistics, Online, July 2020, pp. 7871–7880. Association for Computational Linguistics (2020)
21. Lin, J., Ma, X.: A few brief notes on DeepImpact, coil, and a conceptual framework for information retrieval techniques. CoRR, abs/2106.14807 (2021)
22. Lin, S.-C., Yang, J.-H., Lin, J.: In-batch negatives for knowledge distillation with tightly-coupled teachers for dense retrieval. In: Proceedings of the 6th Workshop on Representation Learning for NLP, RepL4NLP-2021, August 2021, pp. 163–173, Online. Association for Computational Linguistics (2021)
23. Lu, J., Ábrego, G.H., Ma, J., Ni, J., Yang, Y.: Multi-stage training with improved negative contrast for neural passage retrieval. In: Proceedings of the 2021 Conference on Empirical Methods in Natural Language Processing, Punta Cana, Dominican Republic, November 2021, pp. 6091–6103. Association for Computational Linguistics (2021)
24. Luan, Y., Eisenstein, J., Toutanova, K., Collins, M.: Sparse, dense, and attentional representations for text retrieval. Trans. Assoc. Comput. Linguist. 9, 329–345 (2021)
25. Ma, X., Sun, K., Pradeep, R., Lin, J.: A replication study of dense passage retriever. CoRR, abs/2104.05740 (2021)

26. Macdonald, C., McCreadie, R., Santos, R.L.T., Ounis, I.: From puppy to maturity: experiences in developing terrier. In: Proceedings of the OSIR at SIGIR, pp. 60–63 (2012)

27. Mao, Y., et al.: Generation-augmented retrieval for open-domain question answering. In: Proceedings of the 59th Annual Meeting of the Association for Computational Linguistics and the 11th International Joint Conference on Natural Language Processing (Volume 1: Long Papers), Online, August 2021, pp. 4089–4100. Association for Computational Linguistics (2021)

28. Nogueira, R., Lin, J.: From doc2query to docTTTTTquery (2019). Online

29. Nogueira, R., Yang, W., Lin, J., Cho, K.: Document expansion by query prediction. CoRR, abs/1904.08375 (2019)

30. Ponte, J.M., Croft, W.B.: A language modeling approach to information retrieval. In: Proceedings of the 21st Annual International ACM SIGIR Conference on Research and Development in Information Retrieval, SIGIR 1998, pp. 275–281. Association for Computing Machinery, New York (1998)

31. Pradeep, R., Nogueira, R., Lin, J.: The expando-mono-duo design pattern for text ranking with pretrained sequence-to-sequence models. CoRR, abs/2101.05667 (2021)

32. Qu, Y., et al.: RocketQA: an optimized training approach to dense passage retrieval for open-domain question answering. In: Proceedings of the 2021 Conference of the North American Chapter of the Association for Computational Linguistics: Human Language Technologies, Online, June 2021, pp. 5835–5847. Association for Computational Linguistics (2021)

33. Raffel, C., et al.: Exploring the limits of transfer learning with a unified text-to-text transformer. J. Mach. Learn. Res. **21**(140), 1–67 (2020)

34. Roberts, K.: TREC-COVID: rationale and structure of an information retrieval shared task for COVID-19. J. Am. Med. Inf. Assoc. **27**(9), 1431–1436 (2020). https://doi.org/10.1093/jamia/ocaa091

35. Robertson, S.E., Walker, S., Jones, S., Hancock-Beaulieu, M., Gatford, M.: Okapi at TREC-3. In: TREC (1994)

36. Sciavolino, C., Zhong, Z., Lee, J., Chen, D.: Simple entity-centric questions challenge dense retrievers. CoRR, abs/2109.08535 (2021)

37. Tao, T., Wang, X., Mei, Q., Zhai, C.: Language model information retrieval with document expansion. In: Proceedings of the Human Language Technology Conference of the NAACL, Main Conference, June 2006, pp. 407–414. Association for Computational Linguistics, New York (2006)

38. Thakur, N., Reimers, N., Rücklé, A., Srivastava, A., Gurevych, I.: BEIR: a heterogeneous benchmark for zero-shot evaluation of information retrieval models. In: 35th Conference on Neural Information Processing Systems Datasets and Benchmarks Track (2021)

39. Vaswani, A., et al.: Attention is all you need. In: Proceedings of the 31st International Conference on Neural Information Processing Systems, NIPS 2017, Red Hook, NY, USA, pp. 6000–6010. Curran Associates Inc. (2017)

40. Voorhees, E.: Overview of the TREC 2004 robust retrieval track, 01 Aug 2005 (2005)

41. Wang, L.L., et al.: CORD-19: the COVID-19 open research dataset. In: Proceedings of the 1st Workshop on NLP for COVID-19, ACL 2020, Online, July 2020. Association for Computational Linguistics (2020)

42. Wang, S., Zhuang, S., Zuccon, G.: Bert-based dense retrievers require interpolation with BM25 for effective passage retrieval. In: Proceedings of the 2021 ACM SIGIR International Conference on Theory of Information Retrieval, ICTIR 2021, pp. 317–324. Association for Computing Machinery, New York (2021)
43. Xiong, L., et al.: Approximate nearest neighbor negative contrastive learning for dense text retrieval. In: International Conference on Learning Representations (2021)
44. Yang, P., Fang, H., Lin, J.: Anserini: reproducible ranking baselines using Lucene. J. Data Inf. Qual. **10**(4), 1–20 (2018)
45. Zhan, J., Mao, J., Liu, Y., Zhang, M., Ma, S.: RepBERT: contextualized text embeddings for first-stage retrieval. CoRR, abs/2006.15498 (2020)

Incorporating Ranking Context for End-to-End BERT Re-ranking

Xiaoyang Chen[1,2], Kai Hui[3], Ben He[1,2(✉)], Xianpei Han[2], Le Sun[2], and Zheng Ye[4(✉)]

[1] University of Chinese Academy of Sciences, Beijing, China
`chenxiaoyang19@mails.ucas.ac.cn, benhe@ucas.ac.cn`
[2] Institute of Software, Chinese Academy of Sciences, Beijing, China
`{xianpei,sunle}@iscas.ac.cn`
[3] Amazon Alexa, Berlin, Germany
[4] South-Central University for Nationalities, Wuhan, China
`yezheng@scuec.edu.cn`

Abstract. Ranking context has been shown crucial for the performance of learning to rank. Its use for the BERT-based re-rankers, however, has not been fully explored. In this work, an end-to-end BERT-based ranking model has been proposed to incorporate the ranking context by modeling the interactions between a query and multiple documents in the same ranking jointly, using the pseudo relevance feedback to adjust the relevance weightings. Extensive experiments on standard TREC test collections confirm the effectiveness of the proposed model in improving the BERT-based re-ranker with low extra computation cost.

1 Introduction

Recent advances in information retrieval have shown promising performance gain by utilizing large-scale pre-trained transformer-based language models like BERT [12,27,40,57]. Most of these models, however, consider query-document pairs independently. Actually, unlike in ordinal classification, the main goal of a ranking problem is to optimize ranking lists given queries, making the consideration of the context of the ranking important, such as the *local ranking context* in terms of cross-document interactions [2,43,44]. There have been many successful attempts to incorporate the ranking context, mostly in learning-to-rank-based methods. In early works, loss functions have been proposed to optimize on top of a pair or a list of documents [5,28,31,55], modeling the cross-document interactions at loss level, achieving superior performance on L2R benchmark [46]. In addition, a groupwise ranking framework for multivariate scoring functions is proposed [2] to determine the relevance scores of a group of documents jointly, taking handcrafted learning-to-rank features as query-document presentations and using stack of dense layers to evaluate the relevance. More recently, a neural learning-to-rank model named SetRank [43] is proposed to directly learn a

---

K. Hui—Now at Google AI.

M. Hagen et al. (Eds.): ECIR 2022, LNCS 13185, pp. 111–127, 2022.
https://doi.org/10.1007/978-3-030-99736-6_8

ranking model defined on document sets, employing a stack of multi-head self-attention blocks to learn the embedding for all documents jointly, successfully incorporating the local ranking context and leading to promising improvements.

To the best of our knowledge, however, such ranking context has not been successfully used to enhance the state-of-the-art neural ranking models based on pre-trained language models, like BERT. Indeed, as mentioned in [45], using pairwise loss when employing BERT for re-ranking does not lead to improvements. Beyond single query-document pairs, duoBERT [41] concatenates two documents and the query before feeding into BERT layers, and the output from BERT is trained to learn pairwise comparisons between two documents. However, there exists no straightforward extension to incorporate the full local ranking context using duoBERT as BERT model can not encode very long sequence. Inspired by the success of SetRank [43], in this work, we aim to develop a novel model that could incorporate the ranking context on top of the BERT-based contextualized ranking models, advancing the state-of-the-art BERT-ranker. In a nutshell, comparing with SetRank, BERT-based ranker requires the learning of the encoder during the incorporation of the ranking context, requiring novel framework to enable the end-to-end training. Besides, due to the complexity and huge size of the BERT model [13], special designs are desired to enable the joint modeling of hundreds or even thousands of documents.

To bridge this gap, we propose a groupwise BERT-based ranking model, *Co-BERT*, which is equipped to consider the ranking context. In the groupwise scorer, inspired by [43], candidate documents are grouped together and their interaction representations are passed through several BERT layers to model the ranking context, before projecting the outputs into ranking scores. This groupwise scorer and the BERT encoder for individual query-document pairs are trained end-to-end with pointwise loss. Therein, the groupwise scorer should be able to incorporate the ranking context for hundreds or even thousands of documents; however, individual batch can only include a limited number of documents due to the huge amount of parameters in BERT. To mitigate this dilemma, a ranking list is divided into groups of documents from the same ranking, and pseudo-relevance feedback (PRF) is exploited to capture the query-specific information, calibrating the relevance weightings among different groups.

Contributions in this paper are threefold. 1) We propose an end-to-end groupwise BERT-based ranking model, enabling the joint learning of the query-document interactions and the intra-documents ranking context over BERT. 2) A light-weight PRF-based calibration method is proposed to incorporate ranking context for long list of documents, further boosting the groupwise scorer with small extra computational cost. 3) Extensive evaluation demonstrates that Co-BERT can advance the effectiveness of the state-of-the-art BERT re-ranker. Besides, while providing improvements in ranking effectiveness, the extra computation cost of Co-BERT during inference is as least as 0.3% compared with a standard BERT re-ranker. Source code and data are publicly available at https://github.com/VerdureChen/Co-BERT.

2   Related Work

BERT-Based Ranking and Pseudo Relevance Feedback (PRF). Many existing works have attempted to apply BERT for ranking from different aspects, including training models with large amounts of data [40], scoring documents with sentence-level or passage-level information [12,20,27,54,57], multi-stage fine-tuning with BERT [41], pre-training BERT with various external signals [32–34], as well as combining BERT with existing neural models [37] or LTR methods [18]. Beyond that, two-tower retrievers [22,47,56], ColBERT [23], EPIC [36], TK [21], as well as PreTTR [35] pre-compute the passage representations to reduce query-time latency, and are further improved by TAS-Balanced [19], PAIR [48] and JPQ [60]. As can be seen, most of the mentioned BERT-based ranking models consider query-document pairs independently or use time-consuming pairwise loss, ignoring the ranking context based on more than two documents. There are also works that exploit PRF information to boost ranking. Padaki et al. [42] investigate several traditional keyword expansion approaches and find that they are not necessarily beneficial. Zheng et al. [62] propose BERT-QE that expands the original query by text snippets, instead of individual keywords, selected by a fine-tuned BERT ranker. Based on Transformer-XH [61], Yu et al. [58] propose the graph-based PGT model that utilizes a configurable number of feedback documents. PRF mechanism and query expansion approaches are also incorporated with dense retrievers to boost IR performances [50,53,59]. Unlike Co-BERT, the motivation of these works is to expand the queries to mitigate the vocabulary mismatch between queries and documents. Instead, Co-BERT aims to use the PRF signals to calibrate the relevance weightings for documents in different groups but from the same ranking, supplementing the groupwise scorer component in a light-weight fashion.

Incorporating Ranking Context. In early works, pairwise or listwise losses were used to learn from multiple documents [5,28,31,55]. Recently, the cross-document interactions are further incorporated into the ranking models. Ai et al. [1] employ a recurrent neural network to encode the top-ranked results, from which a context model learns to incorporate the query-specific feature distributions. They further develop a general framework for multivariate scoring functions, in which the relevance score of a document is determined by considering multiple other documents in the list [2]. Pasumarthi et al. [44] leverage the cross-document interaction by a self-attention based neural network, showing improved effectiveness and efficiency on several learning to rank (L2R) datasets. Pang et al. [43] propose a transformer-based L2R approach, SetRank, that directly learns a permutation-invariant ranking model defined on document sets. Very recently, Chen et al. [7] propose a listwise learning framework combining four pooling-based losses over three neural retrieval models. Feng et al. [14] apply Bi-LSTM and self-attention mechanism to model the contextual information to guide the generation of the recommendation results. Among these works, evaluation on learning to rank datasets shows performance gain of SetRank [43] over strong

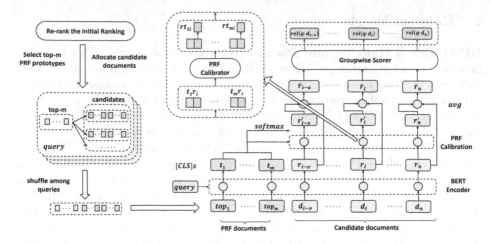

Fig. 1. Model architecture of Co-BERT.

baselines. SetRank uses FNN to encode document features, and feeds the representations into Set-Transformer [26] to capture the local context information from cross-document interactions, jointly scoring retrieved documents. However, when using cross-encoders, it becomes infeasible to put the BERT representations of the entire ranking list in the memory at one time. To the best of our knowledge, these existing models are built upon handcrafted features and do not have straightforward extensions to make uses of the recent pre-trained language models, e.g. BERT. This work extends this research direction for the state-of-the-art BERT-based re-ranker, proposing an end-to-end model that jointly learns the interaction representation and the ranking context.

3   Method

In this section, we present the Co-BERT model for document re-ranking, wherein the query-document interaction representations and the ranking context are learned jointly. The model architecture is summarized in Fig. 1.

3.1   Overview

Given a query q and k ranked documents, e.g., from BM25, a re-ranking method aims to provide each document d a relevance score $rel(q, d)$ that estimates to what degree document d satisfies the query q. As shown in SetRank [43], referring to other candidate documents from the same ranking is important, wherein the query-document representations are in the form of handcrafted feature vectors before modeling the ranking context. Inspired by this, an end-to-end framework is proposed to boost the ranking of these k documents by learning the query-document encoding and the ranking context together. The proposed model is composed of the cross-documents interaction encoder named groupwise scorer, and the components to calibrate the relevance weighting among different batches for the same ranking using pseudo relevance feedback (PRF).

Recall that, when using cross attention to model the relevance with BERT [29], the token sequence from a query q and from a document d are first concatenated into $[CLS]Query[SEP]Document[SEP]$ before passing through multiple self-attention layers, and the interaction representation for $[CLS]$ is used to encode the relevance between the query and the document [13]. In the groupwise scorer, inspired by [2,43], instead of independently evaluating the relevance of individual documents, n documents are considered together using a four-layers BERT model, and the relevance of these n documents are evaluated jointly. Due to the huge amount of the trainable parameters in BERT, a ranking is split into multiple groups and a groupwise scorer models the documents within each group independently. To calibrate the relevance weighting among these groups, similar to [1], the top-m out of the top-k documents are used as pseudo relevance feedback (PRF) set, providing the query-specific context. When evaluating the relevance of a document, beyond directly using the interaction presentation $[CLS]$ between the token sequence from a document-query pair, we first use the m interaction representations from PRF documents to calibrate it. Likewise in [12], since a document could be too long to be encoded using BERT model, we split a document into overlapped passages with the same length. Similar to BERT-QE [62], a BERT checkpoint pre-trained on MS MARCO [40] is used to score each passage relative to the query, and the passage with the highest score in each document is actually used in place of the original document in both training and inference. For brevity, we use the term "document" in the following.

3.2 End-to-End Groupwise Scorer

Given a query-document pair, the $[CLS]$ vector from the last layer of BERT is used as the query-document interaction representation. We denote the k interaction vectors for the top-k documents as r_j where $j \in [1 \cdots k]$, each corresponds to one document to be evaluated, and r_j is a l-dimension dense vector, e.g., $l = 768$ when using BERT-Base. As mentioned in [2], scoring individual documents independently could lead to sub-optimal ranker due to the comparing natural in the ranking problem. Inspired by SetRank [43], we propose a groupwise relevance scorer using BERT, hoping to evaluate the document relevance more effectively by encoding the cross-documents interactions from the same ranking. Due to the size of the BERT encoder, e.g., 110M parameters in BERT-Base, we group n candidate documents ($n \leq k$) together, before modeling the relevance of these n documents jointly. To maintain the cross-references among different groups, we employ a straightforward method by allowing an overlap with o documents, in between neighbouring groups from the initial ranking. For the n documents in a single group, their interaction representations are stacked into a sequence with length n, namely, $r_1, r_2, r_3, \cdots, r_n$. Thereafter, this sequence of interaction representations are passed through multiple layers of BERT, before being projected into n relevance scores, which are used to rank the documents. Herein, a BERT model named $uncased\_L\text{-}4\_H\text{-}768\_A\text{-}12$, with four layers, the hidden size of 768, and, 12 attention heads, is used. We initialise this four-layers BERT model using pre-trained checkpoint from Google [16]. The choice of n is up to the maximum batch size that is allowed by the hardware. Different from the existing ranking model incorporating ranking context, like SetRank [43], groupwise encoder takes the $[CLS]$ from the query-document encoding as input, enabling the end-to-end training of the query-document interaction representation and the ranking context modeling.

Recall that the transformer model [51] relies on the positional embedding to encode the position information. According to our pilot experiments, we do not configure the positional embedding within a group, and simply generate different groups following the initial ranking.

3.3   Light-Weight Pseudo Relevance Feedback

As mentioned, there exist multiple groups when modeling the ranking including many documents, namely, $n \leq k$. In this section, we further introduce a novel building block using PRF information to calibrate relevance weighting among different groups from the same ranking.

The top-m documents are selected as the pseudo relevance feedback (PRF) set, which are used to provide the query-specific context among different groups. We first construct prototype representation for the interaction representations using these m PRF documents. Similar to the computation of each r_j, the m output embedding of the token $[CLS]$ from BERT, each for one of the PRF documents, encode the interaction between the query and the corresponding PRF set. In favor of the description, we denote these m $[CLS]$ vectors as t_i instead of using $r.$, where $i \in [1, \cdots, m]$. Thereafter, the k interaction vectors from Sect. 3.2 are calibrated using these m prototypes t_i with a shallow BERT model of two layers before passing through the groupwise scorer. In particular, the interaction prototype t_i and each interaction representation r_j are stacked into a sequence with two tokens, namely, $t_i r_j$, before passing through the two-layer BERT. The calibrated interaction representation corresponding to r_j using prototype t_i from the two-layers BERT output sequence is denoted as rt_{ij}. Thereby, for each r_j, there are m calibrated representations. Ultimately, we combine these m calibrated presentations into one using a simple weighted average, where the weight is the relevance of the prototype t_i, as in Eq. 1, and W_t and b_t are trainable weights for the projection. Similar to the residual connection in the multi-head attention [51], as shown in Eq. 2, we average the calibrated interaction representation and the origin presentation and use the resulting vector as the inputs for the follow-up scorer. We show that this residual connection is important to the effectiveness in Sect. 5. In this work, for the two-layers BERT model in the calibration, we employ the configuration named *uncased_L-2_H-768_A-12*, which is with two layers, hidden size equaling 768, and 12 attention heads. We use the pre-trained BERT checkpoint from Google [15] to initialise this model.

$$r'_j = \sum_{i \in [1 \cdot m]} softmax(W_t t_i + b_t) \cdot rt_{ij} \tag{1}$$

$$\hat{r}_j = \frac{r_j + r'_j}{2} \tag{2}$$

3.4 End-to-End Training of the Model

Given a query q and k documents, we first select m PRF documents using BERT ranker pre-trained on MS Marco [40]. Thereafter, the batch size is determined based on the constrains of GPU hardware. Therein, in each batch, n candidate documents, together with the m PRF documents are batched together. During training, cross-entropy loss is computed for individual documents as in Eq. 3, where I_{pos} and I_{neg} denote the sets of indexes for relevant and non-relevant documents, respectively, and pr_j is the probability of the document j being relevant according to the model. The probability is computed using a *softmax* function, namely, $pr_j = softmax(rel(q,d))$, where $rel(q,d)$ is the relevance score of d given by Co-BERT.

$$\mathcal{L}(I_{pos}, I_{neg}, q, d_j) = -\sum_{j \in I_{pos}} \log(pr_j) - \sum_{j \in I_{neg}} \log(1 - pr_j) \tag{3}$$

Note that, we use pointwise loss as in Eq. 3 to train the groupwise scorer and leave the study of other losses to future work. The cross-documents interaction is implemented using the four-layers BERT-based groupwise scorer described in Sect. 3.2 and the two-layers BERT-based calibrator in Sect. 3.3.

4   Experiment Setup

4.1   Dataset and Metrics

We experiment on the widely-used Robust04 [52], GOV2 [8], and ClueWeb09-B [9] datasets. We employ 249 title queries for Robust04, 150 title queries for GOV2, and 200 title queries for ClueWeb09-B. Since we have similar observations on NDCG@20 and P@20, we report P@20 to enable the comparisons on the shallow pool; and MAP@1K is reported for deep pool. All statistical tests are based on the paired t-tests at $p < 0.05$ with corrections [6].

4.2   Baselines and Co-BERT Variants

DPH+KL, the unsupervised DPH retrieval model [4] with Rocchio's query expansion using KL divergence [3,49] is used to generate the **initial ranking** of top-1k documents. The implementation from Terrier [38] has been adopted.

BM25+RM3 is another unsupervised ranking model using pseudo relevance feedback signals [25]. We follow the experimental settings from [57], and the implementation from Anserini [30] with default settings is used.

BERT-Base is the BERT-Base ranker boosted by transfer learning. The model is initialised using a checkpoint that has been trained on MS Marco [40], before being fine-tuned on target datasets using the top-1 passage from each relevant document as positive examples as in [62].

BERT-Groupwise is a multi-stage training method. Since there has been no existing work trying to integrate pre-trained language model encoders and groupwise methods, we implement the model by directly combining a SetRank [43] like groupwise model of four-layers BERT with a BERT-based encoder. Text representation of BERT-Base is saved before training groupwise scorer. Due to the pre-storage of text representation, the batch size of training is expanded to 500. Other configurations are similar to Co-BERT.

duoBERT [41] is a pairwise BERT re-ranker initialised using a BERT-Base checkpoint trained on MS Marco, which follows the default setting of the top-30 BERT-Base re-ranking and 512 sequence length.

PGT [58] is a pseudo relevance feedback method that uses a graph-based Transformer. In addition to the results on TREC 19&20 Deep Learning Track [10,11] as in Table 3, we also report our implementation on the other datasets.

BERT-QE [62] is a BERT re-ranking model exploiting the PRF signals. Unlike Co-BERT, BERT-QE is an inference framework and has not been trained end-to-end. In this work, to enable comparisons, we use the BERT-QE variances using three BERT-Base components (namely, BERT-QE-BBB), each for one of its phases. For fair comparisons, we re-implement BERT-QE with the same passage slicing and the same max sequence length as Co-BERT.

The following **variants of Co-BERT** are included for comparisons.

Co-BERT is the model as described in Sect. 3 using BERT-based groupwise scorer on top of the calibrated interaction representations based on PRF.

Co-BERT with PRF calibrator only is a variant of Co-BERT. The relevance of documents are evaluated independently using Eq. 4 without passing the batch of calibrated interaction representations into the groupwise scorer. In particular, we simply project individual $\hat{r}_j$ from Eq. 2 into a relevance score using a shared trainable weights W_{rel} and b_{rel} for each of the k documents, as in Eq. 4.

$$rel(q, \mathcal{R}_m, d_j) = W_{rel}\hat{r}_j + b_{rel} \tag{4}$$

Co-BERT with groupwise scorer only is another variant of Co-BERT without using the PRF calibration, and only use the groupwise scorer described in Sect. 3.2. This means we do not use any feedback signals in the re-ranking, but still use the groupwise scorer for training and inference.

Note that, the efficient design in dense retrieval and contrastive learning [47, 56] are deemed orthogonal to the use of the ranking context, and the dense retrieval models thus have not been included for comparisons. Moreover, the results for the baselines and the Co-BERT variants are based on the standalone ranking models *without* the interpolation with the unsupervised ranking score.

4.3   Model Training and Inference

Data Preparation. Both training and inference are based on the top-1k documents from DPH+KL. Akin to [12], for BERT-Base, PGT, BERT-QE and

Table 1. Effectiveness of Co-BERT relative to baseline models. The gain/loss is reported relative to BERT-Base, on top of which the Co-BERT network architecture is established. The statistical significance at 0.05 relative to (PRF only), (groupwise only), and Co-BERT are denoted as †, ‡ and §, respectively.

Model	Robust04			Gov2			ClueWeb09-B		
	P@20	MAP@1k	FPs	P@20	MAP@1k	FPs	P@20	MAP@1K	FPs
BM25+RM3 [30]	0.3821	0.2903	-	0.5634	0.3350	-	0.2669	0.1819	-
DPH+KL [38]	0.3924	0.3046	-	0.5896	0.3605	-	0.2962	0.2019	-
BERT-Base	$0.4430^{\S}$	$0.3407^{\S}$	+0%	$0.5725^{\ddagger\S}$	$0.3531^{\ddagger\S}$	+0%	$0.3285^{\S}$	$0.2171^{\ddagger\S}$	+0%
BERT-Groupwise	$0.4436^{\S}$	$0.3408^{\S}$	+0.3%	$0.5889^{\ddagger\S}$	$0.3567^{\ddagger\S}$	+0.1%	0.3343	$0.2223^{\ddagger\S}$	+0.5%
duoBERT [41]	$0.4293^{\dagger\ddagger\S}$	$0.3173^{\dagger\ddagger\S}$	+14.6%	$0.5923^{\ddagger\S}$	$0.3553^{\ddagger\S}$	+3.5%	$0.3323^{\S}$	$0.2163^{\ddagger\S}$	+10.3%
PGT [58]	$0.4131^{\dagger\ddagger\S}$	$0.3085^{\dagger\ddagger\S}$	+50.1%	$0.5859^{\ddagger\S}$	$0.3144^{\dagger\ddagger\S}$	+12.0%	$0.2833^{\dagger\ddagger\S}$	$0.1736^{\dagger\ddagger\S}$	+35.4%
BERT-QE [62]	0.4614	0.3555	+86.1%	$0.6198^{\S}$	$0.3662^{\ddagger\S}$	+20.5%	$0.3152^{\ddagger\S}$	0.2131	+60.7%
(PRF only)	0.4526	0.3480	+1.3%	0.5802	0.3550	+0.3%	0.3273	0.2153	+1.0%
	(+2.2%)	(+2.1%)	-	(+1.3%)	(+0.5%)	-	(−0.4%)	(−0.8%)	-
(groupwise only)	0.4500	0.3530	+0.3%	0.6493	0.3993	+0.1%	0.3457	0.2418	+0.5%
	(+1.6%)	(+3.6%)	-	(+13.4%)	(+13.1%)	-	(+5.2%)	(+11.4%)	-
Co-BERT	**0.4629**	**0.3631**	+1.3%	**0.6668**	**0.4022**	+0.3%	**0.3598**	**0.2463**	+1.0%
	(+4.5%)	**(+6.6%)**	-	**(+16.5%)**	**(+13.9%)**	-	**(+9.5%)**	**(+13.5%)**	-

Co-BERT, the documents are chunked using sliding windows of 150 words with an overlap of 75 words. As mentioned in Sect. 3.1, for all four models, the most relevant passage is selected using a BERT ranker pre-trained on MS Marco [40] to represent individual documents. To feed individual query-paragraph (i.e. the text chunk with 150 words) pairs into the model, query and paragraph are concatenated with a maximum sequence length of 256.

Batching and Loss Function. We train BERT-Base and Co-BERT using cross-entropy loss as in Eq. 3 for five epochs with a batch size of 64 on one NVIDIA TITAN RTX 24G. For Co-BERT, according to preliminary results, we configure the number of PRF documents for calibration as $m = 4$, the number of candidate documents in individual group as 60 ($n = 60$), and the overlap between the neighbouring groups is set to four ($o = 4$). During training, we randomly shuffle the batches before feeding them into the model. The Adam optimizer [24] is used with the learning rate schedule from [40]. We configure the initial learning rate as 3e−6, and the warming up steps are set to the 10% of the total training steps.

Cross-Validation. Similar to the configuration in DRMM [17], we use 5-fold cross-validation to report the results with a 3-1-1 split. The query partition on Robust04 follows the settings from [12]. On GOV2 and ClueWeb09-B, queries are partitioned by the order of TREC query id in a round-robin manner. The average performance on the test splits from all folds is reported.

5 Results

In this section, we examine the effectiveness and efficiency of Co-BERT relative to baseline models, before studying how groupwise mechanism and PRF calibrator work with BERT. Finally, we report the results on the TREC Deep Learning track query sets [10,11] for further comparisons.

Table 2. Impacts of the residual connections in Eq. 2. Two alternative feeding orders of batches during training are also investigated. Relative comparison in terms of percentage (in bracket) in comparisons with BERT-Base is also reported. Statistical significance at levels 0.05 is denoted with † and ‡, relative to BERT-Base and Co-BERT, respectively.

Model	Robust04			Gov2		
	P@20	NDCG@20	MAP@1K	P@20	NDCG@20	MAP@1K
BERT-Base	0.4430	0.5109	0.3407	0.5725	0.5040	0.3531
Co-BERT (Random training)	**0.4629**	**0.5213**	**0.3631**	**0.6668**	**0.5781**	**0.4022**
(w/o residual connection in Eq. 2)	0.4554	0.5102	$0.3567^{\dagger}$	$0.6326^{\dagger\ddagger}$	$0.5484^{\dagger\ddagger}$	$0.3951^{\dagger}$
	(−1.6%)	(−2.2%)	(−1.9%)	(−6.0%)	(−5.9%)	(−2.0%)
(Train following initial ranking)	0.4422	0.5029	$0.3457^{\ddagger}$	$0.6211^{\dagger\ddagger}$	$0.5308^{\dagger\ddagger}$	$0.3728^{\dagger\ddagger}$
	(−4.7%)	(−3.6%)	(−5.1%)	(−8.0%)	(−9.4%)	(−8.3%)
(Train reversing initial ranking)	0.4454	0.5026	$0.3429^{\ddagger}$	$0.6322^{\dagger\ddagger}$	$0.5429^{\dagger\ddagger}$	$0.3799^{\dagger\ddagger}$
	(−4.0%)	(−3.7%)	(−5.9%)	(−6.0%)	(−7.0%)	(−6.3%)

5.1 Overall Performance of Co-BERT

Given a query, different BERT-based ranking models, including the variants of Co-BERT model described in Sect. 4, are used to re-rank the top-1k documents from DPH+KL. We also include two classical unsupervised ranking models, namely, BM25+RM3 and DPH+KL, for references. The ranking effectiveness are summarised on both shallow (P@20) and deep pool (MAP@1K) in Table 1.

Effectiveness of Co-BERT. According to Table 1, Co-BERT outperforms all of the unsupervised baselines. As both BERT-Base and Co-BERT have been initialised using the ranking model pre-trained on MS Marco [40], and are fine-tuned in the same way. Thereby, we are assured that the performance difference between Co-BERT and BERT-Base comes from the novel model architecture introduced in Sect. 3. Actually, Co-BERT also achieves better results than the most recent transformer-based ranking models using PRF signals and query expansion such as PGT [58] and BERT-QE [62], confirming the superior effectiveness of the complete Co-BERT, especially on the deep pool.

Efficiency of Co-BERT. The FLOPs, i.e. the number of floating point operations, of various BERT-based models are reported in Table 1, in the form of the relative comparisons to BERT-Base. From Table 1, comparing with BERT-Base, it can be seen that Co-BERT only requires an extra 1.3% computation overheads when significantly boosting the effectiveness on both shallow (4.5%) and deep pool (6.6%) on Robust04; meanwhile, with only 0.3% extra computation cost, Co-BERT could provide more than 13% boosts on both shallow and deep pools on GOV2. Remarkably, though being able to outperform BERT-Base in most cases, the extra computation cost of Co-BERT is actually limited.

5.2   Study of Groupwise Ranking

End-to-End Training Plays an Important Role. As shown in Table 1, although the batch size is large, the detached groupwise architecture of BERT-Groupwise shows little benefit on Robust04 when compared to BERT-Base, and can only achieve marginal improvements on GOV2 and ClueWeb09-B. However, Co-BERT with groupwise scorer only, using the same model component as BERT-Groupwise but with end-to-end training, can significantly improve the performances compared to BERT-Base. On the deep pool in terms of MAP@1K, more than 13% boosts have been observed on both GOV2 and ClueWeb09-B. On shallow pool, the end-to-end training method can also improve P@20 by 1.6%, 13.4% and 5.2%, on the three datasets used, respectively. Recall that BERT-Groupwise attempts to apply groupwise scorer directly to the text representation using a SetRank-like approach. When the query-document representation pre-generated by a fine-tuned BERT ranker is used for groupwise scorer training, the effectiveness of groupwise ranking is limited, demonstrating the importance of the end-to-end training for the BERT-based groupwise ranker.

Impacts of Feeding Order. As mentioned in Sect. 3.4, when the total number of documents for ranking (namely, k) is too large to be fed into single batch, we have to group $n < k$ documents into batches during training and inference, and then feed the data for training after random shuffling. We investigate two alternative ways for the feeding order of training data, namely, training following initial ranking and training reversing initial ranking. Training following initial ranking means when feeding training batches for the same query, the batches are ordered following the initial ranking. On the contrary, when training following the reversed order in initial ranking, the batches are fed in the reversed order of the initial ranking. Note that, among different epochs, the training data is still shuffled among queries to avoid over-fitting. For brevity, we only report results from Robust04 and Gov2, as results obtained on ClueWeb09-B and Gov2 lead to similar observations. According to the results in Table 2, it can be seen that, with the alternative feeding order for the training data, Co-BERT could still outperform BERT-Base on GOV2. Such alternative order, however, leads to at least 3.5% drops among all different metrics on both dataset and the resulting models are significantly worse than Co-BERT trained using fully shuffled batches.

5.3   Study of Light-Weight PRF Calibrator

As can be seen in Table 1, when only using the PRF calibrator without the groupwise scorer, on Robust04, the PRF-calibrator-only variant can outperform BERT-Base with up to 2% margin on two metrics. While on GOV2 and ClueWeb09-B, Co-BERT does not show advantage over BERT-Base. However, when being used with groupwise scorer, namely the full Co-BERT, the PRF calibrator is able to further enhance the effectiveness, although groupwise has already made a significant improvement over BERT-Base. Recall that the purpose of the PRF calibrator is to provide a lightweight performance boost to the

Table 3. Effectiveness of Co-BERT on TREC DL query sets.

Model	TREC DL 19			TREC DL 20		
	MRR@10	NDCG@10	MAP@1K	MRR@10	NDCG@10	MAP@1K
BERT-Base	0.9280	**0.6999**	0.4715	0.7847	0.6776	**0.4553**
PGT [58]	0.9297	0.6938	0.4232	0.8108	0.6818	0.4184
Co-BERT	**0.9581**	0.6996	**0.4838**	**0.8391**	**0.6992**	0.4505

groupwise BERT scorer. The above findings confirm the ability of the PRF calibrator in improving the groupwise BERT ranker with relatively low extra computational overhead as shown in Sect. 5.1. Moreover, as described in Sect. 3.3, the averaging operation in Eq. 2 adds back the origin interaction representation after the PRF calibration, providing more direct connections between early layers and the scorer layers. In Table 2, we report the results of Robust04 and Gov2 without the averaging operation, the performances of Co-BERT drops on all metrics of the results. The same phenomenon is also observed on ClueWeb09-B. This highlights the importance to add this skip connection after calibrating the interaction representation using pseudo relevance feedback.

5.4 Effectiveness on TREC DL

We additionally report the results on the TREC Deep Learning track query sets [10,11] using the MS MARCO passage corpus [39]. TREC DL 19 & 20 contains 43 and 54 queries respectively, which are manually annotated by NIST on a four-point scale. As the MS Marco document set is similar to the passage set in nature, we only report on the latter for brevity. BM25 is used as the initial ranker and the official metrics, MRR@10, NDCG@10 and MAP@1k, are reported. We compare our model with BERT-Base and PGT [58] by re-ranking the top-1000 documents from BM25. According to the results in Table 3, Co-BERT obtains higher scores than PGT in all metrics, however, Co-BERT's performance is overall comparable to BERT-Base. A likely cause for the insignificant difference between Co-BERT and BERT-base is the QA-oriented nature of the MS Marco dataset, which normally has only one prototype answer for a given question. Due to the lack of diversity in the relevant passages for each query, groupwise ranking may not benefit from highlighting different relevant content by the cross-attention.

6   Conclusion

In this paper, we propose an end-to-end BERT-based re-ranking models, named Co-BERT, wherein the relevances of a group of documents are modeled jointly. Evaluation on three standard TREC test collections, namely, Robust04, GOV2, and Clueweb09-B, demonstrates that the proposed Co-BERT could advance the state-of-the-art BERT-based ranking model by a considerable margin. In addition, the results highlight the importance of the end-to-end training of a groupwise BERT ranker, as opposed to the groupwise ranking over the pre-trained text representation using a SetRank-like approach. Finally, the lightweight PRF calibrator is shown to be able to provide a further performance boost over the groupwise ranker with small extra computation overhead.

Acknowledgements. This work is supported by National Key R&D Program of China (2020AAA0105200).

References

1. Ai, Q., Bi, K., Guo, J., Croft, W.B.: Learning a deep listwise context model for ranking refinement. In: SIGIR, pp. 135–144. ACM (2018)
2. Ai, Q., Wang, X., Bruch, S., Golbandi, N., Bendersky, M., Najork, M.: Learning groupwise multivariate scoring functions using deep neural networks. In: ICTIR, pp. 85–92. ACM (2019)
3. Amati, G.: Probability models for information retrieval based on divergence from randomness. Ph.D. thesis, University of Glasgow, UK (2003)
4. Amati, G., Ambrosi, E., Bianchi, M., Gaibisso, C., Gambosi, G.: FUB, IASI-CNR and university of Tor Vergata at TREC 2007 blog track. In: Proceedings of The Sixteenth Text REtrieval Conference. NIST Special Publication, vol. 500-274, pp. 1–10. National Institute of Standards and Technology (2007)
5. Cao, Z., Qin, T., Liu, T.Y., Tsai, M.F., Li, H.: Learning to rank: from pairwise approach to listwise approach. In: Proceedings of the 24th International Conference on Machine Learning, pp. 129–136 (2007)
6. Carterette, B.A.: Multiple testing in statistical analysis of systems-based information retrieval experiments. ACM Trans. Inf. Syst. **30**(1), 4:1–4:34 (2012). https://doi.org/10.1145/2094072.2094076
7. Chen, Z., Eickhoff, C.: PoolRank: max/min pooling-based ranking loss for listwise learning & ranking balance. CoRR abs/2108.03586 (2021). https://arxiv.org/abs/2108.03586
8. Clarke, C.L.A., Craswell, N., Soboroff, I.: Overview of the TREC 2004 terabyte track. In: Proceedings of the Thirteenth Text REtrieval Conference. NIST Special Publication, vol. 500-261, pp. 1–9. National Institute of Standards and Technology (2004)
9. Clarke, C.L.A., Craswell, N., Soboroff, I.: Overview of the TREC 2009 web track. In: Voorhees, E.M., Buckland, L.P. (eds.) Proceedings of The Eighteenth Text REtrieval Conference, TREC 2009, Gaithersburg, Maryland, USA, 17–20 November 2009. NIST Special Publication, vol. 500-278. National Institute of Standards and Technology (NIST) (2009). http://trec.nist.gov/pubs/trec18/papers/WEB09.OVERVIEW.pdf
10. Craswell, N., Mitra, B., Yilmaz, E., Campos, D.: Overview of the TREC 2020 deep learning track. CoRR abs/2102.07662 (2021). https://arxiv.org/abs/2102.07662
11. Craswell, N., Mitra, B., Yilmaz, E., Campos, D., Voorhees, E.M.: Overview of the TREC 2019 deep learning track. CoRR abs/2003.07820 (2020). https://arxiv.org/abs/2003.07820

12. Dai, Z., Callan, J.: Deeper text understanding for IR with contextual neural language modeling. In: Proceedings of the 42nd International ACM SIGIR Conference on Research and Development in Information Retrieval, pp. 985–988. ACM (2019)

13. Devlin, J., Chang, M., Lee, K., Toutanova, K.: BERT: pre-training of deep bidirectional transformers for language understanding. In: Proceedings of the 2019 Conference of the North American Chapter of the Association for Computational Linguistics: Human Language Technologies, pp. 4171–4186. Association for Computational Linguistics (2019)

14. Feng, Y., Hu, B., Gong, Y., Sun, F., Liu, Q., Ou, W.: GRN: generative rerank network for context-wise recommendation. CoRR abs/2104.00860 (2021). https://arxiv.org/abs/2104.00860

15. Google-Research: bert_uncased_L-2_H-768_A-12 (2020). https://storage.google apis.com/bert_models/2020_02_20/uncased_L-2_H-768_A-12.zip

16. Google-Research: bert_uncased_L-4_H-768_A-12 (2020). https://storage.google apis.com/bert_models/2020_02_20/uncased_L-4_H-768_A-12.zip

17. Guo, J., Fan, Y., Ai, Q., Croft, W.B.: A deep relevance matching model for ad-hoc retrieval. In: Proceedings of the 25th ACM International Conference on Information and Knowledge Management, pp. 55–64. ACM (2016)

18. Han, S., Wang, X., Bendersky, M., Najork, M.: Learning-to-rank with BERT in TF-ranking. CoRR abs/2004.08476 (2020)

19. Hofstätter, S., Lin, S., Yang, J., Lin, J., Hanbury, A.: Efficiently teaching an effective dense retriever with balanced topic aware sampling. In: Diaz, F., Shah, C., Suel, T., Castells, P., Jones, R., Sakai, T. (eds.) SIGIR 2021: The 44th International ACM SIGIR Conference on Research and Development in Information Retrieval, Virtual Event, Canada, 11–15 July 2021, pp. 113–122. ACM (2021). https://doi.org/10.1145/3404835.3462891

20. Hofstätter, S., Mitra, B., Zamani, H., Craswell, N., Hanbury, A.: Intra-document cascading: learning to select passages for neural document ranking. In: Diaz, F., Shah, C., Suel, T., Castells, P., Jones, R., Sakai, T. (eds.) SIGIR 2021: The 44th International ACM SIGIR Conference on Research and Development in Information Retrieval, Virtual Event, Canada, 11–15 July 2021, pp. 1349–1358. ACM (2021). https://doi.org/10.1145/3404835.3462889

21. Hofstätter, S., Zlabinger, M., Hanbury, A.: Interpretable & time-budget-constrained contextualization for re-ranking. In: 24th European Conference on Artificial Intelligence. Frontiers in Artificial Intelligence and Applications, vol. 325, pp. 513–520. IOS Press (2020)

22. Karpukhin, V., et al.: Dense passage retrieval for open-domain question answering. In: Proceedings of the 2020 Conference on Empirical Methods in Natural Language Processing (EMNLP), pp. 6769–6781 (2020)

23. Khattab, O., Zaharia, M.: ColBERT: efficient and effective passage search via contextualized late interaction over BERT. In: Proceedings of the 43rd International ACM SIGIR Conference on Research and Development in Information Retrieval, pp. 39–48. ACM (2020)

24. Kingma, D.P., Ba, J.: Adam: a method for stochastic optimization. In: 3rd International Conference on Learning Representations, pp. 1–15 (2015)

25. Lavrenko, V., Croft, W.B.: Relevance-based language models. In: Proceedings of the 24th Annual International ACM SIGIR Conference on Research and Development in Information Retrieval, pp. 120–127. ACM (2001)

26. Lee, J., Lee, Y., Kim, J., Kosiorek, A.R., Choi, S., Teh, Y.W.: Set transformer: a framework for attention-based permutation-invariant neural networks. In: Chaudhuri, K., Salakhutdinov, R. (eds.) Proceedings of the 36th International Conference on Machine Learning, ICML 2019, 9–15 June 2019, Long Beach, California, USA. Proceedings of Machine Learning Research, vol. 97, pp. 3744–3753. PMLR (2019). http://proceedings.mlr.press/v97/lee19d.html

27. Li, C., Yates, A., MacAvaney, S., He, B., Sun, Y.: PARADE: passage representation aggregation for document reranking. CoRR abs/2008.09093 (2020). https://arxiv.org/abs/2008.09093

28. Li, H.: A short introduction to learning to rank. IEICE Trans. Inf. Syst. **94**(10), 1854–1862 (2011)

29. Lin, J., Nogueira, R., Yates, A.: Pretrained transformers for text ranking: BERT and beyond. CoRR abs/2010.06467 (2020)

30. Lin, J., et al.: Toward reproducible baselines: the open-source IR reproducibility challenge. In: Ferro, N., et al. (eds.) ECIR 2016. LNCS, vol. 9626, pp. 408–420. Springer, Cham (2016). https://doi.org/10.1007/978-3-319-30671-1_30

31. Liu, T., Joachims, T., Li, H., Zhai, C.: Introduction to special issue on learning to rank for information retrieval. Inf. Retr. **13**(3), 197–200 (2010)

32. Ma, X., Guo, J., Zhang, R., Fan, Y., Ji, X., Cheng, X.: PROP: pre-training with representative words prediction for ad-hoc retrieval. In: Lewin-Eytan, L., Carmel, D., Yom-Tov, E., Agichtein, E., Gabrilovich, E. (eds.) WSDM 2021, The Fourteenth ACM International Conference on Web Search and Data Mining, Virtual Event, Israel, 8–12 March 2021, pp. 283–291. ACM (2021). https://doi.org/10.1145/3437963.3441777

33. Ma, X., Guo, J., Zhang, R., Fan, Y., Li, Y., Cheng, X.: B-PROP: bootstrapped pre-training with representative words prediction for ad-hoc retrieval. In: Diaz, F., Shah, C., Suel, T., Castells, P., Jones, R., Sakai, T. (eds.) SIGIR 2021: The 44th International ACM SIGIR Conference on Research and Development in Information Retrieval, Virtual Event, Canada, 11–15 July 2021, pp. 1318–1327. ACM (2021). https://doi.org/10.1145/3404835.3462869

34. Ma, Z., et al.: Pre-training for ad-hoc retrieval: hyperlink is also you need. CoRR abs/2108.09346 (2021). https://arxiv.org/abs/2108.09346

35. MacAvaney, S., Nardini, F.M., Perego, R., Tonellotto, N., Goharian, N., Frieder, O.: Efficient document re-ranking for transformers by precomputing term representations. In: Proceedings of the 43rd International ACM SIGIR Conference on Research and Development in Information Retrieval, pp. 49–58. ACM (2020)

36. MacAvaney, S., Nardini, F.M., Perego, R., Tonellotto, N., Goharian, N., Frieder, O.: Expansion via prediction of importance with contextualization. In: Proceedings of the 43rd International ACM SIGIR Conference on Research and Development in Information Retrieval, pp. 1573–1576. ACM (2020)

37. MacAvaney, S., Yates, A., Cohan, A., Goharian, N.: CEDR: contextualized embeddings for document ranking. In: Proceedings of the 42nd International ACM SIGIR Conference on Research and Development in Information Retrieval, pp. 1101–1104. ACM (2019)

38. Macdonald, C., McCreadie, R., Santos, R.L.T., Ounis, I.: From puppy to maturity: experiences in developing terrier. In: Proceedings of the SIGIR 2012 Workshop on Open Source Information Retrieval, pp. 60–63. University of Otago, Dunedin, New Zealand (2012)

39. Nguyen, T., Rosenberg, M., Song, X., Gao, J., T.: MS MARCO: a human generated machine reading comprehension dataset. In: CoCo@ NIPS (2016)

40. Nogueira, R., Cho, K.: Passage re-ranking with BERT. CoRR abs/1901.04085 (2019)
41. Nogueira, R., Yang, W., Cho, K., Lin, J.: Multi-stage document ranking with BERT. CoRR abs/1910.14424 (2019)
42. Padaki, R., Dai, Z., Callan, J.: Rethinking query expansion for BERT reranking. In: Jose, J.M., et al. (eds.) ECIR 2020. LNCS, vol. 12036, pp. 297–304. Springer, Cham (2020). https://doi.org/10.1007/978-3-030-45442-5_37
43. Pang, L., Xu, J., Ai, Q., Lan, Y., Cheng, X., Wen, J.: SetrRank: learning a permutation-invariant ranking model for information retrieval. In: SIGIR, pp. 499–508. ACM (2020)
44. Pasumarthi, R.K., Wang, X., Bendersky, M., Najork, M.: Self-attentive document interaction networks for permutation equivariant ranking. CoRR abs/1910.09676 (2019)
45. Qiao, Y., Xiong, C., Liu, Z., Liu, Z.: Understanding the behaviors of BERT in ranking. arXiv: abs/1904.07531 (2019)
46. Qin, T., Liu, T.Y., Xu, J., Li, H.: LETOR: a benchmark collection for research on learning to rank for information retrieval. Inf. Retrieval **13**(4), 346–374 (2010)
47. Qu, Y., et al.: RocketQA: an optimized training approach to dense passage retrieval for open-domain question answering. In: Proceedings of the 2021 Conference of the North American Chapter of the Association for Computational Linguistics: Human Language Technologies, pp. 5835–5847 (2021)
48. Ren, R., et al.: PAIR: leveraging passage-centric similarity relation for improving dense passage retrieval. In: Zong, C., Xia, F., Li, W., Navigli, R. (eds.) Findings of the Association for Computational Linguistics: ACL/IJCNLP 2021, Online Event, 1–6 August 2021. Findings of ACL, vol. ACL/IJCNLP 2021, pp. 2173–2183. Association for Computational Linguistics (2021). https://doi.org/10.18653/v1/2021.findings-acl.191
49. Rocchio, J.: Relevance feedback in information retrieval. In: The SMART Retrieval System: Experiments in Automatic Document Processing, pp. 313–323. Prentice Hall, Englewood, Cliffs, New Jersey (1971)
50. Tang, H., Sun, X., Jin, B., Wang, J., Zhang, F., Wu, W.: Improving document representations by generating pseudo query embeddings for dense retrieval. In: Zong, C., Xia, F., Li, W., Navigli, R. (eds.) Proceedings of the 59th Annual Meeting of the Association for Computational Linguistics and the 11th International Joint Conference on Natural Language Processing, ACL/IJCNLP 2021, (Volume 1: Long Papers), Virtual Event, 1–6 August 2021, pp. 5054–5064. Association for Computational Linguistics (2021). https://doi.org/10.18653/v1/2021.acl-long.392
51. Vaswani, A., et al.: Attention is all you need. In: NIPS, pp. 5998–6008 (2017)
52. Voorhees, E.M.: Overview of the TREC 2004 robust track. In: Proceedings of the Thirteenth Text REtrieval Conference. NIST Special Publication, vol. 500-261, pp. 1–10. National Institute of Standards and Technology (2004)
53. Wang, X., Macdonald, C., Tonellotto, N., Ounis, I.: Pseudo-relevance feedback for multiple representation dense retrieval. In: Hasibi, F., Fang, Y., Aizawa, A. (eds.) ICTIR 2021: The 2021 ACM SIGIR International Conference on the Theory of Information Retrieval, Virtual Event, Canada, 11 July 2021, pp. 297–306. ACM (2021). https://doi.org/10.1145/3471158.3472250
54. Wu, Z., et al.: Leveraging passage-level cumulative gain for document ranking. In: The Web Conference 2020, pp. 2421–2431. ACM/IW3C2 (2020)
55. Xia, F., Liu, T.Y., Wang, J., Zhang, W., Li, H.: Listwise approach to learning to rank: theory and algorithm. In: Proceedings of the 25th International Conference on Machine Learning, pp. 1192–1199 (2008)

56. Xiong, L., et al.: Approximate nearest neighbor negative contrastive learning for dense text retrieval. arXiv preprint arXiv:2007.00808 (2020)
57. Yilmaz, Z.A., Yang, W., Zhang, H., Lin, J.: Cross-domain modeling of sentence-level evidence for document retrieval. In: Proceedings of the 2019 Conference on Empirical Methods in Natural Language Processing and the 9th International Joint Conference on Natural Language Processing, pp. 3488–3494. Association for Computational Linguistics (2019)
58. Yu, H., Dai, Z., Callan, J.: PGT: pseudo relevance feedback using a graph-based transformer. CoRR abs/2101.07918 (2021)
59. Yu, H., Xiong, C., Callan, J.: Improving query representations for dense retrieval with pseudo relevance feedback. CoRR abs/2108.13454 (2021). https://arxiv.org/abs/2108.13454
60. Zhan, J., Mao, J., Liu, Y., Guo, J., Zhang, M., Ma, S.: Jointly optimizing query encoder and product quantization to improve retrieval performance. CoRR abs/2108.00644 (2021). https://arxiv.org/abs/2108.00644
61. Zhao, C., Xiong, C., Rosset, C., Song, X., Bennett, P.N., Tiwary, S.: Transformer-XH: multi-evidence reasoning with extra hop attention. In: 8th International Conference on Learning Representations, ICLR 2020, Addis Ababa, Ethiopia, 26–30 April 2020. OpenReview.net (2020). https://openreview.net/forum?id=r1eIiCNYwS
62. Zheng, Z., Hui, K., He, B., Han, X., Sun, L., Yates, A.: BERT-QE: contextualized query expansion for document re-ranking. In: Proceedings of the 2020 Conference on Empirical Methods in Natural Language Processing: Findings, pp. 4718–4728 (2020)

WANDS: Dataset for Product Search Relevance Assessment

Yan Chen[✉], Shujian Liu, Zheng Liu, Weiyi Sun, Linas Baltrunas, and Benjamin Schroeder

Search and Recommendation, Wayfair, Boston, USA
{ychen4,sliu1,zliu2,wsun1,lbaltrunas,beschroeder}@wayfair.com

Abstract. Search relevance is an important performance indicator used to evaluate search engines. It measures the relationship between users' queries and products returned in search results. E-commerce sites use search engines to help customers find relevant products among millions of options. The scale of the data makes it difficult to create relevance-focused evaluation datasets manually. As an alternative, user click logs are often mined to create datasets. However, such logs only capture a slice of user behavior in the production environment, and do not provide a complete set of candidates for annotation. To overcome these challenges, we propose a systematic and effective way to build a discriminative, reusable, and fair human-labeled dataset, Wayfair Annotation DataSet (WANDS), for e-commerce scenarios. Our proposal introduces an important cross-referencing step to the annotation process which significantly increases dataset completeness. Experimental results show that this process is effective in improving the scalability of human annotation efforts. We also show that the dataset is effective in evaluating and discriminating between different search models. As part of this contribution, we also released the dataset. To our knowledge, it is the biggest publicly available search relevance dataset in the e-commerce domain.

Keywords: Product search · Search Relevance · Dataset · Evaluation

1 Introduction

Search engines are a big part of our day-to-day lives. They are behind many applications we have come to rely on daily, from web retrieval to e-commerce. Thus, it is hardly a surprise that a lot of research has been poured into improving and evaluating search engines. Search relevance is a measure of the accuracy of the relationship between the search query and the search results. It is commonly used to assess the performance of search engines.

Evaluating search relevance is inherently tricky. It is a common practice to use annotators to indicate the relevancy of a query-result pair. However, on a large scale, it is not possible to ensure the completeness of the evaluation set. The purpose and use case of queries also vary significantly, which makes discerning the intent of the query a challenge. This, in turn, makes it hard to pinpoint the exact

M. Hagen et al. (Eds.): ECIR 2022, LNCS 13185, pp. 128–141, 2022.
https://doi.org/10.1007/978-3-030-99736-6_9

search results that are expected. For example, if a user is interested in finding *induction cooktops*, and attempts to search for them using the query *cooktop*, it poses an interesting annotation challenge - how do we discern between results which include only induction cooktops from those which return all cooktops?

In this paper, we introduce and describe *WANDS*, an open-source e-commerce product dataset that can be used to fairly and accurately evaluate the relevancy of e-commerce product search engines. We will explain our data collection methodology, as well as share experiments that we have conducted to validate the efficacy and value of *WANDS*. The key contributions of this paper include:

– releasing a public dataset, which is built on top of real-world e-commerce production data. To the best of our knowledge, this is the biggest search relevance dataset in the e-commerce domain.
– detailing the methodology used to construct the dataset to allow for transparency and reproducibility.
– proposing an iterative product mining technique called "cross-referencing" to improve the completeness of our annotations while keeping the annotation problem tractable.

2   Related Work

There has been a sizable body of work created on the problem of evaluating search relevance. We partition this prior work into Web Search Relevance and Product Search Relevance.

Web Search Relevance deals with retrieving unstructured search responses from large web-scale datasets. The best-known body of work around web-scale relevance evaluation is from the Text REtrieval Conferences (TREC), a series of evaluation workshops conducted for several years. TREC 2007 and 2008 featured the million query track [5,6] which involved searching over the GOV2 dataset [2]. The dataset used is a collection of web pages from within the .gov domain, and includes around 25 million documents. Part of the track's goal was to investigate whether multiple shallow judgments might be a better alternative to using fewer, more thorough judgments. The 2009 run of this track [9] used a new ClueWeb09 dataset [3] instead of GOV2. This is a much larger dataset of one billion web pages in 10 languages.

Besides academia, multiple enterprises in the tech industry have also shared their research in this space. Google released a sample of their internal annotation guidelines[1]. While it provides a useful peek at how they define relevance, the guidelines do not shed sufficient insights into what Google defines as a *"best"* match. Microsoft Bing made available a package of benchmark dataset *LETOR* [18] for learning to rank, which contains standard features, data, and evaluation tools.

---

[1] https://static.googleusercontent.com/media/guidelines.raterhub.com/en//searchqualityevaluatorguidelines.pdf.

Sogou, a Chinese search engine, released *SOGOU-SRR* (Sogou Search Result Relevance) [27] and *SOGOU-QCL* [29]. These are large and high-quality datasets. However, these datasets would not be the most appropriate for evaluating product search relevance, since their ranking target is web pages instead of products.

Product Search Relevance focuses on retrieving items from datasets of products and merchandise. The community has adopted two main approaches to build product search relevance datasets: mining user click logs and annotating via crowdsourcing.

Mining user click logs is a popular way to build up significantly sized datasets for large enterprises which have ready access to these logs. The rising popularity of embedding-based product retrieval [15, 24–26, 28] is facilitated by datasets assembled from these web-scale search logs. However, these datasets can be noisy as users can click on irrelevant but popular products, and also because non-clicks are difficult to interpret in terms of relevance. Moreover, such datasets are proprietary and have not been released to the public domain.

Datasets in the public domain tend to be crowdsourced datasets that do not leak proprietary and important data. The two following datasets are closely related to *WANDS*.

- Home Depot[2] released the "*Home Depot Product Search Relevance Dataset*" [1,10] on Kaggle. It contains 75K training data samples and 166K evaluation samples. Queries are sampled from Home Depot's search logs. Ground truth labels, between 1 (not relevant) to 3 (highly relevant), are created via crowdsourcing. Each annotation was evaluated by at least three human raters, and the final relevance score is defined to be the average of these human ratings.
- Crowdflower also released a dataset [4] that contains relevance annotations from several e-commerce sites. This is a smaller dataset than the Home Depot one, including 261 search terms and a list of products for each of these terms. Annotations are based on a sliding scale from 1 to 4, where 4 indicates that the product fully satisfies the search query, and 1 indicates that the product does not match a query.

Both of these datasets only provide relevance scores or labels for their training samples, but not for testing samples. This reduces the usability of these datasets for benchmarks and comparison purposes.

Compared to the Home Depot and Crowdflower datasets, *WANDS* is significantly larger in terms of annotated query and product pairs. It includes relevance labels for both training and evaluation datasets to facilitate benchmarking and comparisons. Unlike the two existing datasets, with our *WANDS* dataset we will also release the full annotation guidelines we used, to ensure reproducibility and also to share best practices for future data collectors. *WANDS* also innovates on the annotation process to improve the number of relevant products per query (i.e., the cross-referencing process described in Sect. 4).

---

[2] Major U.S.A. home improvement retailer: http://www.homedepot.com.

3   Annotation Guidelines Design

3.1   Design Principles

In this section, we will detail the design of our annotation guidelines. We design the *WANDS* dataset to meet the following criteria:

Reusable. Our dataset should apply to a wide variety of systems, and provide reproducible results. The most straightforward way to annotate an evaluation dataset is to present a particular Information Retrieval (IR) system's outputs to the annotators and to obtain human judgement specific to the IR system outputs. However, such annotation is not suitable for judging a different IR system. We aim to design a relevance dataset that can be used to evaluate multiple systems.

Fair [17]. It should be agnostic to the systems to be evaluated, and be able to evaluate product search engines fairly and objectively. As discussed in Sect. 2, user behavioral data becomes an increasingly popular choice as relevance evaluation datasets [15, 24–26, 28]. User behavior log data suffer from positional biases and would favor the rankings similar to the production system. We will alleviate positional bias issues by presenting pair-wise query and product information for annotators to judge.

Discriminative. It should have the power to discern the performance of different product search engines given a robust and discriminative evaluation metrics such as nDCG [14, 19–21]. In order to design a dataset that can differentiate great search algorithms from the good ones, we make sure to include hard negatives, the products are almost relevant to a query but not quite. We mined the hard negatives both arithmetically and from user behavior logs.

Completeness. As a core element in the Cranfield paradigm [11], completeness has been a debated quality of a relevance dataset since then [23]. Completeness refers to the property that within a relevance dataset, all relevant documents for a given query are known. Indeed, modern relevance datasets have mostly prioritized dataset size over completeness [22], and various evaluation metrics have been proposed to deal with incompleteness of evaluation datasets [7, 19]. However, as we will show in Sect. 6, incompleteness in Product Search dataset does negatively impact the discriminative power of the evaluation. Incompleteness in Product Search evaluation data also contributes to the problem that offline evaluation results cannot predict online metrics [13]. While we acknowledge that absolute completeness is impossible to achieve for a dataset the size of *WANDS*, we take measures to minimize the impact of incompleteness.

To understand why *completeness* is important, let's assume that we have a target query, which is expected to return 3 products $(p1, p2, p3)$ out of a set of 10. Let's assume that we have two versions of the dataset, A, and B. A includes 2 "relevant" annotations for $p1$ and $p2$. B includes 3 "relevant" annotations for $p1, p2, p3$. Suppose also that we have two search engines that we want to evaluate, α and β. α is able to return two results $(p1, p2)$, while β returns all three relevant products. When evaluated on dataset A, the two search engines

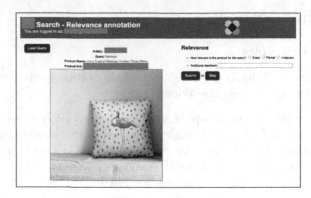

Fig. 1. Screenshot of Annotation Web UI.

will perform identically. It is not possible to tell them apart. However, we will be able to tell using dataset B that β is the better search engine, because it is able to return a better set of results than α.

3.2 Query-Product Annotation

We had three dedicated annotators from the data annotation agent working on this project. Each query, q_k, and its set of candidate products, $\theta_{q_k} = \{i_{k1}, i_{k2}, ..., i_{kn}\}$, are sent to the three annotators. The annotators see one query-product pair at a time and judge each query-product pairing with one of these possible annotations[3]:

- **Exact match:** The surfaced product fully matches the search query.
- **Partial match:** The surfaced product does not fully match the search query. It only matches the target entity of the query, but does not satisfy the modifiers for the query.
- **Irrelevant:** The product is not relevant to the query.

The annotators are given access to a web-based annotation tool to perform the labeling tasks as shown in Fig. 1.

4   Annotation Process

The overview of the annotation process is illustrated in Fig. 2. We started by stratified-sampling of search queries from a pool of historical customer queries stored in the e-commerce customer behavior logs. We then collected the products potentially relevant to one or more of the selected queries and constructed a Product Pool. Once the query and product pools were constructed, we performed Iterative Product Mining to identify the query-product pairs to be annotated. Three annotators then provided independent judgments on the selected

---

[3] Please refer to our Annotation Guidelines released as a supplement to the dataset.

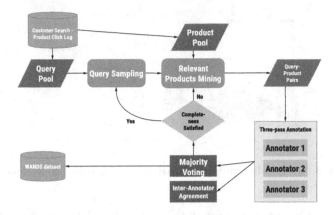

Fig. 2. Overview of the Annotation Process

query-product pairs, according to the Annotation Guidelines. To reduce dataset incompleteness, we introduce the Iterative Product Mining process as described in Sect. 4.3. In the remainder of this section, we will discuss in detail each step of the annotation process.

4.1 Query Sampling

Our e-commerce website serves millions of queries every day. A good search relevance dataset should represent the diversity of real-world queries. To this end, we performed stratified sampling over tens of millions of customer search queries from the 2021 first quarter search log at our U.S. website. This resulted in a total of 480 English search queries.

Specifically, we stratified search queries along the following dimensions: 1) on-site organic searches vs. marketing-redirected searches, 2) searches that resulted in customer engagement (e.g., added products to cart) vs. searches that didn't result in customer engagement, and 3) popularity over the past two years. Within each stratified query group, we picked queries from both the head (frequent) and the tail (infrequent) of the frequency distribution. This approach improves the diversity of the queries in the query pool. Figure 3 illustrated the diverse query distribution over the popularity and engagement dimensions.

4.2 Constructing the Product Pool

Our product catalog contains tens of millions of products. For this annotation task, we need to sample a small subset of our product catalog, such that the resulting relevance data set can differentiate great search models from good ones. This means that for the selected queries, we not only need to include clearly relevant products and clearly irrelevant products, but also need to ensure that there are hard-to-determine, almost-relevant products.

To mimic the real-world difficulty of a product search engine, we adopted two strategies to construct the product pool: using customer engagement data, and using a combination of lexical and neural retrieval systems:

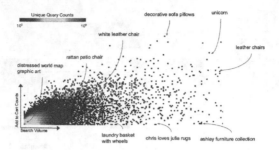

Fig. 3. Query Distributions on Add to Cart and Search Volume.

1. We leveraged user engagement data, and included the products that users clicked on or added to shopping cart during a search experience. Our hypothesis is that the user's added-to-cart products are good approximation of potentially relevant products[4], and their clicked-on (but not added-to-cart) products could be hard negatives, or almost-relevant products.
2. We further mine our product catalog using an open source lexical search engine Solr[5], and a neural product retrieval system inspired by Nigam et al. [16]. The two systems provide different ways to approximate relevant product retrieval. Neither system is perfect thus providing us more chances to include almost-relevant products in additional to relevant products.
3. We didn't attempt to sample easy negative samples (i.e. clearly irrelevant products). We assume that some of the selected products for certain queries will become negative samples for other queries.

4.3   Iterative Product Mining for Dataset Completeness

The query-product pairs resulting from the Product Pool Construction step were sent to the annotators as the first batch of annotation data. Pooling products related to different queries can cause dataset incompleteness [23]. In an ideal world, we would ask our annotators to judge every product and query pairs but that would be intractable - to do so in *WANDS* would require 60 million annotation judgements (480 queries x 42,994 products x 3 passes). To reduce the number of unjudged but relevant query-product pairs, we iteratively mined the entire product pool for unjudged but potentially relevant products for each query as cross-referencing. We presented the mined product-query pairs to the annotators in batches of decreasing likelihood of containing relevant pairs, and monitored the percentage of exact match query-product pairs in the annotation results. Once the percentage of exact match labels dropped to a predetermined level (5%), we would stop the product mining step and assume that the majority of relevant products had been found.

---

[4] Users purchasing irrelevant products in search results is a well documented phenomenon [8], however, it is not a concern in our case.

[5] https://solr.apache.org/.

Table 1. Summaries of *WANDS* and other open-source datasets.

Feature	*WANDS*	Home Depot	Crowdflower
Query			
Counts	480	11,795	261
Predicted Class	✓	✗	✗
Product			
Counts	42,994	54,682	29,790
Primary Class	✓	✗	✗
Title	✓	✓	✓
Description	✓	✓	✓
Attributes	✓	✓	✗
Category Hierarchy	✓	✗	✗
Average Rating	✓	✗	✗
Number of Reviews	✓	✗	✗
Annotated Query-Product Relevance Labels			
Counts	233,448	74,067	22,513

Specifically, we applied the lexical and neural retrieval systems described in Sect. 4.2 to discover more potentially relevant products. We further utilized a proprietary deep learning query classification model, which won during A/B test, to predict the product type that a certain query refers to (e.g. query "textured cotton throw pillow" was classified to "accent pillow" product class), and collected all the items in the product pool that belonged to this product class. After the iterative mining, we have reduced the chance of having unjudged but related query-product pairs in our dataset, and improved dataset completeness.

5   Dataset

The main contribution of this paper is the *WANDS* dataset[6] itself. We collected a total of 480 queries, 42,994 products, and 233K annotated query-product relevance labels. Table 1 shows a summary of *WANDS* relative to the Home Depot and Crowdflower datasets. *WANDS* contains the largest number of relevance labels for query-product pairs. It also contains the richest descriptions of the products and queries in the English language. It includes details such as: product title, product description, primary classes that product belongs to (i.e., chair), product category hierarchy, various product attributes such as size and color, average customer ratings, and review numbers.

Each entry in the dataset maps a (query-product) pair to a single relevance label, which could be one of 1) exact match, 2) partial match, or 3) irrelevant. This label is obtained by aggregating up to 3 entries from our annotators, using the majority vote strategy.

---

[6] https://github.com/wayfair/WANDS.

Table 2. Change in inter-annotator agreement over time.

	Months 1 & 2	Month 3	Month 4
Cohen's Kappa	0.467	0.664	0.826
OPA	0.688	0.812	0.916

Quality Assurance is a common challenge for human-annotated datasets. Without a rigorous quality control strategy, annotators would produce an abundance of poor judgments. To ensure the quality of the annotations, we tracked changes in inter-annotator agreement over time. We do this using two objective quality metrics: 1) Cohen's Kappa [12] and 2) the overlap percentage of agreement (OPA). Both metrics measure the agreement between raters, based on the judgments they make. OPA describes how frequently annotators agree with each other. For example, if 3 annotators all come to the same conclusion, then the inter-annotator agreement is 100%. If 2 out of 3 of them have the same conclusion, then the agreement is 66%. Table 2 shows the changes in inter-annotator agreement over a period of several months. The annotators started with moderate agreement, which steadily increased over the period of 4 months to an almost perfect agreement. Overall, there is a high level of agreement between our annotators leading to the high-quality dataset. We identified four reasons that contribute to significantly improved agreement: 1) Daily routine to discuss the conflicting annotations can help our annotators get calibrated to the annotation guidelines. 2) Regular audits and reviews help to train and align annotators. 3) With the input from annotators, we refine and fine-tune annotation guidelines. 4) Each query-product pair from a new annotator is also labeled by the other two annotators. This ensures data quality and facilitates future alignment. As annotators get trained and are more effective at the task, the overlapped examples are reduced to improve throughput.

Throughput is an important practical aspect of data collection. To determine the initial throughput, we piloted an annotation exercise with four team members. Following our annotation guidelines, we could achieve an initial throughput of 200 query-product pairs per hour, with an OPA of over 90%. The annotators performed consistently for the observed period of time. The throughput after 4 months is at around 190 query-product pairs per hour.

6   Experimental Evaluation

Datasets. Publicly available datasets are used to evaluate performance of search models. We prefer a dataset that provides statistically significant separation between competing search models. We designed an experiment to compare discriminative power of *WANDS* with other two public datasets. Home Depot [1] and Crowdflower [4] are public e-commerce product search datasets consisting of query and product pairs from popular e-commerce websites. Table 1 summarizes the differences between each of these datasets.

Table 3. Mapping of labels across each dataset to a standardized set of scores for metric computation.

Home Depot Dataset	Crowdflower Dataset	*WANDS*	Relevance Score
≥ 2.5	1	exact_match	1.0
≥ 1.5 and < 2.5	2	partial_match	0.5
	3		
≤ 1.5	4	irrelevant	0.0

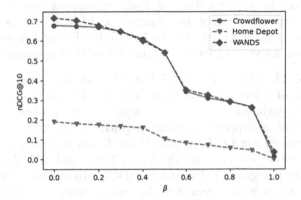

Fig. 4. nDCG@10 with varying β.

Search Models. For the experiment, we also needed to select a set of search models that by design, have known retrieval performance. Note that we are not evaluating the models themselves, but rather how well the dataset can differentiate between two similar models. We chose the following search models with known performance properties:

- **Random ranking (RANDOM).** This is a naive baseline that generates a random list of products as the result to a query.
- **Okapi BM25 (BM25).** That is a probabilistic retrieval ranking model which is integrated into `Apache Solr`. It is based on a bag-of-words representation and uses TF-IDF to estimate the relevance between query and products. This is a widely used and very popular unsupervised search algorithm. In our experiments, we make use of product titles and descriptions in each of the different dataset for ranking.
- **Linear combination of RANDOM and BM25 (LINEAR-β).** This combines both RANDOM and BM25 linearly. The score of a product is computed as follows: $\beta * S_{RANDOM} + (1 - \beta) * S_{BM25}$, where S_x denotes the score assigned to a product by system x and β is a parameter that defines mixing ratio between two base algorithms.

We use nDCG@10 [20] to evaluate the performance of different search models. To compute the metric, we have to resolve the differences in labels used across all three datasets. We map them to the relevance score as shown in Table 3.

Table 4. Label Distribution for different sampling sources.

	exact_match	partial_match	irrelevant
User click logs	8,420 (62.47%)	4,624 (34.31%)	434 (3.22%)
Open-source ranking systems	12,040 (30.95%)	24,797 (63.74%)	2,066 (5.31%)

Results. We ran *RANDOM, BM25* and *LINEAR-β* on three datasets, while varying the values of $β$ from 0.0 to 1.0. Each experiment was repeated 5 times and values averaged. Note that performance of *LINEAR-0.0* is equivalent to *BM25* and *LINEAR-1.0* is equivalent to *RANDOM*. Figure 4 shows a plot of the nDCG@10 scores for varying levels of $β$ on each of the experimental datasets.

Observations. The first observation is that nDCG@10 remains nearly constant across all values of $β$ for the Home Depot dataset. This indicates the dataset is not able to differentiate between search engines in terms of performance, even though *RANDOM* is expected to under-perform *BM25*. In fact, we could not differentiate between any LINEAR-$β$ on Home Depot dataset (one-sided T-test, $p < 0.01$). On the other hand, on the Crowdflower and *WANDS* datasets, we can see an expected gradual decrease in nDCG@10 scores as the value of $β$ increases. The graph is monotonically decreasing, with the highest nDCG@10 score for *BM25* (i.e. $β = 0.0$), and the lowest for *RANDOM* (i.e. $β = 1.0$). When comparing Crowdflower and *WANDS*, we can see that *WANDS* is more discriminative of the two. We can reject the null hypothesis that *LINEAR-0.0* and *LINEAR-0.3* have the same performance (one-sided T-test, $p < 0.01$). However, we cannot statistically separate *LINEAR-0.0* and *LINEAR-0.3* when using the Crowdflower dataset. For Crowdflower, we only see the same level of statistical significance for *LINEAR-0.0* and *LINEAR-0.5*. Thus, we conclude that WANDS has the highest discriminative power as compared to other datasets.

7   Discussion

7.1   Effectiveness of Sampling Sources

Constructing the product pool step uses two candidate sources: user click logs, and open-source ranking systems (e.g. `Solr`). Using user click logs is a popular way to gather query-product pairs and provide a valuable relevance signal. However, relying solely on click logs can lead to an incomplete dataset. This approach misses out on a lot of relevant items that users do not interact with. We augment this with results using open-source ranking systems. While these systems are imperfect, they do greatly expand the possible query-product pairings.

Table 4 shows the breakdown of the distribution of our annotation labels for each of these two sources. We see that the `exact_match` labels mined from user click logs are relatively high at 62.47%, and irrelevant candidates only account for 3.22% of all labels. For open-source ranking systems, we achieve around 31% of `exact_match` labels and 64% `partial_match` labels.

The high proportion of relevant matches from both approaches suggests that our sampling step is working as it was intended - to help narrow down a good list of candidates so that our annotators can pick out relevant matches efficiently.

7.2 Iterative Product Mining for Dataset Completeness Step

We analyzed iterative annotation process to understand how much value it added. After completing the step *Constructing the product pool*, for 49,390 query-product candidates and 480 queries, we obtained a set of 46,875 relevant (e.g. either `exact_match` or `partial_match`) query-product annotations. After iterative mining, the number of relevant matches increased to 172,247 out of a total of 233,448 annotations. This represents a 3x increase and shows that Step 2 is critical to the annotation process.

Table 5. Distribution of labels across annotation steps 1 and 2.

	exact_match	partial_match	irrelevant
Step 1	18,018 (36.48%)	28,857 (58.43%)	2,515 (5.09%)
Step 2	7,596 (4.12%)	117,776 (63.99%)	58,686 (31.88%)

Table 5 lists the differences in distribution of labels we obtained in both steps. We can see that we get a higher proportion of `exact_match` for Step 1 than Step 2 (i.e. 36% vs 4%). Step 2 produces a higher proportion of `partial_match` labels (i.e. 64% vs 58%).

This second step is important, since it can give us more `exact_match` labels. And also we can view these `partial_match` labels as a possible reflection of the harder-to-score/debatable items on the decision boundary. This increases the difficulty of the dataset to further the discriminative power of the dataset.

8 Conclusions and Future Work

Search engines are critical to the success of e-commerce platforms. Much of the work around the evaluation of these systems tends to be proprietary. We hope that the release of *WANDS* will spur continued research in this domain. In this paper, we described the annotation process we have used in detail, as well as shared evaluation results to showcase the discriminative power of *WANDS*. To recap, our key contributions include: 1) making the dataset available in the public domain, 2) introducing the annotation process and releasing the annotation guidelines we used for reproducibility, and 3) sharing our proposal of cross-referencing as a way to improve dataset completeness while keeping the annotation problem tractable. To the best of our knowledge, *WANDS* is the largest search relevance dataset targeted at e-commerce applications.

Looking ahead, we plan to investigate and compare more approaches for cross-referencing. We also want to confirm our hypothesis that the guidelines we have refined through our annotator training process are sufficient to allow less-trained crowdsourced annotators to produce similarly high-quality datasets.

Acknowledgement. We like to thank Elizabeth Yukman and Alex Wolff for their help with editing this manuscript, John Costello and Ariel Nissan for their legal support, and Natali Vlatko and her team for facilitating the open-sourcing of *WANDS*.

References

1. Home depot product search relevance. https://www.kaggle.com/c/home-depot-product-search-relevance/overview
2. Gov2 dataset. http://ir.dcs.gla.ac.uk/test_collections/gov2-summary.htm
3. Clueweb09 dataset. http://boston.lti.cs.cmu.edu/Data/clueweb09/
4. Crowdflower search results relevance. https://www.kaggle.com/c/crowdflower-search-relevance/overview
5. Allan, J., Aslam, J., Carterette, B., Pavlu, V., Kanoulas, E.: Million Query Track 2008 Overview, p. 23. NIST Special Publication, Gaithersburg (2008)
6. Allan, J., Carterette, B., Dachev, B., Aslam, J.A., Pavlu, V., Kanoulas, E.: Million query track 2007 overview. In: Proceedings of the Sixteenth Text REtrieval Conference, TREC 2007 (2007)
7. Buckley, C., Voorhees, E.M.: Retrieval evaluation with incomplete information. In: Proceedings of the 27th Annual International ACM SIGIR Conference on Research and Development in Information Retrieval, pp. 25–32 (2004)
8. Carmel, D., Haramaty, E., Lazerson, A., Lewin-Eytan, L., Maarek, Y.: Why do people buy seemingly irrelevant items in voice product search? On the relation between product relevance and customer satisfaction in ecommerce. In: Proceedings of the 13th International Conference on Web Search and Data Mining, pp. 79–87 (2020)
9. Carterette, B., Pavluy, V., Fang, H., Kanoulas, E.: Million query track 2009 overview. In: TREC, January 2009
10. Choi, J.I., Kallumadi, S., Mitra, B., Agichtein, E., Javed, F.: Semantic product search for matching structured product catalogs in e-commerce (2020). https://arxiv.org/pdf/2008.08180.pdf
11. Cleverdon, C.: The Cranfield tests on index language devices. In: Aslib Proceedings. MCB UP Ltd. (1967)
12. Cohen, J.: A coefficient of agreement for nominal scales. Educ. Psychol. Measure. **20**(1), 37–46 (1960)
13. Deng, A., Shi, X.: Data-driven metric development for online controlled experiments: seven lessons learned. In: Proceedings of the 22nd ACM SIGKDD International Conference on Knowledge Discovery and Data Mining, pp. 77–86 (2016)
14. Lu, X., Moffat, A., Culpepper, J.S.: The effect of pooling and evaluation depth on IR metrics. Inf. Retrieval J. **19**(4), 416–445 (2016)
15. Magnani, A., Liu, F., Xie, M., Banerjee, S.: Neural product retrieval at walmart.com. In: Companion Proceedings of the 2019 World Wide Conference (2019)
16. Nigam, P., et al.: Semantic product search. In: Proceedings of the 25th ACM SIGKDD International Conference on Knowledge Discovery & Data Mining (2019)

17. Pitoura, E., Stefanidis, K., Koutrika, G.: Fairness in rankings and recommendations: an overview. VLDB J. 1–28 (2021). https://doi.org/10.1007/s00778-021-00697-y
18. Qin, T., Liu, T.: Introducing LETOR 4.0 datasets. CoRR abs/1306.2597 (2013). http://arxiv.org/abs/1306.2597
19. Sakai, T., Kando, N.: On information retrieval metrics designed for evaluation with incomplete relevance assessments. Inf. Retrieval **11**(5), 447–470 (2008)
20. Sanderson, M., Zobel, J.: Information retrieval system evaluation: effort, sensitivity, and reliability. In: Proceedings of the 28th ACM SIGIR Conference on Research and Development in Information Retrieval (2005)
21. Valcarce, D., Bellogín, A., Parapar, J., Castells, P.: On the robustness and discriminative power of information retrieval metrics for top-n recommendation. In: Proceedings of the 12th ACM Conference on Recommender Systems, pp. 260–268 (2018)
22. Voorhees, E.: The sixteenth text retrieval conference (TREC 2007) (2007). https://tsapps.nist.gov/publication/get_pdf.cfm?pub_id=890068
23. Voorhees, E.M.: The philosophy of information retrieval evaluation. In: Peters, C., Braschler, M., Gonzalo, J., Kluck, M. (eds.) CLEF 2001. LNCS, vol. 2406, pp. 355–370. Springer, Heidelberg (2002). https://doi.org/10.1007/3-540-45691-0_34
24. Xia, X., et al.: SearchGCN: powering embedding retrieval by graph convolution networks for e-commerce search. In: Proceedings of the 44th International ACM SIGIR Conference on Research and Development in Information Retrieval (2021)
25. Zhang, H., et al.: Towards personalized and semantic retrieval: an end-to-end solution for e-commerce search via embedding learning. In: Proceedings of the 43rd International ACM SIGIR Conference on Research and Development in Information Retrieval (2020)
26. Zhang, H., Wang, T., Meng, X., Hu, Y., Wang, H.: Improving semantic matching via multi-task learning in e-commerce. In: Proceedings of the 42nd International ACM SIGIR on Research and Development in Information and Retrieval, Workshop on eCommerce (2019)
27. Zhang, J., Liu, Y., Ma, S., Tian, Q.: Relevance estimation with multiple information sources on search engine result pages. In: Proceedings of the 2018 ACM on Conference on Information and Knowledge Management (2018)
28. Zhang, J., Liu, Y., Ma, S., Tian, Q.: Que2Search: fast and accurate query and document understanding for search at Facebook. In: Proceedings of the 27th ACM SIGKDD Conference on Knowledge Discovery & Data Mining (2021)
29. Zheng, Y., Fan, Z., Liu, Y., Luo, C., Zhang, M., Ma, S.: Sogou-QCL: a new dataset with click relevance label. In: Proceedings of the 41st International ACM SIGIR Conference on Research and Development in Information Retrieval (2018)

Searching, Learning, and Subtopic Ordering: A Simulation-Based Analysis

Arthur Câmara$^{(\boxtimes)}$, David Maxwell, and Claudia Hauff

Delft University of Technology, Delft, The Netherlands
{a.barbosacamara,d.m.maxwell,c.hauff}@tudelft.nl

Abstract. Complex search tasks—such as those from the *Search as Learning (SAL)* domain—often result in users developing an information need composed of several aspects. However, current models of searcher behaviour assume that individuals have an *atomic* need, regardless of the task. While these models generally work well for simpler informational needs, we argue that searcher models need to be developed further to allow for the decomposition of a complex search task into multiple aspects. As no searcher model yet exists that considers both aspects and the SAL domain, we propose, by augmenting the *Complex Searcher Model* (CSM), the *Subtopic Aware Complex Searcher Model* (SACSM)—modelling aspects as subtopics to the user's need. We then instantiate several agents (i.e., simulated users), with different *subtopic selection* strategies, which can be considered as different prototypical learning strategies (e.g., *should I deeply examine one subtopic at a time, or shallowly cover several subtopics?*). Finally, we report on the first large-scale simulated analysis of user behaviours in the SAL domain. Results demonstrate that the SACSM, under certain conditions, simulates user behaviours accurately.

1 Introduction

Over the years, a series of models[1] that describe searcher behaviour have been defined [7,8,23]. These often provide a post-hoc explanation of—and reasoning behind—the actions of a searcher during information seeking. One of the main drawbacks of such models is their lack of predictive capabilities: we can neither use these models to investigate what is likely to occur in different instantiations of a retrieval system; nor can we use them for simulating user behaviour.[2]

Indeed, models examining searcher behaviours with predictive power [2,3,17] have only recently been explored in the field of *Interactive Information Retrieval (IIR)*. Such models enable us to relate changing costs (e.g., the cost of examining a document) to changing searcher behaviours. Prior works employing these models have investigated how searchers interact with ranked lists [35], the impact of different browsing costs on a searcher's behaviour [5,22], and stopping behaviours

---

[1] In this paper, we refer to a *model* as a ***model of user behaviour***.

[2] This research has been supported by *NWO* projects *SearchX* (639.022.722) and *Aspasia* (015.013.027).

M. Hagen et al. (Eds.): ECIR 2022, LNCS 13185, pp. 142–156, 2022.
https://doi.org/10.1007/978-3-030-99736-6_10

on *Search Engine Results Pages (SERPs)* [31,56]. Search topics are usually considered *atomic* in all of these prior works, with a simple information need. That is, over a *search session*[3], a single topic is considered—with retrieved documents considered to be either relevant or non-relevant to that one topic. These works do not consider the different *aspects* that may constitute a wider topic. In this work, we introduce the first model of user behaviour that incorporates such thinking. More specifically, we take as a starting point the *Complex Searcher Model (CSM)* [30], a model that considers a user's interactions throughout a search session (over multiple queries), and extend it to yield the *Subtopic-Aware Complex Searcher Model (SACSM)*—which, by considering the aspects as subtopics of a larger information need, models: *(i)* subtopic selection; and *(ii)* subtopic switching steps in the search process.

With the SACSM, we explore the effect of different subtopic switching strategies for multiple types of users within a particular domain to ground our work. We consider *Search as Learning (SAL)*, defined by Marchionini [26], as an iterative process whereby learners engage by reading, scanning and processing a large number of documents retrieved by a search system. Here, the goal is to gain knowledge about a specific learning objective. With web search engines having become an essential resource for learners [15], it is therefore vital to provide support to learners (e.g., through the form of novel interface designs [10,45,47] or rankers optimised for human learning [49]) that help improve their learning efficiency while searching. As a learner's complex information needs can often be decomposed into several subtopics, a natural question to ask is *how searchers should tackle the different subtopics to learn efficiently.*

To answer this question, we present an exploratory study of the SACSM where we simulate different types of learners as *agents*[4], and compare these to each other, examining the effect their search behaviour has on their ability to discover documents containing important keywords, as well as how they navigate throughout the subtopic space. We instantiate a series of agents that subscribe to the SACSM—with four tunable parameters that control their simulated searching behaviour: *(i)* **learning speed** (λ), or how fast agents incorporate novel terms into their vocabulary; *(ii)* **exploration** (ξ), or how willing agents are to explore each subtopic; *(iii)* **tolerance** (τ), or how willing an agent is to click on a search result snippet; and *(iv)* **subtopic switching** (φ), the strategy that agents employ to navigate through subtopics. As such, we present the first SAL study that employs simulation to examine the search behaviours of learners. By grounding a series of simulated agents with interaction data from a prior user study, we run extensive simulations of interaction to address the following research question:

RQ *How do **subtopic switching** (φ) strategies for learning-oriented search tasks affect the search behaviour of simulated agents?*

---

[3] We consider a *search session* as interactions with a search interface, which can include the issuing of multiple queries—and the examination of multiple documents.

[4] *Agents* are *simulated users* that are able to make judgements as to the relevancy/attractiveness of information *without* recourse to relevance information [29].

To answer **RQ**, we measure behaviours by tracking how specific measures— the *number of keywords found*, the *order of keywords found*, and *subtopic exploration*—evolve over an agent's search sessions. We argue that to be considered effective, a strategy should allow an agent to: *(i)* discover as many keywords as possible in the early stages of the session; and *(ii)* help the agent to complete the subtopic space exploration in as few steps as possible.

The main findings of our work are: *(i)* subtopic switching strategies that prioritise ordering in the subtopic picking process yield improved discovery of keywords and exploration of subtopics; and *(ii)* the SACSM is enough to instantiate agents that display behaviour similar to real-world learners in a SAL context. Findings suggest that the SACSM is a high-quality model that provides a solid step in approximating searcher behaviours in the SAL domain. This is vital for works that rely on large-scale simulations, such as reinforcement learning for training new rankers optimised for human learning, as well as quickly evaluating new interfaces and algorithmic changes cheaply—all in a simulated environment.

2   Related Work

Models of Searcher Behaviour. Models of searcher behaviour typically fall into one of two categories: *(i)* descriptive models [7,16,20,23,43], allowing us to gain an intuition about the search process; and *(ii)* models that are expressed in more formal (mathematical) language [2,6,11,17,52,53]. The latter category of model provides *predictive power* about why users behave in a certain way. As such, they can be used as the basis of *simulations of interaction* [4]. Here, a model of searcher behaviour that provides a credible approximation of reality can be used to ground simulations to examine what may happen under given circumstances. Despite the advantages that simulations provide, formulating such descriptive models is non-trivial. Contemporary SERPs for example are complex user interfaces, with new components (e.g., *entity cards* [37]) added all the time. In contrast, searcher models typically assume a simple SERP in the format of the traditional *ten blue links* [18]. Numerous studies have been undertaken on this more simplistic design, such as the cost of scrolling [1,5,42], typing [12,40] or response time lag [28,48].

Subtopics. The *Information Retrieval (IR)* community primarily considers the notion of subtopics from a system-centred point of view, with prior works focusing on ranking functions optimised for subtopic retrieval and result diversification [13,21,38,57,58]. Automatic subtopic (structure) extraction has also been investigated, generally based on a given starting query or document [19,51]. The influence of subtopic characteristics on users has not been frequented in IIR. One exception is by Câmara et al. [10], who provided study participants with a list of subtopics and (visual) indicators about the extent of their subtopic exploration. The impact of subtopic ordering on users was not investigated.

SAL. We ground our work in the domain of learning which has attracted considerable attention in recent years. Beyond studies investigating how learning-oriented searches are conducted [15,36] and how to measure learning occurring in search sessions [9,15,55], multiple recent studies have investigated the impact of certain user characteristics and user actions on learning during search sessions—examples include the impact of domain knowledge [39,54], source selection strategies [25], and the cognitive abilities of users [41]. While observational studies are numerous, works proposing novel retrieval algorithms [49] and novel interface elements [10,47] to support learning whilst searching remain sparse.

3 Subtopic-Aware Complex Searcher Model (SACSM)

For our study, we augment the CSM [32,33] to be subtopic-aware, turning it into the SACSM. The CSM is a conceptual model of the IIR process (or a *search session*), describing the flow of activities and decisions that a searcher undertakes when interacting with a search engine. The CSM is built on the work of other conceptual models of the IIR process, such as the models of Baskaya et al. [6] and Thomas et al. [52]. Conceptual models provide us with the necessary scaffolding from which we can expand and develop the model further for a SAL context—and instantiate the model in such a way that we can run our simulations of interaction [4].

The SACSM is illustrated in Fig. 1; it includes a series of additional activities and decision points (compared to CSM) pertaining to the idea of **subtopic selection**, with novel components highlighted in **blue**. Key activities are represented as boxes □, with key decision points undertaken by subscribing agents represented as diamonds ◇. Upon starting at ●, a user (or, in the case of a simulation, an agent) following the SACSM will first examine the given **topic** Ⓐ. SACSM then directs the agent to examine a list of the provided **subtopics** Ⓑ for the given topic, before then deciding **what subtopic** Ⓒ to examine in detail. From here, the agent will **consider a number of potential queries** Ⓓ to issue pertaining to the selected subtopic, before **selecting a query** Ⓔ to issue Ⓕ. The agent will then obtain an 'overview' of the SERP Ⓖ, and decide whether to **enter it** [31] Ⓗ—and if they do, they begin to **examine a snippet** Ⓘ. If the present snippet is **sufficiently attractive** Ⓙ, the agent will **click the associated link** Ⓚ, and **assess the document** Ⓛ for usefulness and/or relevancy, before deciding to **continue on the SERP** Ⓜ (and examining further snippets if so). If not, the decision to **continue with the current subtopic** Ⓝ is then made. If this is the case, further queries are issued Ⓔ—meaning that the snippet and document examination activities are repeated for the results of the new query. This also means that subtopic exploration can entail multiple queries. If the agent instead decides to **abandon the subtopic** Ⓝ, they must then decide whether to **stop the search session** Ⓞ altogether. This process is repeated until all subtopics have been exhausted by the agent Ⓟ, or some other condition is met—such as running out of session time.

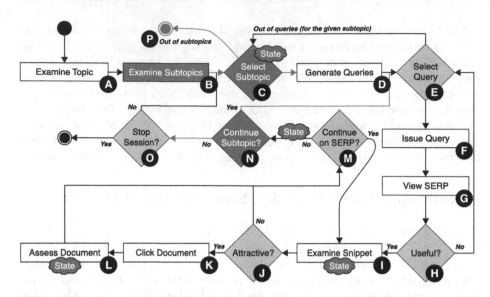

Fig. 1. The *Subtopic-Aware Complex Searcher Model (SACSM)*. Changes from the CSM are highlighted in blue. Refer to Sect. 3 for more about the sequence and shapes.

Note that compared to previous instantiations of the CSM [30–33], we have removed activities and decision points about assessing documents for relevance. Unlike simple search sessions, with *atomic* information needs, a SAL task generally has a more complex and nuanced need [15]. Therefore, we are interested in examining the *content* of retrieved documents (and thus learning from them)— not simply whether the documents themselves are considered relevant, as has been the norm for prior simulations of interaction [27].

In order to keep track of terms/concepts that are examined by agents subscribing to the SACSM (as vocabulary learning is a typical manner to measure learning gains in SAL [10,46,50]), we must also incorporate some type of *state* within it. This state model was considered in the study by Maxwell and Azzopardi [30, Fig. 3] through the *User State Model (USM)*, which *"represents the user's cognitive state"*. Instead of representing the USM as a global, session-based model accumulating state and knowledge of the information examined, we consider a state model for the individual subtopics examined by agents. Each subtopic state consists of a representation of the terms observed by the agent to help them identify fundamental terms about the subtopic, which is used for query generation and determining what snippets (based on the snippet text provided by the underlying retrieval system) should be clicked on, with the corresponding document examined in more detail. Agents following this model only accrue knowledge when examining documents in full, deterministically deducing whether a document is worth examining without recourse to any relevance judgements. The state model is updated at points represented by ● in Fig. 1.

4 Experimental Method

In this section, we describe the details of our instantiation of the SACSM and our simulations. We start by defining how we instantiate each of the components of the SACSM from Fig. 1, dividing them between fixed (i.e., no difference between agents) and variable components (i.e., changes between each agent). By tweaking the variable components, we can instantiate agents that simulate users with different characteristics. For example, an agent with a high λ *(how fast am I at learning new terms?)*, low ξ *(how much content should I explore?)* and low τ *(how liberal am I at clicking links?)* simulates a learner that can quickly absorb new concepts, while only skimming through documents and clicking on almost all documents presented to them. We also outline our search setup, our simulation setup, as well as the datasets, and topics (and subtopics!) used.

4.1 Fixed SACSM Components

We instantiate the SACSM in various ways to evaluate how different **subtopic switching** strategies performed for different types of users. Although the SACSM has many activities and decision points to instantiate, we fixed a number of these to reduce the space that we were required to examine.

Query Generation. We use the $QS3^+$ querying strategy proposed by Maxwell and Azzopardi [30], where three query terms are selected from a language model learned from the documents the agent has already explored (plus the topic description). Previous user studies in the SAL domain [10, 47] have shown that three query terms per query is reasonable, and close to what real-world searchers use.

SERP Examination. Considered by Maxwell and Azzopardi [31], SERP examination strategies provide users with the ability to survey a SERP before committing to examining it in detail. Here, We choose to reduce the complexity of our agents (and explored space) and use the *Always Examine* approach—agents always enter the SERP and examine at least one result snippet.

User Interaction Costs. To realistically mimic how long agents should spend on each phase of their search process, we present in Table 1 the costs (in seconds) from the interaction data of a prior user study [10]. Note the high document examination cost—as participants of the user study were attempting to formulate ideas about concepts, they spent on average longer on documents when compared to other, non-SAL based studies (e.g., [31]). We also note that the total session times influence the stopping behaviours of agents since, when agents reach the time limit of their sessions, they automatically stop—regardless of the number of remaining queries to be issued, as generated by the $QS3^+$ strategy.

Snippet-Level Stopping Strategies. Different *snippet-level stopping strategies* can be employed, generally classified between *fixed* (i.e., the agent will evaluate snippets until a certain depth) or *adaptive* strategies (i.e., the number of snippets evaluated may change depending on factors like agent state, presented snippet

Table 1. Interaction costs grounding our agents, as derived from Câmara et al. [10].

Time required to...	Value (in seconds)
...issue a query	9.42
...examine a SERP	2.00
...examine a result snippet	3.00
...examine a document	80.00
Total session time	2400

content, etc.). For our agents, we use only a fixed snippet-level stopping strategy, where agents examine snippets to a depth of 10. This is a reasonable depth to examine to, and avoids issues with SERP pagination.

4.2 Variable SACSM Components

For this study, agents can be instantiated using four variables, according to the type of user to be simulated. An overview of these variables is presented in Fig. 2.

Subtopic Switching (φ). We propose four different strategies for agents to select and switch between subtopics during their search sessions. These implement the **Select Subtopic** decision point, as shown in Fig. 1. To determine whether agents have explored a subtopic sufficiently, we use a method similar to the approach outlined by Câmara et al. [10] for tracking subtopic exploration. Each clicked document is embedded using SBERT [44] and compared—using the dot product—to pre-computed embeddings for each subtopic of the current topic, as extracted from their *Wikipedia* articles[5]. Therefore, each document clicked by an agent will update an internal state tracker for each subtopic, summing how much the agent *'explored'* each subtopic. We evaluate four strategies.

- **Greedy** For this strategy, an agent examines each subtopic in turn, according to the order provided by the respective Wikipedia article, only deciding to move to the next subtopic when they have achieved a certain level of progress. Intuitively, this would be the most rational type of user, since they follow a subtopic ordering that is optimised for human understanding (i.e., the order comes from a Wikipedia page). In other words, they will attempt to master one subtopic before moving to the next (prescribed) topic.
- **Greedy-Skip** Instead of the above, an agent subscribing to **Greedy-Skip** moves to the next subtopic with the *next lowest completion value*. This instantiated agent attempts to minimise the number of documents to be read by querying in a domain with lesser knowledge.
- **Reverse** This strategy is similar to **Greedy**, but the agent examines the subtopics in reverse order as presented from the corresponding Wikipedia article. The rationale here is that an agent attempts to game the system by first learning the most complex subtopics *before* moving to easier ones.

---

[5] Refer to Sect. 4.3 for more information on the use of Wikipedia articles.

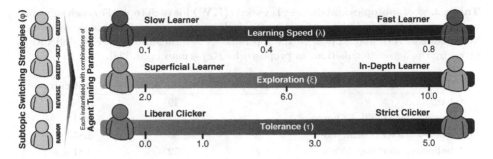

Fig. 2. Overview of the four variable parameters for instantiating simulated agents.

– **Random** This strategy randomly selects a new subtopic after each query, with no predefined order. This strategy models a non-rational learner, and serves as a lower bound for our experiments.

Learning Speed (λ). This parameter is the same λ from the language model proposed by Maxwell and Azzopardi [30]. It controls how much an agent relies on their acquired knowledge (i.e., novel terms) when considering if a given snippet should be clicked or not. The language model is updated every time the agent clicks on a relevant snippet. In addition, a *Maximum Likelihood Estimator* [34] is used for deciding if a given snippet is attractive or not. An agent with a low λ gives lower weights to terms learned during the session, simulating a *slow learner*. An agent with a higher λ, in turn, mimics a user that quickly incorporate new terms, being a *fast learner*. In our simulations, we use $\lambda \in [0.1, 0.4, 0.8]$.

Exploration (ξ). This parameter controls how much the agent should explore a subtopic before being satisfied by what it has *'learnt'*. A lower number implies that such an agent that is only 'skimming' trough the topic, inspecting only a few documents per subtopic. In contrast, a higher value implies that such an agent is willing to explore deeper within each subtopic. For the simulations reported in this paper, we trial $\xi \in [2.0, 6.0, 10.0]$.

Tolerance (τ). Finally, this parameter is the threshold that controls how attractive a snippet should be to be clicked [34]. An agent with low τ is a *strict clicker*, clicking on fewer, *'safer'* snippets, while a higher τ implies a *liberal clicker* agent, more willing to explore. In our simulations, we trial $\tau \in [0.0, 1.0, 3.0, 5.0]$.

4.3 Simulation Setup

The setup of our experiments follows that of the user study presented by Câmara et al. [10]. Here, we use the same eight topics extracted from the TREC CAR 2017 dataset [14], as shown in Table 2. Subtopics were also derived from the TREC CAR dataset: they were extracted from first-level headings of the respective *Wikipedia* articles as they were in December 2017—the dataset's creation.

Table 2. # of subtopics and distinct keywords (**KW**) for each topic. For each subtopic, we determine the ten KWs with higher TF-IDF on the respective subtopic section on Wikipedia. A KW may appear in the top ranks of several subtopics. KW difficulty is given by the age-of-acquisition, as proposed by Kuperman et al. [24].

Topic	#Subtopics	#Unique KWs	KW Difficulty
Ethics	6	49	10.85
Genetically Modified Organism	5	33	9.97
Noise-Induced Hearing Loss	8	56	8.85
Subprime Mortgage Crisis	8	52	9.81
Radiocarbon Dating	4	35	9.77
Business Cycle	4	32	10.70
Irritable Bowel Syndrome	10	72	9.88
Theory of Mind	8	67	9.63

Our study uses the *Bing Search API* to provide a ranking for queries issued by real-world users and our simulated agents. We used a manually curated blocklist[6] of URLs serving *Wiki*-style clones to filter results returned from the Bing API to prevent agents from encountering a single page that would give them all the information on all subtopics at once. This encourages agents to examine multiple documents and issue multiple queries to find information pertinent to their learning task. To match with our stopping strategy (see Sect. 4.1), 10 results per page were presented to agents.

5   Results

By combining all values of ξ, λ, τ and φ, we instantiate 144 unique agents (using a modified version `SimIIR` [30]), and run each agent over all of the topics shown in Table 2. Our version of `SimIIR`—together with the raw outputs of our simulations—are available at https://github.com/ArthurCamara/simiir_subtopics/. With some methods being non-deterministic, each agent was run ten times—with the average reported. In total, we ran a total of $11,520$ simulations. We show representative examples for each set of measures. In all plots, the x axes denote how many documents the agent examined during a search session. While values on y axes may seem low, they are averaged over a large number of simulations with varying degrees of complexity.

Table 3 shows the average value for key measures over all agents of each φ, over the eight topics. In the first row, we also show the measures from the FEEDBACK$_{SC}$ cohort from Câmara et al. [10]. Our simulated agents are similar on these measures compared to how real-world learners would behave, with a similar number of queries, snippets examined, and documents clicked. While a high deviation is expected, recall that this is an average of 288 agents with a large

---

[6] https://github.com/ArthurCamara/CHIIR21-SAL-Scaffolding/blob/master/data/ blocklist.txt (All URLs last accessed January 18th, 2022.).

Table 3. Overview of (average) measures across agents and subtopic switching strategies, and real learners extracted from the FEEDBACK$_{SC}$ cohort from Câmara et al. [10].

Strategy (φ)	#Queries Issued	#Snippets Examined	#Documents Clicked
FEEDBACK$_{SC}$ (N=36)	11.86($\pm$7.60)	152.44($\pm$84.23)	18.50($\pm$9.56)
Greedy	13.05($\pm$14.93)	133.37($\pm$176.29)	21.32($\pm$7.60)
Greedy-Skip	13.05($\pm$15.13)	133.23($\pm$177.26)	21.44($\pm$8.88)
Reverse	12.01($\pm$14.55)	123.28($\pm$173.34)	21.42($\pm$8.78)
Random	13.03($\pm$16.07)	117.61($\pm$155.80)	21.82($\pm$8.30)

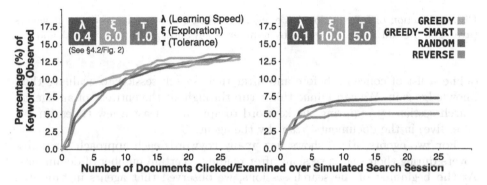

Fig. 3. Accumulated percentage of the keywords seen for two different agents (averaged over all topics, weighted by number of keywords) with varying ξ, λ and τ.

variation on their parameters (compared to only 36 real-world learners). Results show that our agents are indeed similar to real-world learners. To address **RQ**, we break down our analysis further into three sub-questions.

How Many Keywords Can the Agents Find? To measure how well the agents can find documents with a high concentration of potentially valuable keywords[7], we extracted ten keywords for each subtopic from their respective paragraphs from the topic's Wikipedia article.[8] To do this, we begin by ranking all terms from their portions of the articles (excluding stopwords) by their TF-IDF, with the IDF computed over the whole TREC CAR Wikipedia dump— and selecting the top 10 terms as keywords. We use this subtopic-wise approach (instead of extracting keywords from the whole article) to ensure a fair distribution of keywords over all subtopics, providing a less biased overview of how the agent is performing over the topic. Therefore, each topic has a different number of keywords, reflected by the number of subtopics it contains. Table 2 shows how many subtopics and unique keywords each topic contains. This setup is similar to previous SAL user studies [10,46,47], where study participants were asked to

---

[7] As noted in Sect. 3, we do not have explicit relevance judgements.

[8] As an example, the following are extracted keywords for the topic *Ethics*: ethical, ontology, propositions, consequentialism, normative and principles.

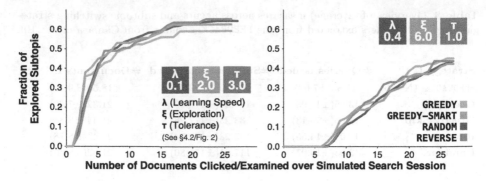

Fig. 4. Fraction of fully explored (i.e. agent reached ξ value) for two different agents (averaged over all topics, weighted by number of keywords) with varying ξ, λ and τ.

define a list of concepts before and after their search session to evaluate their knowledge gain. We can mimic this setup throughout the entirety of an agent's search session by requiring the keyword to appear at least a few times (in our case, five) in the documents *'read'* by the agent.

For two agents, Fig. 3 shows how many keywords each approach for φ discovers during their search sessions after reading a certain number of documents. At the beginning of the search sessions, we observed that agents instantiated with **Greedy** and **Greedy-Skip** strategies found keywords faster than agents with **Random** or **Reverse**. However, this difference diminishes over time. This is expected, since the subtopics ordering comes from Wikipedia articles, which are optimised for human understanding. Therefore, an agent that searches for subtopics in order has a higher probability of encountering documents with a higher number of keywords earlier in the session when compared with one that does not. We can also note that **Random** with higher τ found more unique keywords in total, given their high probability of clicking in any document.

Are the Agents Exploring Enough of the Subtopics? Another way to measure how the agents behave is by investigating how their internal *Subtopic Trackers* evolve during the session (as explained in Sect. 3). If an agent can reach ξ for a given subtopic in a few documents, we can infer that they could quickly find documents related to that subtopic. Figure 4 shows a similar trend to that observed previously, with agents using **Greedy** and **Greedy-Skip** strategies clicking on documents that advance their internal tracking faster. This implies that these strategies effectively lead the agents towards better documents faster.

Are the Agents Following the Order of the Subtopics? While the previous measures show that the agents are indeed effective in finding documents related to the topic, they fail to incorporate another essential learning feature, namely that keywords have dependencies between them. We assume that, for an agent to comprehend what a keyword means entirely, they have to comprehend at least some other, more basic concepts related to the topic at hand. Therefore, an agent that can find documents so that they will encounter more primary

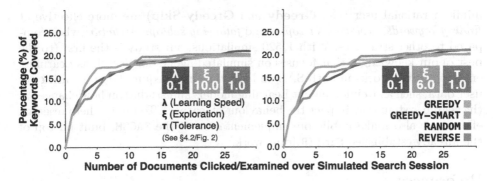

Fig. 5. Fraction of keywords properly *'learned'* by two different agents (averaged over all topics, weighted by number of keywords) with varying ξ, λ and τ.

keywords for the topic (i.e., that appear earlier in the Wikipedia article) earlier in the session before facing more complex keywords (i.e., that appear later in the Wikipedia article) is more desirable for a SAL environment. As an example, consider the keyword `consequentialist` for the topic *Ethics*. Before an agent can adequately understand what it means in this context, they probably need to understand other concepts, like *virtue* and *morality*. Therefore, for this analysis, we consider a keyword to be *'learned'* after the agent has already encountered a certain number of keywords that appear prior to it in the original Wikipedia article. To account for possible noises in our keyword extraction method, we define this number as 50% of the keywords seen prior to the current one (e.g., the keyword `consequentialist` is the 19th out of 49 keywords to appear in the list of extracted terms for the topic *Ethics*). Therefore, we only consider a given keyword as learned after the agent has learned at least 30% of the prior terms.[9] As seen in Fig. 5, we see similar behaviour to the one observed above, with **Greedy** and **Greedy-Skip** outperforming **Random** and **Reverse**, with the difference slowly disappearing throughout the search session. Again, almost all agents repeat this behaviour. These results show that our simulations are close to real users and that there is a clear difference between strategies, with **Greedy** and **Greedy-Skip** following the logical structure of the subtopics, and generally being better strategies for agents exploring the subtopic space. Consequently, these should be taken into account when simulating agents for SAL scenarios.

6   Conclusions

We have proposed a novel user model for simulating agents focused on SAL tasks: the *Subtopic-Aware Complex Searcher Model*, SACSM. Recalling our original research question which considered how different subtopic switching strategies (φ) affected the behaviour or simulated agents, we show that strategies that

---

[9] This number was decided experimentally, as it showed to be the best to distinguish between the different methods trialled.

mimic a rational user (i.e., **Greedy** and **Greedy-Skip**) are more effective at *finding keywords, exploring subtopics* and *following subtopic structure* when compared to other strategies. With $1,520$ simulations, our study is the first (to the best of our knowledge) that focuses on simulated agents for Search as Learning, enabling future works in both SAL and IIR that may require large quantities of user data, such as Reinforcement Learning models and studies on how changes in the search system may impact the behaviour of learners. To further help research efforts, we also make public our implementation of the SACSM, built on top of the already established SimIIR framework.

References

1. Albers, M., Kim, L.: Information design for the small-screen interface: an overview of web design issues for personal digital assistants. Tech. Comm. **49**(1), 45–60 (2002)
2. Azzopardi, L.: The economics in interactive information retrieval. In: Proceedings of 34th ACM SIGIR, pp. 15–24 (2011)
3. Azzopardi, L.: Modelling interaction with economic models of search. In: Proceedings of 37th ACM SIGIR, pp. 3–12 (2014)
4. Azzopardi, L., Järvelin, K., Kamps, J., Smucker, M.D.: Report on the SIGIR 2010 workshop on the simulation of interaction. SIGIR Forum **44**(2), 35–47 (2011)
5. Azzopardi, L., Zuccon, G.: Two scrolls or one click: a cost model for browsing search results. In: Proceedings of 38th ECIR, pp. 696–702 (2016)
6. Baskaya, F., Keskustalo, H., Järvelin, K.: Modeling behavioral factors in interactive information retrieval. In: Proceedings of 22nd ACM CIKM, pp. 2297–2302 (2013)
7. Bates, M.J.: The design of browsing and berrypicking techniques for the online search interface. Online Review (1989)
8. Belkin, N.J.: The cognitive viewpoint in information science. J. Info. Sci. **16**(1), 11–15 (1990)
9. Bhattacharya, N., Gwizdka, J.: Relating eye-tracking measures with changes in knowledge on search tasks. In: Proceedings of 13th ACM ETRA, pp. 1–5 (2018)
10. Câmara, A., Roy, N., Maxwell, D., Hauff, C.: Searching to learn with instructional scaffolding. In: Proceedings of 6th ACM CHIIR, pp. 209–218 (2021)
11. Carterette, B., Kanoulas, E., Yilmaz, E.: Simulating simple user behavior for system effectiveness evaluation. In: Proceedings of the 20th ACM CIKM, pp. 611–620 (2011)
12. Crescenzi, A., Kelly, D., Azzopardi, L.: Time pressure and system delays in information search. In: Proceedings of 38th ACM SIGIR, pp. 767–770 (2015)
13. Dai, W., Srihari, R.: Minimal document set retrieval. In: Proceedings of 14th ACM CIKM, pp. 752–759 (2005)
14. Dietz, L., Verma, M., Radlinski, F., Craswell, N.: TREC complex answer retrieval overview. TREC (2017)
15. Eickhoff, C., Teevan, J., White, R., Dumais, S.: Lessons from the journey: a query log analysis of within-session learning. In: Proceedings of 7th ACM WSDM, pp. 223–232 (2014)
16. Ellis, D.: Modeling the information-seeking patterns of academic researchers: a grounded theory approach. Libr. Q. **63**(4), 469–486 (1993)
17. Fuhr, N.: A probability ranking principle for interactive information retrieval. Inf. Retrieval **11**(3), 251–265 (2008)

18. Hearst, M.: Search User Interfaces. Cambridge University Press (2009)
19. Hearst, M., Plaunt, C.: Subtopic structuring for full-length document access. In: Proceedings of 16th ACM SIGIR, pp. 59–68 (1993)
20. Ingwersen, P., Järvelin, K.: The Turn: Integration of Information Seeking and Retrieval in Context, vol. 18, p. 448. Springer, Dordrecht (2006). https://doi.org/10.1007/1-4020-3851-8
21. Jiang, Z., Wen, J.R., Dou, Z., Zhao, W.X., Nie, J.Y., Yue, M.: Learning to diversify search results via subtopic attention. In: Proceedings of 40th ACM SIGIR, pp. 545–554 (2017)
22. Kashyap, A., Hristidis, V., Petropoulos, M.: FACeTOR: cost-driven exploration of faceted query results. In: Proceedings of 19th ACM CIKM, pp. 719–728 (2010)
23. Kuhlthau, C.C.: Developing a model of the library search process: cognitive and affective aspects, pp. 232–242 (1988)
24. Kuperman, V., Stadthagen-Gonzalez, H., Brysbaert, M.: Age-of-acquisition ratings for 30,000 English words. Behav. Res. Methods 44(4), 978–990 (2012)
25. Liu, C., Song, X.: How do information source selection strategies influence users' learning outcomes. In: Proceedings of 3th ACM CHIIR, pp. 257–260 (2018)
26. Marchionini, G.: Exploratory search: from finding to understanding. Commun. ACM 49(4), 41–46 (2006)
27. Maxwell, D.: Modelling Search and Stopping in Interactive Information Retrieval. Ph.D. thesis, University of Glasgow, Scotland (2019)
28. Maxwell, D., Azzopardi, L.: Stuck in traffic: how temporal delays affect search behaviour. In: Proceedings of 5th IIiX, pp. 155–164 (2014)
29. Maxwell, D., Azzopardi, L.: Agents, simulated users and humans: an analysis of performance and behaviour. In: CIKM, pp. 731–740. ACM (2016)
30. Maxwell, D., Azzopardi, L.: Simulating interactive information retrieval: SimIIR: a framework for the simulation of interaction. In: Proceedings of 39th ACM SIGIR, pp. 1141–1144 (2016)
31. Maxwell, D., Azzopardi, L.: Information scent, searching and stopping. In: Proceedings of 40th ECIR, pp. 210–222 (2018)
32. Maxwell, D., Azzopardi, L., Järvelin, K., Keskustalo, H.: An initial investigation into fixed and adaptive stopping strategies. In: Proceedings of 38th ACM SIGIR, pp. 903–906 (2015)
33. Maxwell, D., Azzopardi, L., Järvelin, K., Keskustalo, H.: Searching and stopping: an analysis of stopping rules and strategies. In: Proceedings of 24th ACM CIKM, pp. 313–322 (2015)
34. Meij, E., Weerkamp, W., de Rijke, M.: A query model based on normalized log-likelihood. In: Proceedings of 18th ACM CIKM, pp. 1903–1906 (2009)
35. Moffat, A., Scholer, F., Thomas, P.: Models and metrics: IR evaluation as a user process. In: Proceedings of 17th ADCS, pp. 47–54 (2012)
36. Moraes, F., Putra, S.R., Hauff, C.: Contrasting search as a learning activity with instructor-designed learning. In: Proceedings of 27th ACM CIKM, pp. 167–176 (2018)
37. Navalpakkam, V., Jentzsch, L., Sayres, R., Ravi, S., Ahmed, A., Smola, A.: Measurement and modeling of eye-mouse behavior in the presence of nonlinear page layouts. In: Proceedings of the 22nd WWW, pp. 953–964 (2013)
38. Nguyen, T.N., Kanhabua, N.: Leveraging dynamic query subtopics for time-aware search result diversification. In: Proceedings of 36th ECIR, pp. 222–234 (2014)
39. O'Brien, H.L., Kampen, A., Cole, A.W., Brennan, K.: The role of domain knowledge in search as learning. In: Proceedings of 5th ACM CHIIR, pp. 313–317 (2020)

40. Ong, K., Järvelin, K., Sanderson, M., Scholer, F.: Qwerty: the effects of typing on web search behavior. In: Proceedings of 3rd ACM CHIIR, pp. 281–284 (2018)
41. Pardi, G., von Hoyer, J., Holtz, P., Kammerer, Y.: The role of cognitive abilities and time spent on texts and videos in a multimodal searching as learning task. In: Proceedings of 5th ACM CHIIR, pp. 378–382 (2020)
42. Peytchev, A., Couper, M.P., McCabe, S.E., Crawford, S.D.: Web survey design: paging versus scrolling. Intl. J. Pub. Opin. Q. **70**(4), 596–607 (2006)
43. Pirolli, P., Card, S.: Information foraging. Psychol. Rev. **106**(4), 643 (1999)
44. Reimers, N., Gurevych, I.: Sentence-BERT: sentence embeddings using siamese BERT-networks. arXiv preprint arXiv:1908.10084 (2019)
45. Roy, N., Câmara, A., Maxwell, D., Hauff, C.: Incorporating widget positioning in interaction models of search behaviour. In: Proceedings of 11th ACM ICTIR, pp. 53–62 (2021)
46. Roy, N., Moraes, F., Hauff, C.: Exploring users' learning gains within search sessions. In: Proceedings of 5th ACM CHIIR, pp. 432–436 (2020)
47. Roy, N., Torre, M.V., Gadiraju, U., Maxwell, D., Hauff, C.: Note the highlight: incorporating active reading tools in a search as learning environment. In: Proceedings of 6th ACM CHIIR, pp. 229–238 (2021)
48. Schurman, E., Brutlag, J.: Performance related changes and their user impact. In: O'Reilly Velocity Conference (2009)
49. Syed, R., Collins-Thompson, K.: Retrieval algorithms optimized for human learning. In: Proceedings of 40th ACM SIGIR, pp. 555–564 (2017)
50. Syed, R., Collins-Thompson, K.: Exploring document retrieval features associated with improved short-and long-term vocabulary learning outcomes. In: Proceedings of 3th ACM CHIIR, pp. 191–200 (2018)
51. Takaki, T., Fujii, A., Ishikawa, T.: Associative document retrieval by query subtopic analysis and its application to invalidity patent search. In: Proceedings of 13th ACM CIKM, pp. 399–405 (2004)
52. Thomas, P., Moffat, A., Bailey, P., Scholer, F.: Modeling decision points in user search behavior. In: Proceedings of 5th IIiX, pp. 239–242 (2014)
53. Wang, J., Zhu, J.: Portfolio theory of information retrieval. In: Proceedings of 32nd ACM SIGIR, pp. 115–122 (2009)
54. Wildemuth, B.M.: The effects of domain knowledge on search tactic formulation. J. Am. Soc. Info. Sci. Tech. **55**(3), 246–258 (2004)
55. Wilson, M.J., Wilson, M.L.: A comparison of techniques for measuring sensemaking and learning within participant-generated summaries. J. Am. Soc. Info. Sci. Tech. **64**(2), 291–306 (2013)
56. Wu, W.C., Kelly, D., Sud, A.: Using information scent and need for cognition to understand online search behavior. In: Proceedings of 37th ACM SIGIR, pp. 557–566 (2014)
57. Zhai, C., Cohen, W.W., Lafferty, J.: Beyond independent relevance: methods and evaluation metrics for subtopic retrieval. SIGIR Forum **49**(1), 2–9 (2015)
58. Zuccon, G., Azzopardi, L., Hauff, C., van Rijsbergen, C.K.: Estimating interference in the QPRP for subtopic retrieval. In: Proceedings of 33rd ACM SIGIR, pp. 741–742 (2010)

Immediate Text Search on Streams Using Apoptosic Indexes

Patrick Eades(✉) ⓘ, Anthony Wirth ⓘ, and Justin Zobel ⓘ

School of Computing and Information Systems, The University of Melbourne,
Parkville, Australia
{patrick.eades,awirth,jzobel}@unimelb.edu.au

Abstract. Applications that involve streams of documents require a mechanism for search over the newest arrivals. In this paper we explore provision of immediate indexing and fast search of recent documents only, in contrast to focus on dynamic construction of an index of all observed material. Our contribution is a new structure, an apoptosic index, that operates in a fixed volume of memory and in which expired index entries vanish without significant overhead; there is neither explicit removal of old data nor explicit memory management. We demonstrate the practicality of apoptosic indexes with a straightforward implementation and experiments on microblog and newswire data, showing dramatically faster performance than observed with alternatives.

Keywords: Streams · Dynamic indexes · Windowed search

1 Introduction

Fast or *streaming* data is ubiquitous, and includes text sources such as microblogs and newswire data. Streams can produce data more quickly than can be reasonably stored or analysed using classical approaches, and in some applications it is attractive to be able to access it immediately [14].

In the context of streaming, it can be desirable to search across the most recent documents [19], which must be available as soon as is feasible. This paper concerns methods for such search, where only recent material is of interest and only fixed resources are available, and in particular indexing must take place in a specified volume of memory. It is also desirable that time spent doing housework – garbage collection and so on – is kept to a minimum so that the structure does not intermittently *stall* while being locked for removal of expired material.

More formally, for a stream of documents the *sliding-window search problem* is to find matches in recent material from a stream and to ensure that new

---

J. Zobel—This research was supported by the Australian Government through the Australian Research Council's Discovery Projects funding scheme (project DP190102078). The views expressed herein are those of the authors and are not necessarily those of the Australian Government or Australian Research Council.

M. Hagen et al. (Eds.): ECIR 2022, LNCS 13185, pp. 157–169, 2022.
https://doi.org/10.1007/978-3-030-99736-6_11

material is immediately available to the search process. Such a process could, for example, be run in parallel with a separate process for building a permanent index, thus allowing new material to be searched while older material is made available through existing indexing processes. In such a context it is only necessary to ensure that sufficient memory is available to keep material until it is available through the longer-term process.

The key challenge is to maintain an index over the most recent observed documents to support similarity search. An *inverted index* associates each word, or *term*, with a *posting list* of documents that contain that term. Similarity queries, which score and rank the documents best matching a list of query terms, can be easily answered by traversing the posting list for each term.

For the sliding-window search problem, a point of comparison is exhaustive search – which is often regarded as an unrealistic reference point because it is so obviously flawed for large collections, but is of relevance here, and moreover has been found to be not excessively slow compared to indexed search in some contexts [20]. Static indexing admits heuristics that reduce the costs of indexed search to times that are not much greater than constant [24], but even without such heuristics, indexed search is far more efficient than is exhaustive search.

However, index construction is in general super-linear in time and size, and a dynamic index needs to be maintained, with non-trivial costs for locating and removing out-of-date material: factors that might overwhelm the advantage over exhaustive search. The core challenge for sliding-window search is to efficiently maintain an index structure that allows expired material to be removed quickly, allowing memory to be reclaimed. An assumption in this work is that memory is bounded and that solutions that make use of external storage are too slow, so that all material must be stored in memory.

Contribution. We have developed an *apoptosic* inverted index, where new documents can be immediately added and old material is expired without any explicit memory or structure management. This transient structure, which is continually rewritten, operates in a fixed volume of memory and, as we show, allows newswire documents to be continually added in less than a tenth of a millisecond – and microblog entries in a microsecond or so – on a basic server (or, indeed, a commodity laptop). While we do not experimentally explore refinements, it is, e.g., straightforward to organise the index as a series of compacted blocks, with the trade-off of a brief stall while compaction takes place.

The basis of this index is our new, general method for efficient management of lists of recent items, which we call *apoptosic lists*. The core element of these lists is that they are stored in a fixed circular array, with material added cyclically so that the address of a pointer corresponds to its age and, while traversing a list, it can be determined implicitly that a pointer is no longer valid. Apoptosis can be applied to any data structure based on a linked list, including for example applications where age of an item implies exponential decay.

As a demonstration, we report experiments on processing streams while supporting simple 'most similar recent document' queries, on newswire and microblogs, across a range of memory volumes. These show that search with apoptosic indexes is significantly faster than is exhaustive search – which was not a certainty in this context – and in particular gives a speed improvement of a factor of twenty on the Twitter data. Encouragingly, search times for the data volumes we are concerned with are about the same as with a static index using a public-domain tool. Although significant refinement is possible, apoptosis is already competitive and suitable for recency searches on text. It could for example be used alongside methods that create enduring indexes by batching.

2 Background

Text search is a key component in a wide range of contexts, from the early applications in collections of library materials to Web and site search, personal collections, and microblogs. Efficiency demands an index, though the requirements are set by context; a key differentiator is static versus dynamic indexing.

Text Indexing. Research on indexes for text search has a long history [24], and continues to yield advances even for the longest-standing challenges. However, the value of past methods to our application is unclear; static indexes are obviously unsuitable. Many dynamic methods [10,21,24] build an index over all observed data, without deletions, which conflicts with our goal of using fixed space. These approaches typically include a buffer of recent data that is accumulated until full and then indexed, and a fresh buffer commenced, meaning that it is not immediately available except by exhaustive search.

Other methods do allow deletions [11], but with some penalty. At the cost of stalling while the index is locked to remove old postings, offline methods can rebuild the index, in full or (if the index is in blocks) partially. Online deletions require a mechanism for identifying the end of each postings list, checking the ends to see which postings have expired, and then freeing the expired postings for future memory allocation and updating pointers that have been invalidated. This implies significant overhead in either space, for additional structures, or repeated traversal of all lists. Deletion has not been widely considered in the text search literature.

Search in Microblogs. The problem of indexing a fast stream of tweets for text search is addressed by Twitter's Earlybird system [3]. In a fixed amount of memory, Earlybird maintains an index over the most recent tweets. However this memory is a *latched*, rather than sliding, window. Once the fixed memory is full, the data structure allocates a new, empty block of memory and compresses the full block to add to an archive – that is, there is a delay before new material is available, which is the essence of the challenge that we wish to address. (This is different from, but as inconvenient as, the stalling caused by housekeeping.) Earlybird provides fast update time and admits querying of data soon – but not immediately – after it has been received; it does not offer a sliding window or give an obvious way to manage the decaying relevance of data points. An alternative solution by Chen *et al.* [6] builds a real-time index only on tweets considered likely to be returned by popular search queries, while the remaining tweets are batch-indexed in the background.

In their approach to microblogs, Wu et al. [21] adapt the log-structure merge tree [16] for real-time search, providing a sequence of indexes with exponentially-increasing capacity. New documents are added to the smallest index; when an index reaches its capacity it is merged into the next level and emptied. This is similar to the grow-by-exponent approach described by Lester et al. [10], and both make use of buffering of new material. Magdy et al. [13] describe a micro-blog search that returns items within a fixed recency and physical location bound. The cost of removing old items is piggybacked onto insertions, and a periodic clean-up phase is needed, creating possible stalls.

Motivating such work, Teevan et al. [19] studied Twitter search logs and found that the vast majority of searches targeted recent posts. Mishne et al. [15] showed that traditional big data approaches are not well suited to solving fast data problems. While they address the issue of maintaining an index over recent data given an infinite stream, none of the data structures above provide an implicit, efficient way to remove old data, or to manage memory fragmentation.

An approach to immediate search is described by Wang and Lin [20], who describe approaches to exhaustive search of microblog streams. With a focus on immediate search, it is the most comparable previous work to our own, though its efficiency relies on the number of query terms being small (making it unsuitable for 'like document' searches) and makes use of specialised hardware. With these optimisations and constraints, they find that exhaustive search can require only a small multiple of the time required for indexed search.

Locality-Sensitive Hashing. The intent of locality-sensitive hashing (LSH) [8] is to hash similar items to common locations. LSH represents each item by a small, fixed number of hash values, with each being a hash of a subpart of the item, such that each pair of sufficiently similar items is likely to have at least one common hash value. The most common hash families for text documents are MinHash [2] and SimHash [5]. LSH methods are well suited to detecting duplicate or very similar short items, but for general text-search problems, the number of hash functions required, or necessary level of similarity, makes LSH ineffective.

Petrovic et al. [17] use a LSH-based index to solve first-story detection over tweet data. They manage space over an infinite stream by setting a maximum posting list length, and removing the oldest posting from any list which reaches its capacity. This means that some lists refer to much older data than do others.

Kraus et al. [9] describe an approach to applying LSH to a general data stream. Their LSH-based index finds similar items over an infinite stream, randomly removing some postings at every insertion. Items deemed significant are inserted multiple times to reduce the probability they are deleted, thus adding to the space consumption, and in effect increasing the cost of deletion, while also creating some likelihood of false misses. Sundaram et al. [18] had a similar method, with a temporary index called a delta table for the most recent data, which then gets merged into a read-only index, and grows indefinitely; query resolution via the delta table requires exhaustive search.

Circular Array. Building a data structure on top of a circular array to maintain sliding window information, while retaining fast update times, appears in a recent paper by Assaf *et al.* [1]. However their goal is to maintain summary statistics of the data, such as item frequency and the number of distinct items; they do not support richer data structures that an index would require.

3 Apoptosic Structures

We now describe an *apoptosic list*, a kind of linked list designed to allow perpetual insertion in a fixed-memory setting, which we have developed to support operations on streams. We then describe how apoptosic lists are a key component of an index for text search over recent data.

Apoptosic Lists. Our fixed-memory setting comprises a circular array, of size n, and a *write pointer*. New data is inserted at the write pointer, overwriting older data, and the write pointer is advanced.

Consider inserting a linked list into memory by prepending a node at a time, possibly with other, unrelated data in memory between each node. We re-index the array such that the write pointer has the largest index, this is just a rotation of the indices modulo n. Then following the linked list from head to tail yields a strictly decreasing sequence of indices. If a node ever points to a next node with a larger index than its own, the linked list must have wrapped around past the write pointer, and hence the pointer refers to data which has been overwritten; we describe such a pointer as *dead*. An illustration is in Fig. 1. The head pointer of a list can become dead if the list is not updated through a complete cycle. Below, we show that these are the only two circumstances under which a pointer may point to overwritten, expired data. Therefore, by ensuring that both kinds of dead pointers are never accessed, an automatically expiring, *apoptosic*, linked list can be maintained in fixed memory with no explicit housework such as garbage collection or deletions.

Apoptosic Indexes. We now describe the construction of our principal index structure, and how the apoptosic technique maintains the data structure's correctness. We allocate a contiguous, circular block of memory, $\mathcal{M}$, comprising n cells, to be used as list nodes: each can store a document id, a term frequency and a pointer to another cell in $\mathcal{M}$. Let w be an index in $\mathcal{M}$ representing the current end of the circular array; we call this the write pointer because it marks where new postings will be inserted. We also allocate a hash table $\mathcal{H}$ of m slots that maps terms to cells in $\mathcal{M}$.

We divide $\mathcal{M}$ into b blocks of equal size; we do this solely to provide a point at which housework (cleaning of the hash table) takes place. The hash table is small relative to total memory allocated and in practice this process is expected be fast, say a few milliseconds; the frequency of clean-up depends on b.

We hash all terms on input to a fixed universe of size m. For simplicity, the term strings are not considered further and each term is represented only by an

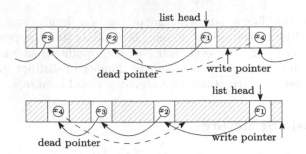

Fig. 1. Above: an apoptosic list inside a fixed-size slice of memory. The pointer out of x_4 is dead because it crosses the write pointer, and refers to data which has been overwritten. Below: the same memory, rotated so the write pointer takes the maximum value. We can see the pointer from x_4 is dead because it points to an index which is larger than its own; all living pointers point to smaller indices.

integer. This introduces some error due to hash collisions, but this is mitigated by choosing sufficiently large m.

We allocate document ids sequentially, modulo d, where d is twice the maximum (estimated) number of documents in memory at any given time. This quantity will only appear in the logarithm, so we can over-estimate without significant cost. If required we may use a separate lookup table mapping our sequential document ids to some other reference (e.g., locations on disk, DoIs).

Operations. To insert a document,

- Hash each distinct term, t, in the document to a value in the range $[0..m-1]$ and record its within-document frequency $f_{t,d}$.
- For each term, create a node with a next-node pointer, the document ID, and $f_{t,d}$; insert at the start of the list for that term's hash value, t.
- When a block is full, the oldest block must be expired. Clean up by making a linear pass through the hash table; if the pointer from a hash-table slot points to a node in the block that is expiring, set that slot to NULL.

Query processing is straightforward. Each term in the query document is hashed – recall that the task in this demonstration implementation is to find similar documents, not to support ad hoc queries – and the list for each hash value is processed in turn. Similarity calculation is described in more detail later.

The list of postings for a term can be accessed by looking up the term in $\mathcal{H}$ and following the pointer stored there to the head of a linked list of term postings. Because the postings are inserted sequentially in memory, each always points to lower memory addresses than its own address (when viewed circularly, with the write pointer as the largest address in the array). As the links are in a consistent direction, the list is somewhat cache-friendly, though not as much so as contiguous storage; this can be improved as noted under 'Compression' below.

A pointer whose target has a larger address than itself is pointing to something on the other side of the write pointer, i.e., something that has fallen out

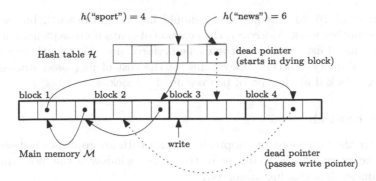

Fig. 2. There are two ways a pointer can become dead. There are currently four postings for "sport", but the pointer from $\mathcal{M}_{15}$ to $\mathcal{M}_6$ is dead because it crosses the write pointer, the posting at $\mathcal{M}_6$ has been overwritten and no longer contains "sport". This can be detected at query time. The pointer for "news" from $\mathcal{H}_6$ to $\mathcal{M}_{11}$ is dead because it is a top-level pointer which points into the block currently being overwritten. However, this pointer would have been removed by the clean-up phase when block 3 was started, and so this situation will never occur.

of the sliding window and has been overwritten. When such a *dead pointer* is encountered during a query, it is not followed. See Fig. 2 for an illustration.

We do not need to explicitly delete old items at update time and, since items are removed in a deterministic way, we can make precise guarantees about when each item will be removed from memory.

Correctness. A critical issue is whether pointers that refer to memory that has been overwritten are always recognised as dead pointers and, therefore, not accessed. Because of the housekeeping phase, applied when a block is expired, a hash table slot cannot contain a dead pointer: blocks are expired in turn, at which step pointers into the expiring block are NULL'ed. Similarly, adjacent postings of a term cannot be more than n insertions apart. Therefore a pointer is living if and only if it begins a linked list, or its value is (circularly) smaller than the pointer before it in the linked list. Since we can detect these situations at query time, we never access a dead pointer and our structure is always correct.

Compression. Our data structure does not naturally compress the index, which risks losing the memory and (in some circumstances) speed improvements that compression can provide. Adding compression is not difficult, however, because of the block arrangement of the data structure. Once b terms have been inserted, that is, a block is full, that block can be compressed and pointers from $\mathcal{H}$ into the block from the hash table updated to the new locations. There are no linked-list pointers into the block because it has only just been created. The next block is then started as before.

Such blockwise compression would not be as effective as would be compression of a complete index, but it could yield significant gains. We believe that achievement of 16–20 bits per node should be straightforward, but we leave details for further work. Moreover, the creation of contiguous sequences of entries from each linked list would improve cache performance. That is, there would be capacity and query-time efficiency gains, at the cost of increased stalling while the index is locked at the end of processing of a block.

3.1   Text Search

We explore the behaviour of apoptosic indexes with an application designed to answer the query 'which document in the sliding window is the most similar to the document that was just observed?'

Lists are processed in turn, and an array of accumulators is used to record per-document partial similarities. Here, the number of query terms means that there is tolerance for hash collisions, and thus we chose to not retain the terms themselves. It would however be straightforward, for example, to replace the hash table with a compact trie with minimal effect on space consumption.

The lists can be processed in increasing order of length, potentially allowing early termination when the largest partial similarity sufficiently exceeds the next-largest (a pair of values that can trivially be maintained during accumulator update), but we do not make use of this optimisation in our experiments.

For our similarity measure we use the form of Cosine similarity as described by Zobel and Moffat [24], but any such measure has similar costs and thus this is an arbitrary choice, as we are not measuring search effectiveness. This formulation uses the frequency $f_{d,t}$ of term t in document d, the number N of documents in memory, and the number f_t of documents in memory containing t.

Maintaining a value for N is straightforward: when a new document is inserted, N is incremented. Whenever a posting is removed from memory by being overwritten by a new posting, its document id is compared against the document id of the next most recently removed posting, which is stored for this purpose; if they are different then a document has been entirely removed and N is decremented. This requires storing only a counter and a document id.

There are several ways of keeping track of f_t. If each posting stores the term for which it is a posting, then the total number of postings for each term can be maintained by increasing a counter when a posting is created and decreasing it when a posting is overwritten. This requires $\mathcal{O}(m \log n)$ space to store the counters and $\mathcal{O}(n \log m)$ space to store the term with each posting. Alternatively the posting list for each term in the query can be traversed at the start of the query to compute each relevant f_t, which approximately doubles the query time in practice. A middle ground is to cache values of f_t using $\mathcal{O}(m \log n)$ space and only occasionally recalculate them, for example cyclically, so that each value is visited once per k document insertions; inaccuracy in values should have only a limited impact on the reliability of calculation of 'most similar'. We store true f_t values in our experiments, noting that this has an impact on update time.

Document lengths are easily calculated on insertion. Similarities can be calculated term-at-a-time, thus using accumulators, or document-at-a-time [4], strategies that have been widely explored in the context of standard inverted indexes.

Table 1. Summary statistics of the evaluation data sets.

Dataset	Raw size (MB)	Parsed size (MB)	Documents	Postings	Postings per doc
NYT	1668	1018	314,452	117,986,501	375
NYT Stopped	1668	899	314,452	99,065,711	315
Twitter	355	259	2,155,912	22,482,484	10

The latter would allow early termination of processing for some query types, such as finding the recent document whose similarity exceeds some threshold.

All the space overheads above are fixed, used during the entire operation of the data structure, except for the document lengths, which we store separately. These are expected to be small relative to the remaining data; the documents themselves need to be retained separately in a production implementation.

4   Experiments

We have implemented our apoptosic index in C. Our experiments are performed on a basic server with a 2.5 GHz Intel Xeon CPU in a single thread. Memory usage is fixed, and approximately n times 16 bytes.

Data Sources. We use two different types of text document data, newswire and microblog data, which represent typical streaming document applications. Except in average document length, the two sources are similar; since we are using the documents for similarity queries this also determines query length.

Our data source for newswire documents comes from the TREC AQUAINT collection [7], consisting of 1.6 GB of New York Times (NYT) articles. With metadata and formatting removed, this totals 1 GB of ASCII-encoded data, 314,452 documents, 117,986,501 postings and an average of 375 postings per document. After stopping, the documents are shortened to 315 words on average.

Our source for microblog data is a 355 MB collection of tweets taken from Twittter's sampled stream API.[1] With metadata removed this leaves 259 MB of data, consisting of 2, 155, 912 documents, 22, 482, 484 postings and an average of 10 postings per document.

These are summarised in Table 1. We note that these are small data sets by the usual standards of text indexing experiments, but our work is focused on small sets of recent data. The critical factor is that these sets are much larger than our sliding windows.

Experimental Design. We use two tasks to evaluate the performance of our data structure. First, we stream 1 GB of data into our data structure and measure the average time to insert a document a term at a time. This measures the

---

[1] https://developer.twitter.com/en/docs/twitter-api/tweets/sampled-stream

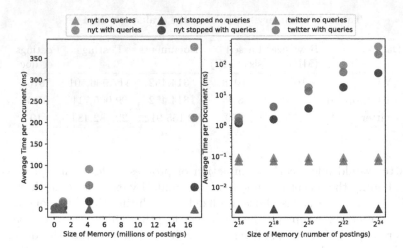

Fig. 3. Left: Memory size plotted against operation time across the datasets for search with apoptosic indexes; as expected, insertion time is unaffected by memory size and query time scales linearly. Right: the same data with log-log axes.

throughput of our data structure and the time delay before data is available to query. Second, by querying the index, we find the most similar recent document to each document as it is inserted, using Cosine as discussed above.

We vary n (the number of nodes) from 2^{16} to 2^{24}. For the NYT data this is an average of about 2,800 documents indexed per million nodes, for Twitter it's approximately 100,000 documents per million nodes. We vary b (block size in nodes) from 2^{12} to 2^{20}, and fix $m = 2^{16}$. By representing each term via its hash, we ensure that the index fits within a fixed memory allocation. Although collisions are now possible, they are of little impact on our experiments. First, this hashed representation is also used for the comparator methods. Second, as each query contains a significant number of terms, the likely number of collisions is low, observing that we have also applied stopping.

Results. Once our data structure has filled and is 'warm', query and insertion times are highly consistent. Average results are shown in Fig. 3. For the NYT data, each insertion takes less than 0.1 ms regardless of n. Query times scale linearly with n, at around 17 ms per query per 2^{20} nodes. For the Twitter data, each insertion takes roughly 0.002 ms, with query times of around 3.5 ms per query per 2^{20} nodes. Similar results were observed on a commodity laptop.

Small block sizes trigger the cleanup (stall) more often, which slows down responsiveness; however, as the per-block cleanup phase only takes 0.3 ms on average – around the time to insert four NYT articles – the cost is small.

We have not explored querying effectiveness, but in principle expect the performance of the query type considered here to be like that of a conventional index because of the number of terms, even though some accuracy is (marginally) lost due to the use of hashing. Some effectiveness would also be lost due to the fact that the volume of material being indexed is small; it is well known that larger collections have smoother statistics. However, in our anticipated use-case – of the apoptosic structure being an ancillary index that holds recent data while an enduring index is built – the statistics held in the enduring index would be available during search, thus ensuring that effectiveness is not degraded.

Standard Indexing Using Anserini. We compare our approach against searching the same data using Anserini,[2] a toolkit built on top of Apache Lucene[3] to support reproducible information retrieval research [12, 22, 23]. We regard Anserini as a reasonable representative of the existing approaches. It is written in a combination of Java and Python, and thus is not directly comparable to our C implementation, but, unlike our code, it is highly optimised.

Using the pyserini interface[4], we index the New York Times collection in 191 seconds, at a rate of 0.60 ms per document, several times slower than our data structure. Indexing the Twitter collection takes 0.02 ms per document, ten times slower than our data structure. Although precise timings are tricky for smaller collections, such as the 2800 NYT documents that require $n = 2^{20}$ postings in our structure, for that value of n time per document is close to 1 ms. However, this does not consider the cost of deleting old items from the Anserini index to maintain fixed memory, which would add considerably to index management time.

As a baseline, we measure query time on a static index built on a window of our data. Since pyserini does not support dynamic index construction and querying while the index is being built, we instead use a separate sample of 500 documents as queries, and take the average query time. For NYT, up to $n = 2^{20}$ query times are stable in the neighbourhood of 20 ms, but grow (somewhat sublinearly) thereafter. By $n = 2^{24}$, Anserini search is significantly faster than within apoptosic indexes, but, noting again that this is a static index, implying a delay (stall) of allowing around 45,000 NYT documents or 1.6 million tweets to arrive before an index is built, or 5 seconds of arrivals at the rate at which the apoptosic index can manage.

Linear Scan Through Documents in Memory. An alternative approach is to keep a sliding window of the most recent documents in memory and find matches via exhaustive search. The expectation is that this method allows fast updates and data availability, at the cost of slower queries. In our implementation, each document is read, one distinct term at a time in sort order, and added as a hash value to a circular array; the hashing is used to provide constant space per term and comparability with our apoptosic-list search demonstrator. Older entries in the array are automatically overwritten, in the same style as apoptosic lists.

At query time, query terms are sorted and then linear-merged at the document level to compute similarities. We use the same Cosine formulation as in our apoptosic-index experiments.

---

[2] http://anserini.io/.
[3] https://lucene.apache.org/.
[4] http://pyserini.io/.

Surprisingly, update – addition of a document to the circular array – is no faster than with the apoptosic index, so for management of a stream of information the lack of an index provides no advantage. Query processing requires about roughly 40 ms per 2^{20} nodes for both NYT and Twitter data, so exhaustive search is slower by a factor of 2 for the long documents (which have a large number of query terms) and a factor of more than 10 for the shorter queries. We note that ad hoc queries are typically only three terms or so, and thus would be a further 3 times faster than even the Twitter queries, showing that the index could give a factor of 30 speed improvement for that task. However, this speed rests on the use of a circular array and apoptosis; a more traditional implementation would be significantly slower.

5   Conclusions

Our apoptosic index addresses the need for low-memory data structures that provide fast search over recent documents without sacrificing throughput or stalling for maintenance. A simple implementation, not significantly optimised, has insertion time competitive with maintaining an unindexed buffer of recent documents, while offering fast query times. Our index cannot answer queries as quickly as an optimised static index, but, for small volumes of indexed data, processes documents faster and does not require rebuilding; and the documents are immediately available for search.

We have demonstrated that apoptosis is a viable principle for design of special-purpose indexes. Combining apoptosis with other heuristics and optimisations from the document search literature, such as pruning or compression, is likely to give significant further improvements in performance. A particular application is to run an apoptosic index for recent data in parallel with a more traditional indexing system that provides ongoing access to historical data.

Similarly, apoptosis may prove to be a useful technique for more general problems on streaming data. A common difficulty in streaming data structure research is removing items from a sliding window without affecting the update time. Apoptosis provides a solution to this problem for any data structure that uses linked list-style pointer chains. It is immediately applicable to maintaining hash maps and exact recency queries, for example.

Several models of data decay are considered in the data stream literature. We have focused on a sliding window because it is conceptually simple. Another common model is *exponential decay*, where the weight, or significance, of an item decays exponentially with its age, and items with weight below a threshold are no longer considered. Our apoptosic index can operate in this model without modification: an exponential decay implies only a fixed window of items need to be considered, and in our index the age of an item is given by its pointer address, so a decay can be applied whenever an item is accessed. Apoptosis can also be applied to related problems such as most recent occurrence and time-weighted occurrence frequency. We plan to explore such applications in future work.

References

1. Assaf, E., Ben-Basat, R., Einziger, G., Friedman, R.: Pay for a sliding bloom filter and get counting, distinct elements, and entropy for free. In: INFOCOM (2018)
2. Broder, A.Z.: On the resemblance and containment of documents. In: SEQUENCES (1997)
3. Busch, M., Gade, K., Larson, B., Lok, P., Luckenbill, S., Lin, J.J.: Earlybird: real-time search at Twitter. In: ICDE (2012)
4. Büttcher, S., Clarke, C.L.A., Cormack, G.V.: Information Retrieval - Implementing and Evaluating Search Engines. MIT Press, Cambridge (2010)
5. Charikar, M.: Similarity estimation techniques from rounding algorithms. In: STOC (2002)
6. Chen, C., Li, F., Ooi, B.C., Wu, S.: TI: an efficient indexing mechanism for real-time search on tweets. In: SIGMOD (2011)
7. Graff, D.: The AQUAINT corpus of English news text LDC2002T31 (2002)
8. Indyk, P., Motwani, R.: Approximate nearest neighbors: towards removing the curse of dimensionality. In: Vitter, J.S. (ed.) STOC (1998)
9. Kraus, N., Carmel, D., Keidar, I.: Fishing in the stream: similarity search over endless data. In: BigData (2017)
10. Lester, N., Moffat, A., Zobel, J.: Fast on-line index construction by geometric partitioning. In: CIKM (2005)
11. Lim, L., Wang, M., Padmanabhan, S., Vitter, J.S., Agarwal, R.C.: Dynamic maintenance of web indexes using landmarks. In: WWW (2003)
12. Lin, J., et al.: Toward reproducible baselines: the open-source IR reproducibility challenge. In: Ferro, N., et al. (eds.) ECIR 2016. LNCS, vol. 9626, pp. 408–420. Springer, Cham (2016). https://doi.org/10.1007/978-3-319-30671-1_30
13. Magdy, A., Mokbel, M.F., Elnikety, S., Nath, S., He, Y.: Mercury: a memory-constrained spatio-temporal real-time search on microblogs. In: ICDE (2014)
14. McCreadie, R., Soboroff, I., Lin, J.J., Macdonald, C., Ounis, I., McCullough, D.: On building a reusable Twitter corpus. In: SIGIR (2012)
15. Mishne, G., Dalton, J., Li, Z., Sharma, A., Lin, J.J.: Fast data in the era of big data: Twitter's real-time related query suggestion architecture. In: SIGMOD (2013)
16. O'Neil, P.E., Cheng, E., Gawlick, D., O'Neil, E.J.: The log-structured merge-tree (LSM-tree). Acta Informatica 33(4), 351–385 (1996)
17. Petrovic, S., Osborne, M., Lavrenko, V.: Streaming first story detection with application to Twitter. In: NAACL-HLT (2010)
18. Sundaram, N., et al.: Streaming similarity search over one billion tweets using parallel locality-sensitive hashing. Proc. VLDB Endow. 6(14), 1930–1941 (2013)
19. Teevan, J., Ramage, D., Morris, M.R.: #TwitterSearch: a comparison of microblog search and web search. In: WSDM (2011)
20. Wang, Y., Lin, J.: The feasibility of brute force scans for real-time tweet search. In: ICTIR (2015)
21. Wu, L., Lin, W., Xiao, X., Xu, Y.: LSII: an indexing structure for exact real-time search on microblogs. In: ICDE (2013)
22. Yang, P., Fang, H., Lin, J.: Anserini: enabling the use of Lucene for information retrieval research. In: SIGIR (2017)
23. Yang, P., Fang, H., Lin, J.: Anserini: reproducible ranking baselines using Lucene. ACM J. Data Inf. Qual. 10(4), 1–20 (2018)
24. Zobel, J., Moffat, A.: Inverted files for text search engines. ACM Comput. Surv. 38(2), 6 (2006)

Influence-Based Deep Network for Next POIs Prediction

Sayda Elmi[✉] and Kian Lee Tan

School of Computing, National University of Singapore, Singapore, Singapore
{saida,tankl}@comp.nus.edu.sg

Abstract. The dramatic proliferation of location-based social networks have resulted in a significant amount of data. This has led to the development of location-based recommendation tools that assist users in discovering attractive Points-of-Interest (POIs). Next POIs recommendation is of great importance for not only individuals but also group of users since group activities have become an integral part of our daily life. However, most existing methods make recommendations through aggregating individual predictive results rather than considering the collective features that govern user preferences made within a group. This insufficiency can directly affect the completeness and semantic accuracy of group features. For this reason, we propose a novel approach which accommodates both individual preferences and group decisions in a joint model. More specifically, based on influencing users in a group, we device a hybrid deep architecture model built with graph convolution networks and attention mechanism to extract connections between group and personal preferences and then capture the impact of each user on the group decision-making, respectively. We conduct extensive experiments to evaluate the performance of our model on two well-known real large-scale datasets, namely, Gowalla and Foursquare. The experimental results show its superiority over the state-of-the-art methods.

Keywords: POIs prediction · Minimum Graph Code · User Influence · Deep Learning · Group Recommendation

1 Introduction

Location-based social networks (LBSNs) such as Yelp, Foursquare, Gowalla, Facebook Place and GeoLife have enabled users to share their experiences and locations via check-ins for Points-Of-Interest (POIs), e.g., restaurants, stores, tourists spots, etc. With the exponential increase in the amount of geo-tagged data collected from the LBSNs, POIs recommendation has attracted wide attention from both academia and industry [15,17]. POIs recommendation is useful for both service providers and users by (i) estimating the number of users that may visit the POI and (ii) helping users discover various POIs for social activities occurring in the near future close to their current locations, respectively. As a natural extension of general POIs recommendation, next POIs recommendation

© The Author(s), under exclusive license to Springer Nature Switzerland AG 2022
M. Hagen et al. (Eds.): ECIR 2022, LNCS 13185, pp. 170–183, 2022.
https://doi.org/10.1007/978-3-030-99736-6_12

aims at predicting the POIs that are most likely to be visited next by analyzing the users' behaviors and mobility given their check-in history. Compared with the general POIs recommendation, next POIs recommendation focuses more on exploiting user movement patterns hidden in the historical check-in data for individuals. However, individuals tend to have several group activities reflecting their social life, e.g. friends often dine out and see a movie and families often watch TV programs, attend parties or travel together. For this reason, it is highly critical to develop group recommender systems to suggest relevant POIs for a group of users, known as group recommendation.

Various heuristic aggregation strategies such as average, least misery and maximum pleasure, have focused on exploring a consensus among group members on an item [1,11]. However, all these predefined aggregation strategies are too simplistic to model the real and complex process of the next POIs problem since they fail to capture the non-linear and complex structure of the related features, leading to sub-optimal group recommendation performance. In recent times, deep learning techniques have shown a great success in capturing non-linear correlations. Meanwhile, several recent studies have shown the utility of deep learning in the area of recommendation systems and information retrieval as well [16]. For group recommendation, attention-based approaches [2,14] have achieved the state of-the-art performance. That is because through an attention mechanism [5,13], the impact of each member in a group is efficiently captured and therefore the final recommendation effectiveness can be improved. Although the above-mentioned attention-based approaches [2,14] have the best performance at present, they ignore the preference interactions between a group and its members and thus fail to consider the correlation between each user influence and the group preferences. These models lack the capability to build a good representation of the group preference as well as group members' preferences, which we believe are crucial to the success of group recommendation systems. This insufficiency can directly affect the completeness and semantic accuracy of group features. To overcome the above-mentioned deficiency, in this paper, we propose an efficient group recommendation model for next POIs prediction called Know Your Destination (KYD) based on a hybrid deep learning model to (1) Build a new representation for the group profile: a user may exhibit different influences and have different weights in different groups. The same user may have a high influence degree in a given group but not in another group, i.e., his personal preferences may influence the preferences of some groups but not all. Thus, considering the preferences of the non influencing members may bias the prediction results. Therefore, we measure the influence of each group member to propose a representative group profile including only its influencing members and (2) Learn connections between group preferences and personal preferences of the members forming the group: for the same POI, each member of a group may have different opinions, so the main challenge is to find correlations between group preferences and the group members with diverse preferences to satisfy each member and the group as a whole.

In this paper, inspired by the success of the attention mechanism architecture in learning group members interactions, we investigate how to apply a hybrid deep learning model by proposing a representation of the different factors and integrate them smoothly in a shared latent space. We propose a high-quality representation for POIs, groups and its members. We then propose our hybrid deep learning model KYD to predict the next POIs for a set of groups exploring and analysing the influence of their members.

2   Problem Definition

Following the convention, we use bold capital letters (e.g., X) to represent both matrices and graphs, and use squiggle capital letters (e.g., $\mathcal{X}$) to denote sets. We use lowercase letters with superscript $\rightarrow$ (e.g., $\vec{x}$) to denote vectors. We employ normal lowercase letters (e.g., x) to denote scalars. Assume a set of POIs $\mathcal{P} = \{p_1, p_2, \ldots, p_{|\mathcal{P}|}\}$ and a set of users $\mathcal{U} = \{u_1, u_2, \ldots, u_{|\mathcal{U}|}\}$ belonging to a LBSN $\mathcal{N}$ and a set of groups $\mathcal{G} = \{g_1, g_2, \ldots, g_{|\mathcal{G}|}\}$. The k-th group $g_k \in \mathcal{G}$ is a set of users and we use $\mathcal{U}_{g_k}$ to denote this set. We make use of the following graphs that reflect the interactions between the different sets $\mathcal{U}$, $\mathcal{P}$ and $\mathcal{G}$. We use bipartite graphs G_{UP} and G_{GP} as shown in Fig. 1, to represent three kinds of interactions: **(i) User-POIs Interaction:** called also a user check-in activity, is a visit of the user u to the POI p at a time t denoted by $x_{u,t}$, i.e., $x_{u,t}$ is the POI visited by the user u at time t. A sequence of historical user check-ins over time (from t-n to t) is called a user check-in profile and represented by the vector $\vec{x}_u$. A given POI p is characterised by the geographical coordinates in terms of longitude and latitude coordinates denoted by l_p and its content (e.g., food, shop, service, etc.) denoted by c_p, **(ii) Group-POIs Interaction:** it is the group check-in activity related to the group g and the POI p at a time t denoted by $x_{g,t}$. The vector $\vec{x}_g$ represents the group profile over time, i.e., a set of POIs visited by the group g over a sequence of historical time-slots and **(iii) User-User Interaction:** it is the relation related to a set of users visiting a given POI at the same time forming an active group $g_k \in \mathcal{G}$.

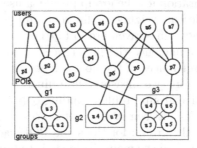

Fig. 1. Illustration of the input data: A set of users visiting a set of POIs and a set of groups interacting with the POIs set.

Problem 1 (Next POIs Prediction). Given a set of groups $\mathcal{G}$, each comprises a set of users, our task is to predict the next POIs of all groups $\mathcal{G}$ in a given LBSN $\mathcal{N}$ that they would be interested in at a future time T where $T = t + z$ and z is the number of the future time-slots to be predicted. More formally, given $\vec{x}_{g_k}$ and $\vec{x}_u / u \in g_k$, we want to predict $x_{g_k, t+i}$ for all $g_k \in \mathcal{G}$ where $1 \leq i \leq z$ and $1 \leq k \leq |\mathcal{G}|$.

3 Methodology

In our framework, the bipartite graphs G_{UP} and G_{GP} and interactions between users in a given LBSN $\mathcal{N}$ are the data containing information about all interactions. A baseline method for predicting the next POIs for a set of users forming a group is to learn the preferences of the group by aggregating the personal preferences of its members. This method may have a high computational cost with (i) larger graphs $\mathcal{N}$ and thus a large number of groups K where $K = |\mathcal{G}|$ and (ii) larger number of users forming the groups $g_k \in \mathcal{G}, \forall k \in [1, K]$. However, when analysing the group behaviour, we have observed that the group decision is related to the preferences of some group members, that we call **influencers**, but not all. Thus, introducing the preferences of the non influencers into the aggregation process may bias the results in terms of accuracy. As shown in Fig. 2, our framework proposes the concept of "the minimum graph code" which replaces the original input graph data considering the preferences of the influencers and neglecting the non acting users. Then, we introduce a representation process which consists of modeling the different interactions between the entities users, groups and POIs sets (as shown in Sects. 3.1 and 3.2). To make the problem learnable, the graph interactions should be either represented as a sequence of visited POIs, or a series of adjacent matrices and vertex features. For sequence inputs we can use CNNs [9], RNNs such as LSTM [19], or Transformers [13] to extract high-level features and learn interactions between the different entities. If the inputs are modeled as series of adjacent matrices and vertex features, we can use convolution GNN to learn vertex representations with message passing from

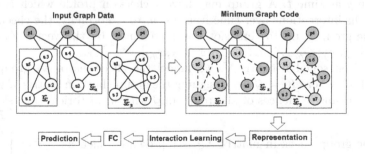

Fig. 2. KYD Framework: it consists of mainly three components: (i) Minimum Graph Code Process, (ii) Representation Process and (iii) Interaction Learning Process.

neighborhoods. After obtaining the representations of our entities, we feed them into an interaction learning module to extract the correlated features from each side. Then we feed the output into fully-connected layers to make predictions.

In the following, we give the representations of both sets Users and Groups. We then model the interactions between these two entities and the POIs set.

3.1   User Representation

The User-POI interaction at a given time denoted by $x_{u,t}$ is the POI visited by the user u at time t. For each user, we create a profile which is a set of User-POIs interactions. A user Profile is a sequence of check-in activities in chronological order from $t-n$ to t. Formally, the check-in sequence of a user u up to time-slot t is represented by the vector $\vec{x}_u$ as follows.

$$\vec{x}_u = [x_{u,t-n}, x_{u,t-(n-1)}, \ldots, x_{u,t-2}, x_{u,t-1}, x_{u,t}] \tag{1}$$

In a given social network $\mathcal{N}$, we model the check-ins of all users in $\mathcal{U}$ by designing the user check-in matrix $X_{\mathcal{U}}$ recording the check-ins sequences of all users in $\mathcal{U}$ with n historical time-slots. Formally, the user check-in matrix $X_{\mathcal{U}}$ is represented as follows.

$$X_{\mathcal{U}} = \begin{bmatrix} x_{u_1,t-n} & \cdots & x_{u_1,t-1} & x_{u_1,t} \\ x_{u_2,t-n} & \cdots & x_{u_2,t-1} & x_{u_2,t} \\ \vdots & \ddots & \vdots & \vdots \\ x_{u_{|\mathcal{U}|-1},t-n} & \cdots & x_{u_{|\mathcal{U}|-1},t-1} & x_{u_{|\mathcal{U}|-1},t} \\ x_{u_{|\mathcal{U}|},t-n} & \cdots & x_{u_{|\mathcal{U}|},t-1} & x_{u_{|\mathcal{U}|},t} \end{bmatrix} \tag{2}$$

Each row vector in $X_{\mathcal{U}}$ contains check-in data for the same user from continuous time slots from $t-n$ up to t, while each column contains check-in data for all users at the same time slot.

3.2   Group Representation

A group is a set of users visiting together one POI at the same time. The Group-POI interaction at a given time denoted by $x_{g,t}$ is the POI visited by the group g at time t. A group may have a check-in profile which is a set of Group-POIs interactions over time from $t-m$ to t. Formally, the check-in profile of the group g up to time-slot t is represented by the vector $\vec{x}_g$ such as $\vec{x}_g = [x_{g,t-m}, x_{g,t-(m-1)}, \ldots, x_{g,t-2}, x_{g,t-1}, x_{g,t}]$.

In a given social network $\mathcal{N}$, we model the check-ins of all groups g_k in $\mathcal{G}, \forall k \in [1, K]$ where $K = |\mathcal{G}|$ by designing the group check-in matrix $X_{\mathcal{G}}$ recording the check-ins sequences of all groups in $\mathcal{G}$ with m historical time-slots. Formally, the group check-in matrix $X_{\mathcal{G}}$ is represented such as $X_{\mathcal{G}} = \begin{bmatrix} \vec{x}_{g_1} \\ \vec{x}_{g_2} \\ \vdots \\ \vec{x}_{g_K} \end{bmatrix}$. In location-based group recommender, we can consider only $X_{\mathcal{G}}$ as an input to learn the mobility behaviour from the historical check-ins of the group as one individual. However, the group preferences may be affected by the personal preferences of its members. Thus, integrating the preferences related to the group members is of great importance for more accurate prediction. However, some of the group members may not be active in the group decision making and some of them, may have more influence than others, i.e., they affect more the group preferences. For this reason, we measure in the following the influence degree of each group member. We then propose the minimum graph code by discarding the non influencing users from the learning module.

The User Influence Modeling. The personal preferences of a user u are represented by the vector $\vec{x}_u$ (See Eq. 1), extracted from the matrix $X_\mathcal{U}$ and reflecting the user check-in profile over time, i.e., all visited POIs, alone or as a group member. Similarly, the vector $\vec{x}_g$ records the group check-ins over time. Let $\phi(\vec{x}_u)$ and $\phi(\vec{x}_g)$ be the sets of distinct check-ins visited by the user u and the group g, respectively.

Definition 1 (Influence Degree). Given a group g_k in $\mathcal{G}$ where $k \in [1, |\mathcal{G}|]$. A set of users forming the group g_k is called $\mathcal{U}_{g_k}$. The influence degree for each user u in $\mathcal{U}_{g_k}$, denoted by $\varphi(u)_{u \in \mathcal{U}_{g_k}}$, is defined as follows

$$\varphi(u)_{u \in \mathcal{U}_{g_k}} = \frac{\phi(\vec{x}_u)_{u \in \mathcal{U}_{g_k}} \cap \phi(\vec{x}_{g_k})}{|\phi(\vec{x}_{g_k})|} \tag{3}$$

The intuition behind the influence degree is to measure the impact of the personal preferences of the user u in $\mathcal{U}_{g_k}$ on the preferences of the group g_k. The more the user u has similar preferences to those of the group g_k, the more the user influenced the group behaviour.

In the following, we model the group representation by proposing the minimum group code.

The Minimum Group Code. Given a group g_k in $\mathcal{G}$ where $k \in [1, |\mathcal{G}|]$. We assume that the users u in $\mathcal{U}_{g_k}$ are either influencers or followers. A group influencers are the group members having the power to affect the decisions of others which are the followers. The set of influencing users i in the group $g_k \in \mathcal{G}$ according to an influence threshold α, is called $\mathcal{I}_{g_k}$ and defined as $\mathcal{I}_{g_k} = \{u \in \mathcal{U}_{g_k} / \varphi(u) \geq \alpha\}$ where $\alpha \in [0, 1]$.

The code of the group g_k contains the check-in profiles of all users u in $\mathcal{U}_{g_k}$. Since personal preferences can affect the group decision and thus improving the prediction accuracy, we integrate into the learning module, the check-in profiles of the users, but not all. The minimum group code is the representation of only

its influencing members i in $\mathcal{I}_{g_k}$ such as $|\mathcal{I}_{g_k}| \leq |\mathcal{U}_{g_k}|$ and $X_{\mathcal{I}_{g_k}} = \begin{bmatrix} \vec{x}_{i_1} \\ \vec{x}_{i_2} \\ \vdots \\ \vec{x}_{|\mathcal{I}_{g_k}|} \end{bmatrix}$

where $\vec{x}_{i_m}/m \in [1, |\mathcal{I}_{g_k}|]$ is the check-in profile of the influencer i_m in $\mathcal{I}_{g_k}$. The minimum group code is the code with the minimum group members with the highest influence according to an influence threshold α. Finally, each group can be represented by the corresponding minimum code, and vice versa. The input of our learning module is the group check-in profile as well as the check-in profiles of all its influencers. More formally, we design the matrix X such as

$$X = X_{\mathcal{G}} \oplus X_{\mathcal{I}_{g_k}/k \in [1,K]} = \begin{bmatrix} \vec{x}_{g_1}, X_{\mathcal{I}_{g_1}} \\ \vec{x}_{g_2}, X_{\mathcal{I}_{g_2}} \\ \vdots \\ \vec{x}_{g_K}, X_{\mathcal{I}_{g_K}} \end{bmatrix} \tag{4}$$

where $K = |\mathcal{G}|$. This representation allows the integration of the group check-in profiles as well as the influencing check-in profiles to be analysed. In the following, given the matrix X, we propose a learning module to extract the hidden features behind the group behaviour.

3.3   The Interaction Learning Module

After obtaining the representations of both users and groups interactions, we propose a hybrid deep learning model which comprises three major components: (1) Learning the group mobility from $X_{\mathcal{G}}$ and then by extracting the hidden patterns of the embedded influencing users using a convolution GNN, (2) a time-series learning using the attention mechanism to model the temporal dependency and (3) the POIs content extraction layer modeling the impact of POIs content.

The learning module proceeds as follows: First, we extract the hidden patterns from the group behaviour over time, i.e., features from historical group check-ins. Then, the influencing users are extracted and embedded into the matrix X. To learn the patterns hidden in the influencing preferences as well as the group preferences, KYD employs GCN as shown in Fig. 3 since it has a very deep structure that can effectively capture the dynamics behind neighbor nodes, i.e., influencing users. Then, we reshape these features to be suitable for time-series learning. Since the group mobility is directly affected by periodicity features, i.e., the mobility behavior during the week-ends may be similar on consecutive week-ends, we feed the group check-in vectors of periodic time intervals into an attention layer, a deep variant sequence modeling, to capture such temporal dynamics. We then employ a fully connected (FC) layers to extract the semantic features that describe the POIs content.

Group Encoding. In graph models, each node has a feature vector, and each edge is used to pass information from its source to its target. GNNs do not need node ids and edge ids explicitly because the adjacency information is included in an adjacent matrix. i.e., interactions between users, groups and POIs. Given that a group is represented as a set of influencing users and thus defined by a minimum graph code, the next encoding step is to transform each influencing

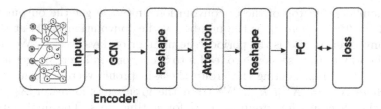

Fig. 3. Interaction Learning Module Architecture.

user into a vector. As explained in Sect. 3.2, we can vectorize node labels into multi-hot vectors represented by the input matrix X as shown in the Eq. 4. In such representation, the adjacent information of the group influencers can be stored in a sparse matrix $(X_{\mathcal{I}_{g_k}/k\in[1,|\mathcal{G}|]})$ to reduce the memory usage and improve the computation speed.

Graph Convolutional Network. After obtaining the input representations, we feed them as inputs of interaction layers to extract the correlated behaviour between the group preferences and those of the group members. Graph Convolutional Network (GCN) [12] is developed specifically to handle multi-relational data in realistic knowledge bases. GCN is acting as an embedding operation or a look-up operation (LK) to integrate the influencing users. A stack of convolutions (CNN) is used to understand the connections between group preferences and user preferences. We employ the Batch Normalization (BN) after the convolution layer for faster training speed. The input of the convolution layer is the check-in profiles of groups as well as the embedded influencing user check-in profiles $\mathcal{X}^{(l)}$. The output layer is to generate the prediction result. At an arbitrary l-th layer, we use $f^{(l)}$ filters to convolve and concatenate all matrices to get $\mathcal{X}^{(l+1)}$. The f-th matrix convolved by the f-th filter can be formulated as follows: $\mathcal{X}^{l,f} = [x_1^{l,f}, \ldots, x_k^{l,f}, \ldots, x_{|\mathcal{G}|}^{l,f}]$ where $x_k^{l,f} = F\big(LK(X_{\mathcal{G}}^{(l-1)}, I) * W^{l,f} + b^{l,f}\big)$. Here, $*$ denotes the convolution operation which uses the f-th filter $W^{l,f}$, F is an activation function, e.g. the rectifier ReLU $F(x) = max(0, x)$ which has achieved a training effectiveness in reducing the problem of gradient vanishing, $W^{l,f}$ and $b^{l,f}$ are the learnable parameters in the l-th layer with the f-th filter. This filter aims at emphasizing the correlated features of influencing users.

Attention Mechanism. In addition to the impact of the influencing users on the group preferences, the group check-in movement may have temporal and sequential patterns, i.e., a visited POI at a given time interval (12:00 pm– 02:00 pm) on Monday can be similar to the next following weekdays for a given group. For learning such patterns, our experimental results show that the attention mechanism outperforms LSTM, GRU and RNN in the benchmark datasets. We thus employ the attention mechanism [13] that has shown great success in capturing such sequential patterns. After C convolution units, we use the attention model to learn the long-term temporal patterns considering the influencing group members. On top of the C convolution units, we get the output tensor $X^C \in \mathbb{R}^{pd \times |\mathcal{G}| \times f_C}$ where f_C is the number of the convolution filters at the last C-th Conv layer and pd is the period (the number of days/weeks). We reshape X^C in the way of time sequence to feed into the attention layer. We get a tensor $X' \in \mathbb{R}^{|\mathcal{G}| \times n \times f_C}$ representing the group check-in profile vectors for all n time-slots, where $X'_{k,t} = X^C[t, k, :]$. We train the group profile vector $X'_{k,t}$ which records the check-ins for group $g_k \in \mathcal{G}$ for n time-slots. The final output of the transformer layers can be represented by the sequential vector Y_S in which the last element is the predicted check-in for the next time iteration such as $Y_S = [Y_{k,t-n}, \ldots, Y_{k,t}, Y_{k,t+1}]$.

Content Learning. Groups at lunchtime tend to go to restaurants for food rather than going to a cinema or any other entertainment points. We thus extract POIs content features by a fully connected layer (FC) as shown in Fig. 3. The output is denoted by Y_C. The predicted POI at the t-th time interval, denoted by $\widetilde{Y}_T$, is defined as $\widetilde{Y}_T = tanh(W_S \circ Y_S + W_C \circ Y_C)$ where $\circ$ is Hadamard product (i.e., element-wise multiplication), W_S and W_C are the learnable parameters in the sequential and the content learning component, respectively. The model output is a probability distribution on all POIs calculated by $\widetilde{Y}_T$. And then we take a gradient step to optimize the loss based on the output and one-hot representations of POIs.

4   Experimental Evaluation

To evaluate the performance of our hybrid deep learning model KYD, we conduct series of experiments on two real large-scale LBSNs datasets[1]: (i) **Foursquare Data-set:** a public data-set, on which we extract social relationships between users forming groups and content features. This data-set contains more than 450,000 check-in records from 2009 to 2011 generated by 2114 users living in USA. For each user, we acquired her POI attendance list and social friend list. For each POI, its visiting time and its content were also collected. Each check-in contains a user, a timestamp and a POI, indicating the user visited the POI at that time; users having relationships visiting the same POI at the same time are considered as a group, and (ii) **Gowalla Data-set:** This data-set contains 6,442,892 check-in historical records from 2009 to 2010 which is much more than the Foursquare data-set. However, it does not contain the content information about POIs. Therefore, each check-in record has the same format with the above Foursquare data-set except for POI-content.

Pre-processing. Given a group profile $\vec{x}_g$ in terms of a collection of group check-in records, we first sort them according to their check-in time-slot order. We then use the 80th percentile as the cut-off point so that check-ins before

---

[1] https://sites.google.com/site/dbhongzhi/.

this point will be used for training and the rest are for testing. For POIs content information, a binary vector is given by one-hot coding to transform and represent the content for the visited POIs. We use Tanh in the output of the KYD model as our final activation. We train our network with the following hyper-parameters setting: mini-batch size (48), learning rate (0.01) with adam optimizer, a variant of Stochastic Gradient Descent (SGD). Afterwards, we continue to train the model on the full training data for a fixed number of epochs (e.g., 20, 50, 100, 200 epochs). A transformer includes L transformer blocks. The hyper-parameters of the transformer are d, k, m, H, and L. The settings of these hyper-parameters are $d = 512$, $k = 64$, $m = 2048$, $H = 8$ and $L = 6$.

Benchmarks. Several prevailing algorithms are chosen for comparisons with our proposed model KYD. **(1) PRME-G** [6]: It uses the metric embedding method to embed users and POIs into the same latent space; **(2) STGCN** [18]: Used to predict next POIs, it is a variant of recurrent neural networks; **(3) RNN** [7]: This method leverages the temporal dependency in user's behavior sequence; **(4) FPMC-LR** [3]: It uses the Markov chains to model the user movement in a given region; **(5) ST-RNN** [10]: Based on the standard RNN model, ST-RNN replaces the single transition matrix in RNN with time-specific transition matrices and distance specific transition matrices to model spatial and temporal contexts of next POIs problem; **(6) LSTM** [19]: This is a variant of RNN model; **(7) HST-LSTM** [8]: It introduces Spatio-Temporal preferences for location prediction into gate mechanism in LSTM and **(8) GRU** [4]: This is a variant of RNN model, which is equipped with two gates to control the information flow. Some of the baselines were used only to predict the next POIs for individuals. To test these methods on groups, we consider a group as a virtual individual. The performance of our KYD model is evaluated by using two metrics Accuracy@K (**Acc@K**) and Mean Average Precision (**MAP**).

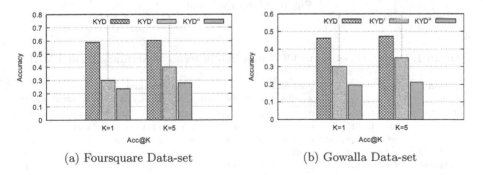

(a) Foursquare Data-set   (b) Gowalla Data-set

Fig. 4. Minimum Graph Code Evaluation.

Effectiveness of the Minimum Graph Code. In addition to KYD, we design two other algorithms KYD'' and KYD' to show the effectiveness of the minimum graph code approach: (i) KYD'' considers the input $X'' = X_{\mathcal{G}}$ to learn the mobility behaviour from the historical check-ins of the group as one individual, (ii) KYD' considers the input X' representing the group check-ins as well as all its members check-ins where $X' = X_{\mathcal{G}} \oplus X_{\mathcal{U}_{g_k}/k\in[1,|\mathcal{G}|]}$ since the group preferences may be affected by the personal preferences of its members. However, KYD employs the minimum graph code and has the input X where $X = X_{\mathcal{G}} \oplus X_{\mathcal{I}_{g_k}/k\in[1,|\mathcal{G}|]}$ as described in the Sect. 3.2. As shown in the Fig. 4, KYD' performs better than KYD'' which proves that including the preferences of the group members captures more patterns and thus improves the prediction accuracy. However, KYD outperforms KYD' since non influencing group members can bias the prediction results. The Fig. 5 shows that the optimal value of α is 0.5. For this reason, we use 0.5 as the default value for all experiments. We can also observe that when $\alpha = 0$ ($\alpha = 1$), the prediction results are similar to those of KYD' (KYD'') which is expected since in this case, we have $\mathcal{I}_{g_k} = \mathcal{U}_{g_k}, \forall k \in [1,|\mathcal{G}|]$ ($\mathcal{G} = \mathcal{I}_{g_k}, \forall k \in [1,|\mathcal{G}|]$, respectively).

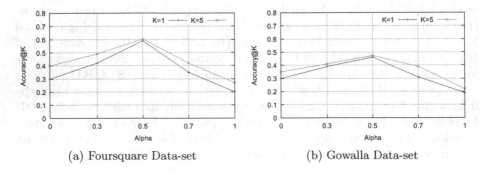

(a) Foursquare Data-set   (b) Gowalla Data-set

Fig. 5. Alpha Evaluation.

Table 1. Evaluation of next POIs Prediction in terms of Accuracy@K and MAP

Method	Ref	Foursquare				Gowalla			
		Acc@1	Acc@5	Acc@10	MAP	Acc@1	Acc@5	Acc@10	MAP
FPMC-LR [3]	IJCAI, 2013	0.216	0.228	0.285	0.251	0.163	0.247	0.276	0.271
PRME-G [6]	IJCAI, 2015	0.181	0.253	0.321	0.301	0.211	0.222	0.251	0.251
RNN [7]	ACM SIGIR, 2018	0.241	0.345	0.395	0.352	0.301	0.342	0.359	0.365
STGCN [18]	AAAI, 2019	0.481	0.501	0.512	0.502	0.391	0.421	0.439	0.442
LSTM [19]	IJCAI, 2017	0.382	0.395	0.421	0.430	0.256	0.278	0.314	0.321
ST-RNN [10]	AAAI, 2016	0.452	0.473	0.491	0.486	0.332	0.351	0.392	0.398
GRU [4]	EMNLP, 2014	0.296	0.325	0.391	0.404	0.218	0.251	0.289	0.312
HST-LSTM [8]	IJCAI, 2018	0.471	0.492	0.519	0.498	0.365	0.398	0.412	0.419
ST-KYD		**0.508**	**0.521**	**0.541**	**0.539**	**0.441**	**0.453**	**0.468**	**0.459**
KYD		**0.589**	**0.603**	**0.657**	**0.695**	**0.461**	**0.473**	**0.498**	**0.490**

Next POIs Prediction on Foursquare. From the experimental results shown in Table 1, we can see the following observations: RNN performs better than Markov chain, i.e., FPMC-LR and PRME-G, thanks to its capability in modeling sequential data and user interests using RNN cell. Both LSTM and GRU slightly improve the performance compared with RNN thanks to their advantages in capturing short-term as well as long-term temporal features. The state-of-the-art method ST-RNN improves the standard RNN method because the single transition matrix in RNN cell is replaced by two matrices representing both time and distance features to model the temporal and spatial contexts of the next POIs problem, respectively. STGCN and HST-LSTM perform better than ST-RNN. It proves the effectiveness of the idea of combining spatial and temporal factors with gates mechanism. Since existing state-of-the-art methods only consider the spatial and temporal factors, we propose for comparative purposes, a simplified version of KYD called ST-KYD which also models the spatial and temporal factors and removes the content learning module. ST-KYD variant performs better than STGCN, HST-LSTM and ST-RNN. This is because these state-of-the-art methods capture the spatial and temporal features by modeling new gates in RNN-based methods. However, ST-KYD uses a new representation to model the mobility of the group members by employing the convolution neural networks which has shown good performance in capturing spatial features. In addition, ST-KYD employs attention mechanism to capture temporal patterns. A combination of these two learning blocks has resulted in a significant improvement in terms of performance comparing to existing state-of-the-art methods. Our proposed KYD, including spatial, temporal and content learning, is significantly better than existing state-of-the-art methods evaluated on the Foursquare dataset considering all metrics. Specifically, KYD outperforms the Markov chain based methods considerably by a large margin. In addition, KYD consistently outperforms five RNN-based methods: RNN, LSTM, GRU, ST-RNN, and HST-LSTM. The significant improvement indicates that the mechanism to model temporal and spatial, in addition to the content learning in KYD can better catch the user's behaviors and are effective for the task of next POIs prediction.

Next POIs Prediction on Gowalla. Table 1 also reports the performance of the prediction models on Gowalla dataset. We can see that the trend of comparison result is similar to that presented for Foursquare data-set. A slight change is reported and is seen when some methods slightly outperform others on Foursquare dataset while they do not on Gowalla. This is because Gowalla dataset has no content information.

5   Conclusion

In this paper, GCN-attention based network named KYD, was proposed for group next POIs prediction. In KYD model, a new representation of the group preferences is introduced and given by the minimum graph code. A hybrid architecture is employed to capture the hidden patterns considering the social relations connecting users in a group. An attention mechanism is then employed to extract the sequential features. We evaluate our model on two real location-based networks, Foursquare and Gowalla, achieving performances which are significantly beyond eight existing methods.

References

1. Amer-Yahia, S., Roy, S.B., Chawlat, A., Das, G., Yu, C.: Group recommendation: semantics and efficiency. Proc. VLDB Endow. **2**(1), 754–765 (2009)
2. Cao, D., He, X., Miao, L., An, Y., Yang, C., Hong, R.: Attentive group recommendation. In: The 41st ACM SIGIR, pp. 645–654 (2018)
3. Cheng, C., Yang, H., Lyu, M.R., King, I.: Where you like to go next: successive point-of-interest recommendation. In: Rossi, F. (ed.) Proceedings of the 23rd IJCAI, 2013, pp. 2605–2611. IJCAI/AAAI (2013)
4. Cho, K., et al.: Learning phrase representations using RNN encoder-decoder for statistical machine translation. In: Moschitti, A., Pang, B., Daelemans, W. (eds.) Empirical Methods in Natural Language Processing, EMNLP, pp. 1724–1734. ACL (2014)
5. Devlin, J., Chang, M.-W., Lee, K., Toutanova, K.: BERT: pre-training of deep bidirectional transformers for language understanding. In: Burstein, J., Doran, C., Solorio, T. (eds.) Proceedings of the 2019 Conference of the North American Chapter of the Association for Computational Linguistics: Human Language Technologies, NAACL-HLT, pp. 4171–4186. Association for Computational Linguistics (2019)
6. Feng, S., Li, X., Zeng, Y., Cong, G., Chee, Y.M., Yuan, Q.: Personalized ranking metric embedding for next new POI recommendation. In: Yang, Q., Wooldridge, M.J. (eds.) Proceedings of the 24th IJCAI, pp. 2069–2075. AAAI Press (2015)
7. Huang, J., Zhao, W.X., Dou, H., Wen, J.-R., Chang, E.Y.: Improving sequential recommendation with knowledge-enhanced memory networks. In: Collins-Thompson, K., Mei, Q., Davison, B.D., Liu, Y., Yilmaz, E. (eds.) The 41st ACM SIGIR Conference, pp. 505–514. ACM (2018)
8. Kong, D., Wu, F.: HST-LSTM: a hierarchical spatial-temporal long-short term memory network for location prediction. In: Lang, J. (ed.) Proceedings of the 27th International Joint Conference on Artificial Intelligence, IJCAI, pp. 2341–2347 (2018). ijcai.org
9. Krizhevsky, A., Sutskever, I., Hinton, G.E.: ImageNet classification with deep convolutional neural networks. In: Bartlett, P.L., Pereira, F.C.N., Burges, C.J.C., Bottou, L., Weinberger, K.Q. (eds.) Advances in NIPS 25: 26th Annual Conference on NIPS, pp. 1106–1114 (2012)
10. Liu, Q., Wu, S., Wang, L., Tan, T.: Predicting the next location: a recurrent model with spatial and temporal contexts. In: Schuurmans, D., Wellman, M.P. (eds.) Proceedings of the Thirtieth AAAI Conference, pp. 194–200. AAAI Press (2016)
11. Salehi-Abari, A., Boutilier, C.: Preference-oriented social networks: group recommendation and inference. In: Proceedings of the 9th ACM Conference on Recommender Systems, RecSys 2015, pp. 35–42, New York, NY, USA, 2015. Association for Computing Machinery (2015)
12. Schlichtkrull, M., Kipf, T.N., Bloem, P., van den Berg, R., Titov, I., Welling, M.: Modeling relational data with graph convolutional networks. In: Gangemi, A., et al. (eds.) ESWC 2018. LNCS, vol. 10843, pp. 593–607. Springer, Cham (2018). https://doi.org/10.1007/978-3-319-93417-4_38

13. Vaswani, A., et al.: Attention is all you need. In: Guyon, I., et al. (eds.) Advances in NIPS 30: Annual Conference on NIPS, pp. 5998–6008 (2017)
14. Tran, L.V., Pham, T.-A.N., Tay, Y., Liu, Y., Cong, G., Li, X.: Interact and decide: medley of sub-attention networks for effective group recommendation. In: Proceedings of the 42nd ACM SIGIR, SIGIR 2019, pp. 255–264, New York, NY, USA. Association for Computing Machinery (2019)
15. Wang, Q., et al.: Next point-of-interest recommendation on resource-constrained mobile devices. In: WWW, pp. 906–916, April 2020
16. Yin, H., Wang, W., Wang, H., Chen, L., Zhou, X.: Spatial-aware hierarchical collaborative deep learning for poi recommendation. IEEE Trans. Knowl. Data Eng. **29**(11), 2537–2551 (2017)
17. Yin, H., Cui, B., Zhou, X., Wang, W., Huang, Z., Sadiq, S.: Joint modeling of user check-in behaviors for real-time point-of-interest recommendation. ACM Trans. Inf. Syst. **35**, 10 (2016)
18. Zhao, P., et al.: Where to go next: a spatio-temporal gated network for next POI recommendation. In: The 33rd AAAI, pp. 5877–5884 (2019)
19. Zhu, Y., et al.: What to do next: modeling user behaviors by time-LSTM. In: Sierra, C. (ed.) Proceedings of the 26th IJCAI, pp. 3602–3608 (2017)

A Dependency-Aware Utterances Permutation Strategy to Improve Conversational Evaluation

Guglielmo Faggioli[1]([✉])(iD), Marco Ferrante[1](iD), Nicola Ferro[1](iD),
Raffaele Perego[2](iD), and Nicola Tonellotto[3](iD)

[1] University of Padova, Padova, Italy
guglielmo.faggioli@phd.unipd.it
[2] ISTI-CNR, Pisa, Italy
[3] University of Pisa, Pisa, Italy

Abstract. The rapid growth in the number and complexity of conversational agents has highlighted the need for suitable evaluation tools to describe their performance. The main evaluation paradigms move from analyzing conversations where the user explores information needs following a scripted dialogue with the agent. We argue that this is not a realistic setting: different users ask different questions (and in a diverse order), obtaining distinct answers and changing the conversation path. We analyze what happens to conversational systems performance when we change the order of the utterances in a scripted conversation while respecting temporal dependencies between them. Our results highlight that the performance of the system widely varies. Our experiments show that diverse orders of utterances determine completely different rankings of systems by performance. The current way of evaluating conversational systems is thus biased. Motivated by these observations, we propose a new evaluation approach based on dependency-aware utterance permutations to increase the power of our evaluation tools.

1 Introduction

The conversational search domain has recently drawn increasing attention from the Information Retrieval (IR) community. A conversational agent, by definition, is expected to interact seamlessly with the user through natural language, either written (i.e. text chat-bots) or spoken (i.e. vocal assistants). Following the development of conversational systems, the evaluation of such systems is receiving a lot of attention. Following the best practices proposed by TREC CAsT [6,7], the principal evaluation campaign in the conversational domain, the evaluation process is very similar to the one used in ad-hoc retrieval. It follows the Cranfield paradigm, with a corpus of passage documents, a set of conversations representing various information needs, and a set of relevance judgements. Each conversation is a sequence of utterances – i.e., phrases issued by the user during the conversation – and the relevance judgements are collected for each utterance. Several

M. Hagen et al. (Eds.): ECIR 2022, LNCS 13185, pp. 184–198, 2022.
https://doi.org/10.1007/978-3-030-99736-6_13

works [1,9,19–21,26,38] have already recognized the drawbacks of using traditional evaluation approaches in a (multi-turn) conversational setup. Among the difficulties that make traditional evaluation techniques hardly applicable in the conversational domain, we can list the following: – *Lack of generalizability* [19]: conversations in the current evaluation collections represent a single interaction between a user and the ideal system. Therefore, when we evaluate using a conversation represented as a sequence of utterances, we consider a snapshot of reality. Since we have a unique sequence of utterances, we cannot generalize to conversations on the same topic that could have happened between the user and the system but are not in the collection. – *Lack of comparability* [19]: conversations have different lengths, they can contain chains of anaphoras or might have multiple self-contained utterances. Evaluation procedures should account for such diversity. – *Interdependency between utterances* [9]: utterances in conversational search are intrinsically dependent, differently from topics in ad-hoc retrieval. Therefore, cannot be treated as independent and identically distributed events. This work aims at providing a new perspective on the first aspect: low generalizability. We show a series of experiments meant to demonstrate the poor generalizability of results obtained using offline evaluation collections. Our work can be formalized with the following research questions:

RQ1 How can we shuffle utterances of a conversation by maintaining their original meaning and inter-dependencies?

RQ2 What is the effect of including dependency-aware permuted conversations in the comparison between systems?

RQ3 Can we improve conversational agents evaluation using permuted dialogues?

By answering the first question, we obtain a sound process to permute utterances of a conversation, producing new conversations to test conversational systems. We, therefore, use such conversations to compare models under the current evaluation paradigm, highlighting and measuring its flaws. Finally, we propose a new strategy to include the permuted conversations in the evaluation methodology. We do not propose a new evaluation measure – as done for example in [9,19] – but show how, by adapting our current instruments, we could partially mitigate the limitations associated with the evaluation of the conversational systems. Our main contributions are the following. We show that:

– Modeling a conversation using a single sequence of utterances only favours some systems, while penalizing others;
– If we consider multiple valid permutations of the conversations, the performance of conversational agents moves from point estimations to distributions of performance (in which the default sequence is an arbitrary point);
– By including multiple permutations in the evaluation, we obtain more reliable and generalizable statistical inference.

Our work is organized as follows: Sect. 2 describes the current state in conversational evaluation. In Sect. 3 we describe our experimental methodology. Section 4 details on the experimental results observed. Finally, Sect. 5 describes the insights of our and outlines the next steps.

2 Related Work

Conversational agents are commonly divided into chit-chat bots [36,37] and task-oriented systems [3,14,26]. Chit-chat bots are used to entertain the users, while the latter guide them to satisfy a goal, such as buy or discover something, through a dialogue. Task-driven conversational systems can be categorized into systems that retrieve and rank answers [14,31,35] and systems that build them through summarization techniques such as T5 [27]. While the latter are traditionally evaluated through traditional NLP and machine translation measures, such as BLEU [25] or METEOR [2], the former still relies on traditional IR evaluation measures such as Precision or Normalized Discounted Cumulated Gain (nDCG) [15], with typically a very small cutoff [6]. Finally, conversational systems can be divided into single-answer systems and multi-turn conversational systems. Among the former, we can list current commercial vocal assistants, handling very short - often scripted - sequences of interactions. The latter should ideally deal with a sequence of interactions of unspecified length. One of the most peculiar aspects related to the multi-turn conversational task is the role played by the concept of "context" [18,23,33]. The context corresponds to the system's internal representation of the conversation state that evolves through time. Correctly maintaining and updating such internal beliefs is essential to approach effectively the multi-turn conversational task. In this work, we focus on the evaluation of *Multi-turn Task-driven Conversational search systems*. Multi-turn conversational search is also the main focus of the TREC Conversational Assistance Track (CAsT) campaign [6,7]. Currently, the track has reached its third edition: a further demonstration of the interest shown by the community. The evaluation aspect of conversational agents is consequently drawing increasing interest [1,9,19–21,26,38]. Even though several efforts aimed at developing proper techniques to evaluate conversational systems [9,38], there is a consensus on the fact that we still lack the properer statistical tools to correctly evaluate such systems. [9] propose to model a conversation through a graph: utterances in a conversation are linked if they concern the same entities. Authors argue that current evaluation approaches introduce biases on systems comparison, by considering utterances as independent events. [10] do not tackle the problem linked to the low generalizability, due to predefined conversations available in current offline collections. [19] start the low generalizability that affects the current offline evaluation of conversational systems. [19] propose to simulate users through a stochastic process, similarly to what done in [38]. In particular, each topic is modelled as a set of subtopics (collected manually and using the available experimental collections). Using crowd assessors, [19] define a Markov chain process that should model how users present utterances to the system when interacting with a conversational agent. This allows producing new simulated conversations. Such a solution partially solves the low generalizability problem. Nevertheless, the need for online data makes it infeasible for purely offline scenarios, where no users are available.

3 Methodology

In this section, we describe the experimental methodology to answer the research questions. In Sect. 3.1 we describe a permutation process capable of preserving the dependencies between utterances (RQ1). Finally, Sect. 3.2 defines a methodology to use ANOVA to evaluate conversational systems, using permuted conversation utterances (RQ3).

3.1 RQ1: A Dependence-Aware Utterance Permutation Strategy

Several works [9,19,26,38] recognize the need of increasing the variety of conversations to improve the generalizability of offline conversational evaluation. As observed by [19], when conversing with a system about a specific topic, distinct users tend to traverse subtopics in different orders. Generalization would ask to observe how distinct users interact with the systems to investigate a specific topic: this is not possible in an offline scenario. A possible approach to simulate how users would experience a system would be permuting the utterances of a given conversation, and measuring how it performs. We cannot however permute utterances completely randomly. In fact, we might lose temporal dependency between the moment the entity is mentioned in an utterance for the first time and referenced later. To solve this limitation, we would have to re-gather the relevance judgements to fit the newly defined anaphoras in the randomly built conversation. This is prohibitive and not suited to an offline evaluation scenario. A better permutation strategy consists in permuting utterances by respecting the temporal dependencies. To this end, we could rely on classification labels (we dub this approach `class-based` permutation) to identify such dependencies. Similarly to what done in [24], we can manually annotate the data using three classes of utterances:

- Self-Explanatory (SE) utterances: utterances that do not contain any semantic omission. Non-contextual retrieval systems can answer such utterances.
- Utterances that depend on the First Topic of the conversation (FT): they contain an - often implicit - reference to the general topic of the conversation, subsumed by the first utterance.
- Utterances that depend on a Previous Topic (PT): the previous SE utterance contains the entity to solve the semantic omission in the current one.

Using this utterance classification, we define a sampling process to randomly permute utterances of a conversation, while preserving temporal dependencies. We define the following rules for the generation of utterance permutations:

- The first utterance in any conversation expresses the main topic of the conversation. It cannot be moved to other positions.
- SE utterances, being independent by definition, can appear in any order inside the conversation.

- PT utterances have to appear immediately after their SE utterance. More in detail, after a SE utterance, in CAsT 2019 conversations, we have an arbitrary number of PT utterances (usually between 0 and 4): such utterances can appear in any order, as long as they occur after the associated SE utterance.
- FT utterances, depending on the global topic of the dialogue can be issued at any moment, since the first utterance cannot be moved.

3.2   RQ3: Exploiting Permuted Conversation Utterances

As a final methodological remark, we show how to embed utterances permutations in the evaluation. To have a common ground with current evaluation strategy, we consider to compare different retrieval models using ANalysis Of VAriance (ANOVA). If we were to apply ANOVA in the current evaluation setup, we would likely rely on the following model:

$$y_{ik} = \mu_{..} + \tau_i + \alpha_k + \varepsilon_{ik} \tag{MD0}$$

Where y_{ik} is the mean performance of all utterances for the conversation i, using the retrieval model k. $\mu_{..}$ is the grand mean, τ_i is the contribution to the performance of the i-th conversation, while α_k is the effect of the k-th system. Finally, ε_{ik} is the unexplained portion of the performance variation using the ANOVA model MD0. This is the traditional two-way ANOVA model used on IR data to recognize statistical differences between systems [5,13,34].

If we also include multiple permutations for each conversation, Model MD0 cannot be applied satisfactorily anymore. The different permutations behave as a nested factor. We need to resort to a three-way ANOVA, that includes the different permutations. A specific permutation is, trivially, a permutation only of one conversation: we cannot treat it as a permutation of others. The variation in the performance due to a permutation should contribute only to the variation in performance of the conversation it represents. Including multiple permutations, which behave as replicates [28], allows computing the interaction factor between retrieval models and conversations in the ANOVA model. In ad-hoc retrieval, such interaction has a medium-to-large size effect [4,12,34] and, if included, allows more powerful inferential analyses. We leave this analysis for future works. We use the following ANOVA model:

$$y_{i(j)k} = \mu_{..} + \tau_i + \nu_{j(i)} + \alpha_k + \varepsilon_{ijk} \tag{MD1}$$

Where, compared to Model MD0, $\nu_{j(i)}$ represent the effect of the j-th permutation of the i-th conversation.

4   Experimental Analysis

In our experimental analysis, we consider the Conversational Assistance Track (CAsT) 2019 [6]. Such collection contains 50 multi-turn conversations, each composed of 9 utterances on average. The utterances in their original formulation contain semantic omissions - anaphoras, ellipsis and co-references. Among the 50 conversations, 30 were used for training and have smaller pools of relevance judgements. The remaining 20 are the test set. In our subsequent analyses, we consider only test conversations, being their relevance judgements much more significant. The corpus is composed of approximately 38 million paragraphs from the TREC Complex Answer Retrieval Paragraph Collection (CAR) [8] and the MS MARCO collection[1]. Regarding the relevance judgements, CAsT 2019 contains graded judgements on a scale from 0 to 4. We adopt nDCG with cutoff at 3, being the most widely diffused evaluation measure for this specific scenario [6]. To ease the reproducibility the code is publicly available[2].

4.1 Conversational Models

As commonly done [10,11,19], we select as baselines a set of models that represent different families of approaches to the multi-turn conversational task. Notice that, for all the rewriting strategies, we used BM25 as ranker.

Non-contextual Baseline Models. We consider three non-contextual baseline models, used as a comparison with other approaches. We compute the runs using the okapi BM25 model with default terrier parameters ($k = 1.2$ and $b = 0.75$). The second baseline is Query Language Model with Bayesian Dirichlet smoothing and $\mu = 2500$. Finally, we include results from a Pseudo-Relevance feedback RM3 rewriting model [22], which considers the 10 most popular terms of the 10 documents ranked the highestd.

Concatenation-Based Models. A simple approach to enrich utterances with context to address the multi-turn conversational challenges consists in concatenating them with one (or more) of the previous ones. We propose three concatenation-based strategies, previously adopted as baselines in the literature [24]:

- First Utterance (FU): each utterance u_j is concatenated with u_1, the first utterance of the conversation.
- Context Utterance (CU): each utterance u_j is concatenated with u_1 and u_{j-1}, the previous utterance.
- Linear Previous (LP): we concatenate u_j with u_{j-1} linearly weighting the terms: $q_j = \lambda * u_j + (1 - \lambda) * u_{j-1}$, with $\lambda \in [0,1]$. In particular, we observed empirically the best results for $\lambda = 0.6$.

Pseudo-Relevance Feedback Based Models. We consider two approaches based on pseudo-relevance feedback (PRF) that account for the "multi-turn" aspect:

- RM3-previous (RM3p): it concatenates the current utterance and the RM3 expansion of the previous one (using BM25 as first stage retrieval model).

---

[1] http://www.msmarco.org/.
[2] https://github.com/guglielmof/utterance_permutations.

Table 1. Number of unique permutations that can be observed for each conversation in CAsT 2019, according to the `class-based` permutation.

Conv. id	31	32	33	34	37	40	49	50	54	56
unique perm.	72	15184	720	720	240	120	5039	120	25676	720
Conv. id	58	59	61	67	68	69	75	77	78	79
unique perm.	720	121	720	289	4996	480	721	48	48	241

- RM3-sequential (`RM3s`): it takes the relevance feedback considering the ranked list retrieved for the previous utterance, and uses it to expand the current.

The difference between the two models is that, for `RM3p`, the ranked list depends only on the previous utterance and the one at hand. Conversely, the latter considers the sequence of utterances observed up to the current one. In both cases, for the first query, we apply directly BM25, without rewriting it.

Language Model-Based Models. Among the neural language models , we consider coref-spanBERT (`anCB`). This method relies on the Higher-order Coreference Resolution model, as defined in [17], but employs the spanBERT [16] embeddings to represent the words. In particular, we use the pre-trained version of the approach available in the AllenNLP framework[3].

4.2 RQ1: Permuting Conversations

Following the sampling process described in Sect. 3.1 we randomly permute the CAsT 2019 conversations. Table 1 reports the number of unique permutations obtained for each of the conversations in CAsT 2019.

The majority of the conversations have the `class-based` permutations in the order of tens to thousands. There are two main exceptions: conversations 54, 32. The larger number of permutations is due to the different structures of such conversations. For example, conversation 54 contains 3 SE utterances plus the first utterance and 5 FT utterances. Given these characteristics, we need to enforce only the first and third constraints to obtain valid `class-based` permutations, producing a larger space of valid permutations[4].

4.3 RQ2: Conversational Systems Performance on Permuted Conversations

Table 2 reports the nDCG@3 observed for the different archetypal conversational retrieval baselines either by considering only the original order of the utterances

---

[3] https://docs.allennlp.org.

[4] If we consider all the random permutations, for an average 9-utterances conversation, we would have approximately 3.6×10^5 permutations: 10 times more than the maximum number of permutations observed using the `class-based` strategy.

Table 2. Performance measured with nDCG@3 for the baselines and PRF conversational models. Baselines results do not depend on the order of the utterances. We report the mean for both standard order of the utterances, and over all permuted conversations. Concerning permuted conversations, we also report the minimum and maximum mean over all conversations that can be observed, using different permutations.

		nDCG@3			
		orig. order	permutations		
	model		min.	mean	max.
baselines	BM25	0.0981	0.0981	0.0981	0.0981
	DLM	0.0794	0.0794	0.0794	0.0794
	RM3	0.1064	0.1064	0.1064	0.1064
concatenation-based	FU	0.1692	0.1692	0.1692	0.1692
	CU	0.1687	0.1185	0.1481	0.1809
	LP	0.1464	0.0906	0.1279	0.1671
PRF-based	RM3p	0.1451	0.1019	0.1353	0.1709
	RM3s	0.1639	0.1108	0.1482	0.1857
neural LM based	anCB	0.1640	0.1410	0.1553	0.1645

as defined in CAsT 2019 or considering the average over multiple permutations for each conversation. To grant a fair comparison between different conversations, since they can have a different number of valid `class-based` permutations, we sample only 100 permutations for each of them. The most interesting insight that Table 2 is that the best performing system is the "First Utterance" (FU). We explain this because the first utterance of the original conversation is often the most generic. If we concatenate it with other utterances it can boost their recall, helping them obtain better results. The FU approach obtains the same results even when we permute conversations. Since we forced the first utterance to remain in its position, the order does not influence this algorithm. Therefore, we do not include it in subsequent analyses that measure the impact of permutations on conversational models. If we consider the result achieved with permuted conversations, we observe a general decrease in the average performance, due to the increased variance caused by the permutations. If we consider the maximum performance achievable, interestingly, all the methods can outperform the results achieved with the original order, indicating that there are situations in which different orders are preferable. The change in performance occurs due to the different information flow. The conversational models selected – as the majority of common conversational strategies – exploit the context to solve the anaphoras and rewrite the utterances. Such context derives from previous turns. By changing the previous turns, we also change the context, and thus the information used by the system. This aims at mimicking a real-world scenario, where we do not know if previous utterances provided good context. Furthermore, such context might change depending on the path followed by the user.

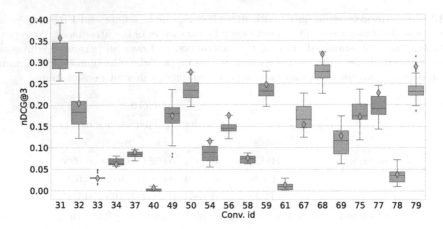

Fig. 1. Distributions of the average systems performance over different permutations of the conversations, considering original CAsT 2019 utterances. The yellow diamond is the average performance achieved using the original order of utterances. Observe that, in most cases the original order of the utterances does not have the best performance. (Color figure online)

Table 3. Maximum distance observed between models, using different permutations. On the diagonal, the maximum average distance from all other systems. The absence of negative numbers indicates that it is always possible to make any model "the best".

model	CU	LP	RM3p	RM3s	anCB
CU	0.0727	0.1030	0.0958	0.0810	0.0882
LP	0.0644	0.0432	0.0460	0.0470	0.0805
RMp	0.0646	0.0577	0.0396	0.0476	0.0803
RMs	0.0955	0.1226	0.1147	0.0937	0.1148
anCB	0.0420	0.0668	0.0593	0.0402	0.0250

Figure 1 plots, for each CAsT 2019 conversation, the distribution over the permutations of the average performance of all systems. The yellow diamond represents the mean performance using the default order of the utterances. It is insightful noticing that the default order rarely gives the best performance: using a different order of utterances strongly influences performance. Such a pattern is also observable for each system singularly[5].

To further investigate the effect of permutations, we select the permutation that maximizes the difference in nDCG@3 between each pair of systems. We repeat this for each conversation. We also select the permutation that maximizes the average difference in performance between a system and all the others. Table 3 reports the results of such analysis. It is always possible to cherry-pick conversations permutations to make any model the best in a pairwise

---

[5] We do not report the figure for each system, to avoid clutter.

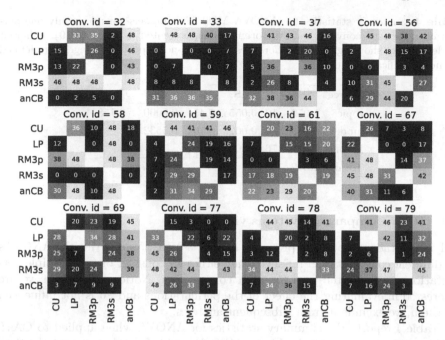

Fig. 2. Conversation-wise comparisons between pairs of systems. Number of times the row system is preferred over the column one, over different permutations of the conversation. Permuting the utterances order changes what system is deemed better: limiting ourselves to only one permutation might lead us to wrong conclusions. 12 out of 20 Conversations have been randomly selected, for the sake of presentation.

comparison. When using a collection with a single sorting of the utterances for each conversation, we need to ask ourselves: is a system better than another or is it an artefact of the collection at hand? Can we trust our results to be generalizable on previously unseen conversations? The difference can be as large as 12%: it is huge if we consider the scale of our performance - see Table 2. Not only it is possible to make any model the best in a pairwise comparison, but we are also able to maximize the distance in terms of performance from any other model, to make an arbitrary system the best in absolute (see diagonal of Table 3). Figure 2 describes how often, conversation by conversation, we would change our opinion over which system is the best, if we present them with different utterances permutations. More in detail, for each conversation, in each cell we report how often the row system is deemed better than column one, over different utterances permutations. When we consider pairwise comparisons between systems, there is seldom a clear winner. For example, consider Conversation 59; in the majority of the pairwise comparisons, there is a 50% chance that one model is better than the other if we select a specific permutation of the utterances. A system wins over another on every permutation only in a few cases.

Table 4. Summary statistics for ANOVA MD0. This models considers only one permutation for each conversation (the original one, presented in CAsT 2019). Different models do not show significant differences. ω^2_{model} is not reported, being ω^2 ill-defined for non-significant factors.

Source	SS	DF	MS	F	p-value	$\hat{\omega}^2_{(fact)}$
topic	1.052	19	0.055	17.454	0.0000	0.758
model	0.010	4	0.002	0.762	0.5532	–
Error	0.241	76	0.003			
Total	1.302	99				

4.4   RQ3: Comparing Systems via ANOVA

Relying on the methodology proposed in Sect. 3.2, we now describe the ANOVAs on different conversational models, when either we consider or not multiple permutations of the utterances for each conversation. Notice that, since we are interested in evaluating the effect of the permutations and FU is not influenced by them, we exclude it from subsequent analyses.

Table 4 reports the summary statistics for ANOVA when applied to CAsT 2019 conversations, using the Model MD0. For each factor, we report the Sum of Squares (SS), the Degrees of Freedom (DF), the Mean Squares (MS), the F statistics, the p-value and the Strength of Association (SOA), measured according to the ω^2 measure.

We observe that the effect of the "conversation" factor is significant and large-sized ($\omega^2 \geq 0.14$). This pattern is often observed in many IR scenarios, such as ad-hoc retrieval [4,13,34] or Query Performance Prediction (QPP) [10]. Conversely, the effect of the Model factor is not significant: none of the models is significantly the best. We are not particularly surprised by that: both Table 3 and Fig. 2 have shown that considering only a single permutation of the utterances, we would likely say something false by saying that a specific system is the best! This indicates the low discriminative power associated with this evaluation approach. If we were to consider state-of-the-art systems, possibly even more complex (and similar) than the ones we used, would we be able to state which system is statistically the best? Being able to discriminate between systems is a fundamental requirement for any evaluation approach [29,30,32]: could we deem ourselves satisfied with what we can achieve with the current evaluation setup in multi-turn conversational search?

Table 5 reports the summary statistics for ANOVA with model MD1. By looking at Table 5 we can see the first huge advantage of including permutations in our evaluation framework: the Model factor is now significant - although small ($0.01 < \omega^2 < 0.06$). As a side note, Tukey's post-hoc analysis shows that anCB is the best model, followed by RM3s which belong to the same tier. Subsequently, we have RM3p and CU, which again are statistically not different from each other, but worse than the previous ones. Finally, LP is the only member of the worst-quality tier. We have moved from having all models equal in Table 4 to a four-tiers

Table 5. Summary statistics for ANOVA MD1. This models considers 100 unique permutations for each conversation plus the original one. Observe that now all the factors have a significant effect.

Source	SS	DF	MS	F	p-value	$\hat{\omega}^2_{\langle fact \rangle}$
topic	38.594	19	2.031	657.983	>1e−3	0.722
perm (topic)	2.438	940	0.003	0.840	0.999	–
model	0.472	4	0.118	38.230	>1e−3	0.030
Error	11.842	3836	0.003			
Total	53.347	4799				

sorting of the models in Table 5. The Permutation factor is not significant. This suggests that there is not a single permutation that allows every system to work better, but rather there is an interaction between the systems and permutations: distinct models behave differently according to the permutation at hand. Table 5 shows that, if we use the permutations as additional evidence of the quality of a model, we discriminate better between them. Furthermore, we do not know in which order the user will pose their utterances. Including permutations allows us to model better the reality: what we observe in our offline experiment is likely to generalize more to a real-world scenario. Permutations allow robust statistical inference, without requiring to gather new conversations, utterances and relevance judgements.

5   Conclusions and Future Works

In this work, we showed that traditional evaluation is seldom reliable when applied to the conversational search. We proposed a methodology to permute the utterances of the conversations used to evaluate conversational systems, enlarging conversational collections. We showed that it is hard to determine the best system when considering multiple conversation permutations. Consequently, any system can be deemed the best, according to specific permutations of the conversations. Finally, we showed how to use permutations of the evaluation dialogues, obtaining by far more reliable and trustworthy systems comparisons.

As future work, we plan to study how to estimate the distribution of systems performance without actually having the permutations and the models at hand. We plan to investigate how to use the performance distributions to compare multi-turn conversational models.

Acknowledgments. Nicola Tonellotto was partially supported by the Italian Ministry of Education and Research (MIUR) in the framework of the CrossLab project (Departments of Excellence).

References

1. Anand, A., Cavedon, L., Joho, H., Sanderson, M., Stein, B.: Conversational search (Dagstuhl Seminar 19461). In: Dagstuhl Reports, vol. 9 (2020)
2. Banerjee, S., Lavie, A.: METEOR: an automatic metric for MT evaluation with improved correlation with human judgments. In: Proceedings of the ACL Workshop on Intrinsic and Extrinsic Evaluation Measures for Machine Translation and/or Summarization, pp. 65–72 (2005)
3. Bangalore, S., Di Fabbrizio, G., Stent, A.: Learning the structure of task-driven human-human dialogs. IEEE Trans. Audio Speech Lang. Process. 16(7), 1249–1259 (2008)
4. Banks, D., Over, P., Zhang, N.F.: Blind men and elephants: six approaches to TREC data. Inf. Retriev. J. 1(1–2), 7–34 (1999)
5. Culpepper, J.S., Faggioli, G., Ferro, N., Kurland, O.: Topic difficulty: collection and query formulation effects. ACM Trans. Inf. Syst. 40(1), 1–36 (2021)
6. Dalton, J., Xiong, C., Callan, J.: TREC CAsT 2019: the conversational assistance track overview. In: Proceedings of TREC (2020)
7. Dalton, J., Xiong, C., Callan, J.: TREC CAsT 2020: the conversational assistance track overview. In: Proceedings of TREC (2021)
8. Dietz, L., Verma, M., Radlinski, F., Craswell, N.: TREC complex answer retrieval overview. In: Proceedings of TREC (2017)
9. Faggioli, G., Ferrante, M., Ferro, N., Perego, R., Tonellotto, N.: Hierarchical dependence-aware evaluation measures for conversational search. In: Proceedings of the 44th International ACM SIGIR Conference on Research and Development in Information Retrieval, pp. 1935–1939 (2021)
10. Faggioli, G., Zendel, O., Culpepper, J.S., Ferro, N., Scholer, F.: An enhanced evaluation framework for query performance prediction. In: Proceedings of the 43rd European Conference on Information Retrieval, pp. 115–129 (2021)
11. Ferro, N., Harman, D.: CLEF 2009: Grid@CLEF pilot track overview. In: Peters, C., et al. (eds.) CLEF 2009. LNCS, vol. 6241, pp. 552–565. Springer, Heidelberg (2010). https://doi.org/10.1007/978-3-642-15754-7_68
12. Ferro, N., Sanderson, M.: Improving the accuracy of system performance estimation by using shards. In: Proceedings of the 42nd International ACM SIGIR Conference on Research and Development in Information Retrieval, pp. 805–814 (2019)
13. Ferro, N., Silvello, G.: A general linear mixed models approach to study system component effects. In: Proceedings of the 39th International ACM SIGIR Conference on Research and Development in Information Retrieval, pp. 25–34 (2016)
14. Gu, J.C., Ling, Z.H., Liu, Q.: Utterance-to-utterance interactive matching network for multi-turn response selection in retrieval-based chatbots. IEEE/ACM Trans. Audio Speech Lang. Proc. 28, 369–379 (2020)
15. Järvelin, K., Kekäläinen, J.: Cumulated gain-based evaluation of IR techniques. ACM Trans. Inf. Syst. 20(4), 422–446 (2002)
16. Joshi, M., Chen, D., Liu, Y., Weld, D.S., Zettlemoyer, L., Levy, O.: Spanbert: improving pre-training by representing and predicting spans. Trans. Assoc. Comput. Linguist. 8, 64–77 (2020)
17. Lee, K., He, L., Zettlemoyer, L.: Higher-order coreference resolution with coarse-to-fine inference. In: Proceedings of the 2018 Conference of the NAACL-HLT, pp. 687–692 (2018)

18. Li, J., et al.: Dialogue history matters! personalized response selection in multi-turn retrieval-based chatbots. ACM Trans. Inf. Syst. **39**(4), 1–25 (2021)
19. Lipani, A., Carterette, B., Yilmaz, E.: How am i doing?: evaluating conversational search systems offline. ACM Trans. Inf. Syst. **39**(4), 1–22 (2021)
20. Liu, C.W., Lowe, R., Serban, I.V., Noseworthy, M., Charlin, L., Pineau, J.: How NOT to Evaluate Your Dialogue System: An Empirical Study of Unsupervised Evaluation Metrics for Dialogue Response Generation (2017)
21. Liu, Z., Zhou, K., Wilson, M.L.: Meta-evaluation of conversational search evaluation metrics. ACM Trans. Inf. Syst. **39**(4), 1–42 (2021)
22. Lv, Y., Zhai, C.: Positional relevance model for pseudo-relevance feedback. In: Proceedings of the 33rd International ACM SIGIR Conference on Research and Development in Information Retrieval, pp. 579–586 (2010)
23. Mele, I., Muntean, C.I., Nardini, F.M., Perego, R., Tonellotto, N., Frieder, O.: Topic propagation in conversational search. In: Proceedings of the 43rd International ACM SIGIR Conference on Research and Development in Information Retrieval, pp. 2057–2060 (2020)
24. Mele, I., Muntean, C.I., Nardini, F.M., Perego, R., Tonellotto, N., Frieder, O.: Adaptive utterance rewriting for conversational search. Inf. Process. Manag. **58**(6), 102682 (2021)
25. Papineni, K., Roukos, S., Ward, T., Zhu, W.J.: Bleu: a method for automatic evaluation of machine translation. In: Proceedings of the 40th Annual Meeting of the Association for Computational Linguistics, pp. 311–318 (2002)
26. Penha, G., Hauff, C.: Challenges in the evaluation of conversational search systems. In: Workshop on Conversational Systems Towards Mainstream Adoption, KDD-Converse (2020)
27. Raffel, C., et al.: Exploring the limits of transfer learning with a unified text-to-text transformer. J. Mach. Learn. Res. **21**(140), 1–67 (2020)
28. Rutherford, A.: ANOVA and ANCOVA, 2nd edn. A GLM Approach. Wiley, New York (2011)
29. Sakai, T.: Evaluating evaluation metrics based on the bootstrap. In: Proceedings of the 29th International ACM SIGIR Conference on Research and Development in Information Retrieval, pp. 525–532 (2006)
30. Smucker, M.D., Allan, J., Carterette, B.A.: A comparison of statistical significance tests for information retrieval evaluation. In: Proceedings of the Sixteenth ACM Conference on Conference on Information and Knowledge Management, pp. 623–632 (2007)
31. Tao, C., Wu, W., Xu, C., Hu, W., Zhao, D., Yan, R.: Multi-representation fusion network for multi-turn response selection in retrieval-based chatbots. In: Proceedings of the 12th ACM International Conference on Web Search and Data Mining, pp. 267–275 (2019)
32. Urbano, J., Lima, H., Hanjalic, A.: Statistical significance testing in information retrieval: an empirical analysis of type I, type II and type III errors. In: Proceedings of the 42nd International ACM SIGIR Conference on Research and Development in Information Retrieval, pp. 505–514 (2019)
33. Vakulenko, S., Longpre, S., Tu, Z., Anantha, R.: Question rewriting for conversational question answering. In: Proceedings of the Fourth ACM International Conference on Web Search and Data Mining (WSDM), pp. 355–363 (2021)
34. Voorhees, E.M., Samarov, D., Soboroff, I.: Using replicates in information retrieval evaluation. ACM Trans. Inf. Syst. **36**(2), 1–21, 102682 (2017)

35. Wu, Y., Wu, W., Xing, C., Zhou, M., Li, Z.: Sequential matching network: a new architecture for multi-turn response selection in retrieval-based chatbots. In: Proceedings of the 55th Annual Meeting of the Association for Computational Linguistics, pp. 496–505 (2017)
36. Yan, R.: "Chitty-chitty-chat bot": deep learning for conversational AI. In: Proceedings of the 27th International Joint Conference on Artificial Intelligence (IJCAI), vol. 18, pp. 5520–5526 (2018)
37. Yu, Z., Xu, Z., Black, A.W., Rudnicky, A.: Strategy and policy learning for non-task-oriented conversational systems. In: Proceedings of the 17th Annual Meeting of the Special Interest Group on Discourse and Dialogue, pp. 404–412 (2016)
38. Zhang, S., Balog, K.: Evaluating conversational recommender systems via user simulation. In: Proceedings of the 26th ACM SIGKDD International Conference on Knowledge Discovery & Data Mining, pp. 1512–1520 (2020)

Sentiment Guided Aspect Conditioned Dialogue Generation in a Multimodal System

Mauajama Firdaus[✉], Nidhi Thakur, and Asif Ekbal

Department of Computer Science and Engineering,
Indian Institute of Technology Patna, Patna, India
{mauajama.pcs16,asif}@iitp.ac.in

Abstract. Multimodality in conversations has become critical for gaining a complete grasp of the user's intention and providing better replies to satisfy the requirements of the customers. Existing multimodal conversational systems suffer from contradictions and generic responses. User sentiments upon the different aspects of a product/service are essential to comprehend the needs of the user and respond in an informative and interactive manner. In this regard, we propose the novel task of sentiment-guided aspect controlled response generation. This task is introduced to ensure consistency and coherence with the sentiments of the users for the aspects mentioned in the ongoing dialogue for generating better responses. In our work, we design a generative framework that utilizes the sentiment information of the previous utterances in a reinforced hierarchical transformer-based network. The decoder is provided the aspect knowledge explicitly for generation. We devise task-specific rewards that guide the generation process in an end-to-end manner. The multi-domain multi-modal conversation (MDMMD) dataset, which includes both text and images, is used to validate our proposed architecture. Quantitative and qualitative analyses show that the proposed network generates consistent and diverse responses, and performs superior to the existing frameworks.

Keywords: Multimodality · Aspect · Sentiment · Generation

1 Introduction

Multimodality in dialogues has helped in bridging the gap between the different fields of artificial intelligence (AI) like natural language processing (NLP) and computer vision (CV). The complementary sources, in the form of video and audio, assists in building robust conversational systems that provide complete understanding to the user. With the advancement in Artificial Intelligence (AI), the dependency of humans on technology has increased exponentially. From booking movie tickets, to reserving tables at restaurant, to online ordering of products has been made easy in today's world. To be able to look at the products and make an informed decision about the things to be bought with the help of multimodality conversational systems [14,35] has been of the running goals of AI.

M. Hagen et al. (Eds.): ECIR 2022, LNCS 13185, pp. 199–214, 2022.
https://doi.org/10.1007/978-3-030-99736-6_14

Fig. 1. Sentiment annotated examples from the Multi-domain Multi-modal Dialogue(MDMMD) dataset

Task-oriented conversational agents are majorly based on the unimodal (textual) information. Growing requirements in various fields such as travel, entertainment, retail, etc. require conversational agents to communicate by incorporating information from the different modalities to build a robust system. In response generation, we take a step ahead by proposing the task of sentiment guided aspect controlled response generation for multimodal dialogue systems. Aspect denotes an attribute or a property of the task-oriented systems. For example, a customer interested in purchasing a laptop considers all the product elements, such as the price, weight, storage, color, etc. Hence, a conversational agent having the capability of generating responses conditioned on different aspect information is essential for building a robust goal-oriented multimodal system. Most of us have had a negative experience and also shared frustration at automated customer service systems. Unfortunately, none of the existing goal-oriented conversational systems can recognise and let alone, act upon the user opinion. Researchers have focused on identifying sentiments in dialogues [23,30,31,51,52] using various deep learning techniques. However, no work has tried to incorporate sentiment information for different aspects for response generation in a multimodal dialogue system in order to create sentiment guided response generation framework. Users are the ultimate evaluators of dialogue systems. Therefore, we think that research on the dialogue framework should aspire for greater user satisfaction.

In Fig. 1, we present example from the MDMMD dataset [14]. It is evident that the data is dependent on images for complete knowledge of the products for assisting the user in a better manner. Simultaneously, it can also be seen that the user's opinion about different aspects of the product is crucial for providing correct and necessary assistance. Therefore, in this work, we not only include user sentiment information as an additional context feature in an end-to-end supervised generation model, but also uses the aspects for which a particular sentiment has been expressed for increasing customer satisfaction and retention. We believe that providing extra feedback from the user in the form of sentiment would guide the model to adapt to user behaviour and assist in generating appropriate and accurate responses according to the user requirements.

The major contributions and/or attributes of our current work are as follows: (i). We propose the task of sentiment guided aspect controlled response generation in a multimodal dialogue system. (ii). We annotate MDMMD dataset with sentiment labels using a semi-supervised approach. (iii). We design hierarchical transformer based end-to-end framework for sentiment-aware aspect conditioned response generation. (iv). The proposed model for both automatic and human evaluation shows its effectiveness over several baselines and the existing approaches.

2   Related Work

Dialogue generation is an essential task in every conversational system. To capture the context of previous queries by the user, the authors in [36–38] proposed a hierarchical framework capable of preserving the past information. Lately, memory networks [22] have been intensely investigated for capturing the contextual information in dialogues for the generation of responses infusing pointer networks. The meta-learning approach [25,32] has been applied to different datasets to increase the domain adaptability for generating the responses. To increase the ability to memorize the dialogue context, authors in [49] used a memory-to-sequence framework along with the pointer generator for response generation. A multi-task framework to enhance the performance of natural language generation (NLG) was investigated in [54].

The research reported in [9,10,12,16,26] has been useful in narrowing the gap between vision and language. With the release of the Multimodal Dialog (MMD) dataset [35], having conversations on the fashion domain incorporating information from both text and images has facilitated response generation in multimodal setup. Several works on the MMD dataset reported in [1,2,13,19] used the hierarchical encoder-decoder model to generate responses by capturing information from text, images, and the knowledge base. Recently, [4] proposed attribute-aware and position-aware attention for generating textual responses. Our present work distinguishes from the prior works of multimodal dialog systems in a sense that we focus here on the task of generating responses conditioned on different aspects of the product or service in accordance with the conversational history and user's sentiments. We use the MDMMD dataset [14] having multi-domain dialogues with both textual and visual information, and extend it by annotating sentiment information for our proposed task.

3   Methodology

We address the task of generating sentiment guided textual responses in a multimodal dialogue system by providing the desired aspect(s) information for constructing informative and interactive responses in our present research. For a given utterance $U_k = (w_{k,1}, w_{k,2}, ..., w_{k,n})$, a set of images $I_k = (i_{k,1}, i_{k,2}, ..., i_{k,j})$, corresponding sentiment labels S_k, conversational history $H_k = ((U_1, I_1), ..., (U_{k-1}, I_{k-1}))$ along with the aspect terms A_k the task is to generate the next textual response $Y = (y_1, y_2, ..., y_{n'})$, where n and n' are the given input utterance and response length, respectively.

The Transformer network [44], as shown in Fig. 2, is the basis of our proposed architecture. We use a hierarchical multimodal transformer network with two encoders: an utterance encoder that transforms the textual utterance $U_i = (w_{k,1}, w_{k,2}, \ldots, w_{k,n})$ and a context encoder at the dialogue level that learns the complete utterance representation (that includes both visual and textual knowledge) provided the prior utterances as context.

Textual Encoder: To learn the representation of a given utterance U_i, we first map $U_i = (w_{k,1}, w_{k,2}, \ldots, w_{k,n})$ into a continuous space by the following equation:

$$E_u = (e_1^i, e_2^i, \ldots, e_{|U_i|}^i); where [e_j^i = e(w_j^i) + p_j] \tag{1}$$

where $e(w_j^i)$ and p_j are the word and positional embedding of every word w_j^i in an utterance, respectively. Glove embeddings are used to represent words along with sine-cosine positional embedding [44] as it achieves better and has fewer trainable parameters. We use the last hidden representation $h_{|U_i|}^i$ (i.e. the representation at the EOS token) as the textual representation of the utterance U_i. The utterance position is also taken into consideration hence, the final textual representation of the utterance U_i is:

$$h_i^{txt} = h_{|U_i|}^i + p_i \tag{2}$$

Note that the positional embedding matrix for words and sentences is the same.

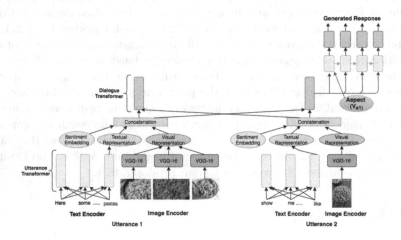

Fig. 2. Architectural diagram of our proposed framework

Visual Encoder: For visual representation, we use the pre-trained VGG-16 [40] having 16-layer deep convolutional network in a similar manner as [14,35]. The global visual context representation is formed by passing the image vector through the linear layer, as shown here:

$$I_{k,i} = VGG(I_{k,i}); T_k = Concat(T_{k,1}, T_{k,2}, \ldots, T_{k,j}); h_{I,k}^{img} = ReLU(W_I T_k + b_I) \tag{3}$$

where, W_I and b_I are the trainable weight matrix and biases, respectively. In every turn, the maximum number of images $i \leq 6$, in the case of only text, vectors of zeros are used instead of image representation.

Context Encoder: For sentiment guided response generation, we concatenate the sentiment label S_k with the final utterance representation $\hat{h}_i$ having both textual and visual representation. The context transformer takes the sequence of utterance representation (having both textual and visual information) $\hat{h}_1, \hat{h}_2, \ldots, \hat{h}_{|D|}$ as input, to obtain the context-sensitive utterance representations $\hat{D} = (\hat{d}_1, \hat{d}_2, \ldots, \hat{d}_{|D|})$.

Aspect Conditioned Decoder: We use an RNN decoder, as illustrated in Fig. 2, to construct the next textual reply using the specified aspect embedding. We use GRU to generate the response in a sequential fashion relying on the hierarchical transformer's contextual representation and the words decoded previously. We use the input feeding decoding along with the attention [21] mechanism for enhancing the performance of the model. Using the decoder state $h_{d,t}^{dec}$ as the query vector, we apply self-attention on the hidden representation of the context-level encoder. The decoder state, the context vector and the aspect term embedding V_a (the aspect embeddings are pre-trained Glove embeddings) are used to calculate a final distribution of the probability over the output tokens.

$$h_{d,t}^{dec} = GRU_d(y_{k,t-1}, [h_{d,t-1}, V_a]); c_t = \sum_{i=1}^{k} \alpha_{t,i} \hat{D}; \alpha_{t,i} = softmax(\hat{D}^T W_f h_{d,t})$$

$$(4)$$

where, W_f are the trainable weight matrices.

Training and Inference: As used in [28], we jointly use reinforcement learning (RL) and machine learning (ML) to train our model in a similar manner as [18]. If $\tilde{y} = \{\tilde{y}_1, \tilde{y}_2, \ldots, \tilde{y}_{n'}\}$ is the gold output tokens for the given generic response tokens h_i^s and conversation history $\hat{D}$, the maximum-likelihood objective using teacher forcing is given by:

$$L_{MLE} = -\sum_{t=1}^{n'} \log p(\tilde{y}_t | \tilde{y}_1, \ldots, \tilde{y}_{t-1}, h_i^s, \hat{D})$$

$$(5)$$

We utilise reinforcement learning to learn by maximising discrete metrics that are task-specific in addition to maximum likelihood error training (which we design as the rewards). We use the self-critical policy gradient algorithm suggested in [34] for training the network. The reward obtained by the inference time algorithm (which performs greedy decoding) is baselined for the REINFORCE [47] algorithm, without the need for training a "critic" network for estimating the value functions. During training, two output sequences are produced: y^s, obtained by sampling $p(y_t^s | y_1^s, \ldots, y_{t-1}^s, x)$ probability distribution, and y^g, the baseline output, obtained by greedily maximizing the output probability distribution at each time step.

$$L_{RL} = (r(y^g) - r(y^s)) \sum_{t=1}^{n'} \log p(y_t^s | y_1^s, \ldots, y_{t-1}^s, h_i^s, \hat{D}) \qquad (6)$$

Our reward function $r(y)$, used for evaluating y against the gold standard output is

$$r(y, \tilde{y}) = \lambda_1 \cdot r1(y, \tilde{y}) + \lambda_2 \cdot r2(y, \tilde{y}) + \lambda_3 \cdot r3(y, \tilde{y}) \qquad (7)$$

The final reward function is the weighted mean of the three terms as given below:

(i). BLEU metric $r1$: Ensures the content matching between the generated response and the ground-truth response to avoid loss of information.

(ii). Sentiment consistency $r2$: Measured by the cosine similarity of the sentiment prediction distribution of the user utterance and generated responses (using pre-trained BERT classifier). It ensures that the sentiment states of the generated response is consistent with the user sentiment.

(iii). Fluency $r3$: The above rewards do not assess if the response content expressed is linguistically fluent. To promote linguistic quality, r3 employs a language model (LM) fine-tuned on a set of utterances to evaluate the language quality of response. To do so, we use the Negative Log-Likelihood (NLL) loss obtained by this LM:

$$r3 = \frac{\alpha - NLL(r)}{\alpha} \qquad (8)$$

where parameter α is used to map any value of NLL that is greater than α to α so that the output of r3 will be between 0 and 1. r3 is not biased to the length of a response as NLL is already normalized by response length.

Finally, we first pre-train using the maximum likelihood (ML) objective and then using a mixed objective function with a reduced learning rate for training the model in an end-to-end manner:

$$\mathcal{L}_{gen} = \eta L_{RL} + (1 - \eta) L_{MLE}, \qquad (9)$$

Model Variations: We compare our proposed hierarchical transformer framework with some of the existing unimodal (i.e., text) and multimodal (text and image) baselines.

The unimodal baselines are: (i) Seq2Seq + Attn: The first framework is the vanilla sequence-to-sequence network with attention [41]. (ii) HRED: We employ a hierarchical encoder-decoder (HRED) RNN framework [37] as one of the baselines. (iii) HVMN: The hierarchical variational memory network (HVMN) [5] has been used to capture the dialogue context to obtain the enhanced representation for generation. (iv) GCN: We also compare our framework with graph convolutional network (GCN) as proposed in [3]. (v) Global-to-local: We similarly employ a pointer memory network to [50] as one of our comparative methods. (vi) Working Memory: The working memory framework investigated in [7] is used as one of the baselines.

The multimodal baselines are as follows: (i) MHRED: We devise the multimodal hierarchical encoder-decoder framework as one of the baselines analogous to [35]. (ii) OAM: We compare our proposed framework to the existing multimodal attention based framework proposed in [4]. (iii) M-GCN + Aspect: We compare our proposed network with multimodal GCN framework as proposed in [14].

4   Dataset

Our work is built upon the Multi Domain Multi Modal Dialogue (MDMMD) dataset [14] that comprises of 130k chat sessions between the customer and sales agent. Domain-specific knowledge having text and image information for the restaurant, electronics and furniture domain was captured during the series of customer-agent interactions. We take a step forward by including the sentiments of the user for providing coherent and consistent responses and in accordance to the user opinions. In order to do so we re-annotate the MDMMD dataset.

Data Annotation: Due to the absence of sentiment labels in the MDMMD dataset, we propose a semi-supervised approach for labeling it with sentiments for which we annotate a portion of the dataset. We create a balanced dataset (MDMMD-annotated) by manually annotating 10k dialogues for all the three domains. We label every utterance in a dialogue with three sentiment labels, *viz.* positive, negative, and neutral. Three annotators were employed to label each utterance of a given dialogue. We observe a multi-rater Kappa [24] agreement ratio of approximately 80%, which can be considered reliable. Majority voting is applied to decide the final label of the utterance.

Sentiment Classifier: We apply a semi-supervised approach for annotating the entire MDMMD dataset with sentiment labels in a similar manner as [13,53]. We first divide the labeled MDMMD-annotated dataset (10k dialogues) into training, validation, and test set in the ratio of 7:1:2. In addition, we also use the BERT based sequence classification model proposed in [48] for sentiment classification. Evaluation results of the various classifiers are demonstrated in Table 1. Finally, for labeling the entire MDMMD dataset, we use the best-performing classifier, RoBERTa.

Table 1. Classification scores of sentiment on the MDMMD dataset. Here, S-F1 denotes the weighted average F1 score of sentiment

Model	S-F1
LSTM	57.06
CNN	64.90
Bi-LSTM	61.87
BERT [11]	78.94
RoBERTa [20]	83.89

5   Experiments

Implementation Details. All the implementations were done using the PyTorch[1] framework. For all the models, including baselines, the batch size is set to 32. We use the dropout [42] with probability 0.45. During decoding, we use a beam search with beam size 10. The model is initialized with the parameters randomly using a Gaussian distribution with the Xavier scheme [17]. The hidden size for all the layers is 512. We employ AMSGrad [33] as the optimizer for model training to mitigate the slow convergence issues. We use uniform label smoothing with $\epsilon = 0.1$ and perform gradient clipping when the gradient norm is above 5. We use 300-dimensional word-embedding initialized with Glove [29] embedding pre-trained on Twitter. For image representation, FC6 (4096 dimension) layer representation of the VGG-19 [40], pre-trained on ImageNet is used. Previous 3 turns are considered for the dialogue history, and maximum utterance length is set to 50. We use $\eta = 0.99$ (similar to [28]) for the joint loss. For the reward function, the values of λ_1, λ_2 and λ_3 are 0.34, 0.33 and 0.33, respectively. We ran for 15 epochs, and the proposed model took about 3 days on a Titan X GPU machine.

Evaluation Metrics: To evaluate the model at relevance and grammatical level, we report the results using the standard metric like perplexity [6]. We also report the results using standard metrics like BLEU-4 [27] to measure the ability of the generated response for capturing the correct information. Success F1 measures the completion of the tasks and is modified from the rate of success in [45, 46]. Instead, we here use Aspect F1 to balance both recall and precision. It is defined as the F1 score of the requested aspects present in the generated response. We also compute sentiment accuracy of the generated responses to check if the user opinion has been expressed in the generated responses using the classifier used for annotating the MDMMD dataset.

We randomly sample 500 responses from the test set for human evaluation. For a given input along with aspect information, six annotators with post-graduate exposure were assigned to evaluate the quality of the generated responses by the different approaches in a similar manner as that of the existing works [4,7,8,39,43]. First, we evaluate the quality of the response on three conventional criteria: *Fluency, Relevance* and *Informativeness*. These are rated on a five-scale, where 1, 3, 5 indicate unacceptable, moderate, and excellent performance, respectively, while 2 and 4 are used for unsure. Secondly, we evaluate the aspect inclusion in a response in terms of *Aspect Consistency* metric and *Sentiment Appropriateness* to judge whether the response generated is in consonance to the specified aspects (e.g., cuisine, color, type, etc.) and is also coherent to the sentiment of the user utterance and the conversational history. In the case of aspect consistency and sentiment appropriateness, 0 indicates irrelevant or contradictory aspects in the response and contradictory sentiment, and 1 represents the consistent response to the specified aspects and user sentiments.

---

[1] https://pytorch.org/.

The Fleiss' kappa [15] score for fluency, relevance and informativeness are 0.75, 0.68 and 0.71, respectively, indicating "reliable agreement" while for aspect consistency and sentiment appropriateness are 0.77 and 0.78, indicating "substantial agreement".

6   Results and Analysis

In Table 2, we present the results of all the existing baselines and the proposed framework for all the automatic metrics discussed in the previous section. From the table, it is clear that the proposed approach outperforms all the existing baseline models (both unimodal and multimodal). As lower perplexity scores establish the fact that the generated response's quality is better, hence from the table, it can be inferred that the proposed $M\text{-}HierTrans + RL + A + S$ model with a remarkably lower perplexity of *1.0043* is effective in generating high-quality responses than all the other baselines. This superior performance of the proposed framework can be attributed to the fact that Transformers provide better-contextualized representation in comparison to other deep learning frameworks such as Recurrent Neural Network (RNN) [37,41], memory network [5,7,50], graph convolutional network [3], etc. Additionally, we compute the BLEU-4 score for all the comparative methods and our proposed approach. It is evident that there is an improvement of more than 12 points from the best performing baseline approach [14]. The proposed network significantly improves the responses generated as opposed to all the unimodal and multimodal baselines. The performance of the unimodal baselines is not at par with the proposed framework. This might be attributed to the additional knowledge available with the images. Also, the Rl training helps in improving the performance of the framework. Despite the fact that the multimodal baselines contain image information, the BLEU-4 scores are indeed lower than our proposed network. This confirms that hierarchical transformers, as opposed to hierarchical RNN frameworks, are better at obtaining context representation (with feed-forward and multi-head attention). The explicit information provided to the decoder aids in the generation of informative replies. This is illustrated by the results in Table 2, as the baseline networks with no aspect information have lower aspect F1 scores in comparison to our proposed framework. Also, sentiment accuracy for the proposed framework with additional sentiment information is higher signifying the fact that sentiment knowledge helps in making the responses better and consistent to the user demands.

The results of the ablation study in case of all measures are presented in Table 2. It is critical to capture both utterance-level and dialogue-level information for better context representation. Hierarchical networks can be used to accomplish this. From the table, it can be shown that the hierarchical *HierTrans* framework outperforms the non-hierarchical *Transformer* network by approximately 3%. In contrast to a single transformer network, it validates the concept that hierarchically combining two transformers gives better dialogue representation for superior response generation (at utterance-level). Because the studies are based

Table 2. Evaluation results with automatic evaluation metrics of different baselines and proposed model on the MDMMD dataset. Here, A: Aspect and S: Sentiment

Model description		Modality		Perplexity	BLEU-4	SA	Aspect F1
		U	M				
Existing baselines	Seq2Seq + Attn [41]	√	–	1.0341	0.4378	45.36	51.59
	HRED [37]	√	–	1.0293	0.5206	48.33	56.83
	HVMN [5]	√	–	1.0189	0.5453	50.14	60.26
	GCN [3]	√	–	1.0119	0.5821	53.75	64.53
	Global-to-local [50]	√	–	1.0169	0.5785	54.66	64.29
	Working Memory [7]	√	–	1.0173	0.5692	55.27	63.78
	MHRED [35]	–	√	1.0142	0.5537	57.48	61.86
	OAM [4]	–	√	1.0136	0.5746	58.35	65.71
	M-GCN + A [14]	–	√	1.0112	0.6014	62.85	67.71
Proposed approach	M-HierTrans + RL + A + S	–	√	**1.0043**	**0.7287**	**72.58**	**80.55**
Ablation study	Transformer	√	–	1.0123	0.6077	54.69	66.18
	Transformer + A	√	–	1.0117	0.6198	57.42	69.34
	Transformer + A + S	√	–	1.0111	0.6295	62.18	71.32
	M-Transformer	–	√	1.0085	0.6534	56.78	68.43
	M-Transformer + A	–	√	1.0077	0.6748	61.39	71.59
	M-Transformer + A + S	–	√	1.0069	0.6927	66.83	73.45
	HierTrans	√	–	1.0116	0.6382	57.23	69.11
	HierTrans + A	√	–	1.0111	0.6577	65.11	73.87
	HierTrans + A + S	√	–	1.0096	0.6825	69.47	76.36
	M-HierTrans	–	√	1.0058	0.6890	59.37	73.69
	M-HierTrans + A	–	√	1.0053	0.7033	67.84	77.51
	M-HierTrans + A + S	–	√	1.0049	0.7101	70.56	78.83

on a multimodal conversation dataset, the integration of another modality (in the form of images) aids in establishing a holistic view of the product and provides a better knowledge of the underlying conversation. The multimodal systems *M-Transformers, M-HierTrans*, which incorporate both text and images information, score higher than the unimodal architectures *Transformers, HierTrans* as shown in the table.

While the most significant objective of our current effort is to incorporate the desired aspect in accordance to the sentiment in the responses, therefore we study different frameworks having no aspect (*Transformers, M-Transformers, HierTrans, M-HierTrans*), aspect (*Transformer + A, M-Transformers + A, HierTrans + A, M-HierTrans + A*) and aspect with sentiment (*Transformers + A + S, M-Transformer + A + S, HierTrans + A + S*) information for response generation. It is evident from the table that the *sentiment aware aspect* models performs superior to the *no aspect* and *aspect only* models, as a result, in the case of all automatic evaluation measures, the responses become more engaging and conversational. The scores of aspect F1 for models with no aspect is obviously lower than the models having aspect information explicitly. In the case of aspect F1, *frameworks with no aspect information*, the scores are lower than the *aspect controlled frameworks*. It shows that aspect information helps in bringing specificity in responses and the ongoing dialogue. While the aspect framework performs considerably better than the frameworks with no aspect information,

Table 3. Evaluation results using human evaluation metrics of different baselines and proposed model on MDMMD dataset. Here, AC: Aspect Consistency; SA: Sentiment Appropriateness; F: Fluency; R: Relevance; I: Informativeness

Model description		Modality		AC	SA	F	R	I
		U	M					
Existing baselines	*Seq2Seq + Attn* [41]	√	–	25.9%	27.1%	2.17	2.21	2.19
	HRED [37]	√	–	34.1%	28.3%	2.54	2.63	2.69
	HVMN [5]	√	–	37.3%	30.1%	2.69	2.71	2.82
	GCN [3]	√	–	44.5%	31.7%	2.75	2.83	3.02
	Global-to-local [50]	√	–	47.1%	33.5%	2.88	2.91	3.19
	Working Memory [7]	√	–	46.5%	33.2%	2.86	2.90	3.15
	MHRED [35]	–	√	53.9%	36.1%	3.15	3.07	3.32
	OAM [4]	–	√	57.8%	38.2%	3.38	3.25	3.51
	M-GCN + A [14]	–	√	58.2%	38.7%	3.74	3.69	3.93
Proposed approach	*M-HierTrans + RL + A + S*	–	√	**68.7%**	**52.2%**	**3.95**	**3.86**	**4.12**
Ablation study	*Transformer*	√	–	49.1%	37.8%	2.91	3.02	3.37
	Transformer + A	√	–	51.3%	38.3%	2.94	3.06	3.41
	Transformer + A + S	√	–	53.5%	40.1%	2.97	3.08	3.47
	M-Transformer	–	√	59.4%	39.5%	3.70	3.59	3.69
	M-Transformer + A	–	√	61.2%	40.9%	3.73	3.63	3.75
	M-Transformer + A + S	–	√	63.7%	43.2%	3.78	3.66	3.83
	HierTrans	√	–	55.3%	41.3%	3.25	3.34	3.51
	HierTrans + A	√	–	57.9%	43.8%	3.28	3.37	3.56
	HierTrans + A + S	√	–	61.1%	46.4%	3.33	3.40	3.63
	M-HierTrans	–	√	65.3%	45.3%	3.80	3.79	3.99
	M-HierTrans + A	–	√	67.2%	48.3%	3.83	3.81	4.05
	M-HierTrans + A + S	–	√	67.7%	50.1%	3.87	3.83	4.07

yet it scores lower than the sentiment guided aspect frameworks. This can be attributed to the fact that sentiment guided aspect frameworks generate more engaging responses.

For complete evaluation, we also present human evaluation results in Table 3. The responses are grammatically fluent compared to the baselines and existing frameworks that ensures the efficacy of the model to generate better responses. With the help of reinforcement learning, the entire framework has been trained to generate interactive and informative responses. The responses are highly relevant ensuring consistency in the responses as the score is the highest for the proposed network having aspect and sentiment information. The ablation study shows that the frameworks with the additional knowledge of sentiment and aspect has better scores for the informativeness metric along with the aspect consistency. This proves the hypothesis of our work to be correct that the attributes help to enhance the performance of the entire network.

We provide the instances of the generated responses in Fig. 3. We provide a few examples to illustrate how the model with aspect and sentiment information generates responses compared to the model without these attributes. The errors detected belong to the following types: (i). Incorrect image: Sometimes incorrect images are selected due to the presence of multiple images. (ii) Aspect inconsistency: The generated responses sometimes generate responses that are

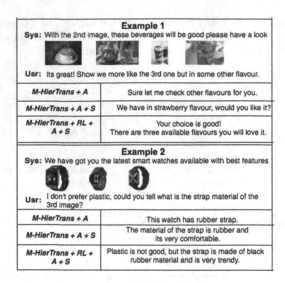

Fig. 3. Generated examples with aspect and sentiment knowledge

inconsistent to the given aspects mainly because of the less number of training samples of such aspects and mostly due to the presence of more than one aspect in the conversational history. (iii) Sentiment disparity: The baseline frameworks and the proposed method sometimes generate responses that are not in accordance to the user's sentiment, thereby, causing breaks/errors in conversations.

7   Conclusion and Future Work

In this paper, we have concentrated upon sentiment guided aspect conditioned response generation using the textual and visual information for conversational systems. We designed a reinforced transformer based framework that captures the contextualized representation for better generation. The sentiment information is provided during encoding while the desired aspect knowledge is explicitly provided during decoding. Task-specific rewards helps to generate informative, fluent and sentiment consistent responses. In this regard, we annotate the MDMMD dataset with sentiments. Quantitative and qualitative analysis show that the proposed framework performs better than the existing baselines.

In the future, we would investigate more efficient fusion techniques to effectively combine different modalities to enhance the generation framework. Also, we would like to focus on aspect based sentiment knowledge in the case of multiple aspects for better dialogue generation.

Acknowledgement. Authors duly acknowledge the support from the Project titled "Sevak-An Intelligent Indian Language Chatbot", Sponsored by SERB, Govt. of India. Asif Ekbal acknowledges the Young Faculty Research Fellowship (YFRF), supported by Visvesvaraya PhD scheme for Electronics and IT, Ministry of Electronics and Information Technology (MeitY), Government of India, being implemented by Digital India Corporation (formerly Media Lab Asia).

References

1. Agarwal, S., Dušek, O., Konstas, I., Rieser, V.: Improving context modelling in multimodal dialogue generation. In: Proceedings of the 11th International Conference on Natural Language Generation, pp. 129–134 (2018)
2. Agarwal, S., Dušek, O., Konstas, I., Rieser, V.: A knowledge-grounded multimodal search-based conversational agent. In: Proceedings of the 2018 EMNLP Workshop SCAI: The 2nd International Workshop on Search-Oriented Conversational AI, pp. 59–66 (2018)
3. Banerjee, S., Khapra, M.M.: Graph convolutional network with sequential attention for goal-oriented dialogue systems. Trans. Assoc. Comput. Linguist. **7**, 485–500 (2019)
4. Chauhan, H., Firdaus, M., Ekbal, A., Bhattacharyya, P.: Ordinal and attribute aware response generation in a multimodal dialogue system. In: Proceedings of the 57th Annual Meeting of the Association for Computational Linguistics, pp. 5437–5447 (2019)
5. Chen, H., Ren, Z., Tang, J., Zhao, Y.E., Yin, D.: Hierarchical variational memory network for dialogue generation. In: Proceedings of the 2018 World Wide Web Conference 2018, Lyon, France, 23–27 April 2018, pp. 1653–1662 (2018)
6. Chen, S., Beeferman, D.H., Rosenfeld, R.: Evaluation metrics for language models (1998)
7. Chen, X., Xu, J., Xu, B.: A working memory model for task-oriented dialog response generation. In: Proceedings of the 57th Conference of the Association for Computational Linguistics, ACL 2019, Florence, Italy, Volume 1: Long Papers, 28 July–2 August 2019, pp. 2687–2693 (2019)
8. Cui, C., Wang, W., Song, X., Huang, M., Xu, X.S., Nie, L.: User attention-guided multimodal dialog systems. In: Proceedings of the 42nd International ACM SIGIR Conference on Research and Development in Information Retrieval, SIGIR 2019, Paris, France, 21–25 July, pp. 445–454 (2019)
9. Das, A., et al.: Visual dialog. In: Proceedings of the IEEE Conference on Computer Vision and Pattern Recognition, pp. 326–335 (2017)
10. De Vries, H., Strub, F., Chandar, S., Pietquin, O., Larochelle, H., Courville, A.: GuessWhat?! Visual object discovery through multi-modal dialogue. In: 2017 IEEE Conference on Computer Vision and Pattern Recognition, CVPR 2017, Honolulu, HI, USA, 21–26 July, pp. 4466–4475 (2017)
11. Devlin, J., Chang, M.W., Lee, K., Toutanova, K.: BERT: pre-training of deep bidirectional transformers for language understanding. In: Proceedings of the 2019 Conference of the North American Chapter of the Association for Computational Linguistics: Human Language Technologies, NAACL-HLT 2019, Volume 1 (Long and Short Papers), Minneapolis, MN, USA, 2–7 June 2019, pp. 4171–4186 (2018)
12. Firdaus, M., Chauhan, H., Ekbal, A., Bhattacharyya, P.: EmoSen: generating sentiment and emotion controlled responses in a multimodal dialogue system. IEEE Trans. Affect. Comput. (2020)
13. Firdaus, M., Pratap Shandeelya, A., Ekbal, A.: More to diverse: generating diversified responses in a task oriented multimodal dialog system. PLoS ONE **15**(11), e0241271 (2020)

14. Firdaus, M., Thakur, N., Ekbal, A.: MultiDM-GCN: aspect-guided response generation in multi-domain multi-modal dialogue system using graph convolution network. In: Proceedings of the 2020 Conference on Empirical Methods in Natural Language Processing: Findings, EMNLP 2020, Online Event, 16–20 November 2020, pp. 2318–2328 (2020)

15. Fleiss, J.L.: Measuring nominal scale agreement among many raters. Psychol. Bull. **76**(5), 378 (1971)

16. Gan, Z., Cheng, Y., Kholy, A.E., Li, L., Liu, J., Gao, J.: Multi-step reasoning via recurrent dual attention for visual dialog. In: Proceedings of the 57th Conference of the Association for Computational Linguistics, ACL 2019, Volume 1: Long Papers, Florence, Italy, 28 July–2 August 2019, pp. 6463–6474 (2019)

17. Glorot, X., Bengio, Y.: Understanding the difficulty of training deep feedforward neural networks. In: Proceedings of the Thirteenth International Conference on Artificial Intelligence and Statistics, AISTATS 2010, Chia Laguna Resort, Sardinia, Italy, 13–15 May, pp. 249–256 (2010)

18. Golchha, H., Firdaus, M., Ekbal, A., Bhattacharyya, P.: Courteously yours: inducing courteous behavior in customer care responses using reinforced pointer generator network. In: Proceedings of the 2019 Conference of the North American Chapter of the Association for Computational Linguistics: Human Language Technologies, Volume 1 (Long and Short Papers), pp. 851–860 (2019)

19. Liao, L., Ma, Y., He, X., Hong, R., Chua, T.S.: Knowledge-aware multimodal dialogue systems. In: 2018 ACM Multimedia Conference on Multimedia Conference, MM 2018, Seoul, Republic of Korea, 22–26 October, pp. 801–809. ACM (2018)

20. Liu, Y., et al.: RoBERTa: a robustly optimized BERT pretraining approach. arXiv preprint arXiv:1907.11692 (2019)

21. Luong, M.T., Pham, H., Manning, C.D.: Effective approaches to attention-based neural machine translation. In: EMNLP (2015)

22. Madotto, A., Wu, C.S., Fung, P.: Mem2Seq: effectively incorporating knowledge bases into end-to-end task-oriented dialog systems. In: Proceedings of the 56th Annual Meeting of the Association for Computational Linguistics, ACL 2018, Volume 1: Long Papers, Melbourne, Australia, 15–20 July 2018, pp. 1468–1478 (2018)

23. Majumder, N., Hazarika, D., Gelbukh, A., Cambria, E., Poria, S.: Multimodal sentiment analysis using hierarchical fusion with context modeling. Knowl. Based Syst. **161**, 124–133 (2018)

24. McHugh, M.L.: Interrater reliability: the kappa statistic. Biochemia Medica **22**(3), 276–282 (2012)

25. Mi, F., Huang, M., Zhang, J., Faltings, B.: Meta-learning for low-resource natural language generation in task-oriented dialogue systems. In: Proceedings of the Twenty-Eighth International Joint Conference on Artificial Intelligence, IJCAI 2019, Macao, China, 10–16 August, pp. 3151–3157 (2019)

26. Mostafazadeh, N., et al.: Image-grounded conversations: multimodal context for natural question and response generation. In: Proceedings of the Eighth International Joint Conference on Natural Language Processing, IJCNLP 2017, Volume 1: Long Papers, Taipei, Taiwan, 27 November–1 December 2017, pp. 462–472 (2017)

27. Papineni, K., Roukos, S., Ward, T., Zhu, W.J.: BLEU: a method for automatic evaluation of machine translation. In: Proceedings of the 40th Annual Meeting of the Association for Computational Linguistics, Philadelphia, PA, USA, 6–12 July 2002, pp. 311–318. Association for Computational Linguistics (2002)

28. Paulus, R., Xiong, C., Socher, R.: A deep reinforced model for abstractive summarization. CoRR arXiv:1705.04304 (2017)

29. Pennington, J., Socher, R., Manning, C.: GloVe: global vectors for word representation. In: Proceedings of the 2014 Conference on Empirical Methods in Natural Language Processing, EMNLP 2014, A Meeting of SIGDAT, a Special Interest Group of the ACL, Doha, Qatar, 25–29 October 2014, pp. 1532–1543 (2014)
30. Pérez-Rosas, V., Mihalcea, R., Morency, L.P.: Utterance-level multimodal sentiment analysis. In: Proceedings of the 51st Annual Meeting of the Association for Computational Linguistics, ACL 2013, Volume 1: Long Papers, Sofia, Bulgaria, 4–9 August 2013, pp. 973–982 (2013)
31. Poria, S., Cambria, E., Hazarika, D., Majumder, N., Zadeh, A., Morency, L.P.: Context-dependent sentiment analysis in user-generated videos. In: Proceedings of the 55th Annual Meeting of the Association for Computational Linguistics, ACL 2017, Volume 1: Long Papers, Vancouver, Canada, 30 July–4 August, pp. 873–883 (2017)
32. Qian, K., Yu, Z.: Domain adaptive dialog generation via meta learning. In: Proceedings of the 57th Conference of the Association for Computational Linguistics, ACL 2019, Volume 1: Long Papers, Florence, Italy, 28 July–2 August 2019, pp. 2639–2649 (2019)
33. Reddi, S.J., Kale, S., Kumar, S.: On the convergence of Adam and beyond. In: 6th International Conference on Learning Representations, ICLR 2018, Conference Track Proceedings, Vancouver, BC, Canada, 30 April–3 May 2018 (2019)
34. Rennie, S.J., Marcheret, E., Mroueh, Y., Ross, J., Goel, V.: Self-critical sequence training for image captioning. In: 2017 IEEE Conference on Computer Vision and Pattern Recognition (CVPR), pp. 1179–1195 (2017)
35. Saha, A., Khapra, M.M., Sankaranarayanan, K.: Towards building large scale multimodal domain-aware conversation systems. In: Proceedings of the Thirty-Second AAAI Conference on Artificial Intelligence, (AAAI 2018), the 30th innovative Applications of Artificial Intelligence (IAAI 2018), and the 8th AAAI Symposium on Educational Advances in Artificial Intelligence (EAAI 2018), New Orleans, Louisiana, USA, 2–7 February, pp. 696–704 (2018)
36. Serban, I.V., Sordoni, A., Bengio, Y., Courville, A., Pineau, J.: Hierarchical neural network generative models for movie dialogues, vol. 7, no. 8. arXiv preprint arXiv:1507.04808 (2015)
37. Serban, I.V., Sordoni, A., Bengio, Y., Courville, A., Pineau, J.: Building end-to-end dialogue systems using generative hierarchical neural network models. In: Proceedings of the Thirtieth AAAI Conference on Artificial Intelligence, Phoenix, Arizona, USA, 12–17 February 2016, pp. 3776–3784 (2016)
38. Serban, I.V., et al.: A hierarchical latent variable encoder-decoder model for generating dialogues. In: Proceedings of the Thirty-First AAAI Conference on Artificial Intelligence, San Francisco, California, USA, 4–9 February 2017, pp. 3295–3301 (2017)
39. Shang, L., Lu, Z., Li, H.: Neural responding machine for short-text conversation. In: Proceedings of the 53rd Annual Meeting of the Association for Computational Linguistics and the 7th International Joint Conference on Natural Language Processing of the Asian Federation of Natural Language Processing, ACL 2015, Volume 1: Long Papers, Beijing, China, 26–31 July 2015, pp. 1577–1586 (2015)
40. Simonyan, K., Zisserman, A.: Very deep convolutional networks for large-scale image recognition. In: 3rd International Conference on Learning Representations, ICLR 2015, Conference Track Proceedings, San Diego, CA, USA, 7–9 May 2015 (2015)
41. Sordoni, A., et al.: A neural network approach to context-sensitive generation of conversational responses. In: NAACL (2015)

42. Srivastava, N., Hinton, G., Krizhevsky, A., Sutskever, I., Salakhutdinov, R.: Dropout: a simple way to prevent neural networks from overfitting. J. Mach. Learn. Res. **15**(1), 1929–1958 (2014)

43. Tian, Z., Bi, W., Li, X., Zhang, N.L.: Learning to abstract for memory-augmented conversational response generation. In: Proceedings of the 57th Conference of the Association for Computational Linguistics, ACL 2019, Florence, Italy, Volume 1: Long Papers, 28 July–2 August 2019, pp. 3816–3825 (2019)

44. Vaswani, A., et al.: Attention is all you need. In: Advances in Neural Information Processing Systems, pp. 5998–6008 (2017)

45. Wen, T.H., et al.: Conditional generation and snapshot learning in neural dialogue systems. In: Proceedings of the 2016 Conference on Empirical Methods in Natural Language Processing, EMNLP 2016, Austin, Texas, USA, 1–4 November 2016, pp. 2153–2162 (2016)

46. Wen, T.H., et al.: A network-based end-to-end trainable task-oriented dialogue system. In: Proceedings of the 15th Conference of the European Chapter of the Association for Computational Linguistics, EACL 2017, Volume 1: Long Papers, Valencia, Spain, 3–7 April 2017, pp. 438–449 (2017)

47. Williams, R.J.: Simple statistical gradient-following algorithms for connectionist reinforcement learning. Mach. Learn. **8**(3–4), 229–256 (1992). https://doi.org/10.1007/BF00992696

48. Wolf, T., et al.: HuggingFace's transformers: state-of-the-art natural language processing. ArXiv arXiv:1910.03771 (2019)

49. Wu, C.S.: Learning to memorize in neural task-oriented dialogue systems. arXiv preprint arXiv:1905.07687 (2019)

50. Wu, C.S., Socher, R., Xiong, C.: Global-to-local memory pointer networks for task-oriented dialogue. In: 7th International Conference on Learning Representations, ICLR 2019, New Orleans, LA, USA, 6–9 May (2019)

51. Zadeh, A., Zellers, R., Pincus, E., Morency, L.P.: Multimodal sentiment intensity analysis in videos: facial gestures and verbal messages. IEEE Intell. Syst. **31**(6), 82–88 (2016)

52. Zhang, Y., Li, Q., Song, D., Zhang, P., Wang, P.: Quantum-inspired interactive networks for conversational sentiment analysis. In: Proceedings of the Twenty-Eighth International Joint Conference on Artificial Intelligence, IJCAI 2019, Macao, China, 10–16 August 2019, pp. 5436–5442 (2019)

53. Zhou, H., Huang, M., Zhang, T., Zhu, X., Liu, B.: Emotional chatting machine: emotional conversation generation with internal and external memory. In: Proceedings of the Thirty-Second AAAI Conference on Artificial Intelligence (AAAI 2018), New Orleans, Louisiana, USA, 2–7 February 2018, pp. 730–739 (2018)

54. Zhu, C., Zeng, M., Huang, X.: Multi-task learning for natural language generation in task-oriented dialogue. In: Proceedings of the 2019 Conference on Empirical Methods in Natural Language Processing and the 9th International Joint Conference on Natural Language Processing, EMNLP-IJCNLP 2019, Hong Kong, China, 3–7 November, pp. 1261–1266 (2019)

An Analysis of Variations in the Effectiveness of Query Performance Prediction

Debasis Ganguly[1][ID], Suchana Datta[2(✉)][ID], Mandar Mitra[3][ID], and Derek Greene[2][ID]

[1] University of Glasgow, Glasgow, UK
debasis.ganguly@glasgow.ac.uk
[2] University College Dublin, Dublin, Ireland
suchana.datta@ucdconnect.ie, derek.greene@ucd.ie
[3] Indian Statistical Institute, Kolkata, India
mandar@isical.ac.in

Abstract. A query performance predictor estimates the retrieval effectiveness of a system for a given query. Query performance prediction (QPP) algorithms are themselves evaluated by measuring the correlation between the predicted effectiveness and the actual effectiveness of a system for a set of queries. This generally accepted framework for judging the usefulness of a QPP method includes a number of sources of variability. For example, "actual effectiveness" can be measured using different metrics, for different rank cut-offs. The objective of this study is to identify some of these sources, and investigate how variations in the framework can affect the outcomes of QPP experiments. We consider this issue not only in terms of the absolute values of the evaluation metrics being reported (e.g., Pearson's r, Kendall's τ), but also with respect to the changes in the ranks of different QPP systems when ordered by the QPP metric scores. Our experiments reveal that the observed QPP outcomes can vary considerably, both in terms of the absolute evaluation metric values and also in terms of the relative system ranks. We report the combinations of QPP evaluation metric and experimental settings that are likely to lead to smaller variations in the observed results.

Keywords: Query Performance Prediction · Variations in QPP Results · QPP Reproducibility

1 Introduction

The problem of *query performance prediction* (QPP) [5, 7–9, 12, 16, 19, 20, 28, 29] has attracted the attention of the Information Retrieval (IR) community over a number of years. QPP involves estimating the retrieval quality of an IR system. A diverse range of pre-retrieval (e.g. avgIDF [9]) and post-retrieval approaches (e.g. WIG [29], NQC [20], UEF [19]) have been proposed for the task of QPP.

The primary use case of QPP can be described as follows: "If we could determine in advance which retrieval approach would work well for a given query,

M. Hagen et al. (Eds.): ECIR 2022, LNCS 13185, pp. 215–229, 2022.
https://doi.org/10.1007/978-3-030-99736-6_15

then hopefully, selecting the appropriate retrieval method on a [per] query basis could improve the retrieval effectiveness significantly" [6]. In other words, the objective of QPP would be to predict how easy or difficult a given query is for an IR system. This prediction could either be a categorical label (e.g., EASY, MODERATE, HARD), or a numerical estimate of a standard IR evaluation metric (which generally lies in $[0, 1]$).

QPP is a challenging problem, however, and this eventual objective has remained elusive thus far. Given a query and an IR system, well-known QPP methods simply compute a real-valued score that is meant to be indicative of the effectiveness of the system for the given query. While this score is typically not interpreted as a statistical estimate of a specific evaluation metric (e.g. AP or nDCG [11]), it is expected to be highly correlated with a standard evaluation measure. Indeed, the quality of a QPP method is usually determined by measuring the correlation between its predicted effectiveness scores and the values of some standard evaluation metric for a set of queries.

Consider a proposed QPP algorithm $\mathcal{P}$. Given an IR system S, and a set of queries $\mathcal{Q} = \{Q_1, Q_2, \ldots, Q_n\}$, S is used to retrieve a ranked list L_i of documents for each $Q_i \in \mathcal{Q}$. For each L_i, $\mathcal{P}$ computes a predicted effectiveness score ϕ_i. Using available relevance assessments as ground-truth, a standard IR metric g_i is also computed for L_i. The correlation between the lists $\{\phi_1, \phi_2, \ldots, \phi_n\}$ and $\{g_1, g_2, \ldots, g_n\}$ is taken to be a measure of how effective $\mathcal{P}$ is as a query performance predictor.

In this study, we analyse the above approach for evaluating and comparing different QPP methods. We identify the sources of variability within this generally accepted framework, and show that these variations can lead to differences in the computed correlations. This, in turn, can lead to differences in

- the absolute values of reported QPP evaluation measures (e.g., the ρ value for NQC [20] measured with AP@100 as the target metric and LM-Dirichlet as the retrieval model can be substantially different from that measured with AP@1000 as the target metric and BM25 as the retrieval model on the *same* set of queries); and also in
- the comparative effectiveness of a number of different QPP measures (e.g., NQC turns out to be better than WIG with AP@100, whereas WIG outperforms NQC when QPP effectiveness is measured using nDCG@10).

Thus, these variations can lead to difficulties in reproducing QPP results, both at the level of the correlation values being reported, and also in terms of the relative performance of different competing methods on standard datasets.

Contributions. We conduct a range of experiments to analyze the potential variations in QPP effectiveness results under different experimental conditions. Specifically, we consider different combinations of IR metrics and IR models (as well as rank cut-off values). The experiments described in Sect. 5 reveal that the results of QPP depend significantly on these settings. Thus, it may be difficult to reproduce QPP experiments without a precise description of the experimental context. While variations in other factors, such as the choice of indexing implementation and set of pre-processing steps, may also matter, we recommend that any empirical study of QPP include a precise description of at least the above experimental settings in order to reduce variations in reported results. More importantly, our findings suggest that it may even be worthwhile to systematically revisit reported comparisons between competing QPP approaches.

2   Related Work

Analyzing the sensitivity of reported results on the experiment settings is important for an empirical discipline such as IR. Buckley and Voorhees while examining the stability of commonly used evaluation measures in IR [3], reported observations, such as P@30 has about twice the average error rate as compared to average precision (AP), or that a stable measurement of P@10 requires an aggregation of over 50 queries etc.

Previous studies have investigated the sensitivity of relative ranks of IR systems to the pooling depth used for relevance assessments. It is reported that smaller samples of the relevance ground-truth obtained with smaller pool depths usually do not lead to significant changes in the relative performance of IR systems [2,4,23,24]. In relation to pooling, Buckley et al. demonstrated that pools created during the TREC 2005 workshop exhibit a specific bias in favor of relevant documents, specifically contain the title words.

The study in [17] analyzed the sensitivity of variations in embeddding vectors used for IR models. The work in [1] stressed the importance of reproducibility in IR research by noting that most of the improvements reported over the years were not statistically significant over their predecessors. Recently, this observation has also been reinforced for neural models by arguing that most of the neural approaches have compared their results against relatively weak baselines [14,22].

Somewhat similar to our investigation of the stability of QPP results relative to IR models and evaluation metrics, an inconsistency in QPP evaluation with respect to IR models and variations in query formulation was shown in [18,21].

3   Anatomy of a QPP Evaluation Framework

In this section, we formally define the various components in a standard QPP evaluation framework. As we demonstrate later, variations in these components can potentially lead to different experimental outcomes.

Definition 1. *The context, $\mathcal{C}(Q)$, of a QPP experiment on a query Q, is a 3-tuple of the form of $(\theta, \mathcal{S}, \kappa)$, where κ is a positive integer; the function $\mathcal{S}$: $Q \times D \mapsto \mathbb{R}$ is a scoring function that computes query-document similarities, and is used to retrieve $L = (D_1, \ldots, D_\kappa)$, the list of κ top-ranked documents for Q from a collection; and $\theta : L \mapsto [0,1]$ is an evaluation metric function that, given a query Q, a list L of top-ranked documents, and $R(Q)$, the relevance assessments for Q, outputs a measure of usefulness of L.*

Definition 2. *The ground-truth or reference value of retrieval effectiveness of a query Q in relation to a QPP context, $\mathcal{C}(Q)$ of Definition 1, is a function of the form $g : \mathcal{C}(Q) \mapsto [0,1]$.*

Definition 3. *A QPP method is a function of the form $\phi(Q, D_1, \ldots, D_k) \mapsto [0, 1]$, which, given a query Q and a list of top-k retrieved documents[1], outputs a number that is indicative of how relevant the retrieved list is. In other words, the output of the predictor $\phi(Q)$ is some measure of the ground-truth retrieval effectiveness measure $g(\mathcal{C}(Q))$ from Definition 2.*

For example, NQC [20] or WIG [29] compute $\phi(Q)$ based on a set of k top-ranked documents[2] and estimating how distinct it is from the rest of the collection. The intuition behind NQC and WIG is that the higher the distinctiveness, the higher the likelihood of finding more relevant documents in the retrieved list.

The next step in QPP evaluation is to measure the correlation between the predicted retrieval effectiveness, $\phi(Q)$, and the ground-truth retrieval effectiveness, $g(\mathcal{C}(Q))$ over a set of benchmark queries $\mathcal{Q}$, using a correlation function, $\chi : (\Phi, \mathcal{G}(\mathcal{C})) \mapsto [0, 1]$, where $\Phi = \bigcup_{Q \in \mathcal{Q}} \phi(Q)$ and $\mathcal{G}(\mathcal{C}) = \bigcup_{Q \in \mathcal{Q}} g(\mathcal{C}(Q)))$. Common choices for χ are Pearson's r, which computes a correlation between the values themselves, and rank correlation measures, such as Spearman's ρ, which compute the correlation between the ordinals of the members of Φ and $\mathcal{G}(\mathcal{C})$.

It is clear from Definitions 1–3 that the QPP outcome, $\chi(\Phi, \mathcal{G})(\mathcal{C})$, depends on the context $\mathcal{C}(Q)$ used for each $Q \in \mathcal{Q}$. Our first objective is to quantify the relative changes in QPP outcomes χ with changes in the context $\mathcal{C}(Q)$. In other words, we wish to compute the relative changes of the form $|\chi(\Phi, \mathcal{G}(\mathcal{C}_i)) - \chi(\Phi, \mathcal{G}(\mathcal{C}_j))|$, for two different instances of QPP contexts $\mathcal{C}_i = (\theta_i, \mathcal{S}_i, \kappa_i)$ and $\mathcal{C}_j = (\theta_j, \mathcal{S}_j, \kappa_j)$. Thus, our first research question is the following:

> **RQ1:** Do **variations in the QPP context**, $\mathcal{C}$, in terms of the IR metric (θ), the IR model ($\mathcal{S}$) and the rank cut-off (κ) used to construct the QPP evaluation ground-truth, $g(\mathcal{C})$, lead to **significant differences in outcome of a QPP method** ϕ?

Next, instead of computing the relative change in the outcome values (correlations) of individual QPP methods, we seek to measure the relative change in the rankings (in terms of effectiveness) of a number of different QPP methods. Formally, given a set of m QPP functions $\{\phi_1, \ldots, \phi_m\}$, we compute the effectiveness of each with respect to a number of different QPP contexts, $\chi(\Phi_i, \mathcal{G}(\mathcal{C}_j))$ for $j = 1, \ldots, n$. The objective is to investigate whether or not the ranking of QPP systems computed with different contexts is relatively stable. For instance, if NQC is the best method for a context that used LM-Dirichlet as retrieval model and AP@100 as evaluation metric, we might wish to investigate whether it remains the best method for a different QPP context, say, BM25 as the retrieval model and nDCG@10 as the evaluation metric. Stated explicitly,

> **RQ2:** Do **variations in the QPP context**, $\mathcal{C}$, in terms of the IR metric (θ), the IR model ($\mathcal{S}$) and the rank cut-off (κ) used to construct the QPP evaluation ground-truth, $g(\mathcal{C})$, lead to **significant differences in the relative ranks of different QPP methods** $\phi_1, \ldots, \phi_m$? (Table 1).

---

[1] For pre-retrieval QPP approaches, $(D_1, \ldots, D_k) = \emptyset$.

[2] k is a parameter of a post-retrieval QPP method, and can be different from κ, the number of top documents used for QPP evaluation.

Table 1. Characteristics of the TREC-Robust dataset used in our QPP experiments. 'Avg.$|Q|$' and 'Avg.#Rel' denote the average number of terms in a query, and the average number of relevant documents for a query, respectively.

Collection	#Docs	Topic Set	#Queries	Avg.$|Q|$	Avg.#Rel
Disks 4,5 (w/o CR)	528,155	TREC-Robust	249	2.68	71.21

4 Experimental Setup

To investigate the research questions from the last section, we conduct QPP experiments[3] on a widely-used dataset, the TREC Robust dataset, which consists of 249 queries. To address RQ1 and RQ2, we first define the set of possible QPP contexts that we explore in our experiments.

IR Evaluation Metrics Investigated. As choices for the IR evaluation metric (i.e., the function θ), we consider 'AP', 'nDCG', 'P@10', and 'recall'. The evaluation functions explored represent a mixture of both precision- and recall-oriented metrics. While AP and nDCG address both the aspects of precision and recall (leaning towards favouring precision), P@10 is a solely precision-oriented metric. To investigate RQ1, we set the cut-off for AP, nDCG, and recall to 100, as is common in the literature on QPP [20,25,26].

IR Models Investigated. IR models represent the second component of a QPP context as per Definition 1. We explore three such models: a) language modeling with Jelinek-Mercer smoothing (LMJM) [10,27], b) language modeling with Dirichlet smoothing (LMDir) [27], and c) Okapi BM25 [15]. The values of the IR model parameters were chosen after a grid search to optimize the MAP values on the TREC-Robust queries. Unless otherwise specified, for LMJM, we used $\lambda = 0.6$, the value of k_1 and b in BM25 were set to 0.7 and 0.3, respectively, and the value of the smoothing parameter μ for LMDir was set to 1000.

QPP Methods Tested. To compare the relative perturbations in preferential ordering of the QPP systems in terms of the evaluated effectiveness, we employ a total of seven different QPP methods, as outlined below:

- **AvgIDF** [9] is a pre-retrieval QPP method that uses the average idfs of the constituent query terms as the predicted query performance estimate.
- **Clarity** [7] estimates a relevance model (RLM) [13] distribution of term weights from a set of top-ranked documents, and then computes its KL divergence with the collection model.
- **WIG** [29] uses the aggregated value of the information gain of each document in the top-retrieved set as a specificity estimate.
- **NQC** [20] or normalized query commitment estimates the specificity of a query as the standard deviation of the RSVs of the top-retrieved documents.
- **UEF** [19] assumes that information from some top-retrieved sets of documents are more reliable than others. A high perturbation of a ranked list

---

[3] Implementation available at: https://github.com/suchanadatta/qpp-eval.git.

after feedback indicates a poor retrieval effectiveness of the initial list. This, in turn, suggests that a smaller confidence should be associated with the QPP estimate of such a query. Formally,

$$\text{UEF}(Q, \phi) = \xi(R_M(Q), R_M(\theta_Q))\phi(Q) \tag{1}$$

where $\phi(Q)$ is the predicted score of a base QPP estimator (e.g. WIG or NQC), $R_M(\theta_Q)$ denotes the re-ranked set of documents post-RLM feedback, the RLM being estimated on $R_M(Q)$ - the top-M documents, and ξ is a rank correlation coefficient of two ordered sets, for which we specifically use Pearson's-ρ, as suggested in [19]. We experiment with three specific instances of the base estimators, namely Clarity, WIG and NQC for UEF, which we denote as UEF(Clarity), UEF(WIG) and UEF(NQC), respectively.

Parameters and Settings. The standard practice in QPP research is to optimize the common hyper-parameter - the number of top documents of post-retrieval QPP approaches (denoted as k in Definition 3). This hyper-parameter is tuned via a grid search on a development set of queries and the optimal setting is used to report the performance on a test set. A common approach is to employ a 50:50 split of the set of queries into development and test sets. This process is usually repeated 30 times and the average results over the test folds are reported [20,26,29].

The focus of our research is different, however, in the sense that we seek to analyze the variations caused due to different settings for constructing the QPP ground-truth, instead of demonstrating that a particular QPP method outperforms others. Moreover, an optimal tuning of the hyper-parameters for each QPP method would require averaging over 30 different experiments for a single way of defining the QPP context for constructing the ground-truth. Hence, to keep the number of experiments tractable, we set $k = 20$, as frequently prescribed in the literature [7,20,26,29]. Another hyper-parameter, specific to UEF, is the number of times a subset of size k is sampled from a set of top-K ($K > k$) documents. We use a total of 10 random samples of $k = 20$ documents from the set of $K = 100$ top documents, as prescribed in [19].

5   Results

5.1   RQ1: Variations in QPP Evaluations

Table 2 reports the standard deviations in the observed values for the QPP experiments[4]. In Tables 2a–d, the value of $\sigma(\theta)$ in each row indicates the standard deviation of the QPP outcome values observed in that row, i.e., these values indicate the standard deviation resulting from the use of different IR metrics for QPP evaluation. Similarly, the value of $\sigma(\mathcal{S})$ in each column is the standard deviation of the r, ρ or τ values reported in that column, i.e., this value denotes

---

[4] Tables 2, 3, 4, 5 and 6 are best viewed in color.

Table 2. Sensitivity of QPP results with variations in the IR evaluation metric (θ) and the IR model (S) for the QPP methods a) AvgIDF, b) NQC, c) WIG and d) UEF(WIG). The metrics - AP, nDCG and recall (R) are measured on the top-100 retrieved documents using retrieval models LMJM($\lambda = 0.6$), BM25($k_1 = 0.7, b = 0.3$) and LMDir($\mu = 1000$) respectively. The lowest (highest) standard deviations for each group of QPP correlation measure are shown in green (red). The lowest and the highest across different correlation measures are shown bold-faced.

(a) AvgIDF

Model(S)	AP	nDCG	R	P@10	$\sigma(\theta)$
r LMJM	0.3795	0.3966	0.3869	0.3311	**0.0291**
BM25	0.5006	0.4879	0.4813	0.2525	**0.1190**
LMDir	0.5208	0.5062	0.4989	0.2851	0.1121
$\sigma(S)$	0.0764	0.0587	0.0602	0.0395	
ρ LMJM	0.4553	0.4697	0.4663	0.3067	0.0788
BM25	0.4526	0.4700	0.4736	0.2842	0.0911
LMDir	0.4695	0.4848	0.4893	0.3017	0.0902
$\sigma(S)$	0.0091	0.0086	0.0118	0.0114	
τ LMJM	0.3175	0.3285	0.3278	0.2193	0.0529
BM25	0.3144	0.3162	0.3319	0.2040	0.0589
LMDir	0.3307	0.3407	0.3440	0.2155	0.0617
$\sigma(S)$	0.0087	0.0123	**0.0084**	0.0120	

(b) NQC

Model(S)	AP	nDCG	R	P@10	$\sigma(\theta)$
r LMJM	0.3652	0.4169	0.4503	0.2548	0.0855
BM25	0.3563	0.4118	0.4495	0.2707	0.0777
LMDir	0.4354	0.4583	0.4854	0.2842	0.0901
$\sigma(S)$	0.0433	0.0255	0.0205	0.0147	
ρ LMJM	0.4545	0.4843	0.5248	0.2918	**0.1022**
BM25	0.4618	0.4887	0.5137	0.3308	0.0814
LMDir	0.5024	0.5260	0.5453	0.3340	0.0969
$\sigma(S)$	0.0258	0.0229	0.0160	0.0235	
τ LMJM	0.3100	0.3319	0.3657	0.2061	0.0688
BM25	0.3170	0.3370	0.3551	0.2374	**0.0519**
LMDir	0.3539	0.3713	0.3828	0.2379	0.0668
$\sigma(S)$	0.0236	0.0214	**0.0140**	0.0182	

(c) WIG

Model(S)	AP	nDCG	R	P@10	$\sigma(\theta)$
r LMJM	0.4056	0.4071	0.3971	0.3054	0.0491
BM25	0.4488	0.4563	0.4386	0.3485	0.0502
LMDir	0.4908	0.4798	0.4632	0.3423	**0.0688**
$\sigma(S)$	0.0426	0.0371	0.0334	0.0233	
ρ LMJM	0.3716	0.3794	0.3790	0.3120	0.0325
BM25	0.4520	0.4601	0.4505	0.3586	0.0480
LMDir	0.4582	0.4688	0.4667	0.3528	0.0561
$\sigma(S)$	0.0483	**0.0493**	0.0467	0.0254	
τ LMJM	0.2514	0.2567	0.2607	0.2209	**0.0181**
BM25	0.3116	0.3181	0.3125	0.2549	0.0297
LMDir	0.3194	0.3267	0.3259	0.2493	0.0375
$\sigma(S)$	0.0372	0.0382	0.0344	**0.0182**	

(d) UEF(WIG)

Model(S)	AP	nDCG	R	P@10	$\sigma(\theta)$
r LMJM	0.4746	0.4763	0.4646	0.3573	0.0575
BM25	0.5386	0.5476	0.5263	0.4182	0.0603
LMDir	0.5693	0.5566	0.5373	0.3971	**0.0797**
$\sigma(S)$	0.0483	0.0440	0.0392	0.0309	
ρ LMJM	0.4385	0.4477	0.4472	0.3682	0.0384
BM25	0.5334	0.5429	0.5316	0.4231	0.0567
LMDir	0.5407	0.5532	0.5507	0.4163	0.0662
$\sigma(S)$	0.0570	**0.0582**	0.0551	0.0300	
τ LMJM	0.3017	0.3080	0.3128	0.2651	**0.0217**
BM25	0.3677	0.3754	0.3688	0.3008	0.0351
LMDir	0.3833	0.3920	0.3911	0.2992	0.0450
$\sigma(S)$	0.0433	0.0445	0.0303	**0.0202**	

the standard deviations in QPP correlations across different IR models. The lowest standard deviations for each QPP correlation type are shown bold-faced. We now discuss the observations that can be made from Table 2.

Table 3. Each cell in the table indicates the correlation (Kendall's τ) between QPP systems ranked in order by their evaluated effectiveness (measured with the help of Pearson's r for the results of this table) for two different IR metrics corresponding to the row and the column name of the cell. A total of 7 QPP systems were used in these experiments, namely AvgIDF, Clarity, WIG, NQC, UEF(Clarity), UEF(WIG) and UEF(NQC). The lowest correlation value for each group is marked in red, and the lowest correlations, overall, are bold-faced.

Model	Metric	AP@100	AP@1000	R@10	R@100	R@1000	nDCG@10	nDCG@100	nDCG@1000
LMJM		0.4286	0.3333	0.9048	0.2381	**−0.1429**	1.0000	0.2381	0.3333
BM25	AP@10	1.0000	0.9048	1.0000	0.9048	0.4286	1.0000	1.0000	0.7143
LMDir		1.0000	0.9048	1.0000	0.9048	0.4286	1.0000	1.0000	0.7143
LMJM			0.9048	0.5238	0.8095	0.4286	0.4286	0.8095	0.9048
BM25	AP@100		0.9048	1.0000	0.9048	0.4286	1.0000	1.0000	0.7143
LMDir			0.9048	1.0000	0.9048	0.4286	1.0000	1.0000	0.7143
LMJM				0.4286	0.8095	0.5238	0.3333	0.9048	1.0000
BM25	AP@1000			0.9048	0.8095	0.3333	0.9048	0.9048	0.8095
LMDir				0.9048	0.8095	0.5238	0.9048	0.9048	0.8095
LMJM					0.3333	−0.0476	0.9048	0.3333	0.4286
BM25	R@10				0.9048	0.4286	1.0000	1.0000	0.7143
LMDir					0.9048	0.4286	1.0000	1.0000	0.7143
LMJM						0.6190	0.2381	1.0000	0.9048
BM25	R@100					0.5238	0.9048	0.9048	0.6190
LMDir						0.5238	0.9048	0.9048	0.6190
LMJM							**−0.1429**	0.6190	0.5238
BM25	R@1000						0.4286	0.4286	0.5238
LMDir							0.4286	0.4286	0.5238
LMJM								0.2381	0.3333
BM25	nDCG@10							1.0000	0.7143
LMDir								1.0000	0.7143
LMJM									0.9048
BM25	nDCG@100								0.7143
LMDir									0.7143

Variations Due to IR Evaluation Metric. The first set of observations, listed below, is in relation to the absolute differences between two different QPP evaluations involving two different QPP contexts.

- **Substantial absolute differences in the QPP outcomes:** Variations in the IR evaluation metric (i.e., the θ component of a QPP context $\mathcal{C}(Q)$ of Definition 1) while keeping the other two components fixed (i.e., retrieval model and cut-off) yields considerable absolute differences in the values. As an example, compare the QPP evaluation of 0.5006 with AP@100 in Table 2a to that of 0.2525 with P@10 obtained with BM25, showing that these absolute differences can be high.
- **Lower variations with τ:** In general, we observe that each QPP method (e.g. NQC, WIG etc.) exhibits considerable differences in measured outcomes specially between AP@100 and P@10. Moreover, the variations, in general,

Table 4. Results of relative changes in the ranks of QPP systems (similar to Table 3), the difference being that the QPP outcomes were measured with τ (instead of r).

Model	Metric	AP@100	AP@1000	R@10	R@100	R@1000	nDCG@10	nDCG@100	nDCG@1000
LMJM		0.5238	0.3333	0.8095	0.4286	0.2381	0.8095	0.4286	0.3333
BM25	AP@10	0.9048	0.7143	0.8095	0.8095	0.5238	1.0000	0.9048	0.5238
LMDir		0.9048	0.8095	1.0000	1.0000	0.8095	1.0000	0.9048	0.7143
LMJM			0.8095	0.5238	0.9048	0.7143	0.3333	0.9048	0.8095
BM25	AP@100		0.8095	0.9048	0.9048	0.6190	0.9048	1.0000	0.6190
LMDir			0.9048	0.9048	0.9048	0.7143	0.9048	1.0000	0.8095
LMJM				0.3333	0.9048	0.7143	0.1429	0.9048	1.0000
BM25	AP@1000			0.7143	0.7143	0.6190	0.7143	0.8095	0.8095
LMDir				0.8095	0.8095	0.8095	0.8095	0.9048	0.9048
LMJM					0.4286	0.2381	0.8095	0.4286	0.3333
BM25	R@10				1.0000	0.7143	0.8095	1.0000	0.5238
LMDir					1.0000	0.8095	1.0000	0.9048	0.7143
LMJM						0.8095	0.2381	1.0000	0.9048
BM25	R@100					0.7143	0.8095	0.9048	0.5238
LMDir						0.8095	1.0000	0.9048	0.7143
LMJM							0.0476	0.8095	0.7143
BM25	R@1000						0.5238	0.6190	0.8095
LMDir							0.8095	0.7143	0.9048
LMJM								0.2381	0.1429
BM25	nDCG@10							0.9048	0.5238
LMDir								0.9048	0.7143
LMJM									0.9048
BM25	nDCG@100								0.6190
LMDir									0.8095

are lower when correlation is measured with the help of Kendall's τ (e.g., compare $\sigma(\theta) = 0.0181$ measured with τ vs. $\sigma(\theta) = 0.0491$ measured with r on documents retrieved with LMJM). The fact that τ exhibits a lower variance in QPP evaluation is likely because the correlation is measured in a pairwise manner (τ being a function of the number of concordant and discordant pairs). As a result, τ depends only on the agreements between the true and the predicted order (of query difficulty) between a query pair, and not on the absolute values of the predicted scores or the reference values of the IR evaluation metric (as in Pearson's r or Spearman's ρ).

– **Lower variances with LMJM:** Similar to our earlier observation that τ should be the preferred QPP evaluation measure (with an objective to minimize the variances in observed results due to changes in IR evaluation metric), we observe from Table 2 that LMJM, in most cases, result in low variances in QPP experiment outcomes.

Variations Due to IR Models. The second set of observations from Table 2 relates to variations in the observed QPP results with respect to variations in IR models. The standard deviations of these values correspond to column-wise

Table 5. Each cell in the table indicates the correlation (Kendall's τ) between QPP systems ranked in order by their evaluated effectiveness (measured with the help of Pearson's r for the results presented in this table) for each pair of IR models for 7 different QPP systems. The lowest correlation value for each group is marked in red. The lowest correlation in the table is bold-faced.

Metric	Model	LMJM (0.6)	BM25 (0.7,0.3)	BM25 (1.0,1.0)	BM25 (0.3,0.7)	LMDir (100)	LMDir (500)	LMDir (1000)
AP@100		1.0000	0.9048	1.0000	0.9048	0.9048	0.9048	0.9048
nDCG@100	LMJM	1.0000	0.8095	0.9048	0.9048	0.9048	0.8095	0.8095
R@100	(0.3)	0.9048	0.8095	0.9048	1.0000	1.0000	0.9048	0.9048
P@10		1.0000	0.8095	1.0000	0.8095	**0.7143**	**0.7143**	1.0000
AP@100			0.9048	1.0000	0.9048	0.9048	0.9048	0.9048
nDCG@100	LMJM		0.8095	0.9048	0.9048	0.9048	0.8095	0.8095
R@100	(0.6)		0.9048	1.0000	0.9048	0.9048	1.0000	1.0000
P@10			0.8095	1.0000	0.8095	**0.7143**	**0.7143**	1.0000
AP@100				0.9048	0.9048	1.0000	1.0000	1.0000
nDCG@100	BM25			0.9048	0.9048	0.9048	1.0000	1.0000
R@100	(0.7,0.3)			0.9048	0.8095	0.8095	0.9048	0.9048
P@10				0.8095	1.0000	0.9048	0.9048	0.8095
AP@100					0.9048	0.9048	0.9048	0.9048
nDCG@100	BM25				1.0000	1.0000	0.9048	0.9048
R@100	(1.0,1.0)				0.9048	0.9048	1.0000	1.0000
P@10					0.8095	**0.7143**	**0.7143**	1.0000
AP@100						1.0000	1.0000	1.0000
nDCG@100	BM25					1.0000	0.9048	0.9048
R@100	(0.3,0.7)					1.0000	0.9048	0.9048
P@10						0.9048	0.9048	0.8095
AP@100							1.0000	1.0000
nDCG@100	LMDir						0.9048	0.9048
R@100	(100)						0.9048	0.9048
P@10							0.8095	**0.7143**
AP@100								1.0000
nDCG@100	LMDir							1.0000
R@100	(500)							1.0000
P@10								**0.7143**

calculation of standard deviations and are shown as the $\sigma(\mathcal{S})$ values. Again, similar to the $\sigma(\theta)$ values, the lowest (highest) values along each row of $\sigma(\mathcal{S})$ are colored in green (red) to reflect the situation of lower the better. The best values across different QPP correlations are bold-faced. We summarise our observations:

– **Lower variations with τ:** Similar to the $\sigma(\theta)$ values it is again observed that mostly measuring QPP outcomes with τ results in the lowest variances in QPP results. Consequently, for better reproducibility it is more useful to report results with Kendall's τ.

Table 6. The difference of this table with Table 5 is that the QPP effectiveness is measured with Kendall's τ (instead of Pearson's r as in Table 5).

Metric	Model	LMJM (0.6)	BM25 (0.7, 0.3)	BM25 (1.0, 1.0)	BM25 (0.3, 0.7)	LMDir (100)	LMDir (500)	LMDir (1000)
AP@100		1.0000	1.0000	1.0000	1.0000	1.0000	1.0000	1.0000
nDCG@100	LMJM	1.0000	1.0000	1.0000	1.0000	1.0000	1.0000	1.0000
R@100	(0.3)	1.0000	1.0000	1.0000	1.0000	1.0000	1.0000	1.0000
P@10		0.9048	1.0000	0.9048	0.8095	0.9095	1.0000	1.0000
AP@100			1.0000	1.0000	1.0000	1.0000	1.0000	1.0000
nDCG@100	LMJM		1.0000	1.0000	1.0000	1.0000	1.0000	1.0000
R@100	(0.6)		1.0000	1.0000	1.0000	1.0000	1.0000	1.0000
P@10			0.9048	1.0000	0.7143	0.7143	0.9048	0.9048
AP@100				1.0000	1.0000	1.0000	1.0000	1.0000
nDCG@100	BM25			1.0000	1.0000	1.0000	1.0000	1.0000
R@100	(0.7, 0.3)			1.0000	1.0000	1.0000	1.0000	1.0000
P@10				0.9048	0.8095	0.8095	1.0000	1.0000
AP@100					1.0000	1.0000	1.0000	1.0000
nDCG@100	BM25				1.0000	1.0000	1.0000	1.0000
R@100	(1.0, 1.0)				1.0000	1.0000	1.0000	1.0000
P@10					0.7143	0.7143	0.9048	0.9048
AP@100						1.0000	1.0000	1.0000
nDCG@100	BM25					1.0000	1.0000	1.0000
R@100	(0.3, 0.7)					1.0000	1.0000	1.0000
P@10						**0.6190**	0.8095	0.8095
AP@100							1.0000	1.0000
nDCG@100	LMDir						1.0000	1.0000
R@100	(100)						1.0000	1.0000
P@10							0.8095	0.8095
AP@100								1.0000
nDCG@100	LMDir							1.0000
R@100	(500)							1.0000
P@10								1.0000

- **Lower variations in the QPP outcomes:** Compared to variations across IR evaluation metrics, we observe that the variations occurring across IR models is lower (compare the bold-faced green $\sigma(\mathcal{S})$ values with those of $\sigma(\theta)$ ones). This entails that experiments need to put more emphasis on a precise description of the IR metrics used for QPP evaluation.
- **Lack of a consistency on which combination of QPP method with IR evaluation context yields least the variance:** While WIG and UEF(WIG) exhibit lowest variances for a precision oriented evaluation of ground-truth retrieval effectiveness, for AvgIDF and NQC methods, the least variations are noted for recall.

5.2   RQ2: Variations in the Relative Ranks of QPP Methods

We now report results in relation to the second research question, where the intention is to measure how stable are QPP system ranks (ordered by their evaluated effectiveness measures) for variations in the QPP context.

Variation Due to IR Metrics. Tables 3 and 4 present the pairwise contingency table for different combinations of IR metrics for three different IR models. The following observations can be made from the results.

- **LMJM leads to the most instability in the relative QPP system ranks**: This behaviour, most likely, can be attributed to the fact that this model has a tendency to favour shorter documents in the top-retrieved in contrast to LMDir or BM25.
- **Some evaluation metrics are more sensitive to rank cut-off values**: For instance, the QPP ground-truth measured with Recall@10 yields considerably different results when the ground-truth corresponds to Recall@1000.
- **Relative ranks of QPP systems more stable with τ**: A comparison between the values of Tables 3 and 4 reveals that a rank correlation measure such as τ leads to better stability of QPP experiments than when r is used to measure the relative effectiveness of QPP models.

Variations Due to IR Models. Tables 5 and 6 present the pairwise contingency between retrieval similarity scores from different evaluation metrics. For this set of experiments, the intention is also to investigate the stability of QPP system ranks with respect to changes, not only to the retrieval model itself, but also for different parameter settings on the same model, e.g. BM25(0.7,0.3)[5] vs. BM25(1,1). We observe the following:

- **Relative ranks of QPP systems are quite stable across IR models**: The correlation values of Tables 5 and 6 are higher than those of Tables 3 and 4, which shows that the QPP experiments are less sensitive to variations in the set of top documents retrieved by different similarity scores.
- **LMJM leads to more instability in the QPP outcomes**: LMJM shows the lowest correlation with other retrieval models. Parameter variations of an IR model usually lead to relatively stable QPP outcomes. For instance, see the correlations between LMDir(500) and LMDir(1000).
- **Relative ranks of QPP systems are more stable with τ**: This observation (a comparison between the values of Tables 5 and 6) is similar to the comparison between Tables 3 and 4. However, the differences between the correlation values are smaller in comparison to those observed between Tables 3 and 4.

---

[5] Values of k_1 and b, respectively, in BM25 [15].

6 Concluding Remarks

We have shown via extensive experiments that QPP outcomes are indeed sensitive to the experimental configuration used. As part of our analysis, we have found that certain factors, such as variations in the IR effectiveness measures, has a greater impact in terms of QPP outcomes than other factors, such as variations in the choice of IR models. An important outcome arising from this study is that further research on QPP should place greater emphasis on a clear specification of the experimental setup to enable better reproducibility. In future we plan to expand our evaluations beyond the TREC Robust dataset. A natural question that we would like to explore concerns the impact of varying Q (the set of benchmark queries) on relative QPP outcomes.

Acknowledgement. The second and the fourth authors were supported by the Science Foundation Ireland (SFI) grant number SFI/12/RC/2289_P2.

References

1. Armstrong, T.G., Moffat, A., Webber, W., Zobel, J.: Improvements that don't add up: ad-hoc retrieval results since 1998. In: Proceedings of the 18th ACM Conference on Information and Knowledge Management, CIKM 2009, pp. 601–610 (2009)
2. Aslam, J.A., Yilmaz, E.: Inferring document relevance from incomplete information. In: Proceedings of the Sixteenth ACM Conference on Information and Knowledge Management, CIKM, pp. 633–642. ACM (2007)
3. Buckley, C., Voorhees, E.M.: Evaluating evaluation measure stability. In: Proceedings of the 23rd Annual International ACM SIGIR Conference on Research and Development in Information Retrieval, SIGIR 2000, pp. 33–40. Association for Computing Machinery (2000)
4. Buckley, C., Voorhees, E.M.: Retrieval evaluation with incomplete information. In: Sanderson, M., Järvelin, K., Allan, J., Bruza, P. (eds.) SIGIR 2004: Proceedings of the 27th Annual International ACM SIGIR Conference on Research and Development in Information Retrieval, Sheffield, UK, 25–29 July 2004, pp. 25–32. ACM (2004)
5. Carmel, D., Yom-Tov, E.: Estimating the query difficulty for information retrieval. In: Proceedings of the 33rd International ACM SIGIR Conference on Research and Development in Information Retrieval. SIGIR 2010, p. 911. ACM, New York (2010)
6. Carmel, D., Yom-Tov, E.: Estimating the query difficulty for information retrieval. Synthesis Lect. Inf. Concepts Retrieval Serv. **2**(1), 1–89 (2010)
7. Cronen-Townsend, S., Zhou, Y., Croft, W.B.: Predicting query performance. In: Proceedings of the 25th Annual International ACM SIGIR Conference on Research and Development in Information Retrieval, SIGIR 2002, pp. 299–306. ACM, New York (2002)
8. Cronen-Townsend, S., Zhou, Y., Croft, W.B.: Precision prediction based on ranked list coherence. Inf. Retr. **9**(6), 723–755 (2006)
9. Hauff, C., Hiemstra, D., de Jong, F.: A survey of pre-retrieval query performance predictors. In: Proceedings of the 17th ACM Conference on Information and Knowledge Management, CIKM 2008, pp. 1419–1420. ACM (2008)

10. Hiemstra, D.: Using language models for information retrieval. University of Twente (2001)
11. Järvelin, K., Kekäläinen, J.: Cumulated gain-based evaluation of IR techniques. ACM Trans. Inf. Syst. **20**(4), 422–446 (2002)
12. Kurland, O., Shtok, A., Carmel, D., Hummel, S.: A unified framework for post-retrieval query-performance prediction. In: Proceedings of the Third International Conference on Advances in Information Retrieval Theory, ICTIR 2011, pp. 15–26 (2011)
13. Lavrenko, V., Croft, W.B.: Relevance based language models. In: Proceedings of SIGIR 2001, pp. 120–127. ACM, New York (2001)
14. Lin, J.: The neural hype, justified! A recantation. In: SIGIR Forum, vol. 53, no. 2, pp. 88–93 (2021)
15. Robertson, S., Walker, S., Beaulieu, M., Gatford, M., Payne, A.: Okapi at TREC-4 (1996)
16. Roitman, H.: An enhanced approach to query performance prediction using reference lists. In: Proceedings of the 37th International ACM SIGIR Conference on Research and Development in Information Retrieval, Proceedings of SIGIR 2017, pp. 869–872. ACM, New York (2017)
17. Roy, D., Ganguly, D., Bhatia, S., Bedathur, S., Mitra, M.: Using word embeddings for information retrieval: how collection and term normalization choices affect performance. In: Proceedings of the 27th ACM International Conference on Information and Knowledge Management, CIKM 2018, pp. 1835–1838 (2018)
18. Scholer, F., Garcia, S.: A case for improved evaluation of query difficulty prediction. In: Proceedings of the 32nd International ACM SIGIR Conference on Research and Development in Information Retrieval, SIGIR 2009, pp. 640–641. Association for Computing Machinery, New York (2009)
19. Shtok, A., Kurland, O., Carmel, D.: Using statistical decision theory and relevance models for query-performance prediction. In: Proceedings of the 33rd International ACM SIGIR Conference on Research and Development in Information Retrieval, SIGIR 2010, pp. 259–266 (2010)
20. Shtok, A., Kurland, O., Carmel, D., Raiber, F., Markovits, G.: Predicting query performance by query-drift estimation. ACM Trans. Inf. Syst. **30**(2), 1–35 (2012)
21. Thomas, P., Scholer, F., Bailey, P., Moffat, A.: Tasks, queries, and rankers in pre-retrieval performance prediction. In: Proceedings of the 22nd Australasian Document Computing Symposium, ADCS 2017. Association for Computing Machinery, New York (2017)
22. Yang, W., Lu, K., Yang, P., Lin, J.: Critically examining the "neural hype": weak baselines and the additivity of effectiveness gains from neural ranking models. In: Proceedings of the 42nd International ACM SIGIR Conference on Research and Development in Information Retrieval, SIGIR 2019, pp. 1129–1132. Association for Computing Machinery (2019)
23. Yilmaz, E., Aslam, J.A.: Estimating average precision with incomplete and imperfect judgments. In: Proceedings of the 15th ACM International Conference on Information and Knowledge Management, CIKM 2006, pp. 102–111. Association for Computing Machinery (2006)
24. Yilmaz, E., Kanoulas, E., Aslam, J.A.: A simple and efficient sampling method for estimating AP and NDCG. In: Proceedings of the 31st Annual International ACM SIGIR Conference on Research and Development in Information Retrieval, SIGIR, pp. 603–610. ACM (2008)

25. Zamani, H., Croft, W.B., Culpepper, J.S.: Neural query performance prediction using weak supervision from multiple signals. In: The 41st International ACM SIGIR Conference on Research and Development in Information Retrieval, SIGIR 2018, pp. 105–114. ACM (2018)
26. Zendel, O., Shtok, A., Raiber, F., Kurland, O., Culpepper, J.S.: Information needs, queries, and query performance prediction. In: Proceedings of SIGIR 2019, pp. 395–404 (2019)
27. Zhai, C., Lafferty, J.: A study of smoothing methods for language models applied to ad hoc information retrieval. In: Proceedings of the 24th Annual International ACM SIGIR Conference on Research and Development in Information Retrieval, SIGIR 2001, pp. 334–342 (2001)
28. Zhou, Y., Croft, W.B.: Ranking robustness: a novel framework to predict query performance. In: Proceedings of the 15th ACM International Conference on Information and Knowledge Management, CIKM 2006, pp. 567–574. ACM, New York (2006)
29. Zhou, Y., Croft, W.B.: Query performance prediction in web search environments. In: Proceedings of 30th Annual International ACM SIGIR Conference on Research and Development in Information Retrieval, SIGIR 2007, pp. 543–550 (2007)

Search Clarification Selection via Query-Intent-Clarification Graph Attention

Chang Gao[✉] and Wai Lam[✉]

The Chinese University of Hong Kong, Shatin, Hong Kong
{gaochang,wlam}@se.cuhk.edu.hk

Abstract. Proactively asking clarifications in response to search queries is a useful technique for revealing the intent of the query. Search clarification is important for both web and conversational search. This paper focuses on the clarification selection task. Inspired by the fact that a good clarification should clarify the query's different intents, we propose a graph attention-based clarification selection model that can exploit the relations among a given query, its intents, and its clarifications via constructing a query-intent-clarification attention graph. The comparison with competitive baselines on large-scale search clarification data demonstrates the effectiveness of our model.

Keywords: Search Clarification · Clarification Selection · Conversational Search

1 Introduction

Search queries are often short, and users' information needs are complex. This makes it challenging for search engines to predict potential user intents and give satisfactory retrieval results. As a consequence, users may need to browse multiple result pages or reformulate their queries. Alternatively, search engines can proactively ask clarifications to the user instead of just giving "ten blue links" [2,4]. Figure 1 shows an example of clarifications in the Bing search engine. Each clarification consists of a clarifying question and a set of candidate answers in response to the query. Users can click on one of the answers to indicate their intents. Zamani et al. [25] show that users enjoy seeing clarifications due to their functional and emotional benefits. Aliannejadi et al. [2] show that asking only one good question can improve the retrieval performance significantly. Moreover, search clarification has been recognized as a critical component of conversational search systems [3,7,15].

Although there is significant progress in exploring the search clarification [2,8,13,19,25,27,28], selecting clarifications is underexplored. In this paper, we

The work described in this paper is substantially supported by a grant from the Research Grant Council of the Hong Kong Special Administrative Region, China (Project Code: 14200620).

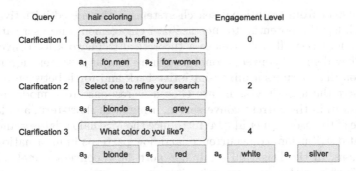

Fig. 1. A query and its three clarifications. Each clarification is associated with an engagement level which is an integer between 0 and 10 based on click-through rates. Higher click-through rates correspond to higher engagement levels.

focus on the clarification selection task, which is a fundamental task in search clarification because: (1) search engines can generate multiple clarifications and then select one of them [2], and (2) search engines can generate new clarifications based on existing clarifications through some operations such as adding or deleting answers, and judge whether the new clarifications are better. The aim of clarification selection is to select the clarification with the highest engagement level for each query.

A good clarification should clarify different intents of the query [27]. There are two challenges for clarification selection: (1) how to estimate the query's intents; (2) how to utilize the query's intents. To address the first challenge, we observe that the candidate answers in the query's clarifications can capture some reasonable intents. For example, as shown in Fig. 1, the answers a_1 to a_7 can reflect the intents of "hair coloring" to a certain extent. To overcome the second challenge, we propose a **G**raph **A**ttention-based **C**larification **S**election (GACS) model, which constructs a query-intent-clarification (QIC) attention graph to exploit the relations among a given query, its intents, and its clarifications. Afterwards, it transforms the graph into a sequence, inputs it to the Transformer [22], and outputs a score to measure how the clarification reflects the query's intents.

We design several different graphs and evaluate the model on two search clarification datasets. Experimental results show that by properly designing the graph structure, GACS can outperform competitive baselines, and the intent can improve the model's performance, especially in scenarios where there are negative clarifications.

2 Related Work

2.1 Conversational Search

The conversational search paradigm aims to satisfy information needs within a conversational format [3]. A key property of conversational search systems that is different from traditional search systems is the mixed-initiative interaction, which has the potential to increase user engagement and user satisfaction. Radlinski and Craswell [15] propose a theoretical framework for conversational search. They define a conversational search system as a system for retrieving information that permits a mixed-initiative back and forth between a user and agent, where the agent's actions are chosen in response to a model of current user needs within the current conversation, using both short-term and long-term knowledge of the user. Qu et al. [14] introduce the MSDialog dataset and analyze user intent distribution, co-occurrence, and flow patterns in information-seeking conversations. In the Dagstuhl seminar report [3], a conversational search system is defined as either an interactive information retrieval system with speech and language processing capabilities, a retrieval-based chatbot with user task modeling, or an information-seeking dialogue system with information retrieval capabilities. Rosset et al. [17] study conversational question suggestion, which aims to proactively engage the user by suggesting interesting, informative, and useful follow-up questions. Ren et al. [16] develop a pipeline for conversation search consisting of six sub-tasks: intent detection, keyphrase extraction, action prediction, query selection, passage selection, and response generation.

In this paper, we focus on search clarification selection, which differs from most existing work in conversational search. Therefore, we review the related work on asking clarifications in the next section.

2.2   Asking Clarifications

Asking clarifications is important in conversational systems since they can only return a limited number of results [2]. Kiesel et al. [9] focus on ambiguous voice queries and conduct a user study for a better understanding of voice query clarifications and their impact on user satisfaction. Aliannejadi et al. [2] propose a workflow for asking clarifying questions in an open-domain conversational search system. Moreover, they build a dataset called Qulac based on the TREC Web Track 2009-2012 collections and develop an offline evaluation protocol. Hashemi et al. [8] propose a Guided Transformer model that can use multiple information sources for document retrieval and next clarifying question selection. Krasakis et al. [10] investigate how different aspects of clarifying questions and user answers affect the quality of document ranking. Previous work ignores the possibility that conversational search systems may generate off-topic clarifying questions that may reduce user satisfaction. Therefore, Wang and Ai [23] propose a risk-aware model to balance the risk of answering user queries and asking clarifying questions. Their system has a risk decision module which can decide whether the system should ask the question or show the documents directly. Tavakoli et al. [21] investigate the characteristics of useful clarifying questions and show that useful clarifying questions are answered by the asker, have an informative answer, and are valuable for the post and the accepted answer. Sekulić et al. [18] propose a facet-driven approach for generating clarifying questions.

For the clarification in web search, Zamani et al. [25] develop a taxonomy of clarification based on large-scale query reformulation data sampled from Bing

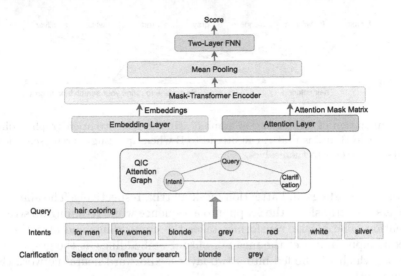

Fig. 2. Overall architecture of GACS.

search logs. Then they propose a rule-based model, a weakly supervised model trained using maximum likelihood, and a reinforcement learning model which maximizes clarification utility for generating clarifications. Later, Zamani et al. [27] analyze user engagements based on different attributes of the clarification and propose a model called RLC for clarification selection. RLC encodes each query-answer-intent triple separately and uses Transformer [22] to capture the interaction between these representations. Compared with RLC, our GACS model exploit the relations among the query, intent, and clarification via constructing a QIC attention graph and is more efficient. To promote research in search clarification, Zamani et al. [26] further introduce a large-scale search clarification dataset called MIMICS, which can be used for training and evaluating various tasks such as generating clarifications, user engagement prediction for clarifications, and clarification selection. In a follow-up study [19], ELBERT is proposed for user engagement prediction. It treats the user engagement prediction task as supervised regression and jointly encodes the query, the clarification, and the SERP elements for predicting the engagement levels. By comparison, we use the intent and its interaction with the query and clarification and focus on clarification selection.

3   Our Proposed Framework

3.1   Model Architecture

Figure 2 depicts the overall structure of GACS. There are four main components, namely, QIC attention graph, embedding layer, attention layer, and Mask-Transformer. The graph is fed into the embedding layer and the attention layer to

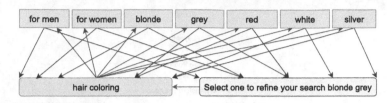

Fig. 3. Illustration of QIC attention graph G_1. In the QIC attention graph, a directed edge from A to B means that A can attend to B when updating its representation. All elements can attend to themselves.

obtain its embeddings and attention mask matrix, respectively. The aim of these two layers is to transform the graph into a sequence while retaining its structural information. The Mask-Transformer takes the embeddings and attention mask matrix as input and uses the mean of output embeddings as the graph's representation, which will be fed into a two-layer feedforward neural network (FNN) to compute the score.

QIC Attention Graph. There are three types of nodes in the QIC attention graph, i.e., query, intent, and clarification. We consider the following four different graphs:

G_1: When updating the representation, the query can attend to itself and its intents, but not the clarification. Thus, the query can focus on its own representation that is independent of a specific clarification. The clarification only needs to attend to itself and the query because the query can absorb its intents' information. For each intent, it can attend to itself, the query, and the clarification, but not other intents. In this way, it can associate itself with the query and determine whether it is reflected in the clarification. Figure 3 provides an illustration of G_1.

G_2: Based on G_1, G_2 considers the fact that each clarification usually only covers some intents. Therefore, G_2 adds edges from the clarification to multiple intents, which allows GACS to model the fact explicitly.

G_3: G_3 is a fully connected graph. It does not impose any special restrictions on the relation among the query, intent, and clarification and has the strongest expressive power.

$G4$: Unlike the previous three graphs, G_4 does not contain the intent, i.e., only the query and clarification are available in this graph. The query and clarification can attend to each other. G_4 is a special case of the QIC attention graph, where intents are masked. The purpose of using this graph is to explore whether the intent is useful or not.

Generally, the more complex the graph is, the stronger its expressiveness is, but the model is more difficult to train. In addition, by designing the graph structure, we can introduce task-related prior knowledge to the model. Therefore, different graphs may be suitable for different situations.

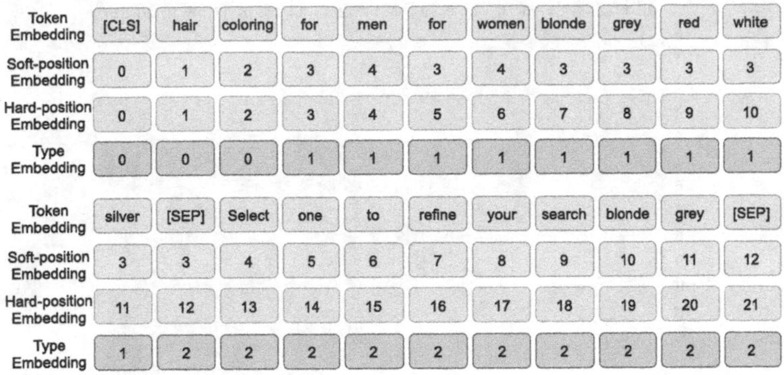

Token Embedding	[CLS]	hair	coloring	for	men	for	women	blonde	grey	red	white
Soft-position Embedding	0	1	2	3	4	3	4	3	3	3	3
Hard-position Embedding	0	1	2	3	4	5	6	7	8	9	10
Type Embedding	0	0	0	1	1	1	1	1	1	1	1

Token Embedding	silver	[SEP]	Select	one	to	refine	your	search	blonde	grey	[SEP]
Soft-position Embedding	3	3	4	5	6	7	8	9	10	11	12
Hard-position Embedding	11	12	13	14	15	16	17	18	19	20	21
Type Embedding	1	2	2	2	2	2	2	2	2	2	2

Fig. 4. Illustration of embedding layer of GACS. (1) The tokens in the QIC attention graph are flattened into a sequence by their hard-position indices. (2) The soft-position embedding is used as position embedding. (3) For the type embedding, type "0", "1", and "2" represent the query, intent, and clarification, respectively.

Embedding Layer. The function of the embedding layer is to convert the QIC attention graph into embeddings that can be fed into the Mask-Transformer. As shown in Fig. 4, the input embedding is the sum of the token embedding, soft-position embedding, and type embedding. The token embedding is consistent with Google BERT [6]. Inspired by [11], we use soft-position embeddings instead of hard-position embeddings because the soft position can reflect the graph structure. As in [20], we use type embeddings to indicate the three types of nodes in the graph. Precisely, type "0", "1", and "2" represent the query, intent, and clarification, respectively.

Attention Layer. The attention layer preserves the structure information of the QIC attention graph via constructing an attention mask matrix. Different graphs have different attention mask matrices. The attention mask matrix of G_1 is shown in Fig. 5. Given the embeddings of the tokens $X \in \mathbb{R}^{n \times d}$, where n is the number of tokens and d is the dimension of embeddings, the attention mask matrix $M \in \mathbb{R}^{n \times n}$ is defined as

$$M_{ij} = \begin{cases} 0 & x_i \ominus x_j \\ -\infty & \text{Otherwise} \end{cases} \tag{1}$$

where $x_i \ominus x_j$ means that x_i and x_j are in the same graph node or there is a directed edge from the node where x_i is to the node where x_j is. i and j are hard-position indices.

Mask-Transformer. The Mask-Transformer follows the Transformer [22] encoder structure. Given the token embeddings X and the attention mask matrix M, the Mask-Transformer uses masked self-attention as shown below:

$$Q, K, V = XW^Q, XW^K, XW^V \tag{2}$$

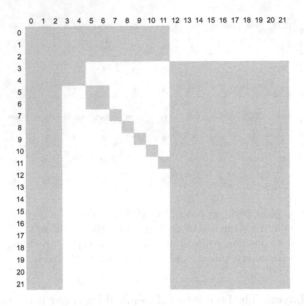

Fig. 5. Attention mask matrix of G_1. The cell at row i, column j is pink means that token i can attend to token j.

$$\text{Attn}(\boldsymbol{Q}, \boldsymbol{K}, \boldsymbol{V}) = \text{Softmax}(\frac{\boldsymbol{Q}\boldsymbol{K}^\top}{\sqrt{d_k}} + \boldsymbol{M})\boldsymbol{V} \tag{3}$$

where $\boldsymbol{W}^Q, \boldsymbol{W}^K, \boldsymbol{W}^V \in \mathbb{R}^{d \times d_k}$ are learnable parameters. The Mask-Transformer can control the information flow according to the attention mask matrix.

3.2 Loss Function

We use a loss function similar to Attention Rank [1]. For each query q and its k clarifications c_1 to c_k, we first compute the best attention allocation $\boldsymbol{a}^e$ and compute the attention allocation $\boldsymbol{a}^s$ with the ranking score $\boldsymbol{s}$:

$$\boldsymbol{a}^e = \text{Softmax}(\boldsymbol{e}), \quad \boldsymbol{a}^s = \text{Softmax}(\boldsymbol{s}) \tag{4}$$

where $\boldsymbol{e}$ and $\boldsymbol{s}$ are k-dimensional vectors. $\boldsymbol{e}_i$ is the engagement level of c_i and $\boldsymbol{s}_i$ is the score of c_i. Then we use the cross entropy between $\boldsymbol{a}^e$ and $\boldsymbol{a}^s$ as the loss:

$$\mathcal{L} = -\sum_{i=1}^{k}(\boldsymbol{a}_i^e \log(\boldsymbol{a}_i^s) + (1 - \boldsymbol{a}_i^e) \log(1 - \boldsymbol{a}_i^s)) \tag{5}$$

The loss function is a listwise function. It does not predict the engagement level of each clarification but focuses on the relative importance of each element in the ranked list.

4   Experiments

4.1   Dataset

We evaluate the model on MIMICS [26], a large-scale dataset collected from the Bing search engine for search clarification. In MIMICS, the engagement level is an integer between 0 and 10 based on click-through rates. The engagement level 0 means that there is no click on the clarification.

MIMICS consists of three subdatasets: MIMICS-Click, MIMICS-ClickExplore and MIMICS-Manual. In this work, we mainly focus on MIMICS-ClickExplore because each query has multiple clarifications in this dataset, but not in the other two datasets. Thus, it can be directly used for evaluating clarification selection models. However, some queries in MIMICS-ClickExplore have several identical clarifications, but their engagement levels are different. We delete such inconsistent data as follows:

- For each query q and its clarifications c_1 to c_n, $1 \leq i, j \leq n, i \neq j$, if c_i and c_j are identical but their engagement levels are different, we delete both of them;
- Afterwards, if the number of clarifications of q is less than 2 or the engagement levels of all its clarifications are 0, we delete the query q.

Finally, 2.44% of queries in MIMICS-ClickExplore are deleted. We call the processed dataset D_1. There are 62446 queries in D_1, and each query is associated with 2.57 clarifications on average. We divide D_1 into training, development, and test sets with a ratio of 8:1:1 based on the query.

The average number of clarifications per query in D_1 is small, we construct a dataset D_2 based on D_1. For each query in D_1, we randomly sample 10 negative clarifications (i.e., clarifications from other queries) from the set to which it belongs. This ensures that the training, validation, and test sets of D_2 have no intersection. Note that there are negative clarifications when testing the model on D_2, which is different from testing the model on D_1. We set the engagement levels of all negative clarifications to -1. Intuitively, it does not make sense to set the click-based engagement level to -1. However, our purpose is to distinguish between positive and negative clarifications. Because the engagement level of a positive clarification may be 0, it is inappropriate to set the engagement level of a negative clarification to 0.

4.2   Baselines

We use the following four baselines:

- **Random.** For each query, it selects one from the query's clarifications randomly.
- **RankNet** [13]. We use the BERT embedding of the query and clarification, the number of characters in the query and question, the number of answers, and the average number of characters in the answers as features and minimize the loss function in Eq. 5.

Table 1. Experimental results on D_1. Best results are in bold. Superscripts 1–4 indicate statistically significant improvements over Random, RankNet, RLC and ELBERT, respectively.

	Hits@1	MRR	nDCG@1	nDCG@2
Random	0.423	0.683	0.494	0.721
RankNet	0.460^1	0.704^1	0.523^1	0.737^1
RLC	0.453^1	0.701^1	0.527^1	0.742^1
ELBERT	0.480^{123}	0.716^{123}	0.547^{123}	0.751^{123}
GACS-G_1	$\mathbf{0.499}^{1234}$	$\mathbf{0.726}^{1234}$	$\mathbf{0.563}^{1234}$	$\mathbf{0.759}^{1234}$
GACS-G_2	0.485^{123}	0.719^{123}	0.555^{123}	0.757^{1234}
GACS-G_3	0.485^{123}	0.719^{123}	0.552^{123}	0.753^{123}
GACS-G_4	0.493^{1234}	0.724^{1234}	0.560^{1234}	$\mathbf{0.759}^{1234}$

- **RLC** [27]. RLC is composed of an intent coverage encoder and an answer consistency encoder. Because the answer consistency encoder requires the answer entity type data which is not available in MIMICS, we implement the intent coverage encoder of RLC as one baseline. In the original paper, they use the query reformulation data and the click data to estimate the intent. However, the two kinds of data are also not available in MIMICS. Therefore, we use the candidate answers to estimate the query's intents as in GACS.
- **ELBERT** [19]. ELBERT shows state-of-the-art performance on the user engagement prediction task. Since we can rank the clarifications according to their predicted engagement levels and select the best one, we implement ELBERT as one baseline. For a fair comparison, we use BERT to encode the query and clarification.

4.3 Evaluation Metrics

For the query q, we denote its clarification with the highest engagement level as c_{best}. We use the following three evaluation metrics:

- **Hits@1:** the percentage of test queries with c_{best} ranking first.
- **MRR:** the mean reciprocal rank of c_{best} of all the test queries.
- **nDCG@p:** the normalized discounted cumulative gain, which is computed as:

$$nDCG@p = \frac{\sum_{i=1}^{p} \frac{e_i}{\log(i+1)}}{\sum_{i=1}^{|E|} \frac{e_i}{\log(i+1)}} \tag{6}$$

where e_i is the engagement level of the i-th clarification in the ranked list and E is the clarification list ordered by their engagement levels up to position p.

Higher Hits@1, MRR, or nDCG@p indicates better performance.

Table 2. Experimental results on D_2. Best results are in bold. Superscripts 1–4 indicate statistically significant improvements over Random, RankNet, RLC and ELBERT, respectively.

	Hits@1	MRR	nDCG@1	nDCG@2
Random	0.075	0.245	−0.119	−0.177
RankNet	0.441^1	0.673^1	0.488^1	0.654^1
RLC	0.445^1	0.694^{12}	0.516^{12}	0.731^{12}
ELBERT	0.465^{123}	0.703^{123}	0.518^{12}	0.717^{12}
GACS-G_1	0.485^{1234}	0.718^{1234}	0.553^{1234}	0.752^{1234}
GACS-G_2	$\mathbf{0.489}^{1234}$	$\mathbf{0.722}^{1234}$	$\mathbf{0.556}^{1234}$	$\mathbf{0.758}^{1234}$
GACS-G_3	$\mathbf{0.489}^{1234}$	0.721^{1234}	$\mathbf{0.556}^{1234}$	0.754^{1234}
GACS-G_4	0.466^{123}	0.703^{123}	0.520^{12}	0.714^{12}

Table 3. Cohen's d and its 95% confidence interval (CI) which indicate the standardized difference between the performance of GACS-G_1 and ELBERT on D_1 and D_2.

		Hits@1	MRR	nDCG@1	nDCG@2
D_1	Cohen's d	0.04	0.04	0.03	0.03
	95% CI	[0.01, 0.07]	[0.01, 0.06]	[0.01, 0.06]	[0.01, 0.05]
D_2	Cohen's d	0.04	0.05	0.07	0.11
	95% CI	[0.01, 0.07]	[0.03, 0.08]	[0.04, 0.10]	[0.08, 0.13]

4.4 Implementation Details

The Transformer encoders of GACS and baselines are initialized with $BERT_{BASE}$ [6]. We use the implementation of HuggingFace's Transformer [24]. For training these models, we use the AdamW [12] optimizer with an initial learning rate of 10^{-5} and a linear learning rate decay scheduler. We fine-tune for 5 epochs and choose the best hyperparameters according to MRR on the development set.

4.5 Experimental Results

Table 1 and Table 2 report the experimental results of our GACS using different graphs and the baselines on D_1 and D_2, respectively. Statistical significance is tested using the paired student's t-test with $p < 0.05$. Due to the multiple comparisons problem [5], we use the Benjamini-Hochberg procedure to adjust p to control the false discovery rate. Moreover, we report the effect size Cohen's d and its 95% confidence interval to compare the performance of GACS-G_1 and the best baseline ELBERT, as shown in Table 3. We have the following observations:

- All other models perform significantly better than the Random model, which shows that the engagement level is a reasonable proxy for the usefulness of the clarification.

Table 4. Performance of GACS-G_3 with different engagement levels of negative clarifications on D_2.

	0	−1	−2	−3	−4
Hits@1	0.477	0.489	0.489	0.483	0.483
MRR	0.715	0.721	0.720	0.717	0.718

Table 5. Experimental results of GACS trained on D_2 and evaluated on D_1.

	Hits@1	MRR	nDCG@1	nDCG@2
GACS-G_1	0.486	0.720	0.555	0.755
GACS-G_2	0.489	0.722	0.556	0.760
GACS-G_3	0.489	0.721	0.556	0.756
GACS-G_4	0.485	0.720	0.552	0.755

- Our framework generally outperforms all the baselines, which is statistically significant and demonstrates the effectiveness of our framework.
- As shown in Table 1, when trained with only positive clarifications, GACS using G_2 or G_3 performs worse than GACS using G_1 or G_4. Although G_2 and G_3 are reasonable and have stronger expressive power, they increase the model's complexity and are more difficult to train, especially considering that there are only 2.57 clarifications per query in D_1. Moreover, GACS-G_1 performs better than GACS-G_4, showing that the intent can improve the model's performance but requires that the graph structure is reasonable and suitable for the dataset.
- As shown in Table 2, when introducing negative clarifications, GACS using G_2 or G_3 can outperform GACS using G_1 or G_4, which is different from the model's performance on D_1. This is because negative clarifications can bring some new information and help train the model. In addition, GACS models that use the intent perform obviously better than those that do not, indicating that the intent can help the model distinguish between positive and negative clarifications.
- Combining the observation from Table 1 and Table 2, we can see that different graphs are suitable for different scenarios. We expect that as the average number of clarifications per query increases, the benefits of the intent will be more obvious.

Effect of Negative Clarifications. According to Eq. 4, the smaller the engagement level of negative clarifications, the lower the importance of them. Table 4 reports the performance of GACS-G_3 with different engagement levels of negative clarifications on D_2. We can see that when the engagement level is −1 or −2, the model performs the best. Moreover, setting the engagement level to 0 performs worse than setting it to a negative value.

Table 6. Experimental results of GACS on D_1.

	Hits@1	MRR	nDCG@1	nDCG@2
GACS-G_1	0.499	0.726	0.563	0.759
w/o st	0.486	0.719	0.553	0.756
GACS-G_2	0.485	0.719	0.555	0.757
w/o st	0.481	0.717	0.546	0.752
GACS-G_3	0.485	0.719	0.552	0.753
w/o st	0.484	0.718	0.549	0.752
GACS-G_4	0.493	0.724	0.560	0.759
w/o st	0.486	0.719	0.554	0.756

To further investigate the effect of negative clarifications, we evaluate the
GACS models trained on D_2 on D_1. The results are shown in Table 5. First, we
can see that models trained on D_2 have very similar performance on D_1 and D_2,
indicating that it is much easier to distinguish between positive and negative
clarifications than to rank the positive clarifications. Thus, although negative
clarifications can help train the model, the benefits they can provide are limited
compared with positive clarifications. Second, after adding negative clarifications
for training, the performance of GACS using G_1 and G_4 does not get better, but
slightly worse. This is because the graphs they use are relatively simple and
negative clarifications may also bring some incorrect information because they
are treated equally.

Effect of Soft-Position Embedding and Type Embedding. In Table 6,
"w/o st" refers to using the hard-position embedding instead of the soft-position
embedding and no type embedding to distinguish between the three types of
nodes in the QIC attention graph. Experimental results show that removing
them will reduce the performance of GACS with different graphs. This indicates
that the soft-position embedding and the type embedding are important for
preserving the structure information in the QIC attention graph, which is crucial,
especially for relatively simple graphs.

5 Conclusion

This paper proposes a graph attention-based model GACS for clarification selection. It can effectively exploit the relations among the query, intent, and clarification by constructing the QIC attention graph and outperform competitive
baselines. The graph structure information is critical to the model's performance.
Moreover, we show that negative clarifications can help train GACS using complex graphs but their benefits are limited compared with positive clarifications.
A better estimation of the intent may further improve the model's performance.
In the future, we will explore how to better estimate the intent.

References

1. Ai, Q., Bi, K., Guo, J., Croft, W.B.: Learning a deep listwise context model for ranking refinement. In: The 41st International ACM SIGIR Conference on Research and Development in Information Retrieval, pp. 135–144 (2018)
2. Aliannejadi, M., Zamani, H., Crestani, F., Croft, W.B.: Asking clarifying questions in open-domain information-seeking conversations. In: Proceedings of the 42nd International ACM SIGIR Conference on Research and Development in Information Retrieval, pp. 475–484 (2019)
3. Anand, A., Cavedon, L., Joho, H., Sanderson, M., Stein, B.: Conversational search (Dagstuhl Seminar 19461). In: Dagstuhl Reports, vol. 9. Schloss Dagstuhl-Leibniz-Zentrum für Informatik (2020)
4. Braslavski, P., Savenkov, D., Agichtein, E., Dubatovka, A.: What do you mean exactly? Analyzing clarification questions in CQA. In: Proceedings of the 2017 Conference on Human Information Interaction and Retrieval, pp. 345–348 (2017)
5. Carterette, B.A.: Multiple testing in statistical analysis of systems-based information retrieval experiments. ACM Trans. Inf. Syst. $30(1)$, 1–34 (2012)
6. Devlin, J., Chang, M.W., Lee, K., Toutanova, K.: BERT: pre-training of deep bidirectional transformers for language understanding. In: Proceedings of the 2019 Conference of the North American Chapter of the Association for Computational Linguistics: Human Language Technologies, Volume 1 (Long and Short Papers), Minneapolis, Minnesota, pp. 4171–4186. Association for Computational Linguistics (2019)
7. Gao, J., Xiong, C., Bennett, P.: Recent advances in conversational information retrieval. In: Proceedings of the 43rd International ACM SIGIR Conference on Research and Development in Information Retrieval, pp. 2421–2424 (2020)
8. Hashemi, H., Zamani, H., Croft, W.B.: Guided transformer: leveraging multiple external sources for representation learning in conversational search. In: Proceedings of the 43rd International ACM SIGIR Conference on Research and Development in Information Retrieval, pp. 1131–1140 (2020)
9. Kiesel, J., Bahrami, A., Stein, B., Anand, A., Hagen, M.: Toward voice query clarification. In: The 41st International ACM SIGIR Conference on Research and Development in Information Retrieval, pp. 1257–1260 (2018)
10. Krasakis, A.M., Aliannejadi, M., Voskarides, N., Kanoulas, E.: Analysing the effect of clarifying questions on document ranking in conversational search. In: Proceedings of the 2020 ACM SIGIR on International Conference on Theory of Information Retrieval, pp. 129–132 (2020)
11. Liu, W., et al.: K-BERT: enabling language representation with knowledge graph. In: Proceedings of the AAAI Conference on Artificial Intelligence, pp. 2901–2908 (2020)
12. Loshchilov, I., Hutter, F.: Decoupled weight decay regularization. In: International Conference on Learning Representations (2017)
13. Lotze, T., Klut, S., Aliannejadi, M., Kanoulas, E.: Ranking clarifying questions based on predicted user engagement. arXiv preprint arXiv:2103.06192 (2021)
14. Qu, C., Yang, L., Croft, W.B., Trippas, J.R., Zhang, Y., Qiu, M.: Analyzing and characterizing user intent in information-seeking conversations. In: The 41st International ACM SIGIR Conference on Research and Development in Information Retrieval, pp. 989–992 (2018)
15. Radlinski, F., Craswell, N.: A theoretical framework for conversational search. In: Proceedings of the 2017 Conference on Human Information Interaction and Retrieval, pp. 117–126 (2017)

16. Ren, P., et al.: Wizard of search engine: access to information through conversations with search engines. In: Proceedings of the 44th International ACM SIGIR Conference on Research and Development in Information Retrieval, pp. 533–543 (2021)

17. Rosset, C., et al.: Leading conversational search by suggesting useful questions. In: Proceedings of the Web Conference 2020, pp. 1160–1170 (2020)

18. Sekulić, I., Aliannejadi, M., Crestani, F.: Towards facet-driven generation of clarifying questions for conversational search. In: Proceedings of the 2021 ACM SIGIR International Conference on Theory of Information Retrieval, pp. 167–175 (2021)

19. Sekulić, I., Aliannejadi, M., Crestani, F.: User engagement prediction for clarification in search. In: Hiemstra, D., Moens, M.-F., Mothe, J., Perego, R., Potthast, M., Sebastiani, F. (eds.) ECIR 2021. LNCS, vol. 12656, pp. 619–633. Springer, Cham (2021). https://doi.org/10.1007/978-3-030-72113-8_41

20. Sun, T., et al.: CoLAKE: contextualized language and knowledge embedding. In: Proceedings of the 28th International Conference on Computational Linguistics, pp. 3660–3670 (2020)

21. Tavakoli, L., Zamani, H., Scholer, F., Croft, W.B., Sanderson, M.: Analyzing clarification in asynchronous information-seeking conversations. J. Assoc. Inf. Sci. Technol. **73**, 449–471 (2021)

22. Vaswani, A., et al.: Attention is all you need. In: Advances in Neural Information Processing Systems, pp. 5998–6008 (2017)

23. Wang, Z., Ai, Q.: Controlling the risk of conversational search via reinforcement learning. In: Proceedings of the Web Conference 2021, pp. 1968–1977 (2021)

24. Wolf, T., et al.: HuggingFace's transformers: state-of-the-art natural language processing. In: Proceedings of the 2020 Conference on Empirical Methods in Natural Language Processing: System Demonstrations, pp. 38–45 (2020)

25. Zamani, H., Dumais, S., Craswell, N., Bennett, P., Lueck, G.: Generating clarifying questions for information retrieval. In: Proceedings of the Web Conference 2020, pp. 418–428 (2020)

26. Zamani, H., Lueck, G., Chen, E., Quispe, R., Luu, F., Craswell, N.: MIMICS: a large-scale data collection for search clarification. In: Proceedings of the 29th ACM International Conference on Information and Knowledge Management, pp. 3189–3196 (2020)

27. Zamani, H., et al.: Analyzing and learning from user interactions for search clarification. In: Proceedings of the 43rd International ACM SIGIR Conference on Research and Development in Information Retrieval, pp. 1181–1190 (2020)

28. Zou, J., Kanoulas, E., Liu, Y.: An empirical study on clarifying question-based systems. In: Proceedings of the 29th ACM International Conference on Information and Knowledge Management, pp. 2361–2364 (2020)

Continual Learning of Long Topic Sequences in Neural Information Retrieval

Thomas Gerald$^{(\boxtimes)}$ and Laure Soulier

CNRS-ISIR, Sorbonne University, Paris, France
{gerald,soulier}@isir.upmc.fr

Abstract. In information retrieval (IR) systems, trends and users' interests may change over time, altering either the distribution of requests or contents to be recommended. Since neural ranking approaches heavily depend on the training data, it is crucial to understand the transfer capacity of recent IR approaches to address new domains in the long term. In this paper, we first propose a dataset based upon the MSMarco corpus aiming at modeling a long stream of topics as well as IR property-driven controlled settings. We then in-depth analyze the ability of recent neural IR models while continually learning those streams. Our empirical study highlights in which particular cases catastrophic forgetting occurs (e.g., level of similarity between tasks, peculiarities on text length, and ways of learning models) to provide future directions in terms of model design.

Keywords: Continual learning · Information retrieval · Neural ranking models

1 Introduction

The information Retrieval (IR) field has seen a keen interest in neural approaches these last years [12, 14, 24, 27] thanks to recent advances in semantic and language understanding. However, these approaches are heavily data dependent, often leading to specialization for a certain type of corpus [28, 32]. If document retrieval remains a core task, many challenges revolve around, such as news detection [37], question answering [43] or conversational search [8]. In all these tasks, users' needs or document content might evolve through time; leading to evolving queries and/or documents and shifting the topic distribution at the inference step [3, 26, 37]. It is, therefore, crucial to understand whether IR models are able to change their ranking abilities to new topics/trends, but also to be still able to perform on previous topics/trends if these ones remain up to date. Accumulating and preserving knowledge is thus an important feature in IR, allowing to continuously adapt to new domains or corpora while still being effective on the old ones. This requirement refers to an emerging research field called *Continual learning* [17, 34, 40]. In practice, continual learning proposes to learn all tasks sequentially by guaranteeing that previous knowledge does not deteriorate through the learning process; this phenomenon is called *catastrophic forgetting*. To solve this issue, one might consider multi-task learning [30] in which models learn together all the sets of tasks.

M. Hagen et al. (Eds.): ECIR 2022, LNCS 13185, pp. 244–259, 2022.
https://doi.org/10.1007/978-3-030-99736-6_17

Another approach would consists in learning a model for each task, but, in this case, the knowledge is not transferred between previous and current tasks. These two last settings are not always realistic in *IR*, since they consider that all tasks are available at the training step. In practice, content and users' needs may evolve throughout the time [3,26].

To the best of our knowledge, only one previous work has addressed the continual learning setting in *IR* [22], highlighting the small weakness of the studied neural models to slightly forget knowledge over time. However this work has two limitations: 1) it only considers few tasks in the stream (2 or 3 successive datasets) and does not allow to exhibit neural model abilities in the more realistic scenario of long-term topic sequences (i.e., a larger number of users and topics implying evolving information needs/trends). 2) Although authors in [22] use datasets of different domains, there is no control of stream properties (e.g., language shift [1,3], information update [26]) allowing to correlate the observed results with *IR* realistic settings, as done in [40] for classification tasks.

The objective of this paper is thus to provide a low-level analysis of the learning behavior of neural ranking models through a continual setting considering long sequences and *IR*-driven controlled topic sequences. In this aim, we propose to study different neural ranking models and to evaluate their abilities to preserve knowledge. To this end, we consider neural rankers successively fined tuned on each task of the sequence. More particularly, our contribution is threefold:

- We design a corpus derived from the MSMarco Passage ranking dataset [29] to address long sequences of topics for continual learning and *IR*-driven controlled topic sequences (Sect. 4).
- We compare the different neural ranking models in a long-term continual *IR* setting (Sect. 5.1) and the controlled settings (Sect. 5.3).
- We in-depth investigate the impact of task similarity level in the continual setting on the learning behavior of neural ranking models (Sect. 5.2).

2 Related Works

Neural Information Retrieval. Deep learning algorithms have been introduced in *IR* to learn representations of tokens/words/texts as vectors and compare query and document representations [5,10,12,13,27,42]. With the advance of sequence-to-sequence models, semantic matching models have grown in popularity, particularly due to the design of new mechanisms such the well-known self-attention in transformer networks [39] or language models such as Bert [6].

Many *IR* approaches benefit from those advances as *CEDR* [24] that combines a Bert language model with relevance matching approaches including KNRM [42] and PACRR [12]. Moreover, recent works addressed ranking with sequence-to-sequence transformers based approach as the *Mono-T5* model [31] for re-ranking documents returned by a BM25 ranker. Using a weak initial ranker such as BM25 may be the bottleneck of reaching higher performances, some approaches are thus reconsidering dense retrieval [7,14,15,44]. All these models are data-dependent, relying on word/topic/query distribution in the training dataset and their application to new domains is not always straightforward

[28,32]. While previous works addressed this issue by leveraging for instance fine-tuning techniques [23,43], one can wonder whether these models are still effective on the word/topic/query distribution of the training dataset. This condition is particularly crucial for open-domain *IR* systems (e.g., public search engines or future conversational search systems) since they should be able to face multiple users and solve both persistent information needs and event-related ones.

Continual Learning. Continual learning generally defines the setting in which a model is trained consecutively on a sequence of tasks and need to adapt itself to new encountered tasks. One main issue of continual learning is that models need to acquire knowledge throughout the sequence without forgetting the knowledge learnt on previous tasks (*catastrophic forgetting*). To solve the catastrophic forgetting issue, three main categories can be outlined [18]. First, regularisation approaches continually learn to address new tasks using soft or hard preservation of weights [17,21,41]. For instance, the *Elastic Weight Consolidation* model [17] softly updates weights for a new task according to their importance in the previous one. Second, replay approaches [2,25,34] (or *rehearsal approaches*), replay examples of previous tasks while training the model on a new one. Third, architecture-based approaches [4,20,40] rely on the decomposition of the inference function. For instance, new approaches leveraging techniques of neural architecture search [20,40] have been proposed.

Recently some works have addressed the continual learning setting for *NLP* tasks. LAMOL [38] for continual language modelling, [19] for conversational systems or [9] for translations tasks. While it exists *IR* approaches to perform on different domains such as using batch balanced topics [11], at the best of our knowledge, only one study addresses *IR* in the continual setting [22], comparing neural ranking models on three successive tasks (MSMarco, TREC-Microblog, and TREC CORD19). Our work follows this line by providing an analysis of the behavior of neural ranking models on longer sequences of topics. We also design *IR*-driven controlled sequences to highlight to what extent neural models face *IR*-specific divergences, such as language drift or documents collection update.

3   Research Design for Continual Learning in *IR*

We address in this paper the following research questions aiming at analyzing the resilience of *IR* models to catastrophic forgetting:

- **RQ1:** How to design a sequence of tasks for continual learning in *IR*?
- **RQ2:** What are the performance of neural ranking models while learning long sequences of topics? Can we perceive signals of catastrophic forgetting?
- **RQ3:** Does the similarity level of tasks in the sequence impact the model effectiveness and their robustness to catastrophic forgetting?
- **RQ4:** How do neural ranking models adapt themselves to queries or documents distribution shifts?

3.1   Continual Learning Setting and Metrics

We propose a continual learning setting based on long sequences. The latter consists in fine-tuning a model on different tasks successively. Following [22], we instantiate tasks by topics/domains, but we rather focus on long sequences of tasks with the perspective that such setting can be connected with long-term trends/changes of user interests. In practice, we consider a sequence of n tasks $S = \{T_1, \ldots, T_i, \ldots, T_n\}$, each task T_i corresponds to a set of queries and their associated relevant documents. We suppose that each task relies on different properties or distributions as in [33]. Neural ranking models are successively fine-tuned over the long sequence S of topics. The objective is to track each task and evaluate each of them at different timestep of the sequence (i.e., after the successive fine-tuning) to measure the model's abilities to adapt to new tasks and their resilience to catastrophic forgetting.

In practice, we propose to track in each sequence a subset of 5 randomly selected tasks (tracking whole tasks throughout the whole sequence is too computationally expensive). For each of these tasks, we will measure at each step of the topic sequence the MRR@K. To measure the catastrophic forgetting mf for a given task T_i at a training step θ_j (associated to task T_j), we identify the maximum value obtained by the model along the sequence S and compare its performances at each training step θ_j with the maximum value:

$$mf(i, \theta_j) = \left(\max_{k \in 1,2,\ldots,|S|} score(i, \theta_k) \right) - score(i, \theta_j) \tag{1}$$

where $score(i, \theta_j)$ refers to a ranking metric for the task T_i using the model obtained training the jth task T_j in the sequence. Looking to $mf(i, \theta_j)$ for all j in the sequence allows observing which tasks have a significant negative transfer impact on T_i (high value) and which have a low negative impact (low value).

3.2   Neural Ranking Models and Learning

We evaluate two different state-of-the-art neural IR models:

- The vanilla Bert [6] (noted **VBert**) estimating a ranking score based on a linear layer applied on the averaged output of the last layer of the Bert language model.
- The *Mono-T5-Ranker* [31] (noted **MonoT5**) based on a *T5-base* model fine-tuning and trained to generate a positive/negative token.

Implementation Details: All models are trained with *Adam* optimizer [16], the optimizer state is not reinitialized for each task of the sequence. Indeed, reinitializing the optimizer will lead to observe a spike in the loss function whether addressing a same or a different task due to the state of *Adam* optimizer parameters. As previous work in *IR* [6,22,31], we perform sparse retrieval by re-ranking top-1000 most relevant documents retrieved by the BM25 model [36].

For MonoT5 we start with the *t5-base*[1] model with a learning rate of $10-3$ and batch size of 16. For the VBert model,[2] the batch size is 16 with a learning rate of 2×10^{-5} for Bert parameters and 10^{-3} for scoring function parameters.

4 MSMarco Continual Learning Corpus

Our continual learning framework is based on learning from a long sequence of tasks. One main difficulty is to create this sequence considering the availability of *IR* datasets. One method would be to build a sequence of datasets of different domains as in [22], but the number of datasets adapted to neural *IR* (with a sufficiently large number of queries and relevance judgments) is not sufficient for long sequences setting. We propose to model the task at a lower granularity level, namely topics, instead of the dataset granularity. In what follows, we present our methodology for creating long sequences of topics using the MSMarco dataset. Once this dataset is validated, it serves as a base for designing controlled settings related to particular *IR* scenarios (all settings and models are open-sourced[3]).

4.1 RQ1: Modeling the Long Topic Sequence

To create the long sequence, we consider the MSMarco dataset [29]. Such dataset is based on real users' questions on Bing. Our intuition is that several queries might deal with the same user's interest (e.g., "what is the largest source of freshwater on earth?" or "what is water shortage mitigation"). These groups of queries denote what we call in the remaining paper *topics*. To extract topics, we propose a two-step method: extracting clusters from randomly sampled queries and populating those clusters with queries from the whole dataset. We use a similarity clustering[4] based on query representations obtained using the sentence-BERT model [35]. The clustering is based on a sample of 50,000 randomly picked queries and estimates the similarity cosine distance according to a threshold t to build clusters of a minimum size of s. We then populate clusters using other queries from the dataset according to threshold t. Finally, we produce the sequence of topics by randomly rearranging clusters to avoid bias of cluster size. Another sequencing method might be envisioned for future work, for instance considering a temporal feature by comparing topic trends in real search logs. In practice, the value of the threshold t differs in each step of clustering and populating, leading to the threshold t_1 and t_2 (with $t_2 < t_1$) to obtain clusters of reasonable size to be used for neural models. Depending on the value of those hyper-parameters (t_1, t_2, s), we obtain three datasets of topic sequences of different sizes (19, 27, and 74), resp. called *MS-TS*, *MS-TM* and *MS-TL* (for small, medium, large).

---

[1] https://huggingface.co/transformers/model_doc/t5.html.
[2] Using bert-base-uncased pretrain.
[3] https://github.com/tgeral68/continual_learning_of_long_topic.
[4] https://www.sbert.net/examples/applications/clustering~(fast~clustering).

Statistics of these three topic sequences are described in Table 1. To build the train/validation/test sets, we constraint the validation and the test set to be composed of approximately 40 queries by topic. Notice that we do not use the original split as it remains difficult to consider enough testing examples falling into the created topics.

4.2   Evaluating the Long Topic Sequence

To verify the relevance of the clusters, we aim at measuring retrieval evidence within and between clusters (i.e., queries within clusters might have similar retrieval evidence and queries between clusters might have different ones). As retrieval evidence, we use the retrieved documents for each query using the *BM25* model with default parameters.[5] Our intuition is that similar queries should share retrieved documents (and vice versa). To compare queries within and between clusters, we randomly select two pools (noted A_i and B_i) of 250

Table 1. Parameters and statistics of the generated dataset and their inter/intra task similarity metric ($c - score$). The intra-score is the mean $c - score$ when comparing a task with itself, and the inter score when comparing different tasks.

Name	t_1	s	t_2	$|\mathcal{T}|$	#queries by topics	inter	intra
MS-TS	0.7	40	0.5	19	$3,650 \pm 1,812$	3.8%	31.4%
MS-TM	0.75	20	0.5	27	$3,030 \pm 1,723$	4.1%	32.1%
MS-TL	0.75	10	0.55	74	$1,260 \pm 633$	3.3%	34.6%
MS-RS	-	-	-	19	$3,650 \pm 1,812$	10.3%	10.2%
MS-RM	-	-	-	27	$3,030 \pm 1,723$	9.9%	9.8%
MS-RL	-	-	-	74	$1,260 \pm 633$	8.7%	8.8%

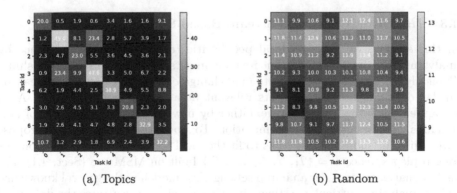

(a) Topics   (b) Random

Fig. 1. Matrix of similarities between topics for 8 tasks of MS-S (1a) and MS-RL (1b) datasets. The c-score ($\times 100$) is processed on all topic pairs, a high value (yellow) denotes the level of retrieved document overlap between queries of topics.

---

[5] Implemented in pyserini: https://github.com/castorini/pyserini.

queries within each cluster associated to task T_i. Let $D^{A_i} = \{D_q | q \in A_i\}$ (resp. $D^{B_i} = \{D_q | q \in B_i\}$) the documents returned by the ranker for the queries in A_i (resp. B_i).

We thus compute the $c - score$ which measures the ratio of common documents between two tasks T_i, T_j (or same task if $i = j$) as follows:

$$c - score(T_i, T_j) = \frac{|D^{A_i} \cap D^{B_j}|}{|D^{A_i}|} \tag{2}$$

This score is then averaged over pairs of topics within the sequence (intra when comparing topics with their-selves and inter when comparing different topics).

To evaluate our topic sequence methodology, for each of the three datasets we create a long topic sequence baseline in which clusters are extracted randomly from the queries of topics based corpora. We obtain three randomized datasets denoted *MS-RS*, *MS-RM* and *MS-RL*.

Table 1 reports for each of the generated datasets the intra and inter c-scores. By comparing the inter metric between both corpus settings (around 3/4% for the clustering-based ones and around 9/10% for randomized ones), one can conclude that our long topic sequence includes clusters that are more different than the ones created in the randomized corpus. The trend is opposite when looking at the intra, meaning that our sequence relies on clusters gathering similar queries but dissimilar from each other. This statement is reinforced in Fig. 1 which depicts the $c - score$ matrix for all couples $(i, j) \in \{1, 2, \ldots, |S|\}^2$ for a subset of 8 tasks (for clarity) of the $MS - S$ and $MS - RS$ corpora. We observe that for the randomized matrix (Fig. 1(b)), the metric value is relatively uniform. In contrast, in the matrix obtained from our long topic sequence based on clustering (Fig. 1(a)), the c-score is very small when computed for different topic clusters (low inter similarity) and higher in the diagonal line (high intra similarity).

4.3 *IR*-Driven Controlled Stream-Based Scenario

In this section, we focus on local peculiarities of the long topic sequence by analyzing *IR*-driven use cases, such as documents or queries distribution shifts. Typically, the available documents may change over time, or even some can be outdated (for instance documents relevant at a certain point in time). Also, it happens that the queries evolve, either by new trends, the emergence of new domains, or shifts in language formulation. To model those scenarios, we propose three different short topic streams to fit the local focus. Topics are based on our long topic sequence $S = \{T_1, \ldots, T_i, \ldots, T_n\}$ built on MSMarco (Sect. 4.1). For each scenario, we consider an initial setting T_{init} modeling the general knowledge before analyzing particular settings. In other words, it constitutes the data used for the pre-training of neural ranking models before fine-tuning on a specific sequence. The proposed controlled settings are:

- **Direct Transfer** [40]: The task sequence is $(T_{init}, T_i^+, T_j, T_i^-)$ where tasks T_i^+ and T_i^- belong to the topic task T_i and have different sizes ($|T_i^-| \ll |T_i^+|$).

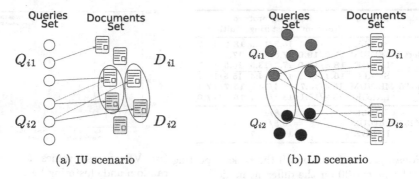

(a) IU scenario   (b) LD scenario

Fig. 2. Both Information Update (IU) and Language Drift (LD) scenarios. The circle of documents or queries represent the pair of documents or queries of different clusters, mapped using the closest neighborhood algorithm. This mapping is used to infer query-relevant documents of different clusters (dotted lines). Solid lines correspond to original query-relevant documents pairs. The red arrows build the training sets of tasks $\mathcal{T}'$ and $\mathcal{T}^*$ while blue arrows compose the one of tasks $\mathcal{T}''$ and $\mathcal{T}^{**}$. (Color figure online)

This setting refers to the case when the same topic comes back in the stream with new available data (new queries and new relevant documents).

- **Information Update:** The task sequence is $(\mathcal{T}_{init}, \mathcal{T}_i', \mathcal{T}_i'')$ where $\mathcal{T}_i'$ and $\mathcal{T}_i''$ have dissimilar document distributions and a similar query distribution. Intuitively, it can be interpreted as a shift in the required documents, such as new trends concerning a topic or an update of the document collection.
- **Language Drift:** The task sequence is $(\mathcal{T}_{init}, \mathcal{T}_i^*, \mathcal{T}_i^{**})$ where $\mathcal{T}_i^*$ and $\mathcal{T}_i^{**}$ have similar document distributions and a dissimilar query distribution. This can correspond to a change of query formulation or focus in a same topic.

To build those sequences, the initial task $\mathcal{T}_{init}$ aggregates k different tasks available in the original sequence topics S. We set $k = 5$ which is a good balance between considering enough tasks for the pre-training and considering not too many tasks to allow an impact of model fine-tuning on our controlled settings.

For the **Direct Transfer**, we randomly select a set of three topics (metrics are then averaged), 75% of the queries are used for $\mathcal{T}_i^+$ and 25% for $\mathcal{T}_i^-$. $\mathcal{T}_j$ is a topic selected randomly.

For **Information Update**, we consider that, for persistent queries, relevant documents might evolve. To do so, we randomly select three topics $\mathcal{T}_i$. For each topic $\mathcal{T}_i$, we cluster the associated relevant documents using a constrained 2-means algorithm[6] based on the cosine similarity metric of Sentence Bert embeddings (used in the Sect. 4.1). We obtain two document sets D_{i1} and D_{i2}: the initial and final information distribution. Since queries in MSMarco passages have in a vast majority one relevant document,[7] we can easily obtain the set of queries Q_{i1} and Q_{i2} associated to document sets D_{i1} and D_{i2} (see Fig. 2(a) -

---

[6] https://pypi.org/project/k-means-constrained/.

[7] If not the case, we sample one document to build the query-relevant document pairs.

Model	Dataset	Learning protocol Random	clustering	Multi-task
VBert	SMALL	18.4/19.6	16.3/17.5	**18.5/19.7**
	MEDIUM	**17.9/19.0**	17.8/18.9	17.5/18.7
	LARGE	**18.8/19.9**	17.3/18.5	18.5/19.7
MonoT5	SMALL	**16.1/17.3**	13.1/14.4	15.5/16.8
	MEDIUM	15.4/16.7	13.4/14.7	**15.7/17.1**
	LARGE	13.9/15.1	13.8/15.1	**15.7/17.0**
BM25	SMALL	10.8/11.7		
	MEDIUM	10.5/11.4		
	LARGE	11.7/12.7		

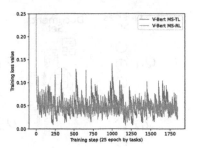

(a) Mean performances on all the tasks reporting $mrr@10/mrr@100$ for the different models.

(b) VBert loss values for both random and clustering-based large corpus.

Fig. 3. General performance of neural ranking models on the long topic sequence.

solid lines being the query-document relevance pairs). To model the information update, we map documents D_{i2} relevant for queries in Q_{i2} (final distribution) to most similar documents in D_{i1} (initial distribution) in the embedding space (circles in Fig. 2(a)). The task T_i' considers the whole set of queries Q_{i1} and Q_{i2} but only the document set D_{i1} as initial information (red arrows in Fig. 2(a)). The task T_i'' corresponds to the update of the information (namely, documents). We thus only consider the query set Q_{i2} for persistent queries with the document set D_{i2} as information update (blue arrows in Fig. 2(a)). We expect that Q_{i1} performs similarly after information update if models do not suffer from catastrophic forgetting and that Q_{i2} improves its performance with the information update. We also consider the reversed setting in which we first consider D_{i2} as the initial information and then update the information with D_{i1}, Q_{i1} (persistent queries).

For the **Language Drift** scenario, we use a similar protocol by clustering queries instead of documents to obtain the sets of queries Q_{i1} and Q_{i2}, and then the associated relevant document sets D_{i1} and D_{i2}. To model the language drift in queries, we consider that one query set will change its query formulation. To do so, let consider that sets Q_{i1} and Q_{i2} reflect resp. the initial and final language distribution of same information needs, and thus, requiring same/similar relevant documents. To observe the language drift, we map pairs of queries $(q_{i1}, q_{i2}) \in Q_{i1} \times Q_{i2}$ according to their similarity in the embedding space (circles in Fig. 2(b)). Thus, we can associate documents of D_{i2} (document relevant for queries of Q_{i2}) to the query set Q_{i1}: q_{i1} has two relevant documents (d_{i1} and d_{i2}) (red arrows in Fig. 2(b)). The T_i^* is composed of the query set Q_{i1} and the associated relevant documents belong to both D_{i1} and D_{i2} (red arrows). The T_i^{**} is based on the query set Q_{i2} (new language for similar information needs) associated to the relevant documents D_{i2} (blue arrows). We also consider the reversed setting in which query sets Q_{i2} and Q_{i1} are resp. used for the initial and final language.

For those two last scenarios (information update and language drift), metrics are respectively averaged over initial and reversed settings.

5 Model Performance and Learning Behavior on Long Topic Sequences

In this section, we report the experiments on the continual settings proposed in Sect. 4. We first analyze the overall retrieval performance of the different models applied on long topic sequences. We then present a fine-grained analysis of the different models with a particular focus on catastrophic forgetting regarding the similarity of topics in the sequence. Finally, we analyze specific *IR* use cases through our controlled settings.

5.1 RQ2: Performances on the MSMarco Long Topic Sequence

We focus here on the global performance of neural ranking models after having successively been fine-tuned on topics in our MSMarco-based long sequence setting (Fig. 3a). For comparison, we use different sequence settings (i.e., the randomized and the topic clustering ones) of different sizes (i.e., small, medium, and large). We also run the multi-task baseline in which models are trained on all the tasks of the sequence jointly (without sequence consideration). At a first glance, we can remark that, in a large majority, neural models after fine-tuning on random sequences or multi-task learning obtain better results than after the fine-tuning on our long topic sequences. This can be explained by the fact that, within our setting, the topic-driven sequence impacts the learning performance: a supplementary effort is needed by the model to adapt to new domains, which is not the case in the random setting. In this latter, the diversity is at the instance level. This trend is depicted in Fig. 3b, highlighting peaks in the clustering-based setting (blue line) referring to topic/cluster changes. This result confirms that catastrophic forgetting might occur with neural ranking models.

5.2 Fine-Grained Analysis

To get a deeper understanding of model behavior, we aim here to analyze the model performance throughout the learning of the sequence. We are particularly interested in explaining the possible behavior of catastrophic forgetting according to the similarity level between tasks in the sequence. For computational reasons, we were not able to track all tasks throughout the whole sequence, we thus considered 5 randomly selected tasks (as described in Sect. 3.1). For each of these 5 tasks T_i, we estimate the catastrophic forgetting using the mf score (Eq. 1) regarding each task T_j of the sequence (with $i \neq j$). For the similarity metric, we use the $c - score$ (Eq. 2) computed between both tasks T_i and T_j. In Fig. 4a, we group together similarity by quartiles and estimate the average of the mf score for tracked tasks in each similarity quartile. We first remark that the mean similarity values of quartiles are relatively small (except the 4th quartile), reinforcing the validation of our dataset building methodology. Also, we observe the following general trends. First, neural ranking models suffer from catastrophic forgetting (positive mf score), particularly the MonoT5 model. The difference in terms of model on both the global effectiveness (Fig. 3(a)) and the similarity analysis suggests that MonoT5 is more sensible to new domains than the VBert model. This can also explain by the difference in the way of updating weights (suggested in the original papers [6,31]). In VBert, two learning rates are used: a small one for the Bert model and a larger for the scorer layer; implying that the gradient descent mainly impacts the scorer. In contrast, the MonoT5 is learnt using a single learning rate leading to modify the whole model. Second, more tasks are similar (high $c - score$), less neural ranking models forget (low mf). In contrast to continual learning in other application domains [17,34] in which fine-tuning models on other tasks always deteriorates task performance, our analysis suggests that tasks might help each other (particularly when they are relatively similar), at least in lowering the catastrophic forgetting. Moreover, as discussed in [10], relevance matching signals play an important role in the model performance, often more than semantic signals. The task sequence may lead to a synergic effect to perceive these relevance signals. Figure 4b shows the VBert performance for three tasks located at different places in the sequence (circle point). To perceive catastrophic forgetting, we look at one part of the curve after the point. One can see that task performances increase after their fine-tuning (higher increase when the task is at the beginning of the sequence), highlighting this synergic effect. In brief, continual learning in IR differs from usual classification/generation lifelong learning setting. It is more likely to have different tasks allowing to "help" each other, either by having closely related topics or by learning a similar structure in the query-document matching.

Model	Dataset	1st	2nd	3rd	4th
Mean Similarity by quartile	MS-S	1.4	2.6	3.9	13.8
	MS-M	1.5	2.8	4.7	15.3
VBert	MS-S	6.3	6.4	5.4	**4.6**
	MS-M	4.2	4.4	5.1	**3.8**
MonoT5	MS-S	9.2	7.0	6.5	**6.3**
	MS-M	6.5	5.3	6.0	**4.5**

(a) Mean mf score grouped by similarities between tasks (mean of 5 selected topics). The results are averaged according to quartile based on the task similarity metric. The mean value of grouped similarity are reported in the head of the table.

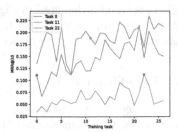

(b) MRR@10 results for three tasks tracked along the training sequence

Fig. 4. Fine-grained analysis of neural ranking model in the long topic sequence.

	DT scenario			B
	T_i^+	T_j	T_i^-	
MonoT5	26.6	24.9	26.6	27.2
VBert	28.5	26.7	27.3	28.9

(a) MRR@10 for task T_i in the Direct Transfer (DT) scenario. See Section 4.3 for building T_i and T_j.

		IU			LD		
		T_i'	T_i''	B	T_i^*	T_i^{**}	B
MonoT5	$Q_{i1}D_{i1}$	28.15	29.6	-	15.6	23.0	-
	$Q_{i2}D_{i2}$	7.75	26.0	-	16.8	26.5	-
	$Q_{i1}D_{i1} \cup Q_{i2}D_{i2}$	18.2	27.8	27.2	15.6	23.8	27.2
VBert	$Q_{i1}D_{i1}$	23.7	30.2	-	28.2	30.1	-
	$Q_{i2}D_{i2}$	14.5	31.4	-	25.5	25.5	-
	$Q_{i1}D_{i1} \cup Q_{i2}D_{i2}$	19.1	30.9	28.9	26.6	27.0	28.9

(b) MRR@10 for the Information Update (IU) and Language Drift (LD) scenarios. See Section 4.3 for the explanation of sets.

Fig. 5. Model performances on *IR*-driven controlled settings. B stands for the baseline.

5.3   RQ3: Behavior on *IR*-Driven Controlled Settings

In this section we review the different scenario described in the Sect. 4.3: **Direct Transfer** (DT), **Information Update** (IU) and **Language Drift** (LD). For all the different settings, we estimate the average metric of the different tracked tasks after each sequence step.

Figure 5a reports the effectiveness of neural models on task T_i (T_i^+ and T_i^- being subsets of T_i) after each fine-tuning step in the **Direct Transfer scenario**. One can see that fine-tuning on a foreign domain (T_2), the performance of both models on task T_i drop, highlighting a behavior towards catastrophic forgetting. However, both models are able to slightly adapt their retrieval performance after the fine-tuning of task T_i^-. This final performance is however lower than the baseline model (training on both T_{init} and T_i) and for the VBert model lower than its initial performance in the beginning of the learning sequence. These two last statements suggest the ability of neural models to quickly reinject a part of the retrained knowledge learnt in the early sequence to adapt to new query/document distributions in the same topic.

Figure 5b reports the average effectiveness metrics for both **Information Update** (IU) and **Language Drift** (LD) scenarios on different sets, $Q_{ik}D_i$ (k = 1,2) denoting the sets used to build relevant pairs of query-document (see Sect. 4.3). In IU scenario, relevant documents of certain queries (Q_{i2}) evolve over time ($D_{i1} \rightarrow D_{i2}$). For both $Q_{i1}D_{i1}$ and particularly $Q_{i2}D_{i2}$ whose queries have encountered the information update, evaluation performances increase throughout the fine-tuning process over the sequence. This denotes the ability of models to adapt to new document distributions (i.e., new information in documents). The adaptation is more important for the MonoT5 model (7.75 vs. 26.0 for the $Q_{i2}D_{i2}$ set), probably explained by its better adaptability to new tasks (as discussed in Sect. 5.2). Interestingly, the performance at the end of the learning sequence overpasses the result of the baseline (fine-tuning on T_i): contrary to the direct transfer scenario, this setting has introduced pseudo-relevant documents in task T_i' which might help in perceiving relevance signals.

For the **Language Drift** LD scenario, the behavior is relatively similar in terms of adaptation: performances increase throughout the sequence and MonoT5 seems more flexible in terms of adaptation. However, it seems more difficult to sufficiently acquire knowledge to reach the baseline performance (although pseudo-relevant documents have also been introduced). This might be due to the length of queries, concerned by the distribution drift: when the vocabulary changes in a short text (i.e., queries), it is more difficult to capture the semantics for the model and to adapt itself in terms of knowledge retention than when the change is carried out on long texts (i.e., documents as in the information update).

6 Conclusion and Future Work

In this paper, we proposed a framework for continual learning based on long topic sequences and carried out a fined-grained evaluation, observing a catastrophic forgetting metric in regards to topic similarity. We also provided specific stream of tasks, each of them addressing a likely scenario in case of *IR* continual learning. Our analysis suggests different design implications for future work: 1) catastrophic forgetting in *IR* exists but is low compared to other domains [17,40], 2) when designing lifelong learning strategy, it is important to care of task similarity, the place of the task in the learning process and of the type of the distribution that needs to be transfered (short vs. long texts). We are aware that results are limited to the experimented models and settings and that much remains to be accomplish for more generalizable results. But, we believe that our in-depth analysis of topic similarity and the controlled settings is a step forward into the understanding of continual *IR* model learning.

Acknowledgements. We thank the ANR JCJC SESAMS project (ANR-18-CE23-0001) for supporting this work. This work was performed using HPC resources from GENCI-IDRIS (Grant 2021-101681).

References

1. Albakour, M.D., Macdonald, C., Ounis, I.: On sparsity and drift for effective real-time filtering in microblogs. In: Proceedings of the 22nd ACM International Conference on Information and Knowledge Management, CIKM 2013, p. 419–428. Association for Computing Machinery, New York (2013). https://doi.org/10.1145/2505515.2505709
2. Asghar, N., Mou, L., Selby, K.A., Pantasdo, K.D., Poupart, P., Jiang, X.: Progressive memory banks for incremental domain adaptation. In: ICLR arXiv:1811.00239 (2020)

3. Cai, F., Liang, S., de Rijke, M.: Time-sensitive personalized query auto-completion. In: Proceedings of the 23rd ACM International Conference on Conference on Information and Knowledge Management, CIKM 2014, pp. 1599–1608. Association for Computing Machinery, New York (2014). https://doi.org/10.1145/2661829.2661921

4. Cai, H., Chen, H., Zhang, C., Song, Y., Zhao, X., Yin, D.: Adaptive parameterization for neural dialogue generation. In: EMNLP-IJCNLP, pp. 1793–1802 (2019)

5. Dai, Z., Xiong, C., Callan, J., Liu, Z.: Convolutional neural networks for soft-matching n-grams in ad-hoc search. In: WSDM, pp. 126–134 (2018)

6. Devlin, J., Chang, M., Lee, K., Toutanova, K.: BERT: pre-training of deep bidirectional transformers for language understanding. In: NAACL-HLT, pp. 4171–4186 (2019)

7. Formal, T., Piwowarski, B., Clinchant, S.: SPLADE: sparse lexical and expansion model for first stage ranking. In: Diaz, F., Shah, C., Suel, T., Castells, P., Jones, R., Sakai, T. (eds.) SIGIR 2021: The 44th International ACM SIGIR Conference on Research and Development in Information Retrieval, Virtual Event, Canada, 11–15 July 2021, pp. 2288–2292. ACM (2021). https://doi.org/10.1145/3404835.3463098

8. Gao, J., Xiong, C., Bennett, P.: Recent advances in conversational information retrieval. In: Proceedings of the 43rd International ACM SIGIR Conference on Research and Development in Information Retrieval, SIGIR 2020, pp. 2421–2424. Association for Computing Machinery, New York (2020). https://doi.org/10.1145/3397271.3401418

9. Garcia, X., Constant, N., Parikh, A.P., Firat, O.: Towards continual learning for multilingual machine translation via vocabulary substitution. In: NAACL-HLT, pp. 1184–1192 (2021)

10. Guo, J., Fan, Y., Ai, Q., Croft, W.B.: A deep relevance matching model for ad-hoc retrieval. In: CIKM, pp. 55–64 (2016)

11. Hofstätter, S., Lin, S., Yang, J., Lin, J., Hanbury, A.: Efficiently teaching an effective dense retriever with balanced topic aware sampling. In: Diaz, F., Shah, C., Suel, T., Castells, P., Jones, R., Sakai, T. (eds.) SIGIR 2021: The 44th International ACM SIGIR Conference on Research and Development in Information Retrieval, Virtual Event, Canada, pp. 113–122. ACM (2021)

12. Hui, K., Yates, A., Berberich, K., de Melo, G.: PACRR: a position-aware neural IR model for relevance matching. In: EMNLP, pp. 1049–1058 (2017)

13. Hui, K., Yates, A., Berberich, K., de Melo, G.: CO-PACRR: a context-aware neural IR model for ad-hoc retrieval. In: WSDM, pp. 279–287 (2018)

14. Karpukhin, V., et al.: Dense passage retrieval for open-domain question answering. In: Webber, B., Cohn, T., He, Y., Liu, Y. (eds.) Proceedings of the 2020 Conference on Empirical Methods in Natural Language Processing, EMNLP 2020, Online, 16–20 November 2020, pp. 6769–6781. Association for Computational Linguistics (2020). https://doi.org/10.18653/v1/2020.emnlp-main.550

15. Khattab, O., Zaharia, M.: ColBERT: efficient and effective passage search via contextualized late interaction over BERT. In: SIGIR, pp. 39–48. ACM (2020)

16. Kingma, D.P., Ba, J.: Adam: a method for stochastic optimization. In: Bengio, Y., LeCun, Y. (eds.) ICLR 2015 (2015)

17. Kirkpatrick, J., et al.: Overcoming catastrophic forgetting in neural networks. CoRR arXiv:1612.00796 (2016)

18. Lange, M.D., et al.: Continual learning: a comparative study on how to defy forgetting in classification tasks. CoRR arXiv:1909.08383 (2019)

19. Lee, S.: Toward continual learning for conversational agents. CoRR arXiv:1712.09943 (2017)
20. Li, X., Zhou, Y., Wu, T., Socher, R., Xiong, C.: Learn to grow: a continual structure learning framework for overcoming catastrophic forgetting. In: ICML, vol. 97, pp. 3925–3934 (2019)
21. Li, Z., Hoiem, D.: Learning without forgetting. IEEE Trans. Pattern Anal. Mach. Intell. **12**, 2935–2947 (2018)
22. Lovón-Melgarejo, J., Soulier, L., Pinel-Sauvagnat, K., Tamine, L.: Studying catastrophic forgetting in neural ranking models. In: Hiemstra, D., Moens, M.-F., Mothe, J., Perego, R., Potthast, M., Sebastiani, F. (eds.) ECIR 2021. LNCS, vol. 12656, pp. 375–390. Springer, Cham (2021). https://doi.org/10.1007/978-3-030-72113-8_25
23. Ma, X., dos Santos, C.N., Arnold, A.O.: Contrastive fine-tuning improves robustness for neural rankers. In: Zong, C., Xia, F., Li, W., Navigli, R. (eds.) Findings of the Association for Computational Linguistics: ACL/IJCNLP 2021, Findings of ACL, pp. 570–582, Online Event, 1–6 August 2021. Association for Computational Linguistics (2021). https://doi.org/10.18653/v1/2021.findings-acl.51
24. MacAvaney, S., Yates, A., Cohan, A., Goharian, N.: CEDR: contextualized embeddings for document ranking. In: SIGIR, pp. 1101–1104 (2019)
25. de Masson d'Autume, C., Ruder, S., Kong, L., Yogatama, D.: Episodic memory in lifelong language learning. CoRR arXiv:1906.01076 (2019)
26. McCreadie, R., et al.: University of Glasgow at TREC 2014: experiments with terrier in contextual suggestion, temporal summarisation and web tracks. In: Voorhees, E.M., Ellis, A. (eds.) Proceedings of The Twenty-Third Text REtrieval Conference, TREC 2014, Gaithersburg, Maryland, USA, vol. 500–308, 19–21 November 2014. National Institute of Standards and Technology (NIST), NIST Special Publication (2014). http://trec.nist.gov/pubs/trec23/papers/pro-uogTr_cs-ts-web.pdf
27. McDonald, R.T., Brokos, G., Androutsopoulos, I.: Deep relevance ranking using enhanced document-query interactions. In: EMNLP, pp. 1849–1860 (2018)
28. Mitra, B., Craswell, N.: An introduction to neural information retrieval. Found. Trends Inf. Retr. **13**(1), 1–126 (2018). https://doi.org/10.1561/1500000061
29. Nguyen, T., et al.: MS MARCO: a human generated machine reading comprehension dataset. In: Besold, T.R., Bordes, A., d'Avila Garcez, A.S., Wayne, G. (eds.) Proceedings of the Workshop on Cognitive Computation: Integrating Neural and Symbolic Approaches 2016 Co-located with the 30th Annual Conference on Neural Information Processing Systems (NIPS 2016), Barcelona, Spain, December 9, 2016, CEUR Workshop Proceedings, vol. 1773. CEUR-WS.org (2016). http://ceur-ws.org/Vol-1773/CoCoNIPS_2016_paper9.pdf
30. Nishida, K., Saito, I., Otsuka, A., Asano, H., Tomita, J.: Retrieve-and-read: multitask learning of information retrieval and reading comprehension. In: Proceedings of the 27th ACM International Conference on Information and Knowledge Management, CIKM 2018, pp. 647–656. Association for Computing Machinery, New York (2018). https://doi.org/10.1145/3269206.3271702
31. Nogueira, R., Jiang, Z., Pradeep, R., Lin, J.: Document ranking with a pretrained sequence-to-sequence model. In: EMNLP, pp. 708–718 (2020)
32. Onal, K.D., et al.: Neural information retrieval: at the end of the early years. Inf. Retr. J. **21**(2–3), 111–182 (2018). https://doi.org/10.1007/s10791-017-9321-y
33. Pan, S.J., Yang, Q.: A survey on transfer learning. IEEE Trans. Knowl. Data Eng. **22**(10), 1345–1359 (2010)
34. Rebuffi, S., Kolesnikov, A., Sperl, G., Lampert, C.H.: iCaRL: incremental classifier and representation learning. In: CVPR, pp. 5533–5542 (2017)

35. Reimers, N., Gurevych, I.: Sentence-BERT: sentence embeddings using Siamese BERT-networks. In: EMNLP (2019). http://arxiv.org/abs/1908.10084
36. Robertson, S.E., Walker, S., Hancock-Beaulieu, M., Gull, A., Lau, M.: Okapi at TREC. In: TREC, vol. 500–207, pp. 21–30 (1992)
37. Sankepally, R.: Event information retrieval from text. In: Proceedings of the 42nd International ACM SIGIR Conference on Research and Development in Information Retrieval, SIGIR 2019, p. 1447. Association for Computing Machinery, New York (2019). https://doi.org/10.1145/3331184.3331415
38. Sun, F., Ho, C., Lee, H.: LAMOL: language modeling for lifelong language learning. In: ICLR (2020)
39. Vaswani, A., et al.: Attention is all you need. In: NeurIPS, pp. 5998–6008 (2017)
40. Veniat, T., Denoyer, L., Ranzato, M.: Efficient continual learning with modular networks and task-driven priors. CoRR arXiv:2012.12631 (2020)
41. Wiese, G., Weissenborn, D., Neves, M.: Neural domain adaptation for biomedical question answering. In: CoNLL 2017, pp. 281–289 (2017)
42. Xiong, C., Dai, Z., Callan, J., Liu, Z., Power, R.: End-to-end neural ad-hoc ranking with kernel pooling. In: SIGIR, pp. 55–64 (2017)
43. Yang, W., Xie, Y., Tan, L., Xiong, K., Li, M., Lin, J.: Data augmentation for BERT fine-tuning in open-domain question answering. CoRR arXiv:1904.06652 (2019)
44. Zhao, T., Lu, X., Lee, K.: SPARTA: efficient open-domain question answering via sparse transformer matching retrieval. In: Toutanova, K., et al. (eds.) Proceedings of the 2021 Conference of the North American Chapter of the Association for Computational Linguistics: Human Language Technologies, NAACL-HLT 2021, Online, 6–11 June 2021, pp. 565–575. Association for Computational Linguistics (2021). https://doi.org/10.18653/v1/2021.naacl-main.47

Ensemble Model Compression for Fast and Energy-Efficient Ranking on FPGAs

Veronica Gil-Costa[1] , Fernando Loor[1] , Romina Molina[1,2,3] ,
Franco Maria Nardini[3] , Raffaele Perego[3] , and Salvatore Trani[3(✉)]

[1] Universidad Nacional de San Luis, San Luis, Argentina
[2] Università degli Studi di Trieste, Trieste, Italy
[3] ISTI-CNR, Pisa, Italy
salvatore.trani@isti.cnr.it

Abstract. We investigate novel SoC-FPGA solutions for fast and energy-efficient ranking based on machine-learned ensembles of decision trees. Since the memory footprint of ranking ensembles limits the effective exploitation of programmable logic for large-scale inference tasks, we investigate binning and quantization techniques to reduce the memory occupation of the learned model and we optimize the state-of-the-art ensemble-traversal algorithm for deployment on low-cost, energy-efficient FPGA devices. The results of the experiments conducted using publicly available Learning-to-Rank datasets, show that our model compression techniques do not impact significantly the accuracy. Moreover, the reduced space requirements allow the models and the logic to be replicated on the FPGA device in order to execute several inference tasks in parallel. We discuss in details the experimental settings and the feasibility of the deployment of the proposed solution in a real setting. The results of the experiments conducted show that our FPGA solution achieves performances at the state of the art and consumes from $9\times$ up to $19.8\times$ less energy than an equivalent multi-threaded CPU implementation.

Keywords: Learning to Rank · Model Compression · Efficient Inference · SoC FPGA

1 Introduction

This work investigates the use of cost-effective SoC-FPGA (System on Chip - Field Programmable Gate Arrays) devices for speeding-up inference tasks based on complex machine-learned ensemble models. Latency and throughput at inference time are critical aspects in many applications of machine learning where the rate of incoming requests is high and tight constraints on prediction quality impose the adoption of computationally-expensive models. In these cases, quality-of-service requirements entail the optimization of the accuracy of the models subject to performing inference in near real-time or within a limited time budget. As a use case where finding the best trade-off between model accuracy and inference time is definitely important and challenging, we consider the

M. Hagen et al. (Eds.): ECIR 2022, LNCS 13185, pp. 260–273, 2022.
https://doi.org/10.1007/978-3-030-99736-6_18

task of ranking documents according to their relevance for user queries. Indeed, ranking for ad-hoc retrieval entangles challenging effectiveness and efficiency constraints in many online services deployed in large-scale Web search engines, e-Commerce platforms and online social networks [6].

We specifically study techniques for performing document ranking with SoC-FPGA devices at a competitive level of quality and speed with respect to the state of the art, but using a fraction of the energy. SoC-FPGA technology provides an energy-efficient alternative to traditional computing due to the possibility of adapting the design of the logic to a specific architecture optimized for the task addressed. The cost and power/performance competitiveness of SoC FPGA makes this technology very attractive for specific tasks such as ranking, where the high cost and power consumption of GPUs make their adoption prohibitive [27]. We claim that SoC-FPGA architectures can provide an efficient and sustainable solution for large-scale query-processing since they can offer efficient ranking capabilities based on state-of-the-art solutions at a fraction of the energy cost incurred by CPU-based or GPU-based solutions. Recently, Molina *et al.* followed the same research line and proposed SoC-FPGA solutions for speeding-up inference based on Learning-to-rank (LtR) ensembles of decision trees [26]. The study identifies in the memory footprint the main issue limiting the computational performance. In this paper, we address this limitation by investigating the use of binning and quantization techniques for reducing the memory occupation of both the ranking model and the feature vectors representing the document-query pairs to be scored. Reducing the memory footprint of the model allows to replicate the ranking logic on the FPGA device to execute several inference tasks in parallel. Furthermore, by compressing the document-query feature vectors, we minimize the transmission costs incurred for transferring them to the FPGA device. We discuss the feasibility of the deployment of the proposed solution in a real setting and evaluate its performance using publicly available LtR datasets. The experiments conducted show that our solution does not impact the quality of the ranking and it provides highly competitive computational performance with very low energy consumption.

The remainder of the paper is organized as follows. Section 2 discusses the related work. Section 3 describes SoC-FPGA technologies. Section 4 introduces ensemble models and the challenges related to their use in the learning to rank scenario. It then discusses the use of binning and quantization for reducing the memory footprint and effectively deploying the ranking process on SoC-FPGA devices. Section 5 discusses the efficiency and effectiveness of the SoC-FPGA deployment compared to the traditional CPU one. It then details an evaluation of the latency introduced by transferring data from the host machine to the SoC-FPGA device. Moreover, it also reports an analysis of the energy consumption provided by both SoC FPGA-based and CPU-based scoring solutions. Finally, Sect. 6 concludes the paper and draws some future work.

2   Related Work

Several effective LtR algorithms and libraries have been proposed in the last years to train complex models able to precisely rank the documents matching a query [9,16,19]. State-of-the-art LtR models include those based on additive ensembles of regression trees learned by Mart [11] and λ-MART [3,31] gradient boosting algorithms. Since such ranking models are made of hundreds of additive regression trees, the tight constraints on query response time require suitable solutions able to provide an optimal trade-off between efficiency and ranking quality [6]. Among the main contributions in the area of efficient ranking, we cite the algorithms for the efficient traversal of tree ensembles [1,10,20,34]. Alternative methods are concerned with: i) strategies for pruning the ensemble during or after the training phase [21,22,24], ii) budget-aware LtR algorithms [1,30], and iii) end-to-end learning of multi-stage LtR pipelines [8,12]. Furthermore, researchers investigated early termination heuristics aimed to reduce, on a document or query-level basis, the cost of the ensemble traversal process without (or minimally) impacting quality [4,5,25]. An analogous strategy was recently proposed to reduce the computational cost of neural re-ranking based on bidirectional transformer networks [32].

Previous work showed that SoC-FPGA devices can handle the complex computation of LtR training algorithms and provide high computing efficiency with low power consumption. Xu et al. describe the design of a FPGA accelerator for a LtR algorithm to reduce training time [33]. Gao and Hsu evaluate a LtR algorithm deployed on a FPGA and explore the design space of the implementation choices [13]. Similar to our work, Qiang et al. present a fixed-point quantization approach for LtR algorithms on FPGA [18]. Experimental results show that the FPGA-based algorithm achieve a 4.42× speedup over a GPU implementation but with 2% accuracy loss. Differently from these previous works focusing on the offline, LtR training phase, we are interested in the online inference phase, where the machine-learned model is deployed in a large infrastructure and used under tight latency constraints. To the best of our knowledge, only Molina et al. previously investigated this important aspect and highlighted the memory usage on the FPGA device as the main issue limiting the computational performance. This work addresses this limitation by exploiting binning and quantization to compress the ranking model and the feature vectors.

3   Using Programmable Logic for Ranking

The features of current SoC-FPGA devices allow their adoption for high-performance computing tasks such as inference under tight time constraints where they can provide an efficient and energy-efficient solution. A SoC-FPGA device integrates on the same chip a general-purpose Processing System (PS) and a Programmable Logic (PL) unit. The PS includes a processor and a memory of greater capacity than the memory available in the PL. The PL includes blocks of memories (BRAM), control and logic components like the Flip Flop (FF) or

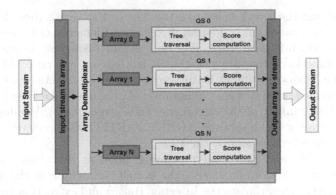

Fig. 1. Hardware design inside the IP block of the FPGA device.

the LookUp Table (LUT). These components are used to implement Intellectual Property (IP) blocks which actually execute the algorithms on the FPGA. The amount of components available in the FPGA is limited and this limitation in turn constrains the design and the deployment of the algorithm which usually involves a trade-off between processing speed and resource utilization. To take full advantage of FPGA logic, we need to process data in parallel, possibly avoiding jumps and recursive calls. To this end, High Level Synthesis (HLS) tools are used to create hardware from a high-level description, using directives to specify concurrency and pipelining opportunities. The HLS tool translates the code to a Register Transfer Level (RTL) specification of the hardware and also returns an estimation of execution latency and resource utilization. In this way, the designer is able to broadly evaluate the performance of different implementation strategies before actually deploying them on the hardware. In this phase the designer is also asked to detail the data communication occurring between the PS and PL. To exploit SoC-FPGA characteristics for ranking, we rely on QUICKSCORER (QS), the state-of-the-art algorithm for the traversal of large tree ensembles [10,20]. QS exploits a representation of the tree ensemble based entirely on linear arrays accessed with high locality. This characteristic permits a very fast traversal of the tree ensemble at inference time by effectively exploiting features and peculiarities of modern processors and memory hierarchies [17,23].

To estimate resource consumption and execution times several directives are inserted in the QS C++ code. Unrolling techniques are used to parallelize the execution of loop constructs such as for, while, do...while, repeat. These loops can be synthesized if and only if the loops bounds and the condition expression can be calculated during compilation time. In other words, the condition expression determining loop exiting cannot dynamically change at run-time. To tackle this problem, the QS algorithm has been modified by removing any dynamic condition expression from the loops and by including additional if-else statements to split the loops into sub-loops of fixed size which can be processed in parallel.

We propose a FPGA hardware design composed of a single Direct Memory Transfer (DMA) and one IP block responsible for accelerating the QS algorithm.

Both the IP and the DMA belong to the PL. In Fig. 1, we show how the QS algorithm is replicated into different Processing Elements (PE) inside the IP block to perform inference in parallel on different instances. To avoid off-chip memory transactions, the input of the IP is a stream composed of batches of instances (i.e., query-document feature vectors) to be predicted and the output is a stream formed by the actual predictions (i.e., the query-document scores producing the final ranking of the documents for a given query). We implement an array demultiplexer and apply the ARRAY_PARTITION directive to distribute the input feature vectors among the PEs.

We also implement task-level pipelining, allowing functions and loops to overlap in their operation, increasing the overall throughput of the design. The PIPELINE directive optimizes the insertion (push) and extraction (pop) of data from the stream. After inference, the final predictions are packed into an output stream adding the corresponding control signals. The PIPELINE directive is also used to speed-up the execution of the function admitting new inputs.

4   Ranking Model Compression

Ensembles of decision trees are among the most successful Machine Learning (ML) models and the winning solutions in many ML competitions.[1] However, inference with ensemble-based models can be computationally expensive since it requires the complete traversal of the tree ensemble, aimed at identifying all the tree leaves contributing to the prediction. To this end, each tree of the ensemble is visited from its root to a leaf by evaluating the splitting conditions (i.e., a test over a single feature with a learned threshold) associated with internal nodes. The contributions of all the leaves reached (i.e., a class label in case of a classification task or a numeric value in case of a regression task) are aggregated to compute the final prediction. This process has a complexity proportional to the number T of trees in the ensemble multiplied by the average depth d of the decision trees. For document ranking, the use case considered in this paper, typical LtR ensembles are made up of hundreds or even thousands of regression trees usually having each from 5 to 9 levels (corresponding to a number of leaves ranging from 32 to 512) [6]. For example, the winning solution of the Yahoo! Learning to Rank challenge used a linear combination of 12 ranking models, 8 of which were λ-MART boosted tree models each composed of about 3,000 trees for a total of 24,000 trees [7]. Since the traversal has to be repeated for each one of the K candidate documents to be scored for a user query, we have that the per-query ranking cost is proportional to $T \cdot d \cdot K$. Also the amount of memory needed for a LtR ensemble is quite large even if we do not consider the data structures needed to support the inference process. We can roughly estimate a lower bound for the space required to store an ensemble by considering the compact representation of each tree obtained by implicitly encoding its structure (parent/child nodes) in a linear array using a breadth-first order. Internal nodes require 8 bits for the feature identifier (assuming to have at most 256 features) and 32 bits for the

---

[1] https://dataaspirant.com/xgboost-algorithm/.

threshold value, summing up to 40 bits per node. Leaf nodes, on the other hand, are usually represented with 32-bit floating-point values. In the LtR case, these values represent the additive contribution of the specific tree to the ranking score predicted for the query-document pair. Let us consider for example an ensemble with $T = 1,000$ and $d = 9$: its compact representation requires about 4.4 MB of memory. Such size surely fits in the memory available on a low-cost FPGA device but reducing the memory requirements provides an interesting opportunity to fully exploit the FPGA logic and increase the number of inference tasks processed concurrently. To this regard, in this work we investigate the use of two popular techniques, namely binning and quantization, to lower the memory occupation of ensemble models and to parallelize the inference on FPGA devices.

- Binning consists in bucketing continuous feature values into discrete bins and representing each value with the index of its bin. It is commonly used to speed up training [16], but, at the best of our knowledge its usage for model compression at inference time has not been previously investigated. Specifically, we used binning to encode each internal node of the trees with only 16 bits: 8 bits for the feature identifier and 8 bits for the identifier of the bin associated with one of the possible 256 threshold values. Another advantage of binning the thresholds is that it allows to represent similarly with only 8 bit instead of 32 also the elements of the feature vectors representing the instances to be predicted. The splitting condition in the internal nodes of each decision trees moves than from $feature[i] <= threshold$ to $binned\_feature[i] <= bin$, with feature vectors values that can now be represented with a single byte storing the bin identifier in place of 32 bits. Let us consider one of the LtR datasets used for the experiments (Istella-S) where each query-document pair to be predicted is represented by 220 real-valued features for a memory occupation of 880 bytes. Binning these values into 256 bins results in a 3/4 reduction of the space needed, thus impacting both the number of instances that can be predicted in parallel on the FPGA and the cost of memory transfers.
- Quantization, on the other hand, consists in mapping continuous real values into a discrete set of finite values. This technique is popular for example to compress deep neural network models [14]. We apply quantization to the leaf values of each tree of the ensemble, so as to further lower the memory footprint. Specifically, we represent the 32-bits real value stored in each leaf of the original model with a 8-bits unsigned integer by mapping the min/max among the actual leaf values to the min/max in the range [0, 255]. This reduces of 3/4 also the space needed for storing the leaves of the ensemble.

The combination of the above binning and quantization techniques permit to represent the ensemble model previously mentioned by using only 1/3 of the original space. In Sect. 5, we will show experimentally that such compression does not introduce significant degradation in the resulting ranking effectiveness. On the other hand, we will show the benefits of compression in lowering the FPGA resources used and the data transmission time.

5   Experiments

We evaluate the impact of the proposed binning and quantization techniques for compressing ensemble models on two publicly available LtR datasets, namely MSLR-WEB30K-F1 (Fold 1)[2] [19], hereinafter simply abbreviated as MSN30K, and Istella-S[3] [22]. Both datasets contain more than $30K$ queries and about $3,5M$ query-document pairs, where each pair is represented with 136 features on MSN30K and 220 features on Istella-S. The query-document pairs in both datasets are labeled by relevance judgments ranging from 0 (irrelevant) to 4 (perfectly relevant). While the two datasets are comparable in size, they differ in the proportion between positive (label > 0) and negative (label $= 0$) examples. Indeed, in the MSN30K dataset about 48% of the documents are labeled with a positive judgement, while in the Istella-S dataset this proportion lower to 11%. The detailed characteristics of the two datasets are listed in Table 1.

Table 1. Characteristics of the two datasets used.

Dataset	MSN30K	Istella-S
queries	31,351	33,018
query-document pairs	3,771,125	3,408,630
features	136	220
positive examples	48.53%	11.39%

Each dataset is split in train, validation and test set according to a 60%-20%-20% scheme. We use training and validation sets to train ensemble models with the λ-MART [3,31] algorithm, while the test set is used for evaluating the performance of the model. The learning process of λ-MART is controlled by several hyper-parameters, some of them controlling the generalization power and the training speed of the learning phase, while others controlling the shape of the trees. Since our objective is to fasten the inference time by exploiting programmable SoC devices, which are limited in the amount of available resources, we start by finding the most compact model providing state-of-the-art performance, i.e., we investigate the optimal trade-off between model size and ranking effectiveness. To this end, we performed several grid searches by varying the hyper-parameters controlling the shape of the final model and allowing each grid exploits the remaining ones. In particular, we varied the number of leaves in $\{64, 128, 256, 512\}$ by keeping fixed the maximum number of trees to 800. We used the implementation of λ-MART available in the LightGBM library [16] for training and the HyperOpt library [2] for tuning the hyper-parameters. When comparing the performance of different models, we also evaluated statistical significance by using the randomization test with 10,000 permutations and p-value ≤ 0.05 [29].

---

[2] http://research.microsoft.com/en-us/projects/mslr/.
[3] http://quickrank.isti.cnr.it/istella-dataset/.

Table 2. Efficiency/effectiveness trade-off: NDCG@10 vs. model size in MB.

Dataset	Model Version	per-tree leaves			
		64	128	256	512
MSN30K	Full Precision	0.524 (0.30)	0.526 (0.87)	0.527 (1.59)	0.528 (2.73)
	Bin. + Quant.	0.524 (0.10)	0.526 (0.29)	0.527 (0.53)	0.528 (0.91)
Istella-S	Full Precision	0.771 (0.44)	0.775 (0.87)	0.779 (1.74)	0.781 (3.38)
	Bin.+ Quant.	0.770 (0.14)	0.776 (0.29)	0.779 (0.58)	0.782 (1.12)

5.1 Effectiveness Assessment

We report on the effectiveness of the resulting fine-tuned models on the MSN30K and Istella-S datasets as a function of the size of the models, with and without model compression. It is worth noting that only quantization can affect effectiveness. Binning in fact, even if not exploited previously for model and feature compression, is natively used by the LightGBM library to speed-up model training and has no impact on the quality of the model trained. Table 2 reports the value of NDCG@10 [15] and the size in MB of the full precision and compressed models trained on the two datasets. In terms of absolute effectiveness, the fine-tuned model with 64 leaves achieves a NDCG@10 equal to 0.524 on MSN30K (0.771 on Istella-S), while the best performing model with 512 leaves reaches 0.528 (0.781). We note that with a quality loss lower than 1% on MSN30K and 1.3% on Istella-S the models with 64 leaves are about 9× smaller than the ones with 512 leaves. Thus, these models largely offer the best effectiveness/space trade-off and are most suited for an efficient FPGA deployment in presence of strict memory constraints. By looking at the NDCG@10 values reported in each column of the table, we see that the impact of quantization on the ranking performance is limited. In most case we do not have differences in the NDCG@10 measured on the test set and in all the cases the differences are not statistically significant. To further investigate the impact of quantization on effectiveness, Fig. 2 reports the NDCG@10 of the ranking models with 64 leaves as a function of the number of trees. Each one of the plots in the figure shows three curves: one for the original, full precision models where the values associated with tree leaves are represented as 32-bit floating point values, and two for models exploiting quantization with 8 and 4-bit representations. The curves plotted confirm that quantization using 8-bit representations does not introduce performance penalty despite it permits to reduce of 4× the space needed for coding the leaves. On the other hand, the models using 4-bit representations perform slightly worse than the full precision ones, showing also a statistically significant difference. Considering the difficulties in working with binary representations smaller than a single byte, hereinafter we will consider models with 64 leaves and 8-bit quantization only. These fine-tuned models have 559 and 649 trees (out of the 800 maximum trees), for the MSN30K and Istella-S datasets, respectively.

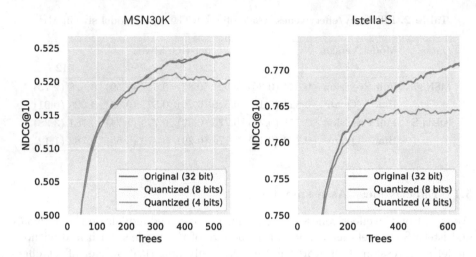

Fig. 2. Impact of quantization on NDCG@10 of λ-MART models.

5.2 Efficiency Assessment

We assess the impact of the proposed binning and quantization techniques on the efficiency of the scoring process by using two versions of QS: the original version [20] and a new one supporting binned and quantized models. Both the versions have been deployed on a high-end multi-core CPU and a SoC-FPGA device. The multi-threaded CPU version [17] runs on a server machine running Ubuntu 20.04 LTS and equipped with two Intel Xeon CPU E5-2630 v3 clocked at 2.40 GHz, 120 GB of RAM. The CPU exploits three levels of cache: 32 KB + 32 KB of L1 cache (data + instructions), 256 KB of L2 cache, and 20,480 KB of L3 cache. The code was compiled with GCC 7.5 with the -O3 optimization flag. We also implemented the same two versions of QS on our SoC-FPGA device by using Vivado HLS 2019.2.1 to directly convert the annotated and optimized C++ code into Register Transfer Level (RTL) code for the FPGA logic.[4] We tested the FPGA implementation on a Zynq UltraScale+MPSoC ZCU102 device with a quad-core ARM CortexTM-A53 processor, dual-core Cortex-R5 real-time processor and Mali-400 MP2 graphics processing unit. The UltraScale FPGA consists of a PS and a PL block integrated on a single die and running independently.

We present the efficiency achieved by our QS deployments on CPU and FPGA for the MSN30K and Istella-S datasets in Table 3. Results are reported in terms of per-instance average inference time measured in μsecs by varying the replication factor (up to 12). In the case of the FPGA this indicates how many times we replicate the PEs inside the IP block (see Fig. 1). For the CPU implementation, it indicates instead the number of parallel threads used to predict the scores. In both cases the execution time is measured on the whole test

---

[4] Code available at https://github.com/hpclab/model_compression_for_ranking_on_fpga.

Table 3. Per-instance average inference time (μsec).

Dataset	QS version	Replication Factor				
		1	2	4	8	12
MSN30K	FPGA-FP	944	502.8	252.4	-	-
	FPGA-BQ	10.8	9.1	4.4	3.3	3.1
	CPU-FP	14.2	7.16	4.15	2.29	1.23
	CPU-BQ	12.8	6.45	3.88	2.15	1.10
Istella-S	FPGA-FP	1072	601.5	338.2	-	-
	FPGA-BQ	9.9	6.7	4.9	4.1	3.8
	CPU-FP	17.5	8.79	5.13	2.87	1.34
	CPU-BQ	16.1	8.11	4.71	2.63	1.24

set and then divided by the number of instances in the test set. The FPGA Full-Precision version (FPGA-FP in Table 3), i.e., the one that do not exploit binning and quantization, obtains an average instance scoring time of 252.4 µs for the MSN30K and 269.2 µs for the Istella-S with 4 replicas. We are not able to increase further the replication factor due to the over-utilization of the resources. The model compression techniques detailed in Sect. 4 allow instead to increase the number of replicas of the scoring logic up to 12. Moreover, the QS version exploiting binning and quantization also improve significantly the inference time due to the many optimizations introduced.

We report the results achieved with the optimized version (FPGA-BQ), which includes loop unrolling and additional if-else statements splitting the QS loops in fixed-size blocks of instructions processed in parallel by the FPGA hardware. By introducing all these optimizations, QS shows a significantly improved average scoring time ranging from 10.8 to 3.1 µs for MSN30K and from 9.9 µs to 3.8 µs for Istella-S. In both cases with 12 replicas we measure a resource utilization exceeding 82%. On the other side, the multi-threaded CPU version of QS that uses binning and quantization (CPU-BQ) techniques also outperforms the version implemented without these techniques (CPU-FP), thus proving that model compression is advantageous even on traditional hardware. Overall, with 8 or more threads running on different cores, the CPU-BQ version obtains lower scoring time than the optimized FPGA-BQ version. However, these slightly higher inference times are counterbalanced by a much lower power consumption as discussed in Sect. 5.4.

5.3 Data Transfer Assessment

The experimental evaluation reported in Table 3 does not include the time needed to transfer data from the host machine performing the retrieval of the candidate documents to the SoC-FPGA device aimed at re-ranking the list of candidates to produce the final results. We experimentally evaluated the impact of the data transfer by conducting additional tests with the MSN30K dataset on an instance of Amazon AWS EC2 F1 node equipped with Xilinx UltraScale+ VU9P FPGAs. These devices are in fact connected to the host machine via a dedicated PCIe Gen3 ×16 bus supporting data transfer with a maximum bandwidth of up to 16 GB/s. We used the AWS instance above for running a simulation measuring the actual bandwidth available when transferring: i) batches of query-document feature vectors from the AWS Amazon EC2 node to the SoC-FPGA, and ii) the resulting scores back from the FPGA to the server.

The result of this simulation shows that the PCIe interface connecting the host machine and the SoC-FPGA device allows to transfer each vector of 136 features, represented with only 136 bytes thanks to our lossless binning technique, with an average latency of 0.24 μs. This latency, although very low, is not an additional overhead to be included in the whole query processing system. Indeed, in a real-world scenario the host device and the SoC-FPGA device operate in pipeline. The host is aimed at retrieving candidate documents from the index and computing query-document feature vectors. The FPGA is instead responsible of inferring the final score to be assigned to each document by using the compressed ensemble model and the QS algorithm. The two operations can be easily pipelined. The resulting scores packed and returned back to the host device can be managed in a similar way. All these query processing operations can be thus overlapped during execution, and the final throughput is given by the slower stage of the pipeline. Since the data transfer time is one order of magnitude lower than the inference on the FPGA (0.24 μs ≪ 3.1 μs) we can conclude that the impact of transferring data from the host to the FPGA and viceversa is negligible on the total latency in a real-world production system.

5.4 Energy Consumption Assessment

We present the results of our energy consumption analysis for running on CPU and FPGA the inference task in Table 4. We report the results as the average energy (in μJoule) spent by the QS algorithm for scoring one single instance. For CPU implementations, we perform the analysis by employing the Mammut[5] library [28]. We use Mammut to read the total energy consumption of the CPU exposed by means of hardware energy counter registries available on Intel CPU architectures. All energy measures obtained are discounted by the energy spent by cores not used by the scoring process. For the FPGA implementation, we employ the Maxim Power Tool USB-to-PMBus Interface Dongle.[6] All energy measures obtained include the energy spent by all components in the PS and in the PL. The PL frequency is set at 200 MHz. Results in Table 4 show that the FPGA-BQ deployment significantly reduces on both datasets the energy consumption measured with respect to the CPU implementations. Specifically, on FPGA we measure an energy consumption for inference that is from 9× up to 19.8× lower than on the CPU. This large difference, consistent with measures

---

[5] https://github.com/DanieleDeSensi/mammut.
[6] https://www.maximintegrated.com/en/products/power/switching-regulators/maxpowertool002.html.

Table 4. Per-instance average energy consumption (μJoule).

Dataset	QS version	Replication Factor				
		1	2	4	8	12
MSN30K	FPGA-BQ	37	35	16	14	13
	CPU-BQ	492	316	239	180	173
Istella-S	FPGA-BQ	31.1	22	17.2	17.6	16.9
	CPU-BQ	616	398	294	224	217

reported in literature for other computing tasks [27], show that an accurate design of the most demanding components of the ranking pipeline with FPGA technology can impact significantly web-scale systems where energy is a major source of cost.

6   Conclusions and Future Work

Modern programmable logic provide an interesting energy-aware alternative to traditional servers for several high-performance tasks. In this paper we tackled the exploitation of SoC-FPGA devices for demanding inference tasks based on complex machine-learned models. We proposed to use binning and quantization to compress additive ensemble of decision trees and increase the number of inference tasks processed in parallel on the FPGA logic. The use case considered was ad-hoc retrieval with fully-optimized LtR models, where finding the best trade-off between accuracy and efficiency is definitely important. Reproducible experiments show that our model compression techniques do not impact the prediction accuracy in a statistically significant measure and that the deployment of a ranking solution based on state-of-the-art algorithms on a low-cost SoC-FPGA device achieves scoring times comparable to those measured on high-end multi-core CPUs. We also showed that the data transfer supplied by the PCIe interface connecting the FPGA to the host machine contribute with a negligible latency with respect to the one of the scoring. On the other hand, the SoC-FPGA solution consumes one order of magnitude less energy in performing the inference. This result can impact significantly large-scale systems where inference is a key task and energy is a major source of cost.

As future work we will investigate the exploitation of SoC-FPGA architectures in the design of optimized algorithms for other inference tasks possibly benefiting from programmable logic. Specifically for the IR domain, we will study the feasibility of this technology for neural ranking.

Acknowledgements. This work was partially supported by the project HAMLET: Hardware Acceleration of Machine LEarning Tasks, funded by CONICET (Argentina) and CNR (Italy) 2017-2018 collaboration program, by the TEACHING project, funded by the EU Horizon 2020 Research and Innovation program (Grant agreement ID: 871385), and by the OK-INSAID project, funded by the Italian Ministry of Education and Research (GA no. ARS01_00917).

References

1. Asadi, N., Lin, J.: Training efficient tree-based models for document ranking. In: Serdyukov, P., et al. (eds.) ECIR 2013. LNCS, vol. 7814, pp. 146–157. Springer, Heidelberg (2013). https://doi.org/10.1007/978-3-642-36973-5_13
2. Bergstra, J., Yamins, D., Cox, D.: Making a science of model search: hyperparameter optimization in hundreds of dimensions for vision architectures. In: Proceedings of ICML, pp. 115–123. PMLR (2013)
3. Burges, C.J.: From ranknet to lambdarank to lambdamart: an overview. Learning 11(23–581), 81 (2010)
4. Busolin, F., Lucchese, C., Nardini, F.M., Orlando, S., Perego, R., Trani, S.: Learning early exit strategies for additive ranking ensembles. In: Proceedings of SIGIR, pp. 2217–2221. ACM (2021)
5. Cambazoglu, B.B., et al.: Early exit optimizations for additive machine learned ranking systems. In: Proceedings of WSDM, pp. 411–420. ACM (2010)
6. Capannini, G., Lucchese, C., Nardini, F.M., Orlando, S., Perego, R., Tonellotto, N.: Quality versus efficiency in document scoring with learning-to-rank models. Inf. Process. Manag. 52(6), 1161–1177 (2016)
7. Chapelle, O., Chang, Y.: Yahoo! Learning to rank challenge overview. J. Mach. Learn. Res. 14, 1–24 (2011). Proceedings Track
8. Chen, R.C., Gallagher, L., Blanco, R., Culpepper, J.S.: Efficient cost-aware cascade ranking in multi-stage retrieval. In: Proceedings of SIGIR, pp. 445–454. ACM (2017)
9. Chen, T., Guestrin, C.: XGBoost: a scalable tree boosting system. In: Proceedings of ACM SIGKDD, pp. 785–794. ACM (2016)
10. Dato, D., et al.: Fast ranking with additive ensembles of oblivious and non-oblivious regression trees. ACM Trans. Inf. Syst. 35(2), 15:1–15:31 (2016)
11. Friedman, J.H.: Greedy function approximation: a gradient boosting machine. Ann. Stat. 29, 1189–1232 (2000)
12. Gallagher, L., Chen, R.C., Blanco, R., Culpepper, J.S.: Joint optimization of cascade ranking models. In: Proceedings of WSDM, pp. 15–23. ACM (2019)
13. Gao, R., Hsu, F.H.: An FPGA-based accelerator for LambdaRank in web search engines. ACM TRETS 4, 1–19 (2011)
14. Han, S., Mao, H., Dally, W.J.: Deep compression: compressing deep neural network with pruning, trained quantization and Huffman coding. In: Proceedings of ICLR (2016)
15. Järvelin, K., Kekäläinen, J.: Cumulated gain-based evaluation of IR techniques. ACM Trans. Inf. Syst. 20(4), 422–446 (2002)
16. Ke, G., et al.: LightGBM: a highly efficient gradient boosting decision tree. In: Proceedings of NIPS, pp. 3149–3157 (2017)
17. Lettich, F., et al.: Parallel traversal of large ensembles of decision trees. IEEE TPDS 30(9), 2075–2089 (2019)
18. Li, Q., Wang, E., Fleming, S.T., Thomas, D., Cheung, P.: Accelerating position-aware top-k ListNet for ranking under custom precision regimes. In: Proceedings of FPL, pp. 81–87 (2019)
19. Liu, T.Y.: Learning to rank for information retrieval. Found. Trends Inf. Retr. 3(3), 225–331 (2009)
20. Lucchese, C., Nardini, F.M., Orlando, S., Perego, R., Tonellotto, N., Venturini, R.: QuickScorer: a fast algorithm to rank documents with additive ensembles of regression trees. In: Proceedings of SIGIR, pp. 73–82. ACM (2015)

21. Lucchese, C., Nardini, F.M., Orlando, S., Perego, R., Silvestri, F., Salvatore, T.: X-CLEaVER: learning ranking ensembles by growing and pruning trees. ACM TIST **9**, 1–26 (2018)
22. Lucchese, C., Nardini, F.M., Orlando, S., Perego, R., Silvestri, F., Trani, S.: Post-learning optimization of tree ensembles for efficient ranking. In: Proceedings of SIGIR, pp. 949–952 (2016)
23. Lucchese, C., Nardini, F.M., Orlando, S., Perego, R., Tonellotto, N., Venturini, R.: Exploiting CPU SIMD extensions to speed-up document scoring with tree ensembles. In: Proceedings of ACM SIGIR, pp. 833–836 (2016)
24. Lucchese, C., Nardini, F.M., Orlando, S., Perego, R., Trani, S.: X-DART: blending dropout and pruning for efficient learning to rank. In: Proceedings of SIGIR, pp. 1077–1080. ACM (2017)
25. Lucchese, C., Nardini, F.M., Orlando, S., Perego, R., Trani, S.: Query-level early exit for additive learning-to-rank ensembles. In: Proceedings of SIGIR, pp. 2033–2036. ACM (2020)
26. Molina, R., Loor, F., Gil-Costa, V., Nardini, F.M., Perego, R., Trani, S.: Efficient traversal of decision tree ensembles with FPGAs. J. Parallel Distrib. Comput. **155**, 38–49 (2021)
27. Qasaimeh, M., Denolf, K., Lo, J., Vissers, K., Zambreno, J., Jones, P.H.: Comparing energy efficiency of CPU, GPU and FPGA implementations for vision kernels. In: Proceedings of IEEE ICESS, pp. 1–8 (2019)
28. Sensi, D.D., Torquati, M., Danelutto, M.: Mammut: high-level management of system knobs and sensors. SoftwareX **6**, 150–154 (2017)
29. Smucker, M.D., Allan, J., Carterette, B.: A comparison of statistical significance tests for information retrieval evaluation. In: Proceedings of CIKM. ACM (2007)
30. Wang, L., Lin, J., Metzler, D.: Learning to efficiently rank. In: Proceedings of SIGIR, pp. 138–145. ACM, New York (2010)
31. Wu, Q., Burges, C., Svore, K., Gao, J.: Adapting boosting for information retrieval measures. Inf. Retrieval **13**, 254–270 (2010). https://doi.org/10.1007/s10791-009-9112-1
32. Xin, J., Tang, R., Yu, Y., Lin, J.: BERxiT: early exiting for BERT with better fine-tuning and extension to regression. In: Proceedings of ACL, pp. 91–104. ACL, April 2021
33. Xu, N.Y., Cai, X.F., Gao, R., Zhang, L., Hsu, F.H.: FPGA acceleration of RankBoost in web search engines. ACM TRETS **1**(4), 1–19 (2009)
34. Ye, T., Zhou, H., Zou, W.Y., Gao, B., Zhang, R.: RapidScorer: fast tree ensemble evaluation by maximizing compactness in data level parallelization. In: Proceedings of SIGKDD, pp. 941–950. ACM (2018)

Local Citation Recommendation with Hierarchical-Attention Text Encoder and SciBERT-Based Reranking

Nianlong Gu[✉], Yingqiang Gao, and Richard H. R. Hahnloser[✉]

Institute of Neuroinformatics, University of Zurich and ETH Zurich,
Zürich, Switzerland
{nianlong,yingqiang.gao,rich}@ini.ethz.ch

Abstract. The goal of local citation recommendation is to recommend a missing reference from the local citation context and optionally also from the global context. To balance the tradeoff between speed and accuracy of citation recommendation in the context of a large-scale paper database, a viable approach is to first prefetch a limited number of relevant documents using efficient ranking methods and then to perform a fine-grained reranking using more sophisticated models. In that vein, BM25 has been found to be a tough-to-beat approach to prefetching, which is why recent work has focused mainly on the reranking step. Even so, we explore prefetching with nearest neighbor search among text embeddings constructed by a hierarchical attention network. When coupled with a SciBERT reranker fine-tuned on local citation recommendation tasks, our hierarchical Attention encoder (HAtten) achieves high prefetch recall for a given number of candidates to be reranked. Consequently, our reranker requires fewer prefetch candidates to rerank, yet still achieves state-of-the-art performance on various local citation recommendation datasets such as ACL-200, FullTextPeerRead, RefSeer, and arXiv.

Keywords: Local citation recommendation · Hierarchical attention · Document reranking

1 Introduction

Literature discovery, such as finding relevant scientific articles, remains challenging in today's age of information overflow, largely arising from the exponential growth in both the publication record [15] and the underlying vocabulary [13]. Assistance to literature discovery can be provided with automatic citation recommendation, whereby a query text without citation serves as the input to a recommendation system and a paper worth citing as its output [9].

Citation recommendation can be dealt with either as a global retrieval problem [2,26,33] or as a local one [12,14,16]. In global citation recommendation, the query text is composed of the title and the abstract of a source paper [2]. In

M. Hagen et al. (Eds.): ECIR 2022, LNCS 13185, pp. 274–288, 2022.
https://doi.org/10.1007/978-3-030-99736-6_19

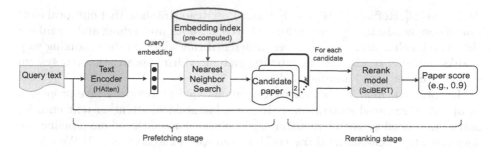

Fig. 1. Overview of our two-stage local citation recommendation pipeline.

contrast, in local citation recommendation, the query consist of two sources of contexts [12,24]: 1) the text surrounding the citation placeholder with the information of the cited paper removed (the **local context**); and 2) the title and abstract of the citing paper as the **global context**. The aim of local citation recommendation is to find the missing paper cited at the placeholder of the local context. In this paper we focus on local citation recommendation.

It is important for a local citation recommendation system to maintain a balance between accuracy (e.g., recall of the target paper among the top K recommended papers) and speed in order to operate efficiently on a large database containing millions of scientific papers. The speed-accuracy tradeoff can be flexibly dealt with using a two-step prefetching-reranking strategy: 1) A fast prefetching model first retrieves a set of candidate papers from the database; 2) a more sophisticated model then performs a fine-grained analysis of scoring candidate papers and reordering them to result in a ranked list of recommendations. In many recent studies [4,6,21,24], either (TF-IDF) [29] or BM25 [30] were used as the prefetching algorithm, which were neither fine-tuned nor taken into consideration when evaluating the recommendation performance.

In this paper, we propose a novel two-stage local citation recommendation system (Fig. 1). In the prefetching stage, we make use of an embedding-based paper retrieval system, in which a siamese text encoder first pre-computes a vector-based embedding for each paper in the database. The query text is then mapped into the same embedding space to retrieve the K nearest neighbors of the query vector. To encode queries and papers of various lengths in a memory-efficient way, we design a two-layer Hierarchical Attention-based text encoder (HAtten) that first computes paragraph embeddings and then computes from the paragraph embeddings the query and document embeddings using a self-attention mechanism [34]. In the reranking step, we fine-tune the SciBERT [1] to rerank the candidates retrieved by the HAtten prefetching model.

In addition, to cope with the scarceness of large-scale training datasets in many domains, we construct a novel dataset that we distilled from 1.7 million arXiv papers. The dataset consist of 3.2 million local citation sentences along with the title and the abstract of both the citing and the cited papers. Extensive experiments on the arXiv dataset as well as on previous datasets including

ACL-200 [24], RefSeer [6,24], and FullTextPeerRead [16] show that our local citation recommendation system performs better on both prefetching and reranking
than the baseline and requires fewer prefetched candidates in the reranking step
thanks to higher recall of our prefetching system, which indicates that our system
strikes a better speed-accuracy balance.

In total, our main contributions are summarized as follows: 1) We propose a
competitive retrieval system consisting of a hierarchical-attention text encoder
and a fine-tuned SciBERT reranker. 2) In evaluations of the whole pipeline, we
demonstrate a well-balanced tradeoff between speed and accuracy. 3) We release
our code and a large-scale scientific paper dataset[1] for training and evaluation
of production-level local citation recommendation systems.

2 Related Work

Local citation recommendation was previously addressed in He et al. [12] in
which a non-parametric probabilistic model was proposed to model the relevance
between the query and each candidate citation. In recent years, embedding-based
approaches [10,19] have been proposed to more flexibly capture the resemblance between the query and the target according to the cosine distance or
the Euclidean distance between their embeddings. Jeong et al. [16] proposed
a BERT-GCN model in which they used Graph Convolutional Networks [18]
(GCN) and BERT [5] to compute for each paper embeddings of the citation
graph and the query context, which they fed into a feed-forward network to estimate relevance. The BERT-GCN model was evaluated on small datasets of only
thousands of papers, partly due to the high cost of computing the GCN, which
limited its scalability for recommending citations from large paper databases.
Although recent studies [4,6,21,24] adopted the prefetching-reranking strategy
to improve the scalability, the prefetch part (BM25 or TF-IDF) only served for
creating datasets for training and evaluating the reranking model, since the target cited paper was added manually if it was not retrieved by the prefetch model,
i.e. the recall of the target among the candidate papers was set to 1. Therefore,
these recommendation systems were evaluated in an artificial situation with an
ideal prefetching model that in reality does not exist.

Supervised methods for citation recommendation rely on the availability of
numerous labeled data for training. It is challenging to assemble a dataset for
local citation recommendation due to the need of parsing the full text of papers
to extract the local contexts and finding citations that are also available in
the dataset, which eliminates a large bulk of data. Therefore, existing datasets
on local citation recommendation are usually limited in size. For example, the
ACL-200 [24] and the FullTextPeerRead [16] contain only thousands of papers.
One of the largest datasets is RefSeer used in Medić and Šnajder [24], which
contains 0.6 million papers in total, but this dataset is not up-to-date as it
only contains papers prior to 2015. Although unarXive [31], a large dataset for

---

[1] Our code and data are available at https://github.com/nianlonggu/Local-Citation-
Recommendation.

citation recommendation, exists, this dataset does not meet the needs of our task because: 1) papers in unarXive are not parsed in a structured manner. For example, the abstract is not separated from the full text, which makes it difficult to construct a global context in our experiments; 2) the citation context is usually a single sentence containing a citation marker, even if the sentence does not contain sufficient contextual information, e.g., "For details, see [#]". These caveats motivate the creation of a novel dataset of high quality.

3   Proposed Dataset

We create a new dataset for local citation recommendation using arXiv papers contained in S2ORC [22], a large-scale scientific paper corpus. Each paper in S2ORC has an identifier of the paper source, such as arXiv or PubMed. Using this identifier, we first obtain all arXiv papers with available titles and abstracts. The title and abstract of each paper are required because they are used as the global context of a query from that paper or as a representation of the paper's content when the latter is a candidate to be ranked. From the papers we then extract the local contexts by parsing those papers for which the full text is available: For each reference in the full text, if the cited paper is also available in the arXiv paper database, we replace the reference marker such as "[#]" or "XXX et al." with a special token such as "CIT", and collect 200 characters surrounding the replaced citation marker as the local context. Note that we "cut off" a word if it lied on the 200-character boundary, following the setting of the ACL-200 and the RefSeer datasets proposed in Medić and Šnajder [24].

Table 1. Statistics of the datasets for local citation recommendation.

Dataset	Number of local contexts			Number of papers	publication years
	Train	Val	Test		
ACL-200	30, 390	9, 381	9, 585	19, 776	2009 – 2015
FullTextPeerRead	9, 363	492	6, 814	4, 837	2007 – 2017
RefSeer	3, 521, 582	124, 911	126, 593	624, 957	– 2014
arXiv (Ours)	2, 988, 030	112, 779	104, 401	1, 661, 201	1991 – 2020

Table 1 shows the statistics of the created arXiv dataset and the comparison with existing datasets used in this paper. As the most recent contexts available in the arXiv dataset is from April 2020, we use the contexts from 1991 to 2019 as the training set, the contexts from January 2020 to February 2020 as the validating set, and the contexts from March 2020 to April 2020 as the test set. The sizes of the arXiv training, validating, and testing sets are comparable to RefSeer, one of the largest existing datasets, whereas our arXiv dataset contains a much larger number of papers, and there are more recently published papers available in the arXiv dataset. These features make the arXiv dataset a more challenging and up-to-date test bench.

4   Approach

Our two-stage telescope citation recommendation system is similar to that of Bhagavatula et al. [2], composed of a fast **prefetching** model and a slower **reranking** model.

4.1   Prefetching Model

The prefetching model scores and ranks all papers in the database to fetch a rough initial subset of candidates. We designed a representation-focused ranking model [11] that computes a query embedding for each input query and ranks each candidate document according to the cosine similarity between the query embedding and the pre-computed document embedding.

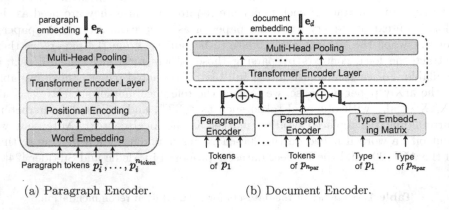

(a) Paragraph Encoder.   (b) Document Encoder.

Fig. 2. The Hierarchical-Attention text encoder (HAtten) used in the prefetching step is composed of a paragraph encoder (a) and a document encoder (b).

The core of the prefetching model is a light-weight text encoder that efficiently computes the embeddings of queries and candidate documents. As shown in Fig. 2, the encoder processes each document or query in a two-level hierarchy, consisting of two components: a paragraph encoder and a document encoder.

Paragraph Encoder. For each paragraph p_i in the document, the paragraph encoder (Fig. 2a) takes as input the token sequence $p_i = [w_1, \ldots, w_{n_i}]$ composed of n_i tokens (words) to output the paragraph embedding e_{p_i} as a single vector. In order to incorporate positional information of the tokens, the paragraph encoder makes use of positional encoding. Contextual information is encoded with a single transformer encoder layer following the configuration in Vaswani et al. [34], Fig. 2a. To obtain a single fixed-size embedding e_p from a variably sized paragraph, the paragraph encoder processes the output of the transformer encoder layer with a multi-head pooling layer [20] with trainable weights. Let

$x_k \in \mathbb{R}^d$ be the output of the transformer encoder layer for token w_k in a paragraph p_i. For each head $j \in \{1, \ldots, n_{\text{head}}\}$ in the multi-head pooling layer, we first compute a value vector $v_k^j \in \mathbb{R}^{d/n_{\text{head}}}$ as well as an attention score $\hat{a}_k^j \in \mathbb{R}$ associated with that value vector:

$$v_k^j = \text{Linear}_v^j(x_k), \quad a_k^j = \text{Linear}_a^j(x_k), \quad \hat{a}_k^j = \frac{\exp a_k^j}{\sum_{m=1}^{n_{\text{token}}} \exp a_m^j}, \tag{1}$$

where Linear() denotes a trainable linear transformation. The weighted value vector $\hat{v}^j$ then results from the sum across all value vectors weighed by their corresponding attention scores: $\hat{v}^j = \sum_{m=1}^{n_{\text{par}}} \hat{a}_m^j v_m^j$. The final paragraph embedding e_p is constructed from the weighted value vectors $\hat{v}^j$ of all heads by a ReLU activation [25] followed by a linear transformation:

$$e_p = \text{Linear}_p(\text{ReLU}(\text{Concat}(\hat{v}^1, \ldots, \hat{v}^{n_{\text{head}}}))). \tag{2}$$

Document Encoder. In order to encode documents with two fields given by the title and the abstract, or to encode queries given by three fields: the local context, the title, and the abstract of the citing paper, we treat each field (local context, title, and abstract) as a paragraph. For a document of n_{par} paragraphs $d = [p_1, \ldots, p_{n_{\text{par}}}]$, we first compute the embeddings of all paragraphs p_i.

Not all fields and paragraphs are treated equally in our document encoder. To allow the document encoder to distinguish between fields, we introduce a *paragraph type* variable, which refers to the field type from which the paragraph originates. We distinguish between three paragraph types: the title, the abstract, and the local context. Each type is associated with a learnable type embedding that has the same dimension as the paragraph embedding. Inspired by the BERT model [5], we produce a type-aware paragraph embedding by adding the type embedding of the given paragraph to the corresponding paragraph embedding (Fig. 2b). All type-aware paragraph embeddings are then fed into a transformer encoder layer followed by a multi-head pooling layer (of identical structures as the ones in the paragraph encoder), which then results in the final document embedding e_d.

Prefetched Document Candidates. The prefetched document candidates are found by identifying the K nearest document embeddings to the query embedding in terms of cosine similarity. The ranking is performed using a brute-force nearest neighbor search among all document embeddings as shown in Fig. 1.

4.2 Reranking Model

The reranking model performs a fine-grained comparison between a query q (consisting of a local and a global context) and each prefetched document candidate (its title and the abstract). The relevance scores of the candidates constitute the final output of our model. We design a reranker based on SciBERT [1], which is

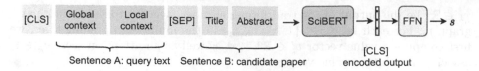

Fig. 3. Structure of our SciBERT Reranker.

a BERT model [5] trained on a large-scale corpus of scientific articles. The input of the SciBERT reranker has the following format: "[CLS] Sentence A [SEP] Sentence B", where sentence A is the concatenation of the global context (title and abstract of the citing paper) and the local context of the query, and sentence B is the concatenation of the title and the abstract of the candidate paper to be scored, Fig. 3. The SciBERT-encoded vector for the "[CLS]" token is then fed into a feed-forward network that outputs the relevance score $s \in [0,1]$ provided via a sigmoid function.

4.3   Loss Function

We use a triplet loss both to train our HAtten text encoder for prefetching and to finetune the SciBERT reranker. The triplet loss is based on the similarity $s(q,d)$ between the query q and a document d. For the prefetching step, $s(q,d)$ is given by the cosine similarity between the query embedding v_q and the document embedding v_d, both computed with the HAtten encoder. For the reranking step, $s(q,d)$ is given by the relevance score computed by the SciBERT reranker. In order to maximize the relevance score between the query q and the cited document d_+ (the positive pair (q,d_+)) and to minimize the score between q and any non-cited document d_- (a negative pair (q,d_-)), we minimize the triplet loss:

$$\mathcal{L} = \max[s(q,d_-) - s(q,d_+) + m, 0] \tag{3}$$

where the margin $m > 0$ sets the span over which the loss is sensitive to the similarity of negative pairs.

For fast convergence during training, it is important to select effective triplets for which $\mathcal{L}$ in Eq. (3) is non-zero [32], which is particularly relevant for the prefetching model, since for each query there is only a single positive document but millions of negative documents (e.g., on the arXiv dataset). Therefore, we employ negative and positive mining strategies to train our HAtten encoder, described as follows.

Negative Mining. Given a query q, we use HAtten's current checkpoint to prefetch the top K_n candidates excluding the cited paper. The HAtten embedding of these prefetched non-cited candidates have high cosine similarity to the HAtten embedding of the query. To increase the similarity between the query and the cited paper while suppressing the similarity between the query and these non-cited candidates, we use the cited paper as the positive document and select the negative document from these K_n overly similar candidates.

Positive Mining. Among the prefetched non-cited candidates, the documents with objectively high textual similarity (e.g. measured by word overlapping, such as the Jaccard index [2]) to the query were considered relevant to the query, even if they were not cited. These textually relevant candidate documents should have a higher cosine similarity to the query than randomly selected documents. Therefore, in parallel with the negative mining strategy, we also select positive documents from the set of textually relevant candidates and select negative documents by random sampling from the entire dataset.

The checkpoint of the HAtten model is updated every N_{iter} training iterations, at which point the prefetched non-cited and the textually relevant candidates for negative and positive mining are updated as well.

In contrast, when fine-tuning SciBERT for reranking, the reranker only needs to rerank the top K_r prefetched candidates. This allows for a simpler triplet mining strategy, which is to select the cited paper as the positive document and randomly selecting a prefetched non-cited papers as the negative document.

5   Experiments

Implementation Details. In the prefetching step, we used as word embeddings of the HAtten text encoder the pre-trained 200-dimensional GloVe embeddings [28], which were kept fixed during training. There are 64 queries in a mini-batch, each of which was accompanied by 1 cited paper, 4 non-cited papers randomly sampled from the top $K_n = 100$ prefetched candidates, and 1 randomly sampled paper from the whole database, which allow us to do negative and positive mining with the mini-batch as described in Sect. 4.3. The HAtten's checkpoint was updated every $N_{\text{iter}} = 5000$ training iterations.

In the reranking step, we initialized the SciBERT reranker with the pre-trained model provided in Beltagy et al. [1]. The feed-forward network in Fig. 3 consisting of a single linear layer was randomly initialized. Within a mini-batch there was 1 query, 1 cited paper (positive sample), and 62 documents (negative samples) randomly sampled from the top $K_r = 2000$ prefetched non-cited documents. In the triplet loss function the margin m was set to 0.1.

We used the Adam optimizer [17] with $\beta_1 = 0.9$ and $\beta_2 = 0.999$. In the prefetching step, the learning rate was set to $\alpha = 1e^{-4}$ and the weight decay to $1e^{-5}$, while in the reranking step these were set to $1e^{-5}$ and to $1e^{-2}$ for fine-tuning SciBERT, respectively. The models were trained on eight NVIDIA GeForce RTX 2080 Ti 11 GB GPUs and tested on two Quadro RTX 8000 GPUs.

Evaluation Metrics. We evaluated the recommendation performance using the Mean Reciprocal Rank (MRR) [35] and the Recall@K (R@K for short), consistent with previous work [8,16,24]. The MRR measures the reciprocal rank of the actually cited paper among the recommended candidates, averaged over multiple queries. The R@K evaluates the percentage of the cited paper appearing in the top K recommendations.

Baselines. In the prefetching step, we compare our HAtten with the following baselines: BM25, Sent2vec [27], and NNSelect [2]. BM25 was used as the prefetching method in previous works [6,21,24]. Sent2vec is an unsupervised text encoder which computes a text embedding by averaging the embeddings of all words in the text. We use the 600-dim Sent2vec pretrained on Wikipedia. NNSelect [2] computes text embeddings also by averaging, and the trainable parameters are the magnitudes of word embeddings that we trained on each dataset using the same training configuration as our HAtten model.

In the reranking step, we compare our fine-tuned SciBERT reranker with the following baselines: 1) a Neural Citation Network (NCN) with an encoder-decoder architecture [6,7]; 2) DualEnh and DualCon [24] that score each candidate using both semantic information and bibliographic information and 3) BERT-GCN [16]. Furthermore, to analyze the influence on ranking performance of diverse pretraining corpuses for BERT, we compared our SciBERT reranker with a BERT reranker that was pretrained on a non-science specific corpus [5] and then fine-tuned on the reranking task.

For a fair performance comparison of our reranker with those of other works, we adopted the prefetching strategies from each of these works. On ACL-200 and RefSeer, we tested our SciBERT reranker on the test sets provided in Medić and Šnajder [24]. For each query in the test set, we prefetched n ($n = 2000$ for ACL-200 and $n = 2048$ for RefSeer) candidates using BM25, and manually added the cited paper as candidate if it was not found by BM25. In other words, we constructed our test set using an "oracle-BM25" with R@$n = 1$. On the FullTextPeerRead dataset, we used our SciBERT reranker to rank all papers in the database without prefetching, in line with the setting in BERT-GCN [16]. On our newly proposed arXiv dataset, we fetched the top 2000 candidates for each query in the test set using the "oracle-BM25" as introduced above.

6 Results and Discussion

In this section, we first present the evaluation results of our prefetching and reranking models separately and compare them with baselines. Then, we evaluate the performance of the entire prefetching-reranking pipeline, and analyze the influence of the number of prefetched candidates to be reranked on the overall recommendation performance.

6.1 Prefetching Results

Our HAtten model significantly outperformed all baselines (including the strong baseline BM25, Table 2) on the ACL-200, RefSeer and the arXiv datasets, evaluated in terms of MRR and R@K. We observed that, first, for larger K, such as $K = 200, 500, 1000, 2000$, the improvement of R@K with respect to the baselines is more pronounced on all four datasets, where the increase is usually larger than 0.1, which means that the theoretical upper bound of the final reranking recall will be higher when using our HAtten prefetching system. Second, the improvements of R@K on large datasets such as RefSeer and arXiv are more prominent than on small datasets such as ACL-200 and FullTextPeerRead, which fits well with the stronger need of a prefetching-reranking pipeline on large datasets due to the speed-accuracy tradeoff.

Table 2. Prefetching performance. For Tables 2, 3 and 4, the asterisks "*" indicate statistical significance ($p < 0.05$) in comparison with the closest baseline in a t-test. The red color indicates a large (> 0.8) Cohen's d effect size [3].

Dataset	Model	avg. prefetch time (ms)	MRR	R@10	R@100	R@200	R@500	R@1000	R@2000
ACL-200	BM25	9.9 ± 20.1	0.138	0.263	0.520	0.604	0.712	0.791	0.859
	Sent2vec	1.8 ± 19.5	0.066	0.127	0.323	0.407	0.533	0.640	0.742
	NNSelect	1.8 ± 3.8	0.076	0.150	0.402	0.498	0.631	0.722	0.797
	HAtten	2.7 ± 3.8	**0.148***	**0.281***	**0.603***	**0.700***	**0.803***	**0.870***	**0.924***
FullText-PeerRead	BM25	5.1 ± 18.6	**0.185***	**0.328***	0.609	0.694	0.802	0.877	0.950
	Sent2vec	1.7 ± 19.6	0.121	0.215	0.462	0.561	0.694	0.794	0.898
	NNSelect	1.7 ± 4.8	0.130	0.255	0.572	0.672	0.790	0.869	0.941
	HAtten	2.6 ± 4.9	0.167	0.306	**0.649***	**0.750***	**0.870***	**0.931***	**0.976***
RefSeer	BM25	216.2 ± 84.9	0.099	0.189	0.398	0.468	0.561	0.631	0.697
	Sent2vec	6.0 ± 20.9	0.061	0.111	0.249	0.306	0.389	0.458	0.529
	NNSelect	4.3 ± 5.5	0.044	0.080	0.197	0.250	0.331	0.403	0.483
	HAtten	6.2 ± 7.3	**0.115***	**0.214***	**0.492***	**0.589***	**0.714***	**0.795***	**0.864***
arXiv	BM25	702.2 ± 104.7	0.118	0.222	0.451	0.529	0.629	0.700	0.763
	Sent2vec	11.3 ± 13.6	0.072	0.131	0.287	0.347	0.435	0.501	0.571
	NNSelect	6.9 ± 4.6	0.042	0.079	0.207	0.266	0.359	0.437	0.520
	HAtten	8.0 ± 4.5	**0.124***	**0.241***	**0.527***	**0.619***	**0.734***	**0.809***	**0.871***

The advantage of our HAtten model is also reflected in the average prefetching time. As shown in Table 2, the HAtten model shows faster prefetching than BM25 on large datasets such as RefSeer and arXiv. This is because for HAtten, both text encoding and embedding-based nearest neighbor search can be accelerated by GPU computing, while BM25[2] benefits little from GPU acceleration because it is not vector-based. Although other embedding-based baselines such as Sent2vec and NNSelect also exhibit fast prefetching, our HAtten prefetcher has advantages in terms of both speed and accuracy.

6.2 Reranking Results

As shown in Table 3, the SciBERT reranker significantly outperformed previous state-of-the-art models on the ACL-200, the RefSeer, and the FullTextPeerRead datasets. We ascribe this improvement to BERT's ability of capturing the semantic relevance between the query text and the candidate text, which is inherited from the "next sentence prediction" pretraining task that aims to predict if two sentences are consecutive. The SciBERT reranker also performed significantly

---

[2] We implemented the Okapi BM25 [23], with $k = 1.2, b = 0.75$.

better than its BERT counterpart, suggesting that large language models pretrained on scientific papers' corpus are advantageous for citation reranking.

Table 3. Comparison of reranking performance on four datasets.

Model	ACL-200		FullTextPeerRead		RefSeer		arXiv	
	MRR	R@10	MRR	R@10	MRR	R@10	MRR	R@10
NCN	-	-	-	-	0.267	0.291	-	-
DualCon	0.335	0.647	-	-	0.206	0.406	-	-
DualEnh	0.366	0.703	-	-	0.280	0.534	-	-
BERT-GCN	-	-	0.418	0.529	-	-	-	-
BERT Reranker	0.482	0.736	0.458	0.706	0.309	0.535	0.226	0.399
SciBERT Reranker	**0.531***	**0.779***	**0.536***	**0.773***	**0.380***	**0.623***	**0.278***	**0.475***

6.3   Performance of Entire Recommendation Pipeline

Table 4. The performance of the entire prefetching-reranking pipeline, measured in terms of R@10 of the final reranked document list. We varied the number of prefetched candidates for reranking. For the RefSeer and arXiv datasets, we evaluated performance on a subset of 10K examples from the test set due to computational resource limitations.

Dataset	Recommendation Pipeline		Number of reranked candidates				
	Prefetch	Rerank	100	200	500	1000	2000
ACL-200	BM25	SciBERT$_{BM25}$	0.457	0.501	0.549	0.577	0.595
	HAtten	SciBERT$_{HAtten}$	**0.513***	**0.560***	**0.599***	**0.619***	**0.633***
FullText-PeerRead	BM25	SciBERT$_{BM25}$	0.527	0.578	0.639	0.680	0.720
	HAtten	SciBERT$_{HAtten}$	**0.586***	**0.651***	**0.713***	**0.739***	**0.757***
RefSeer	BM25	SciBERT$_{BM25}$	0.305	0.332	0.365	0.380	0.383
	HAtten	SciBERT$_{HAtten}$	**0.362***	**0.397***	**0.428***	**0.443***	**0.454***
arXiv	BM25	SciBERT$_{BM25}$	0.333	0.357	0.377	0.389	0.391
	HAtten	SciBERT$_{HAtten}$	**0.374***	**0.397***	**0.425***	**0.435***	**0.439***

The evaluation in Sect. 6.2 only reflects the reranking performance because the prefetched candidates are obtained by an oracle-BM25 that guarantees inclusion of the cited paper among the prefetched candidates, even though such an oracle prefetching model does not exist in reality. Evaluating recommendation systems in this context risks overestimating the performance of the reranking part and underestimating the importance of the prefetching step. To better understand the recommendation performance in real-world scenarios, we compared two pipelines: 1) BM25 prefetching + SciBERT reranker fine-tuned on BM25-prefetched candidates, denoted as SciBERT$_{BM25}$; 2) HAtten prefetching + SciBERT$_{HAtten}$ reranker fine-tuned on HAtten-prefetched candidates. We evaluated recommendation performance by R@10 of the final reranked document list and monitored the dependence of R@10 on the number of prefetched candidates for reranking.

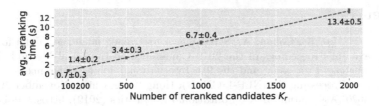

Fig. 4. The reranking time of the SciBERT reranker linearly increases with the number of reranked candidates K_r, tested on arXiv. In comparison, the prefetching time is invariant of K_r, as the prefetcher always scores and ranks all documents in the database to fetch the candidates to be reranked.

As shown in Table 4, the HAtten-based pipeline achieves competitive performance, even when compared with the oracle prefetching model in Sect. 6.2. In particular, on the FullTextPeerRead dataset, using our HAtten-based pipeline, we only need to rerank 100 prefetched candidates to outperform the BERT-GCN model (Table 3) that reranked all 4.8k papers in the database.

Compared to the BM25-based pipeline, our HAtten-based pipeline achieves significantly higher R@10 for any given number of prefetched candidates. Our reranker needs to rerank only 200 to 500 candidates to match the recall score of the BM25-based pipeline needing to rerank 2000 candidates. For large datasets like RefSeer and arXiv, such improvements are even more pronounced. Our pipeline achieves a much higher throughput. For example, on the arXiv dataset, in order to achieve an overall R@10 = 0.39, the BM25-based pipeline takes 0.7 s (Table 2) to prefetch 2000 candidates and it takes another 13.4 s (Fig. 4) to rerank them, which in total amounts to 14.1 s. In contrast, the HAtten-based pipeline only takes 8 ms to prefetch 200 candidates and 1.4 s to rerank them, which amounts to 1.4 s. This results in a 90% reduction of overall recommendation time achieved by our pipeline.

These findings provide clear evidence that a better-performing prefetching model is critical to a large-scale citation recommendation pipeline, as it allows the reranking model to rerank fewer candidates while maintaining recommendation performance, resulting in a better speed-accuracy tradeoff.

7   Conclusion

The speed-accuracy tradeoff is crucial for evaluating recommendation systems in real-world settings. While reranking models have attracted increasing attention for their ability to improve recall and MRR scores, in this paper we show that it is equally important to design an efficient and accurate prefetching system. In this regard, we propose the HAtten-SciBERT recommendation pipeline, in which our HAtten model effectively prefetches a list of candidates with significantly higher recall than the baseline, which allows our fine-tuned SciBERT-based reranker to operate on fewer candidates with better speed-accuracy tradeoff. Furthermore, by releasing our large-scale arXiv-based dataset, we provide a new testbed for research on local citation recommendation in real-world scenarios.

References

1. Beltagy, I., Lo, K., Cohan, A.: SciBERT: a pretrained language model for scientific text. In: Proceedings of the 2019 Conference on Empirical Methods in Natural Language Processing and the 9th International Joint Conference on Natural Language Processing (EMNLP-IJCNLP), Hong Kong, China, November 2019, pp. 3615–3620. Association for Computational Linguistics (2019). https://doi.org/10.18653/v1/D19-1371. https://www.aclweb.org/anthology/D19-1371

2. Bhagavatula, C., Feldman, S., Power, R., Ammar, W.: Content-based citation recommendation. In: Proceedings of the 2018 Conference of the North American Chapter of the Association for Computational Linguistics: Human Language Technologies, Volume 1 (Long Papers), New Orleans, Louisiana, June 2018, pp. 238–251. Association for Computational Linguistics (2018). https://doi.org/10.18653/v1/N18-1022. https://www.aclweb.org/anthology/N18-1022

3. Cohen, J.: Statistical Power Analysis for the Behavioral Sciences. Academic Press, Cambridge (2013)

4. Dai, T., Zhu, L., Wang, Y., Carley, K.M.: Attentive stacked denoising autoencoder with Bi-LSTM for personalized context-aware citation recommendation. IEEE/ACM Trans. Audio Speech Lang. Process. **28**, 553–568 (2020). https://doi.org/10.1109/TASLP.2019.2949925

5. Devlin, J., Chang, M.W., Lee, K., Toutanova, K.: BERT: pre-training of deep bidirectional transformers for language understanding. In: Proceedings of the 2019 Conference of the North American Chapter of the Association for Computational Linguistics: Human Language Technologies, Volume 1 (Long and Short Papers), Minneapolis, Minnesota, June 2019, pp. 4171–4186. Association for Computational Linguistics (2019). https://doi.org/10.18653/v1/N19-1423. https://www.aclweb.org/anthology/N19-1423

6. Ebesu, T., Fang, Y.: Neural citation network for context-aware citation recommendation. In: Proceedings of the 40th International ACM SIGIR Conference on Research and Development in Information Retrieval, SIGIR 2017, New York, NY, USA, pp. 1093–1096. Association for Computing Machinery (2017). https://doi.org/10.1145/3077136.3080730

7. Färber, M., Klein, T., Sigloch, J.: Neural citation recommendation: a reproducibility study. In: BIR@ECIR (2020)

8. Färber, M., Sampath, A.: Hybridcite: a hybrid model for context-aware citation recommendation. In: Proceedings of the ACM/IEEE Joint Conference on Digital Libraries in 2020, JCDL 2020, New York, NY, USA, pp. 117–126. Association for Computing Machinery (2020). https://doi.org/10.1145/3383583.3398534

9. Färber, M., Jatowt, A.: Citation recommendation: approaches and datasets. Int. J. Digit. Libr. **21**(4), 375–405 (2020). https://doi.org/10.1007/s00799-020-00288-2

10. Gökçe, O., Prada, J., Nikolov, N.I., Gu, N., Hahnloser, R.H.: Embedding-based scientific literature discovery in a text editor application. In: Proceedings of the 58th Annual Meeting of the Association for Computational Linguistics: System Demonstrations, Linguistics, pp. 320–326. Association for Computational, July 2020. https://doi.org/10.18653/v1/2020.acl-demos.36. https://www.aclweb.org/anthology/2020.acl-demos.36

11. Guo, J., et al.: A deep look into neural ranking models for information retrieval. Inf. Process. Manag., 102067 (2019)

12. He, Q., Pei, J., Kifer, D., Mitra, P., Giles, L.: Context-aware citation recommendation. In: Proceedings of the 19th International Conference on World Wide Web, pp. 421–430 (2010)

13. Herdan, G.: Type-Token Mathematics, vol. 4. Mouton (1960)
14. Huang, W., Kataria, S., Caragea, C., Mitra, P., Giles, C.L., Rokach, L.: Recommending citations: translating papers into references. In: Proceedings of the 21st ACM International Conference on Information and Knowledge Management, pp. 1910–1914 (2012)
15. Hunter, L., Cohen, K.B.: Biomedical language processing: what's beyond PubMed? Mol. Cell **21**(5), 589–594 (2006)
16. Jeong, C., Jang, S., Park, E.L., Choi, S.: A context-aware citation recommendation model with BERT and graph convolutional networks. Scientometrics **124**(3), 1907–1922 (2020). https://doi.org/10.1007/s11192-020-03561-y
17. Kingma, D.P., Ba, J.: Adam: a method for stochastic optimization. In: Bengio, Y., LeCun, Y. (eds.) 3rd International Conference on Learning Representations, ICLR 2015, San Diego, CA, USA, 7–9 May 2015, Conference Track Proceedings (2015). http://arxiv.org/abs/1412.6980
18. Kipf, T.N., Welling, M.: Semi-supervised classification with graph convolutional networks. In: 5th International Conference on Learning Representations, ICLR 2017, Toulon, France, 24–26 April 2017, Conference Track Proceedings. OpenReview.net (2017). https://openreview.net/forum?id=SJU4ayYgl
19. Kobayashi, Y., Shimbo, M., Matsumoto, Y.: Citation recommendation using distributed representation of discourse facets in scientific articles. In: Proceedings of the 18th ACM/IEEE on Joint Conference on Digital Libraries, JCDL 2018, New York, NY, USA, pp. 243–251. Association for Computing Machinery (2018). https://doi.org/10.1145/3197026.3197059
20. Liu, Y., Lapata, M.: Hierarchical transformers for multi-document summarization. In: Proceedings of the 57th Annual Meeting of the Association for Computational Linguistics, Florence, Italy, July 2019, pp. 5070–5081. Association for Computational Linguistics (2019). https://doi.org/10.18653/v1/P19-1500. https://www.aclweb.org/anthology/P19-1500
21. Livne, A., Gokuladas, V., Teevan, J., Dumais, S.T., Adar, E.: Citesight: supporting contextual citation recommendation using differential search. In: Proceedings of the 37th International ACM SIGIR Conference on Research & Development in Information Retrieval, SIGIR 2014, New York, NY, USA, pp. 807–816. Association for Computing Machinery (2014). https://doi.org/10.1145/2600428.2609585. https://doi.org/10.1145/2600428.2609585
22. Lo, K., Wang, L.L., Neumann, M., Kinney, R., Weld, D.S.: S2orc: the semantic scholar open research corpus. In: Proceedings of the 58th Annual Meeting of the Association for Computational Linguistics, pp. 4969–4983 (2020)
23. Manning, C.D., Raghavan, P., Schütze, H.: Introduction to Information Retrieval. Cambridge University Press, Cambridge (2008). http://nlp.stanford.edu/IR-book/information-retrieval-book.html
24. Medić, Z., Snajder, J.: Improved local citation recommendation based on context enhanced with global information. In: Proceedings of the First Workshop on Scholarly Document Processing, pp. 97–103. Association for Computational Linguistics, November 2020. https://doi.org/10.18653/v1/2020.sdp-1.11. https://aclanthology.org/2020.sdp-1.11
25. Nair, V., Hinton, G.E.: Rectified linear units improve restricted Boltzmann machines. In: ICML (2010)
26. Nallapati, R.M., Ahmed, A., Xing, E.P., Cohen, W.W.: Joint latent topic models for text and citations. In: Proceedings of the 14th ACM SIGKDD International Conference on Knowledge Discovery and Data Mining, pp. 542–550 (2008)

27. Pagliardini, M., Gupta, P., Jaggi, M.: Unsupervised learning of sentence embeddings using compositional n-gram features. In: Proceedings of the 2018 Conference of the North American Chapter of the Association for Computational Linguistics: Human Language Technologies, Volume 1 (Long Papers), New Orleans, Louisiana, June 2018, pp. 528–540. Association for Computational Linguistics (2018). https://doi.org/10.18653/v1/N18-1049. https://www.aclweb.org/anthology/N18-1049

28. Pennington, J., Socher, R., Manning, C.: GloVe: global vectors for word representation. In: Proceedings of the 2014 Conference on Empirical Methods in Natural Language Processing (EMNLP), Doha, Qatar October 2014, pp. 1532–1543. Association for Computational Linguistics (2014). https://doi.org/10.3115/v1/D14-1162. https://aclanthology.org/D14-1162

29. Ramos, J., et al.: Using TF-IDF to determine word relevance in document queries. In: Proceedings of the First Instructional Conference On Machine Learning, New Jersey, USA , vol. 242, pp. 133–142 (2003)

30. Robertson, S., Zaragoza, H.: The Probabilistic Relevance Framework: BM25 And Beyond. Now Publishers Inc. (2009)

31. Saier, T., Färber, M.: unarXive: a large scholarly data set with publications' fulltext, annotated in-text citations, and links to metadata. Scientometrics **125**(3), 3085–3108 (2020). https://doi.org/10.1007/s11192-020-03382-z

32. Schroff, F., Kalenichenko, D., Philbin, J.: FaceNet: a unified embedding for face recognition and clustering. In: 2015 IEEE Conference on Computer Vision and Pattern Recognition (CVPR), pp. 815–823 (2015). https://doi.org/10.1109/CVPR.2015.7298682

33. Strohman, T., Croft, W.B., Jensen, D.: Recommending citations for academic papers. In: Proceedings of the 30th Annual International ACM SIGIR Conference on Research and Development in Information Retrieval, pp. 705–706 (2007)

34. Vaswani, A., et al.: Attention is all you need. In: Advances in neural information processing systems, pp. 5998–6008 (2017)

35. Voorhees, E.M.: The TREC-8 question answering track report. In: Proceedings of TREC-8, pp. 77–82 (1999)

Extending CLIP for Category-to-Image Retrieval in E-Commerce

Mariya Hendriksen[1(✉)], Maurits Bleeker[2], Svitlana Vakulenko[2],
Nanne van Noord[2], Ernst Kuiper[3], and Maarten de Rijke[2]

[1] AIRLab, University of Amsterdam, Amsterdam, The Netherlands
m.hendriksen@uva.nl
[2] University of Amsterdam, Amsterdam, The Netherlands
{m.j.r.bleeker,s.vakulenko,n.j.e.vannoord,m.derijke}@uva.nl
[3] Bol.com, Utrecht, The Netherlands
ekuiper@bol.com

Abstract. E-commerce provides rich multimodal data that is barely
leveraged in practice. One aspect of this data is a category tree that is
being used in search and recommendation. However, in practice, during
a user's session there is often a mismatch between a textual and a visual
representation of a given category. Motivated by the problem, we introduce the task of category-to-image retrieval in e-commerce and propose a
model for the task, CLIP-ITA. The model leverages information from multiple modalities (textual, visual, and attribute modality) to create product
representations. We explore how adding information from multiple modalities (textual, visual, and attribute modality) impacts the model's performance. In particular, we observe that CLIP-ITA significantly outperforms
a comparable model that leverages only the visual modality and a comparable model that leverages the visual and attribute modality.

Keywords: Multimodal retrieval · Category-to-image retrieval ·
E-commerce

1 Introduction

Multimodal retrieval is a major but understudied problem in e-commerce [33].
Even though e-commerce products are associated with rich multi-modal information, research currently focuses mainly on textual and behavioral signals
to support product search and recommendation. The majority of prior work
in multimodal retrieval for e-commerce focuses on applications in the fashion domain, such as recommendation of fashion items [21] and cross-modal
fashion retrieval [6,14]. In the more general e-commerce domain, multimodal
retrieval has not been explored that well yet [10,18]. The multimodal problem
on which we focus is motivated by the importance of category information in
e-commerce. Product category trees are a key component of modern e-commerce
as they assist customers when navigating across large and dynamic product catalogues [13,30,36]. Yet, the ability to retrieve an image for a given product

M. Hagen et al. (Eds.): ECIR 2022, LNCS 13185, pp. 289–303, 2022.
https://doi.org/10.1007/978-3-030-99736-6_20

category remains a challenging task mainly due to noisy category and product data, and the size and dynamic character of product catalogues [17,33].

The Category-to-Image Retrieval Task. We introduce the problem of retrieving a ranked list of relevant images of products that belong to a given category, which we call the *category-to-image* retrieval task. Unlike image classification tasks that operate on a predefined set of classes, in the CtI retrieval task we want to be able not only to understand which images belong to a given category but also to generalize towards unseen categories. Consider the category "Home decor." A CtI retrieval should output a ranked list of k images retrieved from the collection of images that are relevant to the category, which could be anything from images of carpets to an image of a clock or an arrangement of decorative vases. Use cases that motivate the CtI retrieval task include (1) the need to showcase different categories in search and recommendation results [13,30,33]; (2) the task can be used to infer product categories in the cases when product categorical data is unavailable, noisy, or incomplete [39]; and (3) the design of cross-categorical promotions and product category landing pages [24].

The CtI retrieval task has several key characteristics:(1) we operate with categories from non-fixed e-commerce category trees, which range from very general (such as "Automative" or "Home & Kitchen") to very specific ones (such as "Helmet Liners" or "Dehumidifiers"). The category tree is not fixed, therefore, we should be able to generalize towards unseen categories; and (2) product information is highly multimodal in nature; apart from category data, products may come with textual, visual, and attribute information.

A Model for CtI Retrieval. To address the CtI retrieval task, we propose a model that leverages image, text, and attribute information, CLIP-ITA. CLIP-ITA extends upon Contrastive Language-Image Pre-Training (CLIP) [26]. CLIP-ITA extends CLIP with the ability to represent attribute information. Hence, CLIP-ITA is able to use textual, visual, and attribute information for product representation. We compare the performance of CLIP-ITA with several baselines such as unimodal BM25, bimodal zero-shot CLIP, and MPNet [29]. For our experiments, we use the XMarket dataset that contains textual, visual, and attribute information of e-commerce products [2].

Research Questions and Contributions. We address the following research questions: (RQ1) How do baseline models perform on the CtI retrieval task? Specifically, how do unimodal and bi-modal baseline models perform? How does the performance differ w.r.t. category granularity? (RQ2) How does a model, named CLIP-I, that uses product image information for building product representations impact the performance on the CtI retrieval task? (RQ3) How does CLIP-IA, which extends CLIP-I with product attribute information, perform on the CtI retrieval task? (RQ4) And finally, how does CLIP-ITA, which extends CLIP-IA with product text information, perform on the CtI task?

Our main contributions are: (1) We introduce the novel task of CtI retrieval and motivate it in terms of e-commerce applications. (2) We propose CLIP-ITA, the first model specifically designed for this task. CLIP-ITA leverages multimodal product data such as textual, visual, and attribute data. On average, CLIP-ITA outperforms CLIP-I on all categories by 217% and CLIP-IA by 269%. We share our code and experimental settings to facilitate reproducibility of our results.[1]

2 Related Work

Learning Multimodal Embeddings. Contrastive pre-training has been shown to be highly effective in learning joined embeddings across modalities [26]. By predicting the correct pairing of image-text tuples in a batch, the CLIP model can learn strong text and image encoders that project to joint space. This approach to learning multimodal embeddings offers key advantages over approaches that use manually assigned labels as supervision: (1) the training data can be collected without manual annotation; real-world data in which image-text pairs occur can be used; (2) models trained in this manner learn more general representations that allow for zero-shot prediction. These advantages are appealing for e-commerce, as most public multimodal e-commerce datasets primarily focus on fashion only [2]; being able to train from real-world data avoids the need for costly data annotation.

We build on CLIP by extending it to category-product pairs, taking advantage of its ability to perform zero-shot retrieval for a variety semantic concepts.

Multimodal Image Retrieval. Early work in image retrieval grouped images into a restricted set of semantic categories and allowed users to retrieve images by using category labels as queries [28]. Later work allowed for a wider variety of queries ranging from natural language [11,34], to attributes [23], to combinations of multiple modalities (e.g., title, description, and tags) [32]. Across these multimodal image retrieval approaches we find three common components: (1) an image encoder, (2) a query encoder, and (3) a similarity function to match the query to images [7,26]. Depending on the focus of the work some components might be pre-trained, whereas the others are optimized for a specific task.

In our work, we rely on pre-trained image and text encoders but learn a new multimodal composite of the query to perform CtI retrieval.

Multimodal Retrieval in E-Commerce. Prior work on multimodal retrieval in e-commerce has been mainly focused on cross-modal retrieval for fashion [6,16, 42]. Other related examples include outfit recommendation [15,19,21] Some prior work on interpretability for fashion product retrieval proposes to leverage multimodal signals to improve explainability of latent features [20,38]. Tautkute et al. [31] propose a multimodal search engine for fashion items and furniture. When it comes to combining signals for improving product retrieval, Yim et al. [40] propose to combine product images, titles, categories, and descriptions to improve product search, Yamaura et al. [37] propose an algorithm that leverages multimodal product information for predicting a resale price of a second-hand product.

Unlike prior work on multimodal retrieval in e-commerce that mainly focuses on fashion data, we focus on creating multimodal product representations for the general e-commerce domain.

---

[1] https://github.com/mariyahendriksen/ecir2022_category_to_image_retrieval.

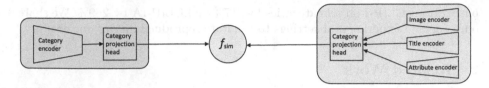

Fig. 1. Overview of CLIP-ITA. The category encoding pipeline is in purple; the category information pipeline in green; f_{sim} is a cosine similarity function. (Color figure online)

3   Approach

Task Definition. We follow the same notation as in [41]. The input dataset can be presented as category-product pairs $(\mathbf{x}_c, \mathbf{x}_p)$, where $\mathbf{x}_c$ represents a product category, and $\mathbf{x}_p$ represents information about product that belong to the category $\mathbf{x}_c$. The product category $\mathbf{x}_c$ is taken from the category tree T and is represented as a category name. The product information comprises titles $\mathbf{x}_t$, images $\mathbf{x}_i$, and attributes $\mathbf{x}_i$, i.e., $\mathbf{x}_p = \{\mathbf{x}_i, \mathbf{x}_t, \mathbf{x}_a\}$.

For the CtI retrieval task, we use the target category name $\mathbf{x}_c$ as a query and we aim to refturn a ranked list of top-k images that belong to the category $\mathbf{x}_c$.

CLIP-ITA. Figure 1 provides a high-level view of CLIP-ITA. CLIP-ITA projects category $\mathbf{x}_c$ and product information $\mathbf{x}_p$ into a d-dimensional multimodal space where the resulting vectors are respectively $\mathbf{c}$ and $\mathbf{p}$. The category and product information is processed by a category encoding pipeline and product information encoding pipeline. The core components of CLIP-ITA are the encoding and projection modules. The model consists out of four encoders: a category encoder, an image encoder, a title encoder, and an attribute encoder. Besides, CLIP-ITA comprises two non-linear projection heads: the category projection head and the multimodal projection head.

While several components of CLIP-ITA are based on CLIP [26], CLIP-ITA differs from CLIP in three important ways: (1) unlike CLIP, which operates on two encoders (textual and visual), CLIP-ITA extends CLIP towards a category encoder, image encoder, textual encoder, and attribute encoder; (2) CLIP-ITA features two projection heads, one for the category encoding pipeline, and one for the product information encoding pipeline; and (3) while CLIP is trained on text-image pairs, CLIP-ITA is trained on category-product pairs, where product representation is multimodal.

Category Encoding Pipeline. The *category encoder* (f_c) takes as input category name $\mathbf{x}_c$ and returns its representation $\mathbf{h}_c$. More specifically, we pass the category name $\mathbf{x}_c$ through the category encoder f_c:

$$\mathbf{h}_c = f_c(\mathbf{x}_c). \tag{1}$$

To obtain this representation, we use pre-trained MPNet model [29]. After passing category information through the category encoder, we feed it to the category projection head. The *category projection head* (g_c) takes as input a query representation $\mathbf{h}_c$ and projects it into d-dimensional multi-modal space:

$$\mathbf{c} = g_c(\mathbf{h}_c), \tag{2}$$

where $\mathbf{c} \in \mathbb{R}^d$.

Product Encoding Pipeline. The product information encoding pipeline represents three encoders, one for every modality, and a product projection head. The *image encoder* (f_i) takes as input a product image $\mathbf{x}_i$ aligned with the category $\mathbf{x}_c$. Similarly to the category processing pipeline, we pass the product image $\mathbf{x}_i$ through the image encoder:

$$\mathbf{h}_i = f_i(\mathbf{x}_i). \tag{3}$$

To obtain the image representation $\mathbf{h}_i$, we use pre-trained Vision Transformer from CLIP model. The *title encoder* (f_t) takes a product title $\mathbf{x}_t$ as input and returns a title representation $\mathbf{h}_t$:

$$\mathbf{h}_t = f_t(\mathbf{x}_t). \tag{4}$$

Similarly to the category encoder f_c, we use pre-trained MPNet to obtain the title representation $\mathbf{h}_t$. The *attribute encoder* (f_a) is a network that takes as input a set of attributes $\mathbf{x}_a = \{a_1, a_2, \ldots, a_n\}$ and returns their joint representation:

$$\mathbf{h}_a = f_a(\mathbf{x}_a) = \frac{1}{n} \sum_{i=1}^{n} f_a(\mathbf{x}_{ai}). \tag{5}$$

Similarly to the category encoder f_c and title encoder f_t, we obtain representation of each attribute with the pre-trained MPNet model. After obtaining title, image and attribute representations, we pass the representations into the product projection head. The *product projection head* (g_p) takes as input a concatenation of the image representation $\mathbf{h}_i$, title representation $\mathbf{h}_t$, and attribute representation $\mathbf{h}_a$ and projects the resulting vector $\mathbf{h}_p = concat(\mathbf{h}_i, \mathbf{h}_t, \mathbf{h}_a)$ into multimodal space:

$$\mathbf{p} = g_p(\mathbf{h}_p) = g_p(concat(\mathbf{h}_i, \mathbf{h}_t, \mathbf{h}_a)), \tag{6}$$

where $\mathbf{p} \in \mathbb{R}^d$.

Loss Function. We train CLIP-ITA using bidirectional contrastive loss [41]. The loss is a weighted combination of two losses: a category-to-product contrastive loss and a product-to-category contrastive loss. In both cases the loss is the InfoNCE loss [25]. Unlike prior work that focuses on a contrastive loss between inputs of the same modality [3,8] and on corresponding inputs of two modalities [41], we use the loss to work with inputs from textual modality (category representation) vs. a combination of multiple modalities (product representation). We train CLIP-ITA on batches of category-product pairs $(\mathbf{x}_c, \mathbf{x}_p)$ with batch size β. For the j-th pair in the batch, the category-to-product contrastive loss is computed as follows:

$$\ell_j^{(c \to p)} = -\log \frac{\exp(f_{sim}(\mathbf{c}_j, \mathbf{p}_j)/\tau)}{\sum_{k=1}^{\beta} \exp(f_{sim}(\mathbf{c}_j, \mathbf{p}_k)/\tau)}, \tag{7}$$

where $f_{sim}(\mathbf{c}_i, \mathbf{p}_i)$ is the cosine similarity, and $\tau \in \mathbb{R}^+$ is a temperature parameter. Similarly, the product-to-category loss is computed as follows:

$$\ell_j^{(p \to c)} = -\log \frac{\exp(f_{sim}(\mathbf{p}_j, \mathbf{c}_j)/\tau)}{\sum_{k=1}^{\beta} \exp(f_{sim}(\mathbf{p}_j, \mathbf{c}_k)/\tau)}. \tag{8}$$

The resulting contrastive loss is a combination of the two above-mentioned losses:

$$\mathcal{L} = \frac{1}{\beta} \sum_{j=1}^{\beta} \left(\lambda \ell_j^{(p \to c)} + (1 - \lambda) \ell_j^{(c \to p)} \right), \tag{9}$$

where β represents the batch size and $\lambda \in [0, 1]$ is a scalar weight.

4   Experimental Setup

Dataset. We use the XMarket dataset recently introduced by Bonab et al. [2] that contains textual, visual, and attribute information of e-commerce products as well as a category tree. For our experiments, we select 38,921 products from the US market. Category information is represented as a category tree and comprises 5,471 unique categories across nine levels. Level one is the most general category level, level nine is the most specific level. Every product belongs to a subtree of categories $t \in T$. In every subtree t, each parent category has only one associated child category. The average subtree depth is 4.63 (minimum: 2, maximum: 9). Because every product belongs to a subtree of categories, the dataset contains 180,094 product-category pairs in total. We use product titles as textual information and one image per product as visual information. The attribute information comprises 228,368 attributes, with 157,049 unique. On average, every product has 5.87 attributes (minimum: 1, maximum: 24).

Evaluation Method. To investigate how model performance changes w.r.t. category granularity, for every product in the dataset, $\mathbf{x}_p$, and the corresponding subtree of categories to which the product belongs, t, we train and evaluate the model performance in three settings: (1) *all categories*, where we randomly select one category from the subtree t; (2) *most general category*, where we use only the most general category of the subtree t, i.e., the root; and (3) *most specific category*, where we use the most specific category of the subtree t. In total, there are 5,471 categories in all categories setup, 34 categories in the most general category, and 4,100 in the most specific category setup. We evaluate every model on category-product pairs $(\mathbf{x}_c, \mathbf{x}_p)$ from the test set. We encode each category and a candidate product data by passing them through category encoding and product information encoding pipelines. For every category $\mathbf{x}_c$ we retrieve the top-k candidates ranked by cosine similarity w.r.t. the target category $\mathbf{x}_c$.

Metrics. To evaluate model performance, we use Precision@K where $K = \{1, 5, 10\}$, mAP@K where $K = \{5, 10\}$, and R-precision.

Baselines. Following [4, 27, 35] we use BM25, MPNet, CLIP as our baselines.

Four Experiments. We run four experiments, corresponding to our research questions as listed at the end of Sect. 1. In *Experiment 1* we evaluate the baselines on the CtI retrieval task (RQ1). We feed BM25 corpora that contain textual product information, i.e., product titles. We use MPNet in a zero-shot manner. For all the products in the dataset, we pass the product title $\mathbf{x}_t$ through the model. During the evaluation, we pass a category $\mathbf{x}_c$ expressed as textual query through MPNet and retrieve top-k candidates ranked by cosine similarity w.r.t. the target category $\mathbf{x}_c$. We compare categories of the top-k retrieved candidates with the target category $\mathbf{x}_c$. Besides, we use pre-trained CLIP in a zero-shot manner with a Text Transformer and a Vision Transformer (ViT) [5] an configuration. We pass the product images $\mathbf{x}_i$ through the image encoder. For evaluation, we pass a category $\mathbf{x}_c$ through the text encoder and retrieve top-k image candidates ranked by cosine similarity w.r.t. the target category $\mathbf{x}_c$. We compare categories of the top-k retrieved images with the target category $\mathbf{x}_c$.

In *Experiment 2* we evaluate image-based product representations (RQ2). After obtaining results with CLIP in a zero-shot setting, we build product representations by training on e-commerce data. First, we investigate how using product image data for building product representations impacts performance on the CtI retrieval task. To introduce visual information, we extend CLIP in two ways: (1) We use ViT from CLIP as image encoder f_i. We add product projection head g_p that takes as an input product visual information $\mathbf{x}_i \in \mathbf{x}_p$. (2) We use the text encoder from MPNet as category encoder f_c; we add a category projection head g_c on top of category encoder f_c thereby completing category encoding pipeline (see Fig. 1). We name the resulting model CLIP-I. We train CLIP-I on category-product pairs $(\mathbf{x}_c, \mathbf{x}_p)$ from the training set. Note that $\mathbf{x}_p = \{\mathbf{x}_i\}$, i.e., we only use visual information for building product representations.

In *Experiment 3*, we evaluate image- and attribute-based product representations (RQ3). We extend CLIP-I by introducing attribute information to the product information encoding pipeline. We add an attribute encoder f_a through which we obtain a representation of product attributes, $\mathbf{h}_a$. We concatenate the resulting attribute representation with image representation $\mathbf{h}_p = concat(\mathbf{h}_i, \mathbf{h}_a)$ and pass the resulting vector to the product projection head g_p. Thus, the resulting product representation $\mathbf{p}$ is based on both visual and attribute product information. We name the resulting model CLIP-IA. We train CLIP-IA on category-product pairs $(\mathbf{x}_c, \mathbf{x}_p)$ where $\mathbf{x}_p = \{\mathbf{x}_i, \mathbf{x}_a\}$, i.e., we use visual and attribute information for building product representation.

In *Experiment 4*, we evaluate image- attribute-, and title-based product representations (RQ4). We investigate how extending the product information processing pipeline with the textual modality impacts performance on the CtI retrieval task. We add title encoder f_t to the product information processing pipeline and

Table 1. Results of Experiments 1–4. The best performance is highligthed in bold.

Model	P@1	P@5	P@10	MAP@5	MAP@10	R-precision
		All categories (5,471)				
BM25 [12]	0.01	0.01	0.01	0.01	0.01	0.01
CLIP [26]	0.01	0.02	0.02	0.03	0.04	0.02
MPNet [29]	0.01	0.06	0.06	0.07	0.09	0.05
CLIP-I (Ours)	3.3	3.8	3.79	6.81	7.25	3.67
CLIP-IA (Ours)	2.5	3.34	3.29	5.95	6.24	3.27
CLIP-ITA (Ours)	**9.9**	**13.27**	**13.43**	**20.3**	**20.53**	**13.42**
		Most general category (34)				
BM25 [12]	2.94	4.71	4.71	8.33	8.28	4.48
CLIP [26]	11.76	12.35	11.76	16.12	15.18	9.47
MPNet [29]	14.70	15.8	15.01	18.44	18.78	9.35
CLIP-I (Ours)	17.85	17.14	16.78	19.88	20.14	13.02
CLIP-IA (Ours)	21.42	21.91	22.78	25.59	26.29	20.74
CLIP-ITA (Ours)	**35.71**	**30.95**	**30.95**	**35.51**	**34.28**	**25.79**
		Most specific category (4,100)				
BM25 [12]	0.02	0.02	0.01	0.01	0.01	0.01
CLIP [26]	11.92	9.81	9.23	15.12	14.95	8.14
MPNet [29]	33.36	28.56	26.93	37.43	36.77	25.29
CLIP-I (Ours)	14.06	12.11	11.53	18.24	17.9	11.22
CLIP-IA (Ours)	35.3	30.21	29.32	39.93	39.27	28.86
CLIP-ITA (Ours)	**45.85**	**41.04**	**40.02**	**50.04**	**49.87**	**39.69**

use it to obtain title representation $\mathbf{h}_t$. We concatenate the resulting representation with product image and attribute representations $\mathbf{h}_p = concat(\mathbf{h}_i, \mathbf{h}_t, \mathbf{h}_a)$. We pass the resulting vector to the product projection head g_p. The resulting model is CLIP-ITA. We train and test CLIP-ITA on category-product pairs $(\mathbf{x}_c, \mathbf{x}_p)$ where $\mathbf{x}_p = \{\mathbf{x}_i, \mathbf{x}_a, \mathbf{x}_t\}$, i.e., we use visual, attribute, and textual information for building product representations.

Implementation Details. We train every model for 30 epochs, with a batch size $\beta = 8$ for most general categories, $\beta = 128$—for most specific categories and all categories. For loss function, we set $\tau = 1$, $\lambda = 0.5$. We implement every projection head as non-linear MLPs with two hidden layers, GELU non-linearities [9] and layer normalization [1]. We optimize both heads with the AdamW optimizer [22].

5 Experimental Results

Experiment 1: Baselines. Following RQ1, we start by investigating how do baselines perform on CtI retrieval task. Besides, we investigate how does the performance on the task differs between the unimodal and the bimodal approach.

The results are shown in Table 1. When evaluating on all categories, all the baselines perform poorly. For the most general category setting, MPNet outperforms CLIP on all metrics except R-precision. The most prominent gain is for Precision@10 where MPNet outperforms CLIP by 28%. CLIP outperforms BM25 on all metrics. For the most specific category setting, MPNet performance is the highest, BM25—the lowest. In particular, MPNet outperforms CLIP by 211% in Precision@10. Overall, MPNet outperforms CLIP and both models significantly outperforms BM25 for both most general and most specific categories. However, when evaluation is done on all categories, the performance of all models is comparable. As an answer to RQ1, the results suggest that using information from multiple modalities is beneficial for performance on the task.

Experiment 2: Image-Based Product Representations. To address RQ2, we compare the performance of CLIP-I with CLIP and MPNet, the best-performing baseline. Table 1, shows the experimental results for Experiment 2. The biggest performance gains are obtained in "all categories" setting. However, there, the performance of the baselines was very poor. For the most general categories, CLIP-I outperforms both CLIP and MPNet. For CLIP-I vs. CLIP, we observe the biggest increase of 51% for Precision@1, for CLIP-I vs. MPNet—39% in R-precision. In the case of the most specific categories, CLIP-I outperforms CLIP but loses to MPNet. Overall, CLIP-I outperforms CLIP in all three settings and outperforms MPNet except the most specific categories. Therefore, we answer RQ2 as follows: the results suggest that extension of CLIP by the introduction of product image data for building product representations has a positive impact on performance on CtI retrieval task.

Experiment 3: Image- and Attribute-Based Product Representations. To answer RQ3, we compare the performance of CLIP-IA with CLIP-I and the baselines. The results are shown in Table 1. When evaluated on all categories, CLIP-IA performs worse than CLIP-I but outperforms MPNet. In particular, CLIP-I obtains the biggest gain relative of 32% on Precision@1 and the lowest gain of 12% on R-precision. For the most general category, CLIP-IA outperforms CLIP-I and MPNet on all metrics. More specifically, we observe the biggest gain of 122% on R-precision over MPNet and the biggest gain of 59% on R-precision for CLIP-I. Similarly, for the most specific category, CLIP-IA outperforms both CLIP-I and MPNet. We observe the biggest relative gain of 138% over CLIP-I. The results suggest that further extension of CLIP by the introduction of the product image and attribute data for building product representations has a positive impact on performance on CtI retrieval task, especially when evaluated on most specific categories. Therefore, we answer RQ4 positively.

Experiment 4: Image-, Attribute-, and Title-Based Product Representations. We compare CLIP-ITA with both CLIP-IA, CLIP-I, and the baselines.

Table 2. Erroneous CLIP-ITA prediction counts for "same tree" vs. " different tree" predictions per evaluation type.

	Same tree	Different tree
All categories	1,655	639
The most general category	2	21
The most specific category	127	1,011
Total	1,786	1,671

The results are shown in Table 1. In general, CLIP-ITA outperforms CLIP-I and CLIP-IA and the baselines in all settings. When evaluated on all categories, the maximum relative increase of CLIP-ITA over CLIP-I is 265% in R-precision, the minimum relative increase is 183% in mAP@10. The biggest relative increase of CLIP-ITA performance over CLIP-IA is 310% in Precision@1, the smallest relative increase is 229% in mAP@10. For the most general categories, CLIP-ITA outperforms CLIP-I by 82% and CLIP-IA by 38%. For most specific categories, we observe the biggest increase of CLIP-ITA over CLIP-I of 254% in R-precision and the smallest relative increase of 172% on mAP@5. At the same time, the biggest relative increase of CLIP-ITA over CLIP-IA is a 38% increase in R-precision and the smallest relative increase is a 27% increase in mAP@5. Overall, CLIP-ITA wins in all three settings. Hence, we answer RQ4 positively.

6   Error Analysis

Distance Between Predicted and Target Categories. We examine the performance of CLIP-ITA by looking at the pairs of the ground-truth and predicted categories (c, c_p) in cases when the model failed to predict the correct category, i.e., $c \neq c_p$. This allows us to quantify how far off the incorrect predictions lie w.r.t. the category tree hierarchy. First, we examine in how many cases target category c and predicted category c_p belong to the same most general category, i.e., belong to the same category tree; see Table 2. In the case of most general categories, the majority of incorrectly predicted categories belong to a tree different from the target category tree. For the most specific categories, about 11% of predicted categories belong to the category tree of the target category. However, when evaluation is done on all categories, 72% of incorrectly predicted cases belong to the same tree as a target category.

Next, we turn to the category-predicted category pairs (c, c_p) where the incorrectly predicted category c_p belongs to the same tree as target category c. We compute the distance d between a category used as a query c and a predicted category c_p. We compute the distance between target category c and a top-1 predicted category c_p as the difference between their respective depths $d(c, c_p) = depth(c_p) - depth(c)$. The distance d is positive if the depth of the predicted category is bigger than the depth of the target category, $depth(c_p) > depth(c)$, i.e., the

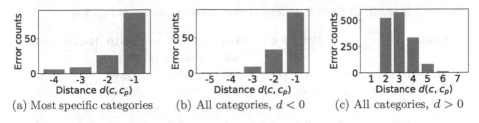

(a) Most specific categories   (b) All categories, $d < 0$ (c) All categories, $d > 0$

Fig. 2. Error analysis for CLIP-ITA. Distance between target category c and a predicted category c_p when c and c_p are in the same tree.

predicted category is more specific than the target category. The setup is mirrored for negative distances. See Fig. 2. We do not plot the results for the most general category because for this setting there are only two cases when target category c and a predicted category c_p were in the same tree. In both cases, predicted category c_p was more general than target category c with distance $d(c, p_c) = 2$. In cases when target category c was sampled from the most specific categories, the wrongly predicted category c_p belonging to the same tree was always more specific than the target category c with the maximum absolute distance between c and c_p, $|d(c, c_p)| = 4$. In 68% of the cases the predicted category was one level above the target category, for 21% $d(c, c_p) = -2$, for 7% $d(c, c_p) = -3$, and for 5% $d(c, c_p) = -4$. For the setting with all categories, in 92% of the cases, the predicted category c_p was more specific than the target category c; for 8% the predicted category was more general.

Overall, for the most general category and the most specific category, the majority of incorrectly predicted categories are located in a category tree different from the one where the target category was located. For the "all categories" setting, it is the other way around. When it comes to the cases when incorrectly predicted categories are in the same tree as a target category, the majority of incorrect predictions are 1 level more general when the target category is sampled from the most specific categories. For the "all categories" setting, the majority of incorrect predictions belonging to the same tree as the target category were more specific than the target category. Our analysis suggests that efforts to improve the performance of CLIP-ITA should focus on minimizing the (tree-based) distance between the target and predicted category in a category tree. This could be incorporated as a suitable extension of the loss function.

Performance on Seen vs. Unseen Categories. Next, we investigate how well CLIP-ITA generalizes to unseen categories. We split the evaluation results into two groups based on whether the category used as a query was seen during training or not; see Table 3. For the most general categories, CLIP-ITA is unable to correctly retrieve an image of the product of the category that was not seen during training at all. For the most specific categories, CLIP-ITA performs better on seen categories than on unseen categories. We observe the biggest relative performance increase of 85% in mAP@10 and the smallest relative increase of 57% in R-precision. When evaluating on all categories, CLIP-ITA performs on unseen

Table 3. CLIP-ITA performance on seen vs. unseen categories.

Model	P@1	P@5	P@10	mAP@5	mAP@10	R-precision
	All categories (5,471)					
CLIP-ITA (unseen cat.)	13.3	18.56	15.55	19.7	19.65	18.52
CLIP-ITA (seen cat.)	10.48	13.95	14.08	21.65	21.65	14.07
	Most general category (34)					
CLIP-ITA (unseen cat.)	0.0	0.0	0.0	0.0	0.0	0.0
CLIP-ITA (seen cat.)	19.23	20.01	17.31	20.41	20.01	15.73
	Most specific category (4,100)					
CLIP-ITA (unseen cat.)	27.27	26.44	26.44	27.92	27.92	26.45
CLIP-ITA (seen cat.)	47.83	43.09	42.14	52.41	51.89	41.58

categories better when evaluated on Precision@k (27% higher in Precision@1, 33% higher in Precision@5, 10% increase in Precision@10) and R-precision (relative increase of 32%). Performance on seen categories is better in terms of mAP@k (10% increase for both mAP@5 and mAP@10).

Overall, for the most general and most specific categories, the model performs much better on categories seen during training. For "all categories" setting, however, CLIP-ITA's performance on unseen categories is better.

7   Conclusion

We introduced the task of category-to-image retrieval and motivated its importance in the e-commerce scenario. In the CtI retrieval task, we aim to retrieve an image of a product that belongs to the target category. We proposed a model specifically designed for this task, CLIP-ITA. CLIP-ITA extends CLIP, one of the best performing text-image retrieval models. CLIP-ITA leverages multimodal product data such as textual, visual, and attribute data to build product representations. In our experiments, we contrasted and evaluated different combinations of signals from modalities, using three settings: on all categories, the most general, and the most specific categories.

We found that combining information from multiple modalities to build product representation produces the best results on the CtI retrieval task. CLIP-ITA gives the best performance both on all categories and on the most specific categories. On the most general categories, CLIP-I, a model where product representation is based on image only, works slightly better. CLIP-I performs worse on the most specific categories and across all categories. For identification of the most general categories, visual information is more relevant. Besides, CLIP-ITA is able to generalize to unseen categories except in the case of most general categories. However, the performance on unseen categories is lower than the performance on seen categories. Even though our work is focused on the e-commerce domain, the findings can be useful for other areas, e.g., digital humanities.

Limitations of our work are due to type of data in the e-commerce domain. In e-commerce, there is typically one object per image and the background is homogeneous, textual information is lengthy and noisy; in the general domain, there is typically more than one object per image, image captions are more informative and shorter. Future work directions can focus on improving the model architecture. It would be interesting to incorporate attention mechanisms into the attribute encoder and explore how it influences performance. Another interesting direction for future work is to evaluate CLIP-ITA on other datasets outside of the e-commerce domain. Future work can also focus on minimizing the distance between the target and predicted category in the category tree.

Acknowledgements. This research was supported by Ahold Delhaize, the Nationale Politie, and the Hybrid Intelligence Center, a 10-year program funded by the Dutch Ministry of Education, Culture and Science through the Netherlands Organisation for Scientific Research, https://hybrid-intelligence-centre.nl.

All content represents the opinion of the authors, which is not necessarily shared or endorsed by their respective employers and/or sponsors.

References

1. Ba, J.L., Kiros, J.R., Hinton, G.E.: Layer normalization arXiv preprint arXiv:160706450 (2016)
2. Bonab, H., Aliannejadi, M., Vardasbi, A., Kanoulas, E., Allan, J.: XMarket: cross-market training for product recommendation. In: CIKM, ACM (2021)
3. Chen, T., Kornblith, S., Norouzi, M., Hinton, G.: A simple framework for contrastive learning of visual representations. In: International Conference on Machine Learning, PMLR, pp. 1597–1607 (2020)
4. Dai, Z., Lai, G., Yang, Y., Le, Q.V.: Funnel-transformer: Filtering out sequential redundancy for efficient language processing. arXiv preprint arXiv:200603236 (2020)
5. Dosovitskiy, A.: An image is worth 16 × 16 words: transformers for image recognition at scale. In: International Conference on Learning Representations (2021)
6. Goei, K., Hendriksen, M., de Rijke, M.: Tackling attribute fine-grainedness in cross-modal fashion search with multi-level features. In: SIGIR 2021 Workshop on eCommerce. ACM (2021)
7. Gupta, T., Vahdat, A., Chechik, G., Yang, X., Kautz, J., Hoiem, D.: Contrastive learning for weakly supervised phrase grounding. In: Vedaldi, A., Bischof, H., Brox, T., Frahm, J.-M. (eds.) ECCV 2020, Part III. LNCS, vol. 12348, pp. 752–768. Springer, Cham (2020). https://doi.org/10.1007/978-3-030-58580-8_44
8. He, K., Fan, H., Wu, Y., Xie, S., Girshick, R: Momentum contrast for unsupervised visual representation learning. In: Proceedings of the IEEE/CVF Conference on Computer Vision and Pattern Recognition, pp. 9729–9738 (2020)
9. Hendrycks, D., Gimpel, K.: Gaussian error linear units (GELUs). arXiv preprint arXiv:160608415 (2016)
10. Hewawalpita, S., Perera, I.: Multimodal user interaction framework for e-commerce. In: 2019 International Research Conference on Smart Computing and Systems Engineering (SCSE), pp 9–16. IEEE (2019)

11. Hu, R., Xu, H., Rohrbach, M., Feng, J., Saenko, K., Darrell, T.: Natural language object retrieval. In: Proceedings of the IEEE Conference on Computer Vision and Pattern Recognition, pp 4555–4564 (2016)
12. Jones, K.S., Walker, S., Robertson, S.E.: A probabilistic model of information retrieval: development and comparative experiments: Part 2. Inf. Process. Manage. **36**(6), 809–840 (2000)
13. Kondylidis, N., Zou, J., Kanoulas, E.: Category aware explainable conversational recommendation. arXiv preprint arXiv:210308733 (2021)
14. Laenen, K., Moens, M.F.: Multimodal neural machine translation of fashion E-commerce descriptions. In: Kalbaska, N., Sádaba, T., Cominelli, F., Cantoni, L. (eds.) FACTUM 2019. Springer, Cham (2019). https://doi.org/10.1007/978-3-030-15436-3_4
15. Laenen, K., Moens, M.F.: A comparative study of outfit recommendation methods with a focus on attention-based fusion. Inf. Process. Manage. **57**(6), 102316 (2020)
16. Laenen, K., Zoghbi, S., Moens, M.F.: Cross-modal search for fashion attributes. In: Proceedings of the KDD 2017 Workshop on Machine Learning Meets Fashion, vol. 2017, pp 1–10, ACM (2017)
17. Laenen, K., Zoghbi, S., Moens, M.F.: Web search of fashion items with multimodal querying. In: Proceedings of the Eleventh ACM International Conference on Web Search and Data Mining, pp. 342–350 (2018)
18. Li, H., Yuan, P., Xu, S., Wu, Y., He, X., Zhou, B.: Aspect-aware multimodal summarization for Chinese e-commerce products. In: Proceedings of the AAAI Conference on Artificial Intelligence, vol. 34, pp. 8188–8195 (2020)
19. Li, X., Wang, X., He, X., Chen, L., Xiao, J., Chua, T.S.: Hierarchical fashion graph network for personalized outfit recommendation. In: Proceedings of the 43rd International ACM SIGIR Conference on Research and Development in Information Retrieval, pp 159–168 (2020)
20. Liao, L., He, X., Zhao, B., Ngo, C.W., Chua, T.S.: Interpretable multimodal retrieval for fashion products. In: Proceedings of the 26th ACM International Conference on Multimedia, pp 1571–1579 (2018)
21. Lin, Y., Ren, P., Chen, Z., Ren, Z., Ma, J., de Rijke, M.: Improving outfit recommendation with co-supervision of fashion generation. In: The World Wide Web Conference, pp. 1095–1105 (2019)
22. Loshchilov, I., Hutter, F.: Decoupled weight decay regularization. arXiv preprint arXiv:171105101 (2017)
23. Nagarajan, T., Grauman, K.: Attributes as operators: factorizing unseen attribute-object compositions. In: Ferrari, V., Hebert, M., Sminchisescu, C., Weiss, Y. (eds.) ECCV 2018, Part I. LNCS, vol. 11205, pp. 172–190. Springer, Cham (2018). https://doi.org/10.1007/978-3-030-01246-5_11
24. Nielsen, J., Molich, R., Snyder, C., Farrell, S.: E-commerce user experience. Nielsen Norman Group (2000)
25. Avd, O., Li, Y., Vinyals, O.: Representation learning with contrastive predictive coding. arXiv preprint arXiv:180703748 (2018)
26. Radford, A., et al.: Learning transferable visual models from natural language supervision. arXiv preprint arXiv:210300020 (2021)
27. Shen, S., et al.: How much can CLIP benefit vision-and-language tasks? arXiv preprint arXiv:210706383 (2021)
28. Smeulders, A., Worring, M., Santini, S., Gupta, A., Jain, R.: Content-based image retrieval at the end of the early years. IEEE Trans. Pattern Anal. Mach. Intell. **22**(12), 1349–1380 (2000). https://doi.org/10.1007/978-3-540-74769-7_81

29. Song, K., Tan, X., Qin, T., Lu, J., Liu, T.Y.: MPNet: Masked and permuted pre-training for language understanding. arXiv preprint arXiv:200409297 (2020)
30. Tagliabue, J., Yu, B., Beaulieu, M.: How to grow a (product) tree: personalized category suggestions for ecommerce type-ahead. arXiv preprint arXiv:200512781 (2020)
31. Tautkute, I., Trzciński, T., Skorupa, A.P., Brocki, Ł, Marasek, K.: DeepStyle: multimodal search engine for fashion and interior design. IEEE Access **7**, 84613–84628 (2019)
32. Thomee, B., et al.: YFCC100M: the new data in multimedia research. Commun. ACM **59**(2), 64–73 (2016)
33. Tsagkias, M., King,T.H., Kallumadi, S., Murdock, V., de Rijke, M.: Challenges and research opportunities in ecommerce search and recommendations. In: SIGIR Forum, vol. 54, issue number 1 (2020)
34. Vo, N., et al.: Composing text and image for image retrieval-an empirical odyssey. In: Proceedings of the IEEE/CVF Conference on Computer Vision and Pattern Recognition, pp. 6439–6448 (2019)
35. Wang, S., Zhuang, S., Zuccon, G.: Bert-based dense retrievers require interpolation with BM25 for effective passage retrieval. In: Proceedings of the 2021 ACM SIGIR International Conference on Theory of Information Retrieval, pp. 317–324 (2021)
36. Wirojwatanakul, P., Wangperawong, A.: Multi-label product categorization using multi-modal fusion models. arXiv preprint arXiv:190700420 (2019)
37. Yamaura, Y., Kanemaki, N., Tsuboshita, Y.: The resale price prediction of second-hand jewelry items using a multi-modal deep model with iterative co-attention. arXiv preprint arXiv:190700661 (2019)
38. Yang, X., et al.: Interpretable fashion matching with rich attributes. In: Proceedings of the 42nd International ACM SIGIR Conference on Research and Development in Information Retrieval, pp. 775–784 (2019)
39. Yashima, T., Okazaki, N., Inui, K., Yamaguchi, K., Okatani, T.: Learning to describe e-commerce images from noisy online data. In: Lai, S.-H., Lepetit, V., Nishino, K., Sato, Y. (eds.) ACCV 2016, Part V. LNCS, vol. 10115, pp. 85–100. Springer, Cham (2017). https://doi.org/10.1007/978-3-319-54193-8_6
40. Yim, J., Kim, J.J., Shin, D.: One-shot item search with multimodal data. arXiv preprint arXiv:181110969 (2018)
41. Zhang, Y., Jiang, H., Miura, Y., Manning, C.D., Langlotz, C.P.: Contrastive learning of medical visual representations from paired images and text. arXiv preprint arXiv:201000747 (2020)
42. Zoghbi, S., Heyman, G., Gomez, J.C., Moens, M.-F.: Cross-modal fashion search. In: Tian, Q., Sebe, N., Qi, G.-J., Huet, B., Hong, R., Liu, X. (eds.) MMM 2016, Part II. LNCS, vol. 9517, pp. 367–373. Springer, Cham (2016). https://doi.org/10.1007/978-3-319-27674-8_35

WIDAR - Weighted Input Document Augmented ROUGE

Raghav Jain[1], Vaibhav Mavi[2], Anubhav Jangra[1(✉)], and Sriparna Saha[1]

[1] Indian Institute of Technology Patna, Patna, India
anubhav0603@gmail.com
[2] New York University, New York, USA

Abstract. The task of automatic text summarization has gained a lot of traction due to the recent advancements in machine learning techniques. However, evaluating the quality of a generated summary remains to be an open problem. The literature has widely adopted Recall-Oriented Understudy for Gisting Evaluation (ROUGE) as the standard evaluation metric for summarization. However, ROUGE has some long-established limitations; a major one being its dependence on the availability of good quality reference summary. In this work, we propose the metric WIDAR which in addition to utilizing the reference summary uses also the input document in order to evaluate the quality of the generated summary. The proposed metric is versatile, since it is designed to adapt the evaluation score according to the quality of the reference summary. The proposed metric correlates better than ROUGE by 26%, 76%, 82%, and 15%, respectively, in coherence, consistency, fluency, and relevance on human judgement scores provided in the SummEval dataset. The proposed metric is able to obtain comparable results with other state-of-the-art metrics while requiring a relatively short computational time (Implementation for WIDAR can be found at - https://github.com/ Raghav10j/WIDAR).

Keywords: Summarization · Evaluation metric · ROUGE

1 Introduction

Accessibility of internet has led to massive increase in content available to a user, making it difficult to obtain the required information. This seemingly perpetual growth of information necessitates the need for automatic text summarization tools. Text summarization can be described as the task of generating fluent and human readable summary while preserving the essence of the original text documents. Evaluation of these automatically generated summaries has been actively explored by the research community for over 5 decades [8]. Since then, various attempts have been made to quantify the effectivenes of the summarization systems; however the evaluation task still remains an open problem till this day.

The most widely adopted evaluation metric for text summarization in the community is *Recall-Oriented Understudy for Gisting Evaluation (ROUGE)* [25]

R. Jain, V. Mavi and A. Jangra—Equal contribution.

which is mainly based on the n-gram overlap between the generated summary and reference summary. However, ROUGE's dependency on a good quality reference summary is one of it's biggest drawback. Fabbri et al. [10] highlighted the inconsistency in quality of some reference summaries in the CNN/DailyMail dataset [26] by describing the summaries consisting of clickbaits instead of being truthful and informative with respect to the input article (refer to Fig. 1). Kryscinski et al. [21] also reported this issue of reference summaries containing irrelevant information such as links to other articles or factual inconsistency in the Newsroom dataset [14]. Even if a reference summary is of satisfactory quality, it is highly unlikely that it is the only acceptable summary of that document as different people tend to produce different summaries for the same document [27,34]. Therefore, all the above-mentioned claims imply that sole dependence on reference summary for an evaluation metric is not optimal. Therefore, we propose an evaluation metric that also considers an input source document while evaluating the quality of its summary.

Input Document 1:

Last week she was barely showing - but Demelza Poldark is now the proud mother to the show's latest addition. Within ten minutes of tomorrow night's episode, fans will see Aidan Turner's dashing Ross Poldark gaze lovingly at his new baby daughter. As Sunday night's latest heartthrob, women across the country have voiced their longing to settle down with the brooding Cornish gentleman - but unfortunately it seems as if his heart is well and truly off the market. Scroll down for video Last week she was barely showing - but Demelza Poldark is now the proud mother to the show's latest addition He may have married his ....

Reference Summary:

SPOILER ALERT: Maid gives birth to baby on Sunday's episode. Only announced she was pregnant with Poldark's baby last week.

Generated Summary:

demelza poldark is now the proud mother to the show's latest addition . fans will see aidan turner's dashing ross poldark gaze lovingly at his new baby daughter . sunday night's latest heartthrob , women across the country have voiced their longing to settle down with the brooding cornish gentleman .

Input Document 2:

Eight Iranian border guards have been killed in clashes with militants near the border with Pakistan, Iranian state media reported. Three of the militants were killed by Iranian forces in the fighting Monday in the southeastern town of Negur, the state-run news agency IRNA reported. The news agency cited Ali Asghar Mirshekari, the deputy governor of Iran's Sistan-Baluchestan province, who said the militants crossed into the country from Pakistan. Iranian officials ....

Reference Summary:

The Pakistani government says its security agencies are investigating. A group believed to be based in Pakistan's Balochistan province claims responsibility.

Generated Summary:

three of the militants were killed by iranian forces in the southeastern town of negur . a militant group called jaish al adal claimed responsibility for the attack . jaish al adal has also claimed responsibility for attacks on iranian territory .

Fig. 1. Examples from DailyMail/CNN dataset [26] where ground truth is unsatisfactory either due to clickbaits (Eg.-1), or information incompleteness (Eg.-2).

In order to design an evaluation metric, it is important to study what comprises of a good summary. Ideally, a summary must be *coherent, non-redundant, fluent, consistent* and *relevant* to the input article [6]. Using these characteristics, recent works have attempted to quantify and compare the performance of existing evaluation metrics [1,10]. These works highlight the limitations of existing metrics and offer various resources for conducting further research on the evaluation task. One such work is the SummEval dataset [10] that provides human annotation scores for - *coherence, consistency, fluency* and *relevance*.

In this paper, we propose an evaluation metric *WIDAR (Weighted Input Document Augmented ROUGE)* in an attempt to overcome the above-mentioned limitations of ROUGE (refer to Fig. 2). The proposed metric utilizes the reference summary and input document to measure the quality of a generated summary. WIDAR introduces the idea of weighted ROUGE that relies on weighting sentences in reference summary based on information coverage and redundancy within the summary. Through experiments, we illustrate that WIDAR is able to outperform ROUGE by a large margin, and is able to obtain comparable results with other state-of-the-art metrics while requiring relatively short computational time.

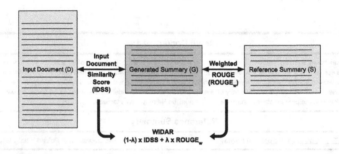

Fig. 2. Model figure for WIDAR.

2   Related Works

The approaches to text summarization can be broadly classified into two categories, extractive methods [22,29,35] and abstractive methods [4,16,38]. Summarization research today has expanded into more complex problems like multilingual summarization [15,36], multi-modal summarization [17–20], across-time summarization [7] etc.

Numerous evaluation metrics have been proposed to assess summarization systems. Some of them are based on text matching between predicted summary and reference summary such as Recall-Oriented Understudy for Gisting Evaluation (ROUGE) [25], ParaEval [46], ROUGE 2.0 [11], Metric for Evaluation of Translation with Explicit ORdering (METEOR) [24], Bilingual Evaluation Understudy (BLEU) score [30], Character n-gram F-score (CHRF) [33], Consensus-based Image Description Evaluation (CIDEr) [43] etc. There are also evaluation metrics that try to capture semantic similarity including word embeddings based techniques such as Word Mover similarity (WMS) [23], MoverScore [45], Sentence Mover Similarity (SMS) [5], ROUGE-WE [28], ELMo-m [41], automated pyramid metric [31] and graph based techniques such as graph based ROUGE (ROUGE-G) [39] and AUTOmatic SUMMary Evaluation based on N-gram Graphs (AutoSummENG) [13]. Other than these, there are also model based learned metrics such as Supervised Summarization Scorer (S^3) [32], BERTScore [44], NeuralTD [2], Support Vector Regression (SVR) [40] and question answering based metrics such as Answering Performance for Evaluation of Summaries (APES) [9] and Semantic QA [3]. In unsupervised settings where evaluation is carried out on the basis of input document rather than depending on a reference summary, SummaQA [37], summarization evaluation with pseudo references and BERT (SUPERT) [12] and BLANC [42] are some of the most recent and state-of-the-art metrics.

3   WIDAR Evaluation Metric

We propose WIDAR (Weighted Input Document Augmented ROUGE), an evaluation metric that utilizes both reference summary (R) and input document (D) to judge the quality of a generated summary (S). For better understanding, we divide our approach into two steps: 1) calculation of Weighted ROUGE (Sect. 3.1), and 2) combination of Weighted ROUGE with similarity score computed between generated summary and input document to obtain WIDAR (Sect. 3.2). Table 1 lists the notations used in the remainder of this paper.

Table 1. Notation of each variable and its corresponding meaning.

Notation	Meaning
D	input document
R	reference summary
S	generated summary
d_i	i^{th} input document's sentence
r_i	i^{th} input reference summary's sentence
s_i	i^{th} input generated summary's sentence
w_{cov_i}	coverage weight assigned to i^{th} generated summary sentence
w_{red_i}	redundancy weight assigned to i^{th} generated summary sentence
w_i	overall weight assigned to i^{th} generated summary sentence

3.1   Weighted ROUGE

As discussed in Sect. 1, ROUGE is highly dependent on the quality of reference summary to perform effectively. However, in real world scenarios, high quality of reference summary is not assured. Therefore, we introduce two special weights for each reference summary sentence to penalize/reward the quality of information present in this sentence. Each reference summary sentence r_i is assigned two scores: 1) Coverage weight (w_{cov_i}) - based on the input document information that is covered by r_i, and 2) Redundancy weight (w_{red_i}) - based on the uniqueness of information presented by r_i in the reference summary. We use Algorithm 1 to compute the redundancy weights and coverage weights for all sentences in the reference summary. We then obtain the overall weight[1] (w_i) for r_i by computing the average of w_{cov_i} and w_{red_i}.

$$w_i = \frac{(w_{cov_i} + w_{red_i})}{2} \times |R| \tag{1}$$

Algorithm 1: Calculating the coverage and redundancy weights. (Here θ_1 and θ_2 are ROUGE-L thresholds for coverage and redundancy respectively.)

Input: $R = \{r_i\}$, $D = \{d_j\}$
Output: $W_{cov} = \{w_{cov_i}\}$ $W_{red} = \{w_{red_i}\}$
$W_{cov}, W_{red} \leftarrow$ emptyList;
for r_i *in* R **do**
 $w_{cov_i} = 0$;
 for d_j *in* D **do**
  if $ROUGE\text{-}L^r (r_i, d_j) \geq \theta_1$ **then**
   w_{cov_i}++;
  end
 end
 $W_{cov} \leftarrow w_{cov_i}/|D|$;
end
for r_i *in* R **do**
 $w_{red_i} = 0$;
 for r_j *in* R **do**
  if $r_i \neq r_j$ & $ROUGE\text{-}L^r (r_i, r_j) \geq \theta_2$ **then**
   w_{red_i}++;
  end
 end
 $W_{red} \leftarrow 1\text{-}(w_{red_i} / |R|)$;
end

We propose sentence-level ROUGE-N ($ROUGE\text{-}N_{SL}$) and sentence-level ROUGE-L ($ROUGE\text{-}L_{SL}$), variations of ROUGE in order to incorporate the sentence-level redundancy and coverage weights (Eq. 1), respectively.

Sentence-Level ROUGE-N: Typically, ROUGE-N measures the number of overlapping n-grams between the reference summary and the generated summary. However, to compute the sentence-level ROUGE-N ($ROUGE\text{-}N_{SL}$) we take into account sentence level n-grams for the overlap count, *viz.* we discard the bridge n-grams (that share words from two or more sentences)[2]. We use the following

---

[1] We multiply the final weights by the number of sentences in the reference summary $|R|$ to ensure that the sum of weights remains the same as in plain ROUGE, i.e., $\sum_i w_i = |R|$.
[2] Note that $ROUGE\text{-}1$ and $ROUGE\text{-}1_{SL}$ denote the same metrics.

equations to measure the precision $(ROUGE\text{-}N_{SL}^p)$, recall $(ROUGE\text{-}N_{SL}^r)$, and f-score $(ROUGE\text{-}N_{SL}^f)$, respectively.

$$ROUGE\text{-}N_{SL}^r = \frac{\sum_{s\text{-}gram_i} \sum_{r\text{-}gram_j} count(s\text{-}gram_i, r\text{-}gram_j)}{\sum_{r\text{-}gram_j} |r\text{-}gram_j|} \tag{2}$$

$$ROUGE\text{-}N_{SL}^p = \frac{\sum_{s\text{-}gram_i} \sum_{r\text{-}gram_j} count(s\text{-}gram_i, r\text{-}gram_j)}{\sum_{s\text{-}gram_j} |s\text{-}gram_j|} \tag{3}$$

$$ROUGE\text{-}N_{SL}^f = \frac{2 \times (ROUGE\text{-}N_{SL}^r) \times (ROUGE\text{-}N_{SL}^p)}{(ROUGE\text{-}N_{SL}^r) + (ROUGE\text{-}N_{SL}^p)} \tag{4}$$

where $s\text{-}gram_i$ and $r\text{-}gram_i$ denote the sentence-level n-grams for i^{th} sentence in the generated summary and in the reference summary, respectively; $count(s\text{-}gram_i, r\text{-}gram_j)$ calculates the number of overlapping n-grams in $s\text{-}gram_i$ and $r\text{-}gram_i$, and $|.|$ denotes the cardinality of a set.

Sentence-Level ROUGE-L: ROUGE-L computes the longest common subsequence of words between the generated summary and the reference summary. Sentence-level ROUGE-L $(ROUGE\text{-}L_{SL})$ is computed as follows:

$$ROUGE\text{-}L_{SL}^r = \frac{\sum_{r_i \in R} UnionLCS(r_i, S)}{|R|} \tag{5}$$

$$ROUGE\text{-}L_{SL}^p = \frac{\sum_{r_i \in R} UnionLCS(r_i, S)}{|S|} \tag{6}$$

$$ROUGE\text{-}L_{SL}^f = \frac{2 \times (ROUGE\text{-}L_{SL}^r) \times (ROUGE\text{-}L_{SL}^p)}{(ROUGE\text{-}L_{SL}^r) + (ROUGE\text{-}L_{SL}^p)} \tag{7}$$

where $UnionLCS(r_i, S)$ is the union of the longest common sub-sequence computed between a reference summary sentence $(r_i \in R)$ and each sentence of generated summary $(s_i \in S)$, and $|R|$ and $|S|$ denote the number of sentences in reference summary and generated summary, respectively.

We integrate into these sentence-level ROUGE metrics the weights (Eq. 1) to obtain Weighted ROUGE-N $(ROUGE\text{-}N_W)$ and Weighted ROUGE-L $(ROUGE\text{-}L_W)$ scores. $ROUGE\text{-}N_W$ is obtained by multiplying w_i in each summation term in Eqs. 2 to 4, and $ROUGE\text{-}L_W$ is obtained by multiplying w_i in each summation term in Eqs. 5 to 7.

3.2 Combining Weighted ROUGE with Input Document Similarity

Input Document Similarity Score (IDSS). We incorporate information overlap of generated summary with input document to make the proposed metric more robust and applicable to the real-world situations where the quality of reference summary might be sometimes inadequate. For simplicity, we use ROUGE-L F-score as the similarity measure, because it performed better than other ROUGE variants in our experiments (refer to Sect. 4.3). Therefore,

$$IDSS = ROUGE\text{-}L^f(S, D) \tag{8}$$

The last step of the evaluation process is to combine the $ROUGE_W$ and $IDSS$ scores in such a way that the final score retains the individual characteristic of both the individual scores. We define $WIDAR$ as follow:

$$WIDAR_K^X = (1 - \lambda) \times IDSS + \lambda \times ROUGE\text{-}K_W^X \tag{9}$$

where $x \in \{r, p, f\}$ and $K \in \{1, 2, L\}$; λ is a hyper-parameter directly proportional to the quality of coverage in reference summary[3].

4   Experiments

4.1   Dataset

For all the experiments conducted in this work, we have used the SummEval dataset [10]. It contains the summaries generated by 23 recent summarization models trained on CNN/DailyMail dataset [26]. The dataset contains human annotation scores for 16 generated summaries of 100 source news articles giving us 1600 summary-text pairs. Each summary is annotated by 3 experts and 5 crowd-source annotators to evaluate the quality of a summary on a range of 1–5 across 4 different characteristics: 1) *Coherence:* measures the quality of smooth transition between different summary sentences such that sentences are not completely unrelated or completely same, 2) *Consistency:* measures the factual correctness of summary with respect to input document, 3) *Fluency:* measures the grammatical correctness and readability of sentences, 4) *Relevance:* measures the ability of a summary to capture important and relevant information from the input document. Apart from human annotation scores, 11 reference summaries for each example, and evaluation scores for generated summaries across different evaluation metrics are also made available in the dataset repository[4].

4.2   Evaluation of Evaluation Metric

In order to measure the performance of the proposed evaluation metric, we calculate the correlation between the scores of that metric and the average annotation

---

[3] λ is a fixed hyper-parameter, which is set to 0.5 in our final experiments. We attempted to make λ a data-driven parameter by setting $\lambda = max(w_{cov_i})$ or $\lambda = mean(w_{cov_i})$, but this setting was not able to outperform the fixed $\lambda = 0.5$ value (refer to Sect. 4.3).

[4] https://github.com/Yale-LILY/SummEval.

scores for each characteristic of each summary for 1600 summary-text examples provided in the SummEval dataset [10] (described in Sect. 4.1). We have used the average of expert annotation scores for our experiments because of the inconsistency between expert and crowd-source scores reported by Fabbri et al. [10]. We use the Kendall's tau correlation coefficient as the correlation metric in our experiments. Kendall's tau correlation between two sequences $X = \{x_i\}$ and $Y = \{y_i\}$ is defined as follows:

$$\tau = \frac{C - D}{C + D} \tag{10}$$

where C is the number of all those pairs that are concordant and D is the number of all those pairs that are discordant in sequences, X and Y.

4.3   Experimental Settings

In this section, we discuss various hyperparameters used in the proposed methodology, along with the tuning experiments carried out to justify them[5].

Weighted sum of $IDSS$ and $ROUGE_W$ (λ) : λ is used to get the weighted sum of information overlap of the generated summary with the input document ($IDSS$) and the reference summary ($ROUGE_W$). We attempted to investigate the optimal value of λ using a data-driven technique. To be more precise, since λ indicates the balance between the degree of attention given to the input document and the reference summary, we hypothesize that making λ adapt to the information shared in reference summary and input document should give us better performance since the higher the overlap, the better the quality of summary, and the higher the λ should be. Hence we perform two different experiments with $\lambda = max(w_{cov_i})$ and $\lambda = mean(w_{cov_i})$. To compare performance of a fixed λ value with the defined data-driven strategy, we plot performance of the proposed technique with fixed values of $\lambda \in \{0.0, 0.1, 0.2, ..., 1.0\}$ (see Fig. 3). Even though both of these λ defining strategies outperform the baseline metric ROUGE, we notice that the d value of $\lambda = 0.5$ is able to outperform these data-driven strategies as well as most of the fixed λ values[6].

---

[5] All the hyperparameter tuning experiments were performed using $ROUGE\text{-}L^f$ unless stated otherwise.

[6] It was also noticed that $\lambda = mean(W_{cov})$ outperforms $\lambda = max(W_{cov})$ in fluency and consistency; while the opposite happens for coherence and relevance. The reason for this can be explained by the fact that $mean(W_{cov}) < max(W_{cov})$; therefore the $\lambda = mean(W_{cov})$ variation always gives more weight to the input document similarity, giving higher fluency and consistency scores because input document consists of all the informationally rich and grammatically correct sentences.

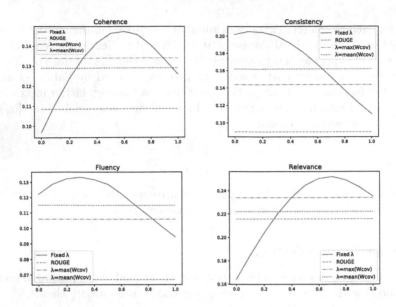

Fig. 3. Correlation plots of WIDAR with human judgement scores (from SummEval dataset [10]) for different λ values.

Thresholds for w_{cov_i} and w_{red_i} : θ_1 and θ_2 are the hyperparameters used in the calculation of coverage weights (w_{cov_i}) and redundancy weights (w_{red_i}) (Algorithm 1), respectively. To obtain the optimal range of these hyperparameters; we first performed an individual search for both θ_1 and θ_2 (see Fig. 4). As per these experiments, $\theta_1 = 0.0$ or 0.1 and $\theta_2 = 0.4$ yielded the best results when analyzed individually. However, on further experimentation, it was found that the best performance was obtained at $\theta_1 = 0.1$ and $\theta_2 = 0.3$.

Similarity Function for $IDSS$: In order to find the most suitable similarity function to compute the information overlap between input document and generated summary, we performed an isolated experiment where correlation coefficient of similarity function candidates was computed with the human judgement scores (Table 2). ROUGE-L^f score was the best performing model, and hence chosen as the similarity function.

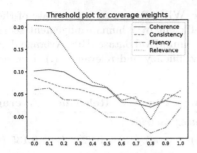 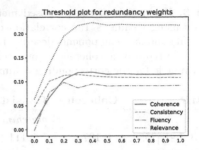

Fig. 4. Correlation plots of WIDAR with human judgement scores (from SummEval dataset [10]) with varying θ_1 values (on left) and θ_2 (on right) values.

Table 2. Various ROUGE-based similarity functions for $IDSS$.

IDSS	Coherence	Consistency	Fluency	Relevance
ROUGE-1^r	0.033	0.101	0.050	0.123
ROUGE-1^f	0.035	0.108	0.055	0.117
ROUGE-2^r	0.066	0.183	0.111	0.149
ROUGE-2^f	0.072	0.194	0.118	0.153
ROUGE-L^r	0.088	0.187	0.112	0.158
ROUGE-L^f	0.097	0.202	0.122	0.164

5   Results and Discussions

We evaluate the performance of our metric with other state-of-the-art techniques using correlation coefficient described in Sect. 4.2. Table 3[7] lists the correlation of WIDAR and other state-of-the art-metric scores available in SummEval with human judgement scores[8]. These scores illustrate the superiority of WIDAR over its predecessor, ROUGE, by a wide margin in all the three variants.

It can be deduced from the results that we need a way to combine these scores to better evaluate the performance of each metric, since a metric like SMS [5] performs well in aspects like consistency and fluency, yet it gives mediocre performance in coherence and relevance. Therefore, we also provide the average of these four scores in an attempt to ascertain the overall performance of each metric. We find out that all three variants of WIDAR are able to perform satisfactory, as they appear as 2nd, 3rd and 4th in the overall rankings; as opposed to their ROUGE counter-parts that end up in the middle-bottom section of the rankings.

---

[7] In case a metric has more than one variation, the version that corresponds to f-score was used.

[8] All the reported metrics in Table 3 have been computed in a multi-reference setting using 11 reference summaries per generated summary.

Table 3. Evaluation of the proposed metric WIDAR against other state-of-the-art methods using Kendall's Tau correlation coefficient over human judgement scores of individual summary components described in SummEval dataset [10]. Average denotes the average score over coherence, consistency, fluency and relevance. (.) denotes the rank of metric for the corresponding column.

Metric	Coherence	Consistency	Fluency	Relevance	Average
Text matching-based metrics					
ROUGE-1 [25]	0.137 (8)	0.111 (14)	0.067 (13)	0.228 (4)	0.135 (9)
ROUGE-2 [25]	0.110 (13)	0.107 (15)	0.054 (15)	0.184 (13)	0.113 (15)
ROUGE-L [25]	0.109 (14)	0.090 (16)	0.067 (13)	0.216 (8)	0.120 (14)
BLEU [30]	0.119 (11)	0.126 (9)	0.104 (8)	0.185 (12)	0.133 (10)
METEOR [24]	0.112 (12)	0.118 (12)	0.079 (12)	0.210 (10)	0.129 (12)
CHRF [33]	0.168 (1)	0.121 (10)	0.086 (11)	0.242 (3)	0.154 (7)
CIDEr [43]	-0.003 (18)	0.006 (18)	0.037 (17)	-0.019 (18)	0.005 (18)
Embedding-based metrics					
MoverScore [45]	0.154 (4)	0.134 (7)	0.117 (4)	0.224 (6)	0.157 (6)
SMS [5]	0.144 (6)	0.188 (2)	0.133 (2)	0.177 (14)	0.160 (5)
ROUGE-WE [28]	0.087 (15)	0.065 (17)	0.012 (18)	0.176 (15)	0.085 (17)
Model-based metrics					
BERTScore [44]	0.126 (10)	0.121 (10)	0.113 (6)	0.213 (9)	0.143 (8)
S^3 [32]	0.166 (2)	0.113 (13)	0.044 (15)	0.227 (5)	0.125 (13)
BlANC [42]	0.084 (16)	0.181 (4)	0.099 (10)	0.168 (16)	0.133 (10)
SUPERT [12]	0.130 (9)	0.259 (1)	0.167 (1)	0.204 (11)	0.190 (1)
SummaQA [37]	0.062 (17)	0.128 (8)	0.101 (9)	0.107 (17)	0.099 (16)
Proposed Metric					
$WIDAR_1$	0.160 (3)	0.178 (5)	0.114 (5)	0.254 (1)	0.176 (2)
$WIDAR_2$	0.138 (7)	0.188 (2)	0.108 (7)	0.221 (7)	0.163 (4)
$WIDAR_L$	0.149 (5)	0.176 (6)	0.119 (3)	0.250 (2)	0.167 (3)

The fact that SUPERT [12] is a model-based metric that evaluates the quality of a summary by taking as input the generated summary and the input document might be the reason for it having high correlation scores with consistency and fluency. Since input document comprises of grammatically correct and factually rich sentences, high performances on fluency and consistency are to be expected. CHRF [33] and S^3 [32] on the other-hand perform well in coherence and relevance; which can be somewhat credited to their evaluation strategy that computes information overlap between generated summary and reference summary. Since reference summary contains only the most significant information from the input document put together in a presentable manner, it results in high relevance and coherence scores. We believe that since WIDAR uses information overlap of generated summary with both the input document and the reference summary efficiently, it performs optimally in all the four characteristics.

5.1   Computational Time Analysis

Table 4 shows the comparison of computational time taken by WIDAR with respect to 5 state-of-the-art models or embedding based metrics computed using a single reference summary. The experiment is conducted for 100 randomly chosen summaries for all the metrics[9]. It is noticed that WIDAR takes about 0.6% of the computational time as compared to the average time taken by all these 5 metrics, while giving similar performance.

Table 4. Computation time taken by WIDAR and various model-based metrics.

Metric	Time-taken
BLANC [42]	1076.35 s
SUPERT [12]	30.40 s
MoverScore [12]	37.60 s
BERTScore [44]	1410.37 s
SummaQA [37]	910.26 s
Average	692.99 s
WIDAR$_L$	3.96 s

Table 5. Ablation Study.

Metric	Coherence	Consistency	Fluency	Relevance
WIDAR$_L$	0.149	0.176	0.119	0.250
ROUGE-L$_W$	0.129	0.108	0.083	0.239
$-W_{red}$	0.105	0.075	0.064	0.201
$-W_{cov}$	0.119	0.115	0.087	0.218
ROUGE-L$_{SL}$	0.102	0.087	0.062	0.204
ROUGE-L	0.109	0.090	0.067	0.216
IDSS	0.097	0.202	0.122	0.164

5.2   Ablation Study

WIDAR comprises of two key-components: (1) weighted ROUGE (ROUGE$_W$) between reference summary and generated summary and (2) similarity overlap (IDSS) between input document and generated summary. In order to establish the necessity of both of these components, we conduct an ablation study. When we consider only ROUGE-L$_W$, we notice a major drop in correlation with consistency (38%) and fluency (30%) (refer to the top two rows in Table 5). We reason

---

[9] This experiment was conducted on a Tyrone machine with Intel's Xeon W-2155 Processor having 196 Gb DDR4 RAM and 11 Gb Nvidia 1080Ti GPU. GPU was only used for BLANC, SUPERT, BERTScore and SummaQA evaluation metrics.

that consistency being the measure of factual correctness in the summary justifies the decrease in consistency scores. An argument can be made regarding fluency that since WIDAR is effectively a string-matching based technique; the input document usually comprises of sentences which are more grammatically sound than ones in reference summary [10,21]) could explain the drop in fluency scores. This argument can be further bolstered when comparing the correlation scores obtained for ROUGE-L and IDSS. IDSS uses ROUGE-L to compute information overlap between generated summary and input document, and ROUGE-L is used to compute information overlap between generated summary and reference summary. We can see that IDSS outperforms ROUGE-L in consistency (by 124%) and fluency (by 82%), supporting the previously mentioned argument.

If we remove W_{red} from weighted ROUGE, we observe drops in coherence (by 18%) and relevance (by 15%) as expected; but we also observe that without these redundancy weights, correlation with consistency and fluency also drop by 30% and 22%, respectively. Removing W_{cov} however yields mixed results in an isolated setting. Yet, together with W_{red}, the weighted ROUGE is able to outperform the sentence-level baseline. This can be noticed from the relevance scores in Table 5; ROUGE-L$_{SL}$ attains 0.204 score, while adding W_{red} yields an increase to 0.218 (shown in row $-W_{cov}$) and adding W_{cov} drops the score to 0.201 (shown in row $-W_{red}$). However, combining these two to obtain ROUGE-L$_W$ attains 0.239 score, better than in the case of the individual components.

5.3   Study of Human Judgement Scores

To analyze how humans have perceived these four characteristics of a summary, we compute and study the Kendall's Tau correlation coefficient between them. The results (refer to Table 6) revealed that coherence and relevance are moderately correlated, while other characteristic pairs do not yield any significant correlation score. This high correlation between coherence and relevance can be attributed to the fact that both relevance and coherence are related to non-redundancy. Coherence explicitly captures the non-redundancy in a summary, since a coherent summary must not have high information overlap across the sentences. Relevance on the other hand implicitly captures the notion of non-redundancy, since a summary that is highly relevant will cover up a major portion of input document, which is not achievable for a redundant summary. This reasoning can also be backed by the results from the ablation study (refer to Table 5), where removing the redundancy weight (W_{red}) from weighted ROUGE affects both the relevance and the coherence scores, implying that humans directly or indirectly consider redundancy of sentences within summary while providing these scores.

Table 6. Kendall's Taus correlation between various summary characteristics.

	Coherence	Consistency	Fluency	Relevance
Coherence	1.00	0.25	0.27	0.53
Consistency	0.25	1.00	0.38	0.27
Fluency	0.27	0.38	1.00	0.23
Relevance	0.53	0.27	0.23	1.00

6   Conclusion

We propose a novel evaluation metric WIDAR that utilizes both input document and reference summary to estimate the quality of the generated summary. We discuss why metrics like ROUGE, METEOR, BLUE etc. that solely depend on reference summary for evaluation do not perform well in real-world situations. We illustrate how the proposed metric is able to outperform its predecessor, ROUGE, by a large margin, and is also able to achieve performance comparable to huge model-based metrics like BERTScore, S^3, SUPERT etc. We also perform an ablation study to establish the necessity of each component in the proposed metric. We believe that the community needs computationally fast and lightweight metrics like WIDAR that can work well in real-world situations.

Acknowledgement. Dr. Sriparna Saha gratefully acknowledges the Young Faculty Research Fellowship (YFRF) Award, supported by Visvesvaraya Ph.D. Scheme for Electronics and IT, Ministry of Electronics and Information Technology (MeitY), Government of India, being implemented by Digital India Corporation (formerly Media Lab Asia) for carrying out this research.

References

1. Bhandari, M., Narayan Gour, P., Ashfaq, A., Liu, P., Neubig, G.: Re-evaluating evaluation in text summarization. In: Proceedings of the 2020 Conference on Empirical Methods in Natural Language Processing (EMNLP) (2020)
2. Böhm, F., Gao, Y., Meyer, C.M., Shapira, O., Dagan, I., Gurevych, I.: Better rewards yield better summaries: learning to summarise without references. arXiv:abs/1909.01214 (2019)
3. Chen, P., Wu, F., Wang, T.: A semantic QA-based approach for text summarization evaluation. In: AAAI (2018)
4. Chopra, S., Auli, M., Rush, A.M.: Abstractive sentence summarization with attentive recurrent neural networks. In: Proceedings of the 2016 Conference of the North American Chapter of the Association for Computational Linguistics: Human Language Technologies, San Diego, California, June 2016, pp. 93–98. Association for Computational Linguistics (2016). https://doi.org/10.18653/v1/N16-1012. https://aclanthology.org/N16-1012

5. Clark, E., Celikyilmaz, A., Smith, N.A.: Sentence mover's similarity: automatic evaluation for multi-sentence texts. In: Proceedings of the 57th Annual Meeting of the Association for Computational Linguistics, Florence, Italy, July 2019. Association for Computational Linguistics (2019). https://doi.org/10.18653/v1/P19-1264. https://aclanthology.org/P19-1264

6. Dang, H.T.: Overview of DUC 2005. In: Proceedings of the Document Understanding Conference, vol. 2005, pp. 1–12 (2005)

7. Duan, Y., Jatowt, A.: Across-time comparative summarization of news articles. In: Proceedings of the Twelfth ACM International Conference on Web Search and Data Mining, pp. 735–743. ACM (2019)

8. Edmundson, H.P.: New methods in automatic extracting. J. ACM **16**, 264–285 (1969)

9. Eyal, M., Baumel, T., Elhadad, M.: Question answering as an automatic evaluation metric for news article summarization. In: Proceedings of the 2019 Conference of the North American Chapter of the Association for Computational Linguistics: Human Language Technologies, Volume 1 (Long and Short Papers), Minneapolis, Minnesota, June 2019, pp. 3938–3948. Association for Computational Linguistics (2019). https://doi.org/10.18653/v1/N19-1395. https://aclanthology.org/N19-1395

10. Fabbri, A.R., Kryscinski, W., McCann, B., Socher, R., Radev, D.: SummEval: re-evaluating summarization evaluation. Trans. Assoc. Comput. Linguist. **9**, 391–409 (2021)

11. Ganesan, K.A.: Rouge 2.0: updated and improved measures for evaluation of summarization tasks. arXiv:abs/1803.01937 (2018)

12. Gao, Y., Zhao, W., Eger, S.: SUPERT: towards new frontiers in unsupervised evaluation metrics for multi-document summarization. In: Proceedings of the 58th Annual Meeting of the Association for Computational Linguistics. Association for Computational Linguistics, July 2020. https://doi.org/10.18653/v1/2020.acl-main.124. https://aclanthology.org/2020.acl-main.124

13. Giannakopoulos, G., Karkaletsis, V.: AutoSumMENG and MeMoG in evaluating guided summaries. Theory Appl. Categ. (2011)

14. Grusky, M., Naaman, M., Artzi, Y.: Newsroom: a dataset of 1.3 million summaries with diverse extractive strategies. In: Proceedings of the 2018 Conference of the North American Chapter of the Association for Computational Linguistics: Human Language Technologies, Volume 1 (Long Papers), New Orleans, Louisiana, June 2018. Association for Computational Linguistics (2018). https://aclanthology.org/N18-1065

15. Hasan, T., et al.: XL-Sum: large-scale multilingual abstractive summarization for 44 languages. In: Findings of the Association for Computational Linguistics: ACL-IJCNLP 2021, pp. 4693–4703 (2021)

16. Jangra, A., Jain, R., Mavi, V., Saha, S., Bhattacharyya, P.: Semantic extractor-paraphraser based abstractive summarization. arXiv preprint arXiv:2105.01296 (2021)

17. Jangra, A., Jatowt, A., Hasanuzzaman, M., Saha, S.: Text-image-video summary generation using joint integer linear programming. In: Jose, J.M., et al. (eds.) ECIR 2020. LNCS, vol. 12036, pp. 190–198. Springer, Cham (2020). https://doi.org/10.1007/978-3-030-45442-5_24

18. Jangra, A., Jatowt, A., Saha, S., Hasanuzzaman, M.: A survey on multi-modal summarization (2021)

19. Jangra, A., Saha, S., Jatowt, A., Hasanuzzaman, M.: Multi-modal summary generation using multi-objective optimization. In: Proceedings of the 43rd International ACM SIGIR Conference on Research and Development in Information Retrieval, pp. 1745–1748 (2020)

20. Jangra, A., Saha, S., Jatowt, A., Hasanuzzaman, M.: Multi-modal supplementary-complementary summarization using multi-objective optimization. In: Proceedings of the 44th International ACM SIGIR Conference on Research and Development in Information Retrieval, pp. 818–828 (2021)

21. Kryscinski, W., Keskar, N.S., McCann, B., Xiong, C., Socher, R.: Neural text summarization: a critical evaluation. In: Proceedings of the 2019 Conference on Empirical Methods in Natural Language Processing and the 9th International Joint Conference on Natural Language Processing (EMNLP-IJCNLP), Hong Kong, China, November 2019. Association for Computational Linguistics (2019). https://doi.org/10.18653/v1/D19-1051. https://aclanthology.org/D19-1051

22. Kupiec, J., Pedersen, J., Chen, F.: A trainable document summarizer. In: Proceedings of the 18th Annual International ACM SIGIR Conference on Research and Development in Information Retrieval, SIGIR 1995, New York, NY, USA, pp. 68–73. Association for Computing Machinery (1995). https://doi.org/10.1145/215206.215333

23. Kusner, M., Sun, Y., Kolkin, N., Weinberger, K.: From word embeddings to document distances. In: Bach, F., Blei, D. (eds.) Proceedings of the 32nd International Conference on Machine Learning. Proceedings of Machine Learning Research, Lille, France, 07–09 July 2015, vol. 37, pp. 957–966. PMLR (2015). https://proceedings.mlr.press/v37/kusnerb15.html

24. Lavie, A., Agarwal, A.: METEOR: an automatic metric for MT evaluation with high levels of correlation with human judgments. In: Proceedings of the Second Workshop on Statistical Machine Translation, Prague, Czech Republic, June 2007, pp. 228–231. Association for Computational Linguistics (2007). https://aclanthology.org/W07-0734

25. Lin, C.Y.: ROUGE: a package for automatic evaluation of summaries. In: Text Summarization Branches Out, Barcelona, Spain, July 2004, pp. 74–81. Association for Computational Linguistics (2004). https://aclanthology.org/W04-1013

26. Nallapati, R., Zhou, B., Santos, C.D., Çaglar Gülçehre, Xiang, B.: Abstractive text summarization using sequence-to-sequence RNNs and beyond. In: CoNLL (2016)

27. Nenkova, A.: Summarization evaluation for text and speech: issues and approaches. In: INTERSPEECH (2006)

28. Ng, J.P., Abrecht, V.: Better summarization evaluation with word embeddings for ROUGE. In: Proceedings of the 2015 Conference on Empirical Methods in Natural Language Processing, Lisbon, Portugal, September 2015, pp. 1925–1930. Association for Computational Linguistics (2015). https://doi.org/10.18653/v1/D15-1222. https://aclanthology.org/D15-1222

29. Paice, C.D.: Constructing literature abstracts by computer: techniques and prospects. Inf. Process. Manag. **26**, 171–186 (1990)

30. Papineni, K., Roukos, S., Ward, T., Zhu, W.J.: BLEU: a method for automatic evaluation of machine translation. In: ACL (2002)

31. Passonneau, R.J., Chen, E., Guo, W., Perin, D.: Automated pyramid scoring of summaries using distributional semantics. In: Proceedings of the 51st Annual Meeting of the Association for Computational Linguistics (Volume 2: Short Papers), Sofia, Bulgaria, August 2013, pp. 143–147. Association for Computational Linguistics (2013). https://aclanthology.org/P13-2026

32. Peyrard, M., Botschen, T., Gurevych, I.: Learning to score system summaries for better content selection evaluation. In: Proceedings of the Workshop on New Frontiers in Summarization, Copenhagen, Denmark, September 2017. Association for Computational Linguistics (2017). https://doi.org/10.18653/v1/W17-4510. https://aclanthology.org/W17-4510

33. Popović, M.: chrF: character n-gram F-score for automatic MT evaluation. In: Proceedings of the Tenth Workshop on Statistical Machine Translation, Lisbon, Portugal, September 2015, pp. 392–395. Association for Computational Linguistics (2015). https://doi.org/10.18653/v1/W15-3049. https://aclanthology.org/W15-3049

34. Rath, G.J., Resnick, S., Savage, T.R.: The formation of abstracts by the selection of sentences (1961)

35. Saini, N., Saha, S., Jangra, A., Bhattacharyya, P.: Extractive single document summarization using multi-objective optimization: exploring self-organized differential evolution, grey wolf optimizer and water cycle algorithm. Knowl. Based Syst. **164**, 45–67 (2019)

36. Scialom, T., Dray, P.A., Lamprier, S., Piwowarski, B., Staiano, J.: MLSUM: the multilingual summarization corpus. In: Proceedings of the 2020 Conference on Empirical Methods in Natural Language Processing (EMNLP), pp. 8051–8067 (2020)

37. Scialom, T., Lamprier, S., Piwowarski, B., Staiano, J.: Answers unite! Unsupervised metrics for reinforced summarization models. In: Proceedings of the 2019 Conference on Empirical Methods in Natural Language Processing and the 9th International Joint Conference on Natural Language Processing (EMNLP-IJCNLP), Hong Kong, China, November 2019. Association for Computational Linguistics (2019). https://doi.org/10.18653/v1/D19-1320. https://aclanthology.org/D19-1320

38. See, A., Liu, P., Manning, C.: Get to the point: summarization with pointer-generator networks. In: Association for Computational Linguistics (2017). https://arxiv.org/abs/1704.04368

39. ShafieiBavani, E., Ebrahimi, M., Wong, R., Chen, F.: A graph-theoretic summary evaluation for ROUGE. In: Proceedings of the 2018 Conference on Empirical Methods in Natural Language Processing, Brussels, Belgium, October–November 2018. Association for Computational Linguistics (2018). https://aclanthology.org/D18-1085

40. ShafieiBavani, E., Ebrahimi, M., Wong, R.K., Chen, F.: Summarization evaluation in the absence of human model summaries using the compositionality of word embeddings. In: COLING (2018)

41. Sun, S., Nenkova, A.: The feasibility of embedding based automatic evaluation for single document summarization. In: Proceedings of the 2019 Conference on Empirical Methods in Natural Language Processing and the 9th International Joint Conference on Natural Language Processing (EMNLP-IJCNLP), Hong Kong, China, November 2019, pp. 1216–1221. Association for Computational Linguistics (2019). https://doi.org/10.18653/v1/D19-1116. https://aclanthology.org/D19-1116

42. Vasilyev, O., Bohannon, J.: Is human scoring the best criteria for summary evaluation? In: Findings of the Association for Computational Linguistics: ACL-IJCNLP 2021. Association for Computational Linguistics, August 2021. https://doi.org/10.18653/v1/2021.findings-acl.192. https://aclanthology.org/2021.findings-acl.192

43. Vedantam, R., Zitnick, C.L., Parikh, D.: CIDER: consensus-based image description evaluation. In: 2015 IEEE Conference on Computer Vision and Pattern Recognition (CVPR), pp. 4566–4575 (2015)

44. Zhang, T., Kishore, V., Wu, F., Weinberger, K.Q., Artzi, Y.: BERTScore: Evaluating text generation with BERT. In: International Conference on Learning Representations (2020). https://openreview.net/forum?id=SkeHuCVFDr

45. Zhao, W., Peyrard, M., Liu, F., Gao, Y., Meyer, C.M., Eger, S.: MoverScore: text generation evaluating with contextualized embeddings and earth mover distance. In: Proceedings of the 2019 Conference on Empirical Methods in Natural Language Processing and the 9th International Joint Conference on Natural Language Processing (EMNLP-IJCNLP), Hong Kong, China, November 2019, pp. 563–578. Association for Computational Linguistics (2019). https://doi.org/10.18653/v1/D19-1053. https://aclanthology.org/D19-1053

46. Zhou, L., Lin, C.Y., Munteanu, D.S., Hovy, E.: ParaEval: using paraphrases to evaluate summaries automatically. In: Proceedings of the Human Language Technology Conference of the NAACL, Main Conference, New York City, USA, June 2006, pp. 447–454. Association for Computational Linguistics (2006). https://aclanthology.org/N06-1057

Bi-granularity Adversarial Training for Non-factoid Answer Retrieval

Zhiling Jin, Yu Hong$^{(\boxtimes)}$, Hongyu Zhu, Jianmin Yao, and Min Zhang

School of Computer Science and Technology, Soochow University, Suzhou, China
tianxianer@gmail.com, jyao@suda.edu.cn

Abstract. Answer Retrieval is a task of automatically retrieving relevant answers towards a specific question. The recent studies, in this field, have witnessed the vast success of non-factoid QA methods which leverage the large pre-trained models. The findings in intensive experiments have shown that the existing large and deep models can be enhanced by the utilization of adversarial examples, the ones which effectively challenge encoders during training and enable them to generalize well during test. However, the majority of adversarial training methods still suffer from two limitations: 1) they separately take into consideration single-granularity adversarial examples (e.g. character, token or sentence-level examples), resulting in a monotonous mode that easily make encoders get accustomed to such examples, and 2) they fail to actively detect and apply the truly challenging adversarial examples for training. In this paper, we propose a Bi-granularity Adversarial Training (BAT) approach. It not only involves multiple perturbation into the generation of adversarial examples, but selectively utilizes them in terms of perturbation strength. A self-adaptive adversarial training method is developed for recognizing perturbative examples. We conduct experiments on the WikiPassageQA and TREC-QA benchmarks. Experimental results show that BAT substantially improve the answer retrieval performance, reaching the MAP score of about 80.05% and MRR of 86.27% for WikiPassageQA, MAP of 93.99% and MRR of 97.55% for TREC-QA.

Keywords: Answer Retrieval · Adversarial Attack · Pre-trained model

1 Introduction

Answer retrieval aims to acquire the most relevant answer from a set of candidates for a specific question. It stands for the crucial component of the current Question Answering (QA) systems [1]. There are mainly two categories of retrieval-based QA systems: factoid and non-factoid QA. Factoid QA pursues short and definite answers, such as the entity-level answer "*William Shakespeare*" of the *WHO*-type question "*Who writes Hamlet?*". By contrast, non-factoid QA is more challenging, which tackles descriptive questions like *WHY*, *WHAT* and *HOW*-type questions. The answers of such questions generally appear as long-winded all-encompassing and well-grounded descriptions. In this paper, we concentrate on the issue of answer retrieval in the scenario of non-factoid QA.

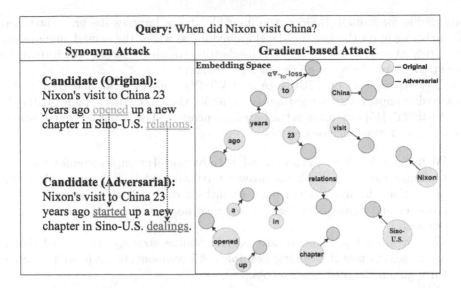

Fig. 1. Fakes that were generated by conducting synonym attack and gradient attack (the gradient one is carried out upon the output of the fine-tuned basic BERT).

Due to the contribution to linguistic representation learning, the pre-trained language models such as BERT [5] have been utilized for answer retrieval. The existing state-of-the-art methods perform perfectly to some extent when deep insights need to be gotten into the semantics of questions and answers. However, the robustness is still less strong. It is not difficult to puzzle the current neural answer retrieval models [31] with some negligible disturbance. Figure 1 shows a couple of fake candidate answers, one of which is obtained by synonymous substitution (titled as "*Synonym Attack*"), while the other results from the slight migration of word embeddings in the semantic space ("*Gradient-based Attack*"). Both barely cause the semantic change, though they result in negative predictions. Dialectically, the previous work proposes to use such fakes as the adversarial examples to challenge neural models during traning (called adversarial training), so as to improve their robustness.

The recent studies of adversarial training concentrate on the generation of adversarial examples, which have been proven effective. However, the resultant adversarial examples are of single granularity, appearing as character [7], token [9] or sentence-level [8] examples, separately. The embedded linguistic units are of less-diverse granularity and, therefore, they fail to bring into severe challenges. As a result, a neural model easily get accustomed to the challenge mode.

To address the issue, we propose a simple but effective method BAT, a self-adaptive method of applying bi-granularity adversarial samples to challenge retrieval models during training. In BAT, the adversarial examples are raised in both the focal visible attack process and general invisible attack. Specifically, the visible attack performs synonyms replacement conditioned on an external knowledge base, generating token-level adversarial examples, where the falsified words are visible for human. Besides, the invisible attack imposes the gradient-based disturbance upon the sentence-level embeddings, where the caused changes are invisible. Moreover, we develop a self-adaptive training strategy which tends to select the most challenging examples for training.

The experiments on TREC-QA and WikiPassageQA show that BAT yields noticeable improvements regardless of whether the basic BERT, large BERT [5] or RoBERTa [15] are taken as the baseline encoders. The potential contributions of this paper are concluded as follows:

1) We utilize token-level and sentence-level adversarial examples to enforce a comprehensive attack towards the answer retrieval models. Experimental results prove that the model enduring such attack during training changes to be stronger and achieves better performance no matter in long or short text scenario.
2) We develop a self-adaptive adversarial training strategy which contributes to the selection of challenging examples. All components are proven effective through our detailed experiments.

2   Related Work

We briefly overview the previous work regarding both answer retrieval and adversarial attack as below.

Answer Retrieval. The previous work on answer retrieval was mainly based on statistical methods such as BM25 and TF-IDF. Recent work has investigated the models based on deep neural networks. According to Rücklé et al.'s survey [26], the models can be classified into two categories. The first kind of models were based on the semantic similarity calculation, such as that of LSTM-DSSM [22], which calculated sentence-level similarity scores between the representations of interrogative sentences and candidate answers. The second ones were focusing on relevance measurement. The most representative model is MatchPyramid [23]. It determines relevance level conditioned on the interaction intensity between questions and candidate answers. Nowadays, the vast success has been achieved by using pre-trained language models, the ones which were obtained through masking mechanism, as well as learning to encode on a large amount of data. From here on, researchers take into account data augmentation. Gary et al. [10] built a large scale dataset, so as to facilitate transfer learning over the dataset. More importantly, they propose a Transfer and Adapt Model (TANDA) to conduct fine-tuning on a larger training data. This method not only improves the answer retrieval performance, but allows the retrieval model to generalize well.

Adversarial Attack. Adversarial attack has recently received keen attention. It is useful for detecting the possible weaknesses of models. More importantly, it generates adversarial examples, so as to enforce neural models to accommodate to pragmatic diversity. Jia and Liang et al. [13] create adversarial examples using manually defined templates. Automatic methods [12] have also been proposed to generate adversarial examples through paraphrasing. These works revealed how neural network systems trained on a large corpus can easily break when facing carefully designed unseen adversarial patterns. However, most previous work on adversarial attack concentrates on a single granularity attack (e.g., character [7], token [9] or sentence [8]). This raises a question of whether multi-granularity linguistic units can be involved into a comprehensive attack. Note that the above remarkable research aims to challenge neural models as competitively as possible. By contrast, we tend to generate and select beneficial adversarial examples for data augmentation, so as to improve performance, instead of reducing it.

3   Methodology

3.1   Task Definition

We define a question set as $Q = \{q_1, q_2, ...q_n\}$. Given a question $q_i \in Q$, there is a candidate set of answers $C_i = \{c_{i1}, c_{i2}, ...c_{im}\}$ for q_i. We suppose to rank the answers in the candidate set C_i for a given question q_i according to their scores. We evaluate an answer retrieval model by verifying whether the answer is highly ranked. We formulate the question-answer pairs with (q_i, c_{ik}, y_{ik}), where y_{ik} is a binary label indicating whether a candidate c_{ik} occurs as the answer for the question q_i. Our goal is to estimate the probability $p(y_{ik}|q_i, c_{ik})$ of y_{ik} in terms of the semantic representations of q_i and c_{ik}, and use it for ranking c_{ik}.

3.2   Basic Answer Retrieval Model (Baseline)

Two standard pre-trained models are employed to build the baseline retrieval model respectively, including BERT [5] and RoBERTa [15]. In our experiments, the pre-trained models are verified on both basic and large configurations. Towards an instance $X = (q, c, p(y))$, we denote $Input_{bert}$=[[CLS], q, [SEP], c, [SEP]] as the input of BERT, while $Input_{roberta}$=[<s>, q, </s>, c, </s>] as that of RoBERTa. [CLS] and [SEP] refer to special tokens in BERT, <s> and </s> are the ones of RoBERTa. The final hidden state obtained for the first token ([CLS] or <s>) is used as the joint representation of the question q and candidate c. It aggregates interactively-attentive latent information of both q and c. We feed the representation into a fully-connected neural layer (i.e., classification layer) with the sigmoid function to compute the probability $p(y)$. Figure 2 (a) shows the baseline architecture in the point-wise ranking scenario, while (b) the pair-wise. We discuss the scenarios in the experimental settings (Sect. 4.2).

Let $Sent = \{x_1, x_2, ...x_n\}$ denotes the input sequence $Input_{bert}$ or $Input_{roberta}$, $F(\cdot)$ is the neural retrieval model mentioned above, and $Loss(\cdot)$ denotes the loss produced by $F(\cdot)$. In our case, the probability of y and $Loss(\cdot)$ is computed as follows (where σ is the sigmoid function and $\hat{y}$ denotes the prediction):

$$p(\hat{y}|q, c) = \sigma(F(Sent)), \tag{1}$$

$$Loss(Sent, y) = -(y * log(p(\hat{y}|q, c)) + (1 - y) * log(1 - p(\hat{y}|q, c))), \tag{2}$$

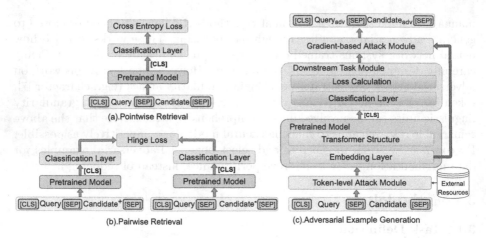

Fig. 2. (a) Point-wise answer retrieval model (b) Pair-wise answer retrieval model (c) Adversarial example generation method for BAT

3.3   Adversarial Examples Generation Strategy

Toward the basic retrieval model, we carry out a bi-granularity adversarial attack during training, which comprises the steps of saliency-based word sorting, token synonym replacement and gradient-based attack. We first falsify the original examples by replacing the top n percent influential words with their synonyms in token-level. On the basis, we keep on disturbing the examples by gradient-based attack in sentence-level. The generating process is shown in Fig. 2(c). From here on, we separately present our methods in the three attacking steps.

Saliency-Based Word Sorting. Echoing the discovery in [21], we suggest that the pre-trained models BERT and RoBERTa overly indulge themselves into (viz., effectively encode) some dominant words (e.g., verbs), the ones which reveal the skeletal semantics of a linguistic sequence. By contrast, they barely embrace other words (e.g., conjunctions). Thus, no matter whether their performance is better or worse, it results from the encoding of the dominant words. Accordingly, falsifying the dominant words helps to produce challenging adversarial examples. To detect the dominant words, we calculate saliency score I_{x_i} for each of them, and pick up the ones holding a higher I_{x_i}. The score I_{x_i} is computed as follow:

$$I_{x_i} = Loss(Sent_{\not{x}_i}, y) - Loss(Sent, y), \tag{3}$$

where $Sent_{\not{x}_i} = \{x_1, \cdots, x_{i-1}, x_{i+1}, \cdots, x_n\}$. In terms of Eq. 3, the saliency score of the candidate word x_i is estimated by the difference of loss which emerges after deleting x_i itself. We verify saliency during training, and select and falsify top-n most salient words by synonym replacement.

Synonym Replacement. As mentioned in [19], the recent adversarial attacking methods pursue the success rate of distracting the neural learning models, regardless of whether the adversarial examples they produced are utterly oppose to the semantic principles or even break the grammatical rules. Retraining with such examples may mislead the learning process thoroughly instead of strengthening it. Different from the previous work, we produce adversarial examples by synonym replacement, where the salient words are considered. This contributes to the preservation of original semantics and grammar, to some extent.

Given a list L of the most salient words, we iteratively replace them with appropriate synonyms. We initialize a synonym candidate set for each word w_i in L. WordNet [17,35] is used to obtain the synonyms. POS-based filtering is applied to filter out the synonym candidates that are of inconsistent POS tags with w_i. Utilizing the rest words in L for replacement, more or less, reduces of the risk of grammatical errors.

We apply a greedy searching method to select a single synonym for substitution, where the nearest neighbor of the considered salient word w_i in the embedding space will be selected. Counter-fitting embedding space [20] is used. In the space, synonyms can be gathered together, while antonyms are anchored apart from each other. Word similarity calculation within the counter-fitting embedding space has achieved the state-of-the art performance on the benchmark dataset SimLex-999 [11]. Given a synonym c_k, we project it and w_i into the counter-fitting space. On the basis, we calculate the semantic distance of their embeddings by cosine similarity. We follow Morris et al. [19] to set the threshold to 0.9. If a synonym has the smallest semantic distance to w_i, and meanwhile the distance is smaller than threshold, it will be selected to substitute the w_i.

Gradient-Based Adversarial Training. Gradient-based sentence-level adversarial attack helps to strengthen the modeling of the continuously distributed representations. Chakraborty et al. [2] utilize it in image processing. Considering that the representations generated by pre-trained models (e.g., BERT) are of successive distribution mode, we utilize gradient-based attack in dealing with the sentence-level BERT-based (and RoBERTa-based) encoding.

The direction of the gradient ascent can lead to the rise of loss quickly and accurately [2]. Therefore, we follow Chakraborty et al. [2] to conduct directional attack, bringing minor perturbations into the sentence-level embeddings along the direction of gradient ascent, which is shown as below:

$$g_{emb} = \nabla_{embedding} Loss(F(X_{adv}), y), \tag{4}$$

$$X_{adv} \leftarrow embedding(X_{adv}) + r * (g_{emb}/\|g_{emb}\|_2), \tag{5}$$

where y denotes the label for adversarial example $X_{adv} = [q'; a']$ generated by us, $embedding(\cdot)$ denotes the input embedding, and r is the intensity of gradient attacking, which is used to determine how far the input embedding changes along the direction of gradient ascent. The gradient g_{emb} of the loss to the embedding layer is calculated via Eq. 4. On this basis, we perturb the embedding layer based on the gradient value by Eq. 5 to enforce a sentence-level attack.

Algorithm 1: Self-Adaptive Adversarial Training Method

Input: original example $X = [q; a]$, adversarial example $X_{adv} = [q'; a']$
generated by X, label y, target model $F(\cdot)$, loss function $Loss(\cdot)$, loss
increase ratio ϵ, loss weight α

Output: None

1 $loss_{clean} = Loss(F(X), y)$;
2 $loss_{adv} = Loss(F(X_{adv}), y)$;
3 **if** $loss_{adv} < \epsilon * loss_{clean}$ **then**
4 | $X_{adv} \leftarrow$ Sentence_Gradient_Attack(X) via Eq. 4 - 5;
5 | $loss_{adv} = Loss(F(X_{adv}), y)$;
6 **end**
7 $loss = \alpha * loss_{clean} + (1 - \alpha) * loss_{adv}$;
8 $loss.backward()$;

3.4 Self-adaptive Adversarial Training

We define the challenging adversarial examples as the ones which not only effectively confuse a learning model but provide novel knowledge to raise the well-generalized learning. Otherwise, the examples are redundant. The use of redundant examples not only reduces the training efficiency but fails to strengthen the learning model. In order to reduce the redundancy, we develop a self-adaptive adversarial training algorithm, where the challenging adversarial examples are recognized and adopted at every step of training (as mentioned in Algorithm 1).

We pursue the challenging adversarial examples at the word-level attack stage. Give an example that is created by synonym replacement, we adopt the increase ratio ϵ of loss as the measure to determine whether the example is challenging enough. If the rate ϵ is high enough, we keep it, otherwise we discard the synonym replacement. Note that the sentence-level gradient-based attack is carried out behind the word-level attack stage, and no matter whether the word-level attack is discarded, the sentence-level attack won't be canceled ever.

4 Experimentation

4.1 Data and Evaluation

We experiment on two publicly-available benchmark datasets, including TREC-QA [30] and WikiPassageQA [4]. TREC-QA is a popular benchmark corpus that is selected from the TREC Question-Answering tracks, containing 1,362 questions. Both the positive and negative answers in TREC-QA are of short sentences. WikiPassageQA are selected from 863 Wikipedia documents by crowd-sourcing, containing 4,165 groups of non-factoid questions and passages which are composed of no less than 6 sentences. The candidate answers of each question derive from the same passage and, in general, the positive and negative answers are highly similar in pragmatics (meanwhile, they often share similar context). The statistics in the sets are shown in Table 1. Mean Average Precision (MAP) and Mean Reciprocal Rank (MRR) are used as the evaluation metrics.

Table 1. Benchmark statistics (Q denotes a question, while A refers to an answer)

Benchmark	WikiPassageQA				TREC-QA			
Properties	Train	Dev	Test	Total	Train	Dev	Test	Total
Questions	3,332	417	416	4,165	1,229	65	68	1,362
Candidate Answer	194,314	25,841	23,981	244,136	53,313	1,117	1,442	55,872
Positive Answer	5,560	707	700	6,967	6299	205	248	6752
Average Length of Q	9.52	9.69	9.44	9.53	8.33	8.01	8.59	8.31
Average Length of A	133.09	134.13	132.65	133.16	28.97	24.89	25.60	26.48

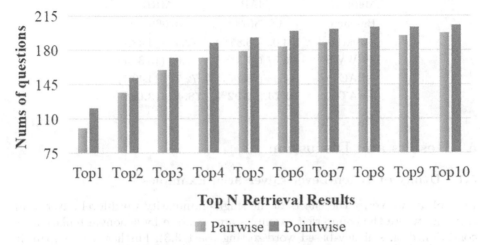

Top N Retrieval Results

■ Pairwise ■ Pointwise

Fig. 3. Comparison between pair-wise and point-wise ranking of candidates (Experiments are conducted on WikiPassageQA, and the BERT-based baseline is used).

4.2 Hyperparameter Settings and Ranking Scenario

We set our experiments on the HuggingFace's Transformers Library [32] with PyTorch version. We use an Adam optimizer with a learning rate of 1.8×10^{-6} and epsilon of 1×10^{-8}, 5 fine-tuning epochs for base configuration and 3 fine-tuning epochs for large configuration. The hyperparameters are set to default values, where set token attack percent n of 1, sentence attack intensity r of 0.25, loss increase ratio ϵ of 1 for base models and 2 for large models, semantic similarity threshold β of 0.9 and loss weight α of 0.5. The number of hidden units is set to 768 for the base configurations of BERT and RoBERTa (1024 for the large configurations). The dropout rate is set to 0.1.

We follow the previous work [16] to examine two sorts of Learning-To-Rank processes, including the point-wise and pair-wise ranking. In the point-wise case, each candidate answer is scored independently of others, which is trained by minimizing the cross-entropy loss via Eq. 2. The BERT-based point-wise ranking architecture is shown in Fig. 2(a). By contrast, the pair-wise approach conducts ranking in terms of contrasting effects among pairs of candidates.

The BERT-based pair-wise ranking architecture is shown in Fig. 2(b). Figure 3 shows that the point-wise ranking approach outperforms the pair-wise, where the vertical axis denotes the number of questions that corresponds to a non-zero recall rate in top-n ($n \in 1,...10$) retrieval results. In all the other experiments, accordingly, we carry out point-wise answer ranking.

Table 2. Test results of the BERT-base answer retrieval (Baseline) and that achieved after using TextFooler, BAE, PWWS and our SAT$_{tkn}$ (Adversarial training models).

Methods	MAP	MRR
Baseline	68.73(0%)	76.06(0%)
+TextFooler	64.02 ($\downarrow$6.85%)	72.38 ($\downarrow$4.48%)
+PWWS	63.57 ($\downarrow$7.51%)	72.00 ($\downarrow$5.34%)
+BAE	64.44 ($\downarrow$6.24%)	72.71 ($\downarrow$4.40%)
+SAT$_{tkn}$	**70.74 ($\uparrow$2.92%)**	**78.40 ($\uparrow$3.08%)**

5   Results and Discussion

5.1   Utility of Token-Level Adversarial Examples

First of all, we verify the effects of the single-granularity word-level adversarial training, where the counterfeit examples are generate by synonym replacement conditioned on saliency-based word sorting (Sect. 3.3). Further, we compare it to other congeneric models, including TextFooler [14], BAE [9], PWWS [25]. All the models generate adversarial examples using TextAttack [18] framework. It is noteworthy that, in this comparison experiment, we merely select 40K question-answer pairs from the WikiPassageQA training set for adversarial training. The involvement of all the training instances into all the models leads to the unaffordable computing resources and time.

In this experiment, we take the BERT-base retrieval model as the baseline (Fig. 2 (a)), which is trained on the original training set. The refined model using our word-level **S**ingle-granularity **A**dversarial **T**raining method is denoted as SAT$_{tkn}$. Table 2 shows the test results. It can be observed that defiantly retraining the baseline by TextFooler, BAE and PWWS causes performance degradation. By contrast, SAT$_{tkn}$ yields substantial performance improvement.

This proves that the generated adversarial examples by SAT$_{tkn}$ are propulsive. In this case, the selective synonym replacement (based on saliency) contributes to the generation of effective adversarial examples. By contrast, the competitors (TextFooler, BAE and PWWS) tend to enlarge the intensity level of adversarial training, at the expense of disordering the canonical grammatics and syntactic rules, with less restrictions and without selective adoption. As a result, the obtained adversarial examples do distract the encoder severely during training, though the cases are far from natural [19] and will be met rarely during test. Therefore, they fail to effectively strengthen the robustness.

5.2   Effect of Bi-granularity Adversarial Training (BAT)

We verify the effect of combining the word-level and sentence-level adversarial training, where synonym replacement and gradient-based directional representation refinement are carried out sequentially for each training instance (one joint attack per instance). We conduct a series of ablation experiments in which the BERT-based answer retrieval model is considered as the baseline, and the variants SAT_{tkn}, SAT_{snt}, BAT_{cmb} and BAT_{all} are obtained by challenging the baseline with 1) single-granularity token-level adversarial examples, 2) single-granularity sentence-level ones (Sect. 3.3), 3) the combined BAT examples as mentioned above, as well as 4) the combined BAT examples that are regulated by self-adaptive filtering (Sect. 3.4). In the experiments, all the training instances are considered for adversarial attacking. That is the reason why SAT_{tkn} has a different performance from that in Table 2.

Table 3. Ablation study on BERT-base retrieval model.

Benchmark	WikiPassageQA		TREC-QA	
Models	MAP	MRR	MAP	MRR
BERT(baseline)	74.90(0%)	82.03(0%)	85.70(0%)	93.70(0%)
+SAT_{tkn}	75.69(↑1.05%)	82.85(↑1.00%)	88.47(↑3.23%)	94.61(↑0.97%)
+SAT_{snt}	75.84(↑1.20%)	82.82(↑0.98%)	88.19(↑2.91%)	94.12(↑0.45%)
+BAT_{cmb}	76.73(↑2.44%)	83.18(↑1.40%)	89.84(↑4.83%)	95.34(↑1.75%)
+BAT_{all}	**77.31(↑3.21%)**	**84.28(↑2.74%)**	**90.99(↑6.17%)**	**96.08(↑2.54%)**

Table 3 shows the ablation experiment results. It can be observed that either SAT_{tkn} or SAT_{snt} obtains slightly better performance than the baseline. By contrast, BAT_{cmb} achieves significant improvements, and outperforms both SAT_{tkn} and SAT_{snt}. This illustrates that different-granularity adversarial examples are mutually complementary, and the sequential utilization of them contributes to the rising of performance. In addition, the redundancy elimination by adaptive filtering (in $_{all}$) improves the performance further.

5.3   Comparison to State of the Art

We compare our answer retrieval models to the previous work that were evaluated on WikiPassageQA and TREC-QA. Let us overview the models separately.

On WikiPassageQA. The recently-reported work on WikiPassageQA comprises different LSTM (Long-Short Term Memory)-based retrieval models (AP-BiLSTM [27], CA-Wang [24] and COALA [26]), as well as BERT-PR [33] and BERTlets[16]. BERT-PR strengthens BERT by a voting mechanism and statistic strategies. BERTlets applies different encoding modes (conditioned on point-wise and pairwise sentence-level representations) with different length constraints.

On TREC-QA. The earlier study on TREC-QA proposes aNMM [34] and KABLSTM [28]. The models develop different kinds of attention-based structures for enhancing the interaction between question and candidate answers. In addition, HyperQA [29] and SD [3] utilize the siamese neural network structure to aggregate the similarity between questions and candidate answers. Besides, TANDA [10] applies the pre-trained transformer for encoding the relationship between questions and candidate answers, and it additionally uses transfer learning that enables pre-finetuning on the external dataset. To the best of our knowledge, TANDA achieves the best performance on TREC-QA.

Table 4. Comparison to the previous work on WikiPassgaeQA and TREC-QA.

WikiPassageQA			TREC-QA		
Models	MAP	MRR	Models	MAP	MRR
AP-BiLSTM [27]	46.98	55.20	aNMM [34]	75.00	81.10
CA-Wang [24]	48.71	56.11	KABLSTM [28]	80.40	88.50
COALA [26]	60.58	69.40	HyperQA [29]	78.40	86.50
BERT-PR [33]	73.55	80.87	SD [3]	78.30	87.80
BERTlets [16]	73.60	81.00	TANDA [10] (SOTA)	**91.20**	**95.10**
BERT-base	74.90	82.03	BERT-base	85.70	93.70
+BAT$_{all}$	**77.31**	**84.28**	+BAT$_{all}$	90.99	96.08
RoBERTa-base	76.01	82.96	RoBERTa-base	88.16	94.12
+BAT$_{all}$	**79.03**	**86.08**	+BAT$_{all}$	91.61	96.32
BERT-large	76.78	82.59	BERT-large	90.40	94.60
+BAT$_{all}$	**79.01**	**84.89**	+BAT$_{all}$	**93.37**	**97.55**
RoBERTa-large	78.13	85.46	RoBERTa-large	91.50	95.22
+BAT$_{all}$	**80.05**	**86.27**	+BAT$_{all}$	**93.99**	**97.55**

Our Models on both Benchmarks. One of our baselines in the comparative experiments is a single BERT-based retrieval model. It couples BERT with a perceptron. Considering that RoBERTa is more robust than BERT, we use RoBERTa to develop another baseline by replacing the Bert-based encoder with RoBERTa. Both the basic and large configurations are considered in the experiments. Accordingly, there are four baselines constructed, including the retrieval models that are based on BERT-base, BERT-large, RoBERTa-base and RoBERTa-large, respectively. In addition, we enhance the baselines by retraining them using both the original and adversarial examples, where the adversarial examples are generated by our BAT$_{all}$. This results in four refined models.

Results of Comparative Experiments. Table 4 shows the test results of all the models mentioned above. It can be observed that all baselines are strengthened by BAT$_{all}$. And the resultant new retrieval models outperform the state-of-the-art models. In particular, BAT$_{all}$ cooperates with RoBERTa better than

BERT, yielding considerable improvements. Over the recently-reported superior models like BERTlets and TANDA, the performance gaps we reached are about 6.5% (MAP) and 5.3% (MRR) for WikiPassageQA, and 2.8% (MAP) and 2.5% (MRR) for TREC-QA. It's noteworthy that, in the case of comparing to TANDA, our models neither perform transfer learning nor utilize external data.

5.4 Statistical and Practical Significance Test

According to Zhu's advice [36], we report statistical significance and practical significance with evaluation metrics of P-value and Cohen's d, respectively. The P-value which is lower than the threshold (usually 0.05) indicates a significant improvement, otherwise insignificant [6]. Besides, Cohen's d is used to measure the effect size of our results, and higher value represents larger effect size.

Table 5. The statistical and practical significance test analysis of all tested models on both benchmarks.

Benchmark	WikiPassageQA		TREC-QA	
T-test	P-value	Cohen's d	P-value	Cohen's d
(Metrics)	(MAP/MRR)	(MAP/MRR)	(MAP/MRR)	(MAP/MRR)
$BERT_{base}$ VS. $BERT_{base}+BAT_{all}$	5.88e−4/2.34e−4	5.08/4.20	1.81e−8 /1.44e−4	4.98/2.75
$BERT_{large}$ VS. $BERT_{large}+BAT_{all}$	4.50e−3/2.25e−3	7.85/4.03	3.13e−4 /8.35e−5	2.29/3.43
$RoBERTa_{base}$ VS. $RoBERTa_{base}+BAT_{all}$	8.19e−4/5.15e−5	3.37/3.35	2.82e−4 /1.21e−3	5.08/4.54
$RoBERTa_{large}$ VS. $RoBERTa_{large}+BAT_{all}$	1.22e−3/1.08e−3	3.76/2.61	1.03e−3 /3.02e−2	4.98/1.97

The P-values and Cohen's d of the t-test for TREC-QA and WikiPassageQA are shown in Table 5. It can be observed that the P-values are far below the threshold and the values of Cohen's d show great effect size on both corpora. This demonstrates that BAT_{all} produces significant improvement and, therefore, it can be utilized as a reliable training method for both long and short text retrieval scenarios.

6 Conclusion

We propose a bi-granularity adversarial attacking method to strengthen the current answer retrieval model. Moreover, a self-adaptive strategy is developed to select informative adversarial examples for retraining. Experiments prove the effectiveness of BAT on both BERT and RoBERTa based retrieval models in both long text and short text scenarios.

Although the proposed method has achieved a higher performance, the efficiency of generating adversarial examples isn't high enough. This reduces the utility over a larger-scale dataset. In order to improve the efficiency, we will explore a transferable multi-task adversarial attacking technique, where all the reliable adversarial behaviors will be shared among the homogeneous data.

Acknowledgement. We thank all reviewers for their insightful comments, as well as the great efforts our colleagues have made so far. This work is supported by National Key R&D Program of China and the national Natural Science Foundation of China (NSFC), via Grant #2020YFB1313601, Nos. 62076174, 61836007.

References

1. Ahmad, A., Constant, N., Yang, Y., Cer, D.: ReQA: an evaluation for end-to-end answer retrieval models. arXiv preprint arXiv:1907.04780 (2019)
2. Chakraborty, A., Alam, M., Dey, V., Chattopadhyay, A., Mukhopadhyay, D.: Adversarial attacks and defences: a survey. arXiv preprint arXiv:1810.00069 (2018)
3. Chen, D., Peng, S., Li, K., Xu, Y., Zhang, J., Xie, X.: Re-ranking answer selection with similarity aggregation. In: SIGIR, pp. 1677–1680 (2020)
4. Cohen, D., Yang, L., Croft, W.B.: WikiPassageQA: a benchmark collection for research on non-factoid answer passage retrieval. In: SIGIR, pp. 1165–1168 (2018)
5. Devlin, J., Chang, M.W., Lee, K., Toutanova, K.: BERT: pre-training of deep bidirectional transformers for language understanding. arXiv:1810.04805 (2018)
6. Dror, R., Baumer, G., Shlomov, S., Reichart, R.: The Hitchhiker's guide to testing statistical significance in natural language processing. In: Proceedings of the 56th Annual Meeting of the Association for Computational Linguistics (Volume 1: Long Papers), pp. 1383–1392 (2018)
7. Eger, S., et al.: Text processing like humans do: visually attacking and shielding NLP systems. arXiv preprint arXiv:1903.11508 (2019)
8. Gan, W.C., Ng, H.T.: Improving the robustness of question answering systems to question paraphrasing. In: ACL, pp. 6065–6075 (2019)
9. Garg, S., Ramakrishnan, G.: BAE: BERT-based adversarial examples for text classification. arXiv preprint arXiv:2004.01970 (2020)
10. Garg, S., Vu, T., Moschitti, A.: TANDA: transfer and adapt pre-trained transformer models for answer sentence selection. In: AAAI, vol. 34, pp. 7780–7788 (2020)
11. Hill, F., Reichart, R., Korhonen, A.: SimLex-999: evaluating semantic models with (genuine) similarity estimation. Comput. Linguist. **41**(4), 665–695 (2015)
12. Iyyer, M., Wieting, J., Gimpel, K., Zettlemoyer, L.: Adversarial example generation with syntactically controlled paraphrase networks. arXiv:1804.06059 (2018)
13. Jia, R., Liang, P.: Adversarial examples for evaluating reading comprehension systems. arXiv preprint arXiv:1707.07328 (2017)
14. Jin, D., Jin, Z., Zhou, J.T., Szolovits, P.: Is BERT really robust? A strong baseline for natural language attack on text classification and entailment. In: AAAI, vol. 34, pp. 8018–8025 (2020)
15. Liu, Y., et al.: RoBERTa: a robustly optimized BERT pretraining approach. arXiv preprint arXiv:1907.11692 (2019)
16. Mass, Y., Roitman, H., Erera, S., Rivlin, O., Weiner, B., Konopnicki, D.: A study of BERT for non-factoid question-answering under passage length constraints. arXiv preprint arXiv:1908.06780 (2019)

17. Miller, G.A.: WordNet: a lexical database for English. Commun. ACM **38**(11), 39–41 (1995)
18. Morris, J., Lifland, E., Yoo, J.Y., Grigsby, J., Jin, D., Qi, Y.: TextAttack: a framework for adversarial attacks, data augmentation, and adversarial training in NLP. In: EMNLP, pp. 119–126 (2020)
19. Morris, J.X., Lifland, E., Lanchantin, J., Ji, Y., Qi, Y.: Reevaluating adversarial examples in natural language. arXiv preprint arXiv:2004.14174 (2020)
20. Mrkšić, et al.: Counter-fitting word vectors to linguistic constraints. arXiv preprint arXiv:1603.00892 (2016)
21. Niven, T., Kao, H.Y.: Probing neural network comprehension of natural language arguments. arXiv preprint arXiv:1907.07355 (2019)
22. Palangi, H., et al.: Semantic modelling with long-short-term memory for information retrieval. arXiv preprint arXiv:1412.6629 (2014)
23. Pang, L., Lan, Y., Guo, J., Xu, J., Wan, S., Cheng, X.: Text matching as image recognition. In: AAAI, vol. 30 (2016)
24. Parikh, A.P., Täckström, O., Das, D., Uszkoreit, J.: A decomposable attention model for natural language inference. arXiv preprint arXiv:1606.01933 (2016)
25. Ren, S., Deng, Y., He, K., Che, W.: Generating natural language adversarial examples through probability weighted word saliency. In: ACL, pp. 1085–1097 (2019)
26. Rücklé, A., Moosavi, N.S., Gurevych, I.: COALA: a neural coverage-based approach for long answer selection with small data. In: AAAI, vol. 33, pp. 6932–6939 (2019)
27. Santos, C.d., Tan, M., Xiang, B., Zhou, B.: Attentive pooling networks. arXiv preprint arXiv:1602.03609 (2016)
28. Shen, Y., et al.: Knowledge-aware attentive neural network for ranking question answer pairs. In: SIGIR, pp. 901–904 (2018)
29. Tay, Y., Tuan, L.A., Hui, S.C.: Hyperbolic representation learning for fast and efficient neural question answering. In: WSDM, pp. 583–591 (2018)
30. Wang, M., Smith, N.A., Mitamura, T.: What is the jeopardy model? A quasi-synchronous grammar for QA. In: EMNLP-CoNLL, pp. 22–32 (2007)
31. Wang, W., Wang, L., Wang, R., Wang, Z., Ye, A.: Towards a robust deep neural network in texts: a survey. arXiv preprint arXiv:1902.07285 (2019)
32. Wolf, T., et al.: HuggingFace's transformers: state-of-the-art natural language processing. arXiv preprint arXiv:1910.03771 (2019)
33. Xu, P., Ma, X., Nallapati, R., Xiang, B.: Passage ranking with weak supervision. arXiv preprint arXiv:1905.05910 (2019)
34. Yang, L., Ai, Q., Guo, J., Croft, W.B.: aNMM: ranking short answer texts with attention-based neural matching model. In: CIKM, pp. 287–296 (2016)
35. Zhang, X., Zhao, J., LeCun, Y.: Character-level convolutional networks for text classification. arXiv preprint arXiv:1509.01626 (2015)
36. Zhu, H., Mak, D., Gioannini, J., Xia, F.: Nlpstattest: a toolkit for comparing NLP system performance. arXiv preprint arXiv:2011.13231 (2020)

RATE: A Reliability-Aware Tester-Based Evaluation Framework of User Simulators

Sahiti Labhishetty$^{(\boxtimes)}$ and ChengXiang Zhai

University of Illinois at Urbana-Champaign, Urbana, IL, USA
{sahitil2,czhai}@illinois.edu

Abstract. Evaluation of user simulators is needed in order to use them for evaluating Interactive Information Retrieval (IIR) Systems. Previous work has proposed a tester-based approach to evaluate user simulators, but it has not addressed the important question about the reliability of the testers themselves, nor has it studied how to generate a single reliability score for a user simulator based on multiple testers. In this paper, we address these two limitations and propose a novel Reliability-Aware Tester-based Evaluation (*RATE*) framework for evaluating the reliability of both User Simulators and testers. In this framework, the reliability of Testers and that of Simulators are jointly learned through unsupervised learning using iterative propagation of reliability. We propose and evaluate two algorithms for unsupervised learning of reliabilities. Evaluation results using TREC data sets show that the proposed RATE framework is effective in measuring the reliability of simulators and testers, thus serving as a foundation for potentially establishing a new paradigm for evaluating IIR systems using user simulation.

Keywords: Reliability of User Simulator · Tester · IIR Systems

1 Introduction

How to evaluate an information retrieval (IR) system has always been an important, yet challenging research topic in IR. Since IR is in general an interactive process, investigation of how to evaluate interactive IR (IIR) systems is especially important. The recent growth of interest in conversational search systems makes it even more urgent to study IIR evaluation. Without an appropriate evaluation methodology, it would be impossible to make solid research progress.

The Cranfield evaluation methodology established in 1960s [9] has so far been the standard methodology for evaluating IR systems [21]. However, because of the static nature of the test collections, such a methodology is inadequate for evaluating IIR systems, which requires involving users in the process of evaluation. Current solutions such as evaluation using user studies [12] or online A/B test [21] are not only expensive but also non-reproducible due to the learning effect of a user. However, reproducible experiments are needed to accurately measure research progress and enable long-term comparison of old methods with new methods.

© The Author(s), under exclusive license to Springer Nature Switzerland AG 2022
M. Hagen et al. (Eds.): ECIR 2022, LNCS 13185, pp. 336–350, 2022.
https://doi.org/10.1007/978-3-030-99736-6_23

A user simulator simulates one or many user actions in a search process such as query formulation, click simulation, reformulating query, ending the session and so on, thus a user simulator can interact with an IIR system and the result is a simulated session with which we can compute the performance of the IIR system throughout the session [27]. User simulators enable reproducible experiments with IIR systems, previous work has proposed to use user simulators to evaluate IIR systems [2,5,8,27]. Many user simulation methods are proposed in literature [1,3–5,13,16] and user simulation has already been applied to evaluate conversational search systems[20,26] and IIR systems using relevance feedback algorithms or query suggestion [11,24].

However, user simulators have variable reliability and utility for evaluating IIR systems [15] and if a user simulator is not reliable, we cannot trust the evaluation of IIR systems based on it. Therefore, before we can confidently use user simulators to evaluate IIR systems, we must first be able to evaluate the reliability of user simulators themselves. Evaluation of user simulation itself has multiple challenges such as variance in user behaviour, unavailability of ground truth, and dependency on the intended use of a user simulator. So far user simulators have been mostly evaluated by assessing how similar a simulator is to a real user using multiple strategies, including comparing simulated and real user session data [16,18], comparing retrieval performance of IR systems with the real user queries and simulated queries [1,3,4,13], comparison of search session data [5,7], log-likelihood and perplexity [7]. However, almost all of these evaluation methods require real user search interaction data which is expensive to acquire. Further, those evaluation methods do not directly assess whether the simulator is reliable for comparing IIR systems.

To address these limitations, a Tester-based evaluation (TBE) framework has been proposed recently by Labhishetty et al. [15] to evaluate the reliability of a simulator, specifically for the purpose of comparing IIR systems, by using multiple Testers. A Tester constitutes a set of IR systems with an expected performance pattern (e.g., one system is expected to be perform better than another), which can be either obtained analytically or empirically using real user data. Thus while TBE also benefits from having real user data available, but it does not necessarily require real user data. A detailed introduction to the TBE framework is given in Sect. 2.

However, the work [15] has not addressed the important question about the reliability of the Testers themselves. Indeed, all Testers are not equally reliable because the expected performance pattern of a Tester may not always be true when generalized to different experiment setups and the variable reliability of a few specific Testers has already been reported in [15]. The reliability issue of Testers must be addressed to make TBE practically useful since without distinguishing reliable Testers from unreliable Testers, there is a risk of using unreliable Testers, making it unable to generate a reliable single reliability score for a user simulator when aggregating its scores on different Testers.

In this work, we study how to quantify the reliability of Testers and extend the previous work [15] by proposing a new Reliability-Aware Tester based Evaluation (*RATE*) framework. In *RATE*, we define the reliability of a Tester as the probability that its expected performance pattern would be satisfied when tested with a population of users. We define and formulate both the reliability of the Tester and the reliability of the Simulator such that they can be jointly learned in an unsupervised manner. We propose two formulations namely *Symmetric formulation* and *Difference formulation* for learning both the reliability of Testers and Simulators as a fixed point solution. These algorithms enable the RATE framework to not only assess the reliabilities of Testers but also directly generate reliability scores for user simulators, facilitating the use of user simulation for evaluating IIR systems.

Evaluation of RATE shows that the framework can meaningfully differentiate Testers and quantify their reliabilities. Furthermore, the results show that the algorithms based on *Symmetric formulation*, the reliability scores and the ranking of Testers or simulators according to reliability are mostly stable against perturbation in the list of simulators and testers, thus indicating the robustness of the framework.

2 Tester-Based Evaluation (TBE)

As our work is an extension of the Tester-based Evaluation (TBE) framework proposed in [15], we first provide a brief overview of TBE in this section and then present our proposed new framework in the next section. The TBE framework was proposed to evaluate the reliability of a user simulator from the perspective of evaluating IIR systems i.e., it is evaluating predictive validity (as described in [25]) of a simulator in differentiating different IIR systems. The framework introduces a component called Tester to evaluate Simulators.

Tester T: A Tester is a set of IR systems along with an expected performance pattern of the IR systems which is generally an expected order of the performance of the IR systems (e.g., the order of the NDCG scores produced by the two IR systems). To evaluate a user simulator using a Tester, the user simulator first interacts with each of the IR systems in the Tester and the retrieval performance (like NDCG score) of each IR system is measured with the simulator. The pattern of the performance of all the systems in a Tester can then be compared with the expected performance pattern. The reliability score of the simulator is computed using a measure quantifying how close the two performance patterns are.

For example, if A and B are two IR systems, where an IR system is defined to include three components: Retrieval method, Document collection, and Topics (Information needs). If we have knowledge that system A performs better than system B, a Tester t can be built as follows: $(A, B, A > B)$ which denotes that A, B are two systems and $A > B$ is the expected performance pattern which means A is expected to perform better than B.

Reliability of Simulator: The reliability score of a simulator S as measured using a Tester T (denoted by $F(S,T)$) is computed as an evaluation measure quantifying how close the obtained performance pattern of S with T and the expected performance pattern of T are. In [15], the reliability score $F(S,T)$ is computed in two ways, Success rate(Sr) and Failure rate(Fr). It is considered as *success* or *failure* if the order of the IR systems given by the simulator is correct or incorrect according to the expected performance pattern of the Tester. Continuing our example, if a simulator s interacts with the IR systems in Tester t, and the resultant performance of the system A and system B with s are computed as p_A, p_B for all the topics. If $p_A > p_B$ then it is considered as success (because the expected pattern is also $A > B$); otherwise it is a failure. Finally, a Success rate (Sr) and a Failure rate (Fr) are computed by aggregating all successes and failures across multiple test cases (multiple information needs). In this work, we continue using Sr and Fr as two different measures for $F(S, T)$.

Limitations of TBE: A major limitation of TBE is that the Testers themselves can be unreliable because the expected performance pattern of a Tester may not be always correct [15]. Indeed, the expected performance pattern is hypothesized using domain knowledge or empirical results of the IR systems, thus the expected performance pattern may not be true when generalized to different information needs, different types of users, or different document collections.

The lack of a way to distinguish reliable from unreliable Testers means that TBE cannot yet be applied to accurately measure the reliability of user simulators. It also makes it hard to generate a reliable single reliability score for a simulator when tested with multiple Testers with different reliability.

In the next section, we address these limitations by proposing a novel Reliability-aware Tester-based Evaluation (RATE) framework.

3   Reliability-Aware Tester-Based Evaluation (RATE)

The main idea of our proposed new framework, named Reliability-Aware Tester-based Evaluation (RATE), is that we evaluate the reliability of the user simulators using Testers which are themselves associated with reliability. We define and formulate both the reliability of the Tester and the reliability of the Simulator such that they can be jointly learned in an unsupervised manner, making it possible to apply the framework to any Testers and any Simulators without requiring any human supervision.

3.1   Reliability of a Tester

How can we compute the reliability of a Tester? Logically, the reliability of a Tester is determined by the certainty of the expected performance pattern. If we have high confidence in the expected pattern of a Tester, we would also have high confidence in saying that a user simulator is not reliable if it fails such a Tester (i.e., fails to show the expected performance pattern), which suggests that the Tester is reliable; in contrast, if our confidence in the expected performance pattern of a Tester is not high, we would be less sure whether a user simulator is unreliable even if it fails such a Tester.

Now, how can we assess the certainty of the expected performance pattern of a Tester? One possibility is to empirically answer this question by running the Tester on real users to see whether the same pattern is always observed. However, an IR system may have different performance when interacting with different types of users or search behaviours. Thus the certainty of the performance pattern, also the reliability of a Tester, inevitably depends on the specific users used to estimate it. Thus the definition of reliability of a Tester is with respect to a group of users. We define the reliability of a Tester as follows.

Reliability of a Tester $R(T)$: *The Reliability of a Tester T is the reliability of the expected performance pattern of that Tester, i.e., the probability that the expected performance pattern is satisfied when tested with a population of users U. Average reliability $R(T)$ is obtained by aggregating reliability scores from different populations of users.*

Formally, if $F(U,T)$ is the probability that the expected performance pattern of T is consistent with the observed performance pattern when the Tester T is applied to a real user U in a population group G. The reliability of T (denoted by $R(T)$) can be defined as the average $F(U,T)$ over all the users in the group.

$$R(T) = \frac{1}{|G|} \sum_{U \in G} F(U,T) \tag{1}$$

From Eq. 1, one of the solutions to compute $F(U,T)$ by testing the Tester using real user experiments. But obtaining $F(U,T)$ using real user experiments for each Tester T is very expensive and generally infeasible. Thus a solution we propose is to approximate a group of real users with a group of user simulators G'. As simulators have variable reliability, we also need to consider each simulator's reliability, i.e., how well a simulator S approximates a real user U, which we denote by $p(U|S)$. Naturally, when S is a real user, this probability would be 1. With this approximation, we have

$$R(T) = \frac{1}{|G|} \sum_{U \in G} \frac{1}{|G'|} \sum_{S \in G'} F(S,T)p(U|S) \tag{2}$$

$$= \frac{1}{|G'|} \sum_{S \in G'} F(S,T) \frac{1}{|G|} \sum_{U \in G} p(U|S) \tag{3}$$

Without additional knowledge of specific users that we are interested in simulating, we can reasonably assume that $p(U|S)$ can simply be approximated by $p(\tilde{U}|S)$, where $\tilde{U}$ denotes an "average" user. With this assumption, $\sum_{U \in G} p(U|S) = |G|p(\tilde{U}|S)$, thus we have

$$R(T) = \frac{1}{G'} \sum_{S \in G'} F(S,T)p(\tilde{U}|S) \tag{4}$$

As the Reliability of the simulator is also evaluating the quality of the simulator in terms of whether it can compare IR systems correctly as real users, we propose that $P(\tilde{U}|S)$ is proportional to the reliability of the simulator S, i.e., $R(S)$. Therefore Reliability of the Tester is dependent on the reliability of the simulators in the following way,

$$P(\tilde{U}|S) \propto R(S), R(T) \propto \sum_{S \in G'} F(S,T)R(S) \tag{5}$$

3.2 Reliability of User Simulator

We now discuss how we can define the reliability of a user simulator $R(S)$. One solution is to compare the behavior of the user simulator with that of a real user and use the similarity to estimate $R(S)$, but we are interested in studying whether it is still possible to define it in some meaningful way even without access to real user data.

One way to achieve the goal is to assume that a simulator is reliable if it gives the expected performance pattern correctly with a reliable Tester, i.e., it has a high success rate, $F(S,T)$, when tested with Tester T. The simulator can be assumed to be even more reliable if it can pass multiple reliable Testers with high success rates. Logically, this is a quite reasonable assumption as it just means that a reliable simulator passes many reliable Testers with high success rates. This heuristic can be potentially implemented in multiple ways. Below we describe two of them.

Symmetric Formulation: In this formulation, we compute $R(S)$ as a weighted mean of success rate $F(S,T)$ over a set of Testers, where the weights are the reliability of the Testers.

$$R(S) = \frac{\sum_T F(S,T) * R(T)}{\sum_T R(T)} \tag{6}$$

We refer to this formulation as symmetric formulation because note that the reliability of Tester and reliability of simulator are both defined in terms of each other as per Eq. 5 and Eq. 6 respectively. Both formulations are similar and symmetric. The symmetric formulation gives a simulator a high reliability score if it passes many highly reliable Testers (high $R(T)$) with high success rates (high $F(S,T)$). Note that $F(S,T)$ and $R(T)$ both range from 0 and 1, therefore $R(S)$ is also between 0 and 1.

Difference Formulation. An alternative method to compute $R(S)$ is by using the definition of reliability of Tester. The reliability of a Tester is the extent to which the Tester expected pattern is satisfied by the real user. We can thus assume that a simulator S is more reliable if it best approximates an average user $(\tilde{U})$ in its behavior on Testers. That is, the score of $F(S,T)$ should be close to $F(\tilde{U},T)$ for all Testers. With this notion of reliability, we define $R(S)$ as being proportional to the difference between $F(S,T)$ and $F(\tilde{U},T)$ where $F(\tilde{U},T)$ is $R(T)$ from Eq. 1 when all users in G are approximated to $\tilde{U}$. We refer to this as Difference formulation. Formally,

$$R(S) \propto 1 / \sum_T (F(S,T) - R(T)) \tag{7}$$

$$R(S) = \frac{2}{\sum_T (F(S,T) - R(T)) + 1} - 1 \tag{8}$$

$R(S)$ in Eq. 8 is scaled such that the value is between *zero* and *one*. Unlike Symmetric formulation, Difference formulation implies that a simulator should have a higher score with highly reliable Testers along with a lower score with less reliable Testers, because the difference $F(S,T) - R(T)$ in such cases will be smaller and $R(S)$ will be higher.

Circular Dependence of Reliability of Tester and Simulator: In the RATE framework, there is a circular dependency between the reliability of a simulator and the reliability of a Tester which is somewhat similar to the HITS algorithm [14]. In Symmetric formulation of $R(S)$ (Eq. 6), the principle is that a simulator has higher reliability when it satisfies reliable Testers and a Tester has high reliability when it is satisfied by reliable simulators obtained from Eq. 5. The principle in the Difference formulation of $R(S)$ (Eq. 8) is that a simulator has higher reliability if it satisfies reliable Testers and does not satisfy unreliable Testers and a Tester has high reliability if it is satisfied by reliable simulators.

The circular dependency between $R(S)$ and $R(T)$ captures their relations intuitively and directly suggests constraints that can be interpreted as defining a fixed point solution for $R(S)$ and $R(T)$ jointly. Thus it enables us to learn both reliabilities from all the test values $F(S,T)$ in an iterative way until they converge. Algorithm 1 and Algorithm 2 are the two iterative algorithms for computing $R(S)$ and $R(T)$ based on Symmetric formulation and Difference formulation respectively. In both algorithms, we start by considering that all Testers are reliable or $R(T) = 1$ which is the assumption of baseline TBE evaluation approach.

Algorithm 1 Iterative algorithm for Symmetric Formulation:

Step1: for each S, for each T, Obtain $F(S,T)$
Step2: Initialize $R(T)^1 = 1$, $n = 1$
Step3: $R(S)^n = \frac{\sum_T F(S,T) * R(T)^n}{\sum_T R(T)^n}$
Step4: $R(T)^{n+1} = \frac{\sum_S F(S,T) * R(S)^n}{\sum_S R(S)^n}$
Step5: $n = n + 1$
Step6: Repeat from Step2 until convergence condition is met.

Algorithm 2 Iterative algorithm for Difference formulation

Step1: for each S, for each T , Obtain $F(S,T)$
Step2: Initialize $R(T)^1 = 1$, $n = 1$
Step3: $R(S)^n = \frac{2}{(\sum_T |F(S,T) - R(T)^n| * \frac{1}{|T|}) + 1} - 1$
Step4: $R(T)^{n+1} = \frac{\sum_S F(S,T) * R(S)^n}{\sum_S R(S)^n}$
Step5: $n = n + 1$
Step6: Repeat from Step2 until convergence condition is met.

---

4   Evaluation

In our experiments, we want to do a preliminary study of the empirical behaviour of the RATE framework and whether RATE can reasonably identify reliable simulators and Testers. Specifically, the following questions are answered through the experiments: RQ1) Can RATE results distinguish unreliable Testers from reliable ones? RQ2) Which of the two formulations, Symmetric or Difference formulation is more effective and robust? RQ3) How sensitive are reliability scores learnt from the iterative algorithm? How sensitive are the rankings of reliable simulators when the list of simulators to evaluate is changed?

4.1 Experiment Design

We used the same experiment setup as in [15]. We used the same Testers and Simulators utilized in [15], which we briefly describe below.

1. Testers:
Query History (QH) Tester: The Tester compares two IR systems where one of them is baseline BM25 and the other is an IIR system that uses previous queries in a session to do query expansion using the method proposed in [22] and improves on the baseline. The expected pattern is that the IIR system should perform better than the baseline. The Tester is denoted by (M, M+QH[α]) where α is the parameter of the query expansion controlling the weight on the current query compared to previous queries. *Click History (CH) Tester*: Similar to QH Tester, this Tester compares the baseline with an IIR system that uses previously clicked documents to do query expansion and improve ranking over the baseline using the method proposed in [19,23]. The Tester is denoted by (M, M+CH[β]) where β controls the weight of the current query compared to the clicked history terms. The expected pattern is similar to QH Tester, $M + CH[\beta] > M$. *BM25 Ablation Testers*: This Tester contains standard BM25 (M) compared with BM25 without TF weighting ($M \backslash TF$) or IDF weighting ($M \backslash IDF$). As the TF and IDF weighting are important components in BM25, the expected pattern for these Testers are $M > M \backslash TF$, $M > M \backslash IDF$ respectively. We used Whoosh [6] library to implement all the Testers.

2. Simulators: We experiment with all the four simulators studied in [15] using SimIIR toolkit [17] to implement these simulators: Smart Ideal TREC user (SIT), Smart Stochastic TREC user (SST), Single Term Stochastic TREC user (STST), and Random User (RU). The detailed configurations of the simulators are not shown to save space, they can be found in [15]. To summarize, Random user(RU) simulates actions randomly and is the worst simulator and SIT and SST are more effective simulators with better query formulations and click simulation techniques.

3. Dataset: The Information needs and document collection are selected from the TREC AQUAINT dataset. A subset of 50000 documents are used as the document collection and the TREC topics are used as Information needs. The simulators use the TREC topic description and topic judgements as input to simulate the user actions to generate a search session.

4. Evaluation Measure: We measure retrieval performance of an IR system with a simulator by computing session-based DCG measure *(sDCG/q)* [10] of the resultant simulated search session. The order of (sDCG/q) scores of two IR systems in the Tester will give whether the simulator has success or not which leads to computing success rate $(F(S, T))$ across multiple topics(Information needs).

5. Experiment Process: We use six Testers which are also used in [15] to conduct our experiments. We first reproduced the results in the paper [15] by using same parameter settings for the Testers. We use two QH Testers $\alpha = 0.5$, $\alpha = 0.01$ denoted by (M, M+QH[0.5]), (M, M+QH[0.01]), and two CH testers with $\beta = 0.5$, $\beta = 0.8$ denoted by (M, M+CH[0.8]), (M, M+CH[0.5]). And the final two testers are (M, M\ TF), (M, M\ IDF). We evaluate the four simulators, by computing success rate $F(S, T)$ using the six Testers, Table 1 shows the results of the success rate scores. From the results in Table 1, it is clear that different Testers give different scores for the simulators. The ablation testers (M, M\TF), (M, M\IDF) do not perform as expected and are argued as potentially unreliable in [15].

By applying the RATE framework, we can now compute the reliability scores of these Testers. Using Table 1 results, we apply the iterative algorithm to compute $R(S)$ and $R(T)$. The convergence condition for the algorithm is when the change in $R(S)$ and $R(T)$ is less than 0.00001. We observe that the convergence condition

Table 1. Success rate of the 4 Simulators with six Testers

F(S,T)	SIT	SST	STST	RU
(M,M+QH[0.5])	0.766	0.745	0.809	0.021
(M,M+QH[0.01])	0.723	0.723	0.872	0.021
(M,M+CH[0.8])	0.787	0.723	0.638	0.021
(M,M+CH[0.5])	0.809	0.638	0.575	0
(M,M\TF)	0.34	0.383	0.362	0
(M,M\IDF)	0.532	0.426	0	0

is always met before 10 iterations and therefore choose $n = 10$ for both Algorithm 1 and Algorithm 2.

4.2   Experiment Results

Tables 2 and 3 shows the resultant R(T) and R(S) scores with Symmetric and Difference formulation respectively. We initialized all $R(T)$ to 1 in *Iteration 0*. We show the reliability scores of Simulators after 10 iterations and similarly for Testers. We can make several interesting observations from the tables.

First, we see that both Symmetric and Difference formulations were able to correctly identify (M, M\TF) and (M, M\IDF) as relatively unreliable Testers, consistent with the conclusions drawn in [15]. Although these two ablation testers are reasonable, depending on the set of information needs and document collection it is possible that the Tester's expected pattern might be incorrect. In the previous work [15], these two BM25 Ablations testers are argued as potentially unreliable; the RATE framework enables us to quantify their reliability with reasonable scores. We also see that the relative order of the Testers according to $R(T)$ scores is the same with both Symmetric and Difference formulations although the absolute values are different, suggesting that the iterative algorithm is generally robust. Moreover, although we started with R(T) as 1.0 for all the testers, the framework learns different reliability scores for each tester and distinguishes unreliable Testers. We can observe that the Testers with different parameters also have different reliabilities i.e., (M, M+QH[0.5]), (M, M+QH[0.01]) have slightly different R(T) scores and (M, M+CH[0.8]), (M, M+CH[0.5]) have a larger difference in their reliability scores. Thus we can answer *RQ1* by concluding that the RATE framework can distinguish reliable Testers from unreliable ones and assign meaningful reliability scores to them.

Table 2. R(S) and R(T) using RATE using Symmetric Formulation of R(S)

Reliability of Simulators		Reliability of Testers	
Simulators	R(S)	Testers	R(T)
SIT	0.70551	(M,M+QH[0.5])	0.76777
SST	0.65107	(M,M+QH[0.01])	0.76519
STST	0.62699	(M,M+CH[0.8])	0.71432
RU	0.01308	(M,M+CH[0.5])	0.67446
		(M,M\TF)	0.3587
		(M,M\IDF)	0.32689

Table 3. R(S) and R(T) using RATE using Difference Formulation of R(S)

Reliability of Simulators		Reliability of Testers	
Simulators	R(S)	Testers	R(T)
SIT	0.77963	(M,M+QH[0.5])	0.68797
SST	0.868	(M,M+QH[0.01])	0.68726
STST	0.80399	(M,M+CH[0.8])	0.63722
RU	0.31133	(M,M+CH[0.5])	0.59603
		(M,M\TF)	0.3216
		(M,M\IDF)	0.28395

Second, from the results in Table 2 and 3, we observe that the reliability scores of Simulators learnt using Symmetric and Difference formulations are very different. Both the formulations correctly identify the random user by giving it the lowest reliability, but the Difference formulation gives a relatively higher score to *RU* which seems undesirable. This is likely because, in Difference formulation, the Random user (RU) is rewarded for lower scores on the two less reliable Testers, (M, M\IDF) and (M, M\TF). Indeed, the difference formulation computes the difference between the simulator score $F(S, T)$ and reliability of Tester $R(T)$ as given in Eq. 8. Since $R(T)$ is a weighted average of success rates of all simulators, the difference algorithm essentially scores a simulator approximately based on how close it is to the average success rates of all simulators. This makes the algorithm more sensitive and less robust to the list of simulators that are considered. As there is a random user (*RU*) in our list of simulators, the *RU* user lowers the average success rates and therefore has a higher score for reliability.

In contrast, Symmetric formulation scores the simulator approximately based on the average success rate of that simulator with the testers, so if a simulator has a higher success rate with all testers, it likely has higher reliability. This principle enables it to identify the random user correctly. It is also more robust with respect to the list of simulators that are evaluated as we will show later.

Overall, compared with the Difference formulation, the Symmetric formulation scores appear to be more meaningful, not only identifying the unreliable random user correctly but also identifying *SIT* as the best simulator, which is quite reasonable as *SIT* uses smart queries and clicks.

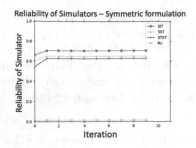

Reliability of Simulators – Symmetric formulation

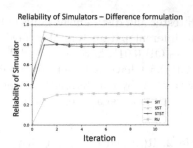

Reliability of Simulators – Difference formulation

Fig. 1. R(S) - Symmetric formulation

Fig. 2. R(S) - Difference formulation

Figures 1 and 2 show the convergence graphs of both algorithms in computing the reliability scores of the four simulators. The values converge fast because there are only a few data points (4 simulators and 6 testers). The convergence graphs further show the difference between both Symmetric and Difference formulations, like the order of the simulators is different in both and RU gains reliability with Difference formulation in Fig. 2.

To further analyze the robustness of Symmetric and Difference formulations and RATE framework overall, we modify the experiment by adding a large number of random testers (RUs). An important question we study is that: *Would simulator ranking be affected much by the set of simulators considered?*. We repeat the experiment with 6 Testers and 50 RU simulators along with SIT, SST, STST simulators. The results are shown in Tables 4 and 5. The Symmetric formulation gives almost the same R(S) scores as in Table 2, but the Difference formulation scores are completely changed where the Random user is now given the highest reliability scores and the other simulators have low reliability. Clearly, the Difference formulation is not robust against changes in the set of simulators especially when more bad user simulators are added. We thus conclude that Symmetric formulation is more effective and robust, answering *RQ2* and partially answering *RQ3*.

Table 4. Reliability scores with multiple Random Testers - Symmetric formulation

Table 5. Reliability scores with multiple Random Testers - Difference formulation

Reliability of Simulators		Reliability of Testers	
Simulators	R(S)	Testers	R(T)
SIT	0.70582	(M,M+QH[0.5])	0.58312
SST	0.65153	(M,M+QH[0.01])	0.58119
STST	0.62784	(M,M+CH[0.8])	0.54287
RU	0.01313	(M,M+CH[0.5])	0.50766
RU	0.01313	(M,M\TF)	0.27001
RU	0.01313	(M,M\IDF)	0.24599
...	0.01313		

Reliability of Simulators		Reliability of Testers	
Simulators	R(S)	Testers	R(T)
SIT	0.2196	(M,M+QH[0.5])	0.03275
SST	0.26046	(M,M+QH[0.01])	0.03284
STST	0.31187	(M,M+CH[0.8])	0.03168
RU	0.98194	(M,M+CH[0.5])	0.01028
RU	0.98194	(M,M\TF)	0.00565
RU	0.98194	(M,M\IDF)	0.00448
...	0.98194		

Finally, to address *RQ3*, we study the behaviour of the RATE framework by altering the list of Simulators or Testers, including removing one simulator (e.g., 3 Simulators (SIT, STST, RU), 6 Testers) and removing two Testers (e.g., (4 Simulators, 4 Testers (QH testers, Ablation testers)) or (4 Simulators, 4 Testers(QH testers, CH testers)). We only use Symmetric formulation for computing R(S) in all these experiments as it is the preferred formulation. The results are shown in Table 6, Table 7 and Table 8. From the results, we observe that the reliability of a tester is not affected much by changing the set of testers, and similarly, the reliability of a simulator is not affected much by changing the set of simulators. For example, R(T) scores of the QH Testers are same in Table 2, Table 6 and Table 7; the R(T) scores of CH Testers are also almost the same in Table 2 and Table 7; and similar results are observed for Ablation testers. Further, the reliability of the simulators SIT, STST and RU is the same as in Table 8 and Table 2. We also experimented with more such configurations and found similar results. Thus we conclude that RATE is stable and robust in the sense that adding or removing unreliable or random simulators does not affect the evaluation of simulators, and adding or removing testers does not affect the evaluation of the testers. That is, the simulator ranking is not affected much by the set of simulators considered in the RATE framework. This implies that once we have a benchmark set of Testers with their learnt reliabilities, we can evaluate any number of simulators using the RATE framework without much worry about the sensitivity of the simulator reliability scores to the number of simulators participating in the evaluation.

However, we observe that changing the list of simulators affects the reliability of Testers, and changing the list of testers affects the reliability of simulators. This is expected as the computation of $R(T)$ itself is proportional to the average success rate of all the simulators when interacting with T, and vice versa. Thus, the average gets affected when the list of simulators or testers change respectively. Interestingly, although the absolute scores are affected, the ranking of Testers is unaffected with the set of simulators which is a positive result indicating the robustness of RATE. Overall, the robustness results indicate that RATE is not highly sensitive to small changes in the list of simulators or Testers.

Table 6. Robustness test - Reliability scores after removing CH Testers

Reliability of Simulators		Reliability of Testers	
Simulators	R(S)	Testers	R(T)
SIT	0.64228	(M,M+QH[0.5])	0.76779
SST	0.63287	(M,M+QH[0.01])	0.7671
STST	0.64228	(M,M\TF)	0.35877
RU	0.0145	(M,M\IDF)	0.31708

Table 7. Robustness test - Reliability scores after removing BM25 Ablation Testers

Reliability of Simulators		Reliability of Testers	
Simulators	R(S)	Testers	R(T)
SIT	0.76973	(M,M+QH[0.5])	0.768
SST	0.70922	(M,M+QH[0.01])	0.76678
STST	0.72997	(M,M+CH_0.8)	0.71215
RU	0.01617	(M,M+CH_0.5)	0.67185

5   Conclusion

In this work, we proposed a new evaluation framework called Reliability-Aware Tester based Evaluation framework(RATE) to evaluate User Simulators for their reliability in comparing IIR systems. Previous work proposed Testers to evaluate simulators but did not consider the reliability of the Testers themselves. In the RATE framework, we defined and formulated

Table 8. Robustness test - Reliability by removing SST simulator

Reliability of Simulators		Reliability of Testers	
Simulators	R(S)	Testers	R(T)
SIT	0.70977	(M,M+QH[0.5])	0.77888
STST	0.63823	(M,M+QH[0.01])	0.786
RU	0.0133	(M,M+CH[0.8])	0.70966
		(M,M+CH[0.5])	0.69139
		(M,M\TF)	0.34699
		(M,M\IDF)	0.27738

reliability of Tester and consequently reliability of simulator where both have circular-dependence between them. We propose two methods to compute $R(S)$, Symmetric and Difference formulation and propose an iterative algorithm to learn both reliability scores. Our experiments show that RATE is effective in distinguishing unreliable Testers from reliable Testers and also can distinguish reliable and unreliable simulators effectively. Between the two formulations, Symmetric formulation is found to be more effective and stable to compute $R(S)$. We further observe that RATE is robust in that the reliability score of a simulator is robust against the list of simulators considered and similarly for testers.

RATE provides a foundation for potentially establishing a new paradigm for evaluating IIR systems using user simulation. As an immediate future work, we envision to leverage RATE to establish a novel open evaluation platform where the research community can regularly add more Testers and simulators (which will naturally happen as researchers develop novel Testers and simulators), and the reliabilities of both the testers and the simulators can then be computed using the Symmetric formulation algorithm. Such a RATE platform will, for the first time, enable the use of potentially many user simulators to evaluate IIR systems with reproducible experiments.

Although we have only used unsupervised learning to learn reliability scores, the framework also allows semi-supervised learning, exploration of which is an interesting future direction. For example, if some of the Testers have the success rate scores when tested with real users $(F(U, T))$, these scores can be used as initial $R(T)$ scores and they can be used to learn the reliability of other Testers and simulators through propagating reliability using the iterative algorithms proposed in the framework. Similarly, if real user data is available to assess the reliability of some simulators, such reliability scores can also be easily incorporated into the framework.

References

1. Azzopardi, L., De Rijke, M., Balog, K.: Building simulated queries for known-item topics: an analysis using six European languages. In: Proceedings of the 30th Annual International ACM SIGIR Conference on Research and Development in Information Retrieval, pp. 455–462. ACM (2007)

2. Azzopardi, L., Järvelin, K., Kamps, J., Smucker, M.D.: Report on the SIGIR 2010 workshop on the simulation of interaction. In: ACM SIGIR Forum, vol. 44, pp. 35–47. ACM New York (2011)
3. Baskaya, F., Keskustalo, H., Järvelin, K.: Simulating simple and fallible relevance feedback. In: Clough, P., et al. (eds.) ECIR 2011. LNCS, vol. 6611, pp. 593–604. Springer, Heidelberg (2011). https://doi.org/10.1007/978-3-642-20161-5_59
4. Baskaya, F., Keskustalo, H., Järvelin, K.: Time drives interaction: Simulating sessions in diverse searching environments. In: Proceedings of the 35th International ACM SIGIR Conference on Research and Development in Information Retrieval, pp. 105–114 (2012)
5. Carterette, B., Bah, A., Zengin, M.: Dynamic test collections for retrieval evaluation. In: Proceedings of the 2015 International Conference on the Theory of Information Retrieval, pp. 91–100. ACM (2015)
6. Chaput, M.: Whoosh. https://whoosh.readthedocs.io/en/latest/#
7. Chuklin, A., Markov, I., Rijke, M.: Click models for web search. Synth. Lect. Inf. Concepts Retr. Serv. **7**(3), 1–115 (2015)
8. Clarke, C.L., Freund, L., Smucker, M.D., Yilmaz, E.: SIGIR 2013 workshop on modeling user behavior for information retrieval evaluation. In: Proceedings of the 36th international ACM SIGIR Conference on Research and Development in Information Retrieval, pp. 1134–1134 (2013)
9. Cleverdon, C.: The Cranfield tests on index language devices. In: Aslib Proceedings. MCB UP Ltd. (1967)
10. Jiang, J., Allan, J.: Correlation between system and user metrics in a session. In: Proceedings of the 2016 ACM on Conference on Human Information Interaction and Retrieval, pp. 285–288 (2016)
11. Jordan, C., Watters, C., Gao, Q.: Using controlled query generation to evaluate blind relevance feedback algorithms. In: Proceedings of the 6th ACM/IEEE-CS Joint Conference on Digital Libraries, pp. 286–295. ACM (2006)
12. Kelly, D.: Methods for Evaluating Interactive Information Retrieval Systems with Users. Now Publishers Inc. (2009)
13. Keskustalo, H., Järvelin, K., Pirkola, A., Sharma, T., Lykke, M.: Test collection-based IR evaluation needs extension toward sessions – a case of extremely short queries. In: Lee, G.G., et al. (eds.) AIRS 2009. LNCS, vol. 5839, pp. 63–74. Springer, Heidelberg (2009). https://doi.org/10.1007/978-3-642-04769-5_6
14. Kleinberg, J.M., Kumar, R., Raghavan, P., Rajagopalan, S., Tomkins, A.S.: The web as a graph: measurements, models, and methods. In: Asano, T., Imai, H., Lee, D.T., Nakano, S., Tokuyama, T. (eds.) COCOON 1999. LNCS, vol. 1627, pp. 1–17. Springer, Heidelberg (1999). https://doi.org/10.1007/3-540-48686-0_1
15. Labhishetty, S., Zhai, C.: An exploration of tester-based evaluation of user simulators for comparing interactive retrieval systems. In: Proceedings of the 44th International ACM SIGIR Conference on Research and Development in Information Retrieval, SIGIR 2021, pp. 1598–1602. Association for Computing Machinery, New York (2021). https://doi.org/10.1145/3404835.3463091
16. Maxwell, D., Azzopardi, L.: Agents, simulated users and humans: an analysis of performance and behaviour. In: Proceedings of the 25th ACM International on Conference on Information and Knowledge Management, pp. 731–740. ACM (2016)
17. Maxwell, D., Azzopardi, L.: Simulating interactive information retrieval: SimIIR: a framework for the simulation of interaction. In: Proceedings of the 39th International ACM SIGIR Conference on Research and Development in Information Retrieval. pp. 1141–1144 (2016)

18. Maxwell, D., Azzopardi, L., Järvelin, K., Keskustalo, H.: Searching and stopping: an analysis of stopping rules and strategies. In: Proceedings of the 24th ACM International on Conference on Information and Knowledge Management, pp. 313–322 (2015)

19. Rocchio, J.: Relevance feedback in information retrieval. The Smart retrieval system-experiments in automatic document processing, pp. 313–323 (1971)

20. Salle, A., Malmasi, S., Rokhlenko, O., Agichtein, E.: Studying the effectiveness of conversational search refinement through user simulation. In: Hiemstra, D., Moens, M.-F., Mothe, J., Perego, R., Potthast, M., Sebastiani, F. (eds.) ECIR 2021. LNCS, vol. 12656, pp. 587–602. Springer, Cham (2021). https://doi.org/10.1007/978-3-030-72113-8_39

21. Sanderson, M.: Test Collection Based Evaluation of Information Retrieval Systems. Now Publishers Inc. (2010)

22. Sriram, S., Shen, X., Zhai, C.: A session-based search engine. In: Proceedings of the 27th Annual International ACM SIGIR Conference on Research and Development in Information Retrieval, pp. 492–493 (2004)

23. Trotman, A., Puurula, A., Burgess, B.: Improvements to BM25 and language models examined. In: Proceedings of the 2014 Australasian Document Computing Symposium, pp. 58–65 (2014)

24. Verberne, S., Sappelli, M., Järvelin, K., Kraaij, W.: User simulations for interactive search: evaluating personalized query suggestion. In: Hanbury, A., Kazai, G., Rauber, A., Fuhr, N. (eds.) ECIR 2015. LNCS, vol. 9022, pp. 678–690. Springer, Cham (2015). https://doi.org/10.1007/978-3-319-16354-3_75

25. Zeigler, B.P., Kim, T.G., Praehofer, H.: Theory of Modeling and Simulation, 2nd edn. Academic Press Inc., USA (2000)

26. Zhang, S., Balog, K.: Evaluating conversational recommender systems via user simulation. In: Proceedings of the 26th ACM SIGKDD International Conference on Knowledge Discovery & Data Mining, pp. 1512–1520 (2020)

27. Zhang, Y., Liu, X., Zhai, C.: Information retrieval evaluation as search simulation: a general formal framework for IR evaluation. In: ACM ICTIR, pp. 193–200. ACM (2017)

HC4: A New Suite of Test Collections for Ad Hoc CLIR

Dawn Lawrie[1]([envelope]) [iD], James Mayfield[1] [iD], Douglas W. Oard[1,2] [iD], and Eugene Yang[1] [iD]

[1] HLTCOE, Johns Hopkins University, Baltimore, MD 21211, USA
{lawrie,mayfield,eugene.yang}@jhu.edu
[2] University of Maryland, College Park, College Park, MD 20742, USA
oard@umd.edu

Abstract. HC4 is a new suite of test collections for ad hoc Cross-Language Information Retrieval (CLIR), with Common Crawl News documents in Chinese, Persian, and Russian, topics in English and in the document languages, and graded relevance judgments. New test collections are needed because existing CLIR test collections built using pooling of traditional CLIR runs have systematic gaps in their relevance judgments when used to evaluate neural CLIR methods. The HC4 collections contain 60 topics and about half a million documents for each of Chinese and Persian, and 54 topics and five million documents for Russian. Active learning was used to determine which documents to annotate after being seeded using interactive search and judgment. Documents were judged on a three-grade relevance scale. This paper describes the design and construction of the new test collections and provides baseline results for demonstrating their utility for evaluating systems.

Keywords: Test Collection · Cross-Language Information Retrieval · CLIR · Evaluation

1 Introduction

Ad hoc Cross-Language Information Retrieval (CLIR) has been studied for decades. Yet until the advent of high-quality machine translation, the usefulness of CLIR has been limited. Easy access to inexpensive or free machine translation has altered this landscape. If one can find a document of interest in a language one cannot read, machine translation is now often sufficient to make the majority of the document's content accessible. Thus, the breadth of the audience for CLIR has increased dramatically in a short period of time.

As machine translation has increased the usefulness of CLIR, recently introduced deep neural methods have improved ranking quality [4,29,43,45,47]. By and large, these techniques appear to provide a large jump in the quality of CLIR output. Yet the evidence for these improvements is based on small, dated test collections [14,15,27,36,37]. Problems with existing collections include:

M. Hagen et al. (Eds.): ECIR 2022, LNCS 13185, pp. 351–366, 2022.
https://doi.org/10.1007/978-3-030-99736-6_24

- Some CLIR test collections are no longer available from any standard source.
- They are typically small, often 100,000 or fewer documents, and some have few known relevant documents per topic.
- Judgment pools were retrieved using older systems. New neural systems are thus more likely to systematically identify relevant unjudged documents [38, 40, 46].
- Many of the early test collections have only binary judgments.

The increased importance of CLIR thus argues for the creation of new ad hoc CLIR collections that ameliorate these problems. A new CLIR collection should contain a large number of recent documents in a standard encoding, with distribution rights that foster broad use, sufficient numbers of relevant documents per topic to allow systems to be distinguished, and graded relevance judgments.

To this end, we have created HC4[1] – the HLTCOE Common Crawl CLIR Collection. In addition to addressing the shortcomings described above and facilitating evaluations of new CLIR systems, this suite of collections has a few unique aspects. First, to mimic well contextualized search sessions, topics are generally inspired by events in the news and written from the perspective of a knowledgeable searcher familiar with the background information on the event. Each topic is associated with a date, and in most cases the topic is linked to Wikipedia page text written immediately prior to that date, generally contemporaneous with the event. This page serves as a proxy for a report that might have written by a searcher prior to their search, reflecting their knowledge at that time. It is included in the collection to enable exploration of contextual search. Second, to maximize recall in the judged set, instead of pooling, active learning identified the documents to be judged [1]. This approach reduces judgment bias toward any specific automated retrieval system.

2   Related Work

The first CLIR test collection was created for Salton's seminal work on CLIR in 1970, in which English queries were manually translated into German [35]. Relevance judgments were exhaustively created for those queries for several hundred abstracts in both languages. In 1995, the first instance of a large-scale CLIR test collection in which documents were selected for assessment using pooling translated Spanish queries from the Fourth Text Retrieval Conference's (TREC-4) Spanish test collection into English for CLIR experimentation [12]. The next year, TREC organizers provided standard English versions of queries for Spanish and Chinese collections [37]. The following year, CLIR became the explicit focus of a TREC track, with collections in German, French, and Italian; that track continued for three years [36]. One enduring contribution from this early work was recognition that to be representative of actual use, translations of topic fields in a test collection should not be made word-by-word, but rather should be re-expressions fluently written in the query language.

---

[1] HC4 can be downloaded from https://github.com/hltcoe/HC4.

With the start of the NACSIS Test Collection Information Retrieval (NTCIR) evaluations in Japan in 1999 [34], the Cross-Language Evaluation Forum (CLEF) in Europe in 2000 [15], and the Forum for Information Retrieval Evaluation (FIRE) in India in 2008 [27], the center of gravity of CLIR evaluation moved away from TREC. Over time, the research in each of these venues has become more specialized, so although CLIR tasks continue, the last large-scale CLIR test collection for ad hoc search of news that was produced in any of the world's four major information retrieval shared-task evaluation venues was created in 2009 for Persian [14]. The decline in test collection production largely reflected a relative stasis in CLIR research, which peaked around the turn of the century and subsequently tailed off. Perhaps the best explanation for the decline is that the field had, by the end of the first decade of the twenty-first century, largely exhausted the potential of the statistical alignment techniques for parallel text that had commanded the attention of researchers in that period.

One consequence of this hiatus is that older test collections do not always age gracefully. As Lin et al. point out, "Since many innovations work differently than techniques that came before, old evaluation instruments may not be capable of accurately quantifying effectiveness improvements associated with later techniques" [25]. The key issue here is that in large test collections, relevance judgments are necessarily sparse. TREC introduced pooling as a way to decide which (typically several hundred) documents should be judged for relevance to each topic, with the remaining documents remaining unjudged. Pools were constructed by merging highly ranked documents from a diverse range of fully automated systems, including some of the best systems of the time, sometimes augmented by documents found using interactive search. Zobel found, using evaluation measures that treat unjudged documents as not relevant, that relevance judgments on such pools result in system comparisons not markedly biased against other systems constructed using similar technology that had not contributed to the pools [48]. Contemporaneously, Voorhees found that comparisons between systems were generally insensitive to substituting judgments from one assessor for those of another [39]. A subsequent line of work found that some newly designed evaluation measures produced system comparisons robust to random ablation of those pools [5,28,33,44]. However, these conclusions do not necessarily hold when new technology finds relevant documents that were not found by earlier methods, as can be the case for neural retrieval methods [25]. In such cases, three approaches might be tried:

1. Re-pool and rejudge an older collection, or create a new collection over newer content using pooling.
2. Select documents to be judged in a manner relatively insensitive to the search technology of the day, without necessarily judging all relevant documents.
3. Use an approach that simply does a better job of finding most of the relevant documents, thus reducing the risk of bias towards any class of system.

We used the third of these approaches to select documents for judgment in HC4. Specifically, we used the HiCAL system [10] to identify documents for judgment using active learning. HiCAL was originally developed to support Technology Assisted Review (TAR) in E-Discovery, where the goal is to identify the largest practical set of relevant documents at a reasonable cost [3,9,31,42]. Similar approaches have been used to evaluate recall-oriented search in the TREC Total Recall and Precision Medicine tracks [17,22,32]. The key idea in HiCAL is to train an initial classifier using a small set of relevance judgments, and then to use active learning with relevance sampling to identify additional documents for review. As Lewis found, relevance sampling can be more effective than the uncertainty sampling approach that is more commonly used with active learning when the prevalence of relevant documents in the collection being searched is low [24]. This low prevalence of relevant documents is often a design goal for information retrieval test collections, both because many real information retrieval tasks exhibit low relevance prevalence, and because (absent an oracle that could fairly sample undiscovered relevant documents) accurately estimating recall requires reasonably complete annotation of the relevant set. One concern that might arise with HiCAL is that if the document space is bifurcated, with little vocabulary overlap between two or more sets of relevant documents, then HiCAL could get stuck in a local optimum, exploiting one part of the document space well but missing relevant documents in another. Experience suggests that this can happen, but that such cases are rare.[2] In particular, we expect such cases to be exceptionally rare in the news stories on which our HC4 test collections are built, since journalists typically go out of their way to contextualize the information that they present.

Early TREC CLIR test collections all included binary relevance judgments, but the introduction of the Discounted Cumulative Gain (DCG) measure in 2000 [20], and the subsequent broad adoption of Normalized DCG (nDCG), increased the demand for relevance judgments with more than two relevance grades (e.g., highly relevant, somewhat relevant, and not relevant). Some of the early CLIR work with graded relevance judgments first binarized those judgments (e.g., either by treating highly and somewhat relevant as relevant, or by treating only highly relevant as relevant) [21]. However, Sakai has noted that using graded relevance in this way can rank systems differently than would more nuanced approaches that award partial credit for finding partially relevant documents [34]. In our baseline runs, we report nDCG using the graded relevance judgments, then binarize those judgments to report Mean Average Precision (MAP) by treating highly and somewhat relevant as relevant.

3   Collection Development Methodology

We adopted several design principles to create HC4. First, to develop a multilingual document collection that was easy to distribute, we chose the Common Crawl News Collection as the basis for the suite of collections. We applied automatic language identification to determine the language of each document.[3] We

---

[2] Personal communication with Gordon Cormack.
[3] https://github.com/bsolomon1124/pycld3

then assembled Chinese, Persian, and Russian documents from August 2016 to August 2019 into ostensibly[4] monolingual document sets. Finally, we automatically identified and eliminated duplicate documents.

The second design principle was to create topics that model the interests of a knowledgeable searcher who writes about world events. Such topics enable CLIR research that addresses complex information needs that cannot be answered by a few facts. Key attributes of a knowledgeable searcher include a relative lack of ambiguity in their information need and an increased interest in named entities. To support this goal, we used events reported in the Wikipedia Current Events Portal (WCEP)[5] as our starting point for topic development. To support exploration of how additional context information could be used to improve retrieval, each topic was associated with a contemporaneous report.

A third design principle was to include topics with relevant documents in multiple languages. Once a topic was developed in one language, it was vetted for possible use with the document sets of other languages.

3.1 Topic Development

Starting from an event summary appearing in WCEP, a topic developer would learn about that event from the English document that was linked to it, and from additional documents about the event that were automatically identified as part of the WCEP multi-document summarization dataset [16]. Topic developers were bilingual, so they could understand how an English topic related to the event being discussed in the news in another language. After learning about the event, the topic developer searched a non-English collection to find documents about the event. After reading a few documents in their language, they were asked to write a sentence or question describing an information need held by the hypothetical knowledgeable searcher. They were then asked to write a three-to-five word summary of the sentence. The summary became the topic title, and the sentence became the topic description. Next, the topic developer would investigate the prevalence of the topic in the collection. To do this they would issue one or more document-language queries and judge ten of the resulting documents. Topic developers answered two questions about each document: (1) How relevant is the most important information on the topic in this document?; and (2) How valuable is the most important information in this document? Relevance was judged as *central*, *tangential*, *not-relevant*, or *unable-to-judge*. The second question was only posed if the answer to the first question was *central*. Allowable answers to the second question were *very-valuable*, *somewhat-valuable*, and *not-valuable*.

To develop topics with relevant documents in more than one language, the title and description, along with the event that inspired the topic, were shown to a topic developer for a different language. The topic developer searched for the presence of the topic in their language. As with the initial topic development,

---

[4] Language ID failure caused some documents in each set to be of the wrong language.
[5] https://en.wikipedia.org/wiki/Portal:Current_events.

ten documents were judged to evaluate whether the document set supported the topic. Topic developers were allowed to modify the topic, which sometimes led to vetting the new topic in the initial language.

3.2   Relevance Judgments

After topic development, some topics were selected for more complete assessment. The titles and descriptions of selected topics were vetted by a committee comprising IR researchers and topic developers. The committee reviewed each topic to ensure that: (a) the title and description were mutually consistent and concise; (b) titles consisted of three to five non-stopwords; (c) descriptions were complete, grammatical sentences with punctuation and correct spelling; and (d) topics were focused and likely to have a manageable number of relevant documents. Corrections were made by having each committee member suggest new phrasing, then a topic developer selecting a preferred alternative.

Given the impracticality of judging millions of documents, and because most documents are not relevant to a given topic, we followed the common practice of assessing as many relevant documents as possible, deferring to the evaluation measure decisions on how unassessed documents should be treated. Because we did not build this collection using a shared task, we did not have diverse systems to contribute to judgment pools. Thus, we could not use *pooling* [41,48]. Instead, we used the active learning system HiCAL [10], to iteratively select documents to be judged. HiCAL builds a classifier based on the known relevant documents using relevance feedback. As the assessor judges documents, the classifier is retrained using the new assessments. To seed HiCAL's classifier, we used ten documents judged during topic development. Because the relevance assessor is likely not the person who developed the topic, and because the topic might have changed during topic vetting, those documents are re-judged. At least one document must be judged relevant to initialize the classifier.

Once assessment was complete, assessors provided a translation of the title and description fields into the language of the documents, and briefly explained (in English) how relevance judgments were made; these explanations were placed in the topic's narrative field. In contrast to the narrative in a typical TREC ad hoc collection, which is written prior to judging documents, these narratives were written after judgments were made; users of these collections must therefore be careful not to use the narrative field as part of a query on the topic.

Our target time for assessing a single topic was four hours. We estimated this would allow us to judge about one hundred documents per topic. According to the designers of HiCAL,[6] one can reasonably infer that almost all findable relevant documents have been found if an assessor judges twenty documents in a row as not relevant. From this, we estimated that topics with twenty or fewer relevant documents were likely to be fully annotated after viewing 100 documents. Treating both *central* and *tangential* documents as relevant would have led to more than twenty relevant documents for most selected topics. Thus,

---

[6] Personal communication with Ian Soboroff.

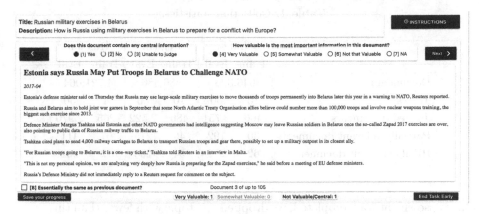

Fig. 1. Annotation interface for relevance judgments.

to support topics that went beyond esoteric facts, we treated only documents deemed central to the topic as relevant.

We established three relevance levels, defined from the perspective of a user writing a report on the topic:

Very-valuable Information in the document would be found in the lead paragraph of a report that is later written on the topic.

Somewhat-valuable The most valuable information in the document would be found in the remainder of such a report.

Not-valuable Information in the document might be included in a report footnote, or omitted entirely.

To map graded relevance values to the binary relevance required by HiCAL, documents judged as *very-valuable* or *somewhat-valuable* were treated as relevant, while documents judged *not-valuable*, and those that were not central to the topic, were considered *not-relevant*. The final collection maps the *not-valuable* category to *not-relevant*. This means that a document can mention a topic without being considered relevant to that topic if it lacks information that would be included in a future report. Because an assessor could judge a topic over multiple days, assessors took copious notes to foster consistency.

To more quickly identify topics too broad to be annotated under our annotation budget, assessors were instructed to end a task early (eliminating the topic from inclusion in the collection) whenever:

- more than five *very-valuable* or *somewhat-valuable* documents were found among the first ten assessed;
- more than fifteen *very-valuable* or *somewhat-valuable* documents were found among the first thirty assessed;
- more than forty *very-valuable* or *somewhat-valuable* documents were found at any point; or
- relevant documents were still being found after assessing 85 or more documents.

Table 1. Collection statistics.

	Chinese		Persian		Russian	
	Train	Eval	Train	Eval	Train	Eval
Documents	646,305		486,486		4,721,064	
Topics	10	50	10	50	4	50
Judged documents	466	2,751	486	2,522	265	2,970
Partially relevant documents	30	192	46	215	67	411
Highly relevant documents	62	282	54	206	12	262

Once assessment was completed, we dropped any topic with fewer than three relevant documents. We subsequently sought to refocus dropped topics to ameliorate the problems encountered during assessment; if this was deemed likely to produce a conforming topic, the refocused topic was added back into the assessment queue. Thus, a few similar but not identical topics are present in different languages.

We used the process described above to develop the topics in each of the three languages. Figure 1 shows the interface used to annotate the collection. Key features include: hot keys to support faster judgment; next document and previous document navigation; identification of near-duplicate documents that were not identified during deduplication; the ability to save progress and return to annotation in another session; counts of how many documents have been judged in different categories; and a button to end the annotation early.

3.3 Contemporaneous Reports

Contemporaneous reports are portions of Wikipedia page text written before a particular date. Each topic was associated with a date, which either came from the date of the event in WCEP that inspired the topic or, if after topic development there was no such event, from the earliest relevant document. The assessor was instructed to find the Wikipedia page most related to the topic and use the edit history of that page to view it as it appeared on the day before the date listed in the topic. The assessor selected text from this page to serve as the contemporaneous report. Because of the date restriction, some contemporaneous reports are less closely related to the topic, since a specific Wikipedia page for the event may not have existed on the day before the event.

Table 2. Multilingual topic counts.

	Chinese+Persian	Chinese+Russian	Persian+Russian	All languages
Train	6	2	2	1
Eval	12	14	10	4

Table 3. Document annotation time in minutes with median of each class and Spearman's ρ correlation between assessment time and the resulting binarized label.

Language	# Doc. (Rel/Not)	Median (Rel/Not)	ρ	Total time
Chinese	1,094/3,863	1.33/0.75	0.1900	8,730.65
Persian	1,576/4,444	1.35/0.80	0.1617	11,807.66
Russian	2,746/5,525	0.79/0.69	0.0584	11,561.20

4   Collection Details

This section introduces collection details, discusses the annotation cost in terms of time, and reports on inter-assessor agreement. Table 1 describes the size of the collection in documents and topics, and presents counts of the number of annotations used in the final collection. Disjoint subsets of Train and Eval topics are defined to encourage consistent choices by users of the test collections. As in most information retrieval collections, the vast majority of the unjudged documents are not relevant. However, because we used active learning to suggest documents for assessment, and because of our desire to create topics with relatively few relevant documents, on average there are only about 50 judged documents per topic. This number ranges from 28 (when no additional relevant documents were discovered during the second phase) to 112 documents (when an assessor used the "Essentially the same" button shown in Fig. 1[7]). Some of the topics have judged documents in multiple languages. Table 2 displays the number of topics with judgments in each pair of languages, and the subset of those with judgments in all three languages. While we sought to maximize the number of multilingual topics, we were constrained by our annotation budget.

The people who performed topic development and relevance assessment were all bilingual. A majority of them were native English speakers, although a few were native speakers in the language of the documents. While some were proficient in more than two languages, none was proficient in more than one of Chinese, Persian or Russian. Highly fluent topic developers verified that the human translations of topics were expressed fluently in the non-English language.

4.1   Development and Annotation Time

As a proxy for the cost of creating these test collections, we report the time spent on topic development and relevance assessment. The total time for developing candidate topics, including those not included in the final collection, is shown in Table 4. A total of about 570 h were spent by 30 developers to create the 559 topics in the three languages. The median time to develop a topic was about 36 min, with an average of about an hour, suggesting a long tail distribution.

---

[7] This button applies the previous relevance judgment without increasing the counter; it was typically used when several news sources picked up the same story, but modified it sufficiently to prevent its being automatically labeled as a near duplicate.

Table 4. Topic development time in minutes.

Language	Topics	Average	Median
Chinese	240	56.49	30.57
Persian	148	52.63	36.02
Russian	181	81.60	46.58

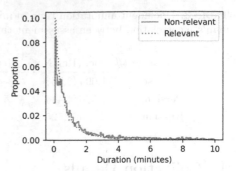

Fig. 2. Document annotation time.

As mentioned in Sect. 3.2, developed topics were filtered before assessment. As shown in Table 3, a total of about 540 h were spent by 33 assessors.[8] These figures include documents rejudged for quality assurance, and topics with incomplete assessments. The median annotation time per document suggests that relevant documents took longer to judge. Here, we aggregated **very-valuable** and **somewhat-valuable** as relevant, and the remaining categories as not relevant. Despite this consistent observation across all three languages, Spearman's ρ suggests only a weak correlation between the judgment time and relevance due to the long tail distribution shown in Fig. 2. There are more **not-relevant** documents that took a shorter time to assess, but as we observe in Fig. 2 the distributions are similar, and the differences are thus not statistically significant by an independent samples t-test.

4.2 Inter-assessor Agreement

Although all topics were assessed by a single assessor for consistency, several were additionally assessed by one or two other assessors for quality assurance. In Table 6 we report the raw agreement (i.e., proportion of the documents all

Table 5. Example for intersection and union agreement

Assessor\Document	D1	D2	D3	D4	D5
A1	✓			✓	✓
A2	✓	✓		✓	
A3	✓			✓	✓
(I)ntersection/(U)nion	I/U	I		I/U	I

---

[8] We replaced the longest 5% of assessment times with the median per language, since these cases likely reflect assessors who left a job unfinished overnight.

Table 6. Inter-assessor agreement on binarized labels.

Language	# of topics	Intersection		Union	
		Agreement	Fleiss' κ	Agreement	Fleiss' κ
Chinese	5	0.85	0.69	0.84	0.62
Persian	4	0.73	0.40	0.69	0.35
Russian	3	0.69	0.33	0.69	0.33

assessors agreed upon) and the Fleiss' κ (i.e., the agreement after chance correction for multiple assessors). Because active learning is path-dependent, each assessor judged a somewhat different set of documents; we thus evaluate agreement on both the intersection and the union of the documents for a complete picture. Unjudged documents were considered ***not-relevant*** for the union agreements. Table 5 shows an example, where only D1 and D4 are in the intersection, judged by all three assessors. D3 was not judged by any assessor, and is thus is not in the union.

All three languages demonstrate at least fair agreement (κ between 0.20 and 0.40 [23]), with Chinese topics having a substantial agreement (κ between 0.60 and 0.80), for both the intersection and the union. The raw agreement indicates that 69% to 85% of the judged documents have the same binarized judgments. The small gap between intersection and union agreements supports our assumption that unjudged documents are not relevant.

5   Baseline Runs

To demonstrate the utility of HC4 for evaluating CLIR systems, we report retrieval evaluation results for a set of baseline CLIR systems on the Eval sets in Table 7. Three retrieval approaches, implemented by Patapsco [11], human query translation, machine query translation, and machine document translation, use BM25 ($k_1 = 0.9$, $b = 0.4$) with RM3 pseudo relevance feedback on title queries. Translation models are trained in-house using the Sockeye toolkit [18].

As examples of neural CLIR models, we evaluated *vanilla* reranking models [26] fine-tuned with MS-MARCO-v1 [2] for at most one epoch with various multi-language pretrained models, including multilingual-BERT (mBERT) [13], XLM-Roberta-large (XLM-R) [8], and infoXLM-large [6]. Model checkpoints were selected by nDCG@100 on HC4 dev sets. Each trained model reranks the top 1000 documents retrieved by the machine query translation BM25 model[9] in a zero-shot fashion [30].

For both nDCG and MAP, human query translation tends to provide the most effective results, usually indistinguishable from machine document translation and from XLM-R (both of which are effective but computationally expensive). In contrast, machine query translation is efficient. Title queries are unlikely

---

[9] Hence, the input of the reranking models is still English queries with documents in the target language.

Table 7. Baseline results of title queries using BM25 with RM3 on Eval sets, QT/DT: query/document translation.

Language	Method	nDCG@100		MAP@100		R@1000		Judged@10	
		Mean	p-value	Mean	p-value	Mean	p-value	Mean	p-value
Chinese	Human QT	0.473	–	0.317	–	0.825	–	**0.544**	–
	Machine QT	0.362	0.009	0.231	0.023	0.708	0.036	0.408	0.010
	Machine DT	0.502	0.566	0.336	0.704	**0.871**	0.345	0.542	0.968
	mBERT	0.348	0.008	0.205	0.011	0.708	0.036	0.302	0.000
	infoXLM	**0.541**	0.155	**0.369**	0.297	0.708	0.036	0.504	0.428
	XLM-R	0.536	0.168	0.368	0.318	0.708	0.036	0.500	0.376
Persian	Human QT	0.428	–	0.277	–	0.858	–	**0.520**	–
	Machine QT	0.355	0.004	0.223	0.006	0.768	0.035	0.460	0.062
	Machine DT	0.411	0.549	0.260	0.489	**0.863**	0.866	0.476	0.319
	mBERT	0.324	0.009	0.179	0.004	0.768	0.035	0.314	0.000
	infoXLM	**0.514**	0.040	**0.366**	0.015	0.768	0.035	**0.520**	1.000
	XLM-R	0.499	0.078	0.349	0.042	0.768	0.035	0.504	0.741
Russian	Human QT	0.373	–	0.239	–	**0.760**	–	**0.448**	–
	Machine QT	0.335	0.237	0.217	0.386	0.710	0.154	0.366	0.285
	Machine DT	0.348	0.533	0.213	0.424	0.756	0.923	0.402	0.324
	mBERT	0.199	0.000	0.087	0.000	0.710	0.154	0.156	0.000
	infoXLM	0.353	0.602	0.233	0.874	0.710	0.154	0.342	0.015
	XLM-R	**0.377**	0.906	**0.249**	0.743	0.710	0.154	0.414	0.384

to be grammatically sound though, so machine translation quality is lower, resulting in lower retrieval effectiveness. We report p-values for two-sided pairwise statistical significance tests. As expected with this number of topics [7], some differences that would be significant at $p < 0.05$ are observed.[10]

The similar levels of *Judged at 10* (the fraction of the top 10 documents that were judged) among the highest-scoring systems by nDCG and MAP suggest that our relevance judgments are not biased toward any of those systems, despite their diverse designs. mBERT yields specifically lower *Judged at 10* due to the significantly worse effectiveness, which has also been found by others [19].

6 Conclusion

Our new HC4 test collections provide a basis for comparing the retrieval effectiveness of both traditional and neural CLIR techniques. HC4 allows for wide distribution since documents are distributed as part of the Common Crawl and the topics and relevance judgments are being made freely available for research use. HC4 is among the first collections in which judged documents are principally identified using active learning. In addition to providing titles and descriptions in

---

[10] Bonferonni correction for 5 tests yields $p < 0.01$ for significance.

English and in the language of the documents, English contemporaneous reports are included to support research into using additional context for retrieval. HC4 will thus help enable development of next generation CLIR algorithms.

References

1. Abualsaud, M., Ghelani, N., Zhang, H., Smucker, M.D., Cormack, G.V., Grossman, M.R.: A system for efficient high-recall retrieval. In: The 41st International ACM SIGIR Conference on Research & Development in Information Retrieval, pp. 1317–1320. ACM (2018)
2. Bajaj, P., et al.: MS MARCO: a human generated machine reading comprehension dataset. arXiv preprint arXiv:1611.09268 (2016)
3. Baron, J., Losey, R., Berman, M.: Perspectives on predictive coding: and other advanced search methods for the legal practitioner. American Bar Association, Section of Litigation (2016). https://books.google.com/books?id=TdJ2AQAACAAJ
4. Bonab, H., Sarwar, S.M., Allan, J.: Training effective neural CLIR by bridging the translation gap. In: Proceedings of the 43rd International ACM SIGIR Conference on Research and Development in Information Retrieval, pp. 9–18 (2020)
5. Buckley, C., Voorhees, E.M.: Retrieval evaluation with incomplete information. In: Sanderson, M., Järvelin, K., Allan, J., Bruza, P. (eds.) SIGIR 2004: Proceedings of the 27th Annual International ACM SIGIR Conference on Research and Development in Information Retrieval, Sheffield, UK, 25–29 July 2004, pp. 25–32. ACM (2004). https://doi.org/10.1145/1008992.1009000
6. Chi, Z., et al.: InfoXLM: an information-theoretic framework for cross-lingual language model pre-training. arXiv preprint arXiv:2007.07834 (2020)
7. Clough, P., Sanderson, M.: Evaluating the performance of information retrieval systems using test collections. Inf. Res. 18(2) (2013)
8. Conneau, A., et al.: Unsupervised cross-lingual representation learning at scale. arXiv preprint arXiv:1911.02116 (2019)
9. Cormack, G.V., Grossman, M.R.: Evaluation of machine-learning protocols for technology-assisted review in electronic discovery. In: Proceedings of the 37th International ACM SIGIR Conference on Research & Development in Information Retrieval, pp. 153–162 (2014)
10. Cormack, G.V., et al.: Dynamic sampling meets pooling. In: Piwowarski, B., Chevalier, M., Gaussier, É., Maarek, Y., Nie, J., Scholer, F. (eds.) Proceedings of the 42nd International ACM SIGIR Conference on Research and Development in Information Retrieval, SIGIR 2019, Paris, France, 21–25 July, pp. 1217–1220 (2019). ACM (2019). https://doi.org/10.1145/3331184.3331354
11. Costello, C., Yang, E., Lawrie, D., Mayfield, J.: Patapasco: a Python framework for cross-language information retrieval experiments. In: Proceedings of the 44th European Conference on Information Retrieval (ECIR) (2022)
12. Davis, M.W., Dunning, T.: A TREC evaluation of query translation methods for multi-lingual text retrieval. In: Harman, D.K. (ed.) Proceedings of The Fourth Text REtrieval Conference, TREC 1995, Gaithersburg, Maryland, USA, 1–3 November 1995. NIST Special Publication, vol. 500–236. National Institute of Standards and Technology (NIST) (1995). http://trec.nist.gov/pubs/trec4/papers/nmsu.ps.gz
13. Devlin, J., Chang, M.W., Lee, K., Toutanova, K.: BERT: pre-training of deep bidirectional transformers for language understanding. arXiv preprint arXiv:1810.04805 (2018)

14. Ferro, N., Peters, C.: CLEF 2009 ad hoc track overview: TEL and persian tasks. In: Peters, C., Di Nunzio, G.M., Kurimo, M., Mandl, T., Mostefa, D., Peñas, A., Roda, G. (eds.) CLEF 2009. LNCS, vol. 6241, pp. 13–35. Springer, Heidelberg (2010). https://doi.org/10.1007/978-3-642-15754-7_2

15. Ferro, N., Peters, C.: Information Retrieval Evaluation in a Changing World: Lessons Learned from 20 Years of CLEF, vol. 41. Springer, Cham (2019). https://doi.org/10.1007/978-3-030-22948-1

16. Ghalandari, D.G., Hokamp, C., The Pham, N., Glover, J., Ifrim, G.: A large-scale multi-document summarization dataset from the Wikipedia current events portal. In: Proceedings of the 58th Annual Meeting of the Association for Computational Linguistics (ACL 2020), pp. 1302–1308 (2020)

17. Grossman, M.R., Cormack, G.V., Roegiest, A.: TREC 2016 total recall track overview. In: Voorhees, E.M., Ellis, A. (eds.) Proceedings of The Twenty-Fifth Text REtrieval Conference, TREC 2016, Gaithersburg, Maryland, USA, 15–18 November 2016. NIST Special Publication, vol. 500–321. National Institute of Standards and Technology (NIST) (2016). http://trec.nist.gov/pubs/trec25/papers/Overview-TR.pdf

18. Hieber, F., Domhan, T., Denkowski, M., Vilar, D., Sokolov, A., Clifton, A., Post, M.: Sockeye: a toolkit for neural machine translation. arXiv preprint arXiv:1712.05690 (2017)

19. Hu, J., Ruder, S., Siddhant, A., Neubig, G., Firat, O., Johnson, M.: XTREME: a massively multilingual multi-task benchmark for evaluating cross-lingual generalisation. In: International Conference on Machine Learning, pp. 4411–4421. PMLR (2020)

20. Järvelin, K., Kekäläinen, J.: IR evaluation methods for retrieving highly relevant documents. In: Yannakoudakis, E.J., Belkin, N.J., Ingwersen, P., Leong, M. (eds.) SIGIR 2000: Proceedings of the 23rd Annual International ACM SIGIR Conference on Research and Development in Information Retrieval, Athens, Greece, 24–28 July 2000, pp. 41–48. ACM (2000). https://doi.org/10.1145/345508.345545

21. Kando, N., Kuriyama, K., Nozue, T., Eguchi, K., Kato, H., Hidaka, S.: Overview of IR tasks. In: Kando, N. (ed.) Proceedings of the First NTCIR Workshop on Research in Japanese Text Retrieval and Term Recognition, NTCIR-1, Tokyo, Japan, 30 August–1 September 1999. National Center for Science Information Systems (NACSIS) (1999). http://research.nii.ac.jp/ntcir/workshop/OnlineProceedings/IR-overview.pdf

22. Kanoulas, E., Li, D., Azzopardi, L., Spijker, R.: CLEF 2019 technology assisted reviews in empirical medicine overview. In: CEUR Workshop Proceedings, vol. 2380 (2019)

23. Landis, J.R., Koch, G.G.: The measurement of observer agreement for categorical data. Biometrics, 159–174 (1977)

24. Lewis, D.D.: A sequential algorithm for training text classifiers: corrigendum and additional data. In: SIGIR Forum, vol. 29, no. 2, pp. 13–19 (1995). https://doi.org/10.1145/219587.219592

25. Lin, J., Nogueira, R., Yates, A.: Pretrained transformers for text ranking: BERT and beyond. Synth. Lect. Hum. Lang. Technol. 14(4), 1–325 (2021)

26. MacAvaney, S., Yates, A., Cohan, A., Goharian, N.: CEDR: contextualized embeddings for document ranking. In: Proceedings of the 42nd International ACM SIGIR Conference on Research and Development in Information Retrieval, pp. 1101–1104 (2019)

27. Majumder, P., et al.: The FIRE 2008 evaluation exercise. ACM Trans. Asian Lang. Inf. Process. 9(3), 10:1–10:24 (2010). https://doi.org/10.1145/1838745.1838747

28. Moffat, A., Zobel, J.: Rank-biased precision for measurement of retrieval effectiveness. ACM Trans. Inf. Syst. **27**(1), 2:1–2:27 (2008). https://doi.org/10.1145/1416950.1416952

29. Nair, S., Galuscakova, P., Oard, D.W.: Combining contextualized and non-contextualized query translations to improve CLIR. In: Proceedings of the 43rd International ACM SIGIR Conference on Research and Development in Information Retrieval, pp. 1581–1584 (2020)

30. Nair, S., et al.: Transfer learning approaches for building cross-language dense retrieval models. In: Proceedings of the 44th European Conference on Information Retrieval (ECIR) (2022)

31. Oard, D.W., Webber, W.: Information retrieval for e-discovery. Inf. Retr. **7**(2–3), 99–237 (2013)

32. Roegiest, A., Cormack, G.V., Clarke, C.L.A., Grossman, M.R.: TREC 2015 total recall track overview. In: Voorhees, E.M., Ellis, A. (eds.) Proceedings of The Twenty-Fourth Text REtrieval Conference, TREC 2015, Gaithersburg, Maryland, USA, 17–20 November 2015. NIST Special Publication, vol. 500–319. National Institute of Standards and Technology (NIST) (2015). https://trec.nist.gov/pubs/trec24/papers/Overview-TR.pdf

33. Sakai, T., Kando, N.: On information retrieval metrics designed for evaluation with incomplete relevance assessments. Inf. Retr. **11**(5), 447–470 (2008). https://doi.org/10.1007/s10791-008-9059-7

34. Sakai, T., Oard, D.W., Kando, N.: Evaluating Information Retrieval and Access Tasks: NTCIR's Legacy of Research Impact. Springer, Singapore (2021). https://doi.org/10.1007/978-981-15-5554-1

35. Salton, G.: Automatic processing of foreign language documents. J. Am. Soc. Inf. Sci. **21**(3), 187–194 (1970)

36. Schäuble, P., Sheridan, P.: Cross-language information retrieval (CLIR) track overview. In: Voorhees, E.M., Harman, D.K. (eds.) Proceedings of The Sixth Text REtrieval Conference, TREC 1997, Gaithersburg, Maryland, USA, 19–21 November 1997. NIST Special Publication, vol. 500–240, pp. 31–43. National Institute of Standards and Technology (NIST) (1997). http://trec.nist.gov/pubs/trec6/papers/clir_track_US.ps

37. Smeaton, A.F.: Spanish and Chinese document retrieval in TREC-5. In: Voorhees, E.M., Harman, D.K. (eds.) Proceedings of the Fifth Text REtrieval Conference, TREC 1996, Gaithersburg, Maryland, USA, 20–22 November 1996. NIST Special Publication, vol. 500–238. National Institute of Standards and Technology (NIST) (1996). http://trec.nist.gov/pubs/trec5/papers/multilingual_track.ps.gz

38. Thakur, N., Reimers, N., Rücklé, A., Srivastava, A., Gurevych, I.: BEIR: a heterogenous benchmark for zero-shot evaluation of information retrieval models. arXiv preprint arXiv:2104.08663 (2021)

39. Voorhees, E.M.: Variations in relevance judgments and the measurement of retrieval effectiveness. Inf. Process. Manag. **36**(5), 697–716 (2000). https://doi.org/10.1016/S0306-4573(00)00010-8

40. Voorhees, E.M.: Coopetition in IR research. In: SIGIR Forum, vol. 54, no. 2, August 2021

41. Webber, W., Moffat, A., Zobel, J.: The effect of pooling and evaluation depth on metric stability. In: EVIA@ NTCIR, pp. 7–15 (2010)

42. Yang, E., Lewis, D.D., Frieder, O.: On minimizing cost in legal document review workflows. In: Proceedings of the 21st ACM Symposium on Document Engineering, August 2021

43. Yarmohammadi, M., et al.: Robust document representations for cross-lingual information retrieval in low-resource settings. In: Proceedings of Machine Translation Summit XVII Volume 1: Research Track, pp. 12–20 (2019)

44. Yilmaz, E., Aslam, J.A.: Estimating average precision with incomplete and imperfect judgments. In: Yu, P.S., Tsotras, V.J., Fox, E.A., Liu, B. (eds.) Proceedings of the 2006 ACM CIKM International Conference on Information and Knowledge Management, Arlington, Virginia, USA, 6–11 November 2006, pp. 102–111. ACM (2006). https://doi.org/10.1145/1183614.1183633

45. Zhang, R., et al.: Improving low-resource cross-lingual document retrieval by reranking with deep bilingual representations. In: Proceedings of the 57th Annual Meeting of the Association for Computational Linguistics, pp. 3173–3179 (2019)

46. Zhang, X., Ma, X., Shi, P., Lin, J.: Mr. TyDi: a multi-lingual benchmark for dense retrieval. arXiv preprint arXiv:2108.08787 (2021)

47. Zhao, L., Zbib, R., Jiang, Z., Karakos, D., Huang, Z.: Weakly supervised attentional model for low resource ad-hoc cross-lingual information retrieval. In: Proceedings of the 2nd Workshop on Deep Learning Approaches for Low-Resource NLP (DeepLo 2019), pp. 259–264 (2019)

48. Zobel, J.: How reliable are the results of large-scale information retrieval experiments? In: Proceedings of the 21st Annual International ACM SIGIR Conference on Research and Development in Information Retrieval, pp. 307–314 (1998)

Did I See It Before? Detecting Previously-Checked Claims over Twitter

Watheq Mansour[1(✉)], Tamer Elsayed[1], and Abdulaziz Al-Ali[1,2]

[1] Computer Science and Engineering Department, Qatar University, Doha, Qatar
{wm1900793,telsayed,a.alali}@qu.edu.qa
[2] KINDI Center for Computing Research, Qatar University, Doha, Qatar

Abstract. With the proliferation of fake news in the last few years, especially during the COVID-19 period, combating the spread of misinformation has become an urgent need. Although automated fact-checking systems were proposed recently, they leave much to be desired in terms of accuracy and explainability. Therefore, involving humans during verification could make the process much easier and more reliable. In this work, we propose an *automated* approach to detect claims that have been already *manually*-verified by professional fact-checkers. Our proposed approach uses recent powerful BERT variants as point-wise rerankers. Additionally, we study the impact of using different fields of the verified claim during training and inference phases. Experimental results show that our proposed pipeline outperforms the state-of-the-art approaches on two English and one Arabic datasets.

Keywords: Claim Retrieval · Fact Checking · Reranking · Verification

1 Introduction

The massive spread of misinformation has a negative impact on many governments, public figures, and organizations, among others [15]. That created an urgent need to combat the spread of misinformation. As a response, many fact-checking organizations, e.g., Politifact[1] and FullFact,[2] arose in the last few years. However, most of these organizations perform fact-checking *manually*, which is indeed time-consuming, and hence cannot cope with the rapid spread of misinformation over social media. To this end, several research directions were pursued to develop automated systems that identify check-worthy claims and investigate their factuality [23,35,39,41]. However, the immature nature of the current automated systems sparked concerns about their credibility. Additionally, fact-checkers and journalists need an explanation for the verdict obtained from automated systems for reliability purposes, which is lacking in current systems.

Many viral claims are indeed dormant news, which show up repeatedly at different time periods, such as those appearing during COVID-19 period. Recognizing that those claims have been already verified by professional fact-checkers

---

[1] https://www.politifact.com/.
[2] http://fullfact.org/.

© The Author(s), under exclusive license to Springer Nature Switzerland AG 2022
M. Hagen et al. (Eds.): ECIR 2022, LNCS 13185, pp. 367–381, 2022.
https://doi.org/10.1007/978-3-030-99736-6_25

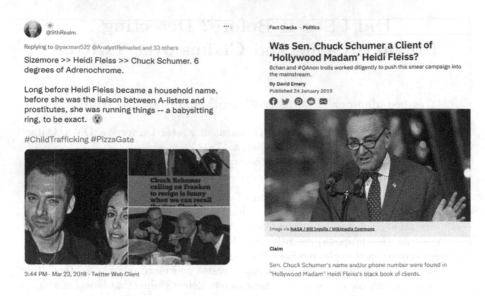

Fig. 1. Example of the claim retrieval problem: tweet (left) and verified claim (right).

has several advantages. It helps mitigate the negative effect of spreading fake news on both society and individuals. Furthermore, it helps journalists put their interviewees on the spot in real-time. It also allows more time for the automated and manual verification systems to focus on verifying unchecked claims.

To that end, in this paper, we tackle the problem of *claim retrieval over Twitter*, defined as follows: given a tweet that includes a claim (denoted as the query) and a collection of previously-checked claims, we aim to retrieve all relevant previously-checked claims with respect to the input tweet. We frame the problem as a *ranking* problem over a collection of previously-verified claims. Figure 1 illustrates an example tweet and corresponding verified claim.

The problem is challenging from two aspects. First, tweets are typically informal and lack context due to the length limitations. Second, claims can be phrased in different forms, calling for semantic matching. To address the problem, we propose a three-step pipeline. First, tweets are preprocessed and expanded with extracted information from embedded URLs, images, and videos. A recall-oriented retrieval step follows, where a classical light-weight retrieval model is used to retrieve an initial set of potentially-relevant claims. Finally, the set is re-ranked by a precision-oriented neural model. In each of the three steps, several alternative models were considered. We conducted our experiments on three datasets that are commonly used in the literature, two English datasets, namely CheckThat! 2020 English (CT2020-En) [5] and CheckThat! 2021 English (CT2021-En) [33], and one in Arabic, namely CheckThat! 2021 Arabic (CT2021-Ar) [33]. Overall, our contributions are three-fold:

- Our proposed approach outperforms the state-of-the-art approaches for both English and Arabic claims.
- We compare several point-wise BERT-based learning-to-rank techniques.
- We examine the effect of using different fields of a given verified claim during both training and inference phases.

The remainder of this paper is organized as follows. Section 2 discusses related work. Section 3 presents our proposed methodology. Section 4 illustrates the experimental setup. Section 5 details our experimental evaluation. Section 6 provides concluding remarks and suggested future directions.

2   Related Work

There are two lines of research that target the claim retrieval problem. The first line is concerned with matching a given query with the body of a fact-checking article [32,34,40]. Shaar et al. [32] used BM25 for initial ranking, then utilized the BM25 scores and sentence-BERT for training a RankSVM reranker. Vo and Lee [40] suggested a framework that uses both text and images to search for fact-checking articles. Sheng et al. [34] proposed a transformer-based reranker that captures the key sentences within a fact-checked article, then exploits them to estimate the relevance score. Although the system proposed by Sheng et al. [34] outperformed proposed systems in [32,40], it is more complex and requires multiple steps to identify the key sentences.

The second line aims to link the given query with previously fact-checked claims. The Verified Claim Retrieval shared tasks in the CheckThat! lab 2020 and 2021 [5,24] are clear representatives of that line. The goal is to detect whether a claim-containing tweet was previously checked among a collection of verified claims. Our focus is directed toward this category of solutions as they are self-explainable and can be used in real-life scenarios. While the task was proposed in English language only in 2020 [5], it was also proposed in Arabic in 2021 [24,33]. Detecting already verified claims for a given claim in political debates is another variation of this task, proposed by Shaar et al. [31]. To capture the context of the claims made in a debate, Shaar et al. [31] used Transformer-XH to model the local and global contexts. They found that modeling the context of the debate is more important than modeling the context of a fact-checking article.

Multiple teams participated in the shared task of the CheckThat! lab 2020 [5] and 2021 [24,33] and followed different strategies in preprocessing and ranking. Bouziane et al. [7], the winning team in CheckThat 2020, utilized multi-modal data and augmented the data with a similar dataset. After that, they fine-tuned the pre-trained RoBERTa model with adversarial hard negative examples to rerank the tweet-verified-claim pairs. We adopt this approach as a strong baseline to compare against for the 2020 dataset.

Passaro et al. [26] adopted a two-phase strategy. In the first phase, they used sentence-BERT [27] to produce a high cosine similarity score to gold pairs. In the second phase, they fine-tuned a sentence-BERT model to achieve the classification task in which the model gives 1 as an output if the pair constitutes a correct match and 0 otherwise. Finally, they rerank the pairs based on the classification score.

McDonald et al. [22] chose the features to be a combination of scores of TF-IDF, BM25, and cosine similarity, similar to [32]. After that, they employed the extracted features in training multiple machine learning models such as Logistic Regression, Random Forest, Gradient Boosted Trees, and Linear SVM. The machine learning models served as scoring models for reranking.

Chernyavskiy et al. [9], the winning team in the CheckThat! lab 2021, made some modifications to the system proposed by Shaar et al. [32]. They utilized the scores of TF.IDF and fine-tuned sentence-BERT as features vector to train LambdaMART reranker. Their system outperformed the other teams and the organizers' baseline by a large margin. We select this system as a competitor baseline for the 2021 dataset.

bigIR team [33] was the first and only team to build a system for Arabic. They exploited AraBERT model as a reranker for an initial set retrieved by BM25. Their proposed system outperformed the organizers' baseline by more than ten points. We consider this work as a baseline for Arabic experiments.

Recently, Kazemi et al. [17] created a dataset of 2343 pairs in five languages (English, Hindi, Bengali, Malayalam, and Tamil) for claim matching. Their work focused on identifying the matching between either pairs of verified claims or pairs of social media posts (namely, WhatsApp messages). However, only 7% of their dataset are pairs of social media content and fact-checked claims.

Most of these methods did not consider the importance of extracting the information of URLs within a tweet, replacing user handles with their usernames, or which BERT variant is the most suitable for this kind of task.

Finally, it is worth mentioning that Google has launched its Fact Check Explorer tool [14] that enables users to search through trusted fact-checking websites for a specific topic or claim. However, this tool cannot address complex claims and has poor performance when used with Arabic text.

3   The Proposed Pipeline

Our proposed approach is a simple three-step pipeline: preprocessing, initial retrieval, and reranking. In this section, we present each of them in detail.

3.1   Preprocessing

Tweets are typically short and usually contain components that are references rather than textual content, such as URLs, user mentions, images, or videos, which might introduce noise to our task. To add more context to the tweets, we expand those components as follows: we convert URL links to their corresponding Web page titles, replace the mentions with their corresponding user names, get the title for the embedded images [7], and add a short description of the embedded videos using reverse image search. As illustrated in Fig. 2, more contextual information is added to the tweet after *expanding* the user mention and embedded image.

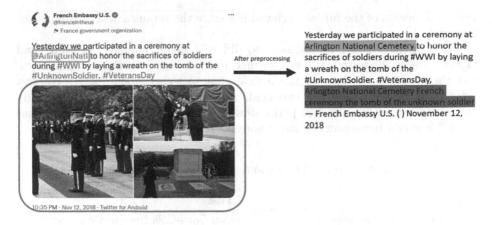

Fig. 2. Example of a tweet before and after preprocessing. Notice the expanded user mention and embedded image.

3.2   Initial Retrieval

The second step in the pipeline retrieves an initial short list of potentially-relevant claims. The goal is to retrieve as many of them, i.e., maximize the recall in the cheapest way, to prepare for the reranking step that aims to push the relevant ones to the top of the list. To achieve that, we leverage the classical retrieval models that are lexical-based, relying on term overlap, such as BM25.

3.3   Reranking

The last step of the pipeline aims to improve the effectiveness of the initial retrieved list using a more expensive reranker that is applied to the initial list. Since transformer-based models, e.g., BERT [11], have shown great success in the information retrieval field, we choose to employ BERT-variants models, e.g., monoBERT [25]. The model takes a query (i.e., the tweet in our context) and a document (i.e., a retrieved verified claim) and classifies the claim based on its relevance to the tweet using a classification layer on top of the neural architecture. The relevance score provided by the classifier is eventually used to rerank the retrieved claims. This is an example of point-wise learning-to-rank models.

The use of BERT as a reranker is not new; however, the novelty in our contribution is mainly characterized by the way we construct the training set, and how the reranker is trained and tested. We choose the negative pairs in the training set so that they cover what the model will be exposed to at inference time. We noticed that many methods in the literature adopt the sampling of "hard" pairs (i.e., the ones that are potential false positives) for tuning the reranker. However, that exposes the model only to hard pairs, with no exposure to other potentially easy ones; which may not necessarily represent what is faced during inference. Alternatively, we choose the negative pairs *randomly* from the top-k documents of the initial retrieved list since the reranker will only consider the top-k documents.

Each of the verified claims has two different fields, namely "VClaim" and "Title", where the former is a well-formed textual version of the original claim, and the latter is the title of the article that fact-checks the claim. We test different combinations of the two fields during training and testing to study their effectiveness on retrieval performance. From Table 1, we can notice that "Title" can add more context and synonyms to the "VClaim".

Table 1. Example VClaim and Title fields of verified claims.

#	VClaim	Title
1	Coca-Cola is an effective contraceptive	Can Coca-Cola Prevent Pregnancy?
2	Coins left on military graves denote visits from living soldiers	Why Are Coins Left on Gravestones in Cemeteries?
3	Officials at Ramapo College forced students to change an "offensive" patriotic barbecue event.	College Threatened to Cancel 'Offensive' America-Themed BBQ Event?

4 Experimental Setup

In this section, we introduce strategies we followed in our experiments. We explain the used datasets, adopted retrieval models and baselines, and the selected BERT variants. Moreover, we elaborate on the applied evaluation measures, tools used for preprocessing, and the method of building the training set.

Datasets. We conducted experiments on three claim retrieval datasets, namely CheckThat! 2020 English (CT2020-En), CheckThat! 2021 English (CT2021-En), and Arabic (CT2021-Ar) datasets. **CT2020-En** [5] is the official dataset for the CheckThat! 2020 lab. The verified claims are collected from Snopes, a well-known fact-checking website, and the queries are crawled tweets that were cited in articles debunking rumors. **CT2021-En** [33] is an extension of CT2020-En, with increasing the number of verified claims and queries. **CT2021-Ar** [33] is the official Arabic dataset for the CheckThat! 2021 lab. The set of verified claims was collected from AraFacts dataset [2] and a translated version of ClaimsKG English dataset [38]. Table 2 shows the statistics of the three datasets. We notice that a query in CT2021-Ar can have more than one relevant verified claim.

Baselines. We chose the top-performing teams in the CheckThat! lab as strong baselines to compare against. For the English models, they are **Buster.AI** [7] and **Aschern** [9] teams from CheckThat! 2020 and 2021, respectively. For Arabic, it is **bigIR** team [33] from CheckThat! 2021.

Table 2. Size of the datasets we used in our experiments. Values in parentheses indicate the average number of relevant verified claims per query.

Dataset	Train	Validation	Test	Verified claims
CT2020-En	800 (1)	197 (1)	200 (1)	10,375
CT2021-En	999 (1)	200 (1)	202 (1)	13,835
CT2021-Ar	512 (1.2)	85 (1.2)	261 (1.3)	30,329

Preprocessing. We employed Twitter API to get usernames out of user handles. Besides, we utilized Meta Reverse Image Search API (MRISA) [36] to perform a reverse image search for pictures and videos.

Initial Retrieval. For initial retrieval, we experiment with multiple classical models, namely, BM25, uni-gram language model with Jelinek-Mercer (JM) smoothing, RM3, and DPH. We used PyTerrier [21] for indexing and retrieval.

BERT Variants and Fine-Tuning. After BERT was introduced [11], many transformer-based pre-trained language models were proposed in the literature to address specific tasks, albeit several were not tried for the claim retrieval task. For English experiments, we studied multiple variants, namely, BERT [11], MPNet [37],[3] RoBERTa [20],[4] Multinligual-MPNet [37], and MiniLM [42].[5] Apart from vanilla BERT, the choice of other models is attributed to their reported performance within SBERT leaderboard[6] on diverse tasks from different domains.

For Arabic experiments, we surveyed several Arabic transformer-based models compared by a recent study [13], namely, Arabic-BERT [30], GigaBERT [18], MARBERT [1], AraBERT [4], QARiB [10], and Arabic-ALBERT [29]. For all BERT variants, we chose the base version due to our limited GPU capacity. For all experiments, we tuned the following hyper-parameters on the dev set: number of epochs (2, 3, or 4), learning rate (2e−5 or 3e−5), and dropout on the classification layer (0.3 or 0.4). We employed one classification layer on top of BERT variants with two output nodes. We fine-tuned each model five times with different random seeds and reported the median performance of those runs.

Evaluation Measures. We follow the evaluation procedure adopted by Check-That! lab, which considers Mean Average Precision at depth 5 (MAP@5) as the main evaluation measure. Furthermore, for the English experiments, we report Precision@1 (P@1) and Mean Reciprocal Rank (MRR) since the average number of relevant documents for a query in these datasets is 1. However, for Arabic experiments, we add the R-Precision (RP) measure to account for queries that have multiple relevant documents. Finally, Recall at depth 100 (R@100) is also reported to show an upper bound performance for the initial retrieval stage.

---

[3] We specifically use STSb-MPNet and Paraphrase-MPNet.
[4] https://huggingface.co/sentence-transformers/msmarco-roberta-base-v2.
[5] https://huggingface.co/sentence-transformers/paraphrase-MiniLM-L12-v2.
[6] https://www.sbert.net/docs/pretrained_models.html.

Significance Test. To establish statistical significance for the reported results, we applied paired t-test significance test on MAP@5 measures. Since we performed multiple tests, we applied Benjamini-Hochberg [6] correction over the p-values to avoid the multiple comparisons problem. We report all corrected p-values and underline the ones that satisfy the 5% significance level. As the computation of the test requires per-query results, we contacted the authors of the adopted baselines to get their submitted runs to CheckThat! lab. We managed to get the runs of the winning teams in CT2021-EN and CT2021-Ar, but missed the winning team run for CT2020-En. To alleviate this problem, we consider the model that is closest in performance to the missing run as a proxy baseline for the purpose of conducting the significance test.

Building the Training Set. To fine-tune BERT-based models, we constructed a balanced training set of positive and negative query-document pairs. The positive pairs are formed by pairing the query with its relevant verified claim and also the title of the verified claim (which constitutes a summary of it). The negative pairs are chosen randomly from the top-k documents of the initial retrieved list, as explained in Sect. 3.3.

5   Experimental Evaluation

In our experiments, we aim to answer the following research questions:

RQ1 What is the effect of the preprocessing steps on the performance of the initial retrieval stage? (Sect. 5.1)

RQ2 Does a monoBERT reranker improve the performance over the initial retrieval stage? What is the best BERT variant for the task? (Sect. 5.2)

RQ3 How can we effectively leverage the title and description of verified claims in training and inference? (Sect. 5.3)

RQ4 What will the performance of the proposed approach be on Arabic data? (Sect. 5.4)

5.1   Initial Retrieval with Preprocessing (RQ1)

To answer RQ1, we apply preprocessing with multiple classical retrieval models, namely BM25 [28], uni-gram language model with Jelinek-Mercer (JM) smoothing [16], DPH [3], and RM3 [19]. We conducted the experiment on both CT2020-En and CT2021-En. Table 3 shows the performance of each of those models before and after applying preprocessing over the dev-set of CT2020-En. We omit the results on CT2021-En as they exhibit very similar performance. The results show that BM25 is superior to the other models for our task. More importantly, the table clearly shows the effectiveness of applying the preprocessing step, as the performance improves for all models over all measures. The p-value for all models except RM3 indicates that preprocessing leads to roughly significant enhancements. Therefore, for the rest of experiments, we adopt BM25 with pre-processing for the initial retrieval. We also notice that the best performance reaches P@1 of about 61%, which leaves a large room for potential improvement for the reranking stage. Furthermore, reaching 95% recall at depth 100 indicates that the improvement is indeed possible by an effective reranker.

Table 3. Performance of the initial retrieval stage on CT2020-En-dev before and after applying preprocessing (PreP).

Model	MAP@5	P@1	MRR	R@100	p-value
BM25	0.710	0.594	0.717	0.949	baseline
BM25+PreP	**0.733**	**0.609**	**0.739**	**0.954**	0.076
DPH	0.685	0.563	0.692	0.939	baseline
DPH+PreP	0.721	0.599	0.727	**0.954**	0.008
JM	0.687	0.558	0.695	0.944	baseline
JM+PreP	0.712	0.579	0.719	0.944	0.066
RM3	0.692	0.594	0.697	0.898	baseline
RM3+PreP	0.713	**0.609**	0.719	0.914	0.141

Table 4. Performance of multiple BERT variants on CT2020-En-dev with varying depth of the initial retrieval stage. We report MAP@5 measure.

Depth	S-MPNet	RoBERTa	P-MPNet	MiniLM
10	0.845	0.850	0.859	**0.846**
20	**0.851**	**0.857**	**0.868**	**0.846**
30	0.844	0.851	0.863	0.837
50	0.845	0.847	0.864	0.832
100	0.842	0.844	0.862	0.824

5.2 MonoBERT for Reranking (RQ2)

As rerankers are more complex, thus expensive, it is critical to first choose a *short* but *effective* depth of the initial retrieved list of claims to be reranked. We conducted an experiment to tune the depth using four variants of BERT, namely, STSb-MPNet (S-MPNet), MSMarco-RoBERTa (RoBERTa), paraphrase-MPNet (P-MPNet), and Paraphrase-MiniLM (MiniLM). Table 4 illustrates the performance of each model as a monoBERT reranker for different depth values of the initial retrieved list (10, 20, 30, 50, and 100). For all models, the experiment shows that depth of 20 exhibits the best performance among the different experimented values; therefore, we stick to it in the rest of experiments. Moreover, we observe that reranking using a BERT-based model gives a considerable improvement over the performance of the initial retrieval models (85% vs. 61% in

P@1), as illustrated earlier in Table 3, which highlights the importance of using contextualized models for this task.

We next turn to answer RQ2 by comparing the performance of multiple BERT-based models with the adopted baselines on the two English datasets. Tables 5 and 6 present the performance of the models over the test sets of CT2020-En, and CT2021-En respectively. The results show that two models, namely S-MPNet and P-MPNet, outperform the best team over CT2020-En, and four models, namely S-MPNet, P-MPNet, RoBERTa, and MiniLM outperform the best team over CT2021-En, as indicated over all measures. Furthermore, the improvement of S-MPNet over the baseline is statistically significant over CT2020-En, but not CT2021-En. We also notice that the vanilla BERT model exhibited poor performance on both datasets compared to other models.

Table 5. Performance of monoBERT models on the test set of CT2020-En.

Model	MAP@5	P@1	MRR	p-value
Buster.AI (Best at CheckThat! 2020)	0.929	0.895	0.927	
Multilingual-MPNet	0.666	0.553	0.650	0.000
BERT	0.741	0.618	0.735	0.000
MiniLM	0.920	0.884	0.920	0.593
RoBERTa	0.926	0.894	0.926	baseline
P-MPNet	0.944	0.925	0.944	0.160
S-MPNet	**0.955**	**0.950**	**0.955**	0.005

Table 6. Performance of monoBERT models on the test set of CT2021-En.

Model	MAP@5	P@1	MRR	p-value
Aschern (Best at CheckThat! 2021)	0.883	0.861	0.884	baseline
Multilingual-MPNet	0.742	0.644	0.749	0.000
BERT	0.834	0.757	0.835	0.100
MiniLM	0.904	0.871	0.906	0.371
RoBERTa	0.916	0.876	0.917	0.147
P-MPNet	0.922	0.886	0.923	0.100
S-MPNet	**0.929**	**0.901**	**0.929**	0.089

5.3   Leveraging Verified Claim Fields (RQ3)

In earlier experiments, we used both VClaim and Title fields of the verified claims separately as training examples, and performed inference using VClaim only, similar to [26]. However, in this experiment, we probe the effect of using other combinations of those fields on the performance. To answer RQ3, we experiment with training using VClaim only, Title only, and both VClaim and Title. We also experiment using both VClaim and Title at inference, where the relevance score of a claim is the average score of tweet-VClaim and tweet-Title pairs.

We conducted such an experiment over CT2021-En. We chose three different models among the top-performing ones in the previous experiment, namely, S-MPNet, RoBERTa, and MiniLM. We excluded P-MPNet as it is the same model of S-MPNet but trained on a different dataset. Table 7 presents the performance using the different combinations. We notice that using both VClaim and Title for training and inference yields the best performance in all models, with a statistically-significant difference with respect to the title-only baselines. We believe training on both increases the model's understanding of the claim context. Moreover, the training set size is doubled when we add Title in training.

Table 7. Performance using different fields of the verified claim on CT2021-En.

Model	Training	Inference	MAP@5	P@1	MRR	p-value
MiniLM	Ttl	Ttl	0.863	0.822	0.867	baseline
	VClaim	VClaim	0.870	0.817	0.871	0.682
	VClaim+Ttl	VClaim	0.904	0.871	0.906	0.050
	VClaim+Ttl	VClaim+Ttl	0.906	0.866	0.907	0.048
RoBERTa	Ttl	Ttl	0.840	0.762	0.841	baseline
	VClaim	VClaim	0.882	0.837	0.883	0.024
	VClaim+Ttl	VClaim	0.916	0.876	0.917	0.000
	VClaim+Ttl	VClaim+Ttl	0.920	0.881	0.920	0.000
S-MPNet	Title	Title	0.884	0.842	0.886	baseline
	VClaim	VClaim	0.908	0.876	0.909	0.095
	VClaim+Ttl	VClaim	0.929	0.901	0.929	0.009
	VClaim+Ttl	VClaim+Ttl	**0.936**	**0.911**	**0.936**	0.005

5.4 Performance on Arabic Data (RQ4)

To answer RQ4, we examine the effectiveness of the proposed pipeline by applying the attained conclusions from English experiments on the Arabic dataset. More specifically, we performed the following steps: (1) We conducted the same preprocessing steps applied to English. We then experimented with multiple classical models for the initial retrieval phase. Here again, BM25 is found to be the best-performing model. (2) We chose AraBERT (as it is the used model by the top team in CheckThat! 2020 [33]) to tune the depth of the initial retrieval set. The best performance is observed when the depth is set to 30. (3) We experimented with the top-performing Arabic BERT-based models as monoBERT rerankers. For all models, we exploited both VClaim and Title fields during training and inference.

Table 8 shows the performance of different BERT-based Arabic models on the test set of CT2021-Ar. Each of AraBERT and GigaBERT outperforms the top team at CheckThat! 2021 in all measures, with statistically-significant improvement in MAP@5. Moreover, two additional models, Arabic-ALBERT and Arabic-BERT, exhibit better performance over the baseline.

Table 8. Evaluation of BERT-based Arabic models on CT2021-Ar.

Model	MAP@5	P@1	MRR	RP	Corrected p-value
bigIR (Best at CheckThat! 2021)	0.908	0.908	0.924	0.895	baseline
MARBERT	0.767	0.743	0.813	0.707	0.000
QARiB	0.903	0.885	0.924	0.861	0.711
Arabic-ALBERT	0.921	0.923	0.948	0.898	0.417
Arabic_BERT	0.932	0.935	0.956	0.910	0.110
GigaBERT-v3	0.939	0.939	0.956	0.918	0.047
AraBERT	**0.940**	**0.946**	**0.959**	**0.927**	0.047

6   Conclusion and Future Work

In this paper, we proposed a pipeline to retrieve previously fact-checked claims with high effectiveness. We converted the ambiguous content in the queries to more informative data. Additionally, we employed a powerful BERT-variant as a point-wise reranker. We also studied the effect of using different fields of the verified claim during the training and testing processes. The experiments showed that the proposed pipeline outperforms the state-of-the-art by a noticeable margin and yet with a simpler approach. Not only does the proposed method outperforms the state-of-the-art in English, but also in Arabic, indicating that it is a promising setup for this task in multiple languages.

One of our future plans is to build a periodically updated collection of fact-checked claims from multiple authorized sources. We also plan to deploy our proposed system in a real-time setup, which in turn utilizes that collection and provides up-to-date predictions. Moreover, we aim to test our methodology on languages other than English and Arabic, and expand the reranking method to pair-wise and list-wise approaches.

Acknowledgments. This work was made possible by NPRP grant# NPRP11S-1204-170060 from the Qatar National Research Fund (a member of Qatar Foundation). The statements made herein are solely the responsibility of the authors.

References

1. Abdul-Mageed, M., Elmadany, A., Nagoudi, E.M.B.: ARBERT & MARBERT: deep bidirectional transformers for Arabic. In: Proceedings of the 59th Annual Meeting of the Association for Computational Linguistics and the 11th International Joint Conference on Natural Language Processing (Volume 1: Long Papers), pp. 7088–7105, August 2021

2. Ali, Z.S., Mansour, W., Elsayed, T., Al-Ali, A.: AraFacts: the first large Arabic dataset of naturally occurring claims. In: Proceedings of the Sixth Arabic Natural Language Processing Workshop, pp. 231–236 (2021)
3. Amati, G., Amodeo, G., Bianchi, M., Gaibisso, C., Gambosi, G.: FUB, IASI-CNR and university of "Tor Vergata" at TREC 2008 blog track. In: The Seventeenth Text REtrieval Conference, TREC 2008, US (2008)
4. Antoun, W., Baly, F., Hajj, H.: AraBERT: Transformer-based model for Arabic language understanding. In: Proceedings of the 4th Workshop on Open-Source Arabic Corpora and Processing Tools, with a Shared Task on Offensive Language Detection, pp. 9–15 (2020)
5. Barrón-Cedeño, A., et al.: Overview of CheckThat! 2020 – automatic identification and verification of claims in social media. In: Proceedings of the 11th International Conference of the CLEF Association: Experimental IR Meets Multilinguality, Multimodality, and Interaction, CLEF 2020, pp. 215–236 (2020)
6. Benjamini, Y., Hochberg, Y.: Controlling the false discovery rate: a practical and powerful approach to multiple testing. J. R. Stat. Soc. Ser. B (Methodol.) **57**(1), 289–300 (1995)
7. Bouziane, M., Perrin, H., Cluzeau, A., Mardas, J., Sadeq, A.: Buster.AI at Check-That! 2020: insights and recommendations to improve fact-checking. In: Cappellato et al. [8] (2020)
8. Cappellato, L., Eickhoff, C., Ferro, N., Névéol, A. (eds.): CLEF 2020 Working Notes. CEUR Workshop Proceedings. CEUR-WS.org (2020)
9. Chernyavskiy, A., Ilvovsky, D., Nakov, P.: Aschern at CheckThat! 2021: lambda-calculus of fact-checked claims. In: Faggioli et al. [12] (2021)
10. Chowdhury, S.A., Abdelali, A., Darwish, K., Soon-Gyo, J., Salminen, J., Jansen, B.J.: Improving Arabic text categorization using transformer training diversification. In: Proceedings of the Fifth Arabic Natural Language Processing Workshop, pp. 226–236 (2020)
11. Devlin, J., Chang, M.W., Lee, K., Toutanova, K.: Bert: Pre-training of deep bidirectional transformers for language understanding. In: Proceedings of the 2019 Conference of the North American Chapter of the Association for Computational Linguistics: Human Language Technologies, Volume 1 (Long and Short Papers). pp. 4171–4186 (2019)
12. Faggioli, G., Ferro, N., Joly, A., Maistro, M., Piroi, F. (eds.): CLEF 2021 Working Notes. Working Notes of CLEF 2021-Conference and Labs of the Evaluation Forum. CEUR Workshop Proceedings. CEUR-WS.org (2021)
13. Farha, I.A., Magdy, W.: Benchmarking transformer-based language models for Arabic sentiment and sarcasm detection. In: Proceedings of the Sixth Arabic Natural Language Processing Workshop, pp. 21–31 (2021)
14. Google: Fact check. https://toolbox.google.com/factcheck/explorer
15. Gunther, R., Beck, P.A., Nisbet, E.C.: Fake news did have a significant impact on the vote in the 2016 election: original full-length version with methodological appendix, Ohio State University, Columbus, OH. Unpublished manuscript (2018)
16. Jelinek, F.: Interpolated estimation of Markov source parameters from sparse data. In: Proceedings of the Workshop on Pattern Recognition in Practice, 1980 (1980)
17. Kazemi, A., Garimella, K., Gaffney, D., Hale, S.: Claim matching beyond English to scale global fact-checking. In: Proceedings of the 59th Annual Meeting of the Association for Computational Linguistics and the 11th International Joint Conference on Natural Language Processing, pp. 4504–4517 (2021)

18. Lan, W., Chen, Y., Xu, W., Ritter, A.: GigaBERT: zero-shot transfer learning from English to Arabic. In: Proceedings of The 2020 Conference on Empirical Methods on Natural Language Processing (EMNLP) (2020)
19. Lavrenko, V., Croft, W.B.: Relevance based language models. In: Proceedings of the 24th Annual International ACM SIGIR Conference on Research and Development in Information Retrieval, pp. 120–127 (2001)
20. Liu, Y., et al.: RoBERTa: a robustly optimized BERT pretraining approach. arXiv preprint arXiv:1907.11692 (2019)
21. Macdonald, C., Tonellotto, N.: Declarative experimentation in information retrieval using PyTerrier. In: Proceedings of ICTIR 2020 (2020)
22. McDonald, T., et al.: The university of sheffield at CheckThat! 2020: claim identification and verification on Twitter. In: Cappellato et al. [8] (2020)
23. Nakov, P., et al.: Automated fact-checking for assisting human fact-checkers. In: International Joint Conferences on Artificial Intelligence (2021)
24. Nakov, P., et al.: Overview of the CLEF–2021 CheckThat! Lab on detecting checkworthy claims, previously fact-checked claims, and fake news. In: Candan, K.S., Ionescu, B., Goeuriot, L., Larsen, B., Müller, H., Joly, A., Maistro, M., Piroi, F., Faggioli, G., Ferro, N. (eds.) CLEF 2021. LNCS, vol. 12880, pp. 264–291. Springer, Cham (2021). https://doi.org/10.1007/978-3-030-85251-1_19
25. Nogueira, R., Yang, W., Cho, K., Lin, J.: Multi-stage document ranking with BERT. arXiv:abs/1910.14424 (2019)
26. Passaro, L., Bondielli, A., Lenci, A., Marcelloni, F.: UNIPI-NLE at CheckThat! 2020: approaching fact checking from a sentence similarity perspective through the lens of transformers. In: Cappellato et al. [8] (2020)
27. Reimers, N., Gurevych, I.: Sentence-BERT: sentence embeddings using siamese BERT-networks. arXiv preprint arXiv:1908.10084 (2019)
28. Robertson, S., Zaragoza, H.: The probabilistic relevance framework: BM25 and beyond. Found. Trends Inf. Retr. **3**(4), 333–389 (2009)
29. Safaya, A.: Arabic-ALBERT, August 2020. https://doi.org/10.5281/zenodo.4718724
30. Safaya, A., Abdullatif, M., Yuret, D.: Kuisail at SemEval-2020 task 12: BERT-CNN for offensive speech identification in social media. In: Proceedings of the Fourteenth Workshop on Semantic Evaluation, pp. 2054–2059 (2020)
31. Shaar, S., Alam, F., Martino, G.D.S., Nakov, P.: The role of context in detecting previously fact-checked claims. arXiv preprint arXiv:2104.07423 (2021)
32. Shaar, S., Babulkov, N., Da San Martino, G., Nakov, P.: That is a known lie: detecting previously fact-checked claims. In: Proceedings of the 58th Annual Meeting of the Association for Computational Linguistics, pp. 3607–3618 (2020)
33. Shaar, S., et al.: Overview of the CLEF-2021 CheckThat! lab task 2 on detecting previously fact-checked claims in tweets and political debates. In: Faggioli et al. [12] (2021)
34. Sheng, Q., Cao, J., Zhang, X., Li, X., Zhong, L.: Article reranking by memory-enhanced key sentence matching for detecting previously fact-checked claims. In: Proceedings of the 59th Annual Meeting of the Association for Computational Linguistics and the 11th International Joint Conference on Natural Language Processing, pp. 5468–5481 (2021)
35. Shu, K., Sliva, A., Wang, S., Tang, J., Liu, H.: Fake news detection on social media: a data mining perspective. ACM SIGKDD Explor. Newsl. **19**(1), 22–36 (2017)
36. Snowdon, R.: Meta reverse image search. https://github.com/vivithemage/mrisa (2017)

37. Song, K., Tan, X., Qin, T., Lu, J., Liu, T.Y.: MPNet: masked and permuted pre-training for language understanding. In: Advances in Neural Information Processing Systems, vol. 33, pp. 16857–16867. Curran Associates, Inc. (2020)

38. Tchechmedjiev, A., et al.: ClaimsKG: a knowledge graph of fact-checked claims. In: Ghidini, C., et al. (eds.) ISWC 2019. LNCS, vol. 11779, pp. 309–324. Springer, Cham (2019). https://doi.org/10.1007/978-3-030-30796-7_20

39. Thorne, J., Vlachos, A.: Automated fact checking: task formulations, methods and future directions. In: Proceedings of the 27th International Conference on Computational Linguistics, pp. 3346–3359 (2018)

40. Vo, N., Lee, K.: Where Are the Facts? Searching for fact-checked information to alleviate the spread of fake news. In: Proceedings of the 2020 Conference on Empirical Methods in Natural Language Processing (EMNLP), pp. 7717–7731 (2020)

41. Vosoughi, S., Roy, D., Aral, S.: The spread of true and false news online. Science **359**(6380), 1146–1151 (2018)

42. Wang, W., Wei, F., Dong, L., Bao, H., Yang, N., Zhou, M.: MiniLM: deep self-attention distillation for task-agnostic compression of pre-trained transformers. In: Advances in Neural Information Processing Systems, vol. 33, pp. 5776–5788. Curran Associates, Inc. (2020)

Transfer Learning Approaches for Building Cross-Language Dense Retrieval Models

Suraj Nair[1,2]([✉]) [iD], Eugene Yang[2] [iD], Dawn Lawrie[2] [iD], Kevin Duh[2] [iD],
Paul McNamee[2] [iD], Kenton Murray[2] [iD], James Mayfield[2] [iD],
and Douglas W. Oard[1,2] [iD]

[1] University of Maryland, College Park, MD 20742, USA
{srnair,oard}@umd.edu
[2] HLTCOE, Johns Hopkins University, Baltimore, MD 21211, USA
{eugene.yang,lawrie,mcnamee,kenton,mayfield}@jhu.edu, kevinduh@cs.jhu.edu

Abstract. The advent of transformer-based models such as BERT has led to the rise of neural ranking models. These models have improved the effectiveness of retrieval systems well beyond that of lexical term matching models such as BM25. While monolingual retrieval tasks have benefited from large-scale training collections such as MS MARCO and advances in neural architectures, cross-language retrieval tasks have fallen behind these advancements. This paper introduces ColBERT-X, a generalization of the ColBERT multi-representation dense retrieval model that uses the XLM-RoBERTa (XLM-R) encoder to support cross-language information retrieval (CLIR). ColBERT-X can be trained in two ways. In *zero-shot* training, the system is trained on the English MS MARCO collection, relying on the XLM-R encoder for cross-language mappings. In *translate-train*, the system is trained on the MS MARCO English queries coupled with machine translations of the associated MS MARCO passages. Results on ad hoc document ranking tasks in several languages demonstrate substantial and statistically significant improvements of these trained dense retrieval models over traditional lexical CLIR baselines.

Keywords: CLIR · ColBERT · ColBERT-X · Dense Retrieval

1 Introduction

BERT-style neural ranking models that use cross-attention between query and document terms [7,16] define the state of the art for monolingual English retrieval. Such models are typically used as rerankers in a retrieve-and-rerank pipeline, due to the quadratic time and space complexity of self-attention in the transformer architecture [30]. Reranking using these models is effective but time-consuming, so the number of documents to be reranked must be tuned to balance the trade-off between effectiveness and efficiency. In contrast to the reranking

M. Hagen et al. (Eds.): ECIR 2022, LNCS 13185, pp. 382–396, 2022.
https://doi.org/10.1007/978-3-030-99736-6_26

approach, dense retrieval models encode query and document representations independently and match them with custom similarity functions (e.g., cosine similarity). Dense retrieval complements the lexical first phase retrieval by using an approximate nearest neighbor search over contextualized representations.

While the retrieve-and-rerank framework has been adapted and explored in cross-language information retrieval (CLIR) [3,11,36–38], most approaches translate queries into the language of the documents and perform monolingual retrieval [28,29]. Dense retrieval models, on the other hand, remain under-explored in CLIR. In this work, we develop an effective dense retrieval model for CLIR.

Dense retrieval models can be broadly categorized into two variants: single-representation and multi-representation [15]. Single-representation models encode queries and documents separately to create a single aggregate representation. However, that can lead to loss of information. Multi-representation models use multiple representations of queries and documents to predict relevance. One such model is ColBERT [13], which computes a similarity between each query term representation and each document term representation. Yet ColBERT is exclusively monolingual. This paper presents *ColBERT-X*, a generalization of the ColBERT approach that supports CLIR. ColBERT-X uses a translate and train fine-tuning approach to exploit existing CLIR training resources.

This generalization poses two challenges: enabling the encoders to process multiple languages, and identifying appropriate resources with which to train the model. To address the former, we adapt XLM-R [5], a multilingual pretrained transformer language model, to initialize the dense retrieval model. For the latter challenge, we use translations of MS MARCO [2], a widely-used passage ranking collection for training monolingual neural retrieval models.

We evaluate ColBERT-X on ad hoc document ranking tasks using English queries to retrieve documents in other languages, exploring two ways to cross the language barrier. In the zero-shot setting, where we lack cross-language training resources, we train the model only on English MS MARCO. In the translate-train setting, the model is trained on machine-generated translations of MS MARCO passages paired with English queries. This paper additionally investigates the effect of machine translation on ColBERT-X retrieval results.

Our main contributions can be summarized as follows:

- We generalize ColBERT to support CLIR and develop a fine-tuning task that leverages translations of existing monolingual retrieval collections.
- We demonstrate significant effectiveness gains over query translation baselines on news in several languages, showing the ability of term-level Approximate Nearest Neighbor (ANN) search to overcome vocabulary mismatch.
- We analyze components of ColBERT-X and techniques to improve effectiveness, including effects of different machine translation models, alternative multilingual encoders, and relevance feedback.
- We release our code to train and evaluate ColBERT-X, and our new machine translations of MS MARCO into Chinese, Persian and Russian.[1]

---

[1] https://github.com/hltcoe/ColBERT-X

2   Related Work

In this section, we briefly review related work on neural retrieval and its extension to cross-lingual settings. For many years, sparse retrieval models such as BM25 [26] and Query Likelihood [24] were the dominant models for ad hoc retrieval tasks. Only in recent years, with the rise of BERT [7] and the availability of large scale retrieval collections such as MSMARCO [2] for training, have neural information retrieval (neural IR) models emerged as the state of the art.

Similar to sparse retrieval models, neural IR models take as input the query and documents, and produce a relevance score. For each query and document pair, matching mechanisms, such as DRMM [9], KNRM [6] or PACCR [10], construct the interaction matrix between the distributed term representations of the query and the documents, and aggregate them into a relevance score. Alternatively, the BERT passage pair classification model [7] considers the query and the document as the input pair, and uses the final classification score as the relevance score [35]. CEDR [18] incorporates contextualized embeddings such as ELMo [23] or BERT [7] into the matching, providing significant effectiveness improvements by taking advantage of contextualization. However, due to the high computational cost, these models are used to rerank top-ranked documents from a sparse retrieval system.

ColBERT [13] further improves efficiency by keeping separate the query-document interaction until the end of the neural architecture. This is called *late interaction*. As opposed to matching mechanisms that require both the query and the document to be present simultaneously, late interaction allows offline encoding of the documents into bags-of-vectors. Document representations are combined with query representations by an efficient *MaxSim* operator, which significantly reduces computation at inference time. This decoupling enables the documents to be encoded offline and indexed to support approximate nearest neighbor search. Further details are discussed in Sect. 3.

Cross-language transfer learning is important for CLIR. Due to the lack of training data for ad hoc neural retrieval models other than in English, prior work explored zero-shot model transfer to other languages, trained with only English retrieval examples [17, 28]. Model initialization with a multilingual language model such as mBERT [7] has been shown to be effective in zero-shot evaluations. However, this approach requires both queries and documents to be in the same language, resulting in evaluation based either on monolingual non-English retrieval [17], or on query translation into the target language [28].

With the availability of translations of the widely-used English ad hoc retrieval resource MS MARCO [4], translate-train (training the retrieval model on a translated collection) using large ad hoc retrieval collections becomes feasible. Prior work explored a dense retrieval approach to translate-train, showing effectiveness gains on monolingual non-English retrieval tasks [29]. However, this approach relied on a single-representation dense retrieval model with an mBERT encoder, combined with sparse retrieval methods such as BM25. Lacking an end-to-end CLIR dense retrieval model that does not require the help of a sparse retrieval system, we bridge the gap by generalizing ColBERT to support directly querying non-English documents with English queries.

3   ColBERT-X

ColBERT is a multi-stage dense retrieval model that uses monolingual BERT [7] to encode both query and document terms. It employs a late-interaction mechanism, MaxSim, that computes the similarity between the encoded query and document term representations. Computing MaxSim for every query and document term pair in the collection is not feasible, so ColBERT has two ways to reduce the number of required similarity comparisons: reranking or end-to-end retrieval. In reranking, a retrieval system such as BM25 generates an initial ranked list, which is then reranked using ColBERT's MaxSim operation. The disadvantage of such a cascaded pipeline is that the overall recall of the system is limited to the recall of the initial ranked list. In the context of CLIR systems, we face the additional complexity of crossing the language barrier that further affects recall. We thus restricted our work to end-to-end (E2E) retrieval.

In the first stage of the E2E setting, a candidate set of documents is generated by ANN search using every query term. Specifically, the k nearest document tokens are retrieved from the ANN index for every query term representation. These tokens are mapped to document IDs, and the union of these IDs creates the final set of candidate documents. In the next stage, these documents are reranked using the late-interaction "MaxSim" operation. For every query term, MaxSim finds the closest document token using the dot product of the encoded query and document term representation. The final score of the document is the summation of individual query term contributions, as shown in Eq. 1. η denotes the monolingual BERT encoder.

$$s_{q,d} = \sum_{i=1}^{|q|} \max_{j=1..|d|} \eta(q_i) \cdot \eta(d_j)^T \tag{1}$$

To generalize ColBERT to CLIR, we replaced monolingual BERT with XLM-R. We call the resulting model ColBERT-X. Initializing the encoder to a multilingual model allows retrieval in any language supported by the embeddings. However, these models must be trained before they can be used for CLIR.

3.1   CLIR Training Strategies

ColBERT was trained using pairwise cross-entropy loss on MS MARCO [2] triples, which consist of an English query, a relevant English passage, and a non-relevant English passage. To train ColBERT-X for CLIR, we explored two strategies from the cross-language transfer learning literature:

1. Zero-Shot: This is a common technique in which a multilingual model (e.g., mBERT or XLM-R) is trained in a high-resource language (usually English)

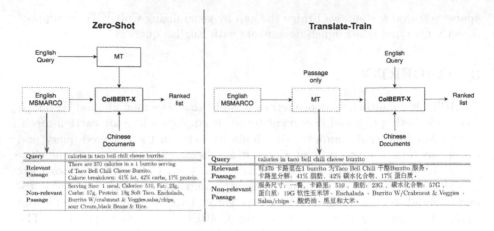

Fig. 1. Two ColBERT-X Transfer Learning Pipelines: Zero-Shot (left) and Translate-Train (right). Dashed boxes denote components used during the training step. In zero-shot, ColBERT-X trained on English MS MARCO is applied on the machine translated queries. With translate-train, the training set consists of translated passages to enable ColBERT-X to cross the language barrier.

and then applied to the document language. In this paper, we first train a ColBERT-X model initialized with an XLM-R encoder on English MS MARCO passage ranking triples. At query time, we use machine translation (MT) to translate the English query to the document language, and use the trained ColBERT-X model to perform retrieval in the document language using Eq. 2. $\hat{q}$ is the translated query. Multilingual language models have demonstrated good cross-language generalization in many other natural language processing tasks; we hypothesized it would also work well for CLIR.

$$s_{\hat{q},d} = \sum_{i=1}^{|\hat{q}|} \max_{j=1..|d|} \eta(\hat{q}_i) * \eta(d_j) \qquad (2)$$

2. Translate-Train: In this setting, an existing high-resource language (e.g., English) collection is translated to the document language. As in zero-shot training, we choose training triples from the MS MARCO passage ranking collection and use a trained MT model to translate them. Since our focus here is using English queries to retrieve content in non-English languages, we pair the original English queries with machine translations of relevant and non-relevant MS MARCO passages to form new triples.[2] We then train ColBERT-X on these newly constructed triples in the same manner as Col-BERT.

---

[2] If we had wanted to experiment with using non-English queries to find English content, we could have instead translated only the MS MARCO queries.

Table 1. Test collection statistics for the CLEF and HC4 newswire collections.

Collection	HC4 Chinese	HC4 Persian	CLEF French	CLEF German	CLEF Italian	CLEF Russian	CLEF Spanish
#documents	646K	486K	129k	294k	157k	16k	454k
#passages	3.6M	3.1M	0.7M	1.6M	0.8M	0.1M	2.7M
#queries	50	50	200	200	200	62	160

Figure 1 shows these two pipelines. The key difference is that in the zero-shot setting we have a single ColBERT-X model for a given query language (in this case English) that is used for retrieval in multiple document languages. In the translate-train setting, we train a ColBERT-X model for each query-document language pair. We might also combine translations in multiple languages to train a single multilingual ColBERT-X model, but we leave that for future work.

3.2 Retrieval

While we train ColBERT-X on passages, our goal is to rank documents. We split large documents into overlapping passages of fixed length with a stride. During indexing, we use the trained ColBERT-X model to generate term representations from these passages. These representations are stored in a FAISS-based ANN index [12] and are saved to disk for subsequent MaxSim computation. At query time, we use the trained ColBERT-X model to generate a ranked list of passages for each query using the approaches discussed in the section above and then use a document's maximum passage score as its document score.

4 Experiments

Collection Statistics. Table 1 provides details for the test collections used in our experiments. We worked with several languages from the 2000 to 2003 Cross-Language Evaluation Forum (CLEF) evaluations [22], using ad hoc news collections for French, German, Italian, Russian and Spanish. We also conducted experiments using the new CLIR Common Crawl Collection (HC4) [14], where the documents are newswire articles from Common Crawl in Chinese or Persian. Throughout, English queries are used to search a collection in a non-English language. We experiment with title and description queries. The MS MARCO [2] passage ranking dataset, which we use for training ColBERT-X, consists of roughly 39M training triples, spanning over 500k queries and 8.8M passages.

ColBERT-X Training and Retrieval. Our two ColBERT-X model strategies, zero-shot (ZS) and translate-train (TT), are trained using mostly the same hyperparameters used to train the original ColBERT model.[3] We replaced the BERT encoder with the XLM-RoBERTa (large) encoder provided by the HuggingFace transformers [33] library (but see Sect. 5.2 for mBERT results). To generate passages from documents, we use a passage length of 180 tokens with a stride of 90 tokens. We index these passages using the trained ColBERT-X model in the same way as the original ColBERT model in the E2E setting.[4]

Machine Translation. For CLEF languages, we use MS MARCO passage translations[5] from Bonifacio *et al.* [4], and the same MT model to translate queries. For the HC4 languages, we use directional MT models built on top of a transformer base architecture (6-layer encoder/decoder) using Sockeye [8]. To produce translations of MS MARCO, the original passages were split using *ersatz* [32], and sentence-level translation was performed using the trained MT model.

Baselines. We compare these strategies with several baselines:

- Human Translation: Monolingual retrieval using Anserini BM25 [34] with the document-language queries provided in the test collection.
- Query Translation: BM25 retrieval using translated queries produced by a specific MT model and original documents in the target language.[6]
- Reranking: We rerank query translation baseline results using the public mT5 reranker[7] trained on translated MS MARCO in 8 languages [4].

Evaluation. We evaluate ranking using Mean Average Precision (MAP). Differences in means are tested for significance using a paired t-test ($p<0.05$) with Holm-Bonferroni multiple test correction.

Results. Table 2 compares the effectiveness of our models to the baselines. Our main finding is that both ColBERT-X variants perform better than BM25 query translation in general. ColBERT-X trained using English MS MARCO alone performs better than query translation and fine-tuning ColBERT-X on translated MS MARCO data helps improve effectiveness further. These gains are statistically significant in both HC4 collections, and for many CLEF collections.

We also compare the ColBERT-X variants to the multilingual T5 reranker that reranks the query translation baseline output. In each collection, ColBERT-X performs consistently and significantly better than the reranker. This is particularly interesting in CLEF collections since both the mT5 reranker and ColBERT-X (TT) were trained on the same MS MARCO translations. However, the reranker was trained on a combined dataset in 8 languages, which might point to the curse of multilinguality [5].

---

[3] We increase our batch size from 32 to 128.

[4] https://github.com/stanford-futuredata/ColBERT#indexing.

[5] https://github.com/unicamp-dl/mMARCO.

[6] To compare the retrieval models fairly, we use the same MT model to translate the queries as the one used to translate the MS MARCO passages.

[7] https://huggingface.co/unicamp-dl/mt5-base-multi-msmarco.

Table 2. Effectiveness results (mean average precision) for CLIR HC4 and CLEF collections using title queries. Statistically significant improvements over the query translation and reranking baselines are marked with * and † respectively. Bold indicates best MAP among the query translation and reranking methods.

Collection(→) Model(↓)	HC4 Chinese	HC4 Persian	CLEF French	CLEF German	CLEF Italian	CLEF Russian	CLEF Spanish
human translation							
BM25	0.301	0.276	0.403	0.304	0.350	0.452	0.452
ColBERT-X (ZS)	0.510	0.343	0.401	0.360	0.328	0.479	0.418
query translation							
BM25	0.237	0.211	0.387	0.263	0.275	0.377	0.405
reranker							
BM25+mT5-multi	0.312	–	0.333	0.297	0.279	0.303	0.370
our methods							
ColBERT-X (ZS)	**0.450***†	0.297*	0.382†	0.328*†	0.272	**0.418**†	0.379
ColBERT-X (TT)	0.408*†	**0.310***	**0.422**†	**0.397***†	**0.339***†	0.410†	**0.415**†

When we compare the two variants of ColBERT-X, we observe that on average translate-train often does better than zero-shot, but these differences are only significant in CLEF collections except Russian and not in HC4 collections. The difference is likely a result of using different MT models in CLEF and HC4 collections, so we conduct further analysis in the next section.

5   Detailed Analysis

This section considers several aspects of ColBERT-X. First, different machine translation models are compared using both MT and CLIR measures. Second, effects of different multilingual encoders are explored. Third, the impact of pseudo-relevance feedback is examined. Then the influence of query length on performance is considered. Finally, ColBERT-X costs in terms of index size are noted.

5.1   Effect of Machine Translation

ColBERT-X utilizes machine translation in two different ways depending on whether it is trained using the zero-shot strategy or the translate-train strategy. In the zero-shot strategy, the queries are translated to the document language at query time, while the translate-train strategy requires an MT system to translate the monolingual training corpus (in this case, the MS MARCO passages) to the document language. The MT systems used to produce translations include:

Table 3. BLEU scores for translation systems using WMT'19 newstest for Chinese and Russian, and TICO-19 (from OPUS[1]) for Persian. These are computed on test sets distinct from the CLIR collections, so absolute BLEU score is not an exact reflection of quality of translations in CLIR experiments. Nevertheless, relative comparison of BLEU scores among MT systems is meaningful.

Language benchmark	Russian newstest'19	Chinese newstest'19	Persian tico-19
OpusMT	26.3	14.6	–
SockeyeMT1	32.1	25.8	4.4
SockeyeMT2	**35.9**	**38.6**	**20.2**

[1]https://opus.nlpl.eu/

- OpusMT – bidirectional MT model(s) with MarianNMT as the base architecture,[8] released by the Helsinki NLP group from Bonifacio *et al.* [4].
- SockeyeMT1 – MT model built on top of a transformer base architecture (6-layer encoder/decoder) trained on bitext. Depending on language, these include publicly available bitext such as OpenSubtitles, UN Corpus, Europarl, and WMT. The model is trained using AWS Sockeye v2 [8].
- SockeyeMT2 – identical model to SockeyeMT1 but trained with 2x–3x more bitext. The number of training sentence pairs for MT1 vs MT2 are 51M vs 120M for Russian, 36M vs 85M for Chinese, and 6M vs 11M for Persian.

Table 3 provides an intrinsic comparison of the systems translating from English on a translation task using BLEU scores [21]. For Russian and Chinese we evaluated using a recent WMT shared task (newstest'19); for Persian we evaluated with a collection of around 3000 sentences about COVID-19, as no WMT test is available. Scores were calculated with *sacrebleu* [25] using the -lc setting. The table reveals that SockeyeMT outputperforms OpusMT and that exposing SockeyeMT to more training data improves the BLEU score.

Table 4 shows that improving BLEU scores likely leads to improvements in CLIR for both training strategies. Table 4a shows the results of translating queries in the zero-shot strategy. While BLEU improvements tend to be realized downstream, this is not seen for HC4 Chinese where OpusMT has better MAP than SockeyeMT1. Note that asking MT to translate keyword queries may not align well with how the systems were trained with complete sentences.

Table 4b shows results for using different translation models on MS MARCO triples, and the effect this has on ColBERT-X retrieval as measured using MAP. Again, we see that the MAP scores tend to improve with improved BLEU; however, in this case the improvement in Russian BLEU from Table 3 between SockeyeMT1 and SockeyeMT2 does not carry over to ColBERT-X, where the performance is essentially the same. Generally, one can expect that improving MT quality will lead to improved effectiveness of ColBERT-X.

---

[8] https://huggingface.co/Helsinki-NLP.

Table 4. MAP using different MT models for ColBERT-X.

MT model	CLEF Russian	HC4 Chinese	HC4 Persian
OpusMT	0.418	0.411	–
SockeyeMT1	0.442	0.391	0.230
SockeyeMT2	**0.461**	**0.450**	**0.297**

(a) ColBERT-X zero-shot

MT model	CLEF Russian	IIC4 Chinese	HC4 Persian
OpusMT	0.410	0.365	–
SockeyeMT1	**0.459**	0.389	0.287
SockeyeMT2	0.456	**0.408**	**0.310**

(b) ColBERT-X translate-train

Table 5. MAP scores for ColBERT-X initialized with the mBERT and XLM-R encoders, and trained on SockeyeMT1 MS MARCO translations.

Multilingual Model	CLEF Russian	HC4 Chinese	HC4 Persian
mBERT	0.341	0.284	0.173
XLM-R	**0.459***	**0.389***	**0.287***

5.2 Effect of Multilingual Language Models

Comparing different multilingual encoders to initialize ColBERT-X, we observe that XLM-R performs significantly better than mBERT, as shown in Table 5. While this might be unsurprising given that the XLM-R model is twice as large and was pretrained on more data than mBERT, tokenization differs across the languages. Considering the case of Chinese, mBERT tokenization produces character-level tokens, whereas the XLM-R tokenizer generates subwords (sentencepieces). This also implies that mBERT indexes are larger than XLM-R indexes, resulting from the term-level storage requirements of ColBERT-X model.

5.3 Pseudo-Relevance Feedback

Pseudo-relevance feedback (PRF) is a form of query expansion that adds discriminative terms extracted from retrieved documents. While PRF has been explored for pre and post translation query expansion [19], here we choose cross-language expansion terms using the ColBERT-X term representation, as suggested by Wang et al. [31]. First, feedback documents (fb-docs) are selected from the top of a ColBERT E2E ranked list. Next, embeddings of terms from the feedback documents are clustered into k clusters. The top ranked centroids of these k clusters[9] by token IDF are used as feedback embeddings (fb-embs). These fb-embs are added to the original query and ColBERT E2E is run again to produce the final ranked list. We extend this approach to the ColBERT-X CLIR setting.

---

[9] Each centroid is mapped to the nearest document token using the ANN index.

Table 6. MAP for query translation BM25 and ColBERT-X translate-train. * or † denote significant improvement over BM25+PRF or ColBERT-X respectively

Retrieval model	CLEF French	CLEF German	CLEF Italian	CLEF Spanish
baseline				
BM25	0.387	0.263	0.275	0.405
ColBERT-X	0.422	0.397	0.339	0.415
with PRF				
BM25	0.410	0.321	0.320	**0.438**
ColBERT-X	**0.459***†	**0.406***†	**0.371***†	0.436†

To better understand the effect of PRF, we compare ColBERT-X translate-train and query translation BM25, with and without PRF. For BM25, we use Anserini's RM3 to perform PRF, with default hyperparameters. For ColBERT-X PRF, we extend Terrier's [20] implementation[10] with default hyperparameters. Table 6 shows the effect of PRF on ColBERT-X translate-train MAP. Except in Spanish, applying PRF to ColBERT-X significantly improves effectiveness compared to ColBERT-X without PRF or compared to BM25 with PRF.

5.4   Effect of Longer Queries

Table 7 analyzes the effect of query type on ColBERT-X translate-train. We compare three representations: *title* (t), which is a short Web-like query; *description* (d), a well-formed sentence describing the information need, and *title+description* (td), the concatenation of the two. Longer queries pose a problem for ColBERT-X, however, since the model only supports queries up to 32 tokens long. To mitigate this problem, we use a list of "stop structures" [1] consisting of phrases (e.g., find documents on, reports of, etc.), which have been shown to work in the past, removing them from the td queries. We observe that td with stop structures removed leads to significant improvements over t or d alone.

5.5   Indexing Footprint

In addition to the FAISS-based ANN index, ColBERT-X requires access to the representation of each term to compute MaxSim. With each term embedded as a 128-dimensional vector and each dimension using 16-bits, that's 256 bytes per term. These are onerous requirements, with index size increasing with collection size. Table 8 provides statistics on storage requirements. ColBERTv2 [27] addresses this issue by clustering token embeddings. That approach could be extended to ColBERT-X for CLIR; we leave it for future work. An artifact

---

[10] https://github.com/terrierteam/pyterrier_colbert.

Table 7. MAP results for ColBERT-X (TT) model using different query representations. * and † denote significant improvements over t and d queries respectively.

Query representation	CLEF French	CLEF German	CLEF Italian	CLEF Spanish
Title	0.422	0.397	0.339	0.415
description	0.434	0.410	0.380	0.456
Title+description	**0.507*†**	**0.466*†**	**0.424*†**	**0.500*†**

Table 8. Collection-specific memory footprint.

Collection	HC4 Chinese	HC4 Persian	CLEF French	CLEF German	CLEF Italian	CLEF Russian	CLEF Spanish
#passages	3.6M	3.1M	0.7M	1.6M	0.8M	0.1M	2.7M
Disk space	154 GB	134 GB	33 GB	70 GB	36 GB	4.7 GB	117 GB

of our design that affects index size is how passages are generated. We use a sliding window of document tokens, so most tokens have two representations. In the future, we will explore the effects of alternative document segmentation approaches.

6   Conclusion

We have developed ColBERT-X, a cross-language generalization of ColBERT that uses a multilingual query and document encoder to improve CLIR beyond what traditional systems such as BM25 can achieve. Using MT systems to translate MS MARCO, we create CLIR collections for training ColBERT-X. We additionally analyze the effect of MT on the CLIR task. In the future, we would like to create a single multilingual model that is trained on the data from many languages, and compare that with a separate models for each language. For pseudo-relevance feedback, it is important to understand which type of queries benefit from it; a per-query comparison could shed some light on that question.

Acknowledgments. This research is based upon work supported in part by the Office of the Director of National Intelligence (ODNI), Intelligence Advanced Research Projects Activity (IARPA), via contract # FA8650-17-C-9117. Views and conclusions contained herein are those of the authors and should not be interpreted as necessarily representing the official policies, expressed or implied, of ODNI, IARPA, or the U.S. Government (USG). The USG is authorized to reproduce and distribute reprints for governmental purposes notwithstanding any copyright annotation therein.

References

1. Allan, J., et al.: INQUERY does battle with TREC-6. NIST Spec. Publ. **500–240**, 169–206 (1998)
2. Bajaj, P., et al.: MS MARCO: a human generated machine reading comprehension dataset. arXiv preprint arXiv:1611.09268v3 (2018)
3. Bonab, H., Sarwar, S.M., Allan, J.: Training effective neural CLIR by bridging the translation gap. In: Proceedings of the 43rd International ACM SIGIR Conference on Research and Development in Information Retrieval, pp. 9–18. Association for Computing Machinery, New York, NY, USA, July 2020
4. Bonifacio, L.H., Campiotti, I., Lotufo, R., Nogueira, R.: mMARCO: a multilingual version of MS MARCO passage ranking dataset. arXiv preprint arXiv:2108.13897 (2021)
5. Conneau, A., et al.: Unsupervised cross-lingual representation learning at scale. In: Proceedings of the 58th Annual Meeting of the Association for Computational Linguistics, pp. 8440–8451. Association for Computational Linguistics, Online, July 2020
6. Dai, Z., Xiong, C., Callan, J., Liu, Z.: Convolutional neural networks for soft-matching n-grams in ad-hoc search. In: Proceedings of the 11th ACM International Conference on Web Search and Data Mining, pp. 126–134 (2018)
7. Devlin, J., Chang, M.W., Lee, K., Toutanova, K.: BERT: pre-training of deep bidirectional transformers for language understanding. In: Proceedings of the 2019 Conference of the North American Chapter of the Association for Computational Linguistics: Human Language Technologies, Volume 1 (Long and Short Papers), pp. 4171–4186. Association for Computational Linguistics, Minneapolis, Minnesota, June 2019
8. Domhan, T., Denkowski, M., Vilar, D., Niu, X., Hieber, F., Heafield, K.: The sockeye 2 neural machine translation toolkit at AMTA 2020. In: Proceedings of the 14th Conference of the Association for Machine Translation in the Americas (Volume 1: Research Track), pp. 110–115. Association for Machine Translation in the Americas, Virtual, October 2020
9. Guo, J., Fan, Y., Ai, Q., Croft, W.B.: A deep relevance matching model for ad-hoc retrieval. In: Proceedings of the 25th ACM International on Conference on Information and Knowledge Management, pp. 55–64 (2016)
10. Hui, K., Yates, A., Berberich, K., de Melo, G.: PACRR: a position-aware neural IR model for relevance matching. In: Proceedings of the 2017 Conference on Empirical Methods in Natural Language Processing, pp. 1049–1058. Association for Computational Linguistics, Copenhagen, Denmark, September 2017
11. Jiang, Z., El-Jaroudi, A., Hartmann, W., Karakos, D., Zhao, L.: Cross-lingual information retrieval with BERT. arXiv preprint arXiv:2004.13005, April 2020
12. Johnson, J., Douze, M., Jégou, H.: Billion-scale similarity search with GPUs. arXiv preprint arXiv:1702.08734 (2017)
13. Khattab, O., Zaharia, M.: ColBERT: efficient and effective passage search via contextualized late interaction over BERT. In: Proceedings of the 43rd International ACM SIGIR Conference on Research and Development in Information Retrieval, pp. 39–48. Association for Computing Machinery, New York, NY, USA, July 2020
14. Lawrie, D., Mayfield, J., Oard, D.W., Yang, E.: HC4: a new suite of test collections for ad hoc CLIR. In: Proceedings of the 44th European Conference on Information Retrieval (2021)

15. Lin, J., Nogueira, R., Yates, A.: Pretrained transformers for text ranking: Bert and beyond. Synt. Lectur. Hum. Lang. Technol. **14**(4), 1–325 (2021)
16. Liu, Y., et al.: Roberta: a robustly optimized BERT pretraining approach. arXiv preprint arXiv:1907.11692 (2019)
17. MacAvaney, S., Soldaini, L., Goharian, N.: Teaching a new dog old tricks: resurrecting multilingual retrieval using zero-shot learning. In: Jose, J.M., Yilmaz, E., Magalhães, J., Castells, P., Ferro, N., Silva, M.J., Martins, F. (eds.) ECIR 2020. LNCS, vol. 12036, pp. 246–254. Springer, Cham (2020). https://doi.org/10.1007/978-3-030-45442-5_31
18. MacAvaney, S., Yates, A., Cohan, A., Goharian, N.: Cedr: contextualized embeddings for document ranking. In: Proceedings of the 42nd International ACM SIGIR Conference on Research and Development in Information Retrieval, pp. 1101–1104 (2019)
19. McNamee, P., Mayfield, J.: Comparing cross-language query expansion techniques by degrading translation resources. In: Proceedings of the 25th Annual International ACM SIGIR Conference on Research and Development in Information Retrieval, pp. 159–166 (2002)
20. Ounis, I., Amati, G., Plachouras, V., He, B., Macdonald, C., Johnson, D.: Terrier information retrieval platform. In: Losada, D.E., Fernández-Luna, J.M. (eds.) ECIR 2005. LNCS, vol. 3408, pp. 517–519. Springer, Heidelberg (2005). https://doi.org/10.1007/978-3-540-31865-1_37
21. Papineni, K., Roukos, S., Ward, T., Zhu, W.J.: BLEU: a method for automatic evaluation of machine translation. In: Proceedings of the 40th Annual Meeting of the Association for Computational Linguistics, pp. 311–318. Association for Computational Linguistics, Philadelphia, Pennsylvania, USA, July 2002
22. Peters, C., Braschler, M.: European research letter: cross-language system evaluation: the CLEF campaigns. J. Am. Soc. Inform. Sci. Technol. **52**(12), 1067–1072 (2001)
23. Peters, M.E., et al.: Deep contextualized word representations. In: Proceedings of NAACL-HLT, pp. 2227–2237 (2018)
24. Ponte, J.M., Croft, W.B.: A language modeling approach to information retrieval. In: Proceedings of the 21st Annual International ACM SIGIR Conference on Research and Development in Information Retrieval, pp. 275–281 (1998)
25. Post, M.: A call for clarity in reporting BLEU scores. In: Proceedings of the Third Conference on Machine Translation: Research Papers, pp. 186–191. Association for Computational Linguistics, Brussels, Belgium, October 2018
26. Robertson, S.E., Walker, S., Jones, S., et al.: Okapi at TREC-3. In: Overview of the Third Text REtrieval Conference (TREC-3) (1995)
27. Santhanam, K., Khattab, O., Saad-Falcon, J., Potts, C., Zaharia, M.: Colbertv2: effective and efficient retrieval via lightweight late interaction. arXiv preprint arXiv:2112.01488 (2021)
28. Shi, P., Lin, J.: Cross-lingual relevance transfer for document retrieval. arXiv preprint arXiv:1911.02989 (2019)
29. Shi, P., Zhang, R., Bai, H., Lin, J.: Cross-lingual training with dense retrieval for document retrieval. arXiv preprint arXiv:2109.01628 (2021)
30. Vaswani, A., et al.: Attention is all you need. arXiv preprint arXiv:1706.03762 (2017)
31. Wang, X., Macdonald, C., Tonellotto, N., Ounis, I.: Pseudo-relevance feedback for multiple representation dense retrieval. arXiv preprint arXiv:2106.11251 (2021)

32. Wicks, R., Post, M.: A unified approach to sentence segmentation of punctuated text in many languages. In: Proceedings of the 59th Annual Meeting of the Association for Computational Linguistics and the 11th International Joint Conference on Natural Language Processing (Volume 1: Long Papers), pp. 3995–4007. Association for Computational Linguistics, Online, August 2021

33. Wolf, T., et al.: Transformers: State-of-the-art natural language processing. In: Proceedings of the 2020 Conference on Empirical Methods in Natural Language Processing: System Demonstrations, pp. 38–45. Association for Computational Linguistics, Online (Oct 2020)

34. Yang, P., Fang, H., Lin, J.: Anserini: enabling the use of Lucene for information retrieval research. In: Proceedings of the 40th International ACM SIGIR Conference on Research and Development in Information Retrieval, pp. 1253–1256. SIGIR 2017, Association for Computing Machinery, New York, NY, USA, August 2017

35. Yang, W., Zhang, H., Lin, J.: Simple applications of BERT for ad hoc document retrieval. arXiv preprint arXiv:1903.10972 (2019)

36. Yu, P., Allan, J.: A study of neural matching models for cross-lingual IR. In: Proceedings of the 43rd International ACM SIGIR Conference on Research and Development in Information Retrieval, pp. 1637–1640. Association for Computing Machinery, New York, NY, USA (2020)

37. Zhang, R., et al.: Improving low-resource cross-lingual document retrieval by reranking with deep bilingual representations. arXiv preprint arXiv:1906.03492 (2019)

38. Zhao, L., Zbib, R., Jiang, Z., Karakos, D., Huang, Z.: Weakly supervised attentional model for low resource ad-hoc cross-lingual information retrieval. In: Proceedings of the 2nd Workshop on Deep Learning Approaches for Low-Resource NLP (DeepLo 2019), pp. 259–264. Association for Computational Linguistics, Hong Kong, China, November 2019

Evaluating the Robustness of Retrieval Pipelines with Query Variation Generators

Gustavo Penha$^{(\boxtimes)}$, Arthur Câmara, and Claudia Hauff

TU Delft, Delft, Netherlands
{g.penha-1,a.barbosacamara,c.hauff}@tudelft.nl

Abstract. Heavily pre-trained transformers for language modeling, such as BERT, have shown to be remarkably effective for Information Retrieval (IR) tasks, typically applied to re-rank the results of a first-stage retrieval model. IR benchmarks evaluate the effectiveness of retrieval pipelines based on the premise that a single query is used to instantiate the underlying information need. However, previous research has shown that (I) queries generated by users for a fixed information need are extremely variable and, in particular, (II) neural models are brittle and often make mistakes when tested with modified inputs. Motivated by those observations we aim to answer the following question: *how robust are retrieval pipelines with respect to different variations in queries that do not change the queries' semantics?* In order to obtain queries that are representative of users' querying variability, we first created a taxonomy based on the manual annotation of transformations occurring in a dataset (UQV100) of user-created query variations. For each syntax-changing category of our taxonomy, we employed different automatic methods that when applied to a query generate a query variation. Our experimental results across two datasets for two IR tasks reveal that retrieval pipelines are not robust to these query variations, with effectiveness drops of $\approx 20\%$ on average. The code and datasets are available at https://github.com/Guzpenha/query_variation_generators.

1 Introduction

Heavily pre-trained transformers for language modeling such as BERT [17] have been shown to be remarkably effective for a wide range of IR tasks [40,43,55]. Commonly, IR benchmarks organized as part of TREC or other evaluation campaigns, evaluate the effectiveness of ranking models—neural or otherwise—based on small sets of topics and their corresponding relevance judgments. Importantly, each topic is typically represented by a single query[1]. However, previous research has shown that queries created by users given a fixed information need may vary widely [6,60]. In the UQV100 [5] dataset for instance, crowd workers on average

---

[1] While TREC topics usually consist of three parts (title, description and narrative), commonly only the TREC topic title is considered as query.

© The Author(s), under exclusive license to Springer Nature Switzerland AG 2022
M. Hagen et al. (Eds.): ECIR 2022, LNCS 13185, pp. 397–412, 2022.
https://doi.org/10.1007/978-3-030-99736-6_27

created 57.7 unique queries for a given information need as instantiated via a backstory.

We thus argue that it is necessary to investigate the robustness of retrieval pipelines in light of *query variations* (i.e., different expressions of the same information need) that are *likely to occur in practice*. That different query variations lead to vastly different ranking qualities is anecdotally shown in Table 1 for a vanilla BERT model for ranking [40]. If, for example, the word order of the original query from TREC-DL-2019 *right pelvic pain causes* is changed to *causes pelvic pain right*, the retrieval effectiveness of the resulting ranking drops by 46%. Similarly, paraphrasing *define visceral* to *what is visceral* reduces the retrieval effectiveness by 38%.

Table 1. Examples of BERT effectiveness drops (nDCG@10 Δ) when we replace the original query from TREC-DL-2019 by an automatic (except for the first two lines that were produced manually) query variation. We focus here on transformations that change the query syntax, but not its semantics.

Original Query	Query Variation	nDCG@10 Δ
popular food in switzerland	popular food in zurich *gen./specialization*	
cost of interior concrete flooring	concrete flooring finishing *aspect change*	
what is theraderm used for	what is **thrraderm** used for *misspelling*	-1.00 (-100%)
anthropological definition of environment	anthropological definition ~~of~~ environment *naturality*	-0.15 (-26%)
right pelvic pain causes	**causes** pelvic pain **right** *ordering*	-0.18 (-46%)
define visceral	**what is** visceral *paraphrasing*	-0.26 (-38%)

In our work, we quantify the extent to which different retrieval models are susceptible to different types of query variations as measured by their drop in retrieval effectiveness. In contrast to prior works that either analyze behaviour of models when faced with modifications to the documents [31], analyze models through the lens of IR axioms [12,47] or analyze NLP models via general natural language text adversarial examples [21,48], we instantiate our *query variations* based on user-created data. Concretely, we manually label a large fraction of UQV100 queries[2] and extract six types of frequently occurring query transitions: *gen./specialization*, *aspect change*, *misspelling*, *naturality*, *ordering* and *paraphrasing*—an example of each is shown in Table 1. The last four of these categories change the query syntax but not its semantics. For each of the four syntax-changing categories, we develop automated approaches that enable us to generate query variations of each category for any input query. With these

---

[2] To our knowledge, UQV100 is the only publicly available dataset that contains a large number of query variations for a set of information needs.

query variation generators in place, we conduct extensive empirical work on the TREC-DL-2019 [15] and ANTIQUE [23] datasets to answer the question: *Are retrieval pipelines robust to different variations in queries that do not change its semantics?* To this end we consider seven ranking approaches: two lexical models (BM25 [49] and RM3 [1]), two neural re-ranking approaches that do not make use of transformers (KNRM [54] and cKNRM [16]) and three transformer-based re-ranking approaches (EPIC [32], BERT [40] and T5 [41]).

We find that the four types of syntax-changing query variations differ in the extent to which they degrade retrieval effectiveness: *misspellings* have the largest effect (with an average drop of 0.25 nDCG@10 points across seven retrieval models for TREC-DL-2019) while the *word ordering* has the least effect (with an average drop of nDCG@10 smaller than 0.01 for TREC-DL-2019).

Our work indicates that more research is required to improve the robustness of retrieval pipelines. Evaluation benchmarks should aim to have multiple query variations for the same information need; we provide here a number of methods to automatically generate such query variations for any dataset.

2   Related Work

Query Variation. A number of studies have argued that evaluation in IR tasks should take into account multiple instantiations of the same information need, i.e. query variations, due to their impact on the effectiveness of ranking models [4–7, 11, 36, 50, 60]. Zuccon et al. [60] proposed a mean-variance framework to explicitly take into account query variations when comparing different IR systems. Bailey et al. [6] argued that a model should be consistent to different query variations, and proposed a measure of consistency which gives additional information to effectiveness measurements.

Besides a better evaluation of models, query variations can also be employed to improve the overall effectiveness of ranking models, for instance by combining the different rankings obtained from them [8, 10] or by modelling relevance of multiple query variations [28]. They have also shown to been helpful for the problem of query performance prediction [57].

Different methods to automatically generate query variations have been proposed. Benham et al. [9] proposed to obtain query expansions through a relevance model which is built by issuing the original query against an external corpora and expanding it with additional terms from the set of external feedback documents. Lu et al. [28] employed a query-url click graph and generated query variations automatically using a two-step backward walk process. Chakraborty et al. [13] generated query variations based on an external knowledge base with a prior term distribution and by building a relevance model in an iterative manner.

Our work differs from previous work in the following ways: (I) our methods do not require access to external corpora, a relevance model or a query-url click graph; (II) we are not concerned with generating queries with the sole purpose of improving effectiveness, but in generating queries that are likely to occur in practice; and (III) each of our generator methods follows a category of our taxonomy of query variations which allows us to *diagnose* ranking models' effectiveness by analyzing what types of variations are more detrimental to what ranking models.

Model Understanding. The success of pre-trained transformer-based language models such as BERT [17] and T5 [46] on several IR benchmarks—a comprehensive account of the effectiveness gains can be found in [27]—has lead to research on understanding their behaviour and the reasons behind their significant gains in ranking effectiveness [12,31,42,45,58].

Câmara and Hauff [12] showed that BERT does not adhere to IR axioms, i.e., heuristics that a reasonable IR model should fulfill, through the use of diagnostic datasets. MacAvaney et al. [31] expanded on the axiomatic diagnostic datasets [47] with ABNIRML, a framework to understand the behaviour of neural ranking models using three different strategies: measure and match (controlling certain measurements such as term frequency and changing another), manipulation of the documents' text (e.g., by shuffling words) and through the transfer of Natural Language Processing (NLP) datasets (e.g., by comparing documents that are more/less formal). We expand on [31] by proposing textual manipulations—unlike previous methods we are inspired by *user-created* variations—to the queries instead of the documents and examine the robustness in terms of effectiveness of ranking models to such manipulations.

A different direction of research in NLP has challenged how well current evaluation schemes are actually evaluating the desired capabilities of the models through the use of held-out test sets. For example, Gardner et al. [21] proposed the manual creation of contrast sets—small perturbations that preserve artifacts but change the true label—in order to evaluate the models' decision boundaries for different NLP tasks. They showed that the model effectiveness on such contrast sets can be up to 25% lower than on the original test sets. Inspired by behavioral testing, i.e. validating input output behaviour without knowledge about internal structure, from software engineering tests, Ribeiro et al. [48] proposed to test NLP models with three different types of tests: minimum functionality tests (simple examples where the model should not fail), label (such as positive, negative and neutral in sentiment analysis) invariant changes to the input, and modifications to the input with known outcomes. With such tests at hand they were able to find actionable failures in different commercial NLP models that had already been extensively tested. It has also been shown that neural models developed for different NLP tasks can be tricked by adversarial examples [2,20,22], i.e. examples with perturbations indiscernible by humans which are misclassified by the model. In terms of query modifications, [53,59] found typos to be detrimental to the effectiveness of neural rankers. Ma et al. [29] showed that contrastive fine-tuning improves the robustness of ranking models to paraphrased and perturbed queries. Wu et al. [53] analyzed the robustness of neural rankers with respect to three dimensions: difficult queries from similar distributions, out-of-domain cases, and defense against adversarial operations. Our work differs from the adversarial line of research by evaluating the robustness of models to query modifications that could be generated by humans, i.e. transformations that naturally occur, and not modifications optimized to trick neural models.

3   Automatic Query Variations

We now first describe how we arrived at our query variation categories in a data-driven manner by annotating a large set of user-created query variations from UQV100. We end up with six categories: four that change the syntax (but not the semantics) and two that change the semantics. **In our work, we focus on the four syntax-changing categories.** We subsequently describe our methods to automatically generate the four types of syntax-changing query variations.

3.1   UQV Taxonomy

In order to better understand how queries differ when we compare different query variations for the same information need, we resort to analyzing variations from the UQV100 dataset. UQV100 contains query variations for 100 (sub)-topics from the TREC 2013 and 2014 web tracks, written by crowd workers who received a "backstory" for each topic as a starting point. On average, UQV100 contains 57.7 spelling corrected (corrected by the UQV100 authors using the spelling service of the Bing search engine) query variations per topic. We consider a query variation pair $\{q_i, q_j\}$ to be a set of two queries q_i and q_j that were provided in UQV100 for the same backstory. In total, 365K such pairs exist; Table 2 (4th column) contains a number of $\{q_i, q_j\}$ examples. We sampled 100 query variation pairs for manual annotation. Three authors of this paper (the "annotators") performed an open card sort [52]. The annotators independently sorted the query variation pairs into different piles and named them, each representing a transformation T that can be applied to q_i and then leads to q_j, i.e. $T(q_i) = q_j$. Multiple transformations might be applied to q_i in order to yield q_j, e.g. $T_2(T_1(q_i)) = q_j$.

After the independent sorting step, the different piles were discussed and merged where necessary, which yielded five categories of transformations. Since the UQV100 data used had already been spelling-corrected by its authors, we added the category *misspellings*. The resulting taxonomy can be found in Table 2. It contains a concrete definition and examples for each of our—in total—six categories: (I) *generalization or specialization*, (II) *aspect change*, (III) *misspelling*, (IV) *naturality*, (V) *word ordering* and (VI) *paraphrasing*. We observed two broad types of transformations: transformations that change the semantics of the query and transformations that do not change the semantics. The *gen./specialization* and *aspect change* transformations fall into the former type, whereas all other categories fall into the latter. We highlight here that unlike previous categorizations that describe how users revise queries in e-commerce [3,24], how to generate better queries to substitute the original query [26], how users reformulate queries in a session [25], we study here how to categorize *query variations* for the same information need which is a related but different problem.

Having arrived at our six categories, our annotators then labeled an additional set of 550 $\{q_i, q_j\}$ randomly sampled pairs from UQV100 in order to determine the distribution of these categories in UQV100. Each $\{q_i, q_j\}$ was labelled

Table 2. Taxonomy of query variations derived from a sample of the UQV100 dataset. Last column is the count of each query variation found on UQV100 based on manual annotation of tuples of queries for the same information need. Categories in grey change the semantics. * typos were already fixed for the UQV100 pairs.

Category	Definition	$\{q_i, q_j\}$ from UQV100		Count
Gen./specialization	Generalizes or specializes within the same information need.	american civil war	↔ number of battles in south carolina during civil war	172
Aspect change	Moves between related but different aspects within the same information need.	what types of spiders can bite you while gardening	↔ signs of spider bite	111
Misspelling	Adds or removes spelling errors.	raspberry pi	↔ raspeberry pi	*
Naturality	Moves between keyword queries and natural language queries.	how does zinc relate to wilson's disease	↔ zinc wilson's disease	118
Ordering	Changes the order of words	carotid cavernous fistula treatment.	↔ treatment carotid cavernous fistula	37
Paraphrasing	Rephrases the query by modifying one or more words.	cures for a bald spot	↔ cures for baldness	215

as belonging to one (or more) of the five categories (with the exception of *misspelling* which, as already stated, had already been corrected by the UQV100 authors). In order to determine the inter-annotator agreement, 25 $\{q_i, q_j\}$ pairs were labelled by all three annotators, and 175 pairs were each labelled by a single annotator. The inter-annotator agreement [14] was moderate (Cohen's $\kappa = 0.42$); the disagreements were highest for the *naturality* and *paraphrasing* categories. We found that a total of 56 $\{q_i, q_j\}$ pairs had more than one category assigned to it[3]. The resulting distribution is shown in Table 2 (right-most column); the categories of query variations that change the query without changing its semantics account for 57% of all the transformations. In contrast, 43% of query variations are semantic changes. Among the syntax-changing categories, we found *naturality* to be the most common with 33% of all transformations falling into this category. Having observed that query variations change the syntax, but not the semantics for the majority of cases, **we focus in the remainder of our work on syntax-changing query variations**. We leave the exploration of query variation generators for *gen./specialization* and *aspect change* as future work.

3.2 Query Generators

For each of the four syntax-changing categories, we explored different methods that generate query variations of the specified category. After an initial

---

[3] For example, the pair { *"what is doctor zhivago all about"*, *"dr zhivago synopsis"*} had both *paraphrasing* and *naturality* labels, as it goes from a natural language question to a keyword-base question and also paraphrases *"doctor [...] all about"* to *"dr [...] synopsis"*.

exploration of different query generator methods for each category, and filtering approaches that did not generate valid variations for the category and approaches that have high correlation with each other, we employed a total of ten different methods. These methods are listed in Table 3, each with an example transformation. We explain each one in more detail in this section. A method M_C receives as input a query q and outputs a query variation $\hat{q}$ for the category C: $M_C(q) = \hat{q}$.

Table 3. Example of applying each query generation method M for the query '*what is durable medical equipment consist of*' from TREC-DL-2019. Rightmost columns indicate the total percentage of valid queries by automatic query variation method based on manual annotation of queries from the test sets of TREC-DL-2019 and ANTIQUE.

C	Method name	M('*what is durable medical equipment consist of*')	TREC	ANT
Misspelling	NeighbCharSwap	*what is durable **medical** equipment consist of*	100.00%	99.50%
	RandomCharSub	*what is durable **medical** equipment consist of*	97.67%	91.00%
	QWERTYCharSub	*what is durable medical equipment **consist** of*	97.67%	98.50%
Naturality	RmvStopWords	*~~what is~~ durable medical equipment consist **~~of~~***	86.05%	99.50%
	T5DescToTitle	*~~what is~~ durable medical equipment ~~consist of~~*	81.40%	68.00%
Ordering	RandOrderSwap	***medical** is durable **what** equipment consist of*	100.00%	100.00%
Paraphrasing	BackTransl	*what is **sustainable** medical equipment ~~consist of~~*	53.49%	46.50%
	T5QQP	*what is durable medical equipment ~~consist of~~*	60.47%	52.50%
	WEmbedSynSwap	*what is durable **medicinal** equipment consist of*	62.79%	62.00%
	WNetSynSwap	*what is **long lasting** medical equipment consist of*	37.21%	35.50%

While most of the methods can generate multiple variations for a single input query (for example by replacing different words of the same query by synonyms or by including several spelling mistakes), for the experiments in the paper we resort to using a single query variation per method which already yields enough data for analysis (see §4). Inspired by adversarial examples, we aim to make minimal perturbations to the input text when possible, e.g. replace only one word by a synonym, thus increasing the chances of obtaining valid variations.

Misspelling. The three methods in this category add one spelling error to the query; the query term an error is introduced in is chosen uniformly at random.

`NeighbCharSwap` Swaps two neighbouring characters from a random query term (excluding stopwords[4]).

`RandomCharSub` Replaces a random character from a random query term (excluding stopwords) with a randomly chosen new ASCII character.

`QWERTYCharSub` Replaces a random character of a random query term (excluding stopwords) with another character from the QWERTY keyboard such that only characters in close proximity are chosen, replicating errors that come from typing too quickly.

Naturality. The two methods in this category transform natural language queries into keyword queries.

`RmvStopWords` Removes all stopwords from the query.

`T5DescToTitle` Applies an encoder-decoder transformer model (here we employ T5 [46]) that we fine-tuned on the task of generating the title of a TREC topic based on the TREC topic description (an example title and description tuple from `trec-robust04` is *'Evidence that rap music has a negative effect on young people.'* → *'Rap and Crime'*). We collect pairs of title and description from eleven datasets available through the IR datasets library [33].[5] Overall, we fine-tuned our model on 1322 such description/title tuples.

Ordering. In this category, we employ only one method to shuffle words as done by previous research on the order of words [31,44].

`RandOrderSwap` Randomly swap two words of the query.

Paraphrasing. The four methods in this category change one or more query terms in the process of paraphrasing.

`BackTransl` Applies a translation method to the query to a pivot language, i.e. an auxiliary language, and from the pivot language back to the original language of the query (in our case: English). In our experiments we employ the M2M100 [18] model, a multilingual model that can translate between any pair of 100 languages, and we use *'German'* as the pivot language, which yielded better results—shown by manual inspection of the generated variations—than the other two languages for which the model had the most data for training (*'Spanish'* and *'French'*). This technique has been used before as a way to generate paraphrases [19,35].

---

[4] We use the NLTK english stopwords list for all the methods; it is available at https://www.nltk.org/.

[5] Concretely, we made use of trec-robust04, trec-tb-2004, aquaint/trec-robust-2005, gov/trec-web-2002, ntcir-www-2, ntcir-www-3, trec-misinfo-2019, cord19/trec-covid, dd-trec-2015, dd-trec-2016 and dd-trec-2017.

T5QQP Applies an encoder-decoder transformer model (T5 [46]) that was fine-tuned on the task of generating a paraphrase question from the original question[6]. The model employs the Quora Question Pairs[7] dataset for fine-tuning, which has 400k pairs of questions like the following: '*How do you start a bakery?*' → '*How can one start a bakery business?*'. We also tested T5 models fine-tuned for PAWS [56] and the combination of PAWS and Quora Question Pairs, but the manual inspection of the generated queries revealed that T5 fine-tuned for Quora Question Pairs generated a higher number of valid variations.

WEmbedSynSwap Replaces a non-stop word by a synonym as defined by the nearest neighbour word in the embedding space according to a counter fitted-Glove embedding which yields better synonyms than standard Glove embeddings [38].

WNetSynSwap Replaces a non-stop word by a the first synonym found on Word-Net[8]. If there are no words with valid synonyms it will not output a variation.

4   Experimental Setup

Datasets. We consider the following datasets: TREC-DL-2019 [15] for the passage retrieval task and ANTIQUE [23] for the non-factoid question answering task. They have 367,013/5,193/43 and 2,426/-/200 instances respectively for training, validation and test. The queries from TREC-DL-2019 are smaller on average: 5.51 terms vs 10.51 from ANTIQUE. For each of the test set queries, we generate one query variation by each generator method, and we use only the valid query variations in our experiments (according to manual annotation), leading to 334 and 1,706 valid query variations for TREC-DL-2019 and ANTIQUE.

Ranking Models. We use different ranking models that range from traditional lexical models, such as BM25, to neural ranking models, such as KNRM and neural ranking models that employ transformer-based language models, such as BERT. For all of our experiments, we apply BM25 as a first stage retriever and re-rank the top 100 results with the neural ranking models, which is an established and efficient approach [27].

For **BM25** [49] and **RM3** [1] we resort to the default hyperparameters and implementation provided by the PyTerrier toolkit [34]. We trained the kernel-based ranking models **KNRM** [54] and **cKNRM** [16] on the training sets of TREC-DL-2019 and ANTIQUE using default settings from the OpenNIR [30] implementation. For the BERT-based methods **EPIC** [32], an efficiency focused model that encodes query and documents separately, and **BERT** [40], also known as monoBERT, which concatenates query and document and makes predictions based on the [CLS] token representation, we fine-tune the `bert-base-uncased` model for the train datasets. For **T5** [46] we use

---

[6] As available here https://huggingface.co/ramsrigouthamg/t5_paraphraser.

[7] https://www.kaggle.com/c/quora-question-pairs.

[8] https://wordnet.princeton.edu/.

the monoT5 [41] implementation of the PyTerrier T5 plugin[9] which has the pre-trained weights for MSMarco [39] by the authors of monoT5.

Query Generators Implementation. As for our methods of generating query variations, for `T5DescToTitle` and `T5QQP` we rely on pre-trained T5 models (`t5-base`) and we fine-tune them using the Huggingface transformers library [51]. For `BackTransl` we use the `facebook/m2m100_418M` pre-trained model from the transformers library[10]. For all other methods, we use the implementations from the TextAttack [37] library.

Quality of Query Generators. Given the automatic nature of the methods we introduced, we need to evaluate their quality. To this end, we consider two properties of the generated queries: (I) $\hat{q}$ maintains the same semantics as q, and (II) the syntax difference between q and $\hat{q}$ can be attributed to the category. All pairs of q and $\hat{q} = M(q)$ from the test sets of TREC-DL-2019 (43 queries) and ANTIQUE (200 queries) for each of the 10 automatic variation methods went through the following process. First, we automatically set the variations from *misspelling*[11] and *ordering* as valid, since they are rule based transformations to the input. Then all transformations that generate a variation that is identical to the input query ($\hat{q} = M(q) = q$) was automatically set to invalid. Three authors then annotated independently the remaining 1,371 pairs of $\{q, \hat{q}\}$ for the two mentioned properties (binary labels). The percentage of queries that are valid (i.e. they have both desired properties) are displayed in the right-most columns of Table 3 for the 10 automatic variation methods used in the paper and all 2,430 combinations of $\{q, \hat{q}\}$. We find the methods in the *paraphrasing* category to yield the largest percentage of invalid query variations: fewer than 38% of query variations generated via `WNetSynSwap` are valid. A manual inspection of the invalid queries reveal the following insights: (I) `T5DescToTitle` at times removes query terms that are important for the query and thus change its semantics (e.g. '*if i had a bad breath what should i do*' → '*if i had a*'), (II) `BackTransl` and `T5QQP` methods can generate an identical copy of the input query which was automatically labelled as invalid and (III) transformations that replace words by their presumed synonyms (`WEmbedSynSwap` and `WNetSynSwap`) at times adds words that are not in fact synonymous in the query context (e.g. '*what is dark energy*' → '*what is blackness energy*' and '*what is a active margin*' → '*what is a active border*').

To evaluate the robustness of the ranking models, we resort to using only the valid queries as defined by the manual annotations. Overall, we have thus 2,040 valid queries for datasets TREC-DL-2019 and ANTIQUE that we employ in the experiments that follow.

---

[9] https://github.com/terrierteam/pyterrier_t5.

[10] https://huggingface.co/facebook/m2m100_418M.

[11] *misspelling* methods can generate invalid queries when all words of the query are stop-words (e.g. '*how is it being you*' from ANTIQUE would generate the same query as output since there is no non stop-words to modify).

5   Results

To explore the robustness of our three types of ranking models we compare the effectiveness of our models when we replace the original query with the respective query variation. The results of this experiment are displayed in Table 4 for both the TREC-DL-2019 and ANTIQUE datasets. Each row shows the effectiveness of the ranking models (columns) when using the queries obtained from each automatic query variation method. The last column ($\#Q$) displays the number of valid queries generated by each query variation method; the invalid queries are replaced with the original ones[12].

The results show that for most of the query variations and ranker combinations we observe a statistical significant effectiveness drop (49 out of 70 times for TREC-DL-2019 and 54 out of 70 times for ANTIQUE), and that no set of query variations improves statistically over using the original query. If we look into the percentage of overall effectiveness decreases considering only the valid queries, we see on average that the models become 20.62% and 19.21% less effective for TREC-DL-2019 and ANTIQUE respectively. **This answers our research question indicating that retrieval pipelines are not robust to query variations.** This confirms previous empirical evidence that query variations induce a big variability effect on different IR systems [6,60]. We show that even with newer large-scale collections such as TREC-DL-2019, retrieval pipelines are not robust to such variations.

There are several potential explanations for this drop in effectiveness besides the lack of robustness of neural rankers. The first-stage ranker may be the point of failure, being unable to retrieve sufficiently many relevant documents for the neural rankers to re-rank. It is also possible that the query variations lead to unjudged documents being ranked highly by the retrieval pipelines, which in the standard retrieval evaluation setup are considered non-relevant. We now present two experiments to show that these alternative explanations are not the cause in drop of retrieval effectiveness.

Let's focus first on the first-stage ranker. We first calculated the average drop in effectiveness when we increase the re-ranking threshold. While the number of documents in the re-ranking set increases[13], neural models still struggle, e.g. for BERT the nDCG@10 decreases on average by 40%, 34% and 31%[14]. This indicates that even if we increase the number of relevant documents to be re-ranked, neural rankers still fail when faced with query variations.

To further isolate the effect of the first-stage retrieval module, we analyzed whether the effectiveness of the pipelines would not degrade in case the first-stage retrieval was performed on the original query. In this experiment only the re-ranker models use the query variations and we check whether the effectiveness drops persist. The results reveal that there are still statistically significant

---

[12] While rows are directly comparable, methods with fewer valid queries are a lower bound of the potential decreases in effectiveness.

[13] BM25 has R@10, R@100 and R@1000 of 0.06, 0.25 and 0.48 for *misspelling*.

[14] Similar results are obtained for other neural rankers.

Table 4. Effectiveness (nDCG@10) of different methods for TREC-DL-2019 and ANTIQUE when faced with different query variations. Bold indicates the highest values observed for each model and ↓/↑ subscripts indicate statistically significant losses/improvements, using two-sided paired Student's T-Test at 95% confidence interval with Bonferroni correction when compared against the model with original queries. #Q is the number of valid query variations (invalid query variations are replaced by the original query).

| | | TREC-DL-2019 | | | | | | | |
Category	Variation	BM25	RM3	KNRM	cKNRM	EPIC	BERT	T5	#Q
–	original query	**0.480**	**0.516**	**0.502**	**0.493**	**0.624**	**0.645**	0.700	43
Misspelling	NeighbCharSwap	0.275$^{\downarrow}$	0.275$^{\downarrow}$	0.316$^{\downarrow}$	0.309$^{\downarrow}$	0.389$^{\downarrow}$	0.416$^{\downarrow}$	0.495$^{\downarrow}$	43
	RandomCharSub	0.231$^{\downarrow}$	0.233$^{\downarrow}$	0.236$^{\downarrow}$	0.226$^{\downarrow}$	0.295$^{\downarrow}$	0.328$^{\downarrow}$	0.396$^{\downarrow}$	42
	QWERTYCharSub	0.244$^{\downarrow}$	0.250$^{\downarrow}$	0.267$^{\downarrow}$	0.297$^{\downarrow}$	0.351$^{\downarrow}$	0.387$^{\downarrow}$	0.446$^{\downarrow}$	42
Naturality	RmvStopWords	0.478	0.511	0.484	0.476	0.621	0.639	0.687	37
	T5DescToTitle	0.421	0.434$^{\downarrow}$	0.392	0.393	0.506$^{\downarrow}$	0.536$^{\downarrow}$	0.571$^{\downarrow}$	35
Ordering	RandOrderSwap	0.480	0.516	0.502	0.471	0.623	0.635	0.697	43
Paraphrasing	BackTransl	0.396	0.420$^{\downarrow}$	0.393	0.361$^{\downarrow}$	0.530	0.547$^{\downarrow}$	0.606	23
	T5QQP	0.472	0.504	0.454	0.461	0.605	0.640	**0.705**	26
	WEmbedSynSwap	0.353$^{\downarrow}$	0.354$^{\downarrow}$	0.382$^{\downarrow}$	0.368$^{\downarrow}$	0.475$^{\downarrow}$	0.472$^{\downarrow}$	0.560$^{\downarrow}$	27
	WNetSynSwap	0.349$^{\downarrow}$	0.365$^{\downarrow}$	0.381$^{\downarrow}$	0.361$^{\downarrow}$	0.449$^{\downarrow}$	0.447$^{\downarrow}$	0.545$^{\downarrow}$	16
		ANTIQUE							
Category	Variation	BM25	RM3	KNRM	cKNRM	EPIC	BERT	T5	#Q
–	original query	**0.229**	**0.217**	0.218	0.207	0.266	**0.421**	**0.334**	200
Misspelling	NeighbCharSwap	0.156$^{\downarrow}$	0.148$^{\downarrow}$	0.159$^{\downarrow}$	0.145$^{\downarrow}$	0.184$^{\downarrow}$	0.287$^{\downarrow}$	0.251$^{\downarrow}$	199
	RandomCharSub	0.162$^{\downarrow}$	0.159$^{\downarrow}$	0.156$^{\downarrow}$	0.148$^{\downarrow}$	0.189$^{\downarrow}$	0.280$^{\downarrow}$	0.249$^{\downarrow}$	182
	QWERTYCharSub	0.161$^{\downarrow}$	0.153$^{\downarrow}$	0.160$^{\downarrow}$	0.155$^{\downarrow}$	0.192$^{\downarrow}$	0.299$^{\downarrow}$	0.266$^{\downarrow}$	197
Naturality	RmvStopWords	0.227	0.216	**0.222**	**0.215**	**0.269**	0.383$^{\downarrow}$	0.320	199
	T5DescToTitle	0.167$^{\downarrow}$	0.165$^{\downarrow}$	0.160$^{\downarrow}$	0.167$^{\downarrow}$	0.200$^{\downarrow}$	0.270$^{\downarrow}$	0.240$^{\downarrow}$	136
Ordering	RandOrderSwap	0.229	0.217	0.218	0.198	0.267	0.413$^{\downarrow}$	0.325$^{\downarrow}$	200
Paraphrasing	BackTransl	0.162$^{\downarrow}$	0.155$^{\downarrow}$	0.160$^{\downarrow}$	0.144$^{\downarrow}$	0.204$^{\downarrow}$	0.305$^{\downarrow}$	0.258$^{\downarrow}$	93
	T5QQP	0.220	0.207	0.210	0.196	0.261	0.393$^{\downarrow}$	0.321	105
	WEmbedSynSwap	0.176$^{\downarrow}$	0.172$^{\downarrow}$	0.190$^{\downarrow}$	0.169$^{\downarrow}$	0.214$^{\downarrow}$	0.325$^{\downarrow}$	0.283$^{\downarrow}$	124
	WNetSynSwap	0.179$^{\downarrow}$	0.175$^{\downarrow}$	0.196$^{\downarrow}$	0.177$^{\downarrow}$	0.212$^{\downarrow}$	0.324$^{\downarrow}$	0.273$^{\downarrow}$	71

effectiveness drops when only the re-ranker models use the query variations, although in smaller magnitude. While the drops in effectiveness of the pipelines when using query variations for the entire pipeline are on average of ≈ 20% in nDCG@10, when using the query variations only for re-ranking they are ≈ 9%. **This indicates that not only the first stage retrieval module is not robust to query variations, but also the neural re-rankers.**

Let's now focus on the matter of unjudged documents. It is possible that we are underestimating the effectiveness of the retrieval pipelines when facing query variations if (I) the number of unjudged documents in the top-10 ranked lists increases and (II) they turn out to be relevant. When counting the amount of judged documents in the top-10 ranked lists of the retrieval pipelines, we find that on average the number actually increases (4.30% for TREC-DL-2019 and 0.36% for ANTIQUE), **meaning that the performance drops of the** retrieval pipelines cannot be attributed to unjudged documents being brought up in the ranking by the query variations.

6 Conclusions

We first described a taxonomy of transformations between two queries for the same information need that characterizes how exactly a query is modified to arrive at one of its variants. We found six different types of transformations, and focused on the ones that do not change the query semantics: *misspelling*, *naturality*, *ordering* and *paraphrasing*. They account for 57% of observed variations in the UQV100 dataset. For each category, we proposed different methods to automatically generate query variations. We studied the quality of the generated query variations, and analyzed how robust retrieval pipelines are to them. Our results on two datasets quantify how much each model is affected by each type of query variation, demonstrating large effectiveness drops of 20% on average when compared to the original queries. As future work, we believe that it is important to study (I) how to automatically generate valid query variation generators for categories that do change the semantics of the query and (II) techniques to improve the robustness of existing ranking pipelines.

Acknowledgements. This research has been supported by NWO projects SearchX (639.022.722) and NWO Aspasia (015.013.027).

References

1. Abdul-Jaleel, N., et al.: UMass at TREC 2004: novelty and hard. Computer Science Department Faculty Publication Series, p. 189 (2004)
2. Alzantot, M., Sharma, Y., Elgohary, A., Ho, B.J., Srivastava, M., Chang, K.W.: Generating natural language adversarial examples. arXiv preprint arXiv:1804.07998 (2018)
3. Amemiya, Y., Manabe, T., Fujita, S., Sakai, T.: How do users revise zero-hit product search queries? In: Hiemstra, D., Moens, M.-F., Mothe, J., Perego, R., Potthast, M., Sebastiani, F. (eds.) ECIR 2021. LNCS, vol. 12657, pp. 185–192. Springer, Cham (2021). https://doi.org/10.1007/978-3-030-72240-1_14
4. Bailey, P., Moffat, A., Scholer, F., Thomas, P.: User variability and IR system evaluation. In: Proceedings of The 38th International ACM SIGIR conference on research and development in Information Retrieval, pp. 625–634 (2015)
5. Bailey, P., Moffat, A., Scholer, F., Thomas, P.: Uqv100: a test collection with query variability. In: Proceedings of the 39th International ACM SIGIR conference on Research and Development in Information Retrieval, pp. 725–728 (2016)
6. Bailey, P., Moffat, A., Scholer, F., Thomas, P.: Retrieval consistency in the presence of query variations. In: Proceedings of the 40th International ACM SIGIR Conference on Research and Development in Information Retrieval, pp. 395–404 (2017)
7. Belkin, N.J., Cool, C., Croft, W.B., Callan, J.P.: The effect multiple query representations on information retrieval system performance. In: Proceedings of the 16th Annual International ACM SIGIR Conference on Research and Development in Information Retrieval, pp. 339–346 (1993)

8. Belkin, N.J., Kantor, P., Fox, E.A., Shaw, J.A.: Combining the evidence of multiple query representations for information retrieval. Inf. Process. Manage. **31**(3), 431–448 (1995)

9. Benham, R., Culpepper, J.S., Gallagher, L., Lu, X., Mackenzie, J.: Towards efficient and effective query variant generation. In: DESIRES, pp. 62–67 (2018)

10. Benham, R., Mackenzie, J., Moffat, A., Culpepper, J.S.: Boosting search performance using query variations. ACM Trans. Inf. Syst. (TOIS) **37**(4), 1–25 (2019)

11. Buckley, C., Walz, J.: The TREC-8 query track. In: TREC (1999)

12. Câmara, A., Hauff, C.: Diagnosing BERT with retrieval heuristics. In: Jose, J.M., et al. (eds.) ECIR 2020. LNCS, vol. 12035, pp. 605–618. Springer, Cham (2020). https://doi.org/10.1007/978-3-030-45439-5_40

13. Chakraborty, A., Ganguly, D., Conlan, O.: Retrievability based document selection for relevance feedback with automatically generated query variants. In: Proceedings of the 29th ACM International Conference on Information & Knowledge Management, pp. 125–134 (2020)

14. Cohen, J.: A coefficient of agreement for nominal scales. Educ. Psychol. Measur. **20**(1), 37–46 (1960)

15. Craswell, N., Mitra, B., Yilmaz, E., Campos, D., Voorhees, E.M.: Overview of the TREC 2019 deep learning track. arXiv preprint arXiv:2003.07820 (2020)

16. Dai, Z., Xiong, C., Callan, J., Liu, Z.: Convolutional neural networks for soft-matching N-grams in ad-hoc search. In: Proceedings of the Eleventh ACM International Conference on Web Search and Data Mining, pp. 126–134 (2018)

17. Devlin, J., Chang, M.W., Lee, K., Toutanova, K.: Bert: pre-training of deep bidirectional transformers for language understanding. In: Proceedings of the 2019 Conference of the North American Chapter of the Association for Computational Linguistics: Human Language Technologies, Volume 1 (Long and Short Papers), pp. 4171–4186 (2019)

18. Fan, A., et al.: Beyond English-centric multilingual machine translation. arXiv preprint arXiv:2010.11125 (2020)

19. Federmann, C., Elachqar, O., Quirk, C.: Multilingual whispers: generating paraphrases with translation. In: Proceedings of the 5th Workshop on Noisy User-generated Text (W-NUT 2019), pp. 17–26 (2019)

20. Gao, J., Lanchantin, J., Soffa, M.L., Qi, Y.: Black-box generation of adversarial text sequences to evade deep learning classifiers. In: 2018 IEEE Security and Privacy Workshops (SPW), pp. 50–56. IEEE (2018)

21. Gardner, M., et al.: Evaluating models' local decision boundaries via contrast sets. In: Proceedings of the 2020 Conference on Empirical Methods in Natural Language Processing: Findings, pp. 1307–1323 (2020)

22. Garg, S., Ramakrishnan, G.: Bae: Bert-based adversarial examples for text classification. arXiv preprint arXiv:2004.01970 (2020)

23. Hashemi, H., Aliannejadi, M., Zamani, H., Croft, W.B.: ANTIQUE: a non-factoid question answering benchmark. In: Jose, J.M., et al. (eds.) ECIR 2020. LNCS, vol. 12036, pp. 166–173. Springer, Cham (2020). https://doi.org/10.1007/978-3-030-45442-5_21

24. Hirsch, S., Guy, I., Nus, A., Dagan, A., Kurland, O.: Query reformulation in e-commerce search. In: Proceedings of the 43rd International ACM SIGIR Conference on Research and Development in Information Retrieval, pp. 1319–1328 (2020)

25. Jansen, B.J., Booth, D.L., Spink, A.: Patterns of query reformulation during web searching. J. Am. Soc. Inf. Sci. Technol. **60**(7), 1358–1371 (2009)

26. Jones, R., Rey, B., Madani, O., Greiner, W.: Generating query substitutions. In: Proceedings of the 15th international conference on World Wide Web, pp. 387–396 (2006)

27. Lin, J., Nogueira, R., Yates, A.: Pretrained transformers for text ranking: Bert and beyond. arXiv preprint arXiv:2010.06467 (2020)

28. Lu, X., Kurland, O., Culpepper, J.S., Craswell, N., Rom, O.: Relevance modeling with multiple query variations. In: Proceedings of the 2019 ACM SIGIR International Conference on Theory of Information Retrieval, pp. 27–34 (2019)

29. Ma, X., Santos, C.N.d., Arnold, A.O.: Contrastive fine-tuning improves robustness for neural rankers. arXiv preprint arXiv:2105.12932 (2021)

30. MacAvaney, S.: OpenNIR: a complete neural ad-hoc ranking pipeline. In: WSDM 2020 (2020)

31. MacAvaney, S., Feldman, S., Goharian, N., Downey, D., Cohan, A.: Abnirml: analyzing the behavior of neural IR models. arXiv preprint arXiv:2011.00696 (2020)

32. MacAvaney, S., Nardini, F.M., Perego, R., Tonellotto, N., Goharian, N., Frieder, O.: Expansion via prediction of importance with contextualization. In: Proceedings of the 43rd International ACM SIGIR Conference on Research and Development in Information Retrieval, pp. 1573–1576 (2020)

33. MacAvaney, S., Yates, A., Feldman, S., Downey, D., Cohan, A., Goharian, N.: Simplified data wrangling with ir_datasets. In: SIGIR (2021)

34. Macdonald, C., Tonellotto, N.: Declarative experimentation in information retrieval using pyterrier. In: Proceedings of ICTIR 2020 (2020)

35. Mallinson, J., Sennrich, R., Lapata, M.: Paraphrasing revisited with neural machine translation. In: Proceedings of the 15th Conference of the European Chapter of the Association for Computational Linguistics: Volume 1, Long Papers, pp. 881–893 (2017)

36. Moffat, A., Scholer, F., Thomas, P., Bailey, P.: Pooled evaluation over query variations: users are as diverse as systems. In: Proceedings of the 24th ACM International on Conference on Information and Knowledge Management, pp. 1759–1762 (2015)

37. Morris, J., Lifland, E., Yoo, J.Y., Grigsby, J., Jin, D., Qi, Y.: Textattack: a framework for adversarial attacks, data augmentation, and adversarial training in NLP. In: Proceedings of the 2020 Conference on Empirical Methods in Natural Language Processing: System Demonstrations, pp. 119–126 (2020)

38. Mrkšić, N., et al.: Counter-fitting word vectors to linguistic constraints. arXiv preprint arXiv:1603.00892 (2016)

39. Nguyen, T., Rosenberg, M., Song, X., Gao, J., Tiwary, S., Majumder, R., Deng, L.: Ms marco: a human generated machine reading comprehension dataset. In: CoCo@ NIPS (2016)

40. Nogueira, R., Cho, K.: Passage re-ranking with bert. arXiv preprint arXiv:1901.04085 (2019)

41. Nogueira, R., Jiang, Z., Lin, J.: Document ranking with a pretrained sequence-to-sequence model. arXiv preprint arXiv:2003.06713 (2020)

42. Padigela, H., Zamani, H., Croft, W.B.: Investigating the successes and failures of Bert for passage re-ranking. arXiv preprint arXiv:1905.01758 (2019)

43. Penha, G., Hauff, C.: Curriculum learning strategies for IR. In: Jose, J.M., Yilmaz, E., Magalhães, J., Castells, P., Ferro, N., Silva, M.J., Martins, F. (eds.) ECIR 2020. LNCS, vol. 12035, pp. 699–713. Springer, Cham (2020). https://doi.org/10.1007/978-3-030-45439-5_46

44. Pham, T.M., Bui, T., Mai, L., Nguyen, A.: Out of order: How important is the sequential order of words in a sentence in natural language understanding tasks? arXiv preprint arXiv:2012.15180 (2020)
45. Qiao, Y., Xiong, C., Liu, Z., Liu, Z.: Understanding the behaviors of Bert in ranking. arXiv preprint arXiv:1904.07531 (2019)
46. Raffel, C., et al.: Exploring the limits of transfer learning with a unified text-to-text transformer. arXiv preprint arXiv:1910.10683 (2019)
47. Rennings, D., Moraes, F., Hauff, C.: An axiomatic approach to diagnosing neural IR models. In: Azzopardi, L., Stein, B., Fuhr, N., Mayr, P., Hauff, C., Hiemstra, D. (eds.) ECIR 2019. LNCS, vol. 11437, pp. 489–503. Springer, Cham (2019). https://doi.org/10.1007/978-3-030-15712-8_32
48. Ribeiro, M.T., Wu, T., Guestrin, C., Singh, S.: Beyond accuracy: behavioral testing of NLP models with checklist. arXiv preprint arXiv:2005.04118 (2020)
49. Robertson, S.E., Walker, S.: Some simple effective approximations to the 2-Poisson model for probabilistic weighted retrieval. In: SIGIR 1994, pp. 232–241. Springer, Cham (1994). https://doi.org/10.1007/978-1-4471-2099-5_24
50. Spark-Jones, K.: Report on the need for and provision of an 'ideal' information retrieval test collection. Computer Laboratory (1975)
51. Wolf, T., et al.: Transformers: state-of-the-art natural language processing. In: Proceedings of the 2020 Conference on Empirical Methods in Natural Language Processing: System Demonstrations, pp. 38–45. Association for Computational Linguistics, Online, October 2020. https://www.aclweb.org/anthology/2020.emnlp-demos.6
52. Wood, J.R., Wood, L.E.: Card sorting: current practices and beyond. J. Usabil. Stud. 4(1), 1–6 (2008)
53. Wu, C., Zhang, R., Guo, J., Fan, Y., Cheng, X.: Are neural ranking models robust? arXiv preprint arXiv:2108.05018 (2021)
54. Xiong, C., Dai, Z., Callan, J., Liu, Z., Power, R.: End-to-end neural ad-hoc ranking with kernel pooling. In: Proceedings of the 40th International ACM SIGIR Conference on Research and Development in Information Retrieval, pp. 55–64 (2017)
55. Yang, W., Zhang, H., Lin, J.: Simple applications of Bert for ad hoc document retrieval. arXiv preprint arXiv:1903.10972 (2019)
56. Yang, Y., Zhang, Y., Tar, C., Baldridge, J.: PAWS-X: a cross-lingual adversarial dataset for paraphrase identification. In: Proceedings of EMNLP (2019)
57. Zendel, O., Shtok, A., Raiber, F., Kurland, O., Culpepper, J.S.: Information needs, queries, and query performance prediction. In: Proceedings of the 42nd International ACM SIGIR Conference on Research and Development in Information Retrieval, pp. 395–404 (2019)
58. Zhan, J., Mao, J., Liu, Y., Zhang, M., Ma, S.: An analysis of Bert in document ranking. In: Proceedings of the 43rd International ACM SIGIR Conference on Research and Development in Information Retrieval, pp. 1941–1944 (2020)
59. Zhuang, S., Zuccon, G.: Dealing with typos for Bert-based passage retrieval and ranking. arXiv preprint arXiv:2108.12139 (2021)
60. Zuccon, G., Palotti, J., Hanbury, A.: Query variations and their effect on comparing information retrieval systems. In: Proceedings of the 25th ACM International on Conference on Information and Knowledge Management, pp. 691–700 (2016)

Exploiting Document-Based Features for Clarification in Conversational Search

Ivan Sekulić[1]([⊠]), Mohammad Aliannejadi[2], and Fabio Crestani[1]

[1] Università della Svizzera Italiana, Lugano, Switzerland
{ivan.sekulic,fabio.crestani}@usi.ch
[2] University of Amsterdam, Amsterdam, Netherlands
m.aliannejadi@uva.nl

Abstract. Asking clarifying questions in order to elicit user's information need is becoming an integral part of modern conversational search systems. Current work heavily relies on pre-collected clarifying questions or large-scale query logs. However, such work is very limited given that collecting all possible clarifying questions on a collection is not feasible. Moreover, modeling clarification based on query reformulation limits a model only to head queries with several occurrences in the log. In this work, we aim to address these limitations by exploiting several document- and ranking-based features to generate clarifying questions. We hypothesise that we can acquire enough evidence about different aspects of a query and extract useful facets to generate clarifying questions about. Specifically, we utilise Part-Of-Speech tagging, entity linking, and topic modelling in order to extract features from the ranked list of documents. Among the extracted features, we then extract potentially useful facets based on three different strategies, aimed to capture feature distinctiveness across documents. We then construct clarifying questions based on the extracted facets that are given to crowdsourcing workers to be evaluated in terms of usefulness. Moreover, our findings show significant improvements (+38% nDCG@3) in document retrieval performance with facet-expanded queries.

Keywords: Conversational search · Facet extraction · Clarifying questions generation

1 Introduction

Users often express their information need using short queries. Whether it is via web interface, smartphone, or conversational assistants, this often leads to incomplete search queries that are open to various interpretations by the search system [11,23]. Search result diversification aimed to address this issue by providing a diverse set of results to cover various aspects of the same query [14]. Given

© The Author(s), under exclusive license to Springer Nature Switzerland AG 2022
M. Hagen et al. (Eds.): ECIR 2022, LNCS 13185, pp. 413–427, 2022.
https://doi.org/10.1007/978-3-030-99736-6_28

that the interface of most conversational systems (e.g., chat bots or voice-based systems) is considerably more limited than web interfaces, these systems can only return a few results to the user which bear additional cost depending on their number [1]. Therefore, more recently, the research on conversational systems has diverged more towards mixed-initiative [27]. A mixed-initiative paradigm of conversational search allows the system to take initiative of the conversation and ask the user clarifying questions [5], or issue other requests. Clarifying the user information need has been shown to be beneficial to both the user and the conversational search system [5, 19, 44], providing a strong motivation for such mixed-initiative systems.

The current research line on search clarification relies either on an available question bank [5] or large-scale query logs [44]. In their proposed offline evaluation methodology, Aliannejadi et al. [4] suggested the use of pre-collected clarifying questions for a limited set of topics. While providing a practical ground for the development and offline evaluation of conversational systems, their assumption of having a set of clarifying questions for all possible topics is not realistic. Zamani et al. [44], on the other hand, proposed a generative model that learns to generate template-based questions, mining a large-scale query logs from Bing.com. Mining query reformulations of various sessions provides crucial insights into how and why a query can be incomplete and what questions can be asked for clarification. However, it is limited to the existence of such large-scale logs. Moreover, it is not applicable to long-tail queries, or queries with very diverse set of reformulations.

Exploiting the top-N retrieved results has proved to be an effective way of approaching several IR tasks, such as pseudo-relevance feedback [10], query performance prediction [15], and query facet extraction [20]. Pseudo-relevance feedback studies lie on the basic assumption of taking the top-k retrieved results in response to a user query as relevant. It has been shown that these documents contain useful domain knowledge that helps the system specialise the user's query and improve the performance [10]. Also, a large body of research on query performance prediction [15, 46] estimates the difficulty of a given query based on the top-k retrieved documents and their predicted relevance scores.

In this work, we aim to experiment and analyse the effectiveness of top-k retrieved documents in generating clarifying questions. As such, we conduct a set of extensive experiments where we systematically study the effectiveness of various feature sets from different aspects, as well as facet extraction techniques. First, we extract three sets of features, namely, part-of-speech (POS) tags, knowledge graph entities, and Latent Dirichlet Allocation (LDA) topics. In the next step, given that each document would have various features, we propose three approaches to analyse the extracted features on a ranking-based manner and extract a few keywords that describe a facet. We call this step facet extraction. We propose the three following techniques for facet extraction: (i) a random selection of features; (ii) selecting the features based on entropy of each extracted features; and (iii) selection based on variance of feature's tf-idf values across the top-N documents.

Regarding the evaluation, we firstly follow [33] in simulating users that provide non-cooperative answers to the systems (i.e., only yes and no with no additional information). By doing so, we are able to test the effectiveness of the predicted facets for each query in terms of document retrieval. Hence, we always assume that a user would respond with a "yes" to the system's posed question. Therefore, following [5], we re-rank the documents by combining the language models of the original query and the extracted facet and evaluate the effectiveness of a facet in terms of how much it improves the document ranking performance. Moreover, we construct template-based questions with the selected facets and evaluate their usefulness with a crowdsourcing study. Furthermore, we discuss how the retrieval and human-annotation results provide new insights on extraction and selection of ranking-based features.

Among the various findings, we find that facet extraction is a critical part of the process of generating clarifying questions. Comparing the results of the random, entropy-based, and TF-IDF-based facet extraction methods, we observe a difference between the three methods and a margin in performance, indicating the significance of the facet extraction technique where we find that the entropy-based method achieves the best performance. Based on the human annotation, we see that LDA-based and entity-based features lead to more useful questions, compared to POS-based questions.

Our contributions can be summarised as follows:

- We perform detailed experiments on clarifying question generation with document-based features;
- We propose novel facet extraction techniques with the aim of generating clarifying questions;
- We conduct an extensive automatic and human evaluation of our approach.

Our findings show that significant improvements in document retrieval performance can be achieved when the retrieval is performed with a facet-expanded query. This finding, combined with the usefulness assessments of the facet-based clarifying questions, support the usefulness of using entity-based and LDA-based facets for clarification in conversational search.

2   Related Work

Recent advances in conversational agents, and in general in automatic voice recognition, have caused an increase interest in the area of conversational search. However, one of the first works in conversational Information Retrieval (IR) dates back to 1987 when Croft and Thomson [12] proposed I^3R, which served as an expert intermediary system by interacting with the user during a search session. Another one of the early works on the topics studies information-seeking strategies for conversational search, utilising case-based reasoning for offering choices in a search session [6]. Conversational search has since then became an integral part of both the field of IR and natural language processing (NLP) [2,16,25,41,43]. More recently, Radlinski and Craswell [27] proposed a theoretical framework for conversational search where they stressed the importance of multi-turn interactions between the system and users, with a goal of narrowing down their specific information needs. Moreover, Trippas et al. [40] studied conversations of real users to identify important aspects of the interactions with a goal of improving the design of a conversational search system.

Clarification in search has been shown to be beneficial and an integral part of conversational search system. For example, Kiesel et al. [19] studied the impact of clarifications on user satisfaction in a voice-only search and found that users like to be asked for clarification. Clarification is even more important in a mixed-initiative paradigm of conversational search, where the system can take initiative and prompt the user for clarification, or other engaging content, at any point of the conversation [5,27]. To foster research in the area of asking clarifying questions to the users, Aliannejadi et al. [5] created an offline evaluation setting where questions are selected from a pre-defined pool, with a goal of elucidating user's need. An extra step was done by Zamani et al. [44], who proposed a supervised, reinforcement learning approach for generating clarifying questions from weak supervision data, which eliminates the shortcoming of pre-defined pool of questions. Moreover, Sekulic et al. [34] proposed GPT-2 based model for generating facet-driven clarifying questions. Researchers further aimed at predicting user engagement on the clarification panes for web search based on various features [21,35]. On the other hand, Ren et al. [31] proposed the task of conversations with search engines, where the system needs to summarise a response to the user, based on the retrieved passages. Other work on the clarification in conversational search includes user simulation for generating answers to the questions [36] and user intent classification [26].

In the area of NLP, studies include various aspects of the topic, such as question ranking [28] and generation [29,42] in conversations. The proposed models often rely on large-scale data from industrial chatbots [42], QA websites [28,29,39], and query logs [30]. Unlike the aforementioned works, we study the task of extracting facets with a goal of construing clarification questions in an IR setting, where the user's request is expressed with a short query and the IR system is expected to return a ranked list of relevant documents.

Facet extraction has previously been studied in the IR field. Notably, Kong and Allan [20] developed a graphical-based model for extracting facets from the set of candidate terms. The candidate set is extracted from the documents retrieved in response to a query. Furthermore, Deveaud et al. [13] proposed a Latent Concept Modeling (LCM) method that aims to understand the conceptual view of user's information need through modelling the search concepts in a latent space. They base their method on LDA model that identifies specific query-related topics from the top K documents retrieved, where the topics are latent variables. In this paper we take ideas from the described LDA-based approach, with a difference that our topic representation needs to become explicit, rather than latent, in order to serve as grounding for clarifying questions. So, we explore various ranking- and document-based features for facet extraction and question generation. Unlike the past work, we rely on neither human-generated

question banks nor large-scale query logs, but generate clarifying questions automatically from the extracted features. Inspired by the related work on facet
extraction, we examine the effect of LDA-based features, as well as others such
as noun phrases and entities that appear on the top of the ranked list.

3   Methodology

3.1   Feature Extraction from Retrieved Documents

In order to extract useful features from the documents and explore their potential
in clarifying question generation, we first retrieve a ranked list of relevant documents for the user's initial queries. For that purpose, we utilise Chatnoir [7] – a
freely accessible Elasticsearch-based search engine with indexes of ClueWeb and
CommonCrawl corpora. The Chatnoir service uses BM25 as its main document
retrieval model.

For each query, we extract several content-based features from the retrieved
list of documents relevant to that query. Formally, for each document D^i in
the ranked list of documents $\mathcal{D} = [D^1, D^2, \ldots, D^N]$, where N is the maximum
number of top ranked documents to consider, we apply a feature extraction
function g, yielding a list of features $\mathcal{F}_g = [F_g^1, F_g^2, \ldots, F_g^N]$:

$$F_g^i = g(D^i) \tag{1}$$

Extracted features for the i-th document in the ranking are represented as feature
terms $F^i = [f_1^i, \ldots, f_j^i, \ldots, f_{M_i}^i]$, where M_i is the number of extracted features
for D^i. Notice that we removed the subscript g from F to simplify the notation.

We experiment with different feature extraction methods, aiming to discover
which content-based features best capture the essence of clarification in conversational search. More specifically, for each document we extract: *1)* nouns; *2)*
noun phrases; *3)* verbs; *4)* named entities; *5)* entities in a knowledge base; *6)*
topics with topic modelling. We utilise *spaCy* [17] for text processing and identifying features *1* through *4*. Moreover, in each document, we link the present
entities to the ones in the English Wikipedia knowledge base (feature *5*) with
Radboud Entity Linker (REL) [18]. The entities acquired by entity linking are
expected to yield more precise and reliable entities than spaCy's named entity
recogniser. Finally, we employ LDA [8] for topic modelling over the document
list. We set the number of topics to extract to 5, as it showed the most promising
results in the initial experiments of the study. We base our LDA implementation
on Stanford's Mallet topic modelling toolkit [22].

3.2   Facet Extraction

Given the extracted feature list for each retrieved document w.r.t. a single query,
the Facet Extraction module aims to select a feature that could be used to
generate a useful clarifying question. Such feature would act as a query facet, and
should ideally be informative and allow us to filter out the retrieved document

list, advancing towards satisfying the user's information need. We find motivation
for constructing the Facet Extraction module by analysing precision and recall
of the extracted raw facets, as described in Sect. 4.2. We experiment with three
different Facet Extraction methods. Namely: *random*, *entropy-based*, and *tf-idf-
based* selection. Each of the methods, given a list of extracted features from the
document list $\mathcal{F}$, selects the most discriminative ones based on different criteria
to act as a facet.

More specifically, the *random* selection method simply randomly selects any
of the features f_i^j from any of the N retrieved documents.

The *entropy-based* selection method computes the entropy for each of the
f_j features by taking into account their term-frequency distribution over the
documents:

$$score_E(f_j) = \mathrm{H}([TF(f_j)^1, TF(f_j)^2, \ldots, TF(F_j)^N])$$

(2)

where $TF(f_j)^i$ is the term frequency count of the feature j in the i-th document.
By computing the entropy of each of the potential facets f_j, we capture their
level of uncertainty and amount of carried information [37]. This means that
features with a balanced probability distribution across the retrieved document
list will have a higher $score_E$, making them more likely to be extracted as facets.

The score based on term frequency-inverse document frequency (tf-idf) is
computed as the standard deviation of tf-idf values for the feature f_j in each of
the documents:

$$score_T(f_j) = \sigma([\frac{TF(f_j)^1}{\log DF(f_j)}, \frac{TF(f_j)^2}{\log DF(f_j)}, \ldots, \frac{TF(f_j)^N}{\log DF(f_j)}]$$

(3)

where the $DF(f_j)$ represents the document frequency of the feature f_j. The
score is based on the proved fact that features with a high tf-idf value in a document are highly discriminative for that document. Thus, features with high
variance between tf-idf scores across documents should capture that notion of
single-document discriminativeness. Intuitively, we want the facet to be discriminative and help in filtering out the result list to tailor it according to the user's
information need, but at the same time be closely related to the topic.

Finally, the facet is selected by taking the one with the maximum score, for
each of the methods separately, that is:

$$facet_{\{E,T\}} = \underset{f_j}{\mathrm{argmax}}\, score_{\{E,T\}}(f_j)$$

(4)

with an additional requirement that the feature f_j must appear in at least 3
different documents, to avoid overestimation by the entropy-based and tf-idf-
based methods.

3.3   Facet-Based Question Generation

As reviewed in the related work section, several methods exist for question generation. However, options are scarce for facet-guided question generation. Recently,

Sekulić et al. [34] proposed a GPT-2 based method for facet-grounded clarifying question generation. Nevertheless, the questions generated by their model are not always about the specific facet, as they can also focus on the query alone. In order to have more control over the experiments, we resort to widely used template-based question construction [44,45]. Specifically, we substitute the slot in the question pattern "Are you interested in {*facet*}?", with *facet* extracted through Eq. 4.

4 Evaluation Setting

In this section, we describe the user data and explain the evaluation framework of our three-step methodology described in the previous section. Specifically, we describe the precision- and recall-based evaluation of extracted features by computing overlap with the human-generated clarifications. Next, we describe the evaluation of facet extraction methods through a document retrieval-based experiment. Finally, we detail the crowdsourcing-based evaluation of the clarifying questions generated based on the extracted facets.

4.1 Evaluation Dataset

For the purpose of evaluating our approach, we focus on the ClariQ dataset [3], an extension of the Qulac dataset [5], aimed at fostering research in the field of asking clarifying questions in open-domain conversational search. ClariQ was created on top of the TREC Web Track 2009-12 collection, which contains ambiguous and faceted queries that often require clarification when addressed in a conversational setting. Given a topic from the dataset, clarifying questions were collected via crowdsourcing. Then, given a topic and a specific facet of the topic, crowdsource workers were employed to gather answers to these clarifying questions. ClariQ contains more than 200 topics, each associated with an initial user query and the relevant document list for each of the query facets. We extract the feature set $\mathcal{F}$ for each of the topics from the ranked list of documents retrieved in response to the initial query for the specific topic.

4.2 Evaluation of Document-Level Features

In order to estimate the effectiveness of various feature extraction methods, we propose a simple, yet informative evaluation procedure based on human-generated clarifying questions in ClariQ. As the human-generated questions capture a large range of query facets, it is justified to evaluate our feature extraction methods by identifying the proportion of the extracted features present in the ClariQ questions. Specifically, for each query, we compute the overlap of the extracted features from the top N ranked documents with the generated questions. Formally, given the feature set $\mathcal{F}$, extracted by one of the methods g from $\mathcal{D}$, we compute the precision and recall with the feature set $\mathcal{F}_{CQ}$ extracted from the set of clarifying questions in the ClariQ dataset, as follows:

$Precision = \frac{|\mathcal{F} \cap \mathcal{F}_{cQ}|}{|\mathcal{F}|}$ and $Recall = \frac{|\mathcal{F} \cap \mathcal{F}_{cQ}|}{|\mathcal{F}_{cQ}|}$. We compute precision and recall for a varying numbers of top documents N, aiming to gain insight into the behaviour of different feature extraction methods depending on the number of considered documents. The results of this study are presented in Sect. 5.1.

4.3 Evaluation of Facet Extraction

In order to evaluate the facet selection methods described in Sect. 3.2, we adopt the document retrieval-based evaluation methodology used in the Conv-AI3 shared task [3]. Specifically, we perform document retrieval with an initial query and evaluate its performance. Then, the initial query is expanded by concatenating the facet terms selected by any of the facet extraction methods, and the retrieval is performed again. We then compare the two retrieval performances, with an assumption that if the selected facet terms are useful for identifying the underlying information need, the retrieval performance will improve with the query expanded with those terms. We analyse the results in terms of traditional IR metrics, namely nDCG@k, precision@1, and MRR. The results of the experiment are presented in Sect. 5.2.

4.4 Evaluation of the Usefulness of Facet-Based Clarifying Questions

The impact of facets selected from various feature extraction methods on the clarifying questions is evaluated in terms of question *usefulness*. Rosset et al. [32] define a conversation-leading clarifying question as *useful*, arguing that questions can be relevant to the user's query, but not necessarily useful. For example, given a query "Tell me about hotels in Las Vegas.", a question such as "Would you like to know about Las Vegas?" is arguably relevant to the query, but not useful, as it is too broad and does not help pin-pointing the underlying user's information need. Notice that *Usefulness* can be related to adequacy [9,38] and informativeness [24], and has previously been used to evaluate clarifying questions in conversational search [34].

We perform a crowdsourcing study to assess the *usefulness* of the generated facet-based clarifying questions. We use Amazon MTurk for acquiring workers, based in the US, with at least 95% task approval rate. The study was done in a pair-wise setting, i.e., each worker was presented with a number of question pairs, where each question in a pair was generated based on different facet extraction methods. Their task was then to provide judgement on which question is more useful, with regard to the context, i.e., the initial query.

We compare clarifying questions based on the three different feature extraction methods in a pairwise setting, namely noun phrases-based, entity-based, and LDA-based features. These features are selected based on their performance in previous experiments in order to reduce the number of pairwise comparisons that grows exponentially with the number of methods to compare. Additionally, we compare the questions based on the facet keywords taken from the ClariQ

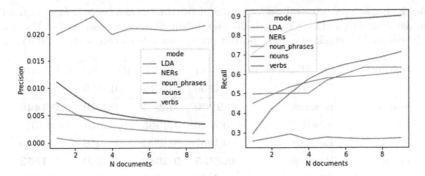

Fig. 1. Precision and recall measured as overlap of the proposed features and the human-generated clarifying questions from ClariQ.

in a similar manner as described in [34], resulting in four methods to compare, i.e., six pairwise experiments. For each pairwise comparison, we annotate 100 question pairs in terms of *usefulness*, with each pair judged by two crowdsource workers. The questions in pairs have been randomly shuffled to mitigate position bias. We define a *win* for model A if both annotators voted the question generated based on model A as more useful, and *loss* for model A if both voted the question generated by model B as more useful. In case the two workers voted differently on a single question pair, it was defined as a *tie*.

5   Results and Discussion

In this section, we aim to answer two main research questions: **RQ1:** Which features yield high-quality facets is useful for clarifying questions?; **RQ2:** Which facet extraction methods result in facets useful for clarification? We discuss RQ1 throughout all of the experiments, while we focus on RQ2 in Sect. 5.2.

5.1   Precision and Recall of Extracted Feature Sets

The results in terms of precision and recall of various feature extraction methods, computed as described in Sect. 4.2, are presented in Fig. 1. The performance is shown as a function of the top N documents the top k features were extracted from. Several observations can be made from the experiment. First, as expected, the precision of all of the methods is declining as the number of documents considered increases. Higher number of documents naturally leads to higher number of extracted features, which harms the precision, but improves the recall, as seen in the figure. Also, it suggests that the higher ranked documents lead to more precise set of features, while considering a deeper ranked list provides a broader set of features that improve the recall.

Moreover, we see that the LDA-based extraction method yields higher-precision features overall. However, the high precision comes from the limited number of predicted topics, leading to a low recall. As the number of topics

Table 1. Impact on retrieval performance of feature extraction and facet extraction methods. Symbol † indicate statistically significant difference compared to the query-only baseline. The significance is reported under two-sided t-test with $p < 0.05$.

Features	FacetExtraction	P@1	MRR	nDCG@3	nDCG@10
–	Query-only	0.1800	0.2761	0.1199	0.1412
Nouns	Random	**0.2200**	0.2957	0.1359	**0.1449**
	Tf-idf	0.2000	**0.2965**	**0.1500†**	0.1341
	Entropy	0.1867	0.2852	0.1416	0.1260
Verbs	Random	**0.2333**	**0.2987**	0.1360	**0.1552**
	Tf-idf	0.2000	0.2907	**0.1384**	0.1302
	Entropy	0.2067	0.2920	0.1401†	0.1394
NEs	Random	0.1800	0.2751	0.1362	0.1460
	Tf-idf	0.2067	0.3073	0.1493†	0.1429
	Entropy	**0.2467†**	**0.3370†**	**0.1519†**	**0.1591†**
NounPhrases	Random	**0.2000**	**0.2853**	0.1153	**0.1343**
	Tf-idf	0.1733	0.2750	0.1326	0.1194
	Entropy	0.1933	0.2816	**0.1419**	0.1244
Entities	Random	0.1733	0.2486	0.1135	0.1266
	Tf-idf	**0.2302†**	0.3205†	**0.1609†**	0.1539
	Entropy	0.2276†	**0.3247†**	0.1574†	**0.1652†**
LDA	Random	0.1667	0.2666	0.1239	0.1342
	Tf-idf	**0.2500†**	**0.3131†**	0.1441	**0.1543**
	Entropy	0.1667	0.2889	**0.1657†**	0.1420

and extracted topic representation terms do not change with the increase of the number of considered documents N, the recall stagnates.

The trend of generally high recall and low precision across all of the methods suggests the need of an additional filtering step in order to identify potentially interesting facets from the large set of features. Obviously, depending on the facet extraction method, one can study the trade-off between recall and precision depending on the depth of the ranked list that a model can take as input, and its effect on the generated questions. These results strongly motivate the facet extraction module described in Sect. 3.2 in our pipeline.

5.2 Facet-Based Retrieval

The results of facet extraction evaluation through document retrieval are presented in Table 1. The first row indicates the performance of the initial query only, while other rows show the performance of queries expanded by various facet extraction methods over the acquired document features. The results reported are for the overall best performing facet for each of the topics. This is to mimic

Table 2. Pairwise comparisons between clarifying questions generated based on different feature extraction methods, as judged by crowdsourcing workers. The * sign indicates statistical significance with $p < 0.05$, calculated by a trinomial test.

Method A	Method B	A Wins	B Wins	Ties	p-value
Entities	HumanF	22%	24%	**44%**	0.1334
Entities*	NounPhrase	**47%**	27%	26%	0.0487
LDA*	Entities	**46%**	14%	40%	0.0076
LDA*	HumanF	**42%**	21%	37%	0.0181
LDA	NounPhrase	30%	26%	**44%**	0.7844
NounPhrase	HumanF*	17%	**52%**	31%	$< 10^{-4}$

the scenario where the extracted facet would indeed be in line with user's information need. We observe several patterns in the results. First, queries expanded by facets from facet extraction-based methods outperform query-only baseline in most cases. The exception is *random*-based facet extraction method, which, perhaps unsurprisingly, often even hurts the performance. Second, entity-based features, where entities are linked to a Wikipedia knowledge graph (*Entities* in the table), and LDA-based features seem to perform the best in terms of several metrics presented. We hypothesis this is due to the fact that these type of features are much less noisy, as they are precision oriented, as opposed to recall oriented like noun- or noun phrases-based features. The *tf-idf*- and *entropy*-based facet extraction methods consistently outperform the baselines for those features. However, we note that, after Bonferroni multiple-comparison corrections, neither the *tf-idf*- nor the *entropy*-based method yielded statistically significant improvements over the *random*-based method (with p of 0.05). Similarly, no statistically significant difference was found between features extracted from the text after the Bonferroni correction. This calls for a future study on the additional filtering step aimed towards informed selection of the appropriate facet terms.

Regarding the RQ1, analysis of the results in this experiment suggests that entity- and topic modelling-based features yield the most useful facets. To address the RQ2, both *tf-idf*- and *entropy*-based facet extraction methods show promising results, but they are heavily dependent on the type of input features. Future work will aim to provide a more detailed analysis of relationships between feature and facet extraction methods, as the presented results do not show which method is consistently the best.

5.3   Usefulness of Clarifying Questions

Table 2 shows the results of the crowdsourcing study aimed at estimating the *usefulness* of clarifying questions generated based on selected facets. All of the facets were extracted with an *entropy*-based method described in Sect. 3.2, but based on different features. The statistical significance is indicated by * symbol and calculated by the trinomial test – a modification of the sign test that takes into account the ties as well. Paired with manual qualitative analysis of the clarifying questions, we make several observations from the acquired results. First, LDA-based facets outperform all of the other methods, even the facets based on human-curated clarifying questions *HumanF*. This phenomenon could be explained by the fact that LDA-based features have the lowest level of noise among all of the considered methods. Second, noun phrase-based features are outperformed by all other facet extraction methods. We attribute the poor performance to the high level of noise in the phrases extracted by spaCy from the retrieved documents. Finally, a large number of ties across all of the comparisons suggests both the challenge of the annotation task and the similarity between the extracted facets. However, it is clear that useful questions should be grounded in low-noise-level facets.

Similarly to the previous experiment, we observe the dominance of entity- and topic modelling-based features over POS-based features. Future work on the topic involves the analysis of larger variety of facet-guided clarifying question generation models.

6   Conclusions

In this paper, we have explored various methods for feature extraction from the list of documents relevant to user's query and their impact on clarifying question generation. We have identified the importance of the facet extraction module, as the experiments have showed low precision of the initial feature extraction methods and can not be considered useful facets without an appropriate filtering. Thus, we experimented with two facet selection methods, namely, entropy- and tf-idf-based. The results suggest improvements in document retrieval performance, when the retrieval is performed with the facet-expanded query. In particular, our proposed method achieved 38% relative improvement in terms of nDCG@3. This finding, combined with the *usefulness* assessments of the facet-based clarifying questions, indicate the adequacy of entity- and LDA-based facets for clarification in conversational search. However, relatively modest improvements call for an additional facet filtering step, in order to help and pin-point the exact terms representing user's underlying information need.

Overall, we observe that document-based feature extraction, together with ranking-based facet extraction can lead to significant improvements when used for generating clarifying questions. Compared to the existing work on question selection [4] and generation based on query logs [44], we conclude that the proposed method for facet extraction and question generation is effective. Our results can be used to inform question generation models with a final goal of an improved end-to-end document retrieval.

Future work encapsulates modelling facet extraction over multi-turn conversations and generating multi-faceted clarifying questions. Also, we plan to evaluate our proposed question generation in a conversational setup with a human-in-the-loop setup and compare with the existing techniques such as question selection.

References

1. Aliannejadi, M., Azzopardi, L., Zamani, H., Kanoulas, E., Thomas, P., Craswel, N.: Analysing mixed initiatives and search strategies during conversational search. CoRR abs/2109.05955 (2021)
2. Aliannejadi, M., Kiaeeha, M., Khadivi, S., Ghidary, S.S.: Graph-based semi-supervised conditional random fields for spoken language understanding using unaligned data. In: ALTA, pp. 98–103 (2014)
3. Aliannejadi, M., Kiseleva, J., Chuklin, A., Dalton, J., Burtsev, M.: ConvAI3: generating clarifying questions for open-domain dialogue systems (ClariQ) (2020)
4. Aliannejadi, M., Kiseleva, J., Chuklin, A., Dalton, J., Burtsev, M.: Building and evaluating open-domain dialogue corpora with clarifying questions. In: EMNLP (2021)
5. Aliannejadi, M., Zamani, H., Crestani, F., Croft, W.B.: Asking clarifying questions in open-domain information-seeking conversations. In: SIGIR, pp. 475–484 (2019)
6. Belkin, N.J., Cool, C., Stein, A., Thiel, U.: Cases, scripts, and information-seeking strategies: on the design of interactive information retrieval systems. Expert Syst. Appl. 9(3), 379–395 (1995)
7. Bevendorff, J., Stein, B., Hagen, M., Potthast, M.: Elastic ChatNoir: search engine for the ClueWeb and the common crawl. In: Pasi, G., Piwowarski, B., Azzopardi, L., Hanbury, A. (eds.) ECIR 2018. LNCS, vol. 10772, pp. 820–824. Springer, Cham (2018). https://doi.org/10.1007/978-3-319-76941-7_83
8. Blei, D.M., Ng, A.Y., Jordan, M.I.: Latent Dirichlet allocation. J. Mach. Learn. Res. 3, 993–1022 (2003)
9. Callison-Burch, C., Osborne, M., Koehn, P.: Re-evaluating the role of bleu in machine translation research. In: ACL (2006)
10. Cao, G., Nie, J., Gao, J., Robertson, S.: Selecting good expansion terms for pseudo-relevance feedback. In: SIGIR, pp. 243–250. ACM (2008)
11. Crestani, F., Mizzaro, S., Scagnetto, I.: Mobile Information Retrieval. Springer Briefs in Computer Science, Springer, Cham (2017). https://doi.org/10.1007/978-3-319-60777-1
12. Croft, W.B., Thompson, R.H.: I^3r: a new approach to the design of document retrieval systems. JASIS 38(6), 389–404 (1987)
13. Deveaud, R., SanJuan, E., Bellot, P.: Accurate and effective latent concept modeling for ad hoc information retrieval. Document numérique 17(1), 61–84 (2014)
14. Drosou, M., Pitoura, E.: Search result diversification. SIGMOD Rec. 39(1), 41–47 (2010)
15. He, B., Ounis, I.: Query performance prediction. Inf. Syst. 31(7), 585–594 (2006)
16. He, Y., Young, S.J.: Semantic processing using the hidden vector state model. Comput. Speech Lang. 19(1), 85–106 (2005)
17. Honnibal, M., Montani, I., Van Landeghem, S., Boyd, A.: spaCy: industrial-strength natural language processing in Python (2020). https://doi.org/10.5281/zenodo.1212303
18. van Hulst, J.M., Hasibi, F., Dercksen, K., Balog, K., de Vries, A.P.: Rel: an entity linker standing on the shoulders of giants. In: Proceedings of the 43rd International ACM SIGIR Conference on Research and Development in Information Retrieval. SIGIR 2020. ACM (2020)
19. Kiesel, J., Bahrami, A., Stein, B., Anand, A., Hagen, M.: Toward voice query clarification. In: SIGIR, pp. 1257–1260 (2018)

20. Kong, W., Allan, J.: Extracting query facets from search results. In: SIGIR, pp. 93–102 (2013)
21. Lotze, T., Klut, S., Aliannejadi, M., Kanoulas, E.: Ranking clarifying questions based on predicted user engagement. CoRR abs/2103.06192 (2021)
22. McCallum, A.K.: Mallet: a machine learning for language toolkit (2002). http:// mallet.cs.umass.edu
23. Mihalkova, L., Mooney, R.: Learning to disambiguate search queries from short sessions. In: Buntine, W., Grobelnik, M., Mladenić, D., Shawe-Taylor, J. (eds.) ECML PKDD 2009. LNCS (LNAI), vol. 5782, pp. 111–127. Springer, Heidelberg (2009). https://doi.org/10.1007/978-3-642-04174-7_8
24. Peng, B., Zhu, C., Li, C., Li, X., Li, J., Zeng, M., Gao, J.: Few-shot natural language generation for task-oriented dialog. arXiv preprint arXiv:2002.12328 (2020)
25. Pieraccini, R., et al.: A speech understanding system based on statistical representation of semantics. In: ICASSP, pp. 193–196 (1992)
26. Qu, C., Yang, L., Croft, W.B., Zhang, Y., Trippas, J.R., Qiu, M.: User intent prediction in information-seeking conversations. In: CHIIR, pp. 25–33 (2019)
27. Radlinski, F., Craswell, N.: A theoretical framework for conversational search. In: CHIIR, pp. 117–126 (2017)
28. Rao, S., Daumé, H.: Learning to ask good questions: Ranking clarification questions using neural expected value of perfect information. In: ACL (1), pp. 2736–2745 (2018)
29. Rao, S., III, H.D.: Answer-based adversarial training for generating clarification questions. arXiv:1904.02281 (2019)
30. Ren, G., Ni, X., Malik, M., Ke, Q.: Conversational query understanding using sequence to sequence modeling. In: WWW, pp. 1715–1724 (2018)
31. Ren, P., Chen, Z., Ren, Z., Kanoulas, E., Monz, C., de Rijke, M.: Conversations with search engines. ACM Trans. Inf. Syst. 1(1) (2020)
32. Rosset, C., Xiong, C., Song, X., Campos, D., Craswell, N., Tiwary, S., Bennett, P.: Leading conversational search by suggesting useful questions. In: TheWebConference, pp. 1160–1170 (2020)
33. Salle, A., Malmasi, S., Rokhlenko, O., Agichtein, E.: Studying the effectiveness of conversational search refinement through user simulation. In: ECIR, pp. 587–602 (2021)
34. Sekulić, I., Aliannejadi, M., Crestani, F.: Towards facet-driven generation of clarifying questions for conversational search. In: Proceedings of the 2021 ACM SIGIR International Conference on Theory of Information Retrieval, pp. 167–175 (2021)
35. Sekulić, I., Aliannejadi, M., Crestani, F.: User engagement prediction for clarification in search. In: Proceedings of the European Conference on Information Retrieval (ECIR) (2021)
36. Sekulić, I., Aliannejadi, M., Crestani, F.: Evaluating mixed-initiative conversational search systems via user simulation. In: Proceedings of the 15th ACM International Conference on Web Search and Data Mining. Association for Computing Machinery (2022)
37. Shannon, C.E.: A mathematical theory of communication. Bell Syst. Tech. J. 27(3), 379–423 (1948)
38. Stent, A., Marge, M., Singhai, M.: Evaluating evaluation methods for generation in the presence of variation. In: CICLing, pp. 341–351 (2005)
39. Tian, Z., Yan, R., Mou, L., Song, Y., Feng, Y., Zhao, D.: How to make context more useful? an empirical study on context-aware neural conversational models. In: ACL (2), pp. 231–236 (2017)

40. Trippas, J.R., Spina, D., Cavedon, L., Joho, H., Sanderson, M.: Informing the design of spoken conversational search: perspective paper. In: CHIIR, pp. 32–41 (2018)
41. Walker, M.A., Passonneau, R.J., Boland, J.E.: Quantitative and qualitative evaluation of Darpa communicator spoken dialogue systems. In: ACL, pp. 515–522 (2001)
42. Wang, Y., Liu, C., Huang, M., Nie, L.: Learning to ask questions in open-domain conversational systems with typed decoders. In: ACL (1), pp. 2193–2203 (2018)
43. Williams, J.D., Raux, A., Ramachandran, D., Black, A.W.: The dialog state tracking challenge. In: SIGDIAL, pp. 404–413 (2013)
44. Zamani, H., Dumais, S., Craswell, N., Bennett, P., Lueck, G.: Generating clarifying questions for information retrieval. In: TheWebConference, pp. 418–428 (2020)
45. Zhang, S., Balog, K.: Evaluating conversational recommender systems via user simulation. In: KDD, pp. 1512–1520 (2020)
46. Zhou, Y., Croft, W.B.: Query performance prediction in web search environments. In: SIGIR, pp. 543–550. ACM (2007)

Adversarial Multi-task Model for Emotion, Sentiment, and Sarcasm Aided Complaint Detection

Apoorva Singh[1], Arousha Nazir[2], and Sriparna Saha[1]

[1] Indian Institute of Technology Patna, Bihta, India
{apoorva_1921cs19,sriparna}@iitp.ac.in
[2] National Institute of Technology Srinagar, Srinagar, India

Abstract. Automatic identification of consumer complaints about products or services purchased can be crucial for businesses and online merchants since they can utilize this knowledge to address the needs of their clients, including handling and resolving complaints. Previous studies on complaint detection do not consider sarcasm, which is often used to express a breach of expectation without directly stating the complaint. Furthermore, since every speech act is influenced by emotions, the customer's emotional state has a considerable impact on the complaint expression. In this paper, we hypothesize that sarcasm, along with two closely related tasks of sentiment and emotion, could aid the process of complaint identification and thereby propose a deep multi-task framework to solve the four problems jointly. We manually annotate the recently released *Complaints* dataset with the emotion, sentiment, and sarcasm classes. We present an attention-based adversarial multi-task deep neural network model for complaint detection. Experimental results on the extended version of the *Complaints* dataset show the effectiveness of our proposed approach for complaint detection over the existing state-of-the-art system. The evaluation also demonstrates that the proposed multi-task system improves performance for the primary task, i.e., complaint detection, with the assistance of the three auxiliary tasks, emotion recognition, sentiment analysis, and sarcasm detection.

Keywords: Complaint Detection · Emotion Recognition · Sentiment Analysis · Sarcasm Detection · Adversarial Multi-task learning · Deep learning

1 Introduction

Complaining is a speech act that generally conveys unpleasant emotions that are caused by a disparity between reality and expectations regarding an entity or event [20]. Complaints are fundamental ways of expressing displeasure in human communication. The expression of complaints varies from person to person depending on the complainers' temperament and specific situations [35]. Presently, social

M. Hagen et al. (Eds.): ECIR 2022, LNCS 13185, pp. 428–442, 2022.
https://doi.org/10.1007/978-3-030-99736-6_29

media platforms and online commercial agencies allow customers to voice their thoughts and observations about a product, service, or event. As a result, commercial and retail organizations perceive product reviews as a valuable source of information that they can utilize to develop their advertising campaigns and address any product-related issues. Additionally, this benefits the customers by providing suggestions on the quality of items or services they plan to purchase. Identifying complaint texts in natural language is crucial for developers of recommendation systems, chatbots [16]; enhancing customer service by acknowledging the level of dissatisfaction [34]; analyzing and solving product-related issues [6].

Sarcasm is frequently employed to convey barely disguised dissatisfaction in a satirical manner. The following example demonstrates this: *"I just loved how this product didn't work at all!"*. The sentence starts on a positive note regarding a product. However, careful observation of the entire context as well as sentiment and emotion of the customer help in comprehending that the customer is dissatisfied with the product and has a negative opinion while expressing the complaint. Here's where emotion and sentiment come into question.

An individual's emotional state and sentiment have a considerable impact on the intended content [17]. In general, sentiment and emotion are regarded as distinct tasks [8,19] although sentiment and emotion are intrinsically linked. Specifically, emotion recognition is a significantly more nuanced and fine-grained analysis than sentiment analysis [15]. Together with emotion and sentiment, sarcasm offers a greater insight into the customer's frame of mind. Emotion, sentiment, and sarcasm are all associated tasks, and each contributes to a clearer understanding of others. We take advantage of these correlations and use sentiment, emotion, and sarcasm tasks to identify complaints in a multi-task setting. The key contributions of our proposed work are outlined as follows:

- *We propose an intra-modal attention-based adversarial multi-task deep neural network model to optimize complaint, emotion, sentiment, and sarcasm tasks jointly. Complaint Identification (CI) is treated as the primary task in our multi-task framework. In contrast, Emotion Recognition (ER), Sentiment Analysis (SA), and Sarcasm Detection (SD) are considered supplementary (i.e., auxiliary) tasks. The proposed multi-task framework for CI shows its efficacy over the existing state-of-the-art system and surpasses the single-task CI considerably.*
- *We manually annotate the recently released Complaints dataset [24] with emotion, sentiment, and sarcasm classes.*
- *We evaluate the proposed model on the extended version of Complaints data. Experimental results indicate that the multi-tasking complaint identification outperforms single task variants and other baselines.*

2 Related Work

In linguistic studies, complaints were classified into different categories based on intensity and directness. The authors in [33] introduced four primary intensity levels: no specific reproach, disapproval, allegation, and blame. Quite recently,

[13] categorized complaints into three levels of directness: very direct, moderately direct, and indirect. Direct or moderately direct complaints include clear violations of expectation. In contrast, indirect complaints either do not explicitly mention or could infer a violation of expectation.

Furthermore, detecting complaints on social media necessitates detecting complaints from fragmented and noisy text snippets with character limits, usage of random acronyms, and colloquialism. In this context, text-based complaints have been previously analyzed based on semi-supervised strategies [31], feature engineering-based machine learning methods [24], feedback likelihood [38], and deep learning models [12], [29]. The authors in [24] developed a logistic regression model with hand-crafted features for identifying complaints. In [29], the authors explored the influence of emotion and sentiment in identifying complaints on a small number of test cases without considering the use of sarcasm when expressing breach of expectation.

In the area of emotion and sentiment recognition in human conversations, attention-based multi-task models have proved to be successful [1,26]. The authors in [2] proposed a multi-task ensemble framework for emotion and sentiment analysis. In another work, [23] employed sentiment and emotion features obtained from pre-trained sentiment, emotion, and personality models on a Twitter-based dataset to predict sarcasm using Convolutional Neural Network. This research has motivated us to investigate the significance of emotion, sentiment, and sarcasm tasks in identifying complaints in tweets.

We first extend the *Complaints* dataset [24] by manually annotating each tweet with emotion, sentiment, and sarcasm labels in our current study. Subsequently, we propose an adversarial multi-task strategy integrated with an intra-modal attention mechanism (A_{intra}) to utilize emotion, sentiment, and sarcasm for predicting complaints. Furthermore, to our knowledge, this is the first attempt to solve the complaint detection problem in an adversarial multi-task framework. Through a detailed experimental analysis, we illustrate that combining knowledge of emotion, sentiment, and sarcasm using an appropriate multi-task framework can help boost complaint detection.

3   Dataset

For our experiments, we used the *Complaints* dataset that consists of 2214 non-complaint and 1235 complaint tweets in English. We chose this dataset since it is publicly available and contains annotated complaints/non-complaints accumulated from Twitter. We manually re-annotated this dataset to introduce emotion, sentiment, and sarcasm labels in addition to complaint labels.

3.1   Data Annotation

For the emotion annotation of the *Complaints* dataset, we consider Ekman's [9] six basic emotions (*anger, disgust, fear, happiness, sadness, and surprise*). Other than these six basic emotions, we add a category out-of-scope (OS) to represent such tweets that do not fall under the scope of above mentioned six emotion classes. For the sentiment annotation, we consider three sentiment classes (*negative, neutral, positive*). The distribution of tweets across the emotion classes is as follows: 844 tweets belong to 'anger', 7 tweets belong to 'disgust', 8 tweets belong to 'fear', 473 tweets belong to 'joy', 1479 tweets belong to 'out-of-scope', 626 tweets belong to 'sadness', and 12 tweets belong to 'surprise'. Furthermore, the distribution of tweets across the sentiment classes is 1041 tweets belong to 'negative', 1198 tweets belong to 'neutral', and 1210 tweets belong to 'positive'. For the annotation of sarcasm task, the tweets were labeled with a specific category, i.e., sarcastic or non-sarcastic. However, on evaluation of the annotation, we found that there were 98% non-sarcastic and 2% sarcastic tweets in the *Complaints* dataset. Due to the highly skewed distribution of these two classes in the dataset, we utilized a more balanced publicly available Twitter-based sarcasm dataset, *SPIRS Sarcasm Dataset* [28] for the training of the sarcasm task. The SPIRS Sarcasm Dataset[1] includes 15,000 non-sarcastic and 15,000 sarcastic tweets. It is a collection of both intended sarcasm and perceived sarcasm.

Table 1. Sample sentences with their complaint, emotion, sentiment and sarcasm labels from the extended version of *Complaints* dataset. Non-Com : Non-Complaint, Com : Complaint

Sentence	Classes	Emotion	Sentiment	Sarcasm
BTW @crantraf @blackanddecker did me right and replaced my two failed batteries under warranty. I'm happy:) Thanks @blackanddecker!	Non-Com	Joy	Positive	Non-sarcastic
Thanks to @NVIDIAQuadro's incompetence, I now can't work till October4th, when the ATI card arrives	Com	Sadness	Negative	Sarcastic
@TommyHilfiger Dramatic shopping exp. Ordered 6 jeans same size (30/32) 2 fits/ 2 too large/2 too slim :same brand & different sizing	Com	Sadness	Neutral	Non-sarcastic

3.2 Annotation Specifications

We assigned three graduate students fluent in English to annotate the reviews with appropriate emotion, sentiment, and sarcasm tags (Table 1). Before commencing the annotation process, the requirements for annotation, as well as some examples, were provided to the annotators. The annotators were asked to initially annotate the tweets with the sarcastic/non-sarcastic tags without knowing the sentiment and emotion labels. The majority voting technique selected the final emotion, sentiment, and sarcasm labels. Reviews with no common emotion

---

[1] https://github.com/bshmueli/SPIRS.

or sentiment label (as determined by the three annotators) were excluded from the final annotated dataset. On the annotated dataset[2], we computed the Cohen-Kappa [5] scores to assess inter-rater agreement among the three annotators. We achieve agreement scores of 0.68, 0.82, and 0.71, respectively, indicating that the annotations are of good quality on the emotion, sentiment, and sarcasm tasks.

4 Proposed Methodology

In this section, we outline our problem and discuss the details of the proposed framework. The major components of the architecture are discussed in the following sub-sections. Figure 1 depicts the overall framework.

4.1 Problem Definition

We intend to learn four closely related tasks at the same time, named as, complaint identification (main task), emotion recognition, sentiment analysis, sarcasm detection (auxiliary tasks). Let $(x_i, e_i, s_i, m_i, c_i)_{i=1}^{X}$ be a set of X instances where e_i, s_i, m_i and c_i represent the matching emotion, sentiment, sarcasm and complaint labels for x_i^{th} tweet, respectively. Here, $x_i \in X$, $e_i \in E$ (emotion classes), $s_i \in S$ (sentiment classes), $m_i \in M$ (sarcasm classes) and $c_i \in C$ (complaint classes).

Our multi-task learning framework's objective is to maximize the function f (Eq. 1) that draws a new instance x_i to its fitting emotion label e_i, sentiment label s_i, sarcasm label m_i and complaint label c_i simultaneously.

$$argmax(\Pi_{k=0}^{R}P(m_i, s_i, e_i, c_i|x_i; \theta)), \tag{1}$$

where x_i is the input sentence whose complaint label (c_i), emotion label (e_i), sentiment label (s_i) and sarcasm label (m_i) are to be predicted. θ denotes the model's parameters we aim to optimize.

4.2 Multi-task Model for Emotion, Sentiment, and Sarcasm Aided Complaint Detection (MTL$_{All}$)

The proposed framework consists of four principal components: (i) Feature extraction with the help of Bidirectional Encoder Representations from Transformers (BERT) [25], (ii) Encoding model, which essentially takes the textual features (extracted above) as input and returns the encoded representations as output, (iii) Intra-modal attention relating the distinct positions of a single instance to compute the final representation of the given instance. (iv) Adversarial Loss ensures that shared layers and task-specific feature space remain mutually exclusive. (v) Output Layer consists of output medium for the four tasks to obtain a generalized representation throughout all the tasks.

---

[2] Dataset available at https://www.iitp.ac.in/~ai-nlp-ml/resources.html#Complaints-ESS.

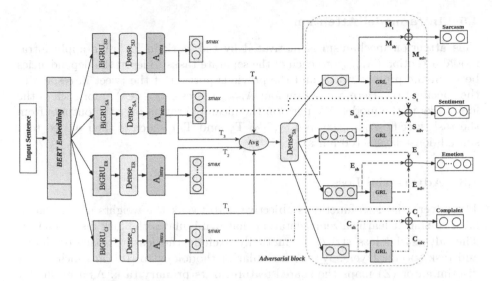

Fig. 1. Adversarial multi-task model for complaint identification. *smax*: softmax activation function, *GRL*: gradient reversal layer, A*intra*: intra-modal attention

4.3   Feature Extraction

We utilized the BERT-BASE pre-trained model for the word embeddings[3] (Uncased: 12-layer, 768-hidden, 12-heads, 110 M parameters). BERT is a multi-layer bidirectional Transformer encoder [7] based on the original work shown in [36].

4.4   Encoding Model

The obtained BERT encoder representation is then passed through four different Bi-directional Gated Recurrent Units (BiGRUs) [3] layers (256 neurons) to sequentially encapsulate these representations into hidden states and learn different contextual dependency-based features associated with various tasks that are CI, ER, SA, and SD. These four BiGRUs learn individual characteristics for each task that are unaffected by features learned from other tasks. The BiGRU layer retains contextual information from both forward ($\overrightarrow{GRU}$) and backward ($\overleftarrow{GRU}$) time steps and produces a hidden representation (h_i) of each word in the sentence. The overall hidden state matrix is represented as: H = [h_1, h_2, h_3,..., h_n], where, $H \in \mathbb{R}^{nX2d}$, n is the dimension of text length, and d is the number of hidden units in each BiGRU. Hence, the final four hidden state matrices correspond to four Bi-GRUs, namely H_1, H_2, H_3, and H_4, respectively. These representations are then forwarded to four fully-connected layers, each of d_i dimensions to learn the attention of different tasks.

---

[3] We also additionally experimented with Sentence Encoding with Multilingual Universal Sentence Encoder (SBERT) [25], but the results were not satisfactory.

4.5 Intra-model Attention

This attention mechanism is motivated by the work in [27]. We apply intra-modal attention (A_{intra}) to each of the separate tasks to learn the dependencies between the present words and the previous portion of the tweet in respect to the specific task. Here, we obtain four A_{intra} scores ($A_i \in \mathbb{R}^{nXd}$) for each of the four tasks in the model. We perform the mean operation over the outputs from the task-specific A_{intra} layers (T_1, T_2, T_3, and T_4) which is then passed to a shared dense layer (100 units).

4.6 Adversarial Loss

The adversarial loss function's objective is to tweak the weights of the shared layer so that it learns a representation that manipulates the task discriminator. The adversarial loss strives to mutually exclude the feature space of shared and task-specific layers. We use a similar methodology to [18], in which a task discriminator (Z) maps the shared feature to its primary task. As a result, for an accurate prediction, as the loss at the shared layer decreases, there is an increment in the adversarial loss and vice versa. Conversely, the shared layer can be adjusted to operate in an adversarial manner, making it impossible for the discriminator to recognize one of the four tasks. The adversarial loss is calculated as: $L_{adv} = min(max(\sum_{u=1}^{U} \sum_{v=1}^{V} (i_v^u * log[I(E(x_v^u))])))$. Where U denotes the type of tasks, i_v^u signifies the actual label amongst U, and x_v^u is the vth example for task u. The gradient reversal layer (GRL) [10] handles the min-max optimization problem.

4.7 Output Layer

As depicted in Fig. 1, the final predictions for complaint, emotion, sentiment and sarcasm tasks are obtained by linearly concatenating the task-specific outputs (C_t, E_t, S_t and M_t), the shared outputs (C_{sh}, E_{sh}, S_{sh} and M_{sh}) and the GRL layer outputs (C_{adv}, E_{adv}, S_{adv} and M_{adv}), respectively.

Calculation of Loss: For the complaint (C), emotion (E), sentiment (S) and sarcasm (M) tasks, we compute the categorical-cross entropy (L_{CE}) losses. The integrated loss function (L) of our proposed MTL$_{All}$ system is realized as follows: $L = p * L_{CE}^C + q * L_{CE}^E + r * L_{CE}^S + t * L_{CE}^M$. We aggregate the weighted sum of the losses from the four tasks to compute the overall loss. Here, p, q, r, and t constants ranging from 0 to 1 determine the loss weights representing the per-task loss-share to the overall loss, respectively.

5 Experiments, Results, and Analysis

In this section, experimental results have been described.

5.1   Baselines

For a comprehensive evaluation of the proposed MTL_{All} model, we compare its performance with the following baselines:

- **Single-task systems:** We develop a single-task deep learning model for CI based on BERT (STL_{CI}). The output of the BiGRU layer goes through the dense layer, followed by the attention and the outer layer.
- **Multi-task systems:** Based on recent works in CI, we develop Baseline$_1$ [29] and Baseline$_2$ [30] models for multi-task baselines. The Baseline$_1$ is a shared private multi-task model for complaint, emotion, and sentiment classification. The authors utilized three different sources of input for training the model. Baseline$_2$ [30] is commonsense knowledge augmented model for simultaneous learning of complaint and sentiment. To generate the sentiment labels, they make use of Valence Aware Dictionary and sEntiment Reasoner (VADER)[4] sentiment analysis tool. We also develop MTL_{glv} model where the textual embeddings are generated from the pre-trained GloVe[5] [22]. The embedding layer's output is forwarded to the word sequence encoder, which analyzes it to extract contextual knowledge from the sentence. The remaining architecture remains the same.
- **Ablation models:** To understand the importance of emotion, sentiment and sarcasm tasks individually and in different combinations on the complaint task, we build dual-task (MTL_{CI+ER}, MTL_{CI+SA}, MTL_{CI+SD}) and tri-task variants ($MTL_{CI+ER+SA}$, $MTL_{CI+ER+SD}$, $MTL_{CI+SA+SD}$) of our proposed framework (MTL_{All}).

5.2   Experimental Setup

To implement our proposed framework and all the baselines we use the Python-based libraries, namely TensorFlow[6], Scikit-learn[7] [21]. We report the accuracy and macro F1-score for the complaint identification task using the metrics module from sklearn. 85% of the *Complaints* dataset was used as training and validation (10%) data, and the rest 15% was used as testing data on all the experimental models. To ensure a fair comparison of the models, a seed value of 42 was chosen, which allowed the models to encounter the same training and testing data. After each of the BiGRU layers (256 units), we apply a *dropout* [32] of 20% each to reduce the chance of overfitting. The output of the task-specific A_{intra} layers are fed to the softmax layers (smax) for shared outputs. In the dense layers (100 units), we employ *ReLU* activation [11]. *Softmax* activation with 2, 7, 3, and 2 neurons are used for the output layers for complaint, emotion, sentiment, and sarcasm classification tasks, respectively. Categorical cross-entropy is used as the loss function to train across all the channels. The epoch size is set to 30.

---

[4] https://github.com/cjhutto/vaderSentiment.
[5] GloVe: http://nlp.stanford.edu/data/wordvecs/glove.840B.300d.zip.
[6] https://www.tensorflow.org/.
[7] https://scikit-learn.org/stable/.

Adam [14] with a learning rate of 0.001 is used as the optimizer. The above values are chosen after thorough hyperparameter tuning using the RandomSearch tuner of the Keras tuner API. The EarlyStopping module was used from Keras[8] [4], with the patience set equal to 10, to restore weights from the epoch that gave the best accuracy on the *Complaints* validation set. Furthermore, for the ablation experiment, the loss weights[9] for the complaints, emotion, sentiment, and sarcasm tasks are set as 1, 0.5, 0.5, and 0.3, respectively.

5.3   Results and Discussion

Please note that the current work aims to improve the performance of CI with the help of the other three supplementary tasks (ER, SA and SD). Therefore, we state the results and analysis with CI strictly serving as the pivotal task in all the task combinations.

Table 2 depicts the classification results from the various experiments. As can be observed, incorporating all the related tasks such as sentiment, emotion, and sarcasm significantly enhance the performance over the single-task, dual-task, and tri-task variants. This enhancement validates the proposed architecture's efficient usage of interaction among the four tasks. This also emphasizes the significance of including the intra-modal attention mechanism in the proposed model. As seen in Table 2, the proposed approach, which includes all four tasks (CI, ER, SA, and SD), outperforms the single-task complaint variant (STL_{CI}). MTL_{CI+ER} outperforms MTL_{CI+SA} and MTL_{CI+SD} in the dual-task variants. This can be driven by the fact that sentiment alone is often insufficient to convey complete information about the user's state of mind. For example, various emotions such as anger, fear, sadness, etc., can lead to negative sentiments about a product. As a result, the distinctive or subtle differences in the state of mind cannot be properly determined and expressed by sentiment alone. In the tri-task variants, $MTL_{CI+ER+SA}$ performs better than the other two tri-task variants. One of the possible reasons for the lower performance of $MTL_{CI+ER+SD}$ and $MTL_{CI+SA+SD}$ could be sarcasm detection being a highly nuanced task that relies heavily on context and does not contribute much in the multi-task architecture as compared to emotion and sentiment, which are more correlated tasks.

Significance of Adversarial Multi-task Architecture. In terms of all the multi-task baselines (Baseline$_1$, Baseline$_2$), these approaches do not take into account the adversarial loss. Whereas the proposed model, MTL_{All}, incorporates the adversarial loss, which enhances the performance of the multi-task model. Moreover, we also report the results by replacing the BERT embedding model with Glove embeddings [22] (MTL_{glv}). The results suggest that each of

---

[8] https://keras.io/.

[9] To fine-tune the loss weights for all of the tasks, we utilize the Grid Search method from the Scikit-learn module.

Table 2. Results of all multi-task baselines, ablation studies conducted and the proposed MTL_{All} model in terms of macro average F1-score and Accuracy value; $\pm$: standard deviation. The maximum scores attained are represented by bold-faced values. The * signifies that these findings are statistically significant.

Model	Accuracy (%)	F1-score (%)
SOTA [12]	$88 \pm .03$	$87.1 \pm .03$
Baseline$_1$ [29]	86.78 ± 1.02	85.44 ± 1.05
Baseline$_2$ [30]	82.45 ± 1.03	81.0 ± 1.02
Single-task baseline		
STL$_{CI}$	78	80.5
Dual-task baselines		
MTL$_{CI+ER}$	83.89 ± 1.07	81.64 ± 1.33
MTL$_{CI+SA}$	83.58 ± 1.12	81.21 ± 1.06
MTL$_{CI+SD}$	83.20 ± 1.04	80.98 ± 1.11
Tri-task baselines		
MTL$_{CI+ER+SA}$	84.56 ± 1.17	82.54 ± 1.05
MTL$_{CI+ER+SD}$	82.62 ± 1.23	81.79 ± 1.18
MTL$_{CI+SA+SD}$	83.39 ± 1.02	80.89 ± 1.06
MTL$_{glv}$	81.21 ± 1.32	$80.55. \pm 1.21$
Proposed approach		
MTL$_{All}$	$\mathbf{88.89^* \pm .06}$	$\mathbf{87.5^* \pm .23}$

these factors considerably boosted the performance of the proposed MTL_{All} framework. *All of the results presented here are statistically significant*[10] [37].

Comparison with State-of-the-art Technique (SOTA): We also compare our proposed approach with the existing state-of-the-art technique [12] for single-task CI. SOTA utilizes an array of neural language models boosted by the use of transformer networks. We state the results in Table 2. The proposed model outperforms the SOTA technique.

Figure 2 depicts the heatmap visualization of the learned weights of individual words for a sample case for the STL_{CI} and the MTL_{ALL} to emphasise the relevance of including emotion, sentiment, and sarcasm as auxiliary tasks. The attention is focused on appropriate disapproval indicating phrases in the multi-task approach; however, in the single-task approach, attention is laid on positive terms such as *thanks*, which has been used sarcastically to express a complaint.

5.4 Error Analysis

Other than the skewed distribution (complaint: 35.8%, non-complaint: 64.2%) of the dataset, the following are some possible explanations for the errors in the complaint prediction:

---

[10] We performed Student's t-test for the test of significance. The results are found to be statistically significant when testing the null hypothesis (p-value < 0.04).

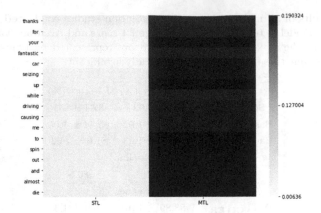

Fig. 2. The visualization of the learned weights for a tweet from Att_{intra} layer- x_1: "thanks @VW for your fantastic car seizing up while driving causing me to spin out and almost die" for STL_{CI} (baseline), and MTL_{All}. The actual label of the tweet is complaint.

Subtle Complaints: Complaints that are conveyed in an implied or subtle manner are wrongly predicted as non-complaint. One of the reasons behind this could be the absence of direct accusation or blame by the complainant. For example, *'Expired beer that we bought today? Whuuuut? @budlight'*. The predicted class is non-complaint whereas, the actual class for the instance is complaint. The tweet does not directly accuse a specific person or company except the negative term (expired) for the beer quality, so the model mispredicts it.

Composite Sentences: Many of the tweets in the dataset consist of multiple short sentences with mixed emotions (more than one emotion) for a single tweet. In such scenarios, learning specific complaint features becomes challenging. For example, *'@BlackandDecker I need a phone number or an email address to receive assistance about a product. Thank you! Sent 2 emails but I did not receive any notifications back.'*; predicted class: non-complaint. The correct class for the preceding example is complaint, but because of the statement's composite nature and contrasting context, it is misclassified as a non-complaint.

Interrogative Complaints: It was also observed that the model misclassifies instances which convey weak dissatisfaction with questioning tone as non-complaints. For example, *'@CBSNews @Dodge @ChryslerCares My driver side air bag has been recalled and replaced, but what about the passenger side?'*. The appropriate class for the preceding example is complaint; because there is no explicit criticism and the statement is interrogative, the model labels it as non-complaint. The system requires contextual understanding and learning the key features in a sentence to accurately classify such cases.

Table 3 shows some sample predictions from the proposed model and the single task (complaint variant) model. For the 1st and 2nd instances in Table 3 the proposed MTL_{All} system correctly predicts the instances as complaints, even

Table 3. Sample sentences with their predicted labels for the best performing multi-task (MTL$_{All}$) model and its single task CI counterpart (STL$_{CI}$). Non-Com: Non-Complaint, Com: Complaint

Sentence	MTL$_{All}$	STL$_{CI}$	Actual label
Very nice @BestBuy was placed on a 48 minute hold just to get hung up on. Great customer service	Com	Non-Com	Com
Wow. Nice work @UPS. Days late AND smashed. Awesome @UPSHelp	Com	Non-Com	Com
@divinecomely @derickmdillard Don't come around here with your facts and common sense. Derick doesn't like that	Non-Com	Com	Non-Com
Who needs a gym when you can work retail during the holidays	Non-Com	Com	Non-Com

though instances express complaint in a sarcastic manner that conveys disapproval humorously. Whereas, in the case of 3rd and 4th instances, even though the cases are satirical in nature, the MTL$_{All}$ model recognizes that there is no breach of expectation. These examples demonstrate the significance of sentiment, emotion, and sarcasm awareness in identifying complaints.

6   Conclusion and Future Work

This work proposes an attention-based, adversarial multi-task framework for simultaneous optimization of complaint classification, emotion recognition, sentiment analysis, and sarcasm detection. We extend an existing *Complaints* dataset by manually annotating emotion, sentiment, and sarcasm labels because no relevant labeled data was available for this problem. Empirical evaluation results on the extended version of the *Complaints* dataset illustrate the effectiveness of our proposed approach for complaint detection over the existing state-of-the-art system. The evaluation also revealed that the proposed adversarial multi-tasking framework yields better performance for the primary task, complaint detection, with the support of emotion recognition, sentiment analysis, and sarcasm detection, the three secondary tasks in our setting.

In the future, we plan to analyze other key contributing factors in complaint identification, such as aspect-based analysis and text politeness markers.

Acknowledgement. This publication is an outcome of the R&D work undertaken in the project under the Visvesvaraya Ph.D. Scheme of Ministry of Electronics & Information Technology, Government of India, being implemented by Digital India Corporation (Formerly Media Lab Asia).

References

1. Akhtar, M.S., Chauhan, D.S., Ghosal, D., Poria, S., Ekbal, A., Bhattacharyya, P.: Multi-task learning for multi-modal emotion recognition and sentiment analysis. arXiv preprint arXiv:1905.05812 (2019)
2. Akhtar, S., Ghosal, D., Ekbal, A., Bhattacharyya, P., Kurohashi, S.: All-in-one: Emotion, sentiment and intensity prediction using a multi-task ensemble framework. IEEE Trans. Affect. Comput. 13(1), 285–297 (2019)
3. Cho, K., van Merrienboer, B., Bahdanau, D., Bengio, Y.: On the properties of neural machine translation: Encoder-decoder approaches. In: Wu, D., Carpuat, M., Carreras, X., Vecchi, E.M. (eds.) Proceedings of SSST@EMNLP 2014, Eighth Workshop on Syntax, Semantics and Structure in Statistical Translation, Doha, Qatar, 25 October 2014, pp. 103–111. Association for Computational Linguistics (2014). https://doi.org/10.3115/v1/W14-4012, https://www.aclweb.org/anthology/W14-4012/
4. Chollet, F., et al.: keras (2015)
5. Cohen, J.: A coefficient of agreement for nominal scales. Educ. Psychol. Meas. 20(1), 37–46 (1960)
6. Coussement, K., Van den Poel, D.: Improving customer complaint management by automatic email classification using linguistic style features as predictors. Decis. Support Syst. 44(4), 870–882 (2008)
7. Devlin, J., Chang, M., Lee, K., Toutanova, K.: BERT: pre-training of deep bidirectional transformers for language understanding. In: Burstein, J., Doran, C., Solorio, T. (eds.) Proceedings of the 2019 Conference of the North American Chapter of the Association for Computational Linguistics: Human Language Technologies, NAACL-HLT 2019, Minneapolis, MN, USA, 2–7 June 2019, vol. 1 (Long and Short Papers), pp. 4171–4186. Association for Computational Linguistics (2019). https://doi.org/10.18653/v1/n19-1423
8. Do, H.H., Prasad, P., Maag, A., Alsadoon, A.: Deep learning for aspect-based sentiment analysis: a comparative review. Expert Syst. Appl. 118, 272–299 (2019)
9. Ekman, P., et al.: Universals and cultural differences in the judgments of facial expressions of emotion. J. Pers. Soc. Psychol. 53(4), 712 (1987)
10. Ganin, Y., Lempitsky, V.: Unsupervised domain adaptation by backpropagation. In: International Conference on Machine Learning, PMLR, pp. 1180–1189 (2015)
11. Glorot, X., Bordes, A., Bengio, Y.: Deep sparse rectifier neural networks. In: Proceedings of the Fourteenth International Conference on Artificial Intelligence and Statistics, pp. 315–323 (2011)
12. Jin, M., Aletras, N.: Complaint identification in social media with transformer networks. In: Scott, D., Bel, N., Zong, C. (eds.) Proceedings of the 28th International Conference on Computational Linguistics, COLING 2020, Barcelona, Spain (Online), 8–13 December 2020, pp. 1765–1771. International Committee on Computational Linguistics (2020). https://doi.org/10.18653/v1/2020.coling-main.157
13. Kakolaki, L.N., Shahrokhi, M.: Gender differences in complaint strategies among Iranian upper intermediate EFL students. Stud. Engl. Lang. Teach. 4(1), 1–15 (2016)
14. Kingma, D.P., Ba, J.: Adam: a method for stochastic optimization. In: Bengio, Y., LeCun, Y. (eds.) 3rd International Conference on Learning Representations, ICLR 2015, San Diego, CA, USA, 7–9 May 2015, Conference Track Proceedings (2015). http://arxiv.org/abs/1412.6980

15. Kumar, A., Ekbal, A., Kawahra, D., Kurohashi, S.: Emotion helps sentiment: a multi-task model for sentiment and emotion analysis. In: 2019 International Joint Conference on Neural Networks (IJCNN), pp. 1–8. IEEE (2019)
16. Lailiyah, M., Sumpeno, S., Purnama, I.E.: Sentiment analysis of public complaints using lexical resources between Indonesian sentiment lexicon and Sentiwordnet. In: 2017 International Seminar on Intelligent Technology and Its Applications (ISI-TIA), pp. 307–312. IEEE (2017)
17. Lewis, M., Haviland-Jones, J.M., Barrett, L.F.: Handbook of Emotions. Guilford Press, New York (2010)
18. Liu, P., Qiu, X., Huang, X.: Adversarial multi-task learning for text classification. arXiv preprint arXiv:1704.05742 (2017)
19. Majumder, N., Poria, S., Hazarika, D., Mihalcea, R., Gelbukh, A., Cambria, E.: DialogueRNN: an attentive RNN for emotion detection in conversations. In: Proceedings of the AAAI Conference on Artificial Intelligence, vol. 33, pp. 6818–6825 (2019)
20. Olshtain, E., Weinbach, L.: 10. complaints: a study of speech act behavior among native and non-native speakers of Hebrew. In: The pragmatic perspective, p. 195. John Benjamins (1987)
21. Pedregosa, F., et al.: Scikit-learn: machine learning in python. J. Mach. Learn. Res. **12**, 2825–2830 (2011)
22. Pennington, J., Socher, R., Manning, C.: GloVe: global vectors for word representation. In: Proceedings of the 2014 Conference on Empirical Methods in Natural Language Processing (EMNLP), pp. 1532–1543 (2014)
23. Poria, S., Cambria, E., Hazarika, D., Vij, P.: A deeper look into sarcastic tweets using deep convolutional neural networks. arXiv preprint arXiv:1610.08815 (2016)
24. Preotiuc-Pietro, D., Gaman, M., Aletras, N.: Automatically identifying complaints in social media. In: Korhonen, A., Traum, D.R., Màrquez, L. (eds.) Proceedings of the 57th Conference of the Association for Computational Linguistics, ACL 2019, Florence, Italy, 28 July – 2 August 2019, vol. 1: Long Papers, pp. 5008–5019. Association for Computational Linguistics (2019). https://doi.org/10.18653/v1/p19-1495
25. Reimers, N., Gurevych, I.: Sentence-bert: Sentence embeddings using siamese bert-networks. CoRR abs/1908.10084 (2019), http://arxiv.org/abs/1908.10084
26. Saha, T., Patra, A., Saha, S., Bhattacharyya, P.: Towards emotion-aided multi-modal dialogue act classification. In: Proceedings of the 58th Annual Meeting of the Association for Computational Linguistics, pp. 4361–4372 (2020)
27. Saha, T., Upadhyaya, A., Saha, S., Bhattacharyya, P.: Towards sentiment and emotion aided multi-modal speech act classification in Twitter. In: Proceedings of the 2021 Conference of the North American Chapter of the Association for Computational Linguistics: Human Language Technologies, pp. 5727–5737 (2021)
28. Shmueli, B., Ku, L.W., Ray, S.: Reactive supervision: a new method for collecting sarcasm data. In: Proceedings of the 2020 Conference on Empirical Methods in Natural Language Processing (EMNLP), pp. 2553–2559. Association for Computational Linguistics, November 2020. https://doi.org/10.18653/v1/2020.emnlp-main.201, https://www.aclweb.org/anthology/2020.emnlp-main.201
29. Singh, A., Saha, S.: Are you really complaining? A multi-task framework for complaint identification, emotion, and sentiment classification. In: Lladós, J., Lopresti, D., Uchida, S. (eds.) ICDAR 2021, Part II. LNCS, vol. 12822, pp. 715–731. Springer, Cham (2021). https://doi.org/10.1007/978-3-030-86331-9_46
30. Singh, A., Saha, S., Hasanuzzaman, M., Dey, K.: Multitask learning for complaint identification and sentiment analysis. Cogn. Comput. **14**(1), 1–16 (2021)

31. Singh, A., Saha, S., Hasanuzzaman, M., Jangra, A.: Identifying complaints based on semi-supervised mincuts. Expert Syst. Appl. **186**, 115668 (2021)
32. Srivastava, N., Hinton, G., Krizhevsky, A., Sutskever, I., Salakhutdinov, R.: Dropout: a simple way to prevent neural networks from overfitting. J. Mach. Learn. Res. **15**(1), 1929–1958 (2014)
33. Trosborg, A.: Interlanguage Pragmatics: Requests, Complaints, and Apologies, vol. 7. Walter de Gruyter, Berlin (2011)
34. Van Noort, G., Willemsen, L.M.: Online damage control: The effects of proactive versus reactive Webcare interventions in consumer-generated and brand-generated platforms. J. Interact. Mark. **26**(3), 131–140 (2012)
35. Vásquez, C.: Complaints online: The case of TripAdvisor. J. Pragmat. **43**(6), 1707–1717 (2011)
36. Vaswani, A., et al.: Attention is all you need. arXiv preprint arXiv:1706.03762 (2017)
37. Welch, B.L.: The generalization of 'student's'problem when several different population varlances are involved. Biometrika **34**(1–2), 28–35 (1947)
38. Yang, W., et al.: Detecting customer complaint escalation with recurrent neural networks and manually-engineered features. In: Proceedings of the 2019 Conference of the North American Chapter of the Association for Computational Linguistics: Human Language Technologies, vol. 2 (Industry Papers), pp. 56–63 (2019)

Joint Personalized Search and Recommendation with Hypergraph Convolutional Networks

Thibaut Thonet[1(✉)], Jean-Michel Renders[1], Mario Choi[2], and Jinho Kim[2]

[1] NAVER LABS Europe, Meylan, France
{thibaut.thonet,jean-michel.renders}@naverlabs.com
[2] NAVER Corporation, Seongnam, South Korea
{mar.io,jinho.kim}@navercorp.com

Abstract. Traditionally, the search and recommendation tasks are performed separately, by distinct models. Having a unique model for the two tasks is however particularly appealing for platforms that offer search and recommendation services to a shared user base over common items. In this paper, we study this unification scenario denoted as Joint Personalized Search and Recommendation (JPSR). To tackle this problem, we introduce HyperSaR, an hypergraph convolutional approach for search and recommendation. From the interaction data, we first build an hypergraph composed of user, item and query keyword nodes in which recommendation instances form user-item edges and search instances define user-item-query hyperedges. We then propagate user, item and query keyword embeddings using hypergraph convolution, and train HyperSaR with the combination of two complementary losses. The first one amounts to assessing the probability of an interaction, while the second one aims at predicting the query of a search interaction given a (user, item) pair. The proposed method is evaluated on the JPSR task using three datasets: a real-world, industrial dataset, and the public MovieLens and Lastfm datasets, which have been adapted to the task. Our experiments demonstrate the superior effectiveness of HyperSaR over competing approaches.

Keywords: Recommendation · Graph Neural Networks · Information retrieval · Personalized search · Hypergraph

1 Introduction

While Information Retrieval (IR) and Recommender Systems (RS) have long been regarded as two distinct facets of the information filtering problem, nowadays this distinction is becoming increasingly blurred. For instance, numerous e-commerce platforms offer users hybrid ways of exploring a large base of heterogeneous items, by deploying a combination of search and recommendation tools to accurately identify what users need or prefer in this vast maze of possibilities. The emergence of mixed-initiative conversational agents and chatbots is another example of applications where IR and RS steps are intimately connected to solve a joint session-based objective [30]. Roughly speaking, IR and RS tasks mainly differ by the presence of a query: IR methods are essentially content-based methods and rely on a

© The Author(s), under exclusive license to Springer Nature Switzerland AG 2022
M. Hagen et al. (Eds.): ECIR 2022, LNCS 13185, pp. 443–456, 2022.
https://doi.org/10.1007/978-3-030-99736-6_30

query understanding or query representation step, in order to effectively match the query with the indexed items. RS methods do not use any query information, but rely on the knowledge of the recommendation context and past user history to score the indexed items. In this paper, we propose to solve a single, unified task, namely "Joint Personalized Search and Recommendation" (JPSR), which consists in ranking a list of items by decreasing order of relevance to user needs or preferences in a given context. For this task, some instances are triggered by a given query (search instances, initiated by the user), while others are not associated with any query (recommendation instances, initiated by the system).

Recently, Graph Neural Networks (GNNs) have been successfully applied to recommendation tasks [9,22,24,26], especially to Collaborative Filtering (CF) problems, where only the interactions between users and items are considered (ratings, clicks, views, purchases, etc.). They basically combine the expressive power of deep neural networks with the exploitation of the structural information given by an underlying graph, by propagating and smoothing node information through convolution operations. This results in graph-contextualized, multihop-aware node embeddings that are then used to solve the final task. Standard CF problems involve two-way interactions, i.e., interactions over (user, item) pairs. Naturally, these interactions can be represented as edges, possibly weighted, in a bipartite graph, and that is the approach followed by most GNN-based recommendation methods. However, when considering n-way interactions with $n > 2$, such as search instances where the outcome of the interaction over a (user, item, query) triplet is observed, bipartite graphs can no longer be adopted. A natural extension is to rely on hypergraphs, where hyperedges precisely capture these n-way interactions, for any possible $n \geq 2$. For that purpose, we introduce in this work HyperSaR, an hypergraph convolution network approach to solve the JPSR problem. The contributions of this paper can be summarized as follows:

- We introduce and formalize the Joint Personalized Search and Recommendation (JPSR) problem.
- We propose HyperSaR, an hypergraph convolutional network approach, to address JPSR.
- We empirically validate HyperSaR and compare it against existing methods on three datasets: a real-world, industrial dataset collected from a commercial search engine widely used in South Korea, and two public datasets, MovieLens and Lastfm, which have been adapted to the JPSR task.
- To foster research on this new problem, we publicly release the code of our approach and baselines, along with the adapted public datasets.[1]

The paper is organized as follows. We first discuss how existing research is related to this work (Sect. 2). We then formally define the JPSR problem (Sect. 3), and describe our proposal HyperSaR to address this problem (Sect. 4). The subsequent sections detail the experimental setup (Sect. 5) and present the results (Sect. 6). Finally, we conclude the paper and propose extra directions to investigate (Sect. 7).

---

[1] https://github.com/naver/hypersar.

2   Related Work

Personalized Search and Recommendation. The research line that is most relevant to this work pertains to unification of search and recommendation through a single model [25,27,28]. [25] investigates the problem of transferring learned knowledge from a recommender system to a search engine defined over a common item base. Although the motivation is similar to ours – i.e., sharing data across search and recommendation tasks – the approach devised in [25] only allows transfer in the recommendation-to-search direction and thus cannot be used for joint personalized search and recommendation. Closer to our work, [27,28] also investigate a joint search and recommendation (JSR) problem. In [28], items are associated with text descriptions that are used along with user-item interactions to train the model. In that regard, the JSR model proposed by [28] is closer to an hybrid recommendation model – combining collaborative filtering and content-based filtering based on item text descriptions – as it was not designed to be trained on both search and recommendation instances. More-over, [27,28] ignore the dependency between users and queries, which is crucial to provide personalized results. On the other hand, existing personalized search works [1–3,11] often rely heavily on text features to match queries and items, or are not explicitly designed to address the empty query case (i.e., recommendation instance). In contrast, the proposed HyperSaR approach links user, item and (possibly empty) query through an hypergraph formulation, thus naturally enabling the integration of both search and recommendation interactions.

Graph Neural Networks for Recommendation. Our work is also grounded in research on graph neural networks which recently received much attention in the recommendation field, due to their ability to capture multihop relationships between users and items. Given the large body of work on this topic, we will only review here selected papers and let the reader refer to [24] for a more exhaustive survey of the literature. One of the early GNN-based recommendation models is NGCF [22], which relies on a message passing architecture and a complex graph propagation operation. Later, the LightGCN [9] model was proposed, advocating for a simpler and more lightweight architecture than the one introduced in NGCF. Such architecture resulted in improved effectiveness over NGCF, thus questioning common GNN practices in recommendation. In the proposed HyperSaR model, we follow the recommendations made in [9] to simplify graph convolutions to their bare minimum: a weighted sum aggregation. While most GNN approaches for recommendation operate on standard graphs, some recent works also devised hypergraph-based approaches to capture item correlation over different time periods [20,21]. These two works are however specifically designed for next-item recommendation and cannot be easily adapted to JPSR.

3   Problem Definition

Before defining the problem we seek to tackle, we first detail the notations used throughout the paper. Let $\mathcal{U}$ and $\mathcal{I}$ be the set of users and items, respectively.

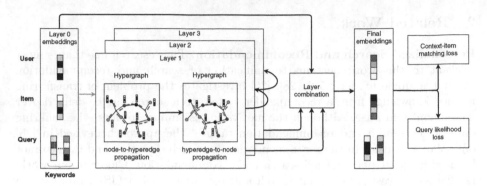

Fig. 1. Illustration of the HyperSaR approach.

We define as $\mathcal{X} = \mathcal{X}_S \cup \mathcal{X}_R$ the set of training interactions, where $\mathcal{X}_S$ and $\mathcal{X}_R$ respectively denote the search instances and recommendation instances. Each search instance $x \in \mathcal{X}_S$ is a tuple composed of a user $u_x \in \mathcal{U}$, a query q_x formulated by this user, and an item $i_x \in \mathcal{I}$ which the user interacted with for this query. A recommendation instance $x \in \mathcal{X}_R$ is a pair made of a user $u_x \in \mathcal{U}$ and an interacted item $i_x \in \mathcal{I}$. Although the nature of interactions considered in this framework is not constrained, we typically consider here implicit feedback (e.g., clicks on items, check-ins for points-of-interest, video views) as this constitutes the most widely available signal in a realistic setting. The problem of Joint Personalized Search and Recommendation (JPSR) then consists in learning a model from the search and recommendation instances in $\mathcal{X}$ which maps items to their relevance scores given a user and a (potentially empty) query.

4   Method

To address the JPSR problem, we propose HyperSaR – an **Hyper**graph convolutional network approach for **S**earch **a**nd **R**ecommendation, illustrated in Fig. 1. As we will detail in the remainder of this section, HyperSaR defines an hypergraph-based formulation of the problem to effectively integrate heterogeneous (search and recommendation) interactions in a unified framework.

4.1   Hypergraph Construction

The HyperSaR approach relies on an undirected hypergraph composed of user nodes, item nodes, and query keyword nodes. In this hypergraph, a search instance gives an hyperedge linking a user, an item and one or several query keywords, while a recommendation instance becomes an hyperedge between a user and an item. Formally, let $\mathcal{H} = (\mathcal{V}, \mathcal{E})$ denote the hypergraph with nodes (or vertices) $\mathcal{V}$ and hyperedges $\mathcal{E}$. The set $\mathcal{V}$ contains user nodes $\mathcal{U}$, item nodes $\mathcal{I}$, and query keyword nodes $\mathcal{W}$, and $\mathcal{E}$ includes the search hyperedges $\mathcal{E}_S \subset \bigcup_{k=1}^{+\infty}(\mathcal{U} \times \mathcal{I} \times \mathcal{W}^k)$ and the recommendation hyperedges $\mathcal{E}_R \subset \mathcal{U} \times \mathcal{I}$. The hyperedges $\mathcal{E}_S$ and $\mathcal{E}_R$ are respectively built from the search instances $\mathcal{X}_S$ and recommendation instances $\mathcal{X}_R$. Given a search instance $x = (u_x, i_x, q_x) \in \mathcal{X}_S$ and the keywords $(w_{x,1}, \ldots, w_{x,k}) \in \mathcal{W}^k$ extracted[2] from the query q_x, we create a $(k+2)$-degree hyperedge $e = \{u_x, i_x, w_{x,1}, \ldots, w_{x,k}\} \in \mathcal{E}_S$. Similarly, a recommendation instance $x = (u_x, i_x) \in \mathcal{X}_R$ yields a 2-degree hyperedge (i.e., ordinary edge) $e = \{u_x, i_x\} \in \mathcal{E}_R$. An example of such hypergraph is illustrated in Fig. 2.

Similarly to the characterization of a graph by an adjacency matrix, an hypergraph is uniquely identified by its *incidence matrix* [4,6,7,20]. The incidence matrix H of the hypergraph $\mathcal{H}$ is a $|\mathcal{V}| \times |\mathcal{E}|$ binary matrix defined as follows:

$$H_{ij} = \begin{cases} 1 & \text{if } v_i \in e_j, \\ 0 & \text{otherwise.} \end{cases}$$

where v_i is the node with index $i \in \{1, \ldots, |\mathcal{V}|\}$ and e_j is the hyperedge with index $j \in \{1, \ldots, |\mathcal{E}|\}$. Intuitively, a row in matrix H indicates all the hyperedges a node is part of, and a column conversely gives all the nodes an hyperedge connects.

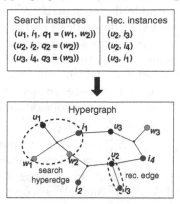

Fig. 2. Example of hypergraph built from search and recommendation instances. Nodes u_*, i_* and w_* denote users, items and query keywords.

4.2 HyperSaR Convolution Operation

We now move on to describing the propagation mechanism of HyperSaR, which aims to smooth node representations based on their neighborhood in the hypergraph detailed in Sect. 4.1. We first associate each node i with a base embedding $E_{\mathcal{V},i}^{(0)}$ – its 'layer-0' embedding. An hypergraph convolution operation can then be decomposed in two steps [6]: (i) the node embeddings are propagated to the hyperedges they belong to and aggregated as the hyperedge embeddings, (ii) the hyperedge embeddings are propagated back to the nodes and aggregated to form the new node embeddings. In this work, we apply to hypergraphs the methodology demonstrated in recent GNN research [23], and in particular on recommendation [5,9], which observed that the GNN aggregation operators need not contain any non-linearity, nor trainable weight matrix but often prove to be more effective as a simple weighted sum.

Formally, we denote as $E_{\mathcal{V},i}^{(k)}$ and $E_{\mathcal{E},j}^{(k)}$ the layer-k embeddings of node i and hyperedge j, respectively. Adopting the sum aggregation operator, along with standard normalization based on node and hyperedge degrees [4,7,20], the HyperSaR convolution operation is defined as follows:

---

[2] In practice, query keywords are obtained by retaining frequent query terms and discarding stop words (see Sect. 5.3). One could alternatively extract intents using query intent understanding techniques (e.g., [29]) but this is beyond the scope of this work.

$$E_{\mathcal{E}}^{(k)} = \left(D_{\mathcal{E}}^{-1/2} H^T D_{\mathcal{V}}^{-1/2}\right) E_{\mathcal{V}}^{(k-1)} \qquad \textit{(node-to-hyperedge propagation)}$$

$$E_{\mathcal{V}}^{(k)} = \left(D_{\mathcal{V}}^{-1/2} H D_{\mathcal{E}}^{-1/2}\right) E_{\mathcal{E}}^{(k)} \qquad \textit{(hyperedge-to-node propagation)} \tag{1}$$

where H is the incidence matrix of the hypergraph, $D_{\mathcal{E}} = \mathrm{diag}\big(|\mathcal{N}_j^{\mathcal{E}}| : j \in \mathcal{E}\big) \in \mathbb{R}^{|\mathcal{E}| \times |\mathcal{E}|}$ is the hyperedge degree matrix, and $D_{\mathcal{V}} = \mathrm{diag}\big(|\mathcal{N}_i^{\mathcal{V}}| : i \in \mathcal{V}\big) \in \mathbb{R}^{|\mathcal{V}| \times |\mathcal{V}|}$ is the node degree matrix. $\mathcal{N}_j^{\mathcal{E}}$ and $\mathcal{N}_i^{\mathcal{V}}$ denote respectively the set of nodes connected by hyperedge j and the set of hyperedges node i is part of. In summary, the core specificities of hypergraph convolution in comparison to standard graph convolution (e.g., that of LightGCN [9]) lie in (i) the use of the incidence matrix instead of the adjacency matrix, and (ii) the two-step propagation mechanism with a node-to-hyperedge step and an hyperedge-to-node step.

After L layers of propagation in the hypergraph, we obtain the node embeddings of each layer $\{E_{\mathcal{V}}^{(0)}, E_{\mathcal{V}}^{(1)}, \dots, E_{\mathcal{V}}^{(L)}\}$. Each $E_{\mathcal{V}}^{(k)}$ can be decomposed into $(E_{\mathcal{U}}^{(k)}, E_{\mathcal{I}}^{(k)}, E_{\mathcal{W}}^{(k)})$ which correspond respectively to the user embeddings, item embeddings and query keyword embeddings at layer k. We aggregate the embeddings at different layers using a convex combination [9] with hyperparameter weights $\alpha^{\mathcal{U}}$, $\alpha^{\mathcal{I}}$ and $\alpha^{\mathcal{W}}$ whose practical choice is further discussed in Sect. 5.3.

$$E_{\mathcal{U}} = \sum_{k=0}^{L} \alpha_k^{\mathcal{U}} E_{\mathcal{U}}^{(k)}, \quad E_{\mathcal{I}} = \sum_{k=0}^{L} \alpha_k^{\mathcal{I}} E_{\mathcal{I}}^{(k)}, \quad E_{\mathcal{W}} = \sum_{k=0}^{L} \alpha_k^{\mathcal{W}} E_{\mathcal{W}}^{(k)}. \tag{2}$$

4.3   Model Training and Inference

Given the final user embeddings $E_{\mathcal{U}}$, item embeddings $E_{\mathcal{I}}$, and query keyword embeddings $E_{\mathcal{W}}$ obtained from the propagation in the hypergraph, we now define the loss used to train the HyperSaR model's parameters – which solely consist of the layer-0 embeddings $E_{\mathcal{V}}^{(0)}$. Our training loss is in fact the combination of two losses: the context-item matching loss and the query likelihood loss.

Context-Item Matching Loss. For a given interaction, the purpose of this loss is to bring the embeddings of the user and query keywords (in the case of search interactions) closer to the embedding of the interacted item. We designate here the user and the query as the *context*. To match the context and the item embeddings, we adopt a dot product because of its demonstrated effectiveness in comparison to more complex scoring operations, and efficiency at inference time [15]. Formally, given an interaction x consisting of a user u_x, an item i_x and a (potentially empty) query q_x, the context-item matching (CIM) score is then computed as the sum of dot products between pairs of embeddings:

$$\hat{y}_{u_x, i_x, q_x} = E_{\mathcal{U}, u_x}{}^T E_{\mathcal{I}, i_x} + E_{\mathcal{U}, u_x}{}^T E_{\mathcal{W}, q_x} + E_{\mathcal{I}, i_x}{}^T E_{\mathcal{W}, q_x}. \tag{3}$$

The query embedding $E_{\mathcal{W}, q_x}$ is defined as the sum of its terms' embeddings $\sum_{i=1}^{k} E_{\mathcal{W}, w_{x,i}}$ if the query $q_x = (w_{x,1}, \dots, w_{x,k})$ is non-empty and 0 otherwise. We adopt here a sum operator instead of a mean operator to use the number of keywords as a 'confidence' factor in the loss by modulating the score magnitude.

The CIM score is finally integrated in the pairwise Bayesian Personalized Ranking (BPR) loss [14], yielding the context-item matching loss:

$$L_{\text{CIM}} = -\frac{1}{|\mathcal{X}|} \sum_{x \in \mathcal{X}} \sum_{i_n \in \mathcal{I} \setminus \{i_x\}} \log \sigma(\hat{y}_{u_x, i_x, q_x} - \hat{y}_{u_x, i_n, q_x}) + \lambda \left\| E_{\mathcal{V}}^{(0)} \right\|_F^2 \qquad (4)$$

where $i_n \in \mathcal{I} \setminus \{i_x\}$ denotes a negative item, σ is the sigmoid function, and λ balances the strength of the L_2 regularization on the model parameters $E_{\mathcal{V}}^{(0)}$.

Query Likelihood Loss. We combine the CIM loss with a query likelihood loss, which was shown to help making the embeddings more suitable for the retrieval task [28]. The proposed loss defined on search instances measures to what extent a user and an item are likely to 'produce' the query keywords. It then complements the CIM loss by considering query keywords in isolation rather than aggregated in the query embedding. First, we define user-specific and item-specific probability distributions over query keywords using a softmax formulation as follows:

$$p(w \mid u) = \frac{\exp(E_{\mathcal{U}, u}{}^T E_{\mathcal{W}, w})}{\sum_{w' \in \mathcal{W}} \exp(E_{\mathcal{U}, u}{}^T E_{\mathcal{W}, w'})}, \; p(w \mid i) = \frac{\exp(E_{\mathcal{I}, i}{}^T E_{\mathcal{W}, w})}{\sum_{w' \in \mathcal{W}} \exp(E_{\mathcal{I}, i}{}^T E_{\mathcal{W}, w'})} \quad (5)$$

Intuitively, the probability $p(w \mid u)$ is large when user u and query keyword w are close in the embedding space, and similarly for $p(w \mid i)$. The query likelihood loss is then obtained from the sum of the likelihood according to $p(w \mid u_x)$ and the likelihood according to $p(w \mid i_x)$ for all search interaction $x \in \mathcal{X}_\mathcal{S}$:

$$L_{\text{QL}} = -\frac{1}{|\mathcal{X}_\mathcal{S}|} \sum_{x \in \mathcal{X}_\mathcal{S}} \frac{1}{|q_x|} \sum_{w \in q_x} \log p(w \mid u_x) - \frac{1}{|\mathcal{X}_\mathcal{S}|} \sum_{x \in \mathcal{X}_\mathcal{S}} \frac{1}{|q_x|} \sum_{w \in q_x} \log p(w \mid i_x) \quad (6)$$

Note that the loss is computed separately from $p(w \mid u_x)$ and $p(w \mid i_x)$, instead of using a single (user, item)-specific distribution $p(w \mid u_x, i_x)$. The motivation for this is to avoid conflating the contribution of the user and that of the item to the query reconstruction, which we found to be detrimental in practice.

Ultimately, the two losses are linearly combined using a balancing hyperparameter η as $L = L_{\text{CIM}} + \eta L_{\text{QL}}$. The model parameters – the layer-0 embeddings $E_{\mathcal{V}}^{(0)}$ – are then learned by minimizing L using stochastic gradient descent.

Model inference. At inference time, an interaction x consists of either a user-query pair (u_x, q_x) (*search case*) or a user u_x and an empty query $q_x = \emptyset$ (*recommendation case*). The goal is for the trained model to produce an item ranking that maximizes utility with respect to the user's information needs and profile. To that end, all items $i \in \mathcal{I}$ are scored based on Eq. 3. This yields the set of context-item matching scores $\{\hat{y}_{u_x, i, q_x}\}_{i \in \mathcal{I}}$, which are then sorted in descending order to form the final item ranking returned to the user.

Table 1. Statistics of the datasets used in the experiments.

Dataset	# users	# items	# search interactions			# recommendation interactions		
			train	valid	test	train	valid	test
POI	22,110	20,376	171,028	50,060	52,132	315,932	97,483	95,411
MovieLens	14,478	18,332	549,517	232,426	223,915	1,328,341	383,884	392,392
Lastfm	1,892	18,022	113,266	36,662	36,551	56,441	18,196	18,197

5   Experimental Setup

5.1   Datasets

Our experiments are performed on three datasets whose statistics are shown in Table 1: an industrial dataset based on a Point-of-Interest search engine (denoted as **POI**), as well as two public datasets, **MovieLens** and **Lastfm**, which have been adapted to our JPSR task due to the lack of existing datasets.[3]

POI. This dataset was collected from the South-Korean search engine Naver[4] between 01/09/2020 and 07/09/2020. Items correspond to South-Korean POIs. Interactions consist of clicks and are either search interactions (user, clicked item, and query) or recommendation interactions (user and clicked item, without query information). Queries are in Korean and were tokenized using KoNLPy [12]. Core-20 filtering was applied to discard low-activity users and items.

MovieLens. The MovieLens-25M[5] dataset contains two types of interactions: *rating interactions* (a user assigns a rating to a movie), and *tagging interactions* (a user adds a tag to a movie). The former are used as recommendation interactions, and the latter as search interactions by assimilating tags and queries. Rating interactions were binarized, considering 4+ ratings as relevant. Users without tagging interactions were discarded, and core-10 filtering was applied.

Lastfm. In Lastfm[6], items are musical artists that users listen to. This dataset consists of *listening interactions* and *tagging interactions*. As on MovieLens, we consider the former as recommendation interactions and the latter as search interactions. No core filtering was needed for this already dense dataset.

5.2   Baselines

In our experiments, we compare the proposed HyperSaR model against several competitive baselines. We first considered recommendation approaches – which

---

[3] Some previous works [1–3,11,27,28] used Amazon datasets for product search. However the common practice [19] is to define synthetic queries from *product categories*, which do not result from a user-specific expression. Therefore, we advocate that such datasets are not suitable for personalized search and, a fortiori, for JPSR.

[4] https://www.naver.com/.

[5] https://grouplens.org/datasets/movielens/25m/.

[6] https://grouplens.org/datasets/hetrec-2011/.

Table 2. Comparison of the model traits for the baselines and proposed approach.

Model trait	MF [10]	LGCN [9]	FM [13]	DeepFM [8]	JSR [28]	BM25 [17]	DREM [2]	HyperSaR
(Hyper)graph propagation	✗	✓	✗	✗	✗	✗	✗	✓
Query exploitation	✗	✗	✓	✓	✓	✓	✓	✓
Result personalization	✓	✓	✓	✓	✓ (rec.) ✗ (search)	✗	✓	✓

do not exploit query information – to show that such approaches are insufficient for the JPSR task:[7] the widely-adopted Matrix Factorization (**MF**) [10]; and the state-of-the-art recommendation model LightGCN (**LGCN**) [9] based on graph neural networks. We also experimented with recommendation approaches that can integrate queries in their modeling: the Factorization Machine (**FM**) [13] which extends matrix factorization by enabling the integration of context features (here, the query keywords); the **DeepFM** model [8] which is a deep learning-based extension of FM; and **JSR** [28][8] which was the first approach to unify search and recommendation in a single model. The word embeddings used in JSR are from FastText[9]. Finally, we include search approaches in the comparison: **BM25** [17][10] is a simple yet strong retrieval baseline; and **DREM** [2] is a recent personalized search approach. Note that BM25 can only be evaluated on search instances and not on recommendation instances, which have no query. The traits of the different baselines are summarized in Table 2.

To ensure a fair comparison with our approach and due to the unavailability of a public implementation for JSR, we re-implemented all baseline methods in PyTorch for our experiments. We publicly release their code with HyperSaR's.[11]

5.3   Hyperparameter Setting

We used the Adam optimizer and we set the embedding size to 64 (as in [9,22]), the batch size to 1024, the learning rate to 0.001 and the number of negative samples to 1. Negative items are sampled uniformly from the whole set of items $\mathcal{I}$. The L_2 regularization weight was simply fixed to 0 due to its limited

---

[7] For these approaches, the search instances are simply considered as user-item pairs by ignoring queries. The same interaction set is then used as for other methods.

[8] Originally, JSR is based on text descriptions attached to items; we adapt the approach to the JPSR task by using the queries in replacement of item descriptions and by linking each query to its interaction instead of its item.

[9] https://fasttext.cc/.

[10] In our setting, items are not associated with text documents, preventing the usage of standard retrieval methods. To apply BM25, we form documents by concatenating training queries pertaining to the same item, and use them for retrieval on test queries.

[11] https://github.com/naver/hypersar.

impact in our preliminary experiments. For the models which exploit queries (FM, DeepFM, JSR, BM25, DREM, and HyperSaR), we use as query keywords the 2000 most frequent terms which occur in less than 10% of all the queries. The impact of the query keyword vocabulary size is further discussed in §6.2. We set the layer weights (see Eq. 2) $\alpha^{\mathcal{U}}$ and $\alpha^{\mathcal{I}}$ to uniform, i.e., $1/(L+1)$, as in [9]. As to $\alpha^{\mathcal{W}}$, we found that allocating more mass to earlier layers was more beneficial in practice and we thus simply set $\alpha_0^{\mathcal{W}} = 1$ and $\alpha_k^{\mathcal{W}} = 0$ for $k > 0$.

Additionally, we selected on the validation set the following hyperparameters based on the HitRate@20 score: edge dropout rate [16] (for LGCN, HyperSaR), weight dropout rate [18] (for JSR, DeepFM), number of layers (for LGCN, DeepFM, JSR, HyperSaR) and loss balancing weight η (for JSR, HyperSaR). To avoid a costly grid search, we first keep $\eta = 0$ for JSR and HyperSaR while searching for the dropout rate and number of layers, and then optimize η using the optimal dropout rate and number of layers. The range for the dropout rate, number of layers and weight η is defined as $\{0.0, 0.1, 0.2, 0.3\}$, $\{0, 1, 2, 3\}$, and $\{0.0, 0.001, 0.01\}$, respectively.

6   Experiment Results

This section presents the results of the evaluation, which is performed on a *per-interaction basis* – at test time models predict an item ranking for each interaction. Based on the groundtruth consisting of the single interacted item, this ranking is then evaluated in terms of HitRate@20 and NDCG@20 (abbreviated as H@20 and N@20), and results are reported separately on search and recommendation instances. In this section, we first compare the proposed HyperSaR against baselines (Sect. 6.1), and then study the impact of its hyperparameters (Sect. 6.2).

6.1   Evaluation on the JPSR Task

The results of the evaluation on JPSR are presented in Table 3. We performed 5 runs per approach on each dataset with different seeds and report the average performance for each metric. To detect significant differences between our approach's and the best baseline's results, we conducted an unpaired Student t-test with a significance level of 0.01. We also report the p-value and effect size.

Overall, we observe that HyperSaR significantly outperforms all baselines on every (dataset, metric) pair and for both recommendation and search instances, with very large effect size. On recommendation instances, the fact that HyperSaR outperforms the state-of-the-art recommendation approach LGCN shows that the query signal from the search data helped our model to learn more accurate user and item representations. Baselines' performance vary across datasets: on POI, FM was the second-best approach, while on MovieLens and Lastfm LGCN obtained the second-best results. This confirms that LGCN is still a strong contender on recommendation instances, and only the proposed HyperSaR was able to outperform it on the three datasets.

Table 3. Results in % on recommendation and search instances (mean over 5 runs). Best (resp. second-best) result is in bold (resp. underlined). The mark † denotes a statistically significant improvement of the proposed HyperSaR over the best baseline according to an unpaired Student t-test ($p < 0.01$).

Model	POI				MovieLens				Lastfm			
	Rec.		Search		Rec.		Search		Rec.		Search	
	H@20	N@20	H@20	N@20	H@20	N@20	H@20	N@20	H@20	N@20	H@20	N@20
MF [10]	8.72	3.30	7.20	2.87	4.88	1.81	2.75	1.02	23.42	10.36	17.03	7.91
LGCN [9]	9.24	3.52	8.21	3.34	<u>5.74</u>	<u>2.17</u>	3.30	1.23	<u>26.32</u>	<u>12.09</u>	20.34	9.94
FM [13]	<u>9.46</u>	<u>3.59</u>	48.23	26.43	5.08	1.89	12.91	5.57	25.07	11.10	<u>27.35</u>	<u>12.80</u>
DeepFM [8]	9.35	3.54	48.71	26.70	5.33	1.99	<u>13.39</u>	<u>5.88</u>	24.56	10.95	26.82	12.56
JSR [28]	9.05	3.42	59.67	<u>36.76</u>	4.95	1.87	12.35	4.82	24.33	10.68	5.01	1.92
BM25 [17]	–	–	<u>60.23</u>	33.89	–	–	11.06	3.99	–	–	3.20	1.16
DREM [2]	9.28	3.53	24.37	11.37	4.09	1.47	6.84	2.63	23.45	10.32	23.58	10.95
HyperSaR	**10.04**†	**3.84**†	**63.17**†	**39.26**†	**5.93**†	**2.27**†	**21.87**†	**11.04**†	**27.11**†	**12.38**†	**30.05**†	**14.76**†
p-value	6e-7	1e-6	2e-14	1e-12	6e-6	5e-7	1e-14	4e-15	4e-5	1e-4	4e-7	4e-8
Effect size	6.30	5.68	56.89	33.12	4.69	6.48	60.83	67.29	3.66	3.04	6.71	8.86

Turning to the results on the search instances, we can see as expected that the approaches which exploit query information (FM, DeepFM, JSR, BM25, DREM, HyperSaR) markedly outperformed the purely collaborative filtering approaches (MF, LGCN) in most cases. This was indeed predictable as having access to the test-set interaction query constitutes a great advantage to identify the relevant item. Here again, HyperSaR obtained the best performance, by a good margin on MovieLens and Lastfm, and followed closely by JSR and BM25 on POI. We note that the search performance of BM25 and JSR dropped slightly on MovieLens and severely on Lastfm. This could be due to their lack of personalization in the search results: unlike the POI dataset which includes actual queries, MovieLens and Lastfm instead contain tags which may be insufficient to identify the relevant items without considering the tastes of the user who wrote the tag.

6.2 Hyperparameter Impact

After validating HyperSaR's recommendation and search performance in comparison to baselines in Sect. 6.1, we turn to analyzing how hyperparameters impact the model's results. In particular, we study here three hyperparameters: (i) the number of layers, (ii) the size of the query keyword vocabulary, and (iii) η which controls the weight of the query likelihood loss over the context-item matching loss. We investigate their effect on the POI and MovieLens datasets – we omit Lastfm for space reasons and due to its similarity with MovieLens. The results are summarized in Fig. 3. Each plot shows the performance of HyperSaR in terms of H@20 (in %) as a function of a given hyperparameter's value, based on one run. The tested values for the number of layers, the query intent vocabulary size, and the loss balancing weight η correspond to $\{0, 1, 2, 3\}$, $\{1000, 2000, 5000, 10000\}$, and $\{0.0, 0.001, 0.005, 0.01, 0.05, 0.1\}$. Unless an hyperparameter is varied, we use the 'default' value identified on the validation set (see Sect. 5.3)

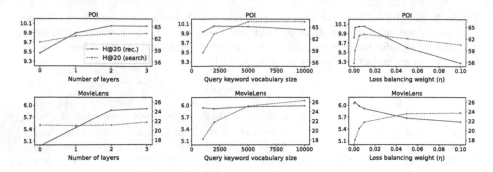

Fig. 3. Results for HyperSaR in H@20 (%) on POI and MovieLens with varied hyperparameters (number of layers, query keyword vocabulary size, and loss balancing weight). Best viewed in color. (Color figure online)

for the number of layers (2 for POI, 3 for MovieLens) and loss balancing weight (0.01 for both), and 2000 for the number of unique query keywords.

We first find that the number of layers is more impactful on recommendation instances than on search instances, although in both cases increasing the number of layers improves performance. On both datasets, a plateau seems to be reached around 2 or 3 layers, which is consistent with the findings of previous works [9, 20]. In contrast, the query keyword vocabulary size has greater impact on search instances than on recommendation instances. This is not surprising, as considering a larger vocabulary implies a greater coverage of the query content. We nonetheless observe a plateau at 5000 or 10000 query keywords, suggesting that further increasing the vocabulary size is not beneficial. Finally, regarding the loss balancing weight η, we note different behaviors across the POI and MovieLens datasets. On POI, a value around 0.01 seems to be optimal for both search and recommendation performance. However, on MovieLens, increasing η overall results in improved search performance and degraded recommendation performance. This disparity could be due to the different nature of the datasets – one is based on 'true' search and recommendation interactions, while the other contains rating and tagging interactions. In a nutshell, these latter results confirm that the query likelihood loss can benefit the HyperSaR model even though it may in some cases imply trading off between search and recommendation performance.

7   Conclusion

This work introduces a novel problem referred to as Joint Personalized Search and Recommendation (JPSR), which consists in learning a unique model from search and recommendation interactions. To address JPSR, we proposed the HyperSaR model based on an hypergraph containing user, item and query keyword nodes built from search and recommendation interactions. Node embeddings are propagated using hypergraph convolution and trained on the combination of a context-item matching objective and a query likelihood objective. HyperSaR is validated on three datasets and shown to significantly outperform baselines.

This work can be extended along several directions. First, we plan to investigate how pre-trained word embeddings could be effectively leveraged in Hyper-SaR to better capture query keyword semantics, similarly to [28]. We also wish to explore how the temporality of the interactions can be integrated in the hypergraph, as in prior work on next-item recommendation [20,21]. Finally, we are interested in exploiting HyperSaR to develop explainable search and recommendation systems by leveraging the intent revealed by the query keywords.

References

1. Ai, Q., Vishwanathan, S.V., Hill, D.N., Bruce Croft, W.: A zero attention model for personalized product search. In: CIKM, pp. 379–388 (2019)
2. Ai, Q., Zhang, Y., Bi, K., Bruce Croft, W.: Explainable product search with a dynamic relation embedding model. ACM Trans. Inf. Syst. **38**(1) (2020)
3. Ai, Q., Zhang, Y., Bi, K., Chen, X., Bruce Croft, W.: Learning a hierarchical embedding model for personalized product search. In: SIGIR, pp. 645–654 (2017)
4. Bai, S., Zhang, F., Torr, P.H.: Hypergraph convolution and hypergraph attention. Pattern Recognit. **110**, 1–30 (2021)
5. Chen, L., Wu, L., Hong, R., Zhang, K., Wang, M.: Revisiting graph based collaborative filtering: a linear residual graph convolutional network approach. In: AAAI, pp. 27–34 (2020)
6. Dong, Y., Sawin, W., Bengio, Y.: HNHN: hypergraph networks with hyperedge neurons. arXiv:2006.12278 (2020)
7. Feng, Y., You, H., Zhang, Z., Ji, R., Gao, Y.: Hypergraph neural networks. In: AAAI, pp. 3558–3565 (2019)
8. Guo, H., Tang, R., Ye, Y., Li, Z., He, X.: DeepFM: a factorization-machine based neural network for CTR prediction. In: IJCAI, pp. 1725–1731 (2017)
9. He, X., Deng, K., Wang, X., Li, Y., Zhang, Y., Wang, M.: LightGCN: simplifying and powering graph convolution network for recommendation. In: SIGIR, pp. 639–648 (2020)
10. Koren, Y., Bell, R.M., Volinsky, C.: Matrix factorization techniques for recommender systems. Computer **42**(8), 30–37 (2009)
11. Liu, S., Gu, W., Cong, G., Zhang, F.: Structural relationship representation learning with graph embedding for personalized product search. In: CIKM, pp. 915–924 (2020)
12. Park, E.L., Cho, S.: KoNLPy: Korean natural language processing in Python. In: HCLT (2014)
13. Rendle, S.: Factorization machines. In: ICDM, pp. 995–1000 (2010)
14. Rendle, S., Freudenthaler, C., Gantner, Z., Schmidt-Thieme, L.: BPR: Bayesian personalized ranking from implicit feedback. In: UAI, pp. 452–461 (2009)
15. Rendle, S., Krichene, W., Zhang, L., Anderson, J.: Neural collaborative filtering vs. matrix factorization revisited. arXiv:2005.09683 (2020)
16. Rong, Y., Huang, W., Xu, T., Huang, J.: DropEdge: towards deep graph convolutional networks on node classification. In: ICLR (2020)
17. Sparck Jones, K., Walker, S., Robertson, S.E.: A probabilistic model of information retrieval: development and comparative experiments - Part 2. Inf. Process. Manage. **36**(6), 809–840 (2000)

18. Srivastava, N., Hinton, G., Krizhevsky, A., Sutskever, I., Salakhutdinov, R.: Dropout: a simple way to prevent neural networks from overfitting. J. Mach. Learn. Res. **15**(1), 1929–1958 (2014)
19. Van Gysel, C., de Rijke, M., Kanoulas, E.: Learning latent vector spaces for product search. In: CIKM, pp. 165–174 (2016)
20. Wang, J., Ding, K., Hong, L., Liu, H., Caverlee, J.: Next-item recommendation with sequential hypergraphs. In: SIGIR, pp. 1101–1110 (2020)
21. Wang, J., Ding, K., Zhu, Z., Caverlee, J.: Session-based recommendation with hypergraph attention networks. In: SDM, pp. 82–90 (2021)
22. Wang, X., He, X., Wang, M., Feng, F., Chua, T.S.: Neural graph collaborative filtering. In: SIGIR, pp. 165–174 (2019)
23. Wu, F., Zhang, T., de Souza, A.H., Fifty, C., Yu, T., Weinberger, K.Q.: Simplifying graph convolutional networks. In: ICML, pp. 11884–11894 (2019)
24. Wu, S., Zhang, W., Sun, F., Cui, B.: Graph neural networks in recommender systems: a survey. arXiv:2011.02260 (2020)
25. Wu, T., et al.: Zero-shot heterogeneous transfer learning from recommender systems to cold-start search retrieval. In: CIKM, pp. 2821–2828 (2020)
26. Ying, R., He, R., Chen, K., Eksombatchai, P., Hamilton, W.L., Leskovec, J.: Graph convolutional neural networks for web-scale recommender systems. In: KDD, pp. 974–983 (2018)
27. Zamani, H., Croft, W.B.: Joint modeling and optimization of search and recommendation. In: DESIRES, pp. 36–41 (2018)
28. Zamani, H., Croft, W.B.: Learning a joint search and recommendation model from user-item interactions. In: WSDM, pp. 717–725 (2020)
29. Zhang, R., Guo, J., Fan, Y., Lan, Y., Cheng, X.: Query understanding via intent description generation. In: CIKM, pp. 1823–1832 (2020)
30. Zhang, Y., Chen, X., Ai, Q., Yang, L., Croft, W.B.: Towards conversational search and recommendation: system ask, user respond. In: CIKM, pp. 177–186 (2018)

Topic Aware Contextualized Embeddings for High Quality Phrase Extraction

V. Venktesh(✉), Mukesh Mohania, and Vikram Goyal

Indraprastha Institute of Information Technology, Delhi, India
{venkteshv,mukesh,vikram}@iiitd.ac.in

Abstract. Keyphrase extraction from a given document is the task of automatically extracting salient phrases that best describe the document. This paper proposes a novel unsupervised graph-based ranking method to extract high-quality phrases from a given document. We obtain the contextualized embeddings from pre-trained language models enriched with topic vectors from Latent Dirichlet Allocation (LDA) to represent the candidate phrases and the document. We introduce a scoring mechanism for the phrases using the information obtained from contextualized embeddings and the topic vectors. The salient phrases are extracted using a ranking algorithm on an undirected graph constructed for the given document. In the undirected graph, the nodes represent the phrases, and the edges between the phrases represent the semantic relatedness between them, weighted by a score obtained from the scoring mechanism. To demonstrate the efficacy of our proposed method, we perform several experiments on open source datasets in the science domain and observe that our novel method outperforms existing unsupervised embedding based keyphrase extraction methods. For instance, on the SemEval2017 dataset, our method advances the F1 score from 0.2195 (EmbedRank) to 0.2819 at the top 10 extracted keyphrases. Several variants of the proposed algorithm are investigated to determine their effect on the quality of keyphrases. We further demonstrate the ability of our proposed method to collect additional high-quality keyphrases that are not present in the document from external knowledge bases like Wikipedia for enriching the document with newly discovered keyphrases. We evaluate this step on a collection of annotated documents. The F1-score at the top 10 expanded keyphrases is 0.60, indicating that our algorithm can also be used for 'concept' expansion using external knowledge.

Keywords: Automatic Keyphrase Extraction (AKE) · contextualized embeddings · unsupervised methods · summarization

1 Introduction

Keyphrases are the salient terms in a document that serve as summaries of the document. They play an important role in many text processing applications like

This research work is supported by Extramarks Education India Pvt. Ltd. (an education technology company), SERB, FICCI (PM fellowship for doctoral research) and TiH Anubhuti (IIITD).

M. Hagen et al. (Eds.): ECIR 2022, LNCS 13185, pp. 457–471, 2022.
https://doi.org/10.1007/978-3-030-99736-6_31

document clustering, classification, information retrieval [13,15] and text generation. Automatic Keyphrase Extraction (AKE) task is a crucial component of these applications to obviate the need for manual extraction of keyphrases. In addition to the above-mentioned applications, we are primarily interested in the applications of AKE in online learning platforms. Learning contents in such learning platforms are tagged at the topic level for accessibility. However, a topic can be further divided into concepts that enable linking of related learning content and easy navigation through the learning contents. In this paper, concepts are characterized as keyphrases as they describe the content of a document. We posit that the Automatic Keyphrase Extraction (AKE) from learning contents can help to index the massive collection of learning contents in online learning platforms enabling better accessibility of the learning contents. Automatic Keyphrase Extraction is a well studied problem [10]. Unlike the supervised methods, the unsupervised methods do not require annotated documents and rely on in-corpus statistical information for extracting keyphrases. In most of the unsupervised keyphrase extraction methods [6,16–18] the candidate phrases are represented by a word graph formed based on a co-occurrence window and then ranked. Recent unsupervised methods like EmbedRank [2] leverage representation learning methods that help to capture the semantic relatedness between the phrases and the document.

We propose a novel unsupervised method to automatically extract keyphrases from a given document. In this method, the candidate phrases and the given document are represented in a continuous vector space by combining the contextual embeddings and the topical information from LDA [3] to strengthen the associations between phrases that occur in similar contexts and also represent similar topics. Then a graph based ranking algorithm where the nodes are represented by phrases than words is employed to rank the phrases. The proposed unsupervised method helps to capture two important characteristics needed for keyphrases: *coherence* and *informativeness*. We posit that the selected phrases are *coherent* if they convey a consistent idea [11] and they are *informative* if they convey the core ideas discussed in the document. In the proposed method, *coherence* is captured as the cosine similarity computed between the embeddings of the candidate phrases. The *informativeness* of a phrase is captured as the cosine similarity computed between the embeddings of the candidate phrase and the document. The proposed algorithm outperforms existing unsupervised AKE methods. For instance, on SemEval2017 dataset our method advances the F1 score from 0.2195 (EmbedRank) to **0.2819**.

Following are the core technical contributions of our paper:

- We propose a new topic aware representation method to represent the phrases and the document. To the best of our knowledge, this representation method has not been applied to the task of keyphrase extraction.
- We propose a graph based ranking method with a new scoring mechanism that captures the *informativeness* and *coherence* measures by using the proposed representation method.

– We apply our algorithm on the task of enriching the set of extracted keyphrases with new keyphrases which are not present in the source document by using external knowledge sources like Wikipedia.

The code and data can be found at https://github.com/VenkteshV/Unsupervised _keyphrase_extraction_CoTagRank_ECIR_2022.

2   Related Work

In this section, we discuss the existing unsupervised AKE methods [21] and also the current advancements in the vector representation methods.

2.1   Unsupervised Keyphrase Extraction

Graph based ranking methods are the most popular among the unsupervised AKE algorithms. Graph based AKE methods were introduced in the seminal work TextRank [18]. It constructs a uniformly weighted graph for the given text where an edge connects word types only if they co-occur within a window of specified size. The SingleRank [26] algorithm was an extension to the TextRank algorithm where the edges were assigned a weight equal to the number of times word types co-occur. The WordAttractionRank [27] algorithm is similar to SingleRank with one difference. It incorporates the distance between the word embeddings into the weighting scheme for the edges between the words. As keyphrases usually appear at the beginning of the document, the Position-Rank [9] algorithm and the MultiPartiteRank [5] algorithm assigns weights to nodes (words and phrases respectively) favouring the terms appearing at initial positions in the text. One of the shortcomings of these approaches (except MultiPartiteRank) is that they rank the phrases by aggregating the scores of the constituent words. This can lead to uninformative candidate phrases being ranked higher just because one of the constituent words has a higher score.

Several existing approaches like TopicRank [6] have leveraged topical information for ranking phrases. Another algorithm that leverages the topical information is the TopicalPagerank (TPR) [16] algorithm. The TPR method runs TextRank for each topic where the topics are obtained using LDA [3]. Another extension to TPR is the Salience Rank algorithm [24] which introduces word salience metric to balance between topic and corpus specificity of words.

In contrast to the graph based methods, the EmbedRank [2] algorithm is an embedding based AKE method that represents both documents and candidate phrases as vectors using document embedding methods like Sent2Vec [20]. The vector representations help to rank candidates by computing cosine similarity between the phrase vectors and the document vector.

2.2   Contextualized Vector Representations

Distributed representations that capture the semantic relationships [19] have helped to advance many NLP tasks. But all the classical embedding methods like fasttext [4], GloVe [22] generate fixed vectors for even polysemous words irrespective of the context in which they occur.

The Bidirectional Encoder Representation from Transformers (BERT) [8] is one of the current state of the art methods that uses a mechanism called *attention* [25]. The attention mechanism helps to encode a word using other positions in the input sequence that would lead to a better representation for the word. The Sentence BERT [23] model was proposed to generate useful sentence embeddings by fine-tuning BERT. Another transformer based sentence encoding model is the Universal Sentence Encoder (USE) [7] that has been specifically trained on semantic textual similarity task. In our experiments, we demonstrate that our novel representation using topic distribution based vectors from LDA (Latent Dirichlet Allocation) and USE embeddings performs better than USE or BERT embeddings in isolation.

3   Methodology

In this section, we describe the proposed extraction algorithm, CoTagRank. First, the candidate phrases are extracted based on the Part Of Speech (POS) tags using the pattern $< NN.* | JJ > * < NN.* >$ [26]. Then the phrases and the document are projected to a continuous vector space and phrases are ranked as discussed in the following sections.

3.1   Vector Representations for the Phrases and the Document

The primary goal of our algorithm is to extract the candidate phrases that best describe the document. We define two measures for achieving this goal namely *coherence* and *informativeness*. The coherence measure can be seen as an indicator that the candidate phrases represent a consistent idea. The informativeness measure can be seen as an indicator as to whether the phrases convey the core ideas discussed in the document. We posit that the above two measures can be captured by leveraging the topical information. Hence, we give a novel vector representation mechanism that combines topical information with the embeddings obtained from the state of the art contextualized embedding methods. We leverage contextualized embeddings to handle polysemous words. An example of polysemy can be seen in the following two sentences : "Consider an **imaginary** box", "An **imaginary number** is a complex number". In the above two sentences, the word "imaginary" has different meanings. Contextualized embeddings capture the context of usage of the word and hence produce different vector representations for the same word depending on the context.

The phrase representations are obtained by combining the contextualized embeddings of the phrases with the topic vectors of their constituent words obtained from LDA. The LDA is a generative probabilistic model in which each word in document d is assumed to be generated by sampling a topic from d's topic distribution θ^d and then sampling a word from the distribution over words denoted by ϕ^t of a topic. We use pre-trained Universal Sentence Encoder (USE) to obtain contextualized embeddings for both the phrases and the sentences as it has been pre-trained on the Semantic Text Similarity (STS) task. This representation method helps in bringing the phrases that are semantically related and having similar topic distributions closer in the vector space. This implies that the phrases that are both semantically related and represent similar topics would have a higher coherence measure (cosine similarity between phrase representations). The phrase representations are obtained in the following manner:

$$LE(CP) = \sum_{w \in CP} [p(w|t_1), p(w|t_2)...]$$

$$CPE = concat(LE(CP), CE(CP)) \tag{1}$$

where LE represents LDA embeddings, CP represents a candidate phrase, CE represents contextualized embeddings, and CPE represents candidate phrase embeddings. The vector $[p(w|t_1), p(w|t_2)...]$ represents the word-topic probabilities that are derived from the word distributions ϕ^t over the topics. Similarly the document representation can be obtained by combining topic distribution of the document with the contextualized embeddings of the document sentences. The document representation is obtained as follows:

$$LE(doc) = [p(t_1|d), p(t_2|d)...]$$

$$DE = concat(LE(doc), CE(doc)) \tag{2}$$

where, the vector $[p(t_1|d), p(t_2|d)...]$ represents the document-topic probabilities and DE represents the document embedding. Latent Dirichlet Allocation (LDA) is run only once on the corpus of documents and not for every document. The vector representations obtained as described are leveraged in the graph based ranking step to compute final scores for the candidate phrases.

3.2   Graph Based Ranking

In this subsection, we explain our graph based candidate phrases ranking method in detail.

It differs from traditional methods like TextRank, SingleRank and Position-Rank. We construct an undirected graph with the candidate phrases as the vertices instead of words. Constructing the graph in this manner circumvents the overgeneration errors that occur when the phrases are ranked by aggregating the scores of the top ranked words. Hence using word graph based AKE methods may result in an uninformative phrase being assigned a high score just because one of the constituent words has a higher score. The edges connecting the phrases are weighted by the semantic relatedness (cosine similarity) computed between the vector representations of the phrases. The vector representations for the phrases are obtained as described in the previous subsection. The edges are formed between the phrases (nodes) that co-occur in the original text within a specified window size (tunable parameter). We demonstrate that when the window size is set to the maximum value for forming a complete graph, we get the maximum performance. The completeness nature of the graph has the benefit of connecting phrases together that may not co-occur together but having similar topic distributions. As mentioned in the previous subsection, our goal is to rank those phrases higher that are coherent and informative for the document. The *coherence* measure is represented by the edge weights of the graph. The *informativeness* measure for each phrase (P_a) is the normalized cosine similarity between the document and the phrase representations computed as follows:

$$n\_Sim(P_a, doc) = \frac{Sim(P_a, doc) - min_{P_b \in P}(Sim(P_b, doc))}{max_{P_b \in P}(Sim(P_b, doc))} \tag{3}$$

where n_Sim is normalized cosine similarity, Sim is the cosine similarity function and doc represents the document. Then the similarity metric is obtained as :

$$F\_Sim(P_a, doc) = \frac{n\_Sim(P_a, doc) - \mu(n\_Sim(P, doc))}{\sigma(n\_Sim(P, doc))} \tag{4}$$

where P is the set of phrases. The F_Sim function returns the final cosine similarity metric obtained after normalization and standardization of cosine similarities. The function n_Sim given a set of embeddings of the phrases and a document embedding as inputs returns a vector of normalized cosine similarities. Each element in the output vector is the normalized similarity between the corresponding embedding in the set and the document embedding. In Eq. (3), $n\_Sim(Pa, doc)$, P_a denotes a set of embeddings having only one element. Whereas in Eq. (4), $n\_Sim(P, doc)$, P has multiple embeddings.

The goal is to find the phrases that maximize the objective:

$$Obj = \lambda S_{coh}(P) + (1 - \lambda)S_{inf}(P, doc)$$

where, S_{inf} denotes the function that returns the *informativeness* measure computed using Eq. 4.

The S_{coh} is the function that computes the *coherence* measure. The S_{coh} denotes a function that takes a set of embeddings of phrases (P) and outputs a vector of cosine similarities computed between embeddings of all possible pairs of distinct phrases in the set. The parameter λ balances the importance given for S_{coh} and S_{inf} factors.

Iteratively optimizing the above objective is similar to random walk based approaches. Hence maximizing the above objective can be done as follows:

Every candidate phrase in the graph is ranked by:

$$R(p_i) = \lambda \sum_{j:p_j -> p_i} \frac{e(p_i, p_j)}{OutDeg(p_j)} R(p_j) + (1 - \lambda)S_{inf}(p_i) \tag{5}$$

where $e(p_i, p_j)$ denotes the weight of the edges between the phrases (p_i and p_j) (*coherence*) and $S_{inf}(p_i)$ is the *informativeness* score that helps in biasing the random jump to phrases (vertices) that are closer to the document in the vector space.

We explore several variants of *CoTagRank* in Sect. 4.

Table 1. Statistics of the datasets used

Dataset	Domain	# of docs	# of tokens/doc	# of gold keys	# of gold keys/doc
Inspec	Science	2000	128.20	29230	14.62
SemEval2017	Science	493	178.22	8969	18.19
SemEval2010	Science	243	8332.34	4002	16.47

4 Experiments and Results

This section, discusses the experimental setup and results.

4.1 Datasets

We evaluate our algorithm on standard datasets like Inspec [12], SemEval 2017 [1] and SemEval 2010 [14] for keyphrase extraction. We choose SemEval2017 and Inspec as they contain documents of short length resembling the learning content in e-learning platforms. We also show the performance of our method on a dataset containing long documents such as SemEval2010. The statistics of the datasets are shown in Table 1. Since our algorithm is completely unsupervised, we evaluate on all the documents in each of these datasets.

4.2 Baselines and Variants of the Proposed Method

In this section, we describe the variants of the proposed CoTagRank algorithm and other baselines. In the proposed CoTagRank algorithm, a complete graph is formed from the phrases. The phrases and the document are represented by combining the contextualized embeddings from Universal Sentence Encoder (USE) (512-dimensional) [7] and topical vectors from Latent Dirichlet Allocation (LDA)[1]. The number of topics K was set at 500 when running LDA.

We compare *CoTagRank* with several variants such as:

- *CoTagRankWindow*: This algorithm is a variant of *CoTagRank* where only the phrases that co-occur in the text within a window of the specified size are connected in the graph. While *CoTagRank* forms a complete graph of phrases, CoTagRankWindow provides a tunable parameter, the window size w, which determines the edges formed between phrases. The vector representation and the ranking method is the same as explained in Sect. 3.2.
- CoTagRanks2v: This algorithm is similar to *CoTagRank* with respect to complete graph formation and ranking using Eq. 5. However, in *CoTagRanks2v* the static sentence representation method like Sent2Vec [20] is used to project the phrases and the document to a continuous vector space.
- *CoTagRankSentenceUSE*: A variant of the *CoTagRank* where the document and phrase are encoded using only Universal Sentence Encoder yielding 512-dimensional representations.

---

[1] We leveraged the sklearn implementation for LDA https://scikit-learn.org/.

Table 2. Performance comparison. [†] indicates significance at 0.01 level (t-test). [‡] indicates that effect size > 0.2.

Dataset	Method	P@10	R@10	F1@10
SemEval2017	TopicalPageRank	0.3523	0.2098	0.2543
	MultiPartiteRank	0.2972	0.1758	0.2133
	SingleRank	0.3428	0.2040	0.2474
	TextRank	0.1848	0.1069	0.1326
	WordAttractionRank	0.2566	0.1482	0.1815
	EmbedRank	0.3061	0.1801	0.2195
	EmbedRankSentenceBERT	0.3329	0.1982	0.2404
	EmbedRankSentenceUSE	0.3286	0.1965	0.2381
	CoTagRank (our algorithm)	**0.3911**[†‡]	**0.2324**[†‡]	**0.2819**[†‡]
	CoTagRankSentenceUSE (our algorithm)	0.3860	0.2290	0.2779
	CoTagRanks2v (our algorithm)	0.3379	0.1990	0.2424
	CoTagRankWindow (w=10) (our algorithm)	0.3797	0.2253	0.2734
	CoTagRankWindow$_{positional}$ (our algorithm)	0.3793	0.2250	0.2731
Inspec	TopicalPageRank	0.2724	0.2056	0.2260
	MultiPartiteRank	0.2210	0.1710	0.1865
	SingleRank	0.2694	0.2044	0.2239
	TextRank	0.1408	0.1020	0.1234
	WordAttractionRank	0.1778	0.1437	0.1516
	EmbedRank	0.2732	0.2034	0.2259
	EmbedRankSentenceBERT	0.2663	0.1970	0.2188
	EmbedRankSentenceUSE	0.2748	0.2049	0.2267
	CoTagRank (our algorithm)	**0.2984**[†‡]	**0.2213**[†‡]	**0.2454**[†‡]
	CoTagRankSentenceUSE (our algorithm)	0.2881	0.2150	0.2377
	CoTagRanks2v (our algorithm)	0.2372	0.1807	0.1983
	CoTagRankWindow (w=10) (our algorithm)	0.2747	0.2062	0.2275
	CoTagRankWindow$_{positional}$ (our algorithm)	0.2750	0.2062	0.2276
SemEval2010	TopicalPageRank	0.0477	0.0293	0.0359
	MultiPartiteRank	**0.1757**[†‡]	**0.1118**[†‡]	**0.1352**[†‡]
	SingleRank	0.0457	0.0277	0.0341
	TextRank	0.0321	0.0199	0.0243
	WordAttractionRank	0.0835	0.0531	0.0641
	EmbedRank	0.0128	0.0082	0.0099
	EmbedRankSentenceBERT	0.0230	0.0137	0.0170
	EmbedRankSentenceUSE	0.0379	0.0241	0.0292
	CoTagRank (our algorithm)	0.0695	0.0434	0.0530
	CoTagRankSentenceUSE (our algorithm)	0.0671	0.0418	0.0511
	CoTagRanks2v (our algorithm)	0.0267	0.0169	0.0204
	CoTagRankWindow (w=10) (our algorithm)	0.1337	0.0867	0.1042
	CoTagRankWindow$_{positional}$ (our algorithm)	0.1494	0.0970	0.1165
SemEval2010 (abstract and intro)	TopicalPageRank	0.1745	0.1100	0.1336
	MultiPartiteRank	0.1646	0.1044	0.1263
	SingleRank	0.1580	0.0998	0.1211
	TextRank	0.1140	0.0719	0.0872
	WordAttractionRank	0.1481	0.0949	0.1145
	EmbedRank	0.0654	0.0407	0.0496
	EmbedRankSentenceBERT	0.0844	0.0521	0.0638
	EmbedRankSentenceUSE	0.1243	0.0760	0.0933
	CoTagRank (our algorithm)	0.1811	0.1134	0.1380
	CoTagRankSentenceUSE (our algorithm)	0.1786	0.1121	0.1363
	CoTagRanks2v (our algorithm)	0.0852	0.0518	0.0636
	CoTagRankWindow (w=10) (our algorithm)	0.1856	0.1170	0.1419
	CoTagRankWindow$_{positional}$ (our algorithm)	**0.1909**[†‡]	**0.1203**[†‡]	**0.1459**[†‡]

We also consider two variants of EmbedRank such as EmbedRankSentence-BERT and EmbedRankSentenceUSE where *bert-base-nli-stsb-mean-tokens* from sentence-transformers[2] and Universal Sentence Encoder are used respectively as vector representation methods. We compare the performance of the proposed algorithms with strong baselines such as EmbedRank (Sent2Vec)[3], SingleRank and other unsupervised AKE methods[4].

4.3   Results and Discussion

The performance comparison of the algorithms are as shown in Table 2. The measures used to evaluate the algorithms are Precision, Recall and F1-score. The metrics were computed using trec-eval[5]. Since the original implementation of EmbedRank did not provide an evaluation script, we use trec-eval to compute the metrics for EmbedRank and we observe different results from those reported in the original EmbedRank paper[6].

As shown in Table 2, the CoTagRank outperforms existing graph based and embedding based unsupervised methods on two of the three datasets and on the third dataset, we get comparable results to the MultiPartiteRank algorithm. The performance gain obtained using CoTagRank over EmbedRankSentence-BERT, EmbedRankSentenceUSE and *CoTagRankSentenceUSE* demonstrates the advantage of fusing topical information with the contextualized embeddings rather than leveraging just contextualized embeddings for phrase and document representations.

However, on long documents, the MultiPartiteRank outperforms all other methods. The MultiPartiteRank algorithm leverages the position of the candidate phrases in the document as a feature that leads to the gain in F1-score on long documents. This result is similar to the result reported in the EmbedRank paper [2]. To overcome this limitation, the authors of EmbedRank propose a variant $EmbedRank_{positional}$, which includes the position of the candidate phrases as a feature to increase the performance on long documents. In contrast to the $EmbedRank_{positional}$ method, we were able to achieve a gain in performance by just tuning the window size in the CoTagRank algorithm. The results in Table 2 show that the performance of a variant of our proposed algorithm, *CoTagRankWindow* with a window size of 10, is close to the performance of MultiPartiteRank. We also verify the intuition of positional bias by proposing $CoTagRankWindow_{positional}$ where we multiply the node weights by the inverse of the start position of the phrase in the input document. We observe that the performance with the positional bias is close to the MultiPartiteRank algorithm. Additionally, we leverage the common knowledge that most keyphrases

---

[2] https://huggingface.co/sentence-transformers/bert-base-nli-stsb-mean-tokens.
[3] https://github.com/swisscom/ai-research-keyphrase-extraction.
[4] https://bit.ly/369Ycg7.
[5] https://github.com/usnistgov/trec_eval.
[6] Our results are close to the implementation in the project https://bit.ly/2IbbyjT which also uses trec-eval and the original EmbedRank implementation.

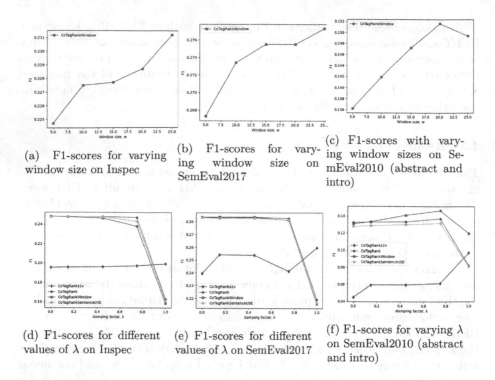

(a)   F1-scores for varying window size on Inspec

(b)   F1-scores for varying window size on SemEval2017

(c) F1-scores with varying window sizes on SemEval2010 (abstract and intro)

(d) F1-scores for different values of λ on Inspec

(e) F1-scores for different values of λ on SemEval2017

(f) F1-scores for varying λ on SemEval2010 (abstract and intro)

Fig. 1. Performance comparison for different hyperparameters

are located at the beginning of the document and hence perform keyphrase extraction only on "Abstract" and "Introduction" sections of every document in SemEval 2010, as shown in Table 2, which advances the F1 scores of CoTagRank and $CoTagRank_{positional}$ to **0.1380** and **0.1459** respectively surpassing Multi-PartiteRank confirming the positional bias intuition. On average, this reduced version of the SemEval 2010 dataset still contains 550 tokens when compared to the number of tokens in Inspec and SemEval 2017, as observed in Table 1.

However, since online learning contents like questions and video transcripts are usually short text documents, the expected performance of $CoTagRank$ on such documents is closer to the results observed for Inspec and SemEval 2017. We also perform statistical significance tests and observe that our results are significant at ($p < 0.01$) with effect sizes of **0.37**, **0.30** and **0.83** for F1 scores on SemEval2017, Inspec and SemEval2010 (abstract and intro) respectively.

4.4   Effects of Different Hyperparameters

In this section, we discuss the effect of varying hyperparameters such as window size (w), damping factor (λ) and number of topics in LDA (LDA embeddings dimension) of the $CoTagRank$ algorithm and its variants. We vary the window size hyperparameter w with values 5, 10, 15, 20 and 25. The graphs in the four students performed an experiment to calculate the density of a stone. while measuring the mass of the stone with the help of spring balance, the first student immersed the stone in water, the second student immersed it in sulphuric acid, the third student immersed it in kerosene and the fourth student allowed it to hang freely in air. the correct value of mass of the stone will be obtained by in this experiment. we need true value of mass that can be measured by suspending the stone freely in air. when we immerse the stone in some liquid, the liquid will exert an upward buoyant force on the stone in upward direction. this changes the reading of the spring balance and we will not get the correct value.

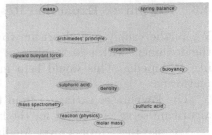

Fig. 2. Keyphrase expansion results for an academic learning content

Fig. 1a and 1b show that the F1-score increases with the increase in window size for the $CoTagRankWindow$ algorithm. The window size can be set to the maximum value encompassing all phrases in the document forming a complete graph. This validates our claim that running the biased PageRank algorithm on a complete graph of phrases helps in producing high quality phrases. However, the same assumption may not hold good for longer documents, as evident from Fig. 1c. We observe that the performance of $CoTagRankWindow$ on SemEval 2010 (abstract and intro) increases with increase in window size drops a little at $w = 25$. This indicates that forming a complete graph may not lead to the highest performance on longer documents. We also vary the damping factor λ in Eq. 5. The values we experiment with are 0, 0.15, 0.45, 0.75, 1.0. The graphs in the Fig. 1d and 1e show that in the proposed CoTagRank algorithm and in the variant $CoTagRankWindow$ the performance declines with an increase in the damping factor. When damping factor is set to 1, the S_{inf} component in Eq. 5 that contributes to *informativeness* of the phrase becomes zero resulting in a drop in F1-score. The decrease in F1-score observed in the plots as the damping factor increases supports our claims of *informativeness* and *coherence* measures. However, we do not observe this trend in CoTagRanks2v. This may be due to the representation method used for the phrases, which do not contribute to the informativeness measure defined in this paper. From Fig. 1f, we can observe that on SemEval2010 (abstract and intro) dataset, when λ is set to 1, there is a drop in F1-score. However, when compared to the previous two graphs, we observe that the relative drop in F1-score is low. This maybe due to the length of the documents in this dataset when compared to short length documents in Inspec and SemEval 2017.

We also vary the number of topics (LDA embeddings dimension) and observe that $CoTagRank$ and $CoTagRankWindow$ achieves the highest performance when number of topics is set to 500. This is similar to the observation made in the TopicalPageRank paper [16] where the authors show that setting the number of topics to 500 gives the highest performance on the Inspec dataset.

4.5 Keyphrase Expansion Results

We further apply our algorithm to the task of keyphrase expansion to enrich the document with new keyphrases with the help of external knowledge sources like Wikipedia. This would help in linking related learning content in online platforms.

The keyphrases extracted from the source document using the CoTagRank algorithm serve as seed set for the keyphrase expansion task.

We use the wrapper over MediaWiki API[7] to extract relevant Wikipedia article titles for each keyphrase in the seed set. Then the expanded phrases are ranked using Eq. 5.

Table 3. Performance comparison at top-10 expanded keyphrases

Dataset	Method	Precision	Recall	F1
Lecture transcripts	CoTagRank (our algorithm)	**0.5448**	**0.7270**	**0.6096**
(Khan academy)	CoTagRanks2v	0.2483	0.3424	0.2956
	$CoTagRankSentenceUSE$	0.5207	0.6950	0.5949

To demonstrate the effectiveness of this algorithm, we applied it for keyphrase expansion on 30 lecture transcripts collected from Khan academy in the science domain. The extracted phrases were given to two annotators who were undergraduate students in the Computer Science department familiar with the concepts. The task was to annotate the phrases as relevant to the document (1) or not relevant to the document (0).

The degree of agreement on relevance of keyphrases between the two annotators was measured using Cohen's kappa κ. We obtained a κ of 0.535 denoting *moderate* agreement between the annotators. A phrase is considered as a ground truth label only if both the annotators consider it to be relevant. We compute the Precision, Recall and F1 metrics as shown in Table 3. The F1 score of **0.6096** indicates that the proposed algorithm was able to retrieve relevant keyphrases from external knowledge sources. We observe that CoTagRanks2v and $CoTagRankSentenceUSE$ do not perform well in this task, indicating that the combination of contextualized embeddings and topic representations help in extracting better keyphrases from external knowledge sources.

Figure 2 shows the results of running the proposed algorithm on an academic content from Khan academy. We observe that our algorithm was able to discover interesting phrases like *Archimedes principle* though it was not present in the source document. The new keyphrases can help in linking related learning content, where the given question in Fig. 2 can be linked with a video explaining *Archimedes principle*. We observed that none of the other algorithms were able to retrieve *Archimedes principle*. This further reinforces the idea that apart

---

[7] https://pypi.org/project/wikipedia/.

from semantic relatedness between phrases that occur in similar contexts, their topic relatedness is also captured through our representation mechanism. The evaluation of the proposed algorithm on this corpus demonstrates that our algorithm could also enrich the existing set of keyphrases with new keyphrases using external knowledge sources like Wikipedia.

5 Conclusions

In this paper, we proposed a novel representation and graph based ranking algorithm, CoTagRank, for keyphrase extraction. The algorithm is currently deployed to extract academic concepts from learning content in an online learning platform. We showed that our method outperforms existing state-of-the-art unsupervised keyphrase extraction methods in shorter texts and comparable performance on longer texts. In addition, forming a complete graph of phrases outperforms window based graph formation methods on short documents. We also demonstrated that including a simple positional bias helps further advance the performance of the algorithm on longer documents. In the future, we aim to incorporate positional embeddings and verify the performance on long texts.

References

1. Augenstein, I., Das, M., Riedel, S., Vikraman, L., McCallum, A.: SemEval 2017 task 10: ScienceIE - extracting keyphrases and relations from scientific publications. In: Proceedings of the 11th International Workshop on Semantic Evaluation (SemEval-2017). Association for Computational Linguistics, Vancouver, Canada, August 2017. https://doi.org/10.18653/v1/S17-2091. https://www.aclweb.org/anthology/S17-2091
2. Bennani-Smires, K., Musat, C., Hossmann, A., Baeriswyl, M., Jaggi, M.: Simple unsupervised keyphrase extraction using sentence embeddings. In: Proceedings of the 22nd Conference on Computational Natural Language Learning, pp. 221–229. Association for Computational Linguistics, Brussels, Belgium, October 2018. https://doi.org/10.18653/v1/K18-1022. https://www.aclweb.org/anthology/K18-1022
3. Blei, D.M., Ng, A.Y., Jordan, M.I.: Latent Dirichlet allocation. J. Mach. Learn. Res. **3**, 993–1022 (2003)
4. Bojanowski, P., Grave, E., Joulin, A., Mikolov, T.: Enriching word vectors with subword information. Trans. Associat. Comput. Linguist. **5**, 135–146 (2017)
5. Boudin, F.: Unsupervised keyphrase extraction with multipartite graphs. In: Proceedings of the 2018 Conference of the North American Chapter of the Association for Computational Linguistics: Human Language Technologies, Volume 2 (Short Papers). Association for Computational Linguistics, New Orleans, Louisiana, June 2018. https://doi.org/10.18653/v1/N18-2105. https://www.aclweb.org/anthology/N18-2105
6. Bougouin, A., Boudin, F., Daille, B.: TopicRank: graph-based topic ranking for keyphrase extraction. In: Proceedings of the Sixth International Joint Conference on Natural Language Processing, pp. 543–551. Asian Federation of Natural Language Processing, Nagoya, Japan, October 2013. https://www.aclweb.org/anthology/I13-1062

7. Cer, D., et al.: Universal sentence encoder for English. In: Proceedings of the 2018 Conference on Empirical Methods in Natural Language Processing: System Demonstrations, pp. 169–174. Association for Computational Linguistics, Brussels, Belgium, November 2018. https://doi.org/10.18653/v1/D18-2029. https://www.aclweb.org/anthology/D18-2029

8. Devlin, J., Chang, M., Lee, K., Toutanova, K.: BERT: pre-training of deep bidirectional transformers for language understanding. CoRR abs/1810.04805 (2018). http://arxiv.org/abs/1810.04805

9. Florescu, C., Caragea, C.: Positionrank: an unsupervised approach to keyphrase extraction from scholarly documents. In: Proceedings of the 55th Annual Meeting of the Association for Computational Linguistics (Volume 1: Long Papers), pp. 1105–1115 (2017)

10. Hasan, K.S., Ng, V.: Automatic keyphrase extraction: a survey of the state of the art. In: Proceedings of the 52nd Annual Meeting of the Association for Computational Linguistics (Volume 1: Long Papers), pp. 1262–1273. Association for Computational Linguistics, Baltimore, Maryland, June 2014. https://doi.org/10.3115/v1/P14-1119

11. He, Y., Xin, D.: Seisa: Set expansion by iterative similarity aggregation. p. 427–436. WWW 2011, Association for Computing Machinery, New York, NY, USA (2011). https://doi.org/10.1145/1963405.1963467

12. Hulth, A.: Improved automatic keyword extraction given more linguistic knowledge. In: Proceedings of the 2003 Conference on Empirical Methods in Natural Language Processing, pp. 216–223 (2003). https://www.aclweb.org/anthology/W03-1028

13. Jones, S., Staveley, M.S.: Phrasier: a system for interactive document retrieval using keyphrases. In: Proceedings of the 22nd Annual International ACM SIGIR Conference on Research and Development in Information Retrieval, pp. 160–167. SIGIR 1999. Association for Computing Machinery, New York, NY, USA (1999). https://doi.org/10.1145/312624.312671,https://doi.org/10.1145/312624.312671

14. Kim, S.N., Medelyan, O., Kan, M.Y., Baldwin, T.: SemEval-2010 task 5 : automatic keyphrase extraction from scientific articles. In: Proceedings of the 5th International Workshop on Semantic Evaluation, pp. 21–26. Association for Computational Linguistics, Uppsala, Sweden, July 2010. https://www.aclweb.org/anthology/S10-1004

15. Kim, Y., Kim, M., Cattle, A., Otmakhova, J., Park, S., Shin, H.: Applying graph-based keyword extraction to document retrieval. In: Proceedings of the Sixth International Joint Conference on Natural Language Processing, pp. 864–868. Asian Federation of Natural Language Processing, Nagoya, Japan, October 2013. https://www.aclweb.org/anthology/I13-1108

16. Liu, Z., Huang, W., Zheng, Y., Sun, M.: Automatic keyphrase extraction via topic decomposition. In: Proceedings of the 2010 Conference on Empirical Methods in Natural Language Processing, pp. 366–376. Association for Computational Linguistics, Cambridge, MA, October 2010. https://www.aclweb.org/anthology/D10-1036

17. Liu, Z., Li, P., Zheng, Y., Sun, M.: Clustering to find exemplar terms for keyphrase extraction. In: Proceedings of the 2009 Conference on Empirical Methods in Natural Language Processing, pp. 257–266. Association for Computational Linguistics, Singapore, August 2009. https://www.aclweb.org/anthology/D09-1027

18. Mihalcea, R., Tarau, P.: TextRank: bringing order into text. In: Proceedings of the 2004 Conference on Empirical Methods in Natural Language Processing, pp. 404–411. Association for Computational Linguistics, Barcelona, Spain, July 2004. https://www.aclweb.org/anthology/W04-3252
19. Mikolov, T., Sutskever, I., Chen, K., Corrado, G., Dean, J.: Distributed representations of words and phrases and their compositionality. In: Proceedings of the 26th International Conference on Neural Information Processing Systems - Volume 2, pp. 3111–3119. NIPS 2013, Curran Associates Inc., Red Hook, NY, USA (2013)
20. Pagliardini, M., Gupta, P., Jaggi, M.: Unsupervised learning of sentence embeddings using compositional n-gram features. In: NAACL-HLT (2018)
21. Papagiannopoulou, E., Tsoumakas, G.: A review of keyphrase extraction (2019)
22. Pennington, J., Socher, R., Manning, C.: Glove: global vectors for word representation. In: Proceedings of the 2014 Conference on Empirical Methods in Natural Language Processing (EMNLP), pp. 1532–1543. Association for Computational Linguistics, Doha, Qatar, October 2014. https://doi.org/10.3115/v1/D14-1162. https://www.aclweb.org/anthology/D14-1162
23. Reimers, N., Gurevych, I.: Sentence-BERT: sentence embeddings using Siamese BERT-networks. In: Proceedings of the 2019 Conference on Empirical Methods in Natural Language Processing and the 9th International Joint Conference on Natural Language Processing (EMNLP-IJCNLP), pp. 3982–3992. Association for Computational Linguistics, Hong Kong, China, November 2019. https://doi.org/10.18653/v1/D19-1410. https://www.aclweb.org/anthology/D19-1410
24. Teneva, N., Cheng, W.: Salience rank: efficient keyphrase extraction with topic modeling. In: Proceedings of the 55th Annual Meeting of the Association for Computational Linguistics (Volume 2: Short Papers), pp. 530–535. Association for Computational Linguistics, Vancouver, Canada, July 2017. https://doi.org/10.18653/v1/P17-2084. https://aclanthology.org/P17-2084
25. Vaswani, A., et al.: Attention is all you need. In: Proceedings of the 31st International Conference on Neural Information Processing Systems, pp. 6000–6010. NIPS 2017. Curran Associates Inc., Red Hook, NY, USA (2017)
26. Wan, X., Xiao, J.: Single document keyphrase extraction using neighborhood knowledge. In: Proceedings of the 23rd National Conference on Artificial Intelligence - Volume 2, pp. 855–860. AAAI 2008. AAAI Press (2008)
27. Wang, R., Liu, W., McDonald, C.: Corpus-independent generic keyphrase extraction using word embedding vectors (2015)

Topic Modeling on Podcast Short-Text Metadata

Francisco B. Valero[ID], Marion Baranes[ID], and Elena V. Epure[(✉)][ID]

Deezer Research, 22-26, rue de Calais, 75009 Paris, France
research@deezer.com

Abstract. Podcasts have emerged as a massively consumed online content, notably due to wider accessibility of production means and scaled distribution through large streaming platforms. Categorization systems and information access technologies typically use topics as the primary way to organize or navigate podcast collections. However, annotating podcasts with topics is still quite problematic because the assigned editorial genres are broad, heterogeneous or misleading, or because of data challenges (e.g. short metadata text, noisy transcripts). Here, we assess the feasibility to discover relevant topics from podcast metadata, titles and descriptions, using topic modeling techniques for short text. We also propose a new strategy to leverage named entities (NEs), often present in podcast metadata, in a Non-negative Matrix Factorization (NMF) topic modeling framework. Our experiments on two existing datasets from Spotify and iTunes and Deezer, a new dataset from an online service providing a catalog of podcasts, show that our proposed document representation, NEiCE, leads to improved topic coherence over the baselines. We release the code for experimental reproducibility of the results (https://github.com/deezer/podcast-topic-modeling).

Keywords: Podcasts · Short-text · Topic modeling · Named entities

1 Introduction

Podcasts is an audio content listened to on-demand for educational, entertainment, or informational purposes. Known as the "spoken" version of blog posts, they have evolved towards a wide variety of formats (e.g. monologues, multiparty conversations, narratives) spanning a wide range of categories (e.g. business, true crime). Podcasts have been massively popularised in the recent years due to increased use of streaming platforms and availability of underlying technology for information access, recording and publishing [11,20,35]. As of summer 2021, the number of available podcasts in the world exceeds $2M$ and over $48M$ episodes [32]. Likewise, the podcast listening audience has grown massively: 50% of the American population has listened to at least a podcast in 2019 and over 32% have listened to podcasts monthly (compared to 14% in 2014) [25].

Given the sharp growing interest in podcasts, researchers and industry players have searched for more suitable ways to enable listeners to effectively navigate these overwhelming collections [3,8,20,35]. Topics are central to any of

M. Hagen et al. (Eds.): ECIR 2022, LNCS 13185, pp. 472–486, 2022.
https://doi.org/10.1007/978-3-030-99736-6_32

the adopted podcast information access technology such as automatic categorization, search engines or recommender systems. However, annotating podcasts with topics is still quite problematic. First, although podcast metadata entails topic-related genres, manually assigned by creators or providers, in reality these are often noisy and unreliable [11,28]: genres could be too heterogeneous or broad (e.g. Kids & Family includes both sleep meditation and parenting advice); and podcast providers may misleadingly label their shows with unrelated genres for broadening exposure. Second, using topic modeling has its limitations inherited from the input text associated with podcasts: 1) metadata, such as title or description, is typically a short text of varying quality; 2) the automatically transcribed speech is noisy having a high word-error-rate especially for NEs and requires the engagement of more resources [3,8,11].

In the current work, we revisit the feasibility of discovering relevant topics from podcast metadata, titles and descriptions, usually documented by podcast providers, creators, or editors. While previous work [3] found podcast metadata less promising for topic-informed search compared to when using podcast transcripts, we hypothesize that it could still be a very useful data source for topic modeling when exploited with appropriate technology. If proven feasible, topic modeling on podcast metadata can be a more economic alternative than automatically extracting and exploiting transcripts of a rapidly growing podcast corpus. Additionally, the discovery of topics directly from metadata offers many opportunities for improved podcast retrieval. Identifying podcast categories at different granularity levels could help editors evolve manually created podcast taxonomies and automatically annotate podcasts with these categories. The discovered topics could also support the consolidation of podcast knowledge graphs [2,11], recently exploited in recommendation, by adding new edges that capture topic-informed podcast similarity based on metadata.

First, we take advantage of advancements in topic modeling, and benchmark multiple algorithms designed for short text on three podcast datasets. Two of these datasets are public: one from Spotify [8] and one from iTunes [23]. We have built a third dataset using Deezer[1], an online service providing a large podcast catalog. This dataset is the largest with both titles and descriptions available at the podcast level. Second, we propose a strategy to leverage NEs, frequently present in podcast titles and descriptions, in a NMF topic modeling framework. As we can see in the following example: *Shields Up! Podcast: Join Chris and Nev as they talk about their favourite Star Trek episodes covering everything from TOS to Lower Decks*, the metadata contains multiple NEs regarding the name of the speakers (*Join Chris* and *Nev*), but also the podcast topic (*Star Trek, TOS* and *Lower Decks*). By injecting cues from NEs in topic modeling, we improve over state-of-the-art (SOTA) methods using plain word embeddings, and show that the data sparsity (very low co-occurrences of semantically related terms) due to short text can be further alleviated.

To sum up the contributions of this work are: a) the most extensive study to date of topic modeling on podcast metadata, covering popular SOTA algorithms

---

[1] https://www.deezer.com/us/.

for short text and datasets from major podcast streaming platforms; b) NEiCE, a new NE-informed document representation for topic modeling with NMF, as an extension of CluWords [31]—our approach improves topic coherence over baselines in most evaluated cases; c) a new podcast dataset entailing English-language titles and descriptions from Deezer, an online service providing a podcast catalog, that is the largest in terms of the number of podcasts/shows.

2   Related Work

Topic modeling on short text faces the challenge of severe data sparsity due to the nature of this type of input [7]. Short text, as it consists of only few words, can be ambiguous and noisy and, in general, has limited context. This means that pairs of words that are topic-related do not or rarely co-occur in the same contexts, leading to conventional topic modeling techniques such as LDA [6] to perform poorly. Various topic modeling techniques have been designed to address this issue. Models can be classified in four groups: pseudo-documents-based [24,38], probabilistic [15], neural [17,33], and NMF-based [29,31]. We further review each group and some representative models.

The principle of pseudo-documents is to aggregate connected short texts in longer documents, which are further used as input to conventional topic modeling [15]. Initial aggregation methods leveraged metadata such as hashtags in tweets [16]. However, this proved limiting for other types of short texts (e.g. search queries) and led to self-aggregation methods, able to aggregate using topic cues based on the corpus only [24,38]. An issue identified with this type of methods is overfitting [38]. Also, they appear overall less competitive than the other groups of topic modeling techniques for short text [29,31], discussed further.

The second group entailing probabilistic models is the most related to conventional topic modeling (LDA) that represents documents and topics as multinomial distributions over topics, respectively words. The adaptation of these models to short text is to assume that each document is sampled only from a single topic, thus restricting document-topic distribution to a mixture of unigrams [22,36,37]. GPU-DMM [15], an effective and fast model in this group, is based on Dirichlet Multinomial Mixture (DMM) model and uses a Generalized Pólya Urna (GPU) as a sampling process to promote topic-related words. The word association is estimated by exploiting pre-trained word embedding [19]. This allows to alleviate data sparsity as it extends the context to words that are semantically related but they do not necessarily co-occur in the same text.

The third group has become popular in the last years with the rise of deep learning. Neural topic modeling is based on Variational Auto-Encoders (VAE) [4,17,30,33]. Typically, an encoder such as a MultiLayer Perceptron (MLP) compresses the Bag-of-Words (BoW) document representation into a continuous vector. Then, a decoder reconstructs the document by generating words independently [4,17]. Negative sampling and Quantization Topic Model (NQTM) [33], the latest topic modeling technique on short texts brings two contributions which yielded the current SOTA results. The first is a new quantification method applied to the encoder's output whose goal is to generate peakier distributions for decoding. The second is to replace the standard decoder with a negative sampling algorithm that proves better at discovering non-repetitive topics.

The NMF-based group learns topics by decomposing the term-document (BoW) matrix representation of the corpus into two low-rank matrices, one corresponding to document representations over topics and the other to topic representations over words [14]. Given the limited contextual information, the Semantics-assisted Non-negative Matrix Factorization (SeaNMF) model [29] adjusts NMF to short texts by integrating into it word-context semantic correlations learnt from the skip-gram view of the input corpus. In contrast to SeaNMF which focuses on the learning part, CluWords [31] enhances the corpus representation before being factorized with standard NMF. The matrix is obtained with a proposed custom TF-IDF strategy that exploits pre-trained word embeddings.

The existing works include in their benchmark, datasets consisting of question or news titles, web snippets, review comments, or tweets [9,15,29,31,33]. Podcast metadata compared to these datasets exhibits a much higher frequency of NEs, which we exploit with the goal to further address data sparsity. To our knowledge, we are the first to assess existing models on podcast metadata and to explicitly consider NE-related cues in short-text topic modeling.

3   Methods

The topic modeling algorithms we benchmark are GPU-DMM [15], NQTM [33], SeaNMF [29] and CluWords [31]. By noticing the high frequency of NEs in podcast titles and descriptions, we also include in the benchmark another standard NMF-based model for which we design a new NE-informed document representation as input. The underlying hypothesis is that NEs convey the main topic information. Thus, we propose to promote vocabulary words related to these NEs by associating them with pseudo-term frequencies as presented in Sect. 3.2. For this, but also to capture word-to-word topic relatedness shown beneficial against data sparsity, we use pre-trained word and NE embeddings [34].

Finally, the rationale behind choosing to explore NE promotion in a NMF framework is twofold. Compared to probabilistic models, NMF-based ones have yielded better results on short text [7,29,31]. Then, the integration of background NE and word information in NMF topic modeling is more straightforward than in deep neural networks. Current autoencoders [30,33] are designed to exploit only the corpus, which we find insufficient by itself to exhibit NE-word relations, especially if these corpora are small or each NE mention is infrequent.

3.1   Notations and Preliminaries

Table 1 summarizes the notations used in the rest of the section. As outlined above, we obtain topics by factorising the short-text corpus representation. Formally, given the corpus $\mathcal{D}$, the vocabulary $\mathcal{V}$ consisting of unique words in $\mathcal{D}$, A the matrix corresponding to BoW representations of each document in $\mathcal{D}$, and the target number of topics K, A can be approximated by the product of two low-rank matrices $A \approx HW$. Each row $W_{j,:}$ represents one of the K topics expressed in terms of words from $\mathcal{V}$ and each row $H_{i,:}$ represents an input document in terms of the learnt K topics.

Table 1. Notations used to present the topic modeling technique.

Name	Description
K, k	Number of topics, the identifier of a single topic
$\mathcal{D}, d$	Short-text documents found in the corpus, a single document
$\mathcal{V}, t, t', v_t$	Vocabulary set, individual terms, the embeddings of term t
$\mathcal{E}, e, v_e$	Set of linked NEs, a NE term, the embedding of a NE e
$A \in \mathbb{N}^{|\mathcal{D}| \times |\mathcal{V}|}$	Term-document matrix with BoW corpus representation
$C \in \mathbb{R}^{|\mathcal{V}| \times |\mathcal{V}|}$	Word similarity matrix computed with pre-trained embeddings
$W \in \mathbb{R}^{K \times |\mathcal{V}|}$	Latent low-rank word-topic matrix
$H \in \mathbb{R}^{|\mathcal{D}| \times K}$	Latent low-rank document-topic matrix
$A^* \in \mathbb{R}^{|\mathcal{D}| \times |\mathcal{V}|}$	Word-document matrix for CluWords corpus representation
$A^{NE} \in \mathbb{R}^{|\mathcal{D}| \times |\mathcal{V}|}$	Word-document matrix for NE-informed corpus representation
$\alpha^{word}, \alpha^{ent} \in [0, 1)$	Minimum cosine similarity between words, or words and NEs

While this is the basic frameworks for NMF-based topic modeling, in practice there are more effective corpus representations than the simple BoW matrix (A), proven to lead to better topics. CluWords [31] is such an example and is based on two components: 1) one that correlates each word, not only with those with which co-occurs in the corpus, but also with other semantically related words, identified with the help of external pre-trained embeddings; 2) another one that derives a novel document representation, inspired by TF-IDF, which is able to incorporate information from the first component regarding word-to-word relatedness. In our work, we choose to extend CluWords document representations to explicitly prioritize NE cues. We further present the original CluWords, followed by the introduced changes in the next Subsect. 3.2.

The first step of CluWords is to compute a matrix C where each element $C_{t,t'}$ is the cosine similarity (cos) of the embeddings corresponding to the pair of terms $t, t' \in \mathcal{V}$. C is constrained to be non-negative as it is used to compute A^*, which is the input to NMF. Thus, a positive cutoff α^{word} is used to select only the most similar term pairs, and nullify the rest of the matrix:

$$C_{t,t'} = \begin{cases} \cos(v_t, v'_t) & \text{if } \cos(v_t, v'_t) > \alpha^{word} \\ 0 & \text{otherwise} \end{cases} \tag{1}$$

Then, the BoW representation is replaced by a TF-IDF-inspired one. Standard TF-IDF uses the corpus statistics to decrease the weight of very frequent terms and give more weight to terms that appear only in some contexts, thus judged more discriminative, while also accounting for term popularity in a document. Equation 2 shows how the TF-IDF score is computed for a term t and a document d, where $\text{tf}(t,d) = A_{d,t}$ is the number of times t appears in d and n_t is the number of documents in $\mathcal{D}$ where t appears:

$$\text{tf\_idf}(t,d) = \text{tf}(t,d) \cdot \log\left(\frac{|\mathcal{D}|}{n_t}\right) \tag{2}$$

CluWords replaces t by $C_{t,:}$ in order to avoid obtaining a very sparse representation matrix due to the limited context of each word in short text. Thus, it redefines the tf and idf (the log ratio) from Eq. 2 to be computed over vector-based term representations instead of individual frequencies. The new tf^* and idf^* in Eq. 3 incorporate information about semantically similar words to the term t of a given document d in order to expand the term's context:

$$A^*_{d,t} = \text{tf}^*(d,t) \cdot \text{idf}^*(t) = (AC)_{d,t} \cdot \log\left(\frac{|\mathcal{D}|}{\sum_{d\in\mathcal{D}} \mu(t,d)}\right) \tag{3}$$

$\mu(t,d)$ is the mean cosine similarity between the term t and its semantically related terms t' in document d denoted $\mathcal{V}^{d,t} = \{t' \in d | C_{t,t'} \neq 0\}$, or 0 when the ratio in the first branch of Eq. 4 is undefined (t is not in d, thus $|\mathcal{V}^{d,t}| = 0$):

$$\mu(t,d) = \begin{cases} \frac{1}{|\mathcal{V}^{d,t}|} \cdot \sum_{t'\in\mathcal{V}^{d,t}} C_{t,t'} & \text{if } |\mathcal{V}^{d,t}| > 0 \\ 0 & \text{otherwise} \end{cases} \tag{4}$$

Let us note that in the limit case where C is the identity matrix, i.e. each term is only similar to itself which can be obtained by taking $\alpha^{word} = \max_{t\neq t'} C_{t,t'}$, Eq. (3) becomes equivalent to Eq. (2).

3.2 NE-informed Corpus Embedding (NEiCE)

Our approach NEiCE consists of a preprocessing step followed by a computation step which creates a new corpus representation matrix A^{NE} leveraging NEs.

Preprocessing Step. We identify NE mentions in podcast titles and descriptions and link them to Wikipedia entities using the Radboud Entity Linker (REL) system [10]. The REL system is based on multiple modules in pipeline specific to different sub-tasks: 1) the detection of NE mentions using Flair [1], a SOTA Named Entity Recognition (NER) framework using contextualized word embeddings; 2) the disambiguation of the identified entity against a list of possible Wikipedia candidates and its linking to the final candidate. In this final linking phase, REL [10] uses Wikipedia2Vec embeddings [34].

The Wikipedia2Vec embeddings that we also leverage in our solution, compared to other embeddings targeting words only [18,19], are learnt jointly for words and NEs from Wikipedia text. Their learning entails the optimization of three skip-gram sub-models [34]: 1) a regular word skip-gram; 2) an anchor context model—for each NE mention appearing as a hyperlink in text its surrounding words become context; and 3) a link graph model—the entities connected to a NE in the Wikipedia graph become context. From all the information REL returns given a specific input, we use: the Wikipedia page of the disambiguated NE and the confidence score that helps us to choose if we treat a span of text as a NE or favour instead to process its words separately.

Finally, when NEs are processed as separate words instead of being linked to Wikipedia entities, we apply an extra vocabulary cleaning step. As we noticed that in podcast metadata mentions of actors, athletes, or celebrities were very common and we want to avoid the extraction of topics focused on names, we remove these concerned words using the package NameDataset[2].

Computation Step. We derive a new corpus representation matrix A^{NE} as explained next. If NEs are identified in a document with high confidence, then we exploit this information as the main topic-related cues. One strategy to achieve this from previous work on regular text [13] is to favour NEs among the top words to describe topics. Specifically, during preprocessing NEs are treated as n-gram terms and included in the vocabulary. Then, re-weighting approaches are applied to these terms before being served as input to a standard or variations of LDA. The idea behind re-weighting is to associate a larger pseudo-frequency (tf) to NEs such that they are more likely to be picked as topic descriptors.

Contrary to the above-mentioned approach, our goal is to take into account NEs without including them in the vocabulary. While indeed humans will find NEs very expressive to convey topics, this only happens if they already know them. For popular NEs which typically appear in news data exploited in [13], this would not necessarily pose a problem. However, the NEs from podcast metadata tend to be less common or very specific to certain domains, hence less informative for humans trying to associate a topic label. For instance, "That Peter Crouch Podcast" requires knowing that Peter Crouch is a footballer before being able to relate this podcast to football or sport.

The approach we propose is to still use re-weighting to boost NEs importance, but, instead of directly targeting NEs, focus on their semantically-related words. Let $\mathcal{E}^e = \{t | \cos(v_e, v_t) \geq \alpha^{ent}, \forall t \in \mathcal{V} - \mathcal{E}\}$ be the set of non-NE words from $\mathcal{V}$ most similar to a NE e. Similar to when we computed C, a threshold α^{ent} is applied to fix a minimum cosine similarity value between a pair of Wikipedia2Vec embeddings involving a NE ($e \in \mathcal{E}$) and a word ($t \in \mathcal{V}$). Then, we still compute A^{NE} with Eq. 3, but replace tf^* with tf^{NE} as follows:

$$\text{tf}_{d,t}^{NE} = \begin{cases} (AC)_{d,t} + \max_{t' \in \mathcal{V}^{d,t}}(AC)_{d,t'} \text{ , if } t \in \mathcal{E}^e, e \text{ in } d \text{ and } |\mathcal{V}^{d,t}| > 0 \\ (AC)_{d,t} \qquad\qquad\qquad\qquad \text{otherwise} \end{cases} \quad (5)$$

We chose to apply the NE-related re-weighting to the tf factor because we wanted to use NE-related words as the main signal for topics and the direct frequencies allowed us to have more control on it, as also emphasized by [13]. Second, there are two branches depending on whether t is a term very similar to a NE e present in d. If that is the case, a pseudo-frequency is computed by taking into account

---

[2] https://github.com/philipperemy/name-dataset.

the maximum in the CluWords tf matrix (tf^*) for a document d. This means that the words related to a NE e become either as important as the term with the largest weight (t') or more important if the word t already appeared in d.

4 Datasets

We start with describing the existing podcast datasets from iTunes [23] and Spotify [8]. Then, we introduce our newly collected dataset, Deezer, which is the largest one among the three as shown in Table 2. All these datasets contain podcast metadata, titles and descriptions, in English-language. Metadata is documented by providers or creators in an RSS feed, used by podcast aggregators and streaming platforms to make podcasts available to listeners. Although metadata exists for both podcasts (shows) and episodes within shows, we currently focus on shows as their information seemed more reliable. By manually analysing episode metadata in the podcast catalog to which we had access, we noticed they often lacked description or inherited show description.

The iTunes dataset [23] consists of 10 155 podcasts, popular at the moment of creation. The Spotify dataset [8] has 105 360 episodes sampled uniformly at random from podcasts proposed by professional creators (about 10%) and by amateur creators (about 90%). The metadata of each episode contains the title and description of the parent show which we extract to create the final dataset used in the experiments. From these two datasets, we keep podcasts with unique titles and with the concatenations of title and description longer than 3 terms. Additionally, for Spotify we select only the podcasts associated with the language identifiers "en" and "en-US".

Table 2. Summary of podcast datasets: the number of podcasts, the vocabulary size, the total number of NE mentions, the total number of podcasts with NEs in metadata, the mean number of words per title, and the mean number of words per description.

Dataset	$|\mathcal{D}|$	$|\mathcal{V}|$	#NE mentions	#podc. with NE	#w/title	#w/descr.
Spotify	17 456	7 336	20 885	9 198	3.5	38.2
iTunes	9 859	7 331	24 973	6 994	4.9	56.4
Deezer	29 539	14 322	67 083	19 969	4.0	62.6

Deezer differs from the others in that it is the largest. It covers 18 genres (Culture & Society, Business, Films & Games, Music & Audio Commentary, Comedian, Sports, Education, Spirituality & Religion, Information & Politics, Health & Fitness, Art, Entertainment, Lifestyle & Entertainment, Stories & Fiction, Science, Child & Family, True Crime, and History), with a minimum of 300 podcasts per genre. Although these categories are related to topics, as we previously discussed in Sect. 1, they tend to be broad and not always reliable. We could notice a significant overlapping (e.g. Entertainment with Lifestyle & Entertainment, Stories & Fiction with True Crime, or Sports with Health & Fitness), but also how a single category gathers multiple topics.

To create the dataset we randomly sampled from the accessed collection, public podcasts which had titles and descriptions, and the language identifier "en". As the language provided in the metadata was not always reliable, we also used two automatic language detectors, fastText [12] and CLD3 [27]. We filtered out podcasts which were not found to be in English by both detectors. Additionally, we also removed podcasts from unpopular genres (<300 shows). Finally, we applied the same preprocessing as for the other two datasets.

Table 2 presents additional statistics of the used datasets. All datasets contain a large number of NEs and we can find NE mentions in 50%–70% of the podcasts per dataset. We can also observe that the average number of words per title is quite similar for all datasets, while the descriptions in Spotify tend to be shorter.

5   Experimental Setup

We describe next the evaluation metric, the detailed preprocessing and experimental setup, and the environment we used for running the models.

We evaluated topic quality by relying on the widely used topic coherence [26]. A set of facts are said to have high coherence if they could support each other. In topic modeling, this translates into mapping terms on facts and measuring the extent to which these terms tend to co-occur in corpora. While the spectrum of word co-occurrence metrics for topic coherence is quite large [21], the exhaustive search performed in [26] shows that C_V correlates best with human judgement of topic ranking. Thus, we decided to report C_V scores in our evaluation. Given a topic k defined by its T top words $t_1, t_2, ..., t_T$, C_V is defined as:

$$C_V(k) = \frac{1}{T} \sum_{i=1}^{T} \cos(v_{NPMI}(t_i), v_{NPMI}(t_{1:T})) \tag{6}$$

$v_{NPMI}(t_i)$ and $v_{NPMI}(t_{1:T})$ yield two vectors computed with the Normalized Pointwise Mutual Information (NPMI) metric as follows:

$$v_{NPMI}(t_i) = \left(\text{NPMI}(t_i, t_j)\right)_{j=1,..,T} \tag{7}$$

$$v_{NPMI}(t_{1:T}) = \left(\sum_{i=1}^{T} \text{NPMI}(t_i, t_j)\right)_{j=1,..,T} \tag{8}$$

$$\text{NPMI}(t_i, t_j) = \frac{\log \frac{p(t_i, t_j)}{p(t_i)p(t_j)}}{-\log(p(t_i, t_j))} \tag{9}$$

where p is the probability of a term occurrence or co-occurrence in an external corpus. We use Palmetto [26] to compute C_V for each topic k on Wikipedia as external corpus, and average over all K topics to obtain an aggregated value.

In all the reported experiments, we fix the number of top words T to 10 and vary the number of topics K between 20, 50, 100 and 200. During preprocessing, we keep all the linked NEs whose REL confidence score is higher than 0.9 even if they only appear once in the corpus. For normal words, same as in [33], we filter out from vocabulary those that appear less than 5 times. We also remove stop words using NLTK [5]. The same preprocessing is applied before each topic modeling baseline. We evaluate GPU-DMM [15], NQTM [33], SeaNMF [29] and CluWords [31] with their default hyper-parameters. We assess the original Clu-Words with both fastText and Wikipedia2Vec embeddings [18].

As discussed in Sect. 3, NEiCE requires two parameters α^{word} and α^{ent}. [31] motivates the choice of α^{word} between 0.35 and 0.4 in CluWords as it allows to select top 2% of most similar pairs of words. Compared to this approach which assumes α^{word} mainly dependent on the pre-trained embeddings, we investigate if it varies per dataset. Thus, we test α^{word} with multiple values (0.2, 0.3, 0.4, 0.5), where larger the value is, fewer words are selected as being semantically-related to a given term. We proceed similarly for α^{ent}. We run the experiments on an Intel Xeon Gold 6134 CPU @ 3.20 GHz with 32 cores and 128 GB RAM.

6   Results and Discussion

The topic coherence scores obtained by the different topic modeling techniques for short text are presented in Table 3. First, we could notice that NMF-based methods (SeaNMF and CluWords) obtain the best scores in most of the cases. Second, when comparing individual techniques, the ranking depends on the case (number of topics and dataset), but few trends emerge. SeaNMF yields best topic coherence for the lowest number of topics (20) on two datasets. Aligned with the previous literature [31,33], the SOTA models, NQTM and CluWords, obtain very often the best or second best scores, with CluWords ranking first in most cases (7/12). These observations support our choices to work in a NMF framework and devise NEiCE as a CluWord extension, but informed by NEs.

Table 3. Topic coherence scores (C_V in %) obtained by baselines on the three podcast datasets for 20, 50, 100 or 200 topics. CluWords is used with fastText embeddings and the default $\alpha^{word} = 0.4$. Best scores are in bold and second best scores are underlined.

Model	Dataset											
	Deezer				**Spotify**				**iTunes**			
	20	50	100	200	20	50	100	200	20	50	100	200
GPU-DMM	39.0	38.3	37.6	40.1	39.5	39.4	<u>39.7</u>	<u>40.1</u>	39.6	38.5	<u>42.0</u>	41.1
NQTM	38.5	<u>42.2</u>	<u>42.9</u>	<u>45.8</u>	<u>42.9</u>	41.6	39.3	**40.2**	**48.4**	**46.6**	38.2	<u>42.8</u>
SeaNMF	**47.7**	40.5	37.3	39.0	**45.5**	36.4	36.6	35.7	42.2	<u>41.8</u>	35.1	36.9
CluWords$_{ft}$	<u>39.7</u>	**44.0**	**46.3**	**54.5**	40.2	**42.3**	**43.4**	39.5	<u>42.7</u>	40.1	**48.6**	**47.9**

Table 4 shows the results for CluWords with Wikipedia2Vec words embeddings for different values of α^{word}. As mentioned in Sect. 5, previously [31] this parameter was fixed depending on the source of embeddings to 0.4 for fast-Text and 0.35 for word2vec. However, no parameter sensitivity analysis was conducted, which we do now per dataset. We can see that the choice of α^{word}:

1) has a significant impact on the results which could vary up to almost 12 percentage points for Spotify, $K = 50$; 2) is dependent on the assessed case (dataset, K) which previously was not considered; and 3) some values appear to emerge as better choices per dataset (e.g. 0.4 for iTunes or 0.5 for Deezer).

Table 4. Topic coherence scores (C_V in %) obtained by CluWords for different α^{word} values (0.2, 0.3, 0.4, 0.5) with Wikipedia2Vec embeddings on the three podcast datasets for $K \in \{20, 50, 100, 200\}$ topics. Best scores are in bold.

Dataset	Deezer				Spotify				iTunes			
	20	50	100	200	20	50	100	200	20	50	100	200
CluWords$_{wk}$(0.2)	41.3	42.8	42.0	**45.9**	43.2	**49.0**	41.9	**43.0**	46.6	46.8	36.6	40.9
CluWords$_{wk}$(0.3)	39.8	41.3	45.6	44.1	42.8	37.8	46.4	37.8	44.6	40.7	39.0	40.3
CluWords$_{wk}$(0.4)	40.2	48.7	42.5	44.4	**48.4**	39.3	41.8	39.9	**52.9**	**48.5**	**49.6**	40.0
CluWords$_{wk}$(0.5)	**43.0**	**49.1**	**47.7**	41.6	47.3	37.2	**49.9**	42.7	45.3	40.4	41.1	**44.9**

Further, we present in Table 5 the topic coherence scores obtained with our proposed document representation, NEiCE, and different values of α^{word} and α^{ent}. First, we could notice that the introduction of NE cues has a positive impact and NEiCE obtains larger coherence scores than the baselines in most cases (datasets and numbers of topics). The average of NEiCE increase over the best baseline scores is of 15.7% for our best choice of parameters α^{word} and α^{ent}, with a maximum increase of 37.7% on Deezer and $K = 50$. Additionally, the underlined scores in Table 5, which represent scores larger than those obtained by the baselines, show that, no matter the choice of α^{word} and α^{ent}, NEiCE still yields better topic coherence in a majority of cases (85.4%). The most challenging case remains Deezer and $K = 200$ in which only $\alpha^{word} = 0.5$ and $\alpha^{ent} = 0.3$ lead to a larger score than the best baseline, although the increase is small so most likely not significant statistically.

Table 5. Topic coherence scores (C_V, in %) obtained by NEiCE, our document embedding strategy, for different values of (α^{word}, α^{ent}) using Wikipedia2Vec embeddings on the three podcast datasets. Best scores per dataset and number of topic are in bold. Scores larger than all baselines presented in Table 3 are underlined.

Dataset	Deezer				Spotify				iTunes			
	20	50	100	200	20	50	100	200	20	50	100	200
NEiCE (0.2, 0.3)	50.2	48.9	51.4	48.4	51.7	49.0	45.2	46.5	49.3	43.3	49.5	47.0
NEiCE (0.2, 0.4)	53.1	49.2	50.8	50.6	48.7	48.7	43.5	41.7	47.2	49.5	**50.7**	**51.3**
NEiCE (0.3, 0.3)	48.5	52.1	51.5	49.8	52.2	49.0	47.5	47.6	50.3	**52.5**	49.0	48.2
NEiCE (0.3, 0.4)	53.3	50.9	**55.3**	51.6	50.1	48.5	51.1	**49.8**	52.5	49.5	49.2	49.8
NEiCE (0.4, 0.3)	53.2	51.5	52.2	50.0	53.2	49.5	**50.5**	45.9	**52.8**	50.1	50.6	51.1
NEiCE (0.4, 0.4)	**56.4**	52.6	48.1	49.0	51.0	48.2	47.3	47.8	52.4	51.9	49.9	47.4
NEiCE (0.5, 0.3)	52.5	56.3	50.8	**55.4**	51.3	47.7	45.6	45.4	50.6	46.5	46.7	49.0
NEiCE (0.5, 0.4)	56.3	**60.6**	54.9	53.3	**55.0**	**49.9**	46.7	45.0	50.5	52.0	48.7	46.1

From Tables 4 and 5, we can notice that the best α^{word} in CluWords is not necessarily the best in NEiCE. For instance, on iTunes, $\alpha^{word} = 0.4$ was the best choice in Table 4, while in Table 5 $\alpha^{word} = 0.2$ appears a better choice. Also, the best pair of values for these parameters seems to depend largely on the case (dataset and K). Thus, a grid search on a hold-out set is advisable with NEiCE.

Table 6. Topics obtained with NEiCE or NQTM on Deezer and $K = 50$.

k	NEiCE	NQTM
1	mindfulness, yoga, meditation, psychotherapy, psychotherapist, hypnotherapy, psychoanalysis, hypnosis, therapist, psychology	psychotherapist, beirut, displays, remixes, weddings, adversity, namaste, kimberly agenda introducing
2	fiction, nonfiction, novel, author, book, novelist, horror, cyberpunk, anthology, fantasy	avenues, werewolf, criminal, pure, imaginative, strategies, demand, agree, oldies, hang
3	republican, senator, senate, libertarian, election, candidate, nonpartisan, conservative, caucus, liberal	hour, sudden, key, genres, keeps, round, neighbor, conservatives, realize, fulfillment

We selected some examples of topics obtained with NEiCE and NQTM[3] for Deezer and $K = 50$ in Table 6. We selected these topics considering the 18 genres introduced in Sect. 4 and assumed them likely related to Health & Fitness (1), Stories & Fiction or True Crime (2), and Information & Politics (3). Although NQTM yields more diverse top words, their association with a topic is less straightforward compared to NEiCE. However, topic 2 in NQTM is clearly about True Crime, while in NEiCE could be also about Stories & Fiction.

Finally, a qualitative analysis of the topics obtained with NEiCE on Deezer also revealed that many topics were related to world regions which, although easy to interpret, may be noisy if too frequent. These results may be related to the podcasts' topics, but a more likely explanation is that region-related NEs are overweighted. Thus, a detailed study of NE weighting in NEiCE is still needed.

7 Conclusion

We presented a detailed study of topic modeling on podcast metadata covering popular SOTA techniques for short text. Moreover, we proposed NEiCE, a new NE-informed document representation exploited in a NMF framework, and we showed it was more effective in terms of topic coherence than the baselines in various evaluation scenarios including three datasets (one of which, the largest, being newly released). Future work aims to extend the study at the episode level, assess the document representation in downstream tasks, gain more insights into NEiCE especially in relation to the pre-trained embeddings and the choices of αs, and conduct expert studies with editors to further validate mined topics.

---

[3] CluWords has similar top words as NEiCE for topics 1&2 and did not find topic 3.

References

1. Akbik, A., Blythe, D., Vollgraf, R.: Contextual string embeddings for sequence labeling. In: COLING 2018, 27th International Conference on Computational Linguistics, pp. 1638–1649 (2018)
2. Benton, G., Fazelnia, G., Wang, A., Carterette, B.: Trajectory based podcast recommendation. arXiv preprint arXiv:2009.03859 (2020)
3. Besser, J., Larson, M., Hofmann, K.: Podcast search: user goals and retrieval technologies. Online Inf. Rev. **43**(3), 395–419 (2010). https://doi.org/10.1108/14684521011054053
4. Bianchi, F., Terragni, S., Hovy, D.: Pre-training is a hot topic: contextualized document embeddings improve topic coherence. In: Proceedings of the 59th Annual Meeting of the Association for Computational Linguistics and the 11th International Joint Conference on Natural Language Processing (Volume 2: Short Papers), pp. 759–766. Association for Computational Linguistics, August 2021. https://doi.org/10.18653/v1/2021.acl-short.96. https://aclanthology.org/2021.acl-short.96
5. Bird, S., Klein, E., Loper, E.: Natural Language Processing with Python: Analyzing Text with the Natural Language Toolkit. O'Reilly Media, Inc., Sebastopol (2009)
6. Blei, D.M., Ng, A.Y., Jordan, M.I.: Latent dirichlet allocation. J. Mach. Learn. Res. **3**(null), 993–1022 (2003)
7. Chen, Y., Zhang, H., Liu, R., Ye, Z., Lin, J.: Experimental explorations on short text topic mining between LDA and NMF based schemes. Knowl. Based Syst. **163**, 1–13 (2019)
8. Clifton, A., et al.: 100,000 podcasts: a spoken English document corpus. In: Proceedings of the 28th International Conference on Computational Linguistics, Barcelona, Spain, December 2020 pp. 5903–5917. International Committee on Computational Linguistics (2020). https://doi.org/10.18653/v1/2020.coling-main.519. https://aclanthology.org/2020.coling-main.519
9. He, R., Zhang, X., Jin, D., Wang, L., Dang, J., Li, X.: Interaction-aware topic model for microblog conversations through network embedding and user attention. In: Proceedings of the 27th International Conference on Computational Linguistics, Santa Fe, New Mexico, USA, August 2018, pp. 1398–1409. Association for Computational Linguistics (2018). https://aclanthology.org/C18-1118
10. van Hulst, J.M., Hasibi, F., Dercksen, K., Balog, K., de Vries, A.P.: REL: an entity linker standing on the shoulders of giants. In: Proceedings of the 43rd International ACM SIGIR Conference on Research and Development in Information Retrieval. SIGIR 2020, New York, NY, USA, pp. 2197–2200. Association for Computing Machinery (2020). https://doi.org/10.1145/3397271.3401416
11. Jones, R., et al.:: Current challenges and future directions in podcast information access. In: Proceedings of the 44th International ACM SIGIR Conference on Research and Development in Information Retrieval. SIGIR 2021, New York, NY, USA, pp. 1554–1565. Association for Computing Machinery (2021). https://doi.org/10.1145/3404835.3462805
12. Joulin, A., Grave, E., Bojanowski, P., Mikolov, T.: Bag of tricks for efficient text classification. In: Proceedings of the 15th Conference of the European Chapter of the Association for Computational Linguistics: Volume 2, Short Papers, Valencia, Spain, April 2017, pp. 427–431. Association for Computational Linguistics (2017). https://aclanthology.org/E17-2068

13. Krasnashchok, K., Jouili, S.: Improving topic quality by promoting named entities in topic modeling. In: Proceedings of the 56th Annual Meeting of the Association for Computational Linguistics (Volume 2: Short Papers), Melbourne, Australia, July 2018. pp. 247–253. Association for Computational Linguistics (2018). https://doi.org/10.18653/v1/P18-2040. https://aclanthology.org/P18-2040

14. Kuang, D., Choo, J., Park, H.: Nonnegative matrix factorization for interactive topic modeling and document clustering. In: Celebi, M.E. (ed.) Partitional Clustering Algorithms, pp. 215–243. Springer, Cham (2015). https://doi.org/10.1007/978-3-319-09259-1_7

15. Li, C., Wang, H., Zhang, Z., Sun, A., Ma, Z.: Topic modeling for short texts with auxiliary word embeddings. In: Proceedings of the 39th International ACM SIGIR Conference on Research and Development in Information Retrieval. SIGIR 2016, New York, NY, USA, pp. 165–174. Association for Computing Machinery (2016). https://doi.org/10.1145/2911451.2911499

16. Mehrotra, R., Sanner, S., Buntine, W., Xie, L.: Improving LDA topic models for microblogs via tweet pooling and automatic labeling. In: Proceedings of the 36th International ACM SIGIR Conference on Research and Development in Information Retrieval. SIGIR 2013, New York, NY, USA, pp. 889–892. Association for Computing Machinery (2013). https://doi.org/10.1145/2484028.2484166

17. Miao, Y., Yu, L., Blunsom, P.: Neural variational inference for text processing. In: Proceedings of the 33rd International Conference on International Conference on Machine Learning - Volume 48. ICML 2016, pp. 1727–1736. JMLR.org (2016)

18. Mikolov, T., Grave, E., Bojanowski, P., Puhrsch, C., Joulin, A.: Advances in pre-training distributed word representations. In: Proceedings of the Eleventh International Conference on Language Resources and Evaluation (LREC 2018), Miyazaki, Japan, May 2018. European Language Resources Association (ELRA) (2018). https://aclanthology.org/L18-1008

19. Mikolov, T., Sutskever, I., Chen, K., Corrado, G., Dean, J.: Distributed representations of words and phrases and their compositionality. In: Proceedings of the 26th International Conference on Neural Information Processing Systems - Volume 2. NIPS 2013, Red Hook, NY, USA, pp. 3111–3119. Curran Associates Inc. (2013)

20. Mizuno, J., Ogata, J., Goto, M.: A similar content retrieval method for podcast episodes. In: 2008 IEEE Spoken Language Technology Workshop, pp. 297 300 (2008). https://doi.org/10.1109/SLT.2008.4777899

21. Newman, D., Lau, J.H., Grieser, K., Baldwin, T.: Automatic evaluation of topic coherence. In: Human Language Technologies: The 2010 Annual Conference of the North American Chapter of the Association for Computational Linguistics. HLT 2010, USA, pp. 100–108. Association for Computational Linguistics (2010)

22. Nigam, K., McCallum, A.K., Thrun, S., Mitchell, T.: Text classification from labeled and unlabeled documents using EM. Mach. Learn. **39**(2), 103–134 (2000)

23. Ozturk, D.G.: Podcasts Data. https://github.com/odenizgiz/Podcasts-Data. Accessed 20 Sept 2021

24. Quan, X., Kit, C., Ge, Y., Pan, S.J.: Short and sparse text topic modeling via self-aggregation. In: Proceedings of the 24th International Conference on Artificial Intelligence. IJCAI 2015, pp. 2270–2276. AAAI Press (2015)

25. Research, E.: The Podcast Consumer 2019 (2019). https://www.edisonresearch.com/the-podcast-consumer-2019/. Accessed 20 Sept 2021

26. Röder, M., Both, A., Hinneburg, A.: Exploring the space of topic coherence measures. In: Proceedings of the Eighth ACM International Conference on Web Search and Data Mining. WSDM 2015, New York, NY, USA, pp. 399–408. Association for Computing Machinery (2015). https://doi.org/10.1145/2684822.2685324

27. Salcianu, A., et al.: Compact Language Detector v3 (CLD3). https://github.com/google/cld3. Accessed 20 Sept 2021
28. Sharpe, M.: A review of metadata fields associated with podcast RSS feeds. arXiv preprint arXiv:2009.12298 (2020)
29. Shi, T., Kang, K., Choo, J., Reddy, C.K.: Short-text topic modeling via non-negative matrix factorization enriched with local word-context correlations. In: Proceedings of the 2018 World Wide Web Conference. WWW 2018, pp. 1105–1114. International World Wide Web Conferences Steering Committee, Republic and Canton of Geneva, CHE (2018). https://doi.org/10.1145/3178876.3186009
30. Srivastava, A., Sutton, C.: Autoencoding variational inference for topic models. In: ICLR (2017)
31. Viegas, F., et al.: CluWords: exploiting semantic word clustering representation for enhanced topic modeling. In: Proceedings of the Twelfth ACM International Conference on Web Search and Data Mining. WSDM 2019, New York, NY, USA, pp. 753–761. Association for Computing Machinery (2019). https://doi.org/10.1145/3289600.3291032
32. Winn, R.: 2021 Podcast Stats & Facts (New Research From April 2021). https://www.podcastinsights.com/podcast-statistics/. Accessed 20 Sept 2021
33. Wu, X., Li, C., Zhu, Y., Miao, Y.: Short text topic modeling with topic distribution quantization and negative sampling decoder. In: Proceedings of the 2020 Conference on Empirical Methods in Natural Language Processing (EMNLP), pp. 1772–1782. Association for Computational Linguistics, November 2020. https://doi.org/10.18653/v1/2020.emnlp-main.138. https://aclanthology.org/2020.emnlp-main.138
34. Yamada, I., et al.: Wikipedia2Vec: an efficient toolkit for learning and visualizing the embeddings of words and entities from Wikipedia. In: Proceedings of the 2020 Conference on Empirical Methods in Natural Language Processing: System Demonstrations, pp. 23–30. Association for Computational Linguistics (2020)
35. Yang, L., Wang, Y., Dunne, D., Sobolev, M., Naaman, M., Estrin, D.: More than just words: modeling non-textual characteristics of podcasts. In: Proceedings of the Twelfth ACM International Conference on Web Search and Data Mining. WSDM 2019, New York, NY, USA, pp. 276–284. Association for Computing Machinery (2019). https://doi.org/10.1145/3289600.3290993
36. Yin, J., Wang, J.: A dirichlet multinomial mixture model-based approach for short text clustering. In: Proceedings of the 20th ACM SIGKDD International Conference on Knowledge Discovery and Data Mining. KDD 2014, New York, NY, USA, pp. 233–242. Association for Computing Machinery (2014). https://doi.org/10.1145/2623330.2623715
37. Zhao, W.X., et al.: Comparing Twitter and traditional media using topic models. In: Clough, P., et al. (eds.) ECIR 2011. LNCS, vol. 6611, pp. 338–349. Springer, Heidelberg (2011). https://doi.org/10.1007/978-3-642-20161-5_34
38. Zuo, Y., et al.: Topic modeling of short texts: a pseudo-document view. In: Proceedings of the 22nd ACM SIGKDD International Conference on Knowledge Discovery and Data Mining. KDD 2016, New York, NY, USA, pp. 2105–2114. Association for Computing Machinery (2016). https://doi.org/10.1145/2939672.2939880

Effective Rating Prediction Using an Attention-Based User Review Sentiment Model

Xi Wang[✉], Iadh Ounis, and Craig Macdonald

Unversity of Glasgow, Glasgow, UK
x.wang.6@research.gla.ac.uk
{iadh.ounis,craig.macdonald}@glasgow.gla.ac.uk

Abstract. We propose a new sentiment information-based attention mechanism that helps to identify user reviews that are more likely to enhance the accuracy of a rating prediction model. We hypothesis that highly polarised reviews (strongly positive or negative) are better indicators of the users' preferences and that this sentiment polarity information helps to identify the usefulness of reviews. Hence, we introduce a novel neural network rating prediction model, called SentiAttn, which includes both the proposed sentiment attention mechanism as well as a global attention mechanism that captures the importance of different parts of the reviews. We show how the concatenation of the positive and negative users' and items' reviews as input to SentiAttn, results in different architectures with various channels. We investigate if we can improve the performance of SentiAttn by fine-tuning different channel setups. We examine the performance of SentiAttn on two well-known datasets from Yelp and Amazon. Our results show that SentiAttn significantly outperforms a classical approach and four state-of-the-art rating prediction models. Moreover, we show the advantages of using the sentiment attention mechanism in the rating prediction task and its effectiveness in addressing the cold-start problem.

1 Introduction

Rating prediction is a classical recommendation task [22], where the recommendation system aims to accurately predict the user rating of an unseen item, so as to better estimate which items to recommend to a user. The predictions are typically based on the existing ratings by users. The rating prediction task remains a challenging and open problem. Indeed, the effectiveness of existing rating prediction-based recommendation systems is still limited, suffering from various types of challenges, including accuracy, data sparsity and the cold-start problem [4,32]. Therefore, many approaches have been proposed to leverage user reviews [16,28] – including the sentiment of the reviews [11,17] – to improve the rating prediction accuracy. Users' reviews can enrich both user and item representations, while sentiment information is often useful for extracting user

© The Author(s), under exclusive license to Springer Nature Switzerland AG 2022
M. Hagen et al. (Eds.): ECIR 2022, LNCS 13185, pp. 487–501, 2022.
https://doi.org/10.1007/978-3-030-99736-6_33

preferences [11]. However, not all reviews are useful to enhance the rating prediction performance, since they may convey varying actionable information about the users' preferences [1]. Recently, a number of approaches have made use of the attention mechanism to estimate the usefulness of reviews [1,24]. Attention mechanism focuses on the parts of review content that contribute to the rating prediction. While these existing approaches demonstrate that the attention mechanism can improve the rating prediction performance, they (i.e. [1,24]) do not leverage the sentiment information actually captured by the reviews.

Given the effectiveness of sentiment information in extracting user preferences, we hypothesise that sentiment information should also be used in estimating the usefulness of reviews, so as to further improve the rating prediction performance. Indeed, reviews with a clear polarised sentiment (i.e. positive or negative) typically convey richer information about items and are more likely to influence the users' decision making when interacting with the corresponding items [11]. In the literature, several approaches focused on leveraging the sentiment information as an additional feature to address the rating prediction task [6,28], while ignoring the potential relationship between the sentiment polarity and the usefulness of reviews in users' decision making. In this study, we propose instead to directly leverage the sentiment scores of reviews to address the aforementioned limitation. Inspired by Wang et al. [28], the sentiment score of a review is estimated as the probability of the review having a clear positive or negative polarity as determined by a sentiment classifier. These sentiment scores are then used in a customised attention mechanism to identify informative reviews with rich user preferences. Hence, SentiAttn assumes that reviews with clearly pronounced user preferences are useful for effective rating prediction. In addition, SentiAttn adds another attention mechanism (i.e. global attention [14]) to capture and model the importance of the parts of reviews that are likely to enhance the rating prediction performances. On the other hand, previous works on *neural architecture search* [7,13] showed that fine-tuning a neural model architecture could have a marked positive impact on the model's performance. To leverage the advantage of fine-tuning the neural models' architectures, in this paper, we propose a strategy where we first concatenate the users' and items' positive and negative reviews as input to SentiAttn, resulting in different SentiAttn architectures with various number of channels (e.g. if we concatenate all reviews for both users and items, then this leads to a single channel-based SentiAttn model). Next, we fine tune the architecture variants of our proposed SentiAttn model with different channel setups on the validation sets of two datasets from Yelp and Amazon, so as to optimise the performances of SentiAttn on different datasets.

Our contributions in this paper are as follows: (1) We propose a new sentiment information-based attention mechanism, which weights the usefulness of reviews by their corresponding sentiment scores. These scores reflect the user preferences since they convey a clear sentiment. To the best of our knowledge, this is the first model to directly encode sentiment information for identifying review usefulness in rating prediction; (2) We examine the impact of the resulting SentiAttn model architectures using different channels on their rating prediction performances. This examination is conducted by fine tuning the architectures on the validation sets of the Yelp and Amazon datasets; **(3)** We show that SentiAttn achieves a significantly better rating prediction accuracy than one classical (NMF [10]) and four existing state-of-the-art rating prediction models (namely, ConvMF [8], DeepCoNN [32], D-Attn [24] and NARRE [1]) over two datasets; **(4)** We show that SentiAttn is particularly effective in addressing the cold-start problem in comparison to the existing baselines.

2   Related Work

In this section, we briefly discuss two bodies of related work.

Review-based Rating Prediction: Several studies have exploited user reviews to improve rating predictions [27,29,32]. Many earlier studies used topic modeling techniques (e.g. Latent Dirichlet Allocation (LDA)) to model user reviews [12,17]. However, with the emergence of word embedding [18] techniques, it has been shown that rating prediction models based on word embeddings can outperform such topic modelling-based approaches. For example, Zheng et al. [32] proposed a deep learning model that initialised both the user and item matrices with word embeddings before jointly modelling the users and items to make rating predictions. However, not all user reviews provide useful information to enhance the rating prediction performance. With this in mind, some previous studies, e.g. [1,24], have applied an attention mechanism to identify useful reviews to improve rating predictions. Seo et al. [24] developed two attention mechanisms to learn review usefulness, i.e. local and global attention mechanisms to generate explainable and better-learned review representation latent vectors. Chen et al. [1] initialised user/item latent vectors with review embedding vectors and the corresponding identification information. The authors used a typical attention mechanism to model the latent vectors. However, although the attention mechanism can be effective for modelling the usefulness of reviews, the attention mechanism does not consider the sentiment information of reviews. Sentiment information has been shown to enhance the rating predictions in many studies [15,26,31] (we discuss in the remainder of this section). In this paper, unlike prior work, we propose to directly leverage the sentiment information within a customised attention mechanism when addressing the rating prediction task.

Sentiment-enhanced Recommendation: Recently, sentiment-enhanced recommendation approaches have benefited from deep-learning techniques. For example, Wang et al. [28] examined the performance of different state-of-the-art sentiment classification approaches (e.g. CNN [9] and LSTM [5]) to generate review sentiment polarity scores, and then validated the usefulness of sentiment information by replacing user ratings with such sentiment scores for making recommendations. Chen et al. [1] used a convolution operation to convert reviews into latent vectors to represent review sentiment information, thereby enhancing the rating prediction performance. These studies validated the usefulness of using sentiment information to identify user preferences in user reviews. Therefore, we postulate that sentiment information can also be useful for identifying useful reviews. To the best of our knowledge, our proposed SentiAttn model is the first sentiment-enhanced recommendation approach to use sentiment information to weight review usefulness in an attention neural network architecture.

3   The SentiAttn Model

In this section, we first state the rating prediction task and the notations used. Next, we illustrate the motivation of using sentiment information to identify useful reviews and describe our proposed SentiAttn rating prediction model.

Table 1. Review examples with sentiment information.

Positive and High Sentiment Score	
Rating: 5 **Sentiment Score:** 0.9726 **Category:** grocery and gourmet food	**Review 1:** This beverage is so delicious. I would like to order more in the future. I drink it to relax
Positive but Low Sentiment Score	
Rating: 5 **Sentiment Score:** 0.1783 **Category:** grocery and gourmet food	**Review 2:** My husband insists on making his own yogurt and won't use any other starter. This assures the same consistency month after month

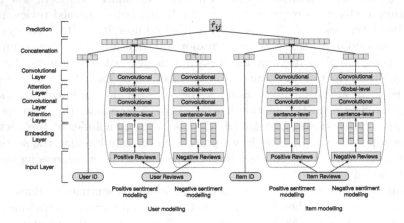

Fig. 1. The architecture of the SentiAttn model

3.1   Task Definition

The rating prediction task aims to predict the ratings of unseen items. Consider a set of users U and items I (of size m and n, respectively). We also have the one-hot embedding vectors E_U and E_I, which map users and items to different randomly initialised vectors. User ratings can be encoded in a rating matrix $R \in \mathbb{R}^{m \times n}$, where entries $r_{u,i} \in R$ represent the previously observed ratings with a range from 1 to 5. In rating prediction, we aim to accurately predict the rating $r_{u,i}$ of an unseen item i for user u. Moreover, each rating $r_{u,i}$ is associated with a textual review $c_{u,i}$. As discussed in Sect. 1, for each review $c_{u,i}$, we also estimate a corresponding sentiment score $s_{u,i}$, which indicates the probability of the review being polarised, i.e. being strongly positive or strongly negative.

3.2   Review Sentiment Information Analysis

To motivate the use of sentiment information in identifying useful reviews, we provide two illustrative review examples in Table 1. The sentiment score corresponds to the probability of a given review being polarised, as further explained in Sect. 3.3. These two reviews are both positive and 5 star-rated. However, when we compare these two reviews, Review 1 better conveys the user's preferences, while Review 2 simply describes a personal event, making it hard for the model to capture the user's preferences. Therefore, Review 1 is deemed more useful than Review 2. In particular, the sentiment scores of Reviews 1 and 2 clearly mirror their usefulness difference (Review 1 is scored 0.9726 as being strongly positive while Review 2 is scored 0.1783 only). Therefore, we propose to leverage the relationship between the sentiment scores and the usefulness of reviews in SentiAttn. Our model identifies useful reviews via a novel sentiment information-based attention mechanism to improve the rating prediction performance.

3.3   Model Architecture

To encode the review usefulness information through their sentiment scores, SentiAttn first uses a customised sentiment attention mechanism to embed the review usefulness information. Next, it integrates another global attention mechanism [14] to capture the parts of reviews that are likely to enhance the rating prediction performance. The architecture of SentiAttn (Fig. 1) comprises eight layers from the input to the rating prediction layer, described further below:

Input and Embedding Layers: In the input layer, users are represented by the reviews they have posted for items while items are represented by the reviews given by users. In particular, the input layer groups reviews into positive and negative reviews according to their corresponding rating values. If the rating $r_{u,i} \geq 4$, the review $c_{u,i}$ is positive, else, if the rating $r_{u,i} \leq 2$, the review $c_{u,i}$ is negative. A review $c_{u,i}$ with a rating of $r_{u,i} = 3$ or with no provided rating is classified as positive or negative according to a CNN-based binary sentiment classifier (described further in Sect. 4.3). Therefore, our SentiAttn model can be divided into four parallel networks (i.e. four channels), which model the positive and negative reviews for users and items. The architecture of our SentiAttn model is flexible and can possibly have two additional variants (i.e. one channel

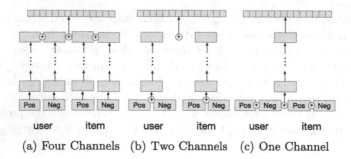

(a) Four Channels   (b) Two Channels   (c) One Channel

Fig. 2. Architectures of the original SentiAttn four channels-based model and its variants (i.e. one and two channels-based) .

or two channels-based). As shown in Fig. 2, instead of modelling the polarised reviews for user and items individually, we can concatenate all the reviews for the user or the item, resulting in the two channels-based SentiAttn model variant. Moreover, if we further concatenate all the reviews of the user and the item together, we can obtain the one channel-based SentiAttn model variant. In particular, for each resulting channel, its review modelling pipeline remains the same as each individual channel depicted in Fig. 1. It is of note that the one channel-based model variant can only be leveraged by a model that uses a value mapping-based predictor (e.g. the factorisation machine and the multi-layer perceptron) and not an interaction-based predictor (e.g. the dot product function), which needs at least two inputs. In this paper, we investigate which SentiAttn model variant exhibits the best performances on the used datasets. Next, following [1,32], in the embedding layer, we convert the reviews text into embedding vectors, denoted as X, which are then given as input to the next layer.

Sentiment Attention Layer: In this layer, we customise a sentiment attention mechanism to encode the usefulness of reviews. Our sentiment attention mechanism is inspired by the dot-product attention function [25], which learns the importance (weight) of different embedding vectors. Then, it multiplies the resulting weighting vectors with the initial word embedding vectors to apply the attention mechanism. Unlike the dot-product attention function, our sentiment attention mechanism obtains the weighting vectors from the sentiment scores of the reviews. These sentiment scores can enrich the user's information and might be helpful in addressing the cold start problem. First, the reviews have been labeled as positive or negative in the previous layer. After that, we process these reviews with a given sentiment classifier and obtain the corresponding probabilities of the positive reviews being positive or the negative reviews being negative (denoted as $p_{u,i}$, which naturally ranges from 0 to 1). The corresponding sentiment scores for the positive reviews are $s_{u,i} = p_{u,i}$. Conversely, we use $s_{u,i} = 1 - p_{u,i}$ for the negative reviews. Hence, the sentiment score indicates the probability of a given review being polarised (positive or negative), and a review is deemed more useful if its sentiment score is closer to 1. Next, with a given review embedding vector X, and its sentiment score vector S, the converted vector X' is calculated as $X' = ((SX^T)^T \oplus X)$, where $\oplus$ is a residual connector. **Convolutional Layer** Our SentiAttn model applies the convolution operation, as in [24,32], on the latent vector X' with g neurons to generate feature vectors for the next layer. Each neuron applies the convolution operation to a sliding window t over latent vectors with width T. The convolution operation of neuron e is obtained as follows: $z_e = f(X'_{1:T} * K_e + b_e)$, where $f(.)$ indicates an activation function to filter the output of the convolution operation, $*$ is the convolution operator of neuron e on the corresponding window of vectors and b_e is a bias parameter [8]. After applying the convolution operation, we apply the max pooling function over the output feature vectors, denoted as Z, and obtain the resulting vector o for each neuron (i.e. $o = max(z_1, z_2, ..., z_e^{T-t+1})$). Next, the outputs of the g neurons are concatenated together into the latent vector X_c.

Global Attention Layer: Apart from the proposed sentiment attention layer, we also use the global attention mechanism from [14]. Accordingly, we add the global attention layer to SentiAttn to model the parts of review content that are likely to contribute to enhancing the rating prediction performances. In particular, the global attention mechanism considers all review embeddings as input and calculates the global attention score vector G of the embedding input X_c: $G = \text{SoftMax}(W_g X_c)$. The embedding input X_c is then further weighted by the global attention score vector G as $X_g = (GX_c^T)^T$. After the global attention layer, we add another convolutional layer, which is the same as the one above the sentiment attention layer, to process the review embeddings. We use the outputs from the convolutional layer as the final latent feature vectors for each channel.

Concatenation and Prediction Layer: In the concatenation layer, we concatenate the latent vectors from two groups of inputs: (1) the resulting latent feature vector from the last convolutional layer of the review modelling channels; (2) the one-hot embedding vectors of each user and item. We refer to the concatenated vector as o. Next, in the prediction layer, we use a two-order factorisation machine [20] as the rating predictor, which is capable of capturing the patterns in data to improve the model's performance [30]. This predictor has also been widely used in the literature to address the rating prediction task [2,3,21]. Each predicted rating $\hat{r}_{u,i}$ is calculated as follows: $\hat{r}_{u,i} = w_0 + b_u + b_i + (\sum_{j=1}^{|o|} w_j o_j) + (\sum_{j=1}^{|o|} \sum_{k=j+1}^{|o|} o_j o_k \mathbf{w}_{j,k})$. This equation has five summands: w_0 is the global bias parameter [20]; next, b_u and b_i correspond to the bias parameters for user u and item i, respectively; in the fourth summand, w_j models the weight of the j_{th} variable in o; the final summand models the interactions between pairs of variable vectors o_j and o_k in o, weighted by a factorised parameter $\mathbf{w}_{j,k} \approx \langle \mathbf{v}_j, \mathbf{v}_k \rangle$ as in [20]. SentiAttn is trained by minimising the prediction error between the true rating value $r_{u,i}$ and the predicted rating value $\hat{r}_{u,i}$ with the MSE function.

4   Experimental Setup

We now examine the performance of SentiAttn through experiments on two real-world datasets, in comparison to a number of classical and state-of-the-art rating prediction models. In particular, we address three research questions: **RQ1:** Which architecture variant of the SentiAttn model (based on 1, 2 or 4 channels) performs the best on the two used datasets? **RQ2:** Does SentiAttn outperform other state-of-the-art models in addressing the rating prediction task and how much does it benefit from (i) the proposed sentiment attention mechanism and (ii) the global attention mechanism? **RQ3:** Does SentiAttn outperform the existing baselines when making rating predictions for cold-start users?

4.1   Datasets

To perform our experiments, we use two popular real-world datasets [17,24]: (i) a Yelp[1] dataset, and (ii) an Amazon Product dataset[2]. The Yelp dataset contains a large number of reviews on venues located in Phoenix, USA. The Amazon dataset contains reviews on products among six categories[3]. The statistics of these two datasets are in Table 2. Following a common setup [1,32], these two datasets are randomly divided into 80% training, 10% validation and 10% testing sets. Moreover, we follow [6] and denote those users with less than 5 reviews in the training dataset as the cold-start users. Table 2 shows that the Yelp dataset is more sparse (i.e. has a lower density[4]) than the Amazon dataset. This observation suggests that the data sparsity's influence might be amplified in a given model's performance when experimenting with the Yelp dataset. Moreover, as per the positive rating percentages in Table 2, most user reviews are positive in both datasets.

Table 2. Statistics of datasets.

Dataset	#Users	#Cold-start Users	#Items	#Reviews	% Density	% Positive ratings
Yelp	45,981	33,306	11,537	229,907	0.043	67.88
Amazon	26,010	7,874	16,514	285,644	0.066	81.24

4.2   Baselines and Evaluation Metrics

We compare our SentiAttn model[5] to the following 5 baselines: **(1) NMF** [10]: NMF is a widely used classical baseline, which characterises users and items with

---

[1] https://kaggle.com/c/yelp-recsys-2013.

[2] http://jmcauley.ucsd.edu/data/amazon/.

[3] 'amazon instant video', 'automotive', 'grocery and gourmet food', 'musical instruments', 'office products' and 'patio lawn and garden'.

[4] % Density = #interactions / (#users × #items)).

[5] Our source code is available at: https://github.com/wangxieric/SentiAttn.

their rating pattern-based latent vectors. **(2) ConvMF** [8]: ConvMF extends the latent feature vectors in NMF with the embedding vector of reviews. **(3) DeepCoNN** [32]: DeepCoNN jointly models reviews to characterise users and items with latent vectors. This approach has been widely used as a strong review-based rating prediction model. **(4) D-Attn** [24]: D-Attn is another review-based rating prediction model that includes two global and local attention mechanisms. D-Attn is another review-based rating prediction model. It includes two global and local attention mechanisms, to improve the explainability and rating prediction accuracy of a rating prediction model. **(5) NARRE** [1]: NARRE is a recent state-of-the-art attention-based rating prediction model. It weights reviews by its learned review usefulness scores. These scores are estimated through the use of an attention mechanism. Moreover, we examine the effectiveness of using our proposed sentiment attention mechanism in comparison to three further baselines derived from SentiAttn as follows: One baseline removes both attention layers in the SentiAttn model (denoted by '**Basic**'), while '**+Glb**' and '**+Sent**' add the global attention layer and the sentiment attention layer to the Basic model, respectively. As for the evaluation metrics, we use Mean Absolute Error (MAE) and Root Mean Squared Error (RMSE) to measure the performances of SentiAttn and the baselines, which are the commonly used metrics to evaluate rating prediction models [1,11,32]. In order to examine the statistical significance of the models' performances, we leverage both the paired t-test, with a significance level of $p < 0.05$, and the post hoc Tukey Honest Significant Difference (HSD) [23] test at $p < 0.05$ to account for the multiple comparisons with the t-tests[6].

4.3 Model Setting

In the input layer, we use an existing CNN-based binary sentiment classifier to group reviews into positive and negative reviews, which has been shown to have a strong accuracy (>95%) for sentiment classification [28]. Other sentiment classifiers could have been used, but the investigation of such classifiers is beyond the scope of this paper. For the used CNN-based binary sentiment classifier, we follow the same experimental setup as [28] and train it on 50,000 positive and 50,000 negative review instances that were sampled from a separate dataset, namely the Yelp Challenge Round 12 dataset[7].

Moreover, the classifier provides each review with its probability $p_{u,i}$ of carrying a strong polarised sentiment, so as to generate a sentiment score $s_{u,i}$ in the sentiment attention layer, as explained in Sect. 3.3. Next, in the embedding layer, we use the pre-trained GloVe [19] word embedding dictionary[8], following [28], and map each word into an embedding vector with 100 dimensions. In

---

[6] Since RMSE is a non-linear aggregation of squared absolute errors, a significance test cannot be conducted with this metric.

[7] https://www.yelp.com/dataset/challenge.

[8] We also apply the pre-trained GloVe word embeddings within the baseline approaches, which ensures fair performance comparisons between approaches.

the convolutional layer, following [24], we set the kernel size to 100 and the activation function to ReLU. In particular, we use the Adam optimiser with a 10^{-4} learning rate. Moreover, it is of note that, to answer RQ1, which investigates the performances of the considered SentiAttn architecture variants, we conduct experiments on the validation sets from the Yelp and Amazon datasets to select the best SentiAttn architecture – thereby mimicking the use of the validation sets for model selection.

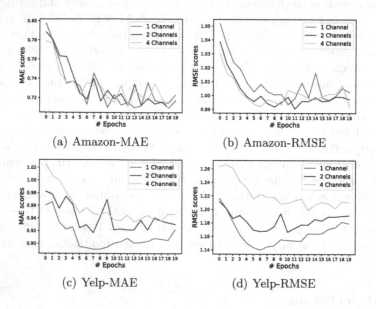

(a) Amazon-MAE   (b) Amazon-RMSE

(c) Yelp-MAE   (d) Yelp-RMSE

Fig. 3. Validation performances of the SentiAttn variants with different #channels.

5   Results

Next, we report and analyse our obtained results:

Performances of Architecture Variants (RQ1). We investigate which of the SentiAttn model architecture variants leads to the best rating prediction performances. In Fig. 3, we report the performances of the three considered architecture variants of the proposed SentiAttn model (namely the one, two and four channels-based SentiAttn architectures). Since we are using the MAE and RMSE error-based evaluation metrics, the lower the metrics' values, the higher is the model's rating prediction performance. First, we compare the performances of the SentiAttn variants on the Amazon dataset in Fig. 3(a) and Fig. 3(b). For both MAE and RMSE, the three variants show similar trends and performances and are overall comparable. However, on the Yelp dataset, Fig. 3(c) and Fig. 3(d) show that the SentiAttn model with one-channel consistently outperforms both the original SentiAttn model with four channels and the two channels-based variant. Using the six components of the Amazon dataset corresponding to each of the categories in Footnote 3, we conducted a further analysis to examine the correlation between a given dataset's statistics and the performances of Senti-Attn with different number of channels. The results of our analysis suggest that the higher the density of interactions in a dataset, the better a variant of Senti-Attn with a larger number of channels performs[9]. Overall, in answer to RQ1, we conclude that a higher number of channels is preferred for datasets with high density of interactions. As per these results, we select the overall best variant model, namely the one channel-based SentiAttn model for the remaining experiments.

Table 3. Rating prediction accuracy; * denotes a significant difference in MAE with SentiAttn with respect to both the paired t-test and the Tukey HSD test, $p < 0.05$.

	All Users					Cold-Start Users			
	Yelp Dataset		Amazon Dataset			Yelp Dataset		Amazon Dataset	
	MAE	RMSE	MAE	RMSE		MAE	RMSE	MAE	RMSE
NMF	0.9866*	1.2630	0.8240*	1.0881	NMF	1.1690*	1.5025	0.9040*	1.1843
ConvMF	0.9748*	1.2329	0.7964*	1.0371	ConvMF	1.0785*	1.3812	0.8565*	1.1154
DeepCoNN	0.9247*	1.1885	0.7233*	0.9929	DeepCoNN	1.0462*	1.3506	0.7882*	1.0749
D-Attn	1.0040*	1.2106	0.8316*	1.0627	D-Attn	1.0154*	1.2394	0.8738*	1.1029
NARRE	0.9163*	1.1781	0.7065*	0.9783	NARRE	1.0289*	1.3481	0.7613*	1.0587
Basic	0.9084*	1.1769	0.7060*	0.9769	Basic	1.0003*	1.3602	0.7451*	1.0520
+Glb	0.8947	1.1734	0.6960	0.9723	+Glb	0.9867	1.3544	0.7253	1.0460
+Sent	0.8932	1.1476	0.6957	0.9685	+Sent	0.9817	1.2408	0.7190	1.0375
SentiAttn	**0.8888**	**1.1463**	**0.6841**	**0.9668**	SentiAttn	**0.9736**	**1.2327**	**0.7090**	**1.0273**

Comparison to the Baselines (RQ2). Table 3 presents the rating prediction errors of both the baseline models and SentiAttn. First, in the obtained results for both the Yelp and Amazon datasets, SentiAttn significantly outperforms the baselines according to both the paired t-test and the Tukey HSD test. In particular, while D-Attn, NARRE and SentiAttn all use an attention mechanism to weight reviews with their estimated usefulness, our SentiAttn model, which relies on a novel sentiment attention and a global attention mechanism returns significantly smaller prediction errors in comparison to competitive baselines on both the Yelp and Amazon datasets. We also evaluate the usefulness of the global attention layer and the proposed sentiment attention layer in Senti-Attn by comparing the performances of SentiAttn with the Basic, +Glb, and +Sent models (introduced in Sect. 4.2). Table 3 shows that SentiAttn significantly (according to both the paired t-test and the Tukey HSD test, $p < 0.05$) outperforms the Basic model on both used datasets, which demonstrates the effectiveness of using the attention mechanisms. Moreover, the results show that the sentiment attention mechanism outperforms the global attention mechanism since it results in lower MAE and RMSE scores (0.8932 vs. 0.8947 (MAE)

---

[9] Due to the page limit, we do not include these experimental results in the paper.

and 1.1476 vs. 1.1734 (RMSE) for +Sent vs. +Glb in Table 3). In particular, we observe that the sentiment attention mechanism is especially effective in decreasing the variance of the rating prediction errors. Indeed, +Sent outperforms both Basic and +Glb providing lower RMSE scores with wide margins on both used datasets.

To further examine the effectiveness of the proposed sentiment attention mechanism, we conducted further analysis on the results of both datasets. We averaged the sentiment scores of the reviews posted by a given user. We group users into two groups: 'sentiment-polarised' vs. 'sentiment-neutral' users. On the Yelp dataset, the sentiment-polarised users have average review scores > 0.88, while the sentiment-neutral users have average scores ≤ 0.88[10]. This leads to 25155 sentiment-polarised and 20826 sentiment-neutral users. We would expect the proposed sentiment attention mechanism to mostly benefit the sentiment-polarised users since these users have more reviews that clearly convey their preferences. Next, we compare performances between the global attention '+Glb' and the sentiment attention '+Sent' models. The results on Yelp show that '+Sent' significantly outperforms '+Glb' for 51.6% of the sentiment-polarised users and 43.3% of the sentiment-neutral users (using a paired t-test on users with the MAE metric). Contrastingly, '+Glb' significantly (paired t-test) outperforms '+Sent' for 38.3% of the sentiment-polarised users and 48.1% of the sentiment-neutral users. These results indicate that the proposed sentiment attention mechanism can indeed help the sentiment-polarised users, but does not exhibit better performances than using the global attention mechanism if most of the users' reviews do not contain highly polarised reviews (i.e. sentiment neutral users). We observed similar conclusions on the Amazon dataset. To answer RQ2, we conclude that the obtained results empirically validate the effectiveness of our SentiAttn model in addressing the rating prediction task in comparison to strong baseline models. The results also show the effectiveness of using the sentiment attention mechanism – which weights the review input according to the corresponding review sentiment scores – thereby outperforming the global attention mechanism.

Cold-Start Users (RQ3). We now evaluate the rating prediction performance of SentiAttn on cold-start users. As introduced in Sect. 4.1, we consider users in the training dataset with less than 5 reviews as cold-start users. Table 3 provides the rating prediction performances of SentiAttn and the various baseline models on both the Yelp and Amazon datasets for cold-start users. The results show that our SentiAttn model obtains a good cold-start performance by significantly outperforming all the strong baseline approaches from the literature on the Yelp and Amazon datasets. Comparing the rating prediction results in Table 3 on the Yelp dataset, we note that as expected from the statistics of this dataset, the rating prediction performances of all models suffer from the cold-start problem. However, the cold-start problem appears to have only a small negative influence on the D-Attn model. To investigate the reasons behind the relative effectiveness of D-Attn in addressing the cold-start problem, we plot the predicted rating

---

[10] The threshold (0.88) is the mean value of the reviews' sentiment score distribution.

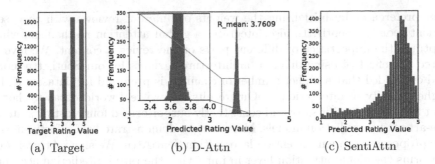

| (a) Target | (b) D-Attn | (c) SentiAttn |

Fig. 4. Cold-start user rating prediction performance comparison (D-Attn vs. Senti-Attn) on the Yelp dataset.

value frequency distribution of the cold-start users on the Yelp dataset using both D-Attn and our SentiAttn model in Figs. 4(b) and 4(c), respectively. These distributions are compared with the target rating distribution in Fig. 4(a). In Fig. 4(b) of the D-Attn model, the predicted rating values shrink between around 3.55 and 3.80, which are all close to the average of the target rating value (i.e. $\overline{R} = 3.7609$). This distribution shows that the performance of D-Attn is less reliable in distinguishing the actual user preferences. On the contrary, in Fig. 4(c), the predicted rating value frequency distribution of our SentiAttn model ranges from 0 to 5 and its shape better aligns with the actual rating distribution of the Yelp dataset in Fig. 4(a).

We also compare the impact of using two attention mechanisms in addressing the cold-start problem. According to the results in Table 3, our sentiment attention mechanism outperforms the global attention mechanism in improving the rating prediction accuracy of the Basic model (e.g. $1.0003 \rightarrow 0.9817$ vs. $1.0003 \rightarrow 0.9867$ on the Yelp dataset) and lowers the variances of the rating prediction errors with a wider margin. For example, on the Yelp dataset, the Basic model benefits from using the global attention mechanism and lowers the RMSE score from 1.3602 to 1.3544. However, when applying the sentiment attention mechanism, the RMSE score of the Basic model is decreased from 1.3602 to 1.2408, indicating a higher improvement than when applying the global attention mechanism. Therefore, in answer to RQ3, our SentiAttn model is particularly effective for the cold-start users compared with the five strong baselines from the literature. Our sentiment attention mechanism also shows its usefulness in improving the rating prediction accuracy, especially lowering the variance of the rating prediction errors for cold-start users. In particular, SentiAttn is more reliable than D-Attn in identifying user preferences, as illustrated by the predicted rating distributions.

6 Conclusions

In this paper, we proposed the SentiAttn model, which leverages user reviews as input and deploys a new sentiment attention mechanism. The latter encodes user preferences by initialising the weights of different reviews with their sentiment scores.   SentiAttn also integrates a global attention mechanism, which captures the importance of different parts of the review's content. We investigated the effect of using different architecture variants for our SentiAttn model and concluded that a higher number of channels is preferred for datasets with a higher density of interactions.   Our results on two real-world datasets showed that SentiAttn significantly and consistently outperformed four existing state-of-the-art rating prediction models. Moreover, we demonstrated the effectiveness of the proposed sentiment attention layer within SentiAttn. We showed that it outperforms the global attention layer in improving the rating prediction accuracy, resulting in a lower variance of the rating prediction errors. Furthermore, we showed that SentiAttn provides a significantly effective rating prediction accuracy and a reliable indication of user preferences for cold-start users. As future work, we plan to consider other review properties (e.g. such as review age) as additional features within SentiAttn to more accurately measure the usefulness of reviews.

References

1. Chen, C., Zhang, M., Liu, Y., Ma, S.: Neural attentional rating regression with review-level explanations. In: Proceedings of WWW (2018)
2. Chen, L., Liu, Y., Zheng, Z., Yu, P.: Heterogeneous neural attentive factorization machine for rating prediction. In: Proceedings of CIKM (2018)
3. Cheng, C., Xia, F., Zhang, T., King, I., Lyu, M.R.: Gradient boosting factorization machines. In: Proceedings of RecSys (2014)
4. Davagdorj, K., Park, K.H., Ryu, K.H.: A collaborative filtering recommendation system for rating prediction. In: Pan, J.-S., Li, J., Tsai, P.-W., Jain, L.C. (eds.) Advances in Intelligent Information Hiding and Multimedia Signal Processing. SIST, vol. 156, pp. 265–271. Springer, Singapore (2020). https://doi.org/10.1007/978-981-13-9714-1_29
5. Hochreiter, S., Schmidhuber, J.: Long short-term memory. Neural Comput. 9(8) (1997)
6. Hyun, D., Park, C., Yang, M.C., Song, I., Lee, J.T., Yu, H.: Review sentiment-guided scalable deep recommender system. In: Proceedings of SIGIR (2018)
7. Jiang, Y., Hu, C., Xiao, T., Zhang, C., Zhu, J.: Improved differentiable architecture search for language modeling and named entity recognition. In: Proceedings of EMNLP (2019)
8. Kim, D., Park, C., Oh, J., Lee, S., Yu, H.: Convolutional matrix factorization for document context-aware recommendation. In: Proceedings of RecSys (2016)
9. Kim, Y.: Convolutional neural networks for sentence classification. In: Proceedings of EMNLP (2014)
10. Koren, Y., Bell, R.M., Volinsky, C.: Matrix factorization techniques for recommender systems. IEEE Comput. 42(8) (2009)
11. Lei, X., Qian, X., Zhao, G.: Rating prediction based on social sentiment from textual reviews. IEEE Trans. Multimedia 18(9), 1910–1921 (2016)
12. Ling, G., Lyu, M.R., King, I.: Ratings meet reviews, a combined approach to recommend. In: Proceedings of RecSys (2014)

13. Liu, H., Simonyan, K., Yang, Y.: DARTS: differentiable architecture search. In: Proceedings of ICLR (2019)
14. Luong, T., Pham, H., Manning, C.D.: Effective approaches to attention-based neural machine translation. In: Proceedings of EMNLP (2015)
15. Ma, X., Lei, X., Zhao, G., Qian, X.: Rating prediction by exploring user's preference and sentiment. Multimedia Tools Applicat. **77**(6) (2018)
16. Manotumruksa, J., Macdonald, C., Ounis, I.: Regularising factorised models for venue recommendation using friends and their comments. In: Proceedings of CIKM (2016)
17. McAuley, J., Leskovec, J.: Hidden factors and hidden topics: understanding rating dimensions with review text. In: Proceedings of RecSys (2013)
18. Mikolov, T., Sutskever, I., Chen, K., Corrado, G.S., Dean, J.: Distributed representations of words and phrases and their compositionality. In: Proceedings of NeurIPS (2013)
19. Pennington, J., Socher, R., Manning, C.D.: GloVe: global vectors for word representation. In: Proceedings of EMNLP (2014)
20. Rendle, S.: Factorization machines with libfm. ACM Trans. Intell. Syst. Technol. (TIST) **3**(3) (2012)
21. Rendle, S., Gantner, Z., Freudenthaler, C., Schmidt-Thieme, L.: Fast context-aware recommendations with factorization machines. In: Proceedings of SIGIR (2011)
22. Ricci, F., Rokach, L., Shapira, B.: Introduction to recommender systems handbook. In: Ricci, F., Rokach, L., Shapira, B., Kantor, P.B. (eds.) Recommender Systems Handbook, pp. 1–35. Springer, Boston, MA (2011). https://doi.org/10.1007/978-0-387-85820-3_1
23. Sakai, T.: Laboratory experiments in information retrieval. Inf. Retrieval Series **40** (2018)
24. Seo, S., Huang, J., Yang, H., Liu, Y.: Interpretable convolutional neural networks with dual local and global attention for review rating prediction. In: Proceedings of RecSys (2017)
25. Vaswani, A., et al.: Attention is all you need. In: Proceedings of NeurIPS (2017)
26. Wang, H., Fu, Y., Wang, Q., Yin, H., Du, C., Xiong, H.: A location-sentiment-aware recommender system for both home-town and out-of-town users. In: Proceedings of SIGKDD (2017)
27. Wang, J., et al.: Human-like decision making: Document-level aspect sentiment classification via hierarchical reinforcement learning. In: Proceedings of EMNLP (2019)
28. Wang, X., Ounis, I., Macdonald, C.: Comparison of sentiment analysis and user ratings in venue recommendation. In: Proceedings of ECIR (2019)
29. Wu, C., Wu, F., Qi, T., Ge, S., Huang, Y., Xie, X.: Reviews meet graphs: enhancing user and item representations for recommendation with hierarchical attentive graph neural network. In: Proceedings of EMNLP (2019)
30. Zhang, W., Du, T., Wang, J.: Deep learning over multi-field categorical data. In: Proceedings of ECIR (2016)
31. Zhao, K., Cong, G., Yuan, Q., Zhu, K.Q.: Sar: a sentiment-aspect-region model for user preference analysis in geo-tagged reviews. In: Proceedings of ICDE (2015)
32. Zheng, L., Noroozi, V., Yu, P.S.: Joint deep modeling of users and items using reviews for recommendation. In: Proceedings of WSDM (2017)

Goldilocks: Just-Right Tuning of BERT for Technology-Assisted Review

Eugene Yang[1(✉)], Sean MacAvaney[2], David D. Lewis[3], and Ophir Frieder[4]

[1] HLTCOE, Johns Hopkins University, Baltimore, USA
eugene.yang@jhu.edu
[2] University of Glasgow, Glasgow, UK
Sean.MacAvaney@glasgow.ac.uk
[3] Reveal-Brainspace, Chicago, USA
ecir2022paper@davelewis.com
[4] IRLab, Georgetown University, Washington, DC, USA
ophir@ir.cs.georgetown.edu

Abstract. Technology-assisted review (TAR) refers to iterative active learning workflows for document review in high recall retrieval (HRR) tasks. TAR research and most commercial TAR software have applied linear models such as logistic regression to lexical features. Transformer-based models with supervised tuning are known to improve effectiveness on many text classification tasks, suggesting their use in TAR. We indeed find that the pre-trained BERT model reduces review cost by 10% to 15% in TAR workflows simulated on the RCV1-v2 newswire collection. In contrast, we likewise determined that linear models outperform BERT for simulated legal discovery topics on the Jeb Bush e-mail collection. This suggests the match between transformer pre-training corpora and the task domain is of greater significance than generally appreciated. Additionally, we show that *just-right* language model fine-tuning on the task collection before starting active learning is critical. Too little or too much fine-tuning hinders performance, worse than that of linear models, even for a favorable corpus such as RCV1-v2.

1 Introduction

High recall retrieval (HRR) tasks (also called *annotation* tasks) involve identifying most or all documents of interest in a large collection. HRR tasks include electronic discovery in the law (eDiscovery) [3], systematic review in medicine [22–24,47], document sensitivity review [34], online content moderation [55], and corpus annotation to support research and development [60].

Technology-assisted review (TAR) refers to the automated methods to reduce the number of documents reviewed in HRR projects [36]. Iterative, pool-based active learning of predictive models for review prioritization is the most commonly applied workflow [9,10]. Linear models such as logistic regression and support vector machines (SVMs) applied to lexical and metadata features are the most common supervised learning approaches. Unlike in classification and

M. Hagen et al. (Eds.): ECIR 2022, LNCS 13185, pp. 502–517, 2022.
https://doi.org/10.1007/978-3-030-99736-6_34

adhoc retrieval tasks, the supervised learning model in TAR is typically discarded after use. This is because each legal case or other project has its own retrieval objective, and because of concerns of leaking confidential information. Therefore, the cost of training the supervised learning model in TAR often cannot be amortized over future data or across tasks.

Pre-trained transformers [46] such as BERT [12], GPT-3 [5], and T5 [38] are effective at a variety of natural language processing tasks. These models learn linguistic patterns from very large corpora in an unsupervised fashion (*pre-training*) and can be tuned to language characteristics of a particular task data set (*LM fine-tuning*) [12,19]. They can then be applied to a task on that data set by zero-shot transfer learning [32,49] or by *task fine-tuning* to labeled training data [20,58]. Transformers have improved effectiveness at tasks related to HRR such as document classification [1], entity extraction [12], and adhoc retrieval [33]. This has inspired initial commercial use of transformers by eDiscovery providers, though not yet in an active learning context.[1]

We are not aware of published studies of transformers in TAR workflows. Several studies have evaluated task fine-tuning using active learning [30,44], including for text classification tasks [13,59]. These studies, however, have evaluated generalization to new data using training/test splits. HRR, like relevance feedback in adhoc search [42], is a transductive setting: evaluation is on the same task corpus from which the training data is selected by active learning.

The transductive setting makes of less importance a key advantage of transformers over traditional methods: their inclusion of language-wide linguistic regularities that might be present in unseen test data. It has already been demonstrated by Gururangan et al. [18] that BERT is more effective when the target task domain is similar to the ones on which BERT was trained (English language books [61] and English Wikipedia). Active learning also reduces transformer advantage, by reducing the labeling cost to learn corpus-specific vocabulary and regularities. Finally, the short useful life of TAR models means limited opportunity to amortize training cost, raising questions about the large computational cost of task fine-tuning for transformers.

The recent TREC-COVID evaluation provides evidence both in favor and against transformers. A SciBERT-based zero-shot reranker of BM25-based text retrieval topped several of the Round 1 evaluation measures [31,32]. On the other hand, another transformer-based effort (which omitted language model fine tuning) struggled [29], a number of other deep learning efforts had mediocre effectiveness, and classic linear models based on lexical features and trained by active learning were highly competitive (leading on one measure) [31,48]. Recently, Ioannidis [21] evaluated BERT and PubMedBERT [17] on CLEF eHealth Technology Assisted Reviews in Empirical Medicine Task [22,23]. Despite the claim, Ioannidis [21] considered a simple ranking and classification setting instead of an iterative task.

Against this context, we provide the first demonstration of fine-tuned transformer-based models in the TAR transductive active learning setting. We

---

[1] https://www.nexlp.com/blog/nexbert-story-engine-cloud.

use BERT [12] as a representative transformer. We fine-tune the language model to each of two (unlabeled) task corpora using a masked language modeling objective, kick off each prioritization task on that corpus with a single positive example, and do task fine-tuning of BERT on each TAR active learning iteration.

Surprisingly, despite the past success stories of BERT in dramatically advancing the retrieval effectiveness, in our work, we found that it only performs on par with the simple logistic regression model due to the transductivity of HRR. On the contrary, under certain scenarios, the BERT model reduces the total reviewing cost, which is the primary objective of HRR tasks. Given its data-hungry property, this cost reduction is counterintuitive but yet very favorable. We highlight our contributions in the following,

- First, we find that language model fine-tuning to the task corpus before active learning is critical, but also that too much of it can be done.
- Second, we find language model fine-tuning is not a cure-all for domain mismatch. Our fine-tuned BERT model beats linear models on a data set (RCV1-v2) similar to the text types on which BERT was trained, but falls short when operating with very different textual characteristics.
- Finally, we provide a running time analysis to demonstrate the computational overhead for applying BERT.

2   Background

HRR projects typically balance thoroughness versus cost by setting a recall target that is high, but below 100%. Targets such as 80% recall are common in eDiscovery [41] and are sometimes encoded in legal agreements [54]. Systematic review often shoots for 95% recall (on smaller and more homogeneous collections) [22,23]. Recall is defined as the number of relevant documents found among the reviewed documents, divided by the number of relevant documents in the defined collection of interest (e.g., all emails from a set of employees relevant to a legal case, or all biomedical research papers that have passed a keyword screening).

TAR workflows reduce costs by using iterative active learning to prioritze batches of documents for review. One-phase TAR workflows continue this process until a stopping rule indicates that the reviewed documents have met the recall target [9]. Two-phase workflows have a training phase followed by a classification phase (on the same data set), with review done in both phases [34,54]. Designing stopping rules that determine as early as possible that a recall target has been reached is an active research area [6,10,11,26,28,43,47,53], but we design our evaluation to avoid the selection and the error incurred by the stopping rule based on the prior studies in TAR cost evaluation [54].

Evaluation for HRR emphasizes a recall/cost tradeoff rather than the related recall/precision tradeoff. In eDiscovery, *Depth for recall (DFR@x)* is the proportion of the collection reviewed to hit a recall target x.[2] Systematic review uses

---

[2] https://www.gibsondunn.com/wp-content/uploads/documents/publications/Evans-Metrics-that-Matter-Inside-Counsel-1.2015.pdf.

Work saved over sampling (WSS@x), which subtracts DFR@x from the expected cost to hit the recall target by random sampling: $WSS@x = x - DFR@x$ [8]. Some early HRR studies also use R-Precision (precision at R where R is the number of relevant documents) [16,41] to capture the effectiveness to lower part of the rank as opposed to Precision at 5 or 10 in adhoc retrieval.

However, these evaluation metrics do not consider the cost of obtaining the labels for training documents. In this study, we adapt the cost evaluation of TAR proposed by Yang et al. [54] to jointly evaluate the effectiveness and the cost of the retrieval results. The total cost of TAR consists of the cost of reviewing (1) the training documents and (2) the minimum number of unreviewed documents ranked by the current classification model for fulfilling the recall target. This cost evaluation approach allows documents in different classes and phases to cost differently, facilitating a more practical HRR evaluation and emphasizing the cost of training the one-time classification model.

Commercial TAR technology relies on traditional text classification approaches such as logistic regression and support vector machines (SVMs) [4, 50], that have been widely studied in both active learning and transductive contexts [9,26,34,53,54]. However, the state of the art in text classification has moved to transformer-based models such as BERT [12] whose properties in these contexts are less well-understood. This gap in understanding motivates the current study.

3   Adapting BERT for TAR

In this section, we describe the adaption of the BERT model to TAR. On a high level, the BERT language model is fine-tuned on the collection of the retrieval interest. At each active learning iteration, we select a set of documents based on the predictions from the model for human review. The acquired labels are fed to the BERT model to perform classification fine-tuning for learning relevancy.

Since the entire task corpus is available before training in TAR, our first step in applying BERT to TAR was language model fine tuning to that corpus. We used the same unsupervised masked language modeling task originally used to train BERT: randomly masking 15% of the tokens in each sequence and tuning BERT's parameters to predict the missing tokens [12]. The key question is how much to move BERT's parameters (encoding the linguistic regularities explicit in a mammoth broad domain corpus) toward the task-specific, but less complete, explicit regularities of the task corpus. Our experiments study this by varying the number of epochs (passes through training set) in language model fine-tuning.

TAR workflows use an active learning method such as relevance feedback [40] or uncertainty sampling [25], where the model trained by supervised learning on iteration $k - 1$ is used to the select the batch of documents to be labeled in iteration k. The union of labeled batches for iterations $1...k - 1$ is the training set for iteration k. One random relevant document was selected at the beginning of the process as the *seed document* to initiate the active learning. All labeled documents are used for classification fine-tuning of the BERT model. Documents labeled in earlier iterations are visited more by the model based on this classification fine-tuning process. However, based on our pilot study on only fine-tuning model on the newly labeled documents at each iteration, the results were far worse than on all labeled documents. We use a cross-entropy loss on the binary class label by adding a dense layer on the [CLS] token on top of BERT. We train for a fixed number of epochs, which previous work on active learning for BERT suggests works as well as choosing epoch number using a validation set [13].

For simple models, training can be done to convergence from scratch on each iteration (as we do for logistic regression and SVMs in our experiments). Classification fine tuning for a transformer is computationally expensive, so we instead use the model trained on iteration $k - 1$ as the starting point for optimization on iteration k. While this potentially gives more influence to examples selected on the first iteration, adaptive example selection by active learning reduces this effect.

4   Experiment Setup

4.1   Data Sets

We simulate TAR reviews on two fully labeled collections widely used in HRR studies [10,11,16,35,41,51,52,56]: RCV1-v2 [27] and the Jeb Bush emails [16, 41].

RCV1-v2 consists of 804,414 news stories with coding for 658 economic news categories. We use the 45 categories subset established by previous high recall retrieval study [54] that spans across three prevalence and three difficulty bins. Text from the title and body was concatenated and tokenized using WordPiece. Documents are truncated with 512 WordPiece tokens as the leading passages of the news documents usually convey the most important aspects of the news articles [7]. The collection is also downsampled to 20% (160,833 documents) for computational efficiency.

The Jeb Bush collection consists of 274,124 unique emails between the former governor of Florida and his colleagues and constituents. The collection was annotated for 44 political topics for the 2015 and 2016 TREC Total Recall Tracks [16,41]. Text from the subject line and body were concatenated. As with RCV1-v2, documents with more than 512 WordPiece tokens were truncated, similar to the preprocessing steps used in prior works in email classification [45]. Since the most recent replies and content are presented at the beginning of the email and the trailing parts are often duplicated from other emails, including only the leading passages are usually sufficient. A 50% random sample of the remainder (137,062 documents) was used. All 44 topics are used in the experiment. For consistency, we refer to these topics as categories in the later sections.

The RCV1-v2 news articles are professionally written texts with topics and vocabulary well covered by the book and encyclopedic text used to train BERT. We view HRR on it as an in-domain task for BERT. The Jeb Bush emails (particularly from constituents) vary wildly in style and formality from message to message, and reference many Florida personalities, places, and issues likely to be poorly covered in the BERT pre-training materials. We therefore view it as an out-of-domain task for BERT.

4.2 Software and Evaluation

We implemented the active learning workflow with `libact` [57], an open-source active learning framework. For each category, a randomly selected positive example formed the sample for the first iteration. On each subsequent iteration, 200 documents were sampled using active learning (either relevance feedback or least-confidence uncertainty sampling) by following experiment settings from prior HRR studies [51, 54].

For BERT runs we used the `BERT-base-cased` model.[3] Masked language model fine-tuning was done with the HuggingFace script `run_mlm.py`,[4] which uses ADAM with no weight decay and warm up period as the optimizer, and a learning rate of 5×10^{-5}. To test the importance of language model fine-tuning, we vary it from no language model fine-tuning to ten iterations over the corpus.

Then on each active learning iteration, we do classification fine-tuning using the ADAM optimizer with a linear weight decay of 0.01 with 50 warm up steps and initial learning rate of 0.001. All reviewed documents (including the ones previously reviewed) are used to fine-tune the model at each active learning iterations with 20 epochs. All hyperparameters were selected based on a pilot study on one selected category of each collection for maximizing the average R-Precision after 20 active learning iterations. The authors also experimented with fine-tuning the model with only the newly queried documents at each iteration, but the results were worse than fine-tuning on all labeled documents by a large margin.

Logistic regression is served as the baseline in our study and is implemented with `scikit-learn` [37] for comparison. It is widely used in HRR research and commercial software [2, 4, 52, 54]. We use the scikit-learn tokenizer and BM25 within document saturated term frequencies as feature values [39, 52]. We use L2 regularization on the logistic losses, with penalty weight 1.0 and fit to convergence with default settings from scikit-learn.

For comparison with prior work, we report R-Precision, which is a metric that often reports in high recall retrieval studies [16, 41, 52]. Despite being an effectiveness measure that jointly considers precision and recall, it does not reflect the actual objective of the retrieval task, which is the reviewing cost.

Therefore, our primary evaluation measure is the total optimal reviewing cost of the TAR run [54], which is the sum of reviewing the training documents and the documents ranked by the current classification model to fulfill the recall target. The latter is referred to as the optimal amount of the second phase review and can be considered as an optimal penalty for the one-phase workflow [26, 54]. We report the minimal total cost that occurs during the 20 active learning

---

[3] https://huggingface.co/bert-base-cased.
[4] https://github.com/huggingface/transformers/blob/master/examples/pytorch/language-modeling/run_mlm.py.

iterations. Without loss of generality, we use 80% recall target as an example, which is a widely used target in eDiscovery study. Higher targets such as 95% yield similar results.

To emphasize the importance of the underlying classification model in the iterative process, we evaluate with both the uniform cost structure (i.e., no reviewing cost difference between documents) and expensive training cost structure. Without loss of generality, we assume the training documents cost ten times more than documents reviewed during the mass reviewing phase as an example [54]. The expensive training cost structure favors classification models that require less training data for optimizing the total cost, enabling us to distinguish the effectiveness of the classification model further.

4.3 Hardware

The active learning experiments are conducted on a cluster of 48 NVIDIA Titan RTX GPUs with 24 GB memory on each. One active learning run (one topic, one sampling strategy, one pretrained BERT model) took on average 18 h. The entire set of experiments $((45 + 44) \times 5 \times 2 = 890$ runs) took around two weeks on our research cluster. The baseline experiments ran on a single CPU. All logistic regression runs $((45 + 44) \times 2 = 178)$ took around one hour. A detailed running time analysis is presented in the next section.

5 Results and Analysis

In this section, we aim to answer the following research questions: does language model fine-tuning improves the retrieval effectiveness? If so, what is the right amount? How much overhead are we paying for applying BERT?

5.1 Language Model Fine-Tuning

Based on our experimental results, BERT with language model (LM) fine-tuning improves the effectiveness only when the domain of the collection aligns with the domain of the pretraining corpora. In Table 1, the reported cost is the average of the proportional relative cost differences between the baseline logistic regression results and the pretrained BERT model. Since the cost varies between categories, averaging the relative differences prevent the naturally harder tasks (with higher baseline cost) from diluting the aggregated values. The paired t-tests are still conducted between the raw cost with a null hypothesis of identical cost between the BERT and the baseline model. In RCV1-v2, BERT models provide roughly the same R-Precision (0.75 to 0.77) as the baseline logistic regression model regardless of the length of LM fine-tuning, suggesting similar quality at the top of the rank list. On the other hand, BERT models reduce the cost, especially when the training documents cost more to review, compared to the baseline model when the amount of LM fine-tuning is *just right* (10% to 15% on average with expensive training cost structure). In our experiments, the *goldilock* amount

Table 1. Averaged evaluation results on the in-domain RCV1-v2 collection and off-domain Jeb Bush collection over categories. Numbers in parentheses are the relative difference between the baseline logistic regression model (LR). Both uniform and expensive training cost (Exp. Train.) values are the relative cost difference between the BERT and the logistic regression models. Values larger than 1.0 indicate higher costs than the baseline. * indicates the statistical significance with 95% confidence between the corresponding pretrained BERT model and the baseline conducted by paired t-test with Bonferroni corrections within each evaluation metric.

Collection	LMFT Epoch	R-Precision (↑)		Uni. Cost (↓)		Exp. Train. (↓)	
		Relevance	Uncertainty	Rel.	Unc.	Rel.	Unc.
In-domain RCV1-v2	LR	0.788 (1.00)	0.760 (1.00)	1.000	1.000	1.000	1.000
	0	0.752 (0.95)	0.756 (0.99)	1.309	1.015	1.178	*0.873
	1	0.757 (0.96)	0.768 (1.01)	1.199	1.039	1.012	0.894
	2	0.759 (0.96)	0.766 (1.01)	1.289	1.028	1.067	0.890
	5	0.756 (0.96)	0.784 (1.03)	1.173	0.893	0.980	*0.844
	10	0.764 (0.97)	0.765 (1.01)	1.192	0.950	1.051	*0.878
Off-domain Jeb Bush	LR	0.904 (1.00)	0.857 (1.00)	1.000	1.000	1.000	1.000
	0	*0.724 (0.80)	*0.719 (0.84)	6.877	5.834	*2.717	*2.194
	1	0.811 (0.90)	0.816 (0.95)	4.678	2.896	1.756	1.413
	2	0.812 (0.90)	0.808 (0.94)	3.257	3.141	1.675	1.446
	5	*0.810 (0.90)	0.813 (0.95)	3.261	2.665	1.583	1.322
	10	0.805 (0.89)	0.815 (0.95)	3.922	2.943	1.601	1.361

is five epochs. However, this amount varies with collection size and other characteristics of the task, which is discussed later in the section. Since reducing the total cost of TAR requires improving the overall rank list [54], these results suggest that the BERT model with five epochs of LM fine-tuning provides a consistent improvement on the entire ranking.

If the target collection is off-domain compared to the original pre-trained corpora, BERT models cannot provide an effective classifier, even worse than simple linear logistic regression. The averaged values in the Jeb Bush collection suggest worse effectiveness (lower R-Precision and higher cost) despite that the differences are not statistically significant. However, the time overhead and computational burden of applying neural models such as BERT are massive compared to linear models. The inability to provide more effective retrieval results is already a failure. Note that the effectiveness of the BERT models could eventually improve over the baseline with more LM fine-tuning despite the decrement from five to ten epochs; the computational cost would be uneconomical. Running time analysis is presented later in this section.

Therefore, applying BERT models to TAR is not guaranteed to lead to more effective retrieval results. The alignment of the domain between the collections and the amount of LM fine-tuning constitutes a considerable variation of the effectiveness, which is counterintuitive to the common wisdom that continuing fine-tuning would result in better results [19]. If *just-right* hyperparameter is not available for the task, which is usually the case for real-world applications, applying BERT models could result in inferior results.

Table 2. Cost of RCV1-v2 categories in each bin under the expensive training cost structure. Values are the relative cost difference between the corresponding BERT and baseline models averaged over the five categories in each bin.

Difficulty	Prevalence	Relevance					Uncertainty				
		0	1	2	5	10	0	1	2	5	10
Hard	Rare	0.918	0.988	1.011	0.997	1.048	1.044	0.843	0.801	0.664	0.870
	Medium	0.774	0.773	0.699	0.622	0.639	0.594	0.670	0.602	0.612	0.613
	Common	0.832	0.856	0.850	0.798	0.755	0.815	0.849	0.842	0.755	0.751
Medium	Rare	0.932	0.916	0.904	0.784	0.951	0.770	0.903	0.868	0.794	0.828
	Medium	1.275	1.311	1.293	1.175	1.229	1.065	1.199	1.203	1.088	1.211
	Common	0.951	0.778	0.830	0.743	0.820	0.946	0.945	0.933	0.845	0.915
Easy	Rare	1.688	1.225	1.362	1.430	1.540	0.587	0.638	0.702	0.632	0.621
	Medium	1.897	1.189	1.182	1.103	1.263	1.073	0.936	1.015	1.069	0.982
	Common	1.336	1.070	1.474	1.165	1.218	0.960	1.061	1.047	1.136	1.112

5.2 *Just-Right* Varies Across Tasks

The 45 categories selected from RCV1-v2 enable further analysis into the effect of the task characteristics. Table 2 demonstrates the averaged relative cost differences compared to the baseline model in each category bin under the expensive training cost structure. Since each bin only contains five runs (five categories), statistical tests are non-indicative; hence omitted.

For relevance feedback where training documents are selected from the top of the rank, BERT models usually perform similarly to logistic regression models with a few exceptions. BERT models are more helpful in hard categories than easy ones since the relevancy is often beyond simple token matching in the hard ones, yielding a 20% to 30% cost reduction. However, when the task is hard and the relevant documents are rare, BERT models are no better than simple linear models, even with more LM fine-tuning.

For uncertainty sampling, where the training documents are ones that the model is the least certain about (with predicted probability around 0.5), BERT models provide a substantial improvement of 20% to 40% cost reduction in both hard and rare categories. These results indicate that BERT models are still more effective in challenging situations – either extremely unbalanced training set or relevancy requires subtle semantic understanding. These are cases where linear models tend to fail if no specific treatments to the collection are made.

However, even in these cases where BERT models demonstrate a clear advantage over the linear models, the amount of LM fine-tuning is still critical. The optimal length of LM fine-tuning varies across difficulty and prevalence bins, which were developed by Yang et al. [54]. For example, the best performing pre-trained model for the *hard-medium* bin is no LM fine-tuning (0.5935, i.e., 41% cost reduction). However, LM fine-tuning for five epochs gives us the lowest cost (0.6637) and seems to be the minimum for *hard-rare*. For *hard-common*, more fine-tuning tends to be consistently improving the model with the lowest cost

(0.7512) occurred at ten epochs in our experiment. The trend is different for medium and easy difficulty bins.

Table 3. Running time in minutes. Running time for LM fine-tuning (LMFT) is agnostic to the categories. Time reported for TAR is the average running time for each category to complete a 20-iteration TAR process, which consists of 20 classification fine-tuning (or training for logistic regression) and scoring the entire collection. Values in parentheses are the standard deviations of the averaged time.

	LMFT	Relevance			Uncertainty		
Collection	Epoch	LMFT	TAR	Total	LMFT	TAR	Total
	0	–	1095	1095 (31.49)	–	1098	1098 (28.97)
	1	98	1094	1192 (17.55)	98	1102	1200 (20.57)
In-domain	2	196	1096	1292 (20.53)	196	1100	1296 (28.33)
RCV1-v2	5	490	1103	1593 (23.57)	490	1103	1593 (19.93)
	10	980	1101	2081 (20.26)	980	1105	2085 (20.90)
	LR	–	0.32	0.34 (0.04)	–	0.38	0.38 (0.05)
	0	–	999	999 (19.02)	–	1008	1008 (19.48)
	1	98	1002	1100 (16.56)	98	1003	1101 (24.78)
Off-domain	2	196	1002	1198 (15.03)	196	1002	1198 (21.80)
Jeb Bush	5	490	1007	1497 (19.34)	490	1004	1494 (27.37)
	10	981	996	1977 (22.48)	981	1006	1987 (26.67)
	LR	–	0.33	0.33 (0.04)	–	0.41	0.41 (0.06)

Beyond minimum cost during the run, the trajectory of cost over the iterations also varies among different numbers of LM fine-tuning epochs. For the *hard-rare* category (I65100) in Fig. 1(a), the transition from the trajectory of 1 epoch of LM fine-tuning to 2 is not smooth and the shape is nowhere similar. The *hard-common* category (I81501 in Fig. 1(c)) also convey no clear relationship between different number of LM fine-tuning epochs.

While BERT models provide significant improvement over the failure cases such as the *medium-rare* category (I42600, Fig. 1(d)) and *hard-medium* category (C182, Fig. 1(b)), the trajectory is nearly identical for the easy categories regardless of the LM fine-tuning epochs, especially with relevance feedback.

Despite making no clear conclusion on the optimal amount of LM fine-tuning, we observe that this hyperparameter is critical and independent of the collection. All TAR runs in Table 2 are based on the same 20% subset of RCV1-v2 collection but with different categories. This poses a challenge for TAR practitioners when applying BERT or potentially other transformer-based classification models to projects: the joint effect of this hyperparameter and the characteristics of the task is so large that it ranges from extremely helpful (50% cost reduction in *hard-medium* categories using uncertainty sampling without LM fine-tuning) to large cost overhead (89% cost overhead in *easy-medium* categories using relevance feedback without LM fine-tuning). Understanding the characteristics of the task

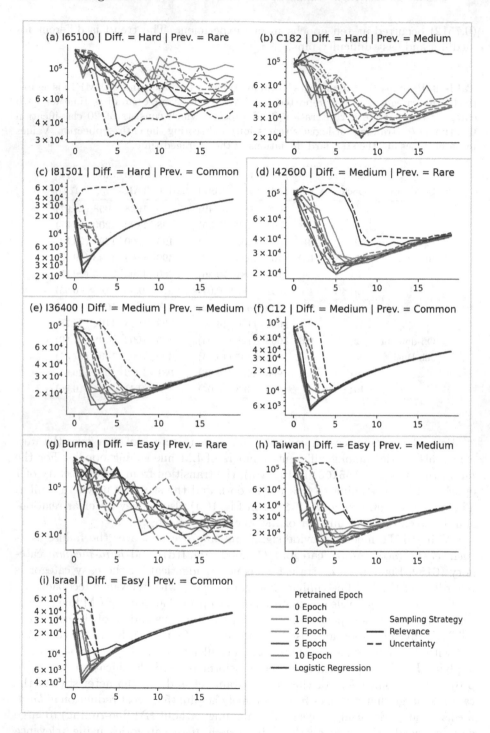

Fig. 1. Example total cost of TAR runs on RCV1-v2 collection over the rounds with expensive training cost structure. The y-axis is the total cost in log-scaled to demonstrate the differences and the x-axis is the number of TAR rounds.

remains crucial but challenging without sufficient annotations, which is one of the purposes for applying TAR.

5.3 Running Time

Finally, we analyze the running time of the TAR runs. In Table 3, the computational overhead of applying BERT is massive. While the training and scoring of the collection during TAR using logistic regression takes on average 20 to 25 s (0.32 to 0.41 min), the BERT model takes around 18 h (1100 min). The majority of the time was spent on scoring the collection, which takes around 40 min at each iteration. The LM fine-tuning is done before the TAR iterative process, taking around 100 min per epoch for the collections we experimented with.

In real high recall retrieval applications where the iterative process spans weeks or even months, each round of reviewing documents takes around half a day. Adding one hour overhead to each iteration is potentially acceptable. However, for smaller projects, this significant time overhead could directly prevent BERT from applying. The computational cost for applying BERT is also not amortized to millions of queries after deployment. Spending 18 h training a single-usage model in exchange for a mild effectiveness improvement could be unnecessary overhead for many HRR projects.

6 Summary and Future Works

We evaluated the effectiveness of TAR with pre-trained BERT as the underlying predictive model. Before entering active learning, the pre-trained BERT model is fine-tuned by the masked language modeling objective with several epochs. Through experiments, we show that the amount of LM fine-tuning is critical even on an in-domain task. For tasks with out-of-domain text, as compared to the BERT model pre-training corpora, LM fine-tuning requires more training, potentially with other similar corpora. Without proper LM fine-tuning, BERT models underperform typical linear models used with TAR. However, our experiments also show that category characteristics also impact how beneficial the BERT models are and the large computational overhead might discourage the application of BERT in real-world HRR projects.

As the first study of applying transformer models to TAR, there is still much to explore in this area. In the future, we will investigate a wider variety of HRR tasks and sampling strategies that are designed for neural models such as Monte Carlo Dropout [14] and Discriminative Active Learning [15]. A comprehensive approach for handling documents with more than 512 tokens should also be studied. Pre-training a transformer model with large email corpora would benefit the community as many eDiscovery tasks are working on emails. Whether the pre-training corpora would carry biases into the final retrieval results in each TAR project is also demanding for future research.

References

1. Adhikari, A., Ram, A., Tang, R., Lin, J.: Docbert: BERT for document classification. arXiv preprint arXiv:1904.08398 (2019)
2. Bannach-Brown, A., et al.: Machine learning algorithms for systematic review: reducing workload in a preclinical review of animal studies and reducing human screening error. Syst. Control Found. Appl. **8**(1), 1–12 (2019)
3. Baron, J., Losey, R., Berman, M.: Perspectives on Predictive Coding: And Other Advanced Search Methods for the Legal Practitioner. American Bar Association, Section of Litigation (2016)
4. Brown, S.: Peeking inside the black box: a preliminary survey of technology assisted review (TAR) and predictive coding algorithms for ediscovery. Suffolk J. Trial App. Advoc. **21**, 221 (2015)
5. Brown, T.B., et al.: Language models are few-shot learners (2020)
6. Callaghan, M.W., Müller-Hansen, F.: Statistical stopping criteria for automated screening in systematic reviews. Syst. Control Found. Appl. **9**(1), 1–14 (2020)
7. Catena, M., Frieder, O., Muntean, C.I., Nardini, F.M., Perego, R., Tonellotto, N.: Enhanced news retrieval: passages lead the way! In: Proceedings of the 42nd International ACM SIGIR Conference on Research and Development in Information Retrieval, pp. 1269–1272 (2019)
8. Cohen, A.M., Hersh, W.R., Peterson, K., Yen, P.Y.: Reducing workload in systematic review preparation using automated citation classification. J. Am. Med. Inform. Assoc. **13**(2), 206–219 (2006)
9. Cormack, G.F., Grossman, M.F.: Evaluation of machine-learning protocols for technology-assisted review in electronic discovery. SIGIR **2014**, 153–162 (2014). https://doi.org/10.1145/2600428.2609601
10. Cormack, G.V., Grossman, M.R.: Engineering quality and reliability in technology-assisted review. In: SIGIR, pp. 75–84, ACM Press, Pisa, Italy (2016), https://doi.org/10.1145/2911451.2911510,http://dl.acm.org/citation.cfm?doid=2911451.2911510, ISBN 978-1-4503-4069-4, 00024
11. Cormack, G.V., Grossman, M.R.: Scalability of continuous active learning for reliable high-recall text classification. In: Proceedings of the 25th ACM International on Conference on Information and Knowledge Management, pp. 1039–1048 (2016)
12. Devlin, J., Chang, M.W., Lee, K., Toutanova, K.: BERT: pre-training of deep bidirectional transformers for language understanding. arXiv preprint arXiv:1810.04805 (2018)
13. Ein-Dor, L., et al.: Active learning for BERT: an empirical study. In: Proceedings of the 2020 Conference on Empirical Methods in Natural Language Processing (EMNLP), pp. 7949–7962 (2020)
14. Gal, Y., Ghahramani, Z.: Dropout as a Bayesian approximation: representing model uncertainty in deep learning. In: International Conference on Machine Learning, pp. 1050–1059. PMLR (2016)
15. Gissin, D., Shalev-Shwartz, S.: Discriminative active learning. arXiv preprint arXiv:1907.06347 (2019)
16. Grossman, M.R., Cormack, G.V., Roegiest, A.: TREC 2016 total recall track overview (2016)
17. Gu, Y., et al.: Domain-specific language model pretraining for biomedical natural language processing. arXiv preprint arXiv:2007.15779 (2020)
18. Gururangan, S., Dang, T., Card, D., Smith, N.A.: Variational pretraining for semi-supervised text classification. arXiv preprint arXiv:1906.02242 (2019)

19. Gururangan, S., et al.: Don't stop pretraining: adapt language models to domains and tasks. In: Proceedings of ACL (2020)
20. Hou, Y., et al.: Few-shot slot tagging with collapsed dependency transfer and label-enhanced task-adaptive projection network. In: Proceedings of the 58th Annual Meeting of the Association for Computational Linguistics, pp. 1381–1393 (2020)
21. Ioannidis, A.: An analysis of a Bert deep learning strategy on a technology assisted review task. arXiv preprint arXiv:2104.08340 (2021)
22. Kanoulas, E., Li, D., Azzopardi, L., Spijker, R.: CLEF 2017 technologically assisted reviews in empirical medicine overview. In: CEUR Workshop Proceedings, vol. 1866, pp. 1–29 (2017)
23. Kanoulas, E., Li, D., Azzopardi, L., Spijker, R.: CLEF 2018 technologically assisted reviews in empirical medicine overview. In: CEUR Workshop Proceedings 2125 (2018). https://strathprints.strath.ac.uk/66446/
24. Kanoulas, E., Li, D., Azzopardi, L., Spijker, R.: Clef 2019 technology assisted reviews in empirical medicine overview. In: CEUR Workshop Proceedings, vol. 2380 (2019)
25. Lewis, D.D., Gale, W.A.: A sequential algorithm for training text classifiers. In: SIGIR 1994, pp. 3–12 (1994)
26. Lewis, D.D., Yang, E., Frieder, O.: Certifying one-phase technology-assisted reviews. In: Proceedings of 30th ACM International Conference on Information and Knowledge Management (2021)
27. Lewis, D.D., Yang, Y., Rose, T.G., Li, F.: RCV1: a new benchmark collection for text categorization research. JMLR 5, 361–397 (2004)
28. Li, D., Kanoulas, E.: When to stop reviewing in technology-assisted reviews: sampling from an adaptive distribution to estimate residual relevant documents. ACM Trans. Inf. Syst. (TOIS) 38(4), 1–36 (2020)
29. Lima, L.C., Het al.: Denmark's participation in the search engine TREC COVID-19 challenge: lessons learned about searching for precise biomedical scientific information on COVID-19. arXiv preprint arXiv:2011.12684 (2020)
30. Liu, M., Tu, Z., Zhang, T., Su, T., Wang, Z.: LTP: a new active learning strategy for CRF-based named entity recognition. arXiv preprint arXiv:2001.02524 (2020)
31. MacAvaney, S., Cohan, A., Goharian, N.: Sledge: a simple yet effective baseline for COVID-19 scientific knowledge search. arXiv e-prints, pp. arXiv-2005 (2020)
32. MacAvaney, S., Cohan, A., Goharian, N.: Sledge-z: a zero-shot baseline for COVID-19 literature search. In: Proceedings of the 2020 Conference on Empirical Methods in Natural Language Processing (2020). https://doi.org/10.18653/v1/2020.emnlp-main.341, https://arxiv.org/abs/2010.05987
33. MacAvaney, S., Yates, A., Cohan, A., Goharian, N.: CEDR: contextualized embeddings for document ranking. In: Proceedings of the 42nd International ACM SIGIR Conference on Research and Development in Information Retrieval, pp. 1101–1104 (2019). https://doi.org/10.1145/3331184.3331317, https://arxiv.org/abs/1904.07094
34. McDonald, G., Macdonald, C., Ounis, I.: Active learning strategies for technology assisted sensitivity review. In: Pasi, G., Piwowarski, B., Azzopardi, L., Hanbury, A. (eds.) ECIR 2018. LNCS, vol. 10772, pp. 439–453. Springer, Cham (2018). https://doi.org/10.1007/978-3-319-76941-7_33
35. Oard, D.W., Sebastiani, F., Vinjumur, J.K.: Jointly minimizing the expected costs of review for responsiveness and privilege in E-Discovery. ACM Trans. Inf. Syst. 37(1), 1–35 (2018). https://doi.org/10.1145/3268928, http://dl.acm.org/citation.cfm?doid=3289475.3268928, ISSN 10468188

36. Oard, D.W., Webber, W.: Information retrieval for e-discovery. Inf. Retrieval **7**(2–3), 99–237 (2013)
37. Pedregosa, F., et al.: Scikit-learn: machine learning in Python. J. Mach. Learn. Res. **12**, 2825–2830 (2011)
38. Raffel, C., et al.: Exploring the limits of transfer learning with a unified text-to-text transformer. J. Mach. Learn. Res. **21**, 1–67 (2020)
39. Robertson, S., Zaragoza, H., et al.: The probabilistic relevance framework: Bm25 and beyond. Found. Trends® Inf. Retrieval **3**(4), 333–389 (2009)
40. Rocchio, J.J.: Relevance feedback in information retrieval (1971)
41. Roegiest, A., Cormack, G.V.: TREC 2015 total recall track overview (2015)
42. Ruthven, I., Lalmas, M.: A survey on the use of relevance feedback for information access systems. Knowl. Eng. Rev. **18**(2), 95–145 (2003)
43. Saha, T.K., Hasan, M.A., Burgess, C., Habib, M.A., Johnson, J.: Batch-mode active learning for technology-assisted review. In: 2015 IEEE International Conference on Big Data (Big Data), pp. 1134–1143 (2015). https://doi.org/10.1109/BigData.2015.7363867, 00003
44. Shelmanov, A., et al.: Active learning with deep pre-trained models for sequence tagging of clinical and biomedical texts. In: 2019 IEEE International Conference on Bioinformatics and Biomedicine (BIBM), pp. 482–489. IEEE (2019)
45. Shu, K., Mukherjee, S., Zheng, G., Awadallah, A.H., Shokouhi, M., Dumais, S.: Learning with weak supervision for email intent detection. In: Proceedings of the 43rd International ACM SIGIR Conference on Research and Development in Information Retrieval, pp. 1051–1060 (2020)
46. Vaswani, A., et al.: Attention is all you need. arXiv abs/1706.03762 (2017)
47. Wallace, B.C., Trikalinos, T.A., Lau, J., Brodley, C., Schmid, C.H.: Semi-automated screening of biomedical citations for systematic reviews. BMC Bioinform. **11**(1), 55 (2010)
48. Wang, X.J., Grossman, M.R., Hyun, S.G.: Participation in TREC 2020 COVID track using continuous active learning. arXiv preprint arXiv:2011.01453 (2020)
49. Wang, Y., Che, W., Guo, J., Liu, Y., Liu, T.: Cross-lingual BERT transformation for zero-shot dependency parsing. In: Proceedings of the 2019 Conference on Empirical Methods in Natural Language Processing and the 9th International Joint Conference on Natural Language Processing (EMNLP-IJCNLP), pp. 5725–5731 (2019)
50. Yang, E., Grossman, D., Frieder, O., Yurchak, R.: Effectiveness results for popular e-discovery algorithms. In: Proceedings of the 16th edition of the International Conference on Artificial Intelligence and Law, pp. 261–264 (2017)
51. Yang, E., Lewis, D.D., Frieder, O.: A regularization approach to combining keywords and training data in technology-assisted review. In: Proceedings of the Seventeenth International Conference on Artificial Intelligence and Law, pp. 153–162 (2019)
52. Yang, E., Lewis, D.D., Frieder, O.: Text retrieval priors for Bayesian logistic regression. In: Proceedings of the 42nd International ACM SIGIR Conference on Research and Development in Information Retrieval, pp. 1045–1048 (2019)
53. Yang, E., Lewis, D.D., Frieder, O.: Heuristic stopping rules for technology-assisted review. In: Proceedings of the 21st ACM Symposium on Document Engineering (2021)
54. Yang, E., Lewis, D.D., Frieder, O.: On minimizing cost in legal document review workflows. In: Proceedings of the 21st ACM Symposium on Document Engineering (2021)

55. Yang, E., Lewis, D.D., Frieder, O.: TAR on social media: a framework for online content moderation. In: 2nd International Conference on Design of Experimental Search & Information REtrieval Systems (2021)
56. Yang, E., Lewis, D.D., Frieder, O., Grossman, D., Yurchak, R.: Retrieval and richness when querying by document. In: International Conference on Design of Experimental Search & Information REtrieval Systems (2018)
57. Yang, Y.Y., Lee, S.C., Chung, Y.A., Wu, T.E., Chen, S.A., Lin, H.T.: libact: pool-based active learning in Python. Technical report, National Taiwan University (2017). https://github.com/ntucllab/libact, available as arXiv preprint https://arxiv.org/abs/1710.00379
58. Yang, Z., Wang, Y., Chen, X., Liu, J., Qiao, Y.: Context-transformer: tackling object confusion for few-shot detection. In: Proceedings of the AAAI Conference on Artificial Intelligence, vol. 34, pp. 12653–12660 (2020)
59. Zhang, L., Zhang, L.: An ensemble deep active learning method for intent classification. In: Proceedings of the 2019 3rd International Conference on Computer Science and Artificial Intelligence, pp. 107–111 (2019)
60. Zhu, J., Wang, H., Hovy, E., Ma, M.: Confidence-based stopping criteria for active learning for data annotation. ACM Trans. Speech Lang. Process. (TSLP) 6(3), 1–24 (2010)
61. Zhu, Y., et al.: Aligning books and movies: Towards story-like visual explanations by watching movies and reading books. In: Proceedings of the IEEE International Conference on Computer Vision, pp. 19–27 (2015)

Multi-modal Sentiment and Emotion Joint Analysis with a Deep Attentive Multi-task Learning Model

Yazhou Zhang[1], Lu Rong[1(✉)], Xiang Li[2], and Rui Chen[1(✉)]

[1] Software Engineering College, Zhengzhou University of Light Industry,
Zhengzhou, China
{lurong,ruichen}@zzuli.edu.cn
[2] Shandong Computer Science Center (National Supercomputing Center in Jinan), Qilu
University of Technology (Shandong Academy of Sciences), Jinan, China

Abstract. Emotion is seen as the external expression of sentiment, while sentiment is the essential nature of emotion. They are tightly entangled with each other in that one helps the understanding of the other, leading to a new research topic, i.e., multi-modal sentiment and emotion joint analysis. There exists two key challenges in this field, i.e., multi-modal fusion and multi-task interaction. Most of the recent approaches treat them as two independent tasks, and fail to model the relationships between them. In this paper, we propose a novel multi-modal multi-task learning model, termed MMT, to generically address such issues. Specially, two attention mechanisms, i.e., cross-modal and cross-task attentions are designed. Cross-modal attention is proposed to model multi-modal feature fusion, while cross-task attention is to capture the interaction between sentiment analysis and emotion recognition. Finally, we empirically show that this method alleviates such problems on two benchmarking datasets, while getting better performance for the main task, i.e., sentiment analysis with the help of the secondary emotion recognition task.

Keywords: Multi-modal sentiment analysis · Emotion recognition · Multi-task learning · Deep learning

1 Introduction

Sentiment is treated as the subjective attitude of a human towards specific target or topic. The recent development of social media platforms, e.g., Instagram, TikTok, etc., has produced a great number of subjective comments of users [10]. Such data has been a rich source of information, including that of attitudes or opinions, providing us a practical and feasible path to understand human intents and affections. Hence, sentiment analysis, as a longevous research focus, has attracted great attention over the past decades [1,31].

Sentiment analysis, also known as opinion mining, usually refers to study, analyze and identify the subjective polarity carried in user generated contents via NLP, statistics knowledge and machine or deep learning approaches. Generally speaking, it could be

M. Hagen et al. (Eds.): ECIR 2022, LNCS 13185, pp. 518–532, 2022.
https://doi.org/10.1007/978-3-030-99736-6_35

considered as a judgment or evaluation on the emotional state (the speaker or author's emotional state) and emotional communication (the emotional effect that the speaker or author wants to convey).

Actually, people's subjective expression is multi-modal, which comprehends a mixture of natural language (text), facial gestures (vision) and vocal actions (audio), instead of text only. Single modality (e.g., text) relying only on basic terms (e.g., words, phrase) and their semantic associations, is insufficient to distinguish complex sentiment while multi-modality could provide vivid description, convey accurate and rich sentimental information and uncover the information that single modality may hide. Multi-modal sentiment analysis is gathering a lot of attention recently considering its potential use. [4, 14].

In this paper, we take a further step towards exploring the use of emotional knowledge to improve sentiment analysis and vice versa. The motivation roots in the difference and similarity between sentiment and emotion. Sentiment is formed on the basis of long term emotional experiences, which is a thought that has been influenced by emotion, e.g., "positive", "negative", etc. Emotion involves a complex psychological state such as happiness, anger, jealousy, grief, etc., which is more inclined to the physical response [17]. Hence, sentiment is tightly coupled with emotion in that one helps the understanding of the other. Hence, jointly analyzing sentiment and emotion would bring benefits to each other where a new research topic, multi-modal sentiment and emotion joint analysis, is brought forth [2].

Different from the traditional single task or text based approaches, there are two intractable challenges in multi-task multi-modal sentiment analysis, i.e., multi-modal fusion and multi-task correlation, which are detailed as follows.

(i) **Multi-modal fusion**. Multi-modal fusion has remained an active research topic in multi-modal sentiment analysis for a long time. Three commonly used fusion strategies, i.e., feature-level fusion (early fusion), decision-level fusion (late fusion) and their combination (hybrid fusion), have been applied in different scenarios [32]. Early fusion targets at grasping the correlations across multi-modal features. Its main obstacle involves data heterogeneity, since each modal feature comes from different semantic spaces. In contrast, late fusion aims to combine the local decisions based on individual features. But it ignores multi-modal feature correlation. To exploit the advantages of both early and late fusion strategies, hybrid fusion is designed with high complexity. Now, researchers fail to reach an agreement on the optimal manner of merging multi-modal information, due to the heterogeneities and augmented diversities across modalities [18, 24].

(ii) **Multi-task correlation**. Multi-task learning paradigm attempts to exploit the inter-relatedness across tasks (e.g., sentiment and emotion, etc.), for improving individual performance. Recent approaches, including the state of the art [1, 2, 4] intuitively argue that one task is tightly intertwined to other tasks and spend considerable effort on implicitly learning a shared representation across multiple tasks, and separately making the prediction for each task. Even though they have achieved impressive results, such implicit way is insufficient to depict the correlations across multiple tasks, e.g., which task reaps the greatest benefit from others, and limits the potential of the model. Hence, multi-task interaction is still an important problem that has not been well addressed.

To generically address the above two issues, we propose a deep attentive multi-modal multi-task learning framework, termed MMT. The key components are two attention mechanisms, i.e., cross-modal (C_m) and cross-task (C_t) attentions. C_m attention is proposed to learn multi-modal feature fusion. The multi-modal representation of the target document is obtained and fed into the decoder to produce its sentimental and emotional results. In contrast, C_t attention is proposed to learn the mutual influence between sentiment analysis and emotion recognition. Extensive experiments are conducted on two benchmarking datasets, i.e., MUStARD and Memotion, in comparison with a wide range of strong baselines, including deep convolutional neural network (CNN), bidirectional encoder representations from transformers (BERT), RCNN-RoBERTa, quantum-like multi-modal framework (QMSA) and two state-of-the-art multi-task learning frameworks (i.e., UPB-MTL and A-MTL). The results show the efficacy of the proposed MMT framework. The evaluation also shows that MMT obtains better performance for the main task, i.e., sentiment analysis with the help of the secondary emotion recognition task. The major innovations of the work presented in this paper can be summarized as follows.

- A novel multi-task multi-modal attentive model is proposed.
- The interaction among related tasks is explicitly captured.
- The correlation between text and image is learned.
- We present the state-of-the-art performance on two datasets.

2 Related Work

2.1 Multi-modal Sentiment Analysis

Multi-modal sentiment analysis aims to identify the polarity expressed in multi-modal documents. Morency [16] first jointly use visual, audio and textual features to solve the problem of tri-modal sentiment analysis. Zhang et al. [32] proposed a quantum-inspired multi-modal sentiment analysis model. Li [13] designed a tensor product based multi-modal representation model for video sentiment analysis. Most recent works are performed from a multi-modal deep learning perspective [3, 12]. Zadeh et al. [29] proposed to use tensor product to perform multi-modal visual and vocal feature fusion. Huang et al. [11] proposed a deep multi-modal attentive fusion approach. Yu et al. [28] introduced a Chinese single- and multi-modal sentiment analysis dataset, called CH-SIMS, and proposed a multi-task learning task. Xu et al. [27] proposed a multi-modal attentive framework for their new task, which is aspect-based multi-modal sentiment analysis.

The above-mentioned works focused on multi-modal feature extraction and multi-modal feature fusion. Few approaches had also attempted to use sentiment or emotion information to enhance the classification performance.

2.2 Multi-modal Emotion Recognition

Multi-modal emotion recognition aims to identify the emotional polarity expressed in multi-modal documents using machine or deep learning approaches. In earlier times, Chuang and Wu [6] constructed a multi-modal emotion recognition framework based on speech signals and textual content. Datcu and Rothkrantz [7] fused early acoustic features with facial expressions for emotion recognition. Zhang et al. [32] proposed a quantum-inspired multi-modal sentiment analysis model. Recently, CNN, RNN and their multifarious variants were commonly used to extract visual and sequential features and built multi-modal emotion recognition framework [21,31].

Emotion recognition in conversation (ERC) has become a popular research task. Majumder [15] described a DialogueRNN model that kept track of the individual party states throughout the conversation and used this information for ERC. Poria et al. [19] created the first multimodal conversational dataset, namely, the multimodal emotion-lines dataset (MELD), to facilitate the development of conversational sentiment analysis. Zhang and Li [30] designed a quantum-inspired interactive network model for textual conversational sentiment analysis and showed its effectiveness. However, they did not take the interactions among different modalities into consideration.

In general, remarkable progress has been made in the current state-of-the-art. However, current approaches still do not explicitly model the interactions between tasks. In this work, we tackle all these two problems in a multi-modal sentiment scenario with a multi-task learning framework.

3 Methodology

In this section, we present the proposed MMT framework step by step.

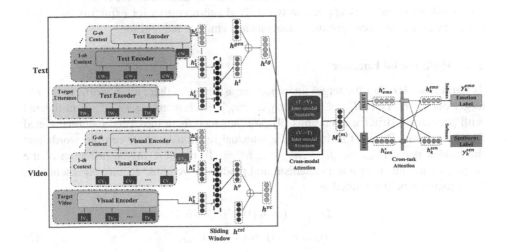

Fig. 1. The overall architecture of the MMT model.

3.1 Task Description and Overall Network

Task Description. Suppose the dataset contains N multi-modal subjective samples, the kth sample X_k could be represented as $\{X_k = (M_k, C_i), Y_k\}$, where M_k, Y_k represent the target multi-modal utterance and the label respectively, and C_i represents the ith context, $k \in [1, 2, ..., N]$, $i \in [1, 2, ..., G]$. In this work, we only consider the textual and vision modalities, e.g., $M_k = (M_k^t, M_k^v)$, $C_i = (C_i^t, C_i^v)$. But we argue that the proposed framework could be extended into the triple-modal task. Given a multi-modal utterance M_k and its context C_G, how to jointly detect the sentiment and emotion polarities, i.e., $Y_k = (Y_k^{sen}, Y_k^{emo})$. We formulate the problem as follows:

$$\zeta = \prod_k p\left(Y_k | M_k, C_G, \Theta\right) \tag{1}$$

where Θ represents the parameter set.

Overall Network. The architecture of the MMT model is shown in Fig. 1. It consists of four core components, i.e., a multi-modal encoder for textual (t) and visual (v) inputs, the external knowledge extraction layer, a multi-head cross-modal attention based multi-modal fusion layer and a cross-task interaction module. (1) We forward the textual utterance and video clip, e.g., M_k^t, M_k^v into multi-modal encoder for obtain their hidden states, denoted as h_k^t and h_k^v respectively. Then, we take the multi-modal contexts appearing in a fixed-window of length L into account, and obtain their hidden representation, h_L^t and h_L^v to merge the target and contextual representation together, i.e., h^t and h^v. (2) We learn the vector representation of the participant gender and that of the overall color as two kinds of external knowledge for augmenting both textual and visual representations. (3) A multi-head cross-modal attention fusion layer is applied to obtain the multi-modal representation of the target document $M_k^{(m)}$. (4) The cross-task attention C_t is applied in two LSTM sub-networks for explicitly modeling the correlations between sentiment analysis and emotion recognition.

3.2   Multi-modal Encoder

Textual Encoder. For the target utterance, we assume that there are n words in the kth target document, i.e., $M_k^t = \{tw_1, tw_2, ..., tw_n\}$. Each word $tw \in \mathcal{R}^{d_t}$ is initialized with pre-trained BERT embeddings [9]. We thus feed them into a bidirectional Gated Recurrent Unit (BiGRU) to learn the contextual relationship between the words and the hidden states $H_{tar} = [h_{tw_1}^t, h_{tw_2}^t, ..., h_{tw_n}^t]$. To measure the contribution of the words, we use the attention mechanism and produce a weighted feature representation h_k^t, which can be formulated as:

$$
\begin{aligned}
D_{tar} &= tanh\left(W_d H_{tar} + b_d\right) \\
\alpha_{tar} &= softmax\left(w^T D_{tar}\right) \\
h_k^t &= \alpha_{tar} H_{tar}
\end{aligned}
\tag{2}
$$

To model the contextual information, we take the multi-modal contexts appearing in a fixed-window of length L into account. We also employ BERT to obtain its pre-trained word embeddings and thus feed them to the attention based BiGRU for learning the contextual representation, h_i^t, where $i \in [1, 2, ..., L]$. In this work, the textual representation is obtained by concatenating the target and the contextual representations, i.e., $h^t = h_k^t \oplus [h_1^t, h_2^t, ..., h_L^t]$.

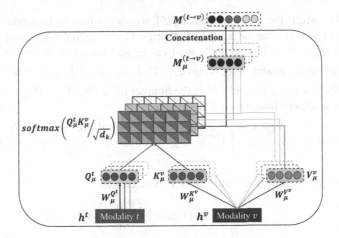

Fig. 2. Cross-modal attention mechanism.

Visual Encoder. For video, suppose that there are n clips in the kth target video, i.e., $M_k^v = \{tv_1, tv_2, ..., tv_n\}$. Each input video clip is scaled to 480×360 and its feature vectors are extracted by the pre-trained EfficientNet network [23]. We use average pooling to get a 768 dimensional feature representation. We thus feed each video clip into the GRU unit to produce its hidden state $H = [h_{tv_1}^v, h_{tv_2}^v, ..., h_{tv_n}^v]$. Based on Eq. 2, the attention mechanism is also adopted to get a weighted visual representation h_k^v. If the target visual document is a static image, we choose to divide the whole image into n visual zones from top to bottom.

Using the same pre-trained model, the visual vectors for each context is represented as $[h_1^v, h_2^v, ..., h_L^v]$. The visual representation is obtained, $h^v = h_k^v \oplus [h_1^v, h_2^v, ..., h_L^v]$.

3.3 External Information Extraction

We also incorporate external information, e.g., gender, color, into the textual and visual vectors to enhance their representation abilities. We represent the gender information using the pre-trained BERT embedding h^{gen} and thus merge it with the textual representation h^t, to construct refined representation, i.e., $h^{tg} = h^t \oplus h^{gen}$.

Similarly, since negative users tend more to deliver darker images, while positive users more likely publish colorful images, we regard the color distribution as another supplement knowledge to improve the visual representation. We extract the HSV space based color histogram of each visual document, denoted as h^{col}, and concatenate it with the visual representation h^v to obtain the refined vector, i.e., $h^{vc} = h^v \oplus h^{col}$.

The final textual and visual vectors h^{tg}, h^{vc} are fed into next hidden layer to perform multi-modal fusion.

3.4 Multi-head Cross-modal Attentive Fusion

We have obtained the textual and visual representations, i.e., h^{tg}, h^{vc}. Then, an cross-modal multi-head self attention based multi-modal fusion layer is designed to obtain the multi-modal representation $M_k^{(m)}$, as depicted in Fig. 2.

Inspired by Multi-modal Transformer [25], we aim to fuse multi-modal information by learning a latent adaptation across modalities. Given textual and visual modalities t and v with their vectors h^{tg} and h^{vc}, we treat textual modality as $Query$, i.e., $Q^t_\mu = W^{Q^t_\mu} h^{tg}$, and visual modality as $Keys$ and $Values$, i.e., $K^v_\mu = W^{K^v_\mu} h^{vc}$ and $V^v_\mu = W^{V^v_\mu} h^{vc}$, where $\mu \in [1, 2, ..., H]$, H is the number of heads, $W^{Q^t}_\mu$, $W^{K^v}_\mu$ and $W^{V^v}_\mu$ are weights. The mapping from t to v is defined as:

$$M^{t \to v}_\mu = softmax\left(\frac{Q^t_\mu K^v_\mu}{\sqrt{d_k}}\right) V^v_\mu = softmax\left(\frac{W^{Q^t_\mu} h^{tg} \cdot W^{K^v_\mu} h^{vc}}{\sqrt{d_k}}\right) W^{V^v_\mu} h^{vc}$$

(3)

Correspondingly, the mapping from t to v is defined as:

$$M^{v \to t}_\mu = softmax\left(\frac{Q^v_\mu K^t_\mu}{\sqrt{d_k}}\right) V^t_\mu$$

(4)

Equation 3 and Eq. 4 will yield H output values respectively. Then, these values are concatenated as: $M^{t \to v} = [M^{t \to v}_1, M^{t \to v}_2, ..., M^{t \to v}_H]$ and $M^{v \to t} = [M^{v \to t}_1, M^{v \to t}_2, ..., M^{v \to t}_H]$. In this work, H is set to eight. Now, we merge them together to obtain the dual-directional representation, as shown in Eq. 5

$$M^{(m)}_k = [M^{t \to v}; M^{v \to t}]$$

(5)

3.5 Cross-task Attention Mechanism

In this section, two attention based sub-networks are proposed to model the correlation between sentiment and emotion labels. Since our focus is sentiment analysis, we regard it as the main task, while emotion recognition is seen as the secondary task, e.g., $(emo \to sen)$. This action indeeds leverage knowledge from other tasks.

To explicitly learn the relationship between the classes of all the tasks, we design a self-attention mechanism. This is:

$$h^*_{sen} = LSTM_s\left(M^{(m)}_k\right), \qquad h^*_{emo} = LSTM_e\left(M^{(m)}_k\right)$$
$$h^{emo}_k = Attention\left(h^*_{emo}, h^*_{sen}\right), \quad h^{sen}_k = Attention\left(h^*_{sen}, h^*_{emo}\right)$$

(6)

where the attention is written as:

$$Attention\left(h_A, h_B\right) = softmax\left(\frac{Q_A K_A}{\sqrt{d}}\right) V_A + softmax\left(\frac{Q_A K_B}{\sqrt{d}}\right) V_B$$

(7)

where A and B denote different sentiment tasks.

3.6 Classification

The outputs, e.g., h^{sen}_k, h^{emo}_k are forwarded through the softmax functions to yield both sentiment and emotion labels. We use cross entropy with L2 regularization as the loss functions ζ_{sen}, ζ_{emo} for training each task, and jointly minimize them with different weights:

$$\zeta = w_{sen}\zeta_{sen} + w_{emo}\zeta_{emo}$$

(8)

where w_{sen} and w_{emo} are weights. We use the backpropagation method to compute the gradients and update all the parameters. To avoid overfitting, we use a dropout strategy.

4   Experimental Setup

We aim to answer the following research questions and experimentally verify the advantages of the MMT model.

RQ1: What is the optimal sliding window size for modeling contextuality?

RQ2: Is the cross-modal attention layer effective for multi-modal feature fusion?

RQ3: Does modeling of human emotion help in sentiment analysis?

4.1   Experiment Settings

Datasets. We choose benchmark datasets that have textual and visual modalities with all sentiment and emotion labels. The extended version of MUStARD (MUStARD$_{ext}$ for short)[1] [5] and Memotion[2] [22] datasets meet the criteria.

(a) **MUStARD$_{ext}$:** MUStARD consists of 690 multi-modal samples from multiple sources, e.g., Big Bang Theory, Friends, etc. The utterance in each conversation is annotated with sarcastic or non-sarcastic labels. As an extended version of MUStARD, MUStARD$_{ext}$ re-annotate sentiment and emotion labels.

(b) **Memotion:** It consists of 6992 training samples and 1879 testing samples. Each memo data has been labelled with semantic dimensions, e.g., sentiment and type of emotion, e.g., sarcasm, humor, etc. The speaker identifiers of all the utterances are also recorded. It is released in the Memotion Analysis 1.0 SemEval 2020 Task.

Evaluation Metrics. We adopt *precision* (P), *recall* (R) and *micro-F1* (M$_i$-F1) as evaluation metrics in our experiments. We also introduce a *balanced accuracy* metric for an ablation test.

Hyper-parameter Setup. The textual and visual inputs are initialized with BERT and EfficientNet. The dimensionality of the embeddings is set to 768. All weight matrices are given their initial values by sampling from a uniform distribution $U(-0.1, 0.1)$, and all biases are set to zeros. We use the Adam algorithm to train the network, and the number of epochs is set to 100. The batch size is set to 64, and the learning rate for sentiment analysis is 0.001, while that for emotion recognition is 0.005. The optimal sliding window size is set to 1 and 2 on two datasets.

4.2   Comparison Models

The state-of-the-art baselines are:

CNN: We apply a deep CNN on the target utterance to extract the textual, visual features respectively, and merge them together. We thus feed this multi-modal feature into the softmax function to make prediction.

SVM+BERT [8]: It uses BERT to produce the textual utterance vector and feeds them into SVM for sentiment and emotion joint analysis. We also concatenate the contextual features.

---

[1] http://www.iitp.ac.in/ai-nlp-ml/resources.html.

[2] https://competitions.codalab.org/competitions/20629.

Table 1. Comparison of different models.

Dataset	Method	Sentiment analysis			Emotion recognition		
		P	R	M_i-F1	P	R	M_i-F1
MUStARD$_{ext}$	CNN	42.14	42.16	42.14	22.54	22.55	22.54
	SVM+BERT	44.08	44.12	44.10	25.64	25.71	25.67
	SVM+BERT (+context)	44.31	44.37	44.34	26.39	26.39	26.39
	RCNN-RoBERTa	46.43	46.53	46.48	31.43	31.47	31.44
	EfficientNet	46.52	46.61	46.59	32.61	32.70	32.65
	UPB-MTL	47.72	47.70	47.70	33.55	33.64	33.60
	QMSA	46.96	46.96	46.96	17.50	17.50	17.50
	A-MTL	49.45	49.48	49.47	33.12	33.07	33.10
	Text-MMT	47.35	47.51	47.39	31.43	31.54	31.40
	Image-MMT	45.14	45.19	45.17	29.11	29.25	29.22
	MMT	**50.47**	**50.32**	**50.37**	**33.72**	**33.80**	**33.75**
	△SOTA	(+1.8%)	(+1.6%)	(+1.7%)	(+0.5%)	(+0.6%)	(+0.5%)
Memotion	CNN	33.97	34.10	34.02	33.64	33.70	33.66
	SVM+BERT	34.57	34.61	34.59	33.87	33.92	33.90
	SVM+BERT (+context)	34.92	34.96	34.93	34.06	34.11	34.07
	RCNN-RoBERTa	36.22	**36.31**	36.27	44.41	44.50	44.45
	EfficientNet	33.04	33.33	33.21	46.35	46.40	46.42
	UPB-MTL	34.41	34.56	34.49	45.17	45.22	45.18
	QMSA	34.88	34.93	34.91	37.78	37.83	37.79
	A-MTL	34.23	34.32	34.27	40.39	40.51	40.44
	Text-MMT	34.57	34.63	34.58	44.33	44.37	44.42
	Image-MMT	30.29	30.36	30.25	36.23	36.45	36.34
	MMT	**36.77**	36.28	**36.55**	**47.04**	**46.85**	**46.97**
	△SOTA	(+1.5%)	(−0.1%)	(+0.7%)	(+1.5%)	(+0.6%)	(+1.1%)

RCNN-RoBERTa [20]: It utilizes pre-trained RoBERTa vectors to represent the utterance and uses a RCNN to obtain its contextual representation. The final classification is performed by the softmax layer.

EfficientNet [23]: It uses a compound scaling method to create different models, which has achieved state-of-the-art performance on the ImageNet challenge.

UPB-MTL [26]: It uses ALBERT to represent the textual utterance and uses VGG-16 to represent the accompanying image.

QMSA [32]: It first extracts visual and textual features using density matrices, and feeds them into the SVM classifier. But we replace the original GloVe embedding with BERT vector.

A-MTL [5]: It designs two attention mechanisms, e.g., intra-segment and inter-segment, attentions to learn the relation between different segments and the relation within the same segment.

4.3   Comparative Analysis

MUStARD$_{ext}$**.** From Table 1, we can notice that in the cases of sentiment analysis, CNN gets the worst results. It also is the second-worst-performing in the case of emotion recognition. One major reason is that its feature representation is not effective to alleviate "the semantic gap". SVM+BERT outperforms CNN by a large margin of 4.7% in term of f1 score. This result is thanks to strong representational capacity of BERT against CNN. By taking the contextual information into consideration, SVM+BERT (+context) achieves slightly improvement for both tasks, which shows the importance of context modeling. Compared with SVM+BERT, RCNN-RoBERTa obtains signifi-cant improvements for the tasks of sentiment analysis and emotion recognition. This because that RoBERTa is trained with longer sequence on a larger dataset, which is capable of better capturing long range dependency. However, RCNN-RoBERTa is only designed for text instead of multi-modality. EfficientNet and UPB-MTL perform bet-ter than RCNN-RoBERTa for the tasks of sentiment analysis and emotion recognition. The performance of QMSA rises and falls steeply for different tasks, which may be due to the instability of quantum density matrix. A-MTL performs very well and achieves the best classification performance among all baselines for the task of sentiment analy-sis, and gets comparable results against UPB-MTL for the task of emotion recognition. Compared with UPB-MTL, the micro f1 scores increase by 3.8%.

Text-MMT and Image-MMT perform not very well, especially the poorly perfor-mance that Image-ATM makes, demonstrating that text and visual modalities cannot be treated independently for multi-modal sentiment and emotion analysis. The proposed MMT model achieves the best micro-F1 scores of 50.37%, 33.75% as compared to micro-F1 of 49.47%, 33.10% of the state-of-the-art system (i.e., A-MTL framework). This shows that MMT successfully leverages the advantages of PLM, multi-task learn-ing and two attention mechanisms in modeling human sentiment.

Memotion. We can see that the performance differences between all models are not as contrasting as they are on MUStARD$_{ext}$. CNN achieves comparable performance against SVM+BERT, because that Memotion contains no conversational context. But SVM+BERT still outperforms deep CNN, which benefits from the outstanding abstract ability of PLM. Meanwhile, since there is no conversational context, the performance of SVM+BERT (+context) is the result of re-running SVM+BERT on Memotion. RCNN-RoBERTa, EfficientNet and UPB-MTL win or lose each other for different tasks. QMSA and A-MTL lose for the tasks of sentiment and emotion joint analysis. One possible interpretation is that QMSA are insufficient to deal with multi-class task. As for A-MTL, the inter-segment attention is useless since there is no context on Memo-tion, and this influences its performance. The proposed MMT model remarkably over-comes all baselines, and achieves the state-of-the-art performance with the micro-f1 of 36.55% and 46.97%. We attribute the main improvements to both PLMs and two attention mechanisms, which ensures that MMT can model inter-modality fusion and multi-task interaction.

4.4   STL v/s MTL Framework

We outline the comparison results between the multi-task (MTL) and single-task (STL) learning frameworks in Table 2. We see that bi-modal (T+V) shows a better performance over unimodal setups.

For sentiment analysis on MUStARD, MTL outperforms STL by a large margin in bi-modality instead of single modalities. The reason is that MUStARD involves many conversational contexts, where both textual and visual information play key roles in multi-modal sentiment and emotion joint analysis. MTL learns more supplementary information when using multi-modal information. For sentiment analysis on Memotion, MTL achieves better performance than STL on text and bi-modal. Because this dataset does not contain context, where textual information contributes more on understanding semantics. The proposed MMT framework could learn the inter-dependence between two related tasks and improves performance.

Table 2. Comparison with single-task learning (STL) and multi-task (MTL) learning frameworks.

Task	Setups	T		V		T+V	
		M_i-F1	Acc	M_i-F1	Acc	M_i-F1	Acc
Sentiment MUStARD	STL	46.55	46.64	45.04	45.05	47.86	48.03
	MTL	47.39	47.51	45.17	45.19	50.37	50.32
Sentiment Memotion	STL	33.34	33.39	30.24	30.26	34.55	34.62
	MTL	34.58	34.63	30.25	30.36	36.55	36.28

Table 3. Ablation experiment results.

Dataset	Models	Metrics	
		M_i-F1	Acc
MUStARD	No C_m Attention	49.26	49.17
	No C_t Attention	48.44	48.51
	No Attention	46.72	46.69
	MMT	50.37	50.32
Memotion	No C_m Attention	35.42	35.47
	No C_t Attention	35.31	35.27
	No Attention	33.29	33.35
	MMT	36.55	36.28

4.5   Ablation Test

We perform the ablation experiments to further analyze the effectiveness of different components of MMT: (1) *No C_m Attention* that replaces the cross-modal attentive fusion with multi-modal feature concatenation; (2) *No C_t Attention* that only handles sentiment analysis without modeling multi-task correlation; (3) *No Attention* that removes both I_e and I_t attentions from MMT.

The results in Table 3 show that the cross-task C_t attention contributes the most to overall performance. Because we treat sentiment analysis as the main task, MMT will share emotional knowledge with sentiment analysis via the connectivity from LSTM, for improving the performance. In addition, *No Attention* achieves the worst results, which shows that cross-modal C_m attention also plays important roles in MMT. In summary, both of them are indispensable components for MMT.

4.6 Misclassification Cases

We check the dataset and show a few misclassification cases (text+image), including the cases that MTL predicts correctly while STL fails, and that both setups fails to predict correctly. These cases are shown in Fig. 3.

For the main task, we notice that misclassification for STL framework happens in the situation where the literal meaning of the text differs from its visual counterpart. The proposed MMT model leverages the depressed emotion to make correct sentiment judgment, and obtains an improvement. Both MTL and STL frameworks fail in the similar situation where the speaker expresses implicit sentiment, e.g., there are no sentiment words or phrases in the utterance. They might require external information, e,g., the speaker's character.

Fig. 3. Wrongly classified multi-modal samples.

5   Conclusions and Future Work

Multi-modal sentiment and emotion joint analysis is an important and challenging NLP task. In this paper, we propose a deep attentive multi-task learning model, termed MMT. The main idea is to use two attention mechanisms, i.e., cross-modal and cross-task attentions, to address the problems of multi-modal fusion and multi-task interaction.

Comprehensive experiments on two benchmark datasets, show that the effectiveness of MMT over state-of-the-art baselines. Since there are closely relationship among sentiment, emotion, sarcasm, humor, etc., our future works will focus on designing an unified multi-task learning model to capture the correlation among triple or more tasks. Those sentiment and emotion conjoint analysis approaches might be applied to emotion dialogue analysis, reviews mining, etc.

Acknowledgment. This work is supported by National Science Foundation of China under grant No. 62006212, the fund of State Key Lab. for Novel Software Technology in Nanjing University under grant No. KFKT2021B41, the Industrial Science and Technology Research Project of Henan Province under Grants 222102210031, 212102210418, 212102310088, the Doctoral Scientific Research Foundation of Zhengzhou Univ. of Light Industry (grant No 2020BSJJ030, 2020BSJJ031).

References

1. Akhtar, M.S., Chauhan, D.S., Ekbal, A.: A deep multi-task contextual attention framework for multi-modal affect analysis. ACM Trans. Knowl. Discov. Data (TKDD) **14**(3), 1–27 (2020)
2. Akhtar, M.S., Chauhan, D.S., Ghosal, D., Poria, S., Ekbal, A., Bhattacharyya, P.: Multi-task learning for multi-modal emotion recognition and sentiment analysis. arXiv preprint arXiv:1905.05812 (2019)
3. Cambria, E., Poria, S., Hussain, A.: Speaker-independent multimodal sentiment analysis for big data. In: Seng, K.P., Ang, L., Liew, A.W.-C., Gao, J. (eds.) Multimodal Analytics for Next-Generation Big Data Technologies and Applications, pp. 13–43. Springer, Cham (2019). https://doi.org/10.1007/978-3-319-97598-6_2
4. Chauhan, D.S., Dhanush, S., Ekbal, A., Bhattacharyya, P.: Sentiment and emotion help sarcasm? A multi-task learning framework for multi-modal sarcasm, sentiment and emotion analysis. In: Proceedings of the 58th Annual Meeting of the Association for Computational Linguistics, pp. 4351–4360 (2020)
5. Chauhan, D.S., Dhanush, S.R., Ekbal, A., Bhattacharyya, P.: Sentiment and emotion help sarcasm? A multi-task learning framework for multi-modal sarcasm, sentiment and emotion analysis. In: Proceedings of the 58th Annual Meeting of the Association for Computational Linguistics, pp. 4351–4360. Association for Computational Linguistics, July 2020. https://doi.org/10.18653/v1/2020.acl-main.401. https://www.aclweb.org/anthology/2020.acl-main.401
6. Chuang, Z.J., Wu, C.H.: Multi-modal emotion recognition from speech and text. Int. J. Comput. Linguist. Chinese Lang. Process. **9**(2), 45–62 (2004). Special Issue on New Trends of Speech and Language Processing
7. Datcu, D., Rothkrantz, L.J.: Semantic audio-visual data fusion for automatic emotion recognition. In: Emotion Recognition: A Pattern Analysis Approach, pp. 411–435 (2014)
8. Devlin, J., Chang, M.W., Lee, K., Toutanova, K.: BERT: pre-training of deep bidirectional transformers for language understanding. In: NAACL-HLT 2019: Annual Conference of the North American Chapter of the Association for Computational Linguistics, pp. 4171–4186 (2019)
9. Devlin, J., Chang, M.W., Lee, K., Toutanova, K.N.: BERT: pre-training of deep bidirectional transformers for language understanding. In: Proceedings of the 2019 Conference of the North American Chapter of the Association for Computational Linguistics: Human Language Technologies, vol. 1 (Long and Short Papers), pp. 4171–4186 (2018)

10. Ghosal, D., Majumder, N., Poria, S., Chhaya, N., Gelbukh, A.F.: DialogueGCN: a graph convolutional neural network for emotion recognition in conversation. In: Proceedings of the 2019 Conference on Empirical Methods in Natural Language Processing and the 9th International Joint Conference on Natural Language Processing (EMNLP-IJCNLP), pp. 154–164 (2019)
11. Huang, F., Zhang, X., Zhao, Z., Xu, J., Li, Z.: Image-text sentiment analysis via deep multimodal attentive fusion. Knowl. Based Syst. **167**, 26–37 (2019)
12. Kumar, A., Garg, G.: Sentiment analysis of multimodal twitter data. Multimedia Tools Appl. **78**, 1–17 (2019)
13. Li, Q., Melucci, M.: Quantum-inspired multimodal representation. In: 10th Italian Information Retrieval Workshop, pp. 1–2 (2019)
14. Liu, Y., Zhang, Y., Li, Q., Wang, B., Song, D.: What does your smile mean? Jointly detecting multi-modal sarcasm and sentiment using quantum probability. In: Findings of the Association for Computational Linguistics: EMNLP 2021, pp. 871–880 (2021)
15. Majumder, N., Poria, S., Hazarika, D., Mihalcea, R., Gelbukh, A., Cambria, E.: DialogueRNN: an attentive RNN for emotion detection in conversations. In: Proceedings of the AAAI Conference on Artificial Intelligence, vol. 33, pp. 6818–6825 (2019)
16. Morency, L.P., Mihalcea, R., Doshi, P.: Towards multimodal sentiment analysis: harvesting opinions from the web. In: Proceedings of the 13th International Conference on Multimodal Interfaces, pp. 169–176. ACM (2011)
17. Munezero, M., Montero, C.S., Sutinen, E., Pajunen, J.: Are they different? Affect, feeling, emotion, sentiment, and opinion detection in text. IEEE Trans. Affect. Comput. **5**(2), 101–111 (2014)
18. Poria, S., Cambria, E., Bajpai, R., Hussain, A.: A review of affective computing: from unimodal analysis to multimodal fusion. Inf. Fusion **37**, 98–125 (2017)
19. Poria, S., Hazarika, D., Majumder, N., Naik, G., Cambria, E., Mihalcea, R.: MELD: a multimodal multi-party dataset for emotion recognition in conversations. In: Proceedings of the 57th Annual Meeting of the Association for Computational Linguistics, vol. 1, pp. 527–536 (2019)
20. Potamias, R.A., Siolas, G., Stafylopatis, A.G.: A transformer-based approach to irony and sarcasm detection. Neural Comput. Appl. **32**(23), 17309–17320 (2020). https://doi.org/10.1007/s00521-020-05102-3
21. Sahu, S., Mitra, V., Seneviratne, N., Espy-Wilson, C.Y.: Multi-modal learning for speech emotion recognition: an analysis and comparison of ASR outputs with ground truth transcription. In: Interspeech, pp. 3302–3306 (2019)
22. Sharma, C., et al.: SemEval-2020 task 8: memotion analysis-the visuo-lingual metaphor! In: Proceedings of the 14th International Workshop on Semantic Evaluation (SemEval 2020), Barcelona, Spain. Association for Computational Linguistics, September 2020
23. Tan, M., Le, Q.V.: EfficientNet: rethinking model scaling for convolutional neural networks. In: International Conference on Machine Learning, pp. 6105–6114 (2019)
24. Tian, D., Zhou, D., Gong, M., Wei, Y.: Interval type-2 fuzzy logic for semisupervised multimodal hashing. IEEE Trans. Cybern. **51**, 3802–3812 (2019)
25. Tsai, Y.H.H., Bai, S., Liang, P.P., Kolter, J.Z., Morency, L.P., Salakhutdinov, R.: Multimodal transformer for unaligned multimodal language sequences. In: Proceedings of the conference. Association for Computational Linguistics. Meeting, vol. 2019, p. 6558. NIH Public Access (2019)
26. Vlad, G.A., Zaharia, G.E., Cercel, D.C., Chiru, C.G., Trausan-Matu, S.: UPB at SemEval-2020 task 8: joint textual and visual modeling in a multi-task learning architecture for memotion analysis. In: Proceedings of the Fourteenth Workshop on Semantic Evaluation, pp. 1208–1214 (2020)

27. Xu, N., Mao, W., Chen, G.: Multi-interactive memory network for aspect based multimodal sentiment analysis. In: Proceedings of the AAAI Conference on Artificial Intelligence, vol. 33, pp. 371–378 (2019)
28. Yu, W., et al.: CH-SIMS: a Chinese multimodal sentiment analysis dataset with fine-grained annotation of modality. In: Proceedings of the 58th Annual Meeting of the Association for Computational Linguistics, pp. 3718–3727. Association for Computational Linguistics, July 2020. https://doi.org/10.18653/v1/2020.acl-main.343. https://www.aclweb.org/anthology/2020.acl-main.343
29. Zadeh, A., Chen, M., Poria, S., Cambria, E., Morency, L.P.: Tensor fusion network for multimodal sentiment analysis. In: Proceedings of the 2017 Conference on Empirical Methods in Natural Language Processing, Copenhagen, Denmark, pp. 1103–1114. Association for Computational Linguistics, September 2017
30. Zhang, Y., Li, Q., Song, D., Zhang, P., Wang, P.: Quantum-inspired interactive networks for conversational sentiment analysis. In: Proceedings of the Twenty-Eighth International Joint Conference on Artificial Intelligence IJCAI 2019, pp. 5436–5442. International Joint Conferences on Artificial Intelligence Organization, July 2019. https://doi.org/10.24963/ijcai.2019/755
31. Zhang, Y., et al.: CFN: a complex-valued fuzzy network for sarcasm detection in conversations. IEEE Trans. Fuzzy Syst. **29**, 3696–3710 (2021)
32. Zhang, Y., et al.: A quantum-inspired multimodal sentiment analysis framework. Theoret. Comput. Sci. **752**, 21–40 (2018)

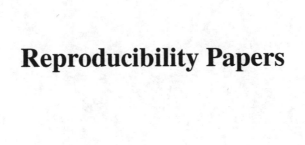
Reproducibility Papers

Do Lessons from Metric Learning Generalize to Image-Caption Retrieval?

Maurits Bleeker[✉] and Maarten de Rijke[✉]

University of Amsterdam, Amsterdam, The Netherlands
{m.j.r.bleeker,m.derijke}@uva.nl

Abstract. The triplet loss with semi-hard negatives has become the de facto choice for image-caption retrieval (ICR) methods that are optimized from scratch. Recent progress in metric learning has given rise to new loss functions that outperform the triplet loss on tasks such as image retrieval and representation learning. We ask whether these findings generalize to the setting of ICR by comparing three loss functions on two ICR methods. We answer this question negatively: the triplet loss with semi-hard negative mining still outperforms newly introduced loss functions from metric learning on the ICR task. To gain a better understanding of these outcomes, we introduce an analysis method to compare loss functions by counting how many samples contribute to the gradient w.r.t. the query representation during optimization. We find that loss functions that result in lower evaluation scores on the ICR task, in general, take too many (non-informative) samples into account when computing a gradient w.r.t. the query representation, which results in sub-optimal performance. The triplet loss with semi-hard negatives is shown to outperform the other loss functions, as it only takes one (hard) negative into account when computing the gradient.

1 Introduction

Given a query item in one modality, *cross-modal retrieval* is the task of retrieving similar items in another modality [41]. We focus on *image-caption retrieval* (ICR) [11,23,24,38]. For the ICR task, given an image or a caption as a query, systems have to retrieve the positive (e.g., matching or similar) item(s) in the other modality. Most ICR methods work with a separate encoder for each modality to map the input data to a representation in a shared latent space [11,12,16,23,24]. The encoders are optimized by using a contrastive-loss criterion, so as to enforce a high degree of similarity between representations of matching items in the latent space. For retrieval, a similarity score between a query and each candidate in a candidate set is computed to produce a ranking with the top-k best matching items. A lot of recent work on ICR relies on (1) pre-training on large amounts of data [16,25,34], and (2) more sophisticated (and data-hungry) model architectures [5,11,12,23,24,30]. However, pre-training on large-scale datasets is not always an option, either due to a lack of compute power, a lack of data, or both. Hence, it is important to continue to develop effective ICR methods that only rely on a modest amount of data.

© The Author(s), under exclusive license to Springer Nature Switzerland AG 2022
M. Hagen et al. (Eds.): ECIR 2022, LNCS 13185, pp. 535–551, 2022.
https://doi.org/10.1007/978-3-030-99736-6_36

To learn the similarity between a query and candidate representations, most ICR work relies on the standard Triplet loss with semi-hard negatives (Triplet SH) [4,5,11,12,23,24,30] or on the cross-entropy based NT-Xent [6,16] loss. In *metric learning*, the focus is on loss functions that result in more accurate item representations (in terms of a given evaluation metric) that can distinguish between similar and dissimilar items in a low-dimensional latent space [31]. There has been important progress in metric learning, with the introduction of new loss functions that result in better evaluation scores on a specific (evaluation) task. For example SmoothAP [1], it is a smooth approximation of the discrete evaluation metric Average Precision. By using SmoothAP, a retrieval method can be optimized with a discrete ranking evaluation metric and can handle multiple positive candidates simultaneously, which is not possible for the standard Triplet loss. Loss functions such as SmoothAP narrow the gap between the training setting and a discrete evaluation objective and thereby improve evaluation scores.

Research Goal. Most metric learning functions work with general representations of similar/dissimilar candidates and, in principle, there is no clear argument why obtained results on a specific task/method should not generalize to other tasks or methods. Hence, *can newly introduced metric learning approaches, that is, alternative loss functions, be used to increase the performance of ICR methods?* We compare three loss function for the ICR task: (1) the Triplet loss [22], including semi-hard negative mining, (2) NT-Xent loss [7], and (3) SmoothAP [1]. We expect SmoothAP to result in the highest performance based on the findings in the context of image retrieval [1] and in representation learning [36].

Main Findings. Following [31], we evaluate the three loss functions on fixed methods, with different datasets, and with a fixed training regime (i.e., training hyper-parameters) to verify which loss function uses the given training data as effectively as possible. Surprisingly, the lessons from metric learning do not generalize to ICR. The Triplet loss with semi-hard negative mining still outperforms the other loss functions that we consider. The promising results obtained by SmoothAP and the NT-Xent loss in other fields do not generalize to the ICR task.

To get a better grasp of this unexpected outcome, we propose *counting contributing samples* (COCOS), a method for analyzing contrastive loss functions. The gradient w.r.t. the query for the Triplet loss, NT-Xent and SmoothAP can be formulated as a sum over the representations of the positive and negative candidates in the training batch. The main difference between the loss functions lies in the number of samples used when computing the gradient w.r.t. the query and how each sample is weighted. We compare loss functions by counting how many samples contribute to the gradient w.r.t. the query representation at their convergence points. This yields an explanation of why one loss function outperforms another on the ICR task.

Main Contributions. (1) We experimentally compare three loss functions from the metric learning domain to determine if promising results from metric learning generalize to the ICR task, and find that the Triplet loss semi-hard (SH) still results in the highest evaluation scores. (2) We propose COCOS, a way of analyzing contrastive loss functions, by defining a count that tells us how many candidates in the batch contribute to the gradient w.r.t. the query. On average, the best performing loss function takes at most one (semi-hard) negative sample into account when computing the gradient.

2 Background and Related Work

Notation. We follow the notation introduced in [1,7,36]. We start with a multi-modal image-caption dataset $\mathcal{D} = \{(\mathbf{x}_I^i, \mathbf{x}_{C_1}^i, \ldots, \mathbf{x}_{C_k}^i)^i, \ldots\}_{i=1}^N$ that contains N image-caption tuples. For each image $\mathbf{x}_I^i$, we have k matching/corresponding captions, $\mathbf{x}_{C_1}^i, \ldots, \mathbf{x}_{C_k}^i$.

In the ICR task, either an image or a caption can function as a query. Given a query $\mathbf{q}$, the task is to rank all candidates in a candidate set $\Omega = \{\mathbf{v}_i \mid i = 0, \ldots, m\}$. A matching candidate is denote as $\mathbf{v}^+$ and a negative candidate(s) as $\mathbf{v}^-$. For each query $\mathbf{q}$, we can split the candidate set Ω into two disjoint subsets: $\mathbf{v}^+ \in \mathcal{P}_\mathbf{q}$ (*positive* candidate set) and $\mathbf{v}^- \in \mathcal{N}_\mathbf{q}$ (*negative* candidate set), where $\mathcal{N}_\mathbf{q} = \{\mathbf{v}^- \mid \mathbf{v}^- \in \Omega, \mathbf{v}^- \notin \mathcal{P}_\mathbf{q}\}$. We assume a binary match between images and captions, they either match or they do not match.

The set with similarity scores for each $\mathbf{v}_i \in \Omega$ w.r.t. query $\mathbf{q}$ is defined as: $S_\Omega^\mathbf{q} = \{s_i = \langle \frac{\mathbf{q}}{\|\mathbf{q}\|}, \frac{\mathbf{v}_i}{\|\mathbf{v}_i\|}\rangle, i = 0, \ldots, m\}$. We use cosine similarity as a similarity scoring function. $S_\Omega^\mathbf{q}$ consists of two disjoint subsets: $S_\mathcal{P}^\mathbf{q}$ and $S_\mathcal{N}^\mathbf{q}$. $S_\mathcal{P}^\mathbf{q}$ contains the similarity scores for the positive candidates and $S_\mathcal{N}^\mathbf{q}$ the similarity scores for the negative candidates. During training, we randomly sample a batch $\mathcal{B}$ with image-caption pairs. Both the images and captions will functions as queries and candidates.

Image-Caption Retrieval. The ICR task can be divided into *image-to text* (i2t) and *text-to-image* (t2i) retrieval. We target specific ICR methods that are optimized for the ICR-task only and satisfy three criteria: (1) The methods we use have solely been trained and evaluated on the same benchmark dataset; (2) the ICR methods we use compute one global representation for both the image and caption; and (3) the methods do not require additional supervision signals besides the contrastive loss for optimization. Below we evaluate two ICR methods with different loss functions: VSE++ [12] and VSRN [24]. In the online appendix of this work,[1] we provide a detailed description of VSE++ and VSRN.

VSE++. The best performing method of VSE++ uses a ResNet-152 [15]) to compute a global image representation. The caption encoder is a single directed GRU-based [10] encoder. Faghri et al. [12] introduce the notion of mining semi hard-negative triplets for the ICR task. By using the hardest negative in the batch for each positive pair (i.e. the negative candidate with the highest similarity score w.r.t. the query), their method outperforms state-of-the-art methods that do not apply this semi-hard negative mining.

---

[1] https://github.com/MauritsBleeker/ecir-2022-reproducibility-bleeker/blob/master/appendix.

VSRN. VSRN takes a set of pre-computed image region features as input. A Graph Convolutional Network [21] is used to enhance the relationships between each region vector. The sequence of region feature vectors is put through an RNN network to encode the global image representation. VSRN uses the same caption encoder and loss as [12].

Other Methods. Following VSE++ and VSRN, the SGRAF [11] and IMRAM [4] methods have been introduced. We do not use these two methods as they either do not outperform VSRN [4] or rely on similar principles as VSRN [11]. The main recent progress in ICR has been characterized by a shift towards transformer-based [37] methods. To the best of our knowledge, TREN/TERAN [29,30] and VisualSparta [28] are the only transformer-based ICR methods that are solely optimized using MS-COCO [26] or Flickr30k [40]. We do not use transformer-based methods, as optimizing them does not scale well for a reproduciblity study with moderately sized datasets. Methods such as OSCAR [25], UNITER [9], Vilbert [27] and ViLT-B [20] use additional data sources and/or loss functions for training. They focus on a wide variety of tasks such as visual QA, image captioning, and image retrieval.

Loss Functions for ICR. In this section we introduce three loss functions for ICR.

Triplet Loss with Semi Hard-Negative Mining. The Triplet loss is commonly used as a loss function for ICR methods [5,11,12,23,24,30]. The *Triplet loss with semi-hard negative mining* (Triplet loss SH), for a query $\mathbf{q}$ is defined as:

$$\mathcal{L}^{\mathbf{q}}_{TripletSH} = \max(\alpha - s^+ + s^-, 0), \tag{1}$$

where α is a margin parameter, $s^- = \max(S^{\mathbf{q}}_{\mathcal{N}})$ and $s^+ = s_0 \in S^{\mathbf{q}}_{\mathcal{P}}$. Here, $S^{\mathbf{q}}_{\mathcal{N}}$ only contains one element per query. The Triplet loss SH over the entire training batch is defined as:

$$\mathcal{L}_{TripletSH} = \sum_{\mathbf{q} \in \mathcal{B}} \mathcal{L}^{\mathbf{q}}_{TripletSH}. \tag{2}$$

Triplet loss SH performs a form of soft-negative mining per query by selecting the negative candidate with the highest similarity score w.r.t. the query, we also refer to this as the maximum violating query. For computational efficiency, this soft-negative mining is executed within the context of the training batch $\mathcal{B}$ and not over the entire training set.

As opposed to the definition above, another possibility is to take the Triplet-loss over all triplets in the batch $\mathcal{B}$. This is the definition of the standard *Triplet-loss* [22]:

$$\mathcal{L}^{\mathbf{q}}_{Triplet} = \sum_{s^- \in S^{\mathbf{q}}_{\mathcal{N}}} \max(\alpha - s^+ + s^-, 0) \tag{3a}$$

$$\mathcal{L}_{Triplet} = \sum_{\mathbf{q} \in \mathcal{B}} \mathcal{L}^{\mathbf{q}}_{Triplet}. \tag{3b}$$

NT-Xent Loss. The *NT-Xent loss* [7] is a loss function commonly used in the field of self-supervised representation learning [7,32]. A similar function has also been proposed by Zhang and Lu [43] in the context of ICR. The NT-Xent loss is defined as:

$$\mathcal{L}_{NT\text{-}Xent} = -\frac{1}{|\mathcal{B}|} \sum_{q \in \mathcal{B}} \log \frac{\exp(s^+/\tau)}{\sum_{s_i \in \mathcal{S}_\Omega^q} \exp(s_i/\tau)}, \tag{4}$$

where τ functions as a temperature parameter. As for the Triplet-loss formulation: $s^+ = s_0 \in \mathcal{S}_\mathcal{P}^q$. The major difference between the Triplet-loss SH is that the NT-Xent loss takes the entire negative candidate set into account.

SmoothAP Loss. The Average Precision metric w.r.t. a query **q** and candidate set Ω is defined as:

$$AP_\mathbf{q} = \frac{1}{|\mathcal{S}_\mathcal{P}^q|} \sum_{i \in \mathcal{S}_\mathcal{P}^q} \frac{\mathcal{R}(i,\mathcal{S}_\mathcal{P}^q)}{\mathcal{R}(i,\mathcal{S}_\Omega^q)}, \tag{5}$$

where $\mathcal{R}(i,\mathcal{S})$ is a function that returns the ranking of candidate $i \in \mathcal{S}$ in the candidate set:

$$\mathcal{R}(i,\mathcal{S}) = 1 + \sum_{j \in \mathcal{S}, i \neq j} \mathbb{1}\{s_i - s_j < 0\}. \tag{6}$$

Let us introduce the $M \times M$ matrix D, where $D_{ij} = s_i - s_j$. By using the matrix D, Eq. 5 can be written as:

$$AP_\mathbf{q} = \frac{1}{|\mathcal{S}_\mathcal{P}^q|} \sum_{i \in \mathcal{S}_\mathcal{P}^q} \frac{1 + \sum_{j \in \mathcal{S}_\mathcal{P}, j \neq i} \mathbb{1}\{D_{ij} > 0\}}{1 + \sum_{j \in \mathcal{S}_\mathcal{P}^q, j \neq i} \mathbb{1}\{D_{ij} > 0\} + \sum_{j \in \mathcal{S}_\mathcal{N}^q} \mathbb{1}\{D_{ij} > 0\}}.$$

The indicator function $\mathbb{1}\{\cdot\}$ is non-differentiable. To overcome this problem, the indicator function can be replaced by a sigmoid function:

$$\mathcal{G}(x;\tau) = \frac{1}{1 + e^{\frac{-x}{\tau}}}. \tag{7}$$

By replacing the indicator function $\mathbb{1}\{\cdot\}$ by $\mathcal{G}$, the Average Precision metric can be approximated with a smooth function:

$$AP_\mathbf{q} \approx \frac{1}{|\mathcal{S}_\mathcal{P}^q|} \sum_{i \in \mathcal{S}_\mathcal{P}^q} \frac{1 + \sum_{j \in \mathcal{S}_\mathcal{P}^q, j \neq i} \mathcal{G}(D_{ij};\tau)}{1 + \sum_{j \in \mathcal{S}_\mathcal{P}^q, j \neq i} \mathcal{G}(D_{ij};\tau) + \sum_{j \in \mathcal{S}_\mathcal{N}^q} \mathcal{G}(D_{ij};\tau)}.$$

This loss function is called *SmoothAP* and has been introduced in the context of image retrieval [1], following similar proposals in document retrieval and learning to rank [2,3,33,39]. The total loss over a batch $\mathcal{B}$ can then be formulated as follows:

$$\mathcal{L}_{AP} = \frac{1}{|\mathcal{B}|} \sum_{\mathbf{q} \in \mathcal{B}} (1 - AP_\mathbf{q}). \tag{8}$$

In the online appendix (see footnote 1), we provide an extended explanation of SmoothAP.

3 Do Findings from Metric Learning Extend to ICR?

In representation learning it was found that NT-Xent loss outperforms the Triplet loss and Triplet loss SH [7]. For both the image retrieval and representation learning task, results show that SmoothAP outperforms both the Triplet loss SH and the NT-Xent loss [1,36]. We examine whether these findings generalize to ICR.

Experimental Setup. We focus on two benchmark datasets for the ICR task: the Flickr30k [40] and MS-COCO [26] datasets. Similar to [12,24], we use the split provided by Karpathy and Fei-Fei [17] for MS-COCO and the Flickr30k. For details of the specific implementations of VSE++ [12][2] and [24][3] we refer to the papers and online implementations. Each method is trained for 30 epochs with a batch size of 128. We start with a learning rate of 0.0002 and after 15 epochs we lower the learning rate to 0.00002.

For VSE++, we do not apply additional fine-tuning of the image encoder after 30 epochs. Our main goal is to have a fair comparison across methods, datasets, and loss functions, not to have the highest overall evaluation scores. For VSE++, we use ResNet50 [15] as image-encoder instead of ResNet152 [15] or VGG [35]. ResNet50 is faster to optimize and the performance differences between ResNet50 and ResNet152 are relatively small.

The VSRN method comes with an additional caption decoder, to decode the original input caption from the latent image representation, this to add additional supervision to the optimization process. We remove the additional image-captioning module, so as to exclude performance gains on the retrieval tasks due to this extra supervision. In [24], the similarity score for a query candidate pair, during evaluation, is based on averaging the predicted similarity scores of (an ensemble of) two trained models. We only take the predicted relevance score of one model. The reason for this is that the evaluation score improvements are marginal when using the scores of two models (instead of one) but optimizing the methods takes twice as long. Therefore, our results are lower than the results published in [24]. For all the remaining details, we refer to our repository.[4] When optimizing with SmoothAP, we take all the k captions into account when sampling a batch, instead of one positive candidate. For this reason, we have to increase the amount of training epochs k times as well to have a fair comparison. For each loss function, we select the best performing hyper-parameter according to its original work.

Experiments. We evaluate each loss function we described in Sect. 2 given a dataset and method. For ease of reference, we refer to each individual evaluation with an experiment number (#) (see Table 1). To reduce the variance in the results we run each experiment five times and report the average score and standard deviation. Similar to [12,24], we evaluate using Recall@k with $k = \{1, 5, 10\}$, for both the image-to text (i2t) and text-to-image (t2i) task. We also report the sum of all the recall scores (rsum) and the average recall value. For the i2t task, we also report the mean average precision at 5 (mAP@5) due to the fact we have k positive captions per image query.

Results. Based on the scores reported in Table 1, we have the following observations:

---

[2] https://github.com/fartashf/vsepp.
[3] https://github.com/KunpengLi1994/VSRN.
[4] https://github.com/MauritsBleeker/ecir-2022-reproducibility-bleeker.

Table 1. Evaluation scores for the Flickr30k and MS-COCO, for the VSE++ and VSRN.

					i2t				t2i				
Loss function	#	hyper param	R@1	R@5	R@10	average recall	mAP@5	R@1	R@5	R@10	average recall	rsum
					Flickr30k								
					VSE++								
Triplet loss	1.1	$\alpha = 0.2$	30.8 ± .7	62.6 ± .3	74.1 ± .8	55.9 ± .3	0.41 ± .00	23.4 ± .3	52.8 ± .1	65.7 ± .3	47.3 ± .1	309.4 ± 0.9
Triplet loss SH	1.2	$\alpha = 0.2$	**42.4 ± .5**	**71.2 ± .7**	**80.7 ± .7**	64.8 ± .6	0.50 ± .01	**30.0 ± .3**	**59.0 ± .2**	**70.4 ± .4**	53.1 ± .2	**353.8 ± 1.6**
NT-Xent	1.3	$\tau = 0.1$	37.5 ± .6	68.4 ± .6	77.8 ± .5	61.2 ± .3	0.47 ± .00	27.0 ± .3	57.3 ± .3	69.1 ± .2	51.1 ± .2	337.1 ± 1.3
SmoothAP	1.4	$\tau = 0.01$	42.1 ± .8	**70.8 ± .6**	**80.6 ± .8**	64.5 ± .4	0.50 ± .00	29.1 ± .3	58.1 ± .1	69.7 ± .2	52.3 ± .2	350.4 ± 1.7
					VSRN								
Triplet loss	1.5	$\alpha = 0.2$	56.4 ± .7	83.6 ± .6	90.1 ± .2	76.7 ± .5	0.63 ± .01	43.1 ± .3	74.4 ± .3	83.1 ± .4	66.9 ± .3	430.7 ± 1.8
Triplet loss SH	1.6	$\alpha = 0.2$	**68.3 ± 1.3**	**89.6 ± .7**	**94.0 ± .5**	84.0 ± .5	0.73 ± .00	**51.2 ± .9**	**78.0 ± .6**	**85.6 ± .5**	71.6 ± .6	**466.6 ± 3.3**
NT-Xent	1.7	$\tau = 0.1$	50.9 ± .5	78.9 ± .7	86.6 ± .4	72.2 ± .4	0.59 ± .00	40.6 ± .6	71.9 ± .2	81.7 ± .3	64.7 ± .2	410.6 ± 1.5
SmoothAP	1.8	$\tau = 0.01$	63.1 ± 1.0	86.6 ± .8	92.4 ± .5	80.7 ± .7	0.69 ± .00	45.8 ± .2	73.7 ± .3	82.3 ± .2	67.3 ± .1	444.0 ± 2.1
					MS-COCO								
					VSE++								
Triplet loss	2.1	$\alpha = 0.2$	22.1 ± .5	48.2 ± .3	61.7 ± .3	44.0 ± .3	0.30 ± .00	15.4 ± .1	39.5 ± .1	53.2 ± .1	36.0 ± .1	240.0 ± 0.9
Triplet loss SH	2.2	$\alpha = 0.2$	**32.5 ± .2**	**61.6 ± .3**	**73.8 ± .3**	56.0 ± .2	0.41 ± .00	**21.3 ± .1**	**48.1 ± .1**	**61.5 ± .0**	43.6 ± .1	**298.8 ± 0.8**
NT-Xent	2.3	$\tau = 0.1$	25.8 ± ± .5	53.6 ± .5	66.1 ± ± .2	48.5 ± .3	0.34 ± .00	18.0 ± .1	43.0 ± .1	56.6 ± .2	39.2 ± .1	263.0 ± 0.9
SmoothAP	2.4	$\tau = 0.01$	30.8 ± .3	60.3 ± .2	**73.6 ± .5**	54.9 ± .3	0.40 ± .00	20.3 ± .2	46.5 ± .2	60.1 ± .2	42.3 ± .2	291.5 ± 1.4
					VSRN								
Triplet loss	2.5	$\alpha = 0.2$	42.9 ± .4	74.3 ± .3	84.9 ± .4	67.4 ± .3	0.52 ± .00	33.5 ± .1	65.1 ± .1	77.1 ± .2	58.6 ± .1	377.8 ± 1.2
Triplet loss SH	2.6	$\alpha = 0.2$	**48.9 ± .6**	**78.1 ± ± .5**	**87.4 ± ± .2**	71.4 ± .4	0.57 ± .01	**37.8 ± .5**	**68.1 ± .5**	**78.9 ± .3**	61.6v.4	**399.0 ± 2.3**
NT-Xent	2.7	$\tau = 0.1$	37.9 ± .4	69.2 ± .2	80.7 ± .3	62.6 ± .1	0.47 ± .00	29.5 ± .1	61.0 ± .2	74.0 ± .2	54.6 ± .1	352.3 ± 0.5
SmoothAP	2.8	$\tau = 0.01$	46.0 ± .6	76.1 ± .3	85.9 ± .3	69.4 ± .3	0.54 ± .00	33.8 ± .2	64.1 ± .1	76.0 ± .2	58.0 ± .2	382.0 ± 1.1

(1) Given a fixed method and default hyper-parameters for each loss function, the Triplet loss SH results in the best evaluation scores, regardless of dataset, method or task.

(2) Similar to [12], we find that the Triplet loss SH consistently outperforms the general Triplet loss, which takes all the negative triplets in the batch into account that violate the margin constraint.

(3) The NT-Xent loss consistently underperforms compared to the Triplet loss SH. This is in contrast with findings in [7], where the NT-Xent loss results in better down-stream evaluation performance on a (augmented image-to-image) representation learning task than the Triplet loss SH. Although the ICR task has different (input) data modalities, the underlying learning object is the same for ICR and augmented image-to-image representation learning (i.e., contrasting positive and negative pairs).

(4) Only for the VSE++ method on the i2t task, SmoothAP performs similar to the Triplet-loss SH.

(5) SmoothAP does not outperform the Triplet loss SH. This is in contrast with the findings in [1], where SmoothAP does outperform Triplet-loss SH and other metric learning functions.

(6) The method with the best Recall@k score also has the highest mAP@k score.

Upshot. Based on our observations concerning Table 1, we conclude the following: (1) The Triplet loss SH should still be the *de facto* choice for optimizing ICR methods. (2) The promising results from the representation learning field that were obtained by using the NT-Xent loss [7], do not generalize to the ICR task. (3) Optimizing an ICR method with a smooth approximation of a ranking metric (SmoothAP) does not result in better Recall@k scores. (4) Optimizing an ICR method by using a pair-wise distance loss between the positive triplet and a semi-hard negative triplet still yields the best evaluation performance. For both methods VSE++ and VSRN, i2t and t2i and for both datasets.

4 A Method for Analyzing the Behavior of Loss Functions

Next, we propose a method for analyzing the behavior of loss functions for ICR. The purpose is to compare loss functions, and explain the difference in performance. If we compare the gradient w.r.t. $\mathbf{q}$ for the Triplet loss and the Triplet loss SH, the only difference is the number of triplets that the two loss functions take into account. If two models are optimized in exactly the same manner, except one model uses the Triplet loss and the other uses Triplet loss SH, the difference in performance can only be explained by the fact that the Triplet loss takes all violating triplets into account. This means that the number of triplets (i.e., candidates) that contribute to the gradient directly relates to the evaluation performance of the model. The same reasoning applies for the NT-Xent and the SmoothAP loss. For example, the gradient w.r.t. $\mathbf{q}$ for the NT-Xent loss also has the form $\mathbf{v}^+ - \mathbf{v}^-$. The major difference between the two functions is that, for the negative candidate the NT-Xent loss computes a weighted sum over all negative to compute a representation of $\mathbf{v}^-$. Therefore, the difference in evaluation performance between the Triplet loss SH and NT-Xent can only be explained by this weighted sum over all negatives. This sum can be turned into a count of negatives, i.e., how many negative approximately contribute to this weighted sum, which can be related to the other losses. By counting the number of candidates that contribute to the gradient, we aim to get a better understanding of why a certain loss function performs better than others. The method we propose is called *counting contributing samples* (COCOS).

First, we provide the form of the derivative of each loss function w.r.t. query $\mathbf{q}$. For each loss function the derivative is a sum over $\mathbf{v}^+ - \mathbf{v}^-$. Loss functions may weight the positive and negative candidate(s) differently, and the number of candidates or triplets that are weighted may differ across loss functions.

Triplet Loss and Triplet Loss SH . The gradient w.r.t. $\mathbf{q}$ for the Triplet loss SH, $\mathcal{L}_{TripletSH}^{\mathbf{q}}$ is the difference between the representation of the positive and negative candidate:

$$\frac{\partial \mathcal{L}_{TripletSH}^{\mathbf{q}}}{\partial \mathbf{q}} = \begin{cases} \mathbf{v}^+ - \mathbf{v}^-, & \text{if } s^+ - s^- < \alpha \\ 0, & \text{otherwise.} \end{cases} \tag{9a}$$

$$\frac{\partial \mathcal{L}_{Triplet}^{\mathbf{q}}}{\partial \mathbf{q}} = \sum_{\mathbf{v}^- \in \mathcal{N}_{\mathbf{q}}} \mathbb{1}\{s^+ - s^- < \alpha\} \left(\mathbf{v}^+ - \mathbf{v}^-\right). \tag{9b}$$

The gradient of Triplet loss $\mathcal{L}_{Triplet}^{\mathbf{q}}$ (Eq. 9b) w.r.t. $\mathbf{q}$ has a similar form. However, there the gradient is a sum over all triplets that violate $s^+ + s^- < \alpha$, and not only the maximum violating one. Based on Eq. 9a we can see that a query $\mathbf{q}$ only has a non-zero gradient when $s^+ - s^- < \alpha$. If this is the case, the gradient always has the form $\mathbf{v}^+ - \mathbf{v}^-$, and this value is independent of the magnitude

$s^+ - s^-$. For this reason, given a batch $\mathcal{B}$, the number of queries $\mathbf{q}$ that have a non-zero gradient is defined by:

$$C_{TripletSH}^{\mathcal{B}} = \sum_{\mathbf{q} \in \mathcal{B}} \mathbb{1}\{s^+ - s^- < \alpha\}, \tag{10}$$

where $s^+ = s^0 \in \mathcal{S}_{\mathcal{P}}^{\mathbf{q}}$ and $s^- = \max(\mathcal{S}_{\mathcal{N}}^{\mathbf{q}})$. We define $C_{TripletSH}^{\mathcal{B}}$ to be *the number of queries $\mathbf{q}$ that have a non-zero gradient given batch $\mathcal{B}$.*

As the Triplet loss takes all the triplets into account that violate the distance margin α, we can count three things: (1) Per query $\mathbf{q}$, we can count how many triplets $\mathbf{v}^+ - \mathbf{v}^-$ contribute to the gradient of $\mathbf{q}$. We define this as $C_{Triplet}^{\mathbf{q}} = \sum_{s^- \in \mathcal{S}_{\mathcal{N}}^{\mathbf{q}}} \mathbb{1}\{s^+ - s^- < \alpha\}$. (2) Given the batch $\mathcal{B}$, we can count how many triplets contribute to the gradient over the entire training batch $\mathcal{B}$. We define this number as $C_{Triplet}^{\mathcal{B}} = \sum_{\mathbf{q} \in \mathcal{B}} C_{Triplet}^{\mathbf{q}}$. (3) Given the entire batch $\mathcal{B}$, we can count how many queries have a gradient value of zero (i.e., no violating triplets). This number is $C_{Triplet}^{0} = \sum_{\mathbf{q} \in \mathcal{B}} \mathbb{1}\{C_{Triplet}^{\mathbf{q}} = 0\}$.

NT-Xent Loss. The gradient w.r.t. $\mathbf{q}$ for the NT-Xent loss is defined as [7]:

$$\frac{\partial \mathcal{L}_{NT-Xent}^{\mathbf{q}}}{\partial \mathbf{q}} = \left(1 - \frac{\exp(s^+/\tau)}{Z(\mathbf{q})}\right)\tau^{-1}\mathbf{v}^+ - \sum_{s^- \in \mathcal{S}_{\mathcal{N}}^{\mathbf{q}}} \left(\frac{\exp(s^-/\tau)}{Z(\mathbf{q})}\right)\tau^{-1}\mathbf{v}^-, \tag{11}$$

where $Z(\mathbf{q}) = \sum_{s_i \in \mathcal{S}_{\Omega}^{\mathbf{q}}} \exp(s_i/\tau)$, a normalization constant depending on $\mathbf{q}$. The gradient w.r.t. $\mathbf{q}$ is the weighted difference of the positive candidate $\mathbf{v}^+$ and the weighted sum over all the negative candidates. The weight for each candidate is based on the similarity with the query, normalized by the sum of the similarities of all candidates. In contrast, for the Triplet-loss (Eq. 9b) all candidates are weighted equally when they violate the margin constraint. The NT-Xent loss performs a natural form of (hard) negative weighting [7]. The more similar a negative sample is to the query, the higher the weight of this negative in the gradient computation. In principle, all the negatives and the positive candidate contribute to the gradient w.r.t. $\mathbf{q}$. In practice, most similarity scores $s^- \in \mathcal{S}_{\mathcal{N}}^{\mathbf{q}}$ have a low value; so the weight of this negative candidate in the gradient computation will be close to 0.

To count the number of negative candidates that contribute to the gradient, we define a threshold value ϵ. If the weight of a negative candidate $\mathbf{v}^-$ is below ϵ, we assume that its contribution is negligible. All candidate vectors are normalized. Hence, there is no additional weighting effect by the magnitude of the vector. For the NT-Xent loss we define three terms: $C_{NTXent}^{\mathbf{qv}^-}$, $W_{NTXent}^{\mathbf{qv}^-}$ and $W_{NT-Xent}^{\mathbf{qv}^+}$: (1) Given a query $\mathbf{q}$, $C_{NT-Xent}^{\mathbf{qv}^-}$ is the number of negative candidates $\mathbf{v}^-$ that contribute to the gradient w.r.t. $\mathbf{q}$: $C_{NT-Xent}^{\mathbf{qv}^-} = \sum_{s^- \in \mathcal{S}_{\mathcal{N}}^{\mathbf{q}}} \mathbb{1}\{\exp(s^-/\tau)Z(\mathbf{q})^{-1} > \epsilon\}$. (2) Given $C_{NT-Xent}^{\mathbf{qv}^-}$, we compute the sum of the weight values of the contributing negative candidates $\mathbf{v}^-$ as $W_{NT-Xent}^{\mathbf{qv}^-} = \sum_{s^- \in \mathcal{S}_{\mathcal{N}}^{\mathbf{q}}} \mathbb{1}\{\exp(s^-/\tau)Z(\mathbf{q})^{-1} > \epsilon\}\exp(s^-/\tau)Z(\mathbf{q})^{-1}$. (3) We define $W_{NT-Xent}^{\mathbf{qv}^+} = \frac{1}{N}\sum_{\mathbf{q} \in \mathcal{B}}(1 - \exp(s^+/\tau)Z(\mathbf{q})^{-1})$, as the mean weight value of the positive candidates in batch $\mathcal{B}$.

We define the two extra terms, $W_{NT-Xent}^{\mathbf{qv}^-}$ and $W_{NT-Xent}^{\mathbf{qv}^+}$ because for the NT-Xent function we have to count the candidates with a weight value above the threshold ϵ. This count on its own does not provide a good picture of the contribution of these candidates to the gradient. Therefore, we compute a mean value of those weight values as well, to provide insight into the number of the samples on which the gradient w.r.t. $\mathbf{q}$ is based.

SmoothAP Loss. A full derivation of the gradient of SmoothAP w.r.t. $\mathbf{q}$ is provided with the implementation of our methods (see footnote 1). We introduce $sim(D_{ij})$, the derivative of (7):

$$
\frac{\partial AP_{\mathbf{q}}}{\partial \mathbf{q}} = \frac{1}{|\mathcal{S}_{\mathcal{P}}^{\mathbf{q}}|} \sum_{i \in \mathcal{S}_{\mathcal{P}}^{\mathbf{q}}} \mathcal{R}(i, \mathcal{S}_{\Omega}^{\mathbf{q}})^{-2} \left(\mathcal{R}(i, \mathcal{S}_{\mathcal{P}}^{\mathbf{q}}) \left(\sum_{j \in \mathcal{S}_{\mathcal{N}}^{\mathbf{q}}} sim(D_{ij})(\mathbf{v}_i - \mathbf{v}_j) \right) \right)
$$
$$
- (\mathcal{R}(i, \mathcal{S}_{\mathcal{N}}^{\mathbf{q}}) - 1) \left(\sum_{j \in \mathcal{S}_{\mathcal{P}}^{\mathbf{q}}, j \neq i} sim(D_{ij})(\mathbf{v}_i - \mathbf{v}_j) \right) \right). \tag{12}
$$

Given Eq. 12, it is less trivial to infer what the update w.r.t. $\mathbf{q}$ looks like in terms of positive candidates $\mathbf{v}_i$ and negative candidates $\mathbf{v}_j$. However, we can derive the following two properties: (1) The lower a positive candidate $\mathbf{v}_i$ is in the total ranking, the less this candidate is taken into account for the gradient computation w.r.t. $\mathbf{q}$, due the inverse quadratic term $\mathcal{R}(i, \mathcal{S}_{\Omega}^{\mathbf{q}})^{-2}$. This is in line with optimizing the AP as a metric; positive candidates that are ranked low contribute less to the total AP score, and therefore are less important to optimize. (2) Each triplet $\mathbf{v}_i - \mathbf{v}_j$ is weighted according to their difference in similarity score D_{ij}. If their difference in similarity score w.r.t. query $\mathbf{q}$ is relatively small (i.e., D_{ij} is close to zero), $sim(D_{ij})$ will have a high value due to the fact that $sim(D_{ij})$ is the derivative of the sigmoid function. Therefore, $sim(D_{ij})$ indicates how close the similarity score (with the query) of candidate $\mathbf{v}_i$ is compared to the similarity score of $\mathbf{v}_j$ This is in line with the SmoothAP loss because we use a sigmoid to approximate the step-function; only triplets of candidates that have a similar similarity score will contribute to the gradient.

Table 2. COCOS w.r.t. query $\mathbf{q}$, for the Triplet loss and the Triplet loss SH.

			#	i2t			t2i		
				$C^{\mathbf{q}}$	$C^{\mathcal{B}}$	C^0	$C^{\mathbf{q}}$	$C^{\mathcal{B}}$	C^0
Flickr30k	VSE++	Triplet loss	1.1	6.79 ± 0.83	768.92 ± 96.87	14.78 ± 3.52	6.11 ± 0.75	774.67 ± 98.05	1.14 ± 1.22
		Triplet loss SH	1.2	1 ± 0.0	98.74 ± 4.83	29.23 ± 4.81	1 ± 0.0	98.22 ± 4.66	29.75 ± 4.62
	VSRN	Triplet loss	1.5	1.39 ± 0.12	60.96 ± 10.30	84.29 ± 5.80	1.28 ± 0.10	61.21 ± 10.01	80.15 ± 6.35
		Triplet loss SH	1.6	1 ± 0.0	45.59 ± 5.93	82.39 ± 5.92	1 ± 0.0	44.98 ± 5.70	82.99 ± 5.70
MS-COCO	VSE++	Triplet loss	2.1	3.51 ± 0.49	353.82 ± 52.71	27.09 ± 4.60	2.94 ± 0.36	341.64 ± 50.80	12.24 ± 4.92
		Triplet loss SH	2.2	1 ± 0.0	88.17 ± 5.25	39.82 ± 5.24	1 ± 0.0	87.24 ± 5.34	40.75 ± 5.33
	VSRN	Triplet loss	2.5	1.21 ± 0.13	29.88 ± 7.46	103.33 ± 5.22	1.15 ± 0.10	30.25 ± 7.49	101.70 ± 5.58
		Triplet loss SH	2.6	1 ± 0.0	33.24 ± 5.39	94.73 ± 5.45	1 ± 0.0	32.90 ± 5.35	95.08 ± 5.4

We define a threshold value ϵ again. If the value of $sim(D_{ij})$ is lower than the threshold value, we consider the contribution of this triplet to be negligible. We have to take into account that all triplets are also weighted by $\mathcal{R}(i, S_\Omega^q)^{-2}$, which is always lower than or equal to 1. We can define C_{Smooth}^q, which *is the number of triplets* $\mathbf{v}^+ - \mathbf{v}^-$ *that contribute to the gradient w.r.t.* $\mathbf{q}$, for SmoothAP as follows:

$$C_{Smooth}^q = \frac{1}{|S_P^q|} \sum_{i \in S_P^q} \left(\sum_{j \in S_N^q} \mathbb{1}\left\{ \frac{sim(D_{ij})}{\mathcal{R}(i,S_\Omega^q)^2} > \epsilon \right\} + \sum_{j \in S_P^q, j \neq i} \mathbb{1}\left\{ \frac{sim(D_{ij})}{\mathcal{R}(i,S_\Omega^q)^2} > \epsilon \right\} \right).$$

(13)

Similar to [1], we use $sim(D_{ij})$ in combination with a threshold value ϵ to indicate which samples have a non-zero gradient in the training batch. We ignore the terms $\mathcal{R}(i, S_P^q)$ and $1 - \mathcal{R}(i, S_N^q)$ for this gradient computation. We also count all queries $\mathbf{q}$ within batch $\mathcal{B}$ that do not have a gradient value. We define this number as $C_{Smooth}^0 = \sum_{q \in \mathcal{B}} \mathbb{1}\{C_{Smooth}^q = 0\}$. This completes the definition of COCOS: for every loss function that we consider, it counts the number of candidates that contribute to the gradient w.r.t. $\mathbf{q}$.

5 Analyzing the Behavior of Loss Functions for ICR

Experimental Setup. To use COCOS, we introduce the following experimental setup. For each loss function, we take the checkpoint of one of the five optimized models. We refer to this checkpoint as *the optimal convergence point* for this loss function. This is not the point with the lowest loss value, but the model checkpoint that results in the highest evaluation scores on the validation set. We freeze all model parameters and do not apply dropout. We iterate over the entire training set by sampling random batches $\mathcal{B}$ (with batch size $|\mathcal{B}| = 128$, similar to the training set-up). For each batch we compute the COCOS and weight values defined in Sect. 4. We report the mean value and standard deviation over the entire training set for both VSE++ and VSRN, for both datasets and for each loss function. The only hyper-parameter for this experiment is ϵ. We use $\epsilon = 0.01$ for both the NT-Xent and SmoothAP loss.

Experimental Outcomes. For each of the loss functions that we consider, we analyze its performance using COCOS.

Table 3. COCOS w.r.t. query $\mathbf{q}$, for the NT-Xent loss [7].

			i2t			t2i	
		#	$C_{NT-Xent}^{qv^-}$	$W_{NT-Xent}^{qv^-}$	$W_{NT-Xent}^{qv^+}$	$C_{NT-Xent}^{qv^-}$	$W_{NT-Xent}^{qv^-}$	$W_{NT-Xent}^{qv^+}$
Flickr30k	VSE++	1.3	9.88 ± 0.51	0.42 ± 0.02	0.56 ± 0.02	9.65 ± 0.51	0.42 ± 0.02	0.56 ± 0.02
	VSRN	1.7	2.45 ± 0.23	0.13 ± 0.02	0.20 ± 0.02	2.46 ± 0.23	0.13 ± 0.02	0.20 ± 0.02
MS-COCO	VSE++	2.3	5.59 ± 0.40	0.36 ± 0.02	0.46 ± 0.02	5.33 ± 0.38	0.36 ± 0.02	0.46 ± 0.02
	VSRN	2.7	1.10 ± 0.14	0.10 ± 0.02	0.14 ± 0.02	1.11 ± 0.14	0.09 ± 0.02	0.14 ± 0.02

Triplet Loss. Our goal is not to show that the Triplet loss SH outperforms the Triplet loss, which has already been shown [12], but to explain this behavior based on COCOS w.r.t. $\mathbf{q}$ and also relate this to the NT-Xent and SmoothAP loss.

Based on Table 1 (row 1.1/1.2 and 1.5/1.6, row 2.1/2.2 and 2.5/2.6) it is clear that the Triplet loss SH always outperforms the general Triplet loss with a large margin. If we look at Table 2, row 1.1/1.2 and 2.1/2.2, respectively, there is a clear relation between $\mathcal{C}^q$ and the final evaluation score for the VSE++ model for both sub-tasks i2t and t2i (Table 1). $\mathcal{C}^q_{Triplet}$ and $\mathcal{C}^{\mathcal{B}}_{Triplet}$ are both much greater than $\mathcal{C}^q_{TripletSH}$ and $\mathcal{C}^{\mathcal{B}}_{TripletSH}$, for both dataset and both the i2t and t2i task. When multiple negatives with small margin violation are combined into a gradient, the gradient is dominated by easy or non-informative negative samples, which results in convergence of the model into a sub-optimal point [12]. Clearly, the loss function with the lowest evaluation score takes into account the most negatives when computing the gradient w.r.t. $\mathbf{q}$. Based on [12] and the COCOS results in Table 2 we conclude that, at the optimal convergence point, the Triplet loss takes too many negatives into account (i.e., too many triplets still violate the margin constraint), leading to lower evaluation scores.

For VSRN the relation between $\mathcal{C}^q_{Triplet}$, $\mathcal{C}^q_{TripletSH}$ and the final evaluation score is less clear. If we look at Table 2, row 1.5/1.6 and 2.5/2.6, respectively, we see that $\mathcal{C}^q_{Triplet} \approx \mathcal{C}^q_{TripletSH} = 1$. This means that at the optimal convergence point, for VSRN, the Triplet loss and the Triplet loss SH (approximately) are a similar to each other and both functions only take one negative triplet into account when computing the gradient w.r.t. $\mathbf{q}$. Thus, both functions should result in approximately the same gradient value while the Triplet loss SH still outperforms the Triplet loss with a large margin. This can be explained as follows: At the start of training, for each query $\mathbf{q}$ (almost) all triplets violate the margin constraint (because all candidate representations are random). Therefore, the gradient(s) computation w.r.t. $\mathbf{q}$ for the Triplet loss is based on all triplets in the batch and therefore this gradient is dominated by a majority of non-informative samples in the beginning of the training, which leads to convergence at a sub-optimal point.

NT-Xent. Based on Table 3, we can see that $\mathcal{C}^{qv^-}_{NT-Xent}$ is higher than 1 for both VSE++ and VSRN, for i2t and t2i, on both datasets. If we relate the evaluation performances of the NT-Xent loss (row 1.3, 1.7, 2.3, 2.7) to the Triplet loss SH (row 1.2, 1.6, 2.2, 2.6) in Table 1, we can see that the Triplet loss SH consistently outperforms the NT-Xent loss, regardless of the method, dataset or sub-task. We therefore can conclude that taking only the most violating negative into account when computing the gradient w.r.t. $\mathbf{q}$ results in better evaluation performances than computing a weighted sum over all negative candidates. We can apply the same reasoning used to explain the performance difference between the Triplet loss and Triplet loss SH. The gradient w.r.t. $\mathbf{q}$ for the NT-Xent is dominated by too many non-informative negatives, which have a weight value bigger than ϵ.

Looking at Table 1, we see that NT-Xent loss outperforms the Triplet loss for the VSE++ method (1.3/1.1 and 2.3/2.1), while taking more negative samples into account when computing the gradient (based on our definition of COCOS). This in contrast with the previous observation for the Triplet loss of the more (non-informative) samples a loss function takes into account when computing the gradient w.r.t. $\mathbf{q}$, the lower the evaluation score. Solely counting the number

Table 4. COCOS w.r.t. query **q**, for the SmoothAP [1] loss.

			i2t		t2i	
		#	$C^q_{SmoothAP}$	$C^0_{SmoothAP}$	$C^q_{SmoothAP}$	$C^0_{SmoothAP}$
Flickr30k	VSE++	1.4	1.27 ± 0.06	2.15 ± 1.51	1.47 ± 0.83	636.72 ± 18.72
	VSRN	1.8	2.33 ± 0.07	0.00 ± 0.00	1.62 ± 0.95	636.49 ± 18.65
MS-COCO	VSE++	2.4	1.48 ± 0.07	0.80 ± 0.90	1.41 ± 0.74	637.10 ± 20.28
	VSRN	2.8	1.67 ± 0.07	0.14 ± 0.37	1.42 ± 0.76	637.23 ± 20.35

of negative examples that contribute to the gradient does not the provide the full picture for the NT-Xent loss; the weight value of each individual sample (including the positive) plays a more important role than initially was assumed. We have tried different values for ϵ, with little impact.

SmoothAP. The observations in Table 4 are in line with the observations in Table 2 and the evaluation performance in Table 1. At the optimal convergence point SmoothAP takes approximately one triplet into account when computing the gradient w.r.t. **q**, which results in close-to or similar performances as the Triplet loss SH. We also observe the following: the only experiment where the Triplet loss SH outperforms SmoothAP with a large margin (Table 1, row 1.5 and 1.8), is also the experiment where the SmoothAP function takes the highest number of negatives into account when computing the gradient w.r.t. **q** (Table 4, row 1.8). This supports the general observation that the more samples that contribute to the gradient, the lower the final evaluation score.

For the t2i task, we also see that $C^0_{SmoothAP}$ is almost as big as the number of samples ($640 = (k = 5) \times (|\mathcal{B}| = 128)$) in the candidate set, for both datasets and methods. Hence, barely any query has a gradient value anymore at the optimal convergence point. However, this is not the case for the i2t task. We conclude that optimizing a ranking metric (i.e., AP) with only one positive candidate (as is the case for the t2i task), might be too easy to optimize and could result in over-fitting. Therefore, it is not useful to optimize a ranking task like ICR with a ranking-based loss function when there is only one positive candidate per query, which is the case for the i2t task. For the i2t task, however, there are barely any queries without a gradient value; here we have k positive candidates per query.

Upshot. In summary, (1) it is important to focus on only one (or a limited) number of (hard) negatives per query during the entire training for the gradient computation, so as to prevent the gradient from being dominated by non-informative or easy negative samples. (2) Weighting each negative candidate by its score (as is done in NT-Xent) as opposed to weighting all negative equally (as is done in the Triplet loss) can beneficial for the gradient computation and therefore for the final evaluation score. However, this weighted sum of negatives does not result in the fact that the NT-Xent loss outperforms the Triplet loss SH, which implies that the gradient computation for the NT-Xent is still based on too many non-informative samples.

6   Conclusion

We have examined three loss functions from the metric learning field to question if the promising results obtained in metric learning generalize to the image-caption retrieval (ICR) task. In contrast with the findings from metric learning, we find that the Triplet loss with semi-hard negative mining still outperforms the NT-Xent and SmoothAP loss. Hence, the Triplet loss should still be the de facto choice as a loss function for ICR; results from metric learning do not generalize directly to ICR. To gain a better understanding of why a loss function results in better performance than others, we have introduced the notion of counting contributing samples (COCOS). We have shown that the best performing loss function only focuses on one (hard) negative sample when computing the gradient w.r.t. the query and therefore results in the most informative gradient. COCOS suggests that the underperforming loss functions take too many (non-informative) negatives into account, and therefore converge to a sub-optimal point.

The definition of COCOS uses a threshold value. The idea that a candidate contributes to the gradient if its weight value is above a certain threshold is insightful but does not provide the complete picture of how strong the influence of this sample is. We encourage two directions for future work: (1) Work on more sophisticated methods to determine the influence of (the number of) samples on the gradient w.r.t. a query. (2) Design new loss functions for the ICR task by taking the lessons from COCOS into account, i.e., loss functions that only take one, or a limited number of, hard negative(s) into account. Additionally, we want to investigate if our findings generalize to fields such as Dense Passage Retrieval (DPR) [18]. DPR methods are also mainly optimized by using two data encoders [18,19], for the query and for documents, and the main learning objective is contrasting positive and negative candidates with a query [8,13,14,18,19,42], similar to ICR.

Acknowledgements. We thank Gabriel Benedict, Mariya Hendriksen, Maria Heuss, Sarah Ibrahimi, and Ana Lucic for feedback and discussions. This research was supported by the Nationale Politie and the Hybrid Intelligence Center through the Netherlands Organisation for Scientific Research. All content represents the opinion of the authors, which is not necessarily shared or endorsed by their respective employers and/or sponsors.

References

1. Brown, A., Xie, W., Kalogeiton, V., Zisserman, A.: Smooth-AP: smoothing the path towards large-scale image retrieval. In: Vedaldi, A., Bischof, H., Brox, T., Frahm, J.-M. (eds.) ECCV 2020. LNCS, vol. 12354, pp. 677–694. Springer, Cham (2020). https://doi.org/10.1007/978-3-030-58545-7_39
2. Bruch, S., Wang, X., Bendersky, M., Najork, M.: An analysis of the softmax cross entropy loss for learning-to-rank with binary relevance. In: International Conference on Theory of Information Retrieval (ICTIR), pp. 75–78. ACM (2019)

3. Bruch, S., Zoghi, M., Bendersky, M., Najork, M.: Revisiting approximate metric optimization in the age of deep neural networks. In: SIGIR Conference on Research and Development in Information Retrieval (SIGIR), pp. 1241–1244. ACM (2019)
4. Chen, H., Ding, G., Liu, X., Lin, Z., Liu, J., Han, J.: IMRAM: iterative matching with recurrent attention memory for cross-modal image-text retrieval. In: Proceedings of the IEEE/CVF Conference on Computer Vision and Pattern Recognition (CVPR), pp. 12655–12663 (2020)
5. Chen, T., Li, L.: Intriguing properties of contrastive losses. arXiv preprint arXiv:2011.02803 (2020)
6. Chen, T., Deng, J., Luo, J.: Adaptive offline quintuplet loss for image-text matching. In: Vedaldi, A., Bischof, H., Brox, T., Frahm, J.-M. (eds.) ECCV 2020. LNCS, vol. 12358, pp. 549–565. Springer, Cham (2020). https://doi.org/10.1007/978-3-030-58601-0_33
7. Chen, T., Kornblith, S., Norouzi, M., Hinton, G.: A simple framework for contrastive learning of visual representations. In: International Conference on Machine Learning, PMLR, pp. 1597–1607 (2020)
8. Chen, X., He, B., Hui, K., Sun, L., Sun, Y.: Simplified TinyBERT: knowledge distillation for document retrieval. In: Hiemstra, D., Moens, M.-F., Mothe, J., Perego, R., Potthast, M., Sebastiani, F. (eds.) ECIR 2021. LNCS, vol. 12657, pp. 241–248. Springer, Cham (2021). https://doi.org/10.1007/978-3-030-72240-1_21
9. Chen, Y.C., et al.: UNITER: UNiversal Image-TExt Representation learning. In: Vedaldi, A., Bischof, H., Brox, T., Frahm, J.-M. (eds.) ECCV 2020. LNCS, vol. 12375, pp. 104–120. Springer, Cham (2020). https://doi.org/10.1007/978-3-030-58577-8_7
10. Cho, K., Van Merriënboer, B., Bahdanau, D., Bengio, Y.: On the properties of neural machine translation: encoder-decoder approaches. arXiv preprint arXiv:1409.1259 (2014)
11. Diao, H., Zhang, Y., Ma, L., Lu, H.: Similarity reasoning and filtration for image-text matching. arXiv preprint arXiv:2101.01368 (2021)
12. Faghri, F., Fleet, D.J., Kiros, J.R., Fidler, S.: VSE++: improving visual-semantic embeddings with hard negatives. In: Proceedings of the British Machine Vision Conference (BMVC) (2018)
13. Formal, T., Piwowarski, B., Clinchant, S.: SPLADE: sparse lexical and expansion model for first stage ranking. In: Proceedings of the 44th International ACM SIGIR Conference on Research and Development in Information Retrieval, pp. 2288–2292 (2021)
14. Gao, L., Dai, Z., Callan, J.: Rethink training of BERT rerankers in multi-stage retrieval pipeline. In: Hiemstra, D., Moens, M.-F., Mothe, J., Perego, R., Potthast, M., Sebastiani, F. (eds.) ECIR 2021. LNCS, vol. 12657, pp. 280–286. Springer, Cham (2021). https://doi.org/10.1007/978-3-030-72240-1_26
15. He, K., Zhang, X., Ren, S., Sun, J.: Deep residual learning for image recognition. In: Proceedings of the IEEE Conference on Computer Vision and Pattern Recognition (CVPR), pp. 770–778 (2016)
16. Jia, C., et al.: Scaling up visual and vision-language representation learning with noisy text supervision. arXiv preprint arXiv:2102.05918 (2021)
17. Karpathy, A., Fei-Fei, L.: Deep visual-semantic alignments for generating image descriptions. In: Proceedings of the IEEE Conference on Computer Vision and Pattern Recognition (CVPR), pp. 3128–3137 (2015)
18. Karpukhin, V., et al.: Dense passage retrieval for open-domain question answering. In: Proceedings of the 2020 Conference on Empirical Methods in Natural Language

Processing (EMNLP), pp. 6769–6781. Association for Computational Linguistics (2020)

19. Khattab, O., Zaharia, M.: ColBERT: efficient and effective passage search via contextualized late interaction over BERT. In: Proceedings of the 43rd International ACM SIGIR Conference on Research and Development in Information Retrieval, pp. 39–48 (2020)

20. Kim, W., Son, B., Kim, I.: ViLT: vision-and-language transformer without convolution or region supervision. arXiv preprint arXiv:2102.03334 (2021)

21. Kipf, T.N., Welling, M.: Semi-supervised classification with graph convolutional networks. arXiv preprint arXiv:1609.02907 (2016)

22. Kiros, R., Salakhutdinov, R., Zemel, R.S.: Unifying visual-semantic embeddings with multimodal neural language models. arXiv preprint arXiv:1411.2539 (2014)

23. Lee, K.H., Chen, X., Hua, G., Hu, H., He, X.: Stacked cross attention for image-text matching. In: Proceedings of the European Conference on Computer Vision (ECCV), pp. 201–216 (2018)

24. Li, K., Zhang, Y., Li, K., Li, Y., Fu, Y.: Visual semantic reasoning for image-text matching. In: Proceedings of the IEEE International Conference on Computer Vision (ICCV), pp. 4654–4662 (2019)

25. Li, X., et al.: OSCAR: object-semantics aligned pre-training for vision-language tasks. In: Vedaldi, A., Bischof, H., Brox, T., Frahm, J.-M. (eds.) ECCV 2020. LNCS, vol. 12375, pp. 121–137. Springer, Cham (2020). https://doi.org/10.1007/978-3-030-58577-8_8

26. Lin, T.Y., et al.: Microsoft COCO: common objects in context. In: Fleet, D., Pajdla, T., Schiele, B., Tuytelaars, T. (eds.) ECCV 2014. LNCS, vol. 8693, pp. 740–755. Springer, Cham (2014). https://doi.org/10.1007/978-3-319-10602-1_48

27. Lu, J., Batra, D., Parikh, D., Lee, S.: ViLBERT: pretraining task-agnostic visiolinguistic representations for vision-and-language tasks. arXiv preprint arXiv:1908.02265 (2019)

28. Lu, X., Zhao, T., Lee, K.: VisualSparta: an embarrassingly simple approach to large-scale text-to-image search with weighted bag-of-words. arXiv preprint arXiv:2101.00265 (2021)

29. Messina, N., Amato, G., Esuli, A., Falchi, F., Gennaro, C., Marchand-Maillet, S.: Fine-grained visual textual alignment for cross-modal retrieval using transformer encoders. arXiv preprint arXiv:2008.05231 (2020)

30. Messina, N., Falchi, F., Esuli, A., Amato, G.: Transformer reasoning network for image-text matching and retrieval. In: International Conference on Pattern Recognition (ICPR), pp. 5222–5229. IEEE (2021)

31. Musgrave, K., Belongie, S., Lim, S.-N.: A metric learning reality check. In: Vedaldi, A., Bischof, H., Brox, T., Frahm, J.-M. (eds.) ECCV 2020. LNCS, vol. 12370, pp. 681–699. Springer, Cham (2020). https://doi.org/10.1007/978-3-030-58595-2_41

32. van den Oord, A., Li, Y., Vinyals, O.: Representation learning with contrastive predictive coding. arXiv preprint arXiv:1807.03748 (2018)

33. Oosterhuis, H., de Rijke, M.: Differentiable unbiased online learning to rank. In: International Conference on Information and Knowledge Management (CIKM), pp. 1293–1302. ACM (2018)

34. Radford, A., et al.: Learning transferable visual models from natural language supervision. arXiv preprint arXiv:2103.00020 (2021)

35. Simonyan, K., Zisserman, A.: Very deep convolutional networks for large-scale image recognition. arXiv preprint arXiv:1409.1556 (2014)

36. Varamesh, A., Diba, A., Tuytelaars, T., Van Gool, L.: Self-supervised ranking for representation learning. arXiv preprint arXiv:2010.07258 (2020)

37. Vaswani, A., et al.: Attention is all you need. In: Advances in Neural Information Processing Systems, pp. 5998–6008 (2017)
38. Verma, G., Vinay, V., Bansal, S., Oberoi, S., Sharma, M., Gupta, P.: Using image captions and multitask learning for recommending query reformulations. In: Jose, J.M., et al. (eds.) ECIR 2020. LNCS, vol. 12035, pp. 681–696. Springer, Cham (2020). https://doi.org/10.1007/978-3-030-45439-5_45
39. Wang, X., Li, C., Golbandi, N., Bendersky, M., Najork, M.: The LambdaLoss framework for ranking metric optimization. In: Conference on Information and Knowledge Management (CIKM), pp. 1313–1322. ACM (2018)
40. Young, P., Lai, A., Hodosh, M., Hockenmaier, J.: From image descriptions to visual denotations: new similarity metrics for semantic inference over event descriptions. Trans. Assoc. Comput. Linguist. **2**, 67–78 (2014)
41. Zeng, D., Yu, Y., Oyama, K.: Deep triplet neural networks with cluster-CCA for audio-visual cross-modal retrieval. ACM Trans. Multimedia Comput. Commun. Appl. (TOMM) **16**(3), 1–23 (2020)
42. Zhan, J., Mao, J., Liu, Y., Guo, J., Zhang, M., Ma, S.: Optimizing dense retrieval model training with hard negatives. In: Proceedings of the 44th International ACM SIGIR Conference on Research and Development in Information Retrieval (SIGIR), pp. 1503–1512. Association for Computing Machinery (2021)
43. Zhang, Y., Lu, H.: Deep cross-modal projection learning for image-text matching. In: Proceedings of the European Conference on Computer Vision (ECCV), pp. 686–701 (2018)

Consumer Fairness in Recommender Systems: Contextualizing Definitions and Mitigations

Ludovico Boratto, Gianni Fenu, Mirko Marras[(✉)], and Giacomo Medda

Department of Mathematics and Computer Science,
University of Cagliari, Cagliari, Italy
{ludovico.boratto,mirko.marras}@acm.org, {fenu,giacomo.medda}@unica.it

Abstract. Enabling non-discrimination for end-users of recommender systems by introducing consumer fairness is a key problem, widely studied in both academia and industry. Current research has led to a variety of notions, metrics, and unfairness mitigation procedures. The evaluation of each procedure has been heterogeneous and limited to a mere comparison with models not accounting for fairness. It is hence hard to contextualize the impact of each mitigation procedure w.r.t. the others. In this paper, we conduct a systematic analysis of mitigation procedures against consumer unfairness in rating prediction and top-n recommendation tasks. To this end, we collected 15 procedures proposed in recent top-tier conferences and journals. Only 8 of them could be reproduced. Under a common evaluation protocol, based on two public data sets, we then studied the extent to which recommendation utility and consumer fairness are impacted by these procedures, the interplay between two primary fairness notions based on equity and independence, and the demographic groups harmed by the disparate impact. Our study finally highlights open challenges and future directions in this field. The source code is available at https://github.com/jackmedda/C-Fairness-RecSys.

Keywords: Recommender Systems · Fairness · Bias · Consumers

1 Introduction

Recommender systems help us make decisions, from selecting books to choosing friends [24]. Their wide adoption has spurred investigations into possibly unfair practices in the systems' mechanisms [5,9,11,12,21]. Fairness is a concept of *non-discrimination* on the basis of the membership to *protected groups*, identified by a *protected feature*, e.g., gender and age in anti-discrimination legislation[1]. *Group fairness* avoids the discrimination of a given group, assessed as the absence of a *disparate impact* in the outcomes generated for them [22]. Despite involving different stakeholders (e.g., providers and sellers), fairness in recommender systems may particularly affect those who receive the recommendations (*consumers*) [2].

---

[1] Please refer to Art. 21 of the EU Charter of Fundamental Rights, Art. 14 of European Convention on Human Rights, Art. 18-25 of the Treaty on the Functioning of EU.

© The Author(s), under exclusive license to Springer Nature Switzerland AG 2022
M. Hagen et al. (Eds.): ECIR 2022, LNCS 13185, pp. 552–566, 2022.
https://doi.org/10.1007/978-3-030-99736-6_37

Hence, group consumer fairness should account for no disparate impact of rec-ommendations on protected groups of consumers. Providing guarantees on this property is a key strategic objective for the responsible advancement of the field.

As fairness is an abstract concept, an abundance of consumer fairness notions have been proposed, along with algorithmic procedures for mitigating unfairness in recommendations according to the proposed notions. Examples of mitigation procedures have been applied in *pre-processing* [13], by transforming the input data, *in-processing* [6,15,18,28], by constraining the training process of state-of-the-art models, and *post-processing* [3,19,23], by ranking again the originally recommended items. Moreover, the evaluation protocol adopted to assess their impact has been often *heterogeneous* (e.g., different data sets, train-test splits) and limited to showing that the proposed mitigation is *better than doing nothing*, making the landscape convoluted. To shape recommender systems that account for consumer fairness, we need a common understanding and practical bench-marks on how and when each procedure can be used in comparison to the others. As a response, with this research work, we address *three research questions*:

RQ1 *Is recommendation utility affected by the mitigation procedures?*
RQ2 *Do the selected mitigation procedures reduce the unfairness estimates?*
RQ3 *Is disparate impact systematically harming the minority group?*

To answer these questions, in a first step (Sect. 2), we conducted a system-atic study on algorithmic procedures for mitigating consumer unfairness in rating prediction or top-n recommendation tasks. To this end, we scanned the proceed-ings of top-tier conferences and journals, identifying *15 relevant papers*. We tried to reproduce the procedures reported in the paper in case the source code was made available by the authors (*only 8 papers*). Our first contribution is hence an assessment of the reproducibility level of mitigations in the area.

In a second step (Sect. 3), we defined a common evaluation protocol, includ-ing two public data sets (MovieLens 1M; LastFM 1K), two sensitive attributes (gender; age) and two fairness notions (equity; independence); we evaluated the recommendation models reported in the papers, with/out the proposed miti-gation procedure, under this common protocol. Our results revealed that, the mitigation procedures did not consistently reduce the utility of the recommen-dations (*RQ1*). We however found that only a minor subset of procedures sub-stantially reduce unfairness, and rarely for the two fairness notions at the same time (*RQ2*). Moreover, disparate impact does not always harm minority groups (*RQ3*). Our second contribution lies in evaluating mitigation procedures under a common protocol and identifying challenges in the area (Sect. 4).

2   Research Methodology

In this section, we describe the collection process for mitigation procedures, the steps for their reproduction, and the common evaluation protocol (Fig. 1).

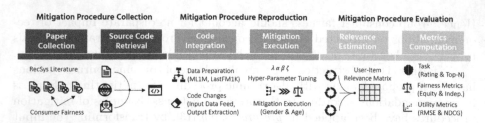

Fig. 1. Method. We systematically collected papers and retrieved their source code. We processed the data sets used in our evaluation protocol, formatted them as per each mitigation requirements, and made the format of the mitigation results uniform. We trained the recommendation models included in the original papers, with/out mitigation, and computed fairness and utility metrics for the target recommendation task.

2.1 Mitigation Procedures Collection

To collect existing mitigation procedures against consumer fairness, we systematically scanned the recent proceedings of top-tier Information Retrieval conferences and workshops, namely CIKM, ECIR, ECML-PKDD, FAccT, KDD, RecSys, SIGIR, WSDM, WWW, and journals edited by top-tier publishers, namely ACM, Elsevier, IEEE, and Springer. The keywords for our manual research were composed by a technical term, *"Recommender System"* or *"Recommendation"*, and a non-technical term, *"Consumer Fairness"* or *"User Fairness"*. We marked a paper to be relevant if (a) it focused on recommender systems, (b) it proposed a mitigation procedure, and (c) that procedure targeted the end users receiving the recommendations. Papers on other domains, e.g., non-personalized rankings, other stakeholders, e.g., providers only, and on pure conceptualization only, e.g., proposing a fairness notion without any mitigation, were excluded. Papers addressing both consumer and provider fairness were included, since they also target the end users. Finally, *15 relevant papers* were considered in our study.

We then attempted to reproduce the mitigation procedure proposed in each relevant paper, relying as much as possible on the source code provided by the authors themselves. We hence tried to obtain the source code for each relevant paper, by searching for the link into the paper, browsing for the official repository on the Web , and sending an e-mail to the authors as a last resort. We considered a mitigation procedure to be reproducible if a working version of the source code was obtained, and required minimal changes to accept another data set and extract the final recommendations. Otherwise, we considered a paper to be non-reproducible given our reproduction approach. We also considered works to be non-reproducible when the source code was obtained but included only a skeleton version of the procedure with many parts and details missing. At the end, *8 out of 15 relevant papers* could be reproduced with a reasonable effort.

In Table 1, for each reproducible paper, we identified the recommendation task (RP: Rating Prediction; TR: Top-N Recommendation), the notion of consumer fairness (EQ: equity of the error/utility score across demographic groups; IND: independence of the predicted relevance scores or recommendations from the demographic group), the consumers' grouping (G: Gender, A: Age, O: Occupation,

Table 1. The considered reproducible mitigation procedures for consumer fairness.

Task	Paper	Year	Mitigation			Evaluation		
			Notion	Groups	Type	Data Sets	Utility Metrics	Fairness Metrics
TR	Burke et al. [6]	2018	EQ	G	IN	ML	NDCG	CES
	Frisch et al. [15]	2021	IND	G-A	IN	ML	NDCG	EPS-CHI
	Li et al. [19]	2021	EQ	B	POST	AM	NDCG-F1	DP
TR + RP	Ekstrand et al. [13]	2018	EQ	G	PRE	ML-LFM	NDCG-MRR	DP
RP	Kamishima et al. [18]	2018	IND	G-A	IN	ML-SS	MAE	KS
	Rastegarpanah et al. [23]	2019	EQ	B	POST	ML	RMSE	GLV
	Ashokan & Haas [3]	2021	EQ	G	POST	ML-SY	RMSE-MAE	GEI-TI
	Wu et al. [28]	2021	IND	G-A-O	IN	ML-LFM	RMSE	AUC-F1

B: Behavioral), the mitigation type (PRE-, IN- or POST-Processing), the evaluation data sets (ML: MovieLens 1M or 10M, LFM: LastFM 1K or 360K, AM: Amazon, SS: Sushi, SY: Synthetic), the utility/accuracy metrics (NDCG: Normalized Discounted Cumulative Gain; F1: F1 Score; AUC: Area Under Curve; MRR: Mean Reciprocal Rank; RMSE: Root Mean-Square Error; MAE: Mean Absolute Error), and fairness metrics (EPS: ϵ-fairness; CHI: Chi-Square Test; KS: Kolmogorov-Smirnov Test; GEI: Generalized Entropy Index; TI: Theil Index; DP: Demographic Parity; EP: Equal Opportunity; CES: Category Equity Score; GLV: Group Loss Variance). The reproducibility ratio was of 53% (8/15) in total: 50% (4/8) for top-n recommendation and 57% (4/7) for rating prediction. We identified [20, 25–27] and [4, 14, 17] as non-reproducible procedures according to our criteria for top-n recommendation and rating prediction, respectively.

2.2 Mitigation Procedures Reproduction

For each reproducible paper, we delve into the core idea and the characteristics reported in Table 1. Our source code includes a directory for each paper, documented with the changes on the original code and the steps to get our results.

Burke et al. [6] proposed to generate recommendations for a user from a neighborhood having an equal number of peers from each group, to reduce unfairness. SLIM, a collaborative filtering method, was extended with a regularization aimed to achieve balance between protected and non-protected neighbors. Fairness was measured with a variant of what is known in statistics as risk ratio; this score is less (greater) than 1 when the protected group is recommended fewer (more) movies of the desired genre, on average (1 means perfect equity). Recommendation utility was measured via NDCG@10. An evaluation on ML 1M (5-fold cross-validation, no train-test split specified) showed that the treated models led to an equity score closer to 1 than the original models. The source code was not included in the paper, but shared by the authors during a scientific tutorial [7].

Frisch et al. [15] aimed at producing fair recommendations using a co-clustering of users and items that respects statistical parity w.r.t. some sensitive attributes. To this end, the authors introduced a co-clustering model based on the Latent

Block Model (LBM), that relies on an ordinal regression model taking the sensitive attributes as inputs. Fairness was measured by monitoring that, for any two items, the proportion of users with the same preference was similar across demographic groups. NDCG monitored the recommendation utility. An evaluation on ML 1M showed that their procedure led to lower unfairness. No source code was linked in the paper; we contacted the authors to get a copy of it.

Li et al. [19] investigated consumer unfairness across user groups based on the level of activity in the platform (more or less active). As a mitigation, the authors adopted a re-ranking method, whose objective function was to select items out of the baseline top-n list of each user so that the overall recommendation utility could be maximized, constrained to the fact that the model should minimize the difference in average recommendation performance between the groups of users. F1@10 and NDCG@10 were used to assess recommendation utility. The difference in NDCG between the groups estimated the unfairness of the model. An evaluation on Amazon data sets showed that their procedure could reduce unfairness between groups significantly, and also improve the overall recommendation utility. The original source code in the paper included only the re-ranking method. We contacted the authors for the complete source code, which was provided in a public repository (https://github.com/rutgerswiselab/NLR).

Ekstrand et al. [13] re-sampled user interactions (random sampling without replacement), such that the representation of user interactions across groups in the training set was balanced, and re-trained the recommendation models with the balanced training set. Recommendation utility was measured with NDCG@10, and fairness was assessed by visually comparing the averaged NDCG scores for the different demographic groups. An evaluation on ML 1M and LFM 1K and 360K showed that their re-sampling procedure led to unfairness mitigation for gender groups. The link to the source code was not reported in the paper. The authors promptly provided us the publicly available archive stored in the authors' university website (https://scholarworks.boisestate.edu/cs_scripts/4/).

Kamishima et al. [18] delved into the concept of recommendation independence, achieved when a recommendation outcome (predicted ratings) is statistically independent from a specified sensitive attribute. The mitigation consisted of optimizing a recommendation model by minimizing the dissimilarity between true ratings and predicted ratings and jointly maximizing the degree of independence between the predicted ratings and sensitive labels. Prediction errors were measured by the MAE. Independence was checked by measuring the equality of the predicted rating distributions between groups (Kolmogorov-Smirnov test; a smaller KS indicates that predicted ratings are more independent). An evaluation on ML 1M, Flixster, and Sushi, and three independence terms (mean-m, bdist-m, and mi-normal), showed that the sensitive information could be removed at the cost of a small loss in MAE. The source code linked in the paper included two complementary repositories (https://github.com/tkamishima/kamrecsys) and (https://github.com/tkamishima/kamiers).

Rastegarpanah et al. [23] investigated whether augmenting the training input with additional data can improve the fairness of the resulting predictions. Given a pretrained matrix factorization model, the mitigation required to add fake users who provided ratings on existing items to the training set; the fake users' ratings were chosen to improve the fairness of the final model for the real users. Recommendation utility was measured via RMSE. Fairness was measured through the variance of the loss across demographic groups, with the loss being the mean squared estimation error over all ratings of users in the group. An evaluation on ML 1M (no train-test split specified) showed that their mitigation could efficiently improve fairness of the considered recommender systems. The source code was found in a public repository (https://github.com/rastegarpanah/antidote-data-framework), whose link was sent to us by the authors.

Ashokan & Haas [3] adjusted the relevance scores predicted by the original model such that a given fairness metric increased. The authors experimented with value-based fairness (given a user, the difference in predicted and actual ratings in the training set for the group the user belongs to was added to the predicted ratings of the user) and parity-based fairness (the overall difference between predicted ratings for two groups on the training set was added to the predicted rating of a user for an item in the test set). Recommendation utility was measured via RMSE and MAE. Fairness was measured, among others, via the Generalized Entropy and the Theil indexes, which estimate inequality of errors across users. An evaluation on the ALS and ItemKNN recommendation models, trained on a synthetic data set and on ML 1M (5-fold cross-validation), showed that increasing fairness can even lead to lower RMSE and MAE in certain cases. No source code was linked in the paper; the authors sent it to us by e-mail.

Wu et al. [28] focused on mitigating unfairness in latent factor models. To this end, their procedure took the user and item embeddings from the original recommendation model as input and learned a filter space where any sensitive information was obfuscated and recommendation utility was preserved. The filters were learnt through a graph-based adversarial training process, where a discriminator tried to predict the sensitive label, and the filters were trained to remove sensitive information exposed in the supporting graph structure. RMSE measured recommendation utility. Fairness was monitored by checking the performance in terms of AUC (binary attributes) and F1 (multi-class attributes) of a classifier that predicts the sensitive attribute, given the user embedding (smaller values denote better fairness). An evaluation on ML 1M (training and test ratio of 9:1) and LFM 360K (training, validation, test ratio of 7:1:2) showed that fairness could be improved without significantly impacting on recommendation utility. The source code linked in the paper omitted important components. The authors provided us with an updated public repository (https://github.com/newlei/LR-GCCF).

2.3 Mitigation Procedures Evaluation

To ensure evaluation consistency and uniformity across mitigation procedures, given the heterogeneity of the original experimental evaluations, we mixed replication and reproduction [1,10]. For readability, we use the term *"reproducibility"*. So, we used the source code provided by the original authors to run their models and mitigation procedures, and our own artifacts (data and source code) to (a) pre-process the input data sets as per their requirements and (b) compute evaluation metrics based on the relevance scores or recommendations they returned.

Data Sets. The assessment of consumer fairness is challenging due to the lack of public data sets with ratings and *sensitive attributes* of the consumers. In our analysis, we considered all the public data sets that (a) were used in at least one reproduced paper, (b) reported at least one sensitive attribute, and (c) included enough ratings to reasonably train a recommender system ($\geq 200,000$ ratings). We hence evaluated the reproduced mitigation procedures on two public data sets on the movies and music domains (Table 2). Each data set was downloaded from the original website and pre-processed according to our common evaluation protocol, in response also to some limitations of the reproduced mitigations. For instance, given that the existing mitigation procedures are often tailored to binary groups only, we grouped users in two groups in case of data sets with multi-class sensitive attributes (*while attributes like gender and age are by no means a binary construct, what we are considering is a binary feature*).

Gender labels were already binary in ML 1M. We binarized age labels, organized in seven age ranges, such that the two groups included consecutive age ranges and had the most similar representation possible. For LFM 1K, we considered only users reporting both their gender and age and filtered those with wrong ages (≤ 0 or ≥ 125). Interactions of a user for the same artist were aggregated, using the number of plays of a user for an artist as a proxy of the rating. We filtered users interacting with less than 20 artists (as in ML 1M), and ratings were log-normalized and scaled in $[1,5]$. Gender labels were already binary. We binarized age labels (integer) with the same criteria used in ML 1M.

Table 2. The data sets with consumer's sensitive attributes included in our study.

Data set	#Users	#Items	#Ratings	Sensitive attributes
ML 1M [16]	6,040	3,952	1,000,209	Gender (M: 71.7%; F: 28.3%) Age (< 35 : 56.6%; ≥ 35 : 43.4%)
LFM 1K [8]	268	51,609	200,586	Gender (M: 57.8%; F: 42.2%) Age (< 25 : 57.8%; ≥ 25 : 42.2%)

Protocol. Each reproduced paper applied the corresponding mitigation procedure to a set of state-of-the-art recommendation models, which was quite heterogeneous across papers due to authors' arbitrary choices or the focus on a specific type of model. These models covered several families, including non-personalized (`TopPopular` [13] and `AvgRating` [13]), memory (`ItemKNN` [3,13], `UserKNN` [13]), matrix factorization (`BiasedMF` [3,19], `PMF` [18,19,28], `FunkSVD` [13]), learning-to-rank (`NCF` [19], `LBM` [15], `SLIM-U` [6], `ALS` [23], `LMaFit` [23]), graph (`GCN` [28]), and session-based (STAMP [19]). In line with our reproduction approach, we applied a given mitigation on the same models considered by the original authors[2].

Specifically, given a data set, a sensitive attribute, and a reproducible paper, we considered the following evaluation protocol. We first performed a train-test split per user, with 20% of the interactions (the most recent if a timestamp was available, randomly selected otherwise) being in the test set and the remaining interactions being in the train set. In case a validation set was needed for best model selection, 10% of interactions (selected in the same way) of each user from the train set were considered as a validation set and the other ones included in the final train set. To fit with the original source code, the format of the considered sets and the sensitive attribute's labels per user were adapted. No changes on the source code specific for the mitigation procedure were applied.

Using the prepared sets and an appropriate hyper-parameters grid, we ran a grid search for each recommendation model, with and without mitigation. For each paper, our source code includes the scripts to format a data set as per the original source code requirements and to compute evaluation metrics as well as the details of models hyper-parameter tuning. For each setup, we obtained the predicted relevance scores and the recommendations, and computed utility and fairness metrics. Utility metrics included NDCG for top-n recommendation (using binary relevances) and RMSE for rating prediction, selected due to their popularity (see Table 1). Consumer fairness metrics monitored equity through Demographic Parity (DP), computed as the difference on utility for the corresponding task between groups, and independence through Kolmogorov-Smirnov (KS), computed on predicted relevance scores, covering two well-known perspectives and steps of the pipeline. Mainly due to space constraints, we left analyses on other fairness notions and implementations of the same fairness notions as a future work. Experiments ran on a Ryzen7 machine with 32 GB RAM.

3   Experimental Results

We now analyze the extent to which the mitigation procedures impact on recommendation utility (*RQ1*), reduce unfairness (*RQ2*), and possibly affect groups differently (*RQ3*). To this end, we report recommendation utility and fairness scores obtained under the above evaluation protocol, for TR (Table 3, gender; Table 4, age) and RP tasks (Table 5, gender; Table 6, age). DP was tested for statistical significance via a Mann-Whitney test. For KS, we used its own score. Note that * and ∧ meant significance at p-values 0.05 and 0.01, respectively.

3.1   Impact on Recommendation Utility (RQ1)

In a first analysis, we assess the impact of mitigation on recommendation utility, focusing on the NDCG/RMSE columns provided in the aforementioned tables.

---

[2] Though some procedures might be applied across models, their transfer often requires arbitrary design choices and core changes that mine our rigorous reproduction.

Table 3. Top-n recommendation (TR) considering *gender* groups.

Paper	Model	ML 1M						LFM 1K					
		NDCG ↑		DP ↓		KS ↓		NDCG ↑		DP ↓		KS ↓	
		Base	Mit	Base	Mit	Base	Mit	Base	Mit	Base	Mit	Base	Mit
Burke et al.	SLIM-U	0.084	0.084	-0.022	-0.028	-0.032	-0.115	0.348	0.301	^−0.128	-0.072	-0.010	-0.142
Frisch et al.	LBM	0.044	0.021	-0.006	-0.004	-0.013	-0.025	0.144	0.212	*−0.035	*−0.058	-0.120	-0.126
Li et al.	BiasedMF	0.112	0.112	-0.016	-0.013	-0.033	**-0.006**	0.246	0.245	^-0.076	*−0.049	-0.026	**-0.001**
	NCF	0.120	0.120	-0.018	-0.015	-0.024	**-0.006**	0.204	0.202	-0.046	-0.023	-0.017	**-0.001**
	PMF	0.123	0.123	-0.020	-0.015	-0.026	**-0.006**	0.163	0.164	-0.069	*−0.049	-0.035	**-0.001**
	STAMP	0.068	0.067	-0.013	-0.009	**-0.007**	**-0.006**	0.110	0.110	-0.024	-0.018	-0.002	**-0.001**
Ekstrand et al.	FunkSVD	0.018	0.015	**-0.004**	**0.002**	-0.027	-0.018	0.010	0.013	**-0.006**	**−0.003**	-0.107	-0.119
	ItemKNN	**0.140**	**0.134**	-0.038	-0.030	-0.030	-0.031	0.287	0.286	-0.127	*-0.116	-0.019	-0.022
	TopPopular	0.110	0.104	-0.035	-0.030	**-0.007**	-0.007	0.312	0.321	*-0.085	*-0.102	**-0.001**	-0.002
	UserKNN	0.137	0.131	-0.031	-0.024	-0.074	-0.052	**0.406**	**0.411**	-0.110	-0.106	-0.067	-0.067

Table 4. Top-n recommendation (TR) considering *age* groups.

Paper	Model	ML 1M						LFM 1K					
		NDCG ↑		DP ↓		KS ↓		NDCG ↑		DP ↓		KS ↓	
		Base	Mit	Base	Mit	Base	Mit	Base	Mit	Base	Mit	Base	Mit
Burke et al.	SLIM-U	0.084	0.048	-0.022	-0.014	-0.009	-0.095	0.348	0.207	*−0.065	−0.145	-0.021	-0.082
Frisch et al.	LBM	0.044	0.042	**-0.005**	**-0.006**	-0.021	-0.027	0.144	0.213	−0.011	−0.021	-0.125	-0.152
Li et al.	BiasedMF	0.112	0.112	-0.018	-0.017	-0.042	**-0.006**	0.246	0.247	−0.044	*−0.060	-0.015	**-0.005**
	NCF	0.120	0.120	-0.022	-0.019	-0.031	**-0.006**	0.204	0.203	−0.035	−0.048	-0.008	**-0.005**
	PMF	0.123	0.123	-0.027	-0.021	-0.027	**-0.006**	0.163	0.164	−0.033	^−0.044	-0.018	**-0.005**
	STAMP	0.068	0.068	**0.005**	**-0.006**	**-0.006**	**-0.006**	0.110	0.110	*−0.030	^−0.034	**-0.005**	**-0.005**
Ekstrand et al.	FunkSVD	0.018	0.016	-0.008	**-0.006**	-0.029	-0.021	0.010	0.016	**0.002**	**−0.004**	-0.054	-0.047
	ItemKNN	**0.140**	**0.138**	-0.027	-0.024	-0.029	-0.033	0.287	0.269	0.010	0.020	-0.133	-0.118
	TopPopular	0.110	0.107	-0.038	-0.034	**-0.006**	**-0.006**	0.312	0.315	−0.044	−0.050	-0.006	-0.007
	UserKNN	0.137	0.137	-0.028	-0.023	-0.060	-0.051	**0.406**	**0.397**	−0.023	−0.031	-0.036	-0.031

Table 5. Rating prediction (RP) considering gender groups.

Paper	Model	ML 1M						LFM 1K					
		RMSE ↓		DP ↓		KS ↓		RMSE ↓		DP ↓		KS ↓	
		Base	Mit	Base	Mit	Base	Mit	Base	Mit	Base	Mit	Base	Mit
Ekstrand et al.	AvgRating	0.905	0.914	^−0.032	*−0.027	-0.047	-0.045	1.239	1.246	0.025	0.024	-0.060	-0.070
	FunkSVD	0.881	0.894	^−0.032	**−0.023**	-0.052	-0.051	1.255	1.268	*0.039	0.039	-0.040	-0.052
	ItemKNN	0.865	0.882	^−0.034	*−0.026	-0.055	-0.056	1.218	1.230	*0.037	*0.035	-0.064	-0.072
	UserKNN	0.896	0.911	^−0.035	−0.025	-0.056	-0.058	1.226	1.239	*−0.047	*0.054	**-0.036**	-0.045
Kamishima et al.	PMF BDist	**0.863**	0.870	^−0.029	−0.046	-0.056	-0.032	1.172	1.179	**0.014**	*0.029	-0.067	-0.029
	PMF Mean	**0.863**	0.870	^−0.029	^−0.048	-0.056	-0.056	1.172	1.179	**0.014**	*0.025	-0.067	-0.054
	PMF Mi	**0.863**	0.870	^−0.029	^−0.046	-0.056	-0.032	1.172	1.179	**0.014**	*0.029	-0.067	-0.029
Rastegarpanah et al.	ALS	0.894	0.890	^−0.034	^−0.034	**0.035**	-0.033	1.490	1.189	-0.145	0.029	**-0.036**	-0.114
Ashokan & Haas	ALS Par	0.867	0.868	^−0.030	−0.029	-0.056	-0.034	**1.145**	1.146	0.016	**0.018**	-0.047	**-0.017**
	ALS Val	0.867	0.867	^−0.030	−0.030	-0.056	-0.057	**1.145**	1.150	0.016	**0.018**	-0.047	-0.050
	ItemKNN Par	0.865	0.866	^−0.034	^−0.033	-0.055	-0.036	1.176	1.183	*0.033	*0.045	v0.061	-0.058
	ItemKNN Val	0.865	**0.865**	^−0.034	^−0.034	-0.055	-0.052	1.176	1.173	*0.033	*0.036	-0.061	-0.046
Wu et al.	FairGo GCN	0.895	0.892	^−0.038	^−0.034	-0.048	-0.045	1.609	1.283	-0.151	0.038	-0.113	-0.113

In a TR task, we observed that the NDCG achieved by the untreated models (Base) in ML 1M was in the range [0.110, 0.140], except for SLIM-U, FunkSVD, LBM, and STAMP, whose NDCG was lower (≤ 0.084). Mitigating unfairness (Mit) in ML 1M did not generally result in a substantial change in utility (±0.006 gender; ±0.003 age). Higher changes were observed in two cases: SLIM-U treated

Table 6. Rating prediction (RP) considering age groups.

Paper	Model	ML 1M RMSE ↓ Base	Mit	DP ↓ Base	Mit	KS ↓ Base	Mit	LFM 1K RMSE ↓ Base	Mit	DP ↓ Base	Mit	KS ↓ Base	Mit
Ekstrand et al.	AvgRating	0.905	0.904	-0.051	-0.056	-0.071	-0.072	1.239	1.248	0.040	0.048	-0.080	-0.092
	FunkSVD	0.881	0.886	-0.042	-0.045	-0.073	-0.081	1.255	1.264	0.032	0.035	-0.083	-0.086
	ItemKNN	0.865	0.875	-0.039	-0.042	-0.074	-0.079	1.218	1.226	**0.019**	**0.028**	-0.088	-0.092
	UserKNN	0.896	0.902	-0.047	-0.050	-0.092	-0.103	1.226	1.233	0.034	0.031	-0.087	-0.095
Kamishima et al.	PMF BDist	**0.863**	0.872	-0.039	-0.031	-0.084	**-0.018**	1.172	1.183	0.045	-0.065	-0.124	-0.047
	PMF Mean	**0.863**	0.872	-0.039	**-0.027**	-0.084	-0.045	1.172	1.184	0.045	-0.069	-0.124	-0.042
	PMF Mi	**0.863**	0.872	-0.039	-0.031	-0.084	**-0.018**	1.172	1.183	0.045	-0.064	-0.124	-0.047
Rastegarpanah et al.	ALS	0.894	0.892	**-0.034**	-0.040	**-0.034**	-0.037	1.490	1.185	0.033	-0.052	**-0.017**	-0.064
Ashokan & Haas	ALS Par	0.867	0.871	-0.041	-0.048	-0.074	-0.026	**1.145**	**1.146**	0.043	-0.046	-0.082	**-0.015**
	ALS Val	0.867	0.866	-0.041	-0.042	-0.074	-0.079	**1.145**	1.149	0.043	-0.046	-0.082	-0.077
	ItemKNN Par	0.865	0.870	-0.040	-0.048	-0.074	-0.031	1.176	1.177	0.029	0.031	-0.085	-0.029
	ItemKNN Val	0.865	**0.864**	-0.040	-0.042	-0.074	-0.071	1.176	1.172	0.029	0.032	-0.085	-0.083
Wu et al.	FairGo GCN	0.895	0.908	-0.040	-0.044	-0.070	-0.074	1.609	1.277	0.043	-0.056	-0.079	-0.120

with Burke et al.'s mitigation (stable for gender; −0.036 age) and LBM treated with Frisch et al.'s (−0.023 gender; stable for age). In LFM 1K, the untreated models (Base) got an NDCG in [0.204, 0.406], overall higher than ML 1M. The models ranking based on NDCG differs for several models from ML 1M. Though their utility was relatively high, PMF, FunkSVD, LBM, and STAMP were still under-performing in LFM 1K. The treated models (Mit) showed changes in NDCG (±0.009 gender; ±0.018 age) larger in magnitude than ML 1M. SLIM-U with Burke et al.'s mitigation (−0.047 gender; −0.141 age) and LBM with Frisch et al.'s mitigation (+0.068 gender; +0.069 age) led to higher changes in NDCG.

Considering an RP task, the untreated models (Base) achieved an RMSE in the range [0.863, 0.905] in ML 1M. By mitigating (Mit) in ML 1M, no substantial changes were observed (±0.017 gender; ±0.013 age). In LFM 1K, the untreated models (Base) achieved a higher RMSE, in the range [1.145, 1.255]. ALS and GCN are the lowest performers (1.490 and 1.609, respectively). The treated models (Mit) showed minimal (±0.0135 gender; ±0.012 age) which are similar to the changes in ML 1M. ALS under Rastegarpanah et al.'s mitigation lowered RMSE (−0.301 gender; −0.305 age), as well as GCN under Wu et al.'s mitigation (−0.326 gender; −0.332 age).

Observation 1. *In general, the mitigation procedures did not substantially impact on recommendation utility, regardless of the sensitive attribute, data set, task. The impact is larger in LFM 1K than ML 1M.*

3.2 Impact on Group Unfairness (RQ2)

In a second analysis, we investigated the impact of mitigation on unfairness. For each table and data set, we consider the DP and KS columns.

We start from a TR task, focusing our presentation on the subset of models that achieved a reasonable NDCG (≤ 0.110 for ML 1M; ≤ 204 for LFM 1K). In ML 1M, the DP and KS achieved by the untreated models (Base) laid in the ranges ([0.016, 0.038] gender; [0.018, 0038] age) and ([0.007, 0.074] gender; [0.006, 0.060] age), respectively. Without any mitigation, in terms of DP, BiasedMF, NCF, and PMF (≤ 0.020 gender; ≤ 0.027 age) were fairer than Top-Popular, UserKNN, and ItemKNN (≥ 0.031 gender; ≥ 0.027 age). To some surprise, when KS was considered, we observed a different pattern. TopPopular was the fairest model (0.007 gender; 0.006 age), followed by NCF and PMF (0.024 and 0.026 gender; 0.031 and 0.027 age), ItemKNN and BiasedMF (0.030 and 0.033 gender; 0.029 and 0.042 age), and UserKNN (0.074 gender; 0.060 age). By mitigating (Mit), DP went down to the range ([0.013, 0.030] gender; [0.017, 0.034] age), while KS laid in the range ([0.006, 0.052] gender; [0.006, 0.051] age). In LFM 1K, models were less fair than in ML 1M. The untreated models (Base) achieved a DP in the ranges ([$-0.046, -0.127$] gender; [0.010, -0.044] age) and a KS in the ranges ([0.001, 0.067] gender; [0.006, 0.133] age). The models ranking in terms of DP and KS was similar between LFM 1K and ML 1M. Once mitigated (Mit), interestingly, we observed that re-sampling by Ekstrand et al. resulted in a decrease of fairness for TopPopular in terms of DP on gender groups (0.017), and for TopPopular, ItemKNN and UserKNN on age groups (≥ 0.06). These findings are replicated for ItemKNN in terms of KS on gender groups (0.03), while, for age groups KS was substantially lowered (0.015). Other cases did not lead to substantial changes.

In a RP task, in ML 1M, untreated models (Base) achieved a DP in [$-0.038, -0.025$] (gender) and [0.034, 0.051] (age), and a KS in [0.035, 0.056] (gender) and [0.034, 0.092] (age). With no mitigation, there were minimal differences in terms of DP between models for the attribute gender (avg. 0.033, std. dev. 0.003). For the attribute age, the untreated models had similar DP (avg. 0.041, std. dev. 0.005). Considering KS, comparable estimates across models were observed (avg. 0.053, std. dev. 0.003 gender; avg. 0.076, std. dev. 0.007 age). ALS (0.035 gender; 0.034 age) resulted in fairer outcomes in terms of KS. Treated models (Mit) showed stable fairness (± 0.010 gender; ± 0.008 age) in all cases, except for Kamishima et al. (± 0.019 gender; ± 0.012 age) when DP was considered. In terms of KS, models treated with Kamishima et al.'s mitigation (for gender only PMF BDist and PMF Mi) and Ashokan et al.'s mitigation (parity setting) were substantially fairer (≥ 0.019 gender; ≥ 0.039 age), while other treated models did not benefit from the mitigation (± 0.003 gender; ± 0.011 age). In LFM 1K, untreated models (Base) achieved a DP in [0.014, 0.151] (gender) and [0.019, 0.045] (age), and a KS in [0.036, 0.113] (gender) and [0.017, 0.124] (age). Without mitigating, findings in ML 1M held in LFM 1K, except for the high DP (0.151) and KS (0.113) of GCN for gender. Treated models (Mit) instead showed stable fairness (≤ 0.015 gender; ≤ 0.009 age) except for Kamishima et al. (≥ 0.019 age), ALS (0.116 gender; 0.019 age), GCN (0.113 gender; 0.013 age), in terms of DP (opposite to ML 1M). In terms of KS, except the mitigations of Kamishima et al. and Ashokan et al. (parity), treated models did not benefit from mitigation (≤ 0.015 gender; ≤ 0.005 age).

Observation 2. *Unfairness depends on the mitigation, model, and fairness notion. Often the mitigation impact is small. Lowering DP does not imply lowering KS, and viceversa. Unfairness is higher in LFM than ML.*

3.3 Relationships Between Representation and Unfairness (RQ3)

In a third analysis, we analyzed whether the disparate impact always harms minority groups (see group representations in Table 2), based on the sign of DP.

In a TR task, positive values of DP mean that models advantage the majority (majority group's NDCG higher than minority's group NDCG; the higher the NDCG the higher the utility). Conversely, negative values show an advantage for the minority. From our results, untreated models (Base) negatively impacted on the minority for both gender (in all cases significantly) and age (9/10 times significantly) in ML 1M. Though unfairness was reduced through mitigation, the same observations were still valid on treated models (Mit). To some surprise, the majority groups were negatively impacted for both attributes in LFM 1K (7/10 times significantly for gender and 2/10 times significantly for age) by untreated models (Base). By mitigating (Mit), 7 out of 10 treated models were significantly unfair for gender. For age, observations were similar.

Considering a RP task, positive values of DP mean that models advantage the minority (majority group's RMSE higher than minority's group RMSE; the higher the RMSE the lower the utility). Conversely, negative values show an advantage for the majority. The results showed that the minority age group was advantaged in both data sets (in all cases significantly in ML 1M) by untreated models (Base). The minority group was also advantaged in LFM 1K for the gender attribute, significantly 7/13 times. Conversely, the majority gender group was advantaged in ML 1M, significantly in all cases. Similarly to the TR task, treated models (Mit) were still significantly unfair against the group disadvantaged in the untreated model.

Observation 3. *The disparate impact does not always harm the minority group. The latter was advantaged for both attributes in LFM 1K (TR), in both data sets for age and in LFM 1K for gender (RP).*

4   Discussion and Conclusions

In this last section, we connect our findings and present the resulting conclusions.

Reproducibility. Several challenges emerged while reproducing existing procedures. For instance, the code base modularity should be improved to easily accommodate different data sets as an input. Moreover, many procedures required extensive computational resources to treat the recommendation models. This issue prevented us from using larger data sets, e.g., LFM 360K, and questions scalability. Future works should account for modularity and efficiency.

Optimization. Mitigating unfairness adds additional hyper-parameters and often requires to deal with a trade-off between recommendation utility and unfairness. It is hence challenging to tune the hyper-parameters. While we provide results for an arbitrary optimal setup, it is up to the stakeholders to select the trade-offs most suitable for their goals. One of the future directions should be to find novel mitigation procedures that embed the constraint on recommendation utility more strictly, to avoid convoluted decisions on the mentioned trade-off.

Comparability. Our study showed that there is an abundance of evaluation metrics to assess fairness and that, despite several papers using similar data sets (e.g., ML 1M), the evaluation setting was often different. Our paper shows the first attempt of comparing a wide range of mitigation procedures under the same evaluation protocol, considering two relevant yet transferable fairness notions. Despite the common protocol, we however could not conclude whether a mitigation is better than another in a given context, given that many of them could not be easily transferred across models. In the future, a mitigation procedure should be tested across recommendation models, data sets and sensitive attributes.

Impact. Our results showed that the impact of the mitigation procedure on utility is often negligible. However, depending on the recommendation model, the data set, and the task, mitigation procedures do not always substantially reduce unfairness. Moreover, being fair in terms of independence does not imply higher fairness in terms of equity. Future work should study the friction across fairness notions, and ensure that the unfairness reduction achieved offline can then provide tangible impacts when moved online.

Overall, our analyses showed that reproducing research in this area is still a challenging task hence and call for more rigor and shared practices in this area. Motivated by our findings, we will extend our analyses to papers published in other outlets and to other notions of consumer fairness. We also plan to devise novel mitigation procedures, following the lessons learnt from this study (e.g., modularity, efficiency, optimization, comparability, impact).

References

1. ACM artifact review and badging. https://www.acm.org/publications/policies/artifact-review-and-badging-current (2021). Accessed 25 Sept 2021
2. Abdollahpouri, H., et al.: Multistakeholder recommendation: survey and research directions. User Model. User Adapt. Interact. **30**(1), 127–158 (2020)
3. Ashokan, A., Haas, C.: Fairness metrics and Bias mitigation strategies for rating predictions. Inf. Process. Manag. **58**(5), 102646 (2021)
4. Bobadilla, J., Lara-Cabrera, R., González-Prieto, Á., Ortega, F.: DeepFair: deep learning for improving fairness in recommender systems. CoRR abs/2006.05255 (2020). https://arxiv.org/abs/2006.05255
5. Boratto, L., Fenu, G., Marras, M.: Interplay between upsampling and regularization for provider fairness in recommender systems. User Model. User Adapt. Interact. **31**(3), 421–455 (2021)

6. Burke, R., Sonboli, N., Ordonez-Gauger, A.: Balanced neighborhoods for multi-sided fairness in recommendation. In: Conference on Fairness, Accountability and Transparency, FAT 2018, 23–24 February 2018, New York, NY, USA. Proceedings of Machine Learning Research, vol. 81, pp. 202–214. PMLR (2018). http://proceedings.mlr.press/v81/burke18a.html

7. Burke, R.D., Mansoury, M., Sonboli, N.: Experimentation with fairness-aware recommendation using librec-auto: hands-on tutorial. In: FAT* 2020: Conference on Fairness, Accountability, and Transparency, Barcelona, Spain, 27–30 January 2020, p. 700. ACM (2020). https://doi.org/10.1145/3351095.3375670

8. Celma, Ò.: Music Recommendation and Discovery - The Long Tail, Long Fail, and Long Play in the Digital Music Space. Springer, Heidelberg (2010). https://doi.org/10.1007/978-3-642-13287-2

9. Chen, J., Dong, H., Wang, X., Feng, F., Wang, M., He, X.: Bias and debias in recommender system: a survey and future directions. CoRR abs/2010.03240 (2020). https://arxiv.org/abs/2010.03240

10. Dacrema, M.F., Cremonesi, P., Jannach, D.: Are we really making much progress? A worrying analysis of recent neural recommendation approaches. In: Proceedings of the 13th ACM Conference on Recommender Systems, RecSys 2019, Copenhagen, Denmark, 16–20 September 2019, pp. 101–109. ACM (2019). https://doi.org/10.1145/3298689.3347058

11. Deldjoo, Y., Bellogín, A., Noia, T.D.: Explaining recommender systems fairness and accuracy through the lens of data characteristics. Inf. Process. Manag. 58(5), 102662 (2021)

12. Ekstrand, M.D., Das, A., Burke, R., Diaz, F.: Fairness and discrimination in information access systems. CoRR abs/2105.05779 (2021). https://arxiv.org/abs/2105.05779

13. Ekstrand, M.D., Tet al.: All the cool kids, how do they fit in?: Popularity and demographic biases in recommender evaluation and effectiveness. In: Conference on Fairness, Accountability and Transparency, FAT 2018. vol. 81, pp. 172–186. PMLR (2018). http://proceedings.mlr.press/v81/ekstrand18b.html

14. Farnadi, G., Kouki, P., Thompson, S.K., Srinivasan, S., Getoor, L.: A fairness-aware hybrid recommender system. CoRR abs/1809.09030 (2018). http://arxiv.org/abs/1809.09030

15. Frisch, G., Leger, J.B., Grandvalet, Y.: Co-clustering for fair recommendation. In: Proceedings of the European Conference on Machine Learning and Principles and Practice of Knowledge Discovery in Databases (2021). https://hal.archives-ouvertes.fr/hal-03239856

16. Harper, F.M., Konstan, J.A.: The movielens datasets: history and context. ACM Trans. Interact. Intell. Syst. 5(4), 19:1–19:19 (2016). https://doi.org/10.1145/2827872

17. Huang, W., Labille, K., Wu, X., Lee, D., Heffernan, N.: Achieving user-side fairness in contextual bandits. CoRR abs/2010.12102 (2020). https://arxiv.org/abs/2010.12102

18. Kamishima, T., Akaho, S., Asoh, H., Sakuma, J.: Recommendation independence. In: Conference on Fairness, Accountability and Transparency, FAT 2018, 23–24 February 2018, New York, NY, USA. Proceedings of Machine Learning Research, vol. 81, pp. 187–201. PMLR (2018). http://proceedings.mlr.press/v81/kamishima18a.html

19. Li, Y., Chen, H., Fu, Z., Ge, Y., Zhang, Y.: User-oriented fairness in recommendation. In: WWW 2021: The Web Conference 2021, pp. 624–632. ACM / IW3C2 (2021). https://doi.org/10.1145/3442381.3449866, https://doi.org/10.1145/3442381.3449866

20. Li, Y., Chen, H., Xu, S., Ge, Y., Zhang, Y.: Towards personalized fairness based on causal notion, pp. 1054–1063. Association for Computing Machinery (2021). https://doi.org/10.1145/3404835.3462966

21. Marras, M., Boratto, L., Ramos, G., Fenu, G.: Equality of learning opportunity via individual fairness in personalized recommendations. Int. J. Artif. Intell. Educ. 1–49 (2021). https://doi.org/10.1007/s40593-021-00271-1

22. Mehrabi, N., Morstatter, F., Saxena, N., Lerman, K., Galstyan, A.: A survey on bias and fairness in machine learning. ACM Comput. Surv. 54(6), 115:1–115:35 (2021). https://doi.org/10.1145/3457607

23. Rastegarpanah, B., Gummadi, K.P., Crovella, M.: Fighting fire with fire: using antidote data to improve polarization and fairness of recommender systems. In: Proceedings of the Twelfth ACM International Conference on Web Search and Data Mining, WSDM 2019, Melbourne, VIC, Australia, 11–15 February 2019, pp. 231–239. ACM (2019). https://doi.org/10.1145/3289600.3291002

24. Ricci, F., Rokach, L., Shapira, B. (eds.): Recommender Systems Handbook. Springer, Heidelberg (2015). https://doi.org/10.1007/978-1-4899-7637-6

25. Tsintzou, V., Pitoura, E., Tsaparas, P.: Bias disparity in recommendation systems. arXiv e-prints. arXiv:1811.01461 (2018)

26. Wu, C., Wu, F., Wang, X., Huang, Y., Xie, X.: Fairness-aware news recommendation with decomposed adversarial learning. In: Proceedings of the AAAI Conference on Artificial Intelligence, vol. 35, pp. 4462–4469 (2021). https://ojs.aaai.org/index.php/AAAI/article/view/16573

27. Wu, H., Ma, C., Mitra, B., Diaz, F., Liu, X.: Multi-FR: a multi-objective optimization method for achieving two-sided fairness in e-commerce recommendation. CoRR abs/2105.02951 (2021). https://arxiv.org/abs/2105.02951

28. Wu, L., Chen, L., Shao, P., Hong, R., Wang, X., Wang, M.: Learning fair representations for recommendation: a graph-based perspective. In: WWW 2021: The Web Conference 2021, Virtual Event/Ljubljana, Slovenia, 19–23 April 2021, pp. 2198–2208. ACM/IW3C2 (2021). https://doi.org/10.1145/3442381.3450015

The Power of Anchor Text in the Neural Retrieval Era

Maik Fröbe[1][(✉)], Sebastian Günther[1], Maximilian Probst[1], Martin Potthast[2],
and Matthias Hagen[1]

[1] Martin-Luther-Universität Halle-Wittenberg, Halle, Germany
maik.froebe@informatik.uni-halle.de
[2] Leipzig University, Leipzig, Germany

Abstract. In the early days of web search, a study by Craswell et al. [11] showed that anchor texts are particularly helpful ranking features for navigational queries and a study by Eiron and McCurley [24] showed that anchor texts closely resemble the characteristics of queries and that retrieval against anchor texts yields more homogeneous results than against documents. In this reproducibility study, we analyze to what extent these observations still hold in the web search scenario of the current MS MARCO dataset, including the paradigm shift caused by pre-trained transformers. Our results show that anchor texts still are particularly helpful for navigational queries, but also that they only very roughly resemble the characteristics of queries and that they now yield less homogeneous results than the content of documents. As for retrieval effectiveness, we also evaluate anchor texts from different time frames and include modern baselines in a comparison on the TREC 2019 and 2020 Deep Learning tracks. Our code and the newly created Webis MS MARCO Anchor Texts 2022 datasets are freely available.

Keywords: Anchor text · MS MARCO · ORCAS · TREC Deep Learning track

1 Introduction

Almost from the beginning, search engines have exploited the Web's link structure to improve their result rankings. But besides the actual links, also the anchor texts (i.e., the clickable texts of the links) were an important ranking feature, since they "often provide more accurate descriptions of web pages than the pages themselves" [2].

The seminal works of Craswell et al. [11] and Eiron and McCurley [24] from 2001 and 2003 examined two important aspects of anchor text. Craswell et al. showed that anchor text especially helps for navigational queries (i.e., queries to find a specific document [3]). This result explained why commercial search engines heavily used anchor text even though no positive effect was observed in TREC scenarios [27,49]: more than 20% of the traffic of commercial search engines were navigational queries [3], but hardly any TREC topic was navigational. Eiron and McCurley showed that retrieval against anchor texts yields more homogeneous results than against documents and that anchor texts closely resemble the characteristics of queries. This result later inspired others to use anchor texts as a replacement for proprietary query logs [7,20,36,38].

© The Author(s), under exclusive license to Springer Nature Switzerland AG 2022
M. Hagen et al. (Eds.): ECIR 2022, LNCS 13185, pp. 567–583, 2022.
https://doi.org/10.1007/978-3-030-99736-6_38

In the two decades since the studies of Craswell et al. and Eiron and McCurley were published, the Web and the search behavior of users have changed. We thus analyze to what extent the original findings can be reproduced on current web crawls and query logs. Additionally, given the recent success of pre-trained transformers [52], we also analyze whether anchor text is still a valuable ranking feature or whether it might be "obsolete" for retrieval pipelines using BERT [43], MonoT5 [44], or DeepCT [18].

As reproducibility scenario for our study, we employ the two available versions of the MS MARCO datasets (3.2 and 12 million documents, 367,013 queries with relevance judgments) [15], the ORCAS query log (18.8 million query–click entries related to MS MARCO documents) [8], and extract anchor texts from Common Crawl snapshots of the last six years to construct the Webis MS MARCO Anchor Texts 2022 dataset: it contains billions of anchor texts for about 1.7 million documents from MS MARCO Version 1 (about 53% of all documents), and for about 4.82 million documents from MS MARCO Version 2 (about 40% of all documents).

The results of our reproducibility study are dichotomous. While we can reproduce Craswell et al.'s observation that anchor text is particularly helpful for navigational queries (details in Sect. 5), we find substantial differences for the results of Eiron and McCurley. In the MS MARCO scenario, the anchor texts are pretty different to queries (e.g., number of distinct terms) and retrieval against them yields less (not more) homogeneous results than against the content of documents (details in Sect. 4). We attribute both changes to the fact that Eiron and McCurley conducted their study in the corporate IBM intranet with queries and anchor texts both formulated by employees of IBM, whereas, in our reproducibility scenario, we have "arbitrary" searchers and anchor text authors from the Web. In the reproducibility experiments for the study of Craswell et al., we also evaluate the effectiveness of anchor text from different time frames and include modern baselines in a comparison on the topics of the TREC 2019 and 2020 Deep Learning tracks. The results still confirm the observation that anchor text only slightly improves the effectiveness in TREC scenarios [11,27,49]. All our code and data is published under a permissible open-source license.[1]

2 Related Work

Exploiting link structure has a long tradition in IR [16]. Already in 1993, Dunlop and van Rijsbergen [23] used text referring to non-textual objects like images to retrieve those non-textual objects for text queries. McBryan [41] refined this process by only including terms from the clickable texts of links: the *anchor texts*. Anchor texts were later reported to be heavily used by commercial search engines [2,24] but had no positive effect in TREC scenarios [1,26,27,49]. Craswell et al. [11] resolved this dichotomy by showing that anchor text is particularly useful for navigational queries (i.e., queries to find a specific document [3]) while hardly any TREC topics were navigational.

After Craswell et al.s result, dedicated shared tasks like homepage finding or named page finding evolved [9,10,12] and more and more systems incorporated anchor text for navigational queries. For instance, Westerveld et al. [49] combined anchor text with a

---

[1] Code and data: https://github.com/webis-de/ECIR-22. Data is integrated in ir_datasets [39]. Data on Zenodo: https://zenodo.org/record/5883456.

document's content, URL, and link count, and Ogilvie and Callan [45] showed that anchor text can also be combined with poor-performing features without harming the overall effectiveness for navigational queries. Since links may "rot" over time [34]—resulting in possibly outdated anchor texts—, several approaches used historical information [17] or importance estimation [22,42] to weight anchor text. Finally, the anchor text source and quantity were shown to be very important. Kamps et al. [29] found that anchor text from the Wikipedia is more effective than anchor text from the general Web while Koolen and Kamps [35] showed that more anchor text led to higher early precision on the TREC 2009 Web track [6], which includes 66 navigational subtopics.

Anchor text became an important retrieval feature also used in lieu of query logs [7,20,24,36,38]. But with the recent paradigm shift due to transformers [52], the IR community's main focus changed from feature engineering to neural re-ranking and dense retrieval models [30]. The MS MARCO datasets, utilized by the TREC Deep Learning tracks [8,14], particularly enabled this shift, but since they lack anchor texts, our goal of reproducing the seminal anchor text studies by Craswell et al. [11] and Eiron and McCurley [24] requires the extra effort of collecting anchor texts for its documents.

3   The Webis MS MARCO Anchor Text 2022 Dataset

MS MARCO does not feature anchor texts, and its documents are only sparsely linked. To overcome this shortcoming for the reproduction of the results of Craswell et al. and Eiron and McCurley on MS MARCO, we compile the Webis MS MARCO Anchor Text 2022 dataset by extracting anchor texts from web pages linking to MS MARCO documents found in Common Crawl snapshots. A high recall has been achieved by processing one randomly selected snapshot from each year between 2016 and 2021 (between 1.7–3.4 billion documents each). Unlike Craswell et al. and Eiron and McCurley, we applied the three filtering steps developed by Chen et al. [5] to remove low-quality anchor texts. An anchor text has been omitted, if it consisted of (1) one or more of the manually selected "stop words" 'click', 'read', 'link', 'mail', 'here', and 'open'; (2) more than 10 words, since these are often due to parsing errors; or, if it (2) originated from an intra-site link (i.e., same source and target domain), since anchor texts of inter-site links are usually more descriptive [42]. These filtering steps removed about 50% of all anchor texts pointing to MS MARCO documents.

Processing the total 17.12 billion Common Crawl documents (343 TiB compressed WARC files) on our 3000 CPU Hadoop cluster [48] yielded 8.16 billion anchor texts for MS MARCO documents. A first data analysis revealed that most links point to only a few very popular documents. To obtain a sensible dataset size both for our experiments and future users, we applied min-wise sampling of 1,000 anchor texts for documents that are targeted by more links than that. This stratified sampling still ensured the inclusion of all anchor texts for most of the documents (94% for MS MARCO version 1; 97% for version 2), downsampling only the most popular documents.

Table 1 shows an overview of all extracted anchor texts (column group 'Anchors') and the downsampled subsets for the two MS MARCO versions ('Sample@V1' and 'Sample@V2'). Overall, the combined samples cover 1.70 million documents of Version 1 (53% of all documents) and 4.82 million documents of Version 2 (40%). For

Table 1. The Webis MS MARCO Anchor Text 2022 dataset at a glance. The samples for Versions 1 and 2 (Sample@V1/V2) include at most 1,000 anchor texts per MS MARCO document.

Common Crawl snapshot			Anchors		Sample@V1		Sample@V2	
Snapshot	Docs	Size	V1	V2	Anchors	Docs cov.	Anchors	Docs cov.
2016-07	1.73 b	28.57 TiB	1.05 b	0.75 b	54.05 m	0.83 m	65.04 m	1.49 m
2017-04	3.14 b	53.95 TiB	0.95 b	0.91 b	61.19 m	1.18 m	94.35 m	2.34 m
2018-13	3.20 b	67.66 TiB	0.83 b	0.68 b	81.24 m	1.27 m	116.59 m	2.45 m
2019-47	2.55 b	53.95 TiB	0.55 b	0.41 b	65.60 m	1.16 m	90.18 m	2.83 m
2020-05	3.10 b	59.94 TiB	0.67 b	0.48 b	78.46 m	1.24 m	108.16 m	3.10 m
2021-04	3.40 b	78.98 TiB	0.52 b	0.36 b	60.62 m	1.14 m	84.93 m	3.18 m
$\sum$	17.12 b	343.05 TiB	4.57 b	3.59 b	207.28 m	1.70 m	341.17 m	4.82 m

each anchor text, our datasets also contain the source URL, the target URL, and the MS MARCO ID of the target document. Besides releasing the dataset to the community, we employ it to reproduce the main findings of Eiron and McCurley [24] (next section) and the retrieval effectiveness results of Craswell et al. [11] (Sect. 5).

4 Properties of Anchor Texts, Queries, and Documents

In 2003, Eiron and McCurley [24] studied properties of anchor texts, queries, and documents on the IBM intranet (2.95 million documents, 2.57 million anchor texts, and 1.27 million queries). They found that anchor texts closely resembled query length, that terms in document titles/bodies and in anchor texts often have different meanings, and that retrieval against anchor text yielded more homogeneous results than against document content. Eiron and McCurley also conducted a study on retrieval effectiveness but we do not reproduce their setup (without relevance judgments) but instead reproduce the retrieval experiments of Craswell et al. [11] with relevance judgments (cf. Sect. 5).

Analyzing to what extent the similarity of anchor texts and queries that Eiron and McCurley observed can be reproduced in a current retrieval scenario is particularly important, since the observation had inspired others to replace proprietary query logs by anchor texts [7,20,38]. We repeat the study of Eiron and McCurley on the MS MARCO Version 1 dataset and the ORCAS query log [8] linked to it. Interestingly, in our "modern" web search scenario with about 27 times more anchor texts (81.24 million in the 2018 subset matching the MS MARCO Version 1 crawling date) and 15 times more queries (18.82 million from ORCAS), we obtain some substantially different results.

Number of Distinct Terms. The plots in Fig. 1 show the distributions of the number of distinct terms per anchor text, query, or document title as reported by Eiron and McCurley for their IBM dataset (left plot) and what we observe for MS MARCO (right). While Eiron and McCurley reported the distributions for anchor texts and queries as highly similar, we find them to be rather dissimilar on MS MARCO.

To assess the similarity of the distributions, we calculate the symmetric Jensen-Shannon distance [25] for all pairs (right plot of Fig. 2; a distance of 0 indicates equal

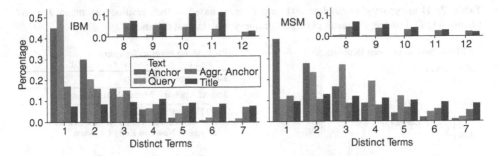

Fig. 1. Distributions of the number of distinct terms in anchor texts, queries, document titles, and aggregated anchor texts (all anchors combined that point to a document) on the IBM data (left) and MS MARCO (MSM; right).

Queries w/ significant differences			
p-value	More	Less	Equally
0.05	6,770	1,121	71
0.01	6,764	1,113	85
0.001	6,748	1,098	116

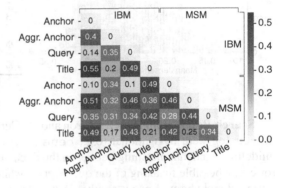

Fig. 2. Left: Number of queries with significantly more, less, or equally homogeneous content-based results. Right: Jensen-Shannon distance of all pairs (0 = identical).

distributions). The anchor text distributions are very similar for the MS MARCO and the IBM data (distance of 0.10) as are the distributions of anchor texts and queries for the IBM data (0.14). However, on the MS MARCO data, anchor texts and queries are more dissimilar (0.28), probably mainly due to the more "web-like" query distribution: the IBM query distribution is pretty different to the ORCAS queries (distance of 0.34; most IBM queries have one term, most ORCAS queries have three terms, etc.).

Frequent Terms. Eiron and McCurley also compared the 16 most frequent terms in document titles, queries, and anchor texts and found that these are rather different. Some terms like 'of' are frequent in all types but most terms frequent in one type are rare in the other types. Eiron and McCurley then argued that the different frequencies indicate that anchor texts should be kept separate and not mixed with document content such that methods depending on term frequencies could better exploit the different contexts of a term's frequencies. We can confirm the observed substantial differences also for the MS MARCO scenario. For example, the frequent terms 'you', 'it', and 'are' for titles, 'meaning', 'online', and 'free' for ORCAS queries, as well as 'home', 'university', or 'website' for anchor texts very rarely occur in the other types.

Table 2. Homogeneity of anchor text and content-based search results: (a) mean Jensen-Shannon (JS) distance, (b) result excerpts for query with largest distance

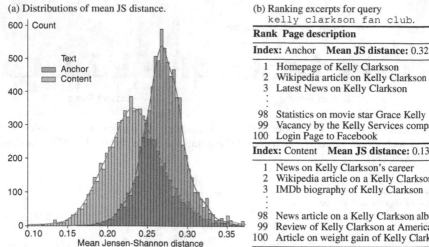

(a) Distributions of mean JS distance.

(b) Ranking excerpts for query `kelly clarkson fan club.`

Rank	Page description
Index: Anchor	**Mean JS distance: 0.32**
1	Homepage of Kelly Clarkson
2	Wikipedia article on Kelly Clarkson
3	Latest News on Kelly Clarkson
⋮	
98	Statistics on movie star Grace Kelly
99	Vacancy by the Kelly Services company
100	Login Page to Facebook
Index: Content	**Mean JS distance: 0.13**
1	News on Kelly Clarkson's career
2	Wikipedia article on a Kelly Clarkson single
3	IMDb biography of Kelly Clarkson
⋮	
98	News article on a Kelly Clarkson album
99	Review of Kelly Clarkson at American Idol
100	Article on weight gain of Kelly Clarkson

Search Result Homogeneity. Eiron and McCurley reported that most of the queries in their log were navigational (e.g., `benefits` or `travel` to find respective IBM guidelines) and that matching queries in the document content tended to retrieve results for every possible meaning of the query terms while matching only in the anchor texts retrieved more homogeneous results—but in an experiment with only 14 queries.

On 10,000 randomly sampled ORCAS queries, we follow the setup of Eiron and McCurley: we rank the MS MARCO documents by either matching their anchor texts or their content, we remove queries with less than 800 results (7,962 queries remain), and we measure the results' homogeneity using the method of Kilgarriff and Rose [33] to compute the mean Jensen-Shannon distances; distributions shown in Table 2a.

In contrast to Eiron and McCurley, we observe that retrieval against document content yields more homogeneous results than against anchor text (cf. Fig. 2 (left table); content yields more homogeneous results for more than 6,700 queries). For example, the top-100 content-based results for the query `kelly clarkson fan club` all refer to Kelly Clarkson while the anchor text-based results are more "diverse" (cf. the excerpts in Table 2b). An explanation for the difference to the observation of Eiron and McCurley probably is twofold: (1) our large-scale dataset has rather diverse authors and queries from different searchers while in the IBM data anchor text writers and searchers probably were IBM employees with experience in intranet search, and, probably more importantly, (2) Eiron and McCurley have experimented with 14 queries only.

5 Anchor Text and Retrieval Effectiveness

To reproduce the result of Craswell et al. [11] (that anchor text helps for navigational queries), we compare the effectiveness of traditional and modern content-based retrieval for navigational queries to the effectiveness of focused retrieval in the MS MARCO anchor text datasets. We also further extend the experiment to the queries with judgments from the TREC Deep Learning tracks [13–15]—all of them informational queries.

5.1 Navigational Queries for MS MARCO

Craswell et al. [11] experimented with three sets of navigational queries to demonstrate the effectiveness of anchor text for web search. For a web crawl with 18.5 million pages, they created 100 navigational queries for random entry pages and 100 navigational queries for random popular entry pages (selected from a manually maintained Yahoo! list of popular entry pages). Additionally, for a crawl of 0.4 million documents from the domain of the Australian National University, they created 100 navigational queries pointing to academic persons or institutions—we omit those academic queries from our reproduction to focus on general web search.

Following Craswell et al. [11], we created 100 navigational queries for random entry pages and 100 for popular entry pages in the MS MARCO document sets as follows. We extracted all MS MARCO Version 1 documents that potentially are entry pages by applying the respective rules of Westerveld et al. [49] (URL-path must be empty or must be index.html). From the resulting 92,562 candidates, we selected 100 pages at random and 100 documents at random with domains listed in the Alexa top-1000 ranking of 2018 (probable crawl date of the MS MARCO Version 1 document set). To actually create the 200 navigational queries, we manually inspected each of the 200 target pages and formulated a query that searchers would probably use to search for that page. We then also checked whether the page is still present in MS MARCO Version 2 and whether the same navigational query still applies. For 194 query–document pairs, the transfer was easily possible while for the 6 remaining ones we manually had to correct changed URLs (e.g., calendar.live.com → outlook.live.com).

5.2 Retrieval Models and Training

For navigational queries, Craswell et al. [11] compared the effectiveness of BM25-based retrieval using document content to BM25-based retrieval using anchor texts. In our reproducibility study, we substantially extend this setup by employing 18 different retrieval systems. We use different anchor text sets to evaluate the effectiveness of anchor text over time and include novel retrieval models that did not exist during the evaluation of Craswell et al. back in 2001.

Seven of the systems in our study retrieve results only against anchor texts using BM25 as the retrieval model; six systems for six different Common Crawl versions of our anchor text dataset and a seventh system that uses all the combined anchor texts. From the other eleven systems that we use for comparison, six solely use the documents' content (one is BM25-based), while the remaining five systems use combinations of document content, anchor text, and ORCAS query–click information [8]. Nine of the eleven comparison systems employ approaches that did not exist during the evaluation of Craswell et al.: DeepCT [18, 19], MonoBERT [43], MonoT5 [44], and LambdaMART [4] (cf. left column of Table 4 for a list of all the 18 systems). For DeepCT, we use different training setups (with or without access to query log information and anchor texts), and for LambdaMART, we use different sets of features (with or without access to query log information and anchor texts) such that we can assess the importance of anchor texts in such models as an additional case study.

We use the Anserini toolkit [51] in our experiments and follow Craswell et al. [11] by not tuning the parameters of BM25—keeping them at Anserini's defaults of $k = 0.9$ and $b = 0.4$. In general, we preprocess queries and the indexed texts via Porter stemming and stopword removal using Lucene's default stopwords for English but for re-ranking documents using MonoT5 and MonoBERT, we follow Nogueira et al. [44] and omit stemming and stopword removal. For all rankers, we break score ties within a ranking via alphanumeric ordering by document ID as implemented in Anserini (given random document IDs, this leads to a random distribution with respect to other document properties such as text length [37]).

BM25 on Anchor Text. Following Craswell et al. [11], we concatenate all anchor texts pointing to the same target page and index these aggregated anchor text "documents" in dedicated Anserini BM25 indexes for all 14 anchor text samples (6 individual Common Crawl versions and their combination for MS MARCO Version 1 and Version 2; see Table 1). At query time, the actual documents are returned in the order of their retrieved aggregated anchor text "documents". With this setup, we mimic the corresponding baseline of Craswell et al. with the novel aspect that we can compare the retrieval effectiveness for the individual anchor text subsets and their combination.

BM25 on Content. Mimicking the baseline of Craswell et al. [11], we concatenate the title and body of the documents and create a respective Anserini BM25 index.

DeepCT on Content. DeepCT [18, 19] estimates the importance of terms in their context, removing unimportant terms while including multiple copies of important terms. With its focus on precision, DeepCT could be particularly suited for navigational queries. We train three DeepCT models: on the training data of MS MARCO Version 1, on the ORCAS data, and on our combined anchor texts. Interestingly, Dai and Callan [18] designed DeepCT to use anchor text as training data but had not tried it for MS MARCO since no anchor text dataset existed—a gap that we now close with the release of our anchor text data and our respective results for DeepCT.

Following Dai and Callan [18], we compute the importance of a term t in a document d as the fraction of queries with clicks on d that contain t as a query term or the fraction of anchor texts pointing to d that contain t. The three different DeepCT-based systems in our comparison are trained on the queries in the official MS MARCO Version 1 training data, on the queries in the ORCAS data, and on our new anchor text data. To avoid any train/test leakage, we remove the 270,511 MS MARCO documents from the training for which any query or anchor text in the training data contains a term from any of the 200 navigational queries used in our evaluation. The DeepCT systems thus are trained on 249,046 documents for the official MS MARCO training data, on

Table 3. (a) Characteristics of the train/test leakage filtered term importance datasets for DeepCT: MS MARCO training data (MARCO), ORCAS data (ORCAS), and the combined anchor texts from the Common Crawls (Anchor). (b) Pairwise comparison of the importance scores' correlations (Kendall's τ, Pearson's ρ) and the Jaccard similarity (J) of terms with non-zero weights.

(a) Term importance training datasets.				(b) Comparison of importance scores.			
Dataset	Docs	Passages	w/o imp. term	Compared datasets	τ	ρ	J
MARCO	0.25 m	2.08 m	0.29 m	Anchor vs. ORCAS	**0.39**	**0.61**	**0.53**
ORCAS	0.88 m	8.17 m	0.92 m	ORCAS vs. MARCO	0.35	0.46	0.51
Anchor	1.43 m	11.64 m	2.02 m	Anchor vs. MARCO	0.26	0.41	0.45

876,950 documents for the ORCAS data, and on 1,432,621 documents for the combined anchor texts. Following a suggestion of Dai and Callan [19], each document is split into fixed-length passages of 250 terms since working with fixed-length passages is more effective than variable-length original passages [31] (passage splitting done with the TREC CAsT tools[2]). Table 3a shows the characteristics of the training datasets including the number of passages that do not contain any important term.

Table 3b shows the correlations (Kendalls τ and Pearsons ρ) of the term importance scores derived from the three training datasets and also the Jaccard similarity of the term sets with non-zero importance scores. Interestingly, anchor texts and the ORCAS queries lead to more similar scores than the two query sets. Still, the differences for any pair are large enough so that we decided to train and compare three individual models. For the training, we use the implementation of Dai and Callan [18] and follow their suggestions: each DeepCT model is trained with a maximum input length of 512 tokens for 100,000 steps with a batch size of 16 and a learning rate of 2e-5. For inference, we process all passages with PyTerrier [40] and index the documents (processed passages concatenated again) in an Anserini BM25 index.

MonoBERT and MonoT5 on Content. Since Transformer-based re-rankers recently caused a paradigm shift in information retrieval [52], we include two such systems in our experiments: MonoBERT [43], the first re-ranker based on BERT [21], and MonoT5 [44] that outperforms MonoBERT on MS MARCO and Robust04 [52] by classifying the relevance of a document to a given query using the sequence-to-sequence Transformer T5 [46]. For both, MonoBERT and MonoT5, we use the implementations of PyGaggle[3] and let the default trained castorini/monobert-large-msmarco model and the castorini/monot5-base-msmarco model re-rank the top-100 BM25 results via the maximum score of a passage as the document score.

BM25 on ORCAS. For each document d, we concatenate all queries that have clicks on d in the ORCAS data and index these aggregated query "documents" with Anserini's BM25 implementation. At query time, the actual documents are returned in the order of their retrieved aggregated query "documents". Note that in the TREC 2021 Deep Learning track that uses MS MARCO Version 2 the ORCAS query log should

---

[2] https://github.com/grill-lab/trec-cast-tools.
[3] https://github.com/castorini/pygaggle.

Table 4. Effectiveness of the 18 retrieval systems in our comparison as mean reciprocal rank (MRR), recall at 3 (R@3), and recall at 10 (R@10) on 100 navigational queries for random entry pages and 100 navigational queries for popular entry pages in MS MARCO version 1 (V1) and version 2 (V2). Bold: highest scores per group.

Retrieval system		Random@V1			Popular@V1			Random@V2			Popular@V2		
		MRR	R@3	R@10	MRR	R@3	R@10	MRR	R@3	R@10	MRR	R@3	R@10
Anchor	BM25@2016-07	0.61	0.63	0.68	**0.62**	**0.72**	0.83	0.56	0.61	0.64	**0.57**	**0.64**	**0.80**
	BM25@2017-04	0.63	0.70	0.73	0.59	0.67	0.84	0.59	0.68	0.70	0.48	0.56	0.73
	BM25@2018-13	0.70	0.76	0.82	0.54	0.65	0.81	0.62	0.68	0.77	0.47	0.54	0.77
	BM25@2019-47	0.63	0.74	0.78	0.58	0.69	0.84	0.59	0.62	0.76	0.49	0.57	0.78
	BM25@2020-05	0.63	0.72	0.79	0.55	0.66	**0.86**	0.56	0.64	0.71	0.45	0.53	0.74
	BM25@2021-04	0.63	0.73	0.77	0.54	0.66	0.80	0.50	0.54	0.64	0.46	0.55	0.73
	BM25@Anchor	**0.74**	**0.83**	**0.89**	0.55	0.66	0.84	**0.67**	**0.73**	**0.85**	0.39	0.48	0.70
Content	BM25@Content	0.21	0.24	0.36	0.02	0.02	0.03	0.21	0.22	0.42	0.02	0.01	0.04
	DeepCT@Anchor	**0.43**	**0.46**	**0.58**	**0.03**	**0.03**	0.08	**0.43**	**0.49**	**0.66**	0.04	0.03	**0.13**
	DeepCT@ORCAS	0.38	0.42	0.57	0.02	0.00	**0.09**	0.36	0.40	0.60	**0.05**	**0.04**	0.10
	DeepCT@Train	0.27	0.28	0.44	0.02	0.01	0.05	0.32	0.34	0.49	0.03	0.02	0.08
	MonoT5	0.39	0.43	0.53	0.02	0.01	0.05	0.38	0.43	0.57	0.04	**0.04**	0.08
	MonoBERT	0.35	0.37	0.51	0.02	0.01	0.05	0.36	0.41	0.56	0.01	0.01	0.02
Other	BM25@ORCAS	**0.60**	**0.64**	**0.70**	0.28	0.32	0.43	0.56	0.59	0.66	**0.28**	**0.33**	**0.44**
	λ-MART@BTOA	0.48	0.55	0.63	0.08	0.07	0.18	0.52	0.57	**0.77**	0.12	0.12	0.21
	λ-MART@BTO	0.41	0.49	0.57	0.07	0.06	0.17	0.49	0.55	0.65	0.08	0.10	0.14
	λ-MART@BTA	0.43	0.51	0.61	0.06	0.06	0.19	0.55	**0.62**	0.75	0.14	0.15	0.24
	λ-MART@BT	0.27	0.31	0.46	0.04	0.03	0.09	0.40	0.44	0.60	0.05	0.05	0.08

not be used since it might cause train/test leakage.[4] However, since we do not evaluate the effectiveness of retrieval models on the topics of the TREC 2021 Deep Learning track, this potential train/test leakage can not occur in our situation and we can use the ORCAS query log also for MS MARCO Version 2 in our navigational query scenario without the risk of train/test leakage.

LambdaMART. To study the effectiveness of anchor text in combination with other features and to analyze whether the observation still holds that anchor text adds only small or no effectiveness in TREC scenarios [24], we train four LambdaMART [4] models—the state-of-the-art for feature-based learning to rank [4,28,50]—on the training and validation labels of MS MARCO Version 1. Again, since we removed the MS MARCO documents from the training for which any query or anchor text contains a term from any of the 200 navigational queries used in our evaluation, there is no risk of train/test leakage. In our setup, we distinguish four feature sources: anchor texts, ORCAS queries, document titles, and document bodies. For each of the four sources, we calculate the following eight feature types using Anserini: TF, TF·IDF, BM25, F2exp, QL, QLJM, PL2, and SPL. Four LambdaMART models are trained with LightGBM [32] on different feature subsets: (1) using all 32 feature types (λ-MART@BTOA), (2) using body, title, and ORCAS (λ-MART@BTO), (3) using body, title, and anchor text (λ-MART@BTA), and (4) using body and title (λ-MART@BT).

---

[4] https://microsoft.github.io/msmarco/TREC-Deep-Learning.html.

5.3   Evaluation

We experimentally compare the effectiveness of the 18 retrieval models on Version 1 and Version 2 of MS MARCO. In a first experiment, we use the above described 200 navigational queries created for MS MARCO to try to reproduce the result of Craswell et al. [11] that anchor text helps to improve MRR for navigational queries. We extend the original study by adding novel aspects like modern neural baselines and by evaluating the effectiveness of anchor text over time. In a second experiment, we then also evaluate the 18 retrieval models on the 88 informational topics from the TREC 2019 and 2020 Deep Learning tracks. Any reported significance test uses $p \leq 0.05$ and includes a Bonferroni correction in case of multiple comparisons.

Retrieval Effectiveness for Navigational Queries. Table 4 shows the retrieval effectiveness for the 200 navigational topics on MS MARCO Version 1 and Version 2.

For queries pointing to random entry pages (columns 'Random@V1' and 'Random@V2'), BM25 retrieval against the combined anchor texts (BM25@Anchor) achieves the best effectiveness scores. While the scores for BM25 on single anchor text snapshots are a little lower (the combination on average has 450 anchor texts per random entry page, each individual snapshot less than 250), the MRR differences from any anchor text-based BM25 retrieval to the best content-based retrieval, DeepCT with importance scores trained on the anchor texts (DeepCT@Anchor), are significant. Within the content-based approaches, the recent improvements of neural approaches are also visible for our navigational queries: the score differences of DeepCT trained on anchor texts or ORCAS, of MonoT5, and of MonoBERT to 'the BM25 content-based retrieval all are statistically significant—as are the differences of the three better LambdaMART models to content-based BM25. Interestingly, also BM25 retrieval on ORCAS queries improves upon all content-only models (all MRR differences are significant), even reaching the effectiveness of some anchor text models. Still, BM25 against the combined anchor texts or the ones from 2018 significantly improves upon BM25 against ORCAS.

For queries pointing to popular entry pages (columns 'Popular@V1' and 'Popular@V2'), all anchor text-based BM25 models are statistically significantly more effective than any other model. Also BM25 on ORCAS queries is significantly better than all non-anchor-based models, again highlighting some similarity of anchor texts to queries.

Altogether, our results confirm the result of Craswell et al. [11] that retrieval against anchor texts is better than retrieval against document content for navigational queries— in our experiments now even including modern neural content-based approaches. However, in almost all of our experimental cases, retrieval for queries pointing to popular entry pages is less effective than for random entry pages. This contradicts an observation of Craswell et al. [11] who reported lower MRR scores for queries pointing to random entry pages than for queries pointing to popular entry pages. For content-based retrieval, the problem is that many other pages "talk" about popular entry pages and mention the respective query terms more often than the actual popular page does.

Retrieval Effectiveness of Anchor Text over Time. To further inspect the impact of crawling time on anchor text effectiveness, we look more deeply into navigational queries that yield at least 100 results against any anchor text snapshot. From the 200 queries, this filtering removes 47 for MS MARCO Version 1 (27 random, 20 popu-

Table 5. (a) Overview of the effectiveness of anchor text on our navigational topics over the crawling period between 2016 and 2021. (b) Overview of the retrieval effectiveness on the TREC Deep Learning topics from 2019 and 2020 where we report nDCG@10 and nDCG@20.

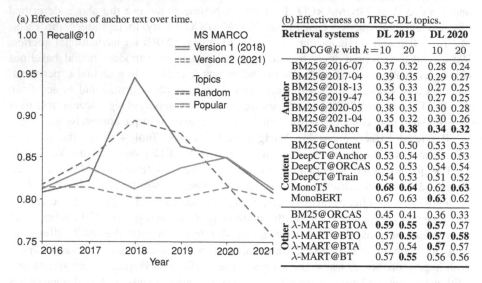

(a) Effectiveness of anchor text over time.

(b) Effectiveness on TREC-DL topics.

Retrieval systems	DL 2019		DL 2020	
nDCG@k with $k=$	10	20	10	20
BM25@2016-07	0.37	0.32	0.28	0.24
BM25@2017-04	0.39	0.35	0.29	0.27
BM25@2018-13	0.35	0.33	0.27	0.25
BM25@2019-47	0.34	0.31	0.27	0.25
BM25@2020-05	0.38	0.35	0.30	0.28
BM25@2021-04	0.35	0.32	0.30	0.26
BM25@Anchor	**0.41**	**0.38**	**0.34**	**0.32**
BM25@Content	0.51	0.50	0.53	0.53
DeepCT@Anchor	0.53	0.54	0.55	0.53
DeepCT@ORCAS	0.52	0.53	0.54	0.54
DeepCT@Train	0.54	0.53	0.51	0.52
MonoT5	**0.68**	**0.64**	0.62	**0.63**
MonoBERT	0.67	0.63	**0.63**	0.62
BM25@ORCAS	0.45	0.41	0.36	0.33
λ-MART@BTOA	**0.59**	**0.55**	**0.57**	0.57
λ-MART@BTO	0.57	**0.55**	**0.57**	**0.58**
λ-MART@BTA	0.57	0.54	**0.57**	0.57
λ-MART@BT	0.57	**0.55**	0.56	0.56

lar) and 53 for Version 2 (34 random, 19 popular). Table 5a shows the Recall@10 over time for the remaining queries. For popular pages, there are only slight changes since they always have many anchors pointing to them. As for the random pages, the anchor text crawling time has a larger impact. In particular, the effectiveness peaks at 2018, reflecting the creation date of MS MARCO Version 1. We also observe this peak for Version 2 (crawled in 2021) since we use the same queries that we originally created by sampling pages from Version 1. Not surprisingly, anchor text indexes should thus be refreshed from time to time to match the temporal changes of navigational queries.

Retrieval Effectiveness for Informational Queries. In a final experiment, we evaluate the effectiveness of the 18 retrieval systems on the TREC Deep Learning tracks of 2019 [14] and 2020 [13] on MS MARCO Version 1 (judgments for Version 2 were not yet available)—the respective 88 topics all are informational. Since not all of the 18 systems did contribute to the judgment pools, we removed all unjudged documents from the rankings to mitigate bias as suggested by Sakai [47]. Table 5b shows the resulting nDCG@10 and nDCG@20 scores. Unsurprisingly, the modern Transformer-based MonoT5 and MonoBERT models achieve the overall best scores. For these informational queries, all models solely based on anchor texts or queries are less effective than BM25 on the content of the documents. Still, more anchor text is more effective (BM25@Anchor). Still, the LambdaMART results show that combining content-based retrieval with anchor texts and queries can very slightly improve the effectiveness. Overall, our experiments confirm the earlier observation [24] that anchor text alone is not effective in TREC-style scenarios with a focus on informational queries.

6 Conclusion

In the scenario of the MS MARCO dataset, we have successfully reproduced the result of Craswell et al. [11] that anchor text is very effective for navigational queries. Trying to also reproduce the other seminal anchor text study of Eiron and McCurley [24] we obtained rather different results. We found that the term distributions of anchor texts and queries today are rather dissimilar and that retrieval against anchor text now yields less homogeneous results than retrieval against the document content.

Besides the above positive and negative reproducibility results, another important result of our study is that Transformer-based approaches, be it in re-ranking scenarios or in the DeepCT context of estimating term importance, are less effective for navigational queries than a "basic" anchor text-oriented BM25 retrieval. Identifying navigational queries and switching to anchor text-based retrieval for them instead of neural models might thus improve the retrieval effectiveness of a general retrieval system. However, in the popular TREC Deep Learning tracks, the impact will be rather limited since the Deep Learning tracks do not involve navigational queries. Our code and the newly created Webis MS MARCO Anchor Texts 2022 datasets are freely available (see footnote (1)).

References

1. Bailey, P., Craswell, N., Hawking, D.: Engineering a multi-purpose test collection for web retrieval experiments. Inf. Process. Manag. **39**(6), 853–871 (2003)
2. Brin, S., Page, L.: The anatomy of a large-scale hypertextual web search engine. Comput. Netw. **30**(1–7), 107–117 (1998)
3. Broder, A.Z.: A taxonomy of web search. SIGIR Forum **36**(2), 3–10 (2002)
4. Burges, C.J.: From RankNet to LambdaRank to LambdaMART: an overview. Learning **11**(23–581), 81 (2010)
5. Chen, W.F., Syed, S., Stein, B., Hagen, M., Potthast, M.: Abstractive snippet generation. In: Huang, Y., King, I., Liu, T., van Steen, M. (eds.) Proceedings of the World Wide Web Conference, WWW 2020, San Francisco, CA, USA, 20–24 April 2020, pp. 1309–1319. ACM (2020). ISBN 978-1-4503-7023-3
6. Clarke, C.L.A., Craswell, N., Soboroff, I.: Overview of the TREC 2009 web track. In: Voorhees, E.M., Buckland, L.P. (eds.) Proceedings of the 18th Text REtrieval Conference, TREC 2009, Gaithersburg, MD, USA, 17–20 November, 2009. NIST Special Publication, vol. 500–278, National Institute of Standards and Technology (NIST) (2009)
7. Craswell, N., Billerbeck, B., Fetterly, D., Najork, M.: Robust query rewriting using anchor data. In: Leonardi, S., Panconesi, A., Ferragina, P., Gionis, A. (eds.) Proceedings of the 6th ACM International Conference on Web Search and Data Mining, WSDM 2013, Rome, Italy, 4–8 February 2013, pp. 335–344. ACM (2013)
8. Craswell, N., Campos, D., Mitra, B., Yilmaz, E., Billerbeck, B.: ORCAS: 20 million clicked query-document pairs for analyzing search. In: d'Aquin, M., Dietze, S., Hauff, C., Curry, E., Cudré-Mauroux, P. (eds.) Proceedings of the 29th ACM International Conference on Information and Knowledge Management, CIKM 2020, Virtual Event, Ireland, 19–23 October 2020, pp. 2983–2989. ACM (2020)

9. Craswell, N., Hawking, D.: Overview of the TREC-2002 web track. In: Voorhees, E.M., Buckland, L.P. (eds.) Proceedings of the 11th Text REtrieval Conference, TREC 2002, Gaithersburg, MD, USA, 19–22 November 2002. NIST Special Publication, National Institute of Standards and Technology (NIST), vol. 500–251 (2002)

10. Craswell, N., Hawking, D.: Overview of the TREC 2004 web track. In: Voorhees, E.M., Buckland, L.P. (eds.) Proceedings of the 13th Text REtrieval Conference, TREC 2004, Gaithersburg, MD, USA, 16–19 November 2004. NIST Special Publication, National Institute of Standards and Technology (NIST), vol. 500–261 (2004)

11. Craswell, N., Hawking, D., Robertson, S.E.: Effective site finding using link anchor information. In: Croft, W.B., Harper, D.J., Kraft, D.H., Zobel, J. (eds.) Proceedings of the 24th Annual International ACM SIGIR Conference on Research and Development in Information Retrieval, SIGIR 2001, New Orleans, LA, USA, 9–13 September 2001, pp. 250–257. ACM (2001)

12. Craswell, N., Hawking, D., Wilkinson, R., Wu, M.: Overview of the TREC 2003 web track. In: Voorhees, E.M., Buckland, L.P. (eds.) Proceedings of the 12th Text REtrieval Conference, TREC 2003, Gaithersburg, MD, USA, 18–21 November 2003. NIST Special Publication, National Institute of Standards and Technology (NIST), vol. 500–255, pp. 78–92 (2003)

13. Craswell, N., Mitra, B., Yilmaz, E., Campos, D.: Overview of the TREC 2020 deep learning track. In: Voorhees, E.M., Ellis, A. (eds.) Proceedings of the 29th Text REtrieval Conference, TREC 2020, Virtual Event, Gaithersburg, MD, USA, 16–20 November 2020. NIST Special Publication, National Institute of Standards and Technology (NIST), vol. 1266 (2020)

14. Craswell, N., Mitra, B., Yilmaz, E., Campos, D., Voorhees, E.M.: Overview of the TREC 2019 deep learning track. In: Voorhees, E., Ellis, A. (eds.) 28th International Text Retrieval Conference, TREC 2019. Maryland, USA, NIST Special Publication, National Institute of Standards and Technology (NIST) (2019)

15. Craswell, N., Mitra, B., Yilmaz, E., Campos, D., Voorhees, E.M., Soboroff, I.: TREC Deep learning track: reusable test collections in the large data regime. In: Diaz, F., Shah, C., Suel, T., Castells, P., Jones, R., Sakai, T. (eds.) Proceedings of the 44th International ACM SIGIR Conference on Research and Development in Information Retrieval, SIGIR 2021, Virtual Event, Canada, 11–15 July 2021, pp. 2369–2375. ACM (2021)

16. Croft, W.B., Metzler, D., Strohman, T.: Search Engines - Information Retrieval in Practice. Pearson Education (2009). ISBN 978-0-13-136489-9

17. Dai, N., Davison, B.D.: Mining anchor text trends for retrieval. In: Gurrin, C., et al. (eds.) ECIR 2010. LNCS, vol. 5993, pp. 127–139. Springer, Heidelberg (2010). https://doi.org/10.1007/978-3-642-12275-0_14

18. Dai, Z., Callan, J.: Context-aware sentence/passage term importance estimation for first stage retrieval. CoRR abs/1910.10687 (2019)

19. Dai, Z., Callan, J.: Context-aware document term weighting for ad-hoc search. In: Huang, Y., King, I., Liu, T., van Steen, M. (eds.) Proceedings of the World Wide Web Conference, WWW 2020, Taipei, Taiwan, 20–24 April 2020, pp. 1897–1907. ACM/IW3C2 (2020)

20. Dang, V., Croft, W.B.: Query reformulation using anchor text. In: Davison, B.D., Suel, T., Craswell, N., Liu, B. (eds.) Proceedings of the 3rd ACM International Conference on Web Search and Web Data Mining, WSDM 2010, New York, NY, USA, 4–6 February 2010, pp. 41–50, ACM (2010)

21. Devlin, J., Chang, M., Lee, K., Toutanova, K.: BERT: pre-training of deep bidirectional transformers for language understanding. In: Burstein, J., Doran, C., Solorio, T. (eds.) Proceedings of the 2019 Conference of the North American Chapter of the Association for Computational Linguistics: Human Language Technologies, NAACL-HLT 2019, Minneapolis, MN, USA, June 2–7, 2019, Volume 1, pp. 4171–4186. Association for Computational Linguistics (2019)

22. Dou, Z., Song, R., Nie, J., Wen, J.: Using anchor texts with their hyperlink structure for web search. In: Allan, J., Aslam, J.A., Sanderson, M., Zhai, C., Zobel, J. (eds.) Proceedings of the 32nd Annual International ACM SIGIR Conference on Research and Development in Information Retrieval, SIGIR 2009, Boston, MA, USA, 19–23 July 2009, pp. 227–234. ACM (2009)

23. Dunlop, M.D., van Rijsbergen, C.J.: Hypermedia and free text retrieval. Inf. Process. Manag. **29**(3), 287–298 (1993)

24. Eiron, N., McCurley, K.S.: Analysis of Anchor text for web search. In: Clarke, C.L.A., Cormack, G.V., Callan, J., Hawking, D., Smeaton, A.F. (eds.) Proceedings of the 26th Annual International ACM SIGIR Conference on Research and Development in Information Retrieval, SIGIR 2003, Toronto, Canada, 28 July–1 August 2003, pp. 459–460. ACM (2003)

25. Fuglede, B., Topsøe, F.: Jensen-Shannon divergence and Hilbert space embedding. In: Proceedings of the 2004 IEEE International Symposium on Information Theory, ISIT 2004, Chicago Downtown Marriott, Chicago, IL, USA, 27 June–2 July 2004, p. 31. IEEE (2004)

26. Hawking, D.: Overview of the TREC-9 web track. In: Voorhees, E.M., Harman, D.K. (eds.) Proceedings of the 9th Text REtrieval Conference, TREC 2000, Gaithersburg, MD, USA, 13–16 November 2000. NIST Special Publication, National Institute of Standards and Technology (NIST), vol. 500–249 (2000)

27. Hawking, D., Voorhees, E.M., Craswell, N., Bailey, P.: Overview of the TREC-8 web track. In: Voorhees, E.M., Harman, D.K. (eds.) Proceedings of the 8th Text REtrieval Conference, TREC 1999, Gaithersburg, MD, USA, 17–19 November 1999. NIST Special Publication, National Institute of Standards and Technology (NIST), vol. 500–246 (1999)

28. Hu, Z., Wang, Y., Peng, Q., Li, H.: Unbiased LambdaMART: an unbiased pairwise learning-to-rank algorithm. In: Liu, L., et al. (eds.) Proceedings of the World Wide Web Conference, WWW 2019, San Francisco, CA, USA, 13–17 May 2019, pp. 2830–2836. ACM (2019)

29. Kamps, J., Kaptein, R., Koolen, M.: Using anchor text, spam filtering and Wikipedia for web search and entity ranking. In: Voorhees, E.M., Buckland, L.P. (eds.) Proceedings of the 19th Text REtrieval Conference, TREC 2010, Gaithersburg, MD, USA, 16–19 November 2010. NIST Special Publication, National Institute of Standards and Technology (NIST), vol. 500–294 (2010)

30. Karpukhin, V., et al.: Dense passage retrieval for open-domain question answering. In: Webber, B., Cohn, T., He, Y., Liu, Y. (eds.) Proceedings of the 2020 Conference on Empirical Methods in Natural Language Processing, EMNLP 2020, Virtual Event, 16–20 November 2020, pp. 6769–6781. Association for Computational Linguistics (2020)

31. Kaszkiel, M., Zobel, J.: Passage retrieval revisited. In: Belkin, N.J., Narasimhalu, A.D., Willett, P., Hersh, W.R., Can, F., Voorhees, E.M. (eds.) Proceedings of the 20th Annual International ACM SIGIR Conference on Research and Development in Information Retrieval, SIGIR 1997, Philadelphia, PA, USA, 27–31 July 1997, pp. 178–185. ACM (1997)

32. Ke, G., et al.: LightGBM: a highly efficient gradient boosting decision tree. In: Guyon, I., et al. (eds.) Advances in Neural Information Processing Systems 30: Annual Conference on Neural Information Processing Systems 2017, 4–9 December 2017, Long Beach, CA, USA, pp. 3146–3154 (2017)

33. Kilgarriff, A., Rose, T.: Measures for Corpus similarity and homogeneity. In: Ide, N., Voutilainen, A. (eds.) Proceedings of the 3rd Conference on Empirical Methods for Natural Language Processing, Palacio de Exposiciones y Congresos, Granada, Spain, 2 June 1998, pp. 46–52. ACL (1998)

34. Kobayashi, M., Takeda, K.: Information retrieval on the web. ACM Comput. Surv. **32**(2), 144–173 (2000)

35. Koolen, M., Kamps, J.: The importance of anchor text for ad hoc search revisited. In: Crestani, F., Marchand-Maillet, S., Chen, H., Efthimiadis, E.N., Savoy, J. (eds.) Proceedings of the 33rd International ACM SIGIR Conference on Research and Development in Information Retrieval, SIGIR 2010, Geneva, Switzerland, 19–23 July 2010, pp. 122–129. ACM (2010)

36. Kraft, R., Zien, J.Y.: Mining anchor text for query refinement. In: Feldman, S.I., Uretsky, M., Najork, M., Wills, C.E. (eds.) Proceedings of the 13th International World Wide Web Conference, WWW 2004, New York, USA, 17–20 May 2004, pp. 666–674. ACM (2004)

37. Lin, J., Yang, P.: The impact of score ties on repeatability in document ranking. In: Piwowarski, B., Chevalier, M., Gaussier, É., Maarek, Y., Nie, J., Scholer, F. (eds.) Proceedings of the 42nd International ACM SIGIR Conference on Research and Development in Information Retrieval, SIGIR 2019, Paris, France, 21–25 July 2019, pp. 1125–1128. ACM (2019)

38. Ma, Z., Dou, Z., Xu, W., Zhang, X., Jiang, H., Cao, Z., Wen, J.: Pre-training for ad-hoc retrieval: hyperlink is also you need. In: 30th ACM International Conference on Information and Knowledge Management (CIKM 2021). ACM (2021)

39. MacAvaney, S., Yates, A., Feldman, S., Downey, D., Cohan, A., Goharian, N.: Simplified data wrangling with ir_datasets. In: Diaz, F., Shah, C., Suel, T., Castells, P., Jones, R., Sakai, T. (eds.) SIGIR 2021: The 44th International ACM SIGIR Conference on Research and Development in Information Retrieval, Virtual Event, Canada, 11–15 July, 2021, pp. 2429–2436. ACM (2021)

40. Macdonald, C., Tonellotto, N.: Declarative experimentation in information retrieval using PyTerrier. In: Balog, K., Setty, V., Lioma, C., Liu, Y., Zhang, M., Berberich, K. (eds.) ICTIR 2020: The 2020 ACM SIGIR International Conference on the Theory of Information Retrieval, Virtual Event, Norway, 14–17 September 2020, pp. 161–168. ACM (2020)

41. McBryan, O.A.: GENVL and WWWW: tools for taming the web. In: Proceedings of the 1st International World Wide Web Conference, WWW 1994, Geneva, Switzerland, 25–27 May 1994, vol. 341 (1994)

42. Metzler, D., Novak, J., Cui, H., Reddy, S.: Building enriched document representations using aggregated anchor text. In: Allan, J., Aslam, J.A., Sanderson, M., Zhai, C., Zobel, J. (eds.) Proceedings of the 32nd Annual International ACM SIGIR Conference on Research and Development in Information Retrieval, SIGIR 2009, Boston, MA, USA, 19–23 July 2009, pp. 219–226. ACM (2009)

43. Nogueira, R., Cho, K.: Passage Re-ranking with BERT. CoRR abs/1901.04085 (2019)

44. Nogueira, R., Jiang, Z., Pradeep, R., Lin, J.: Document ranking with a pretrained sequence-to-sequence model. In: Cohn, T., He, Y., Liu, Y. (eds.) Proceedings of the 2020 Conference on Empirical Methods in Natural Language Processing, EMNLP 2020, Virtual Event, 16–20 November 2020, Findings of ACL, vol. EMNLP 2020, pp. 708–718. Association for Computational Linguistics (2020)

45. Ogilvie, P., Callan, J.P.: Combining document representations for known-item search. In: Clarke, C.L.A., Cormack, G.V., Callan, J., Hawking, D., Smeaton, A.F. (eds.) Proceedings of the 26th Annual International ACM SIGIR Conference on Research and Development in Information Retrieval, SIGIR 2003, 28 July–1 August 2003, Toronto, ON, Canada, pp. 143–150. ACM (2003)

46. Raffel, C., et al.: Exploring the limits of transfer learning with a unified text-to-text transformer. J. Mach. Learn. Res. **21**, 140:1–140:67 (2020)

47. Sakai, T.: Alternatives to Bpref. In: Kraaij, W., de Vries, A.P., Clarke, C.L.A., Fuhr, N., Kando, N. (eds.) Proceedings of the 30th Annual International ACM SIGIR Conference on Research and Development in Information Retrieval, SIGIR 2007, Amsterdam, The Netherlands, 23–27 July 2007, pp. 71–78. ACM (2007)

48. Völske, M., Bevendorff, J., Kiesel, J., Stein, B., Fröbe, M., Hagen, M., Potthast, M.: Web archive analytics. In: Reussner, R.H., Koziolek, A., Heinrich, R. (eds.) 50. Jahrestagung der Gesellschaft für Informatik, INFORMATIK 2020 - Back to the Future, Karlsruhe, Germany, 28 September–2 Oktober 2020, LNI, vol. P-307, pp. 61–72, GI (2020). https://doi.org/10. 18420/inf2020_05
49. Westerveld, T., Kraaij, W., Hiemstra, D.: Retrieving web pages using content, links, URLs and anchors. In: Voorhees, E.M., Harman, D.K. (eds.) Proceedings of the 10th Text REtrieval Conference, TREC 2001, Gaithersburg, MD, USA, 13–16 November 2001. NIST Special Publication, National Institute of Standards and Technology (NIST), vol. 500–250 (2001)
50. Wu, Q., Burges, C.J.C., Svore, K.M., Gao, J.: Adapting boosting for information retrieval measures. Inf. Retrieval 13(3), 254–270 (2010)
51. Yang, P., Fang, H., Lin, J.: Anserini: Enabling the use of Lucene for information retrieval research. In: Kando, N., Sakai, T., Joho, H., Li, H., de Vries, A.P., White, R.W. (eds.) Proceedings of the 40th International ACM SIGIR Conference on Research and Development in Information Retrieval, SIGIR 2017, Shinjuku, Tokyo, Japan, 7–11 August 2017, pp. 1253–1256. ACM (2017)
52. Yates, A., Nogueira, R., Lin, J.: Pretrained transformers for text ranking: BERT and beyond. In: Diaz, F., Shah, C., Suel, T., Castells, P., Jones, R., Sakai, T. (eds.) Proceedings of the 44th International ACM SIGIR Conference on Research and Development in Information Retrieval, SIGIR 2021, Virtual Event, Canada, 11–15 July 2021, pp. 2666–2668. ACM (2021)

Automation of Citation Screening for Systematic Literature Reviews Using Neural Networks: A Replicability Study

Wojciech Kusa[1]([✉])(iD), Allan Hanbury[1,2](iD), and Petr Knoth[3,4](iD)

[1] TU Wien, Vienna, Austria
wojciech.kusa@tuwien.ac.at
[2] Complexity Science Hub, Vienna, Austria
[3] Research Studios Austria, Vienna, Austria
[4] The Open University, Milton Keynes, UK

Abstract. In the process of Systematic Literature Review, citation screening is estimated to be one of the most time-consuming steps. Multiple approaches to automate it using various machine learning techniques have been proposed. The first research papers that apply deep neural networks to this problem were published in the last two years. In this work, we conduct a replicability study of the first two deep learning papers for citation screening [8,16] and evaluate their performance on 23 publicly available datasets. While we succeeded in replicating the results of one of the papers, we were unable to replicate the results of the other. We summarise the challenges involved in the replication, including difficulties in obtaining the datasets to match the experimental setup of the original papers and problems with executing the original source code. Motivated by this experience, we subsequently present a simpler model based on averaging word embeddings that outperforms one of the models on 18 out of 23 datasets and is, on average, 72 times faster than the second replicated approach. Finally, we measure the training time and the invariance of the models when exposed to a variety of input features and random initialisations, demonstrating differences in the robustness of these approaches.

Keywords: Citation screening · Study selection · Systematic literature review (SLR) · Document retrieval · Replicability

1 Introduction

A systematic literature review is a type of secondary study that summarises all available data fitting pre-specified criteria to answer precise research questions. It uses rigorous scientific methods to minimise bias and generate clear, solid conclusions that health practitioners frequently use to make decisions [12].

Unfortunately, conducting systematic reviews is slow, labour intensive and time-consuming as this relies primarily on human effort. A recent estimate shows

M. Hagen et al. (Eds.): ECIR 2022, LNCS 13185, pp. 584–598, 2022.
https://doi.org/10.1007/978-3-030-99736-6_39

that conducting a full systematic review takes, on average, 67 weeks [4], although another past study reports that the median time to publication was 2.4 years [22]. Furthermore, according to [21], 23% of published systematic reviews need updating within two years after completion.

Citation screening (also known as selection of primary studies) is a crucial part of the systematic literature review process [23]. During this stage, reviewers need to read and comprehend hundreds (or thousands) of documents and decide whether or not they should be included in the systematic review. This decision is made on the basis of comparing each article content with predefined exclusion and inclusion criteria. Traditionally it consists of two stages, the first round of screening titles and abstracts, which is supposed to narrow down the list of potentially relevant items. It is followed by a task appraising the full texts, a more detailed (but also more time-consuming) revision of all included papers from the first stage based on the full text of articles.

Multiple previous studies tried to decrease the completion time of systematic reviews by using text mining methods to semi-automate the citation screening process (see a recent systematic review on this topic: [9]). Using the machine learning paradigm, citation screening could be reduced to a binary classification problem. Then, the task is to train a model using the seed of manually labelled citations that can distinguish between documents to be included (includes) and those to be excluded (excludes). One of the challenges is a significant class imbalance (for 23 benchmark datasets, the maximum percentage of included documents is 27%, and on average, it is only 7%). Additionally, existing approaches require training a separate model for each new systematic review.

In this work, we replicate two recent papers related to automated citation screening for systematic literature reviews using neural networks [8,16]. We chose these studies since, to our knowledge, they are the first ones to address this problem using deep neural networks. Both papers represent citation screening as a binary classification task and train an independent model for each dataset. We evaluate the models on 23 publicly available benchmark datasets. We present our challenges regarding replicability in terms of datasets, models and evaluation. In the remaining sections of this article, we will use the name **Paper A** to refer to the study by Kontonatsios et al. [16] and **Paper B** to indicate work by van Dinter et al. [8].

Moreover, we investigate if the models are invariant to different data features and random initialisations. 18 out of 23 datasets are available as a list of Pubmed IDs of the input papers with assigned categories (included or excluded). As we needed to recreate data collection stages for both papers, we wanted to measure if the choice of the document features would influence the final results of the replicated models.

Both papers utilise deep learning due to their claimed substantial superiority over traditional (including shallow neural network) models. We compare the models with previous benchmarks and assess to what extent do these models improve performance over simpler and more traditional models. Finally, we make our data collection and experiment scripts and detailed results publicly available on GitHub[1].

2 Related Work

Out of all stages of the systematic review process, the selection of primary studies is known as the most time-consuming step [2,20,24]. It was also automated the most often in the past using text mining methods. According to a recent survey on the topic of automation of systematic literature reviews [9], 25 out of 41 analysed primary studies published between 2006 and 2020 addressed (semi-)automation of the citation screening process. Another, older systematic review from 2014 found in total 44 studies dealing implicitly or explicitly with the problem of screening workload [18].

Existing approaches to automation of the citation screening process can be categorised into two main groups. The first one uses text classification models [17, 25] and the second one screening prioritisation or ranking techniques that exclude items falling below some threshold [6,10]. Both groups follow a similar approach. They train a supervised binary classification algorithm to solve this problem, e.g. Support Vector Machines (SVMs) [6,25], Naïve Bayes [17] or Random Forest [15]. A significant limitation of these approaches is the need for a large number of human decisions (annotations) that must be completed before developing a reliable model [24].

Kontonatsios et al. [16] (**Paper A**) was the first one to apply deep learning algorithms to automate the citation screening process. They have used three neural network-based denoising autoencoders to create a feature representation of the documents. This representation was fed into a feed-forward network with a linear SVM classifier trained in a supervised manner to re-order the citations. Van Dinter et al. [8] (**Paper B**) presented the first end-to-end solution to citation screening with a deep neural network. They developed a binary text classification model with the usage of a multi-channel convolutional neural network. Both models claim to yield significant workload savings of at least 10% on most benchmark review datasets.

A different procedure to automating systematic reviews was presented during the CLEF 2017 eHealth Lab Technology Assisted Reviews in Empirical Medicine task [13,14]. Here, the user needs to find all relevant documents from a set of PubMed articles given a Boolean query. It overcomes the need for creating an annotated dataset first but makes it harder to incorporate reviewers' feedback.

The recently published BERT model [7] and its variants have pushed the state of the art for many NLP tasks. Ioannidis [11] used BERT-based models to work on document screening within the Technology Assisted Review task achieving better results than the traditional IR baseline models. To our knowledge, this was the first use of a generative neural network model in a document screening task.

---

[1] https://github.com/ProjectDoSSIER/CitationScreeningReplicability.

3   Experiment Setup

3.1   Models

DAE-FF. Paper A presents a neural network-based, supervised feature extraction method combined with a linear Support Vector Machine (SVM) trained to prioritise eligible documents. The data preprocessing pipeline contains stopword removal and stemming with a Porter stemmer. The feature extraction part is implemented as three independent denoising autoencoders (DAE) that learn to reconstruct corrupted Bag-of-Words input vectors. Their concatenated output is used to initialise a supervised feed-forward neural network (FF). These extracted document vectors are subsequently used as an input to an L2-regularised linear SVM classifier. Class imbalance is handled by setting the regularisation parameter $C = 1 \times 10^{-6}$.

Multi-channel CNN. Paper B presents a multi-channel convolutional neural network (CNN) to discriminate between includes and excludes. It uses static, pre-trained GloVe word embeddings [19] to create an input embedding matrix. This embedding is inserted into a series of parallel CNN blocks consisting of a single-dimensional CNN layer followed by global max pooling. Outputs from the layers are concatenated after global pooling and fed into a feed-forward network. The authors experimented with a different number of channels and Conv1D output shapes. Input documents are tokenised and lowercased, punctuation and non-alphabetic tokens are removed. Documents are padded and truncated to a maximum length of 600 tokens. Class imbalance is handled with oversampling. For our replicability study, we have chosen the best performing Model_2.

fastText. We also test a shallow neural network model which is based on fastText word embeddings [3]. This model is still comparable to more complex deep learning models in many classification tasks. At the same time, it is orders of magnitude faster for training and prediction, making it more suitable for active learning scenarios where reviewers could alter the model's predictions by annotating more documents. To make it even simpler, we do not use pre-trained word embeddings to vectorise documents. Data preprocessing is kept minimal as we only lowercase the text and remove all non-alphanumerical characters.

Hyperparameters. Paper A optimised only the number of training epochs for their DAE model. In order to do so, they used two datasets: Statins and BPA reviews and justified this choice with differences between smaller datasets from Clinical and Drug reviews and SWIFT reviews. Other hyperparameters (including the minibatch size and the number of epochs for the feed-forward model) are constant across all datasets. Paper B used the Statins review dataset to tune a set of hyperparameters, including the number of epochs, batch size, dropout, and dense units.

3.2   Data

All 23 datasets are summarised in Table 1, including the dataset source, number of citations, number and percentage of eligible citations, maximum WSS@95% score (Sect. 3.3) and the availability of additional bibliographic metadata. Every citation consists of a title, an abstract, and an eligibility label (included or excluded). Moreover, 18 datasets contain also bibliographic metadata. The percentage of eligible citations (includes) varies between datasets, from 0.55% to 27.04%, but on average, it is about 7%, meaning that the datasets are highly imbalanced.

Table 1. Statistics of 23 publicly available datasets used in the experiments on automated citation screening for Systematic Literature Reviews.

	Dataset name	Introduced in	# Citations	Included citations	Excluded citations	Maximum WSS@95%	Bibliographic metadata
1	ACEInhibitors		2544	41 (1.6%)	2503 (98.4%)	93.47%	Yes
2	ADHD		851	20 (2.4%)	831 (97.6%)	92.77%	Yes
3	Antihistamines		310	16 (5.2%)	294 (94.8%)	89.84%	Yes
4	Atypical Antipsychotics		1120	146 (13.0%)	974 (87.0%)	82.59%	Yes
5	Beta Blockers		2072	42 (2.0%)	2030 (98.0%)	93.07%	Yes
6	Calcium Channel Blockers		1218	100 (8.2%)	1118 (91.8%)	87.20%	Yes
7	Estrogens	Drug	368	80 (21.7%)	288 (78.3%)	74.35%	Yes
8	NSAIDs	(Cohen et al.,	393	41 (10.4%)	352 (89.6%)	85.08%	Yes
9	Opioids	2006)	1915	15 (0.8%)	1900 (99.2%)	94.22%	Yes
10	Oral Hypoglycemics		503	136 (27.0%)	367 (73.0%)	69.16%	Yes
11	Proton PumpInhibitors		1333	51 (3.8%)	1282 (96.2%)	91.32%	Yes
12	Skeletal Muscle Relaxants		1643	9 (0.5%)	1634 (99.5%)	94.45%	Yes
13	Statins		3465	85 (2.5%)	3380 (97.5%)	92.66%	Yes
14	Triptans		671	24 (3.6%)	647 (96.4%)	91.57%	Yes
15	Urinary Incontinence		327	40 (12.2%)	287 (87.8%)	83.38%	Yes
	Average Drug		1249	56 (7.7%)	1192 (92.3%)	87.67%	15/15
16	COPD	Clinical	1606	196 (12.2%)	1410 (87.8%)	83.36%	No
17	Proton Beam	(Wallace et al.,	4751	243 (5.1%)	4508 (94.9%)	90.14%	No
18	Micro Nutrients	2010)	4010	258 (6.4%)	3752 (93.6%)	88.87%	No
	Average Clinical		3456	232 (7.9%)	3223 (92.1%)	87.45%	0/3
19	PFOA/PFOS		6331	95 (1.5%)	6236 (98.5%)	93.56%	Yes
20	Bisphenol A (BPA)	SWIFT	7700	111 (1.4%)	7589 (98.6%)	93.62%	Yes
21	Transgenerational	(Howard et al.,	48638	765 (1.6%)	47873 (98.4%)	93.51%	Yes
22	Fluoride and neurotoxicity	2016)	4479	51 (1.1%)	4428 (98.9%)	93.91%	No
23	Neuropathic pain — CAMRADES		29207	5011 (17.2%)	24196 (82.8%)	78.70%	No
	Average SWIFT		19271	1206 (4.6%)	18064 (95.4%)	90.66%	3/5
	Average (All datasets)		5454	329 (7.0%)	5125 (93.0%)	88.29%	18/23

Cohen et al. [5] was the first one to introduce datasets for training and evaluation of citation screening. They constructed a test collection for 15 different systematic review topics produced by the Oregon Evidence-based Practice Centre (EPC) related to the efficacy of medications in several drug classes.

Another three datasets for evaluation of automated citation screening were released by Wallace et al. [25]. These systematic reviews are related to the clinical outcomes of various treatments. Both drug and clinical reviews contain a small number of citations (varying from 310 to 4751).

The third group of datasets was introduced by Howard et al. [10] and consists of five substantially larger reviews (from 4479 to 48 638 citations) that have been used to assess the performance of the SWIFT-review tool. They were created using broader search strategies which justifies a higher number of citations.

Paper A trained and evaluated their model on all 23 datasets coming from three categories. Paper B used 20 datasets from the Clinical and SWIFT categories. Paper B states that, on average, 5.2% of abstracts are missing in all 20 datasets, varying between 0% for *Neuropathic Pain* and 20.82% for *Statins*. Compared to previous papers, Paper B reports fewer citations for three datasets: *Statins*, *PFOA/PFOS* and *Neuropathic Pain*. This difference is insignificant compared to the dataset size, e.g. 29207 versus 29202 for *Neuropathic Pain*, so it should not influence the model evaluation.

3.3   Evaluation

Evaluation of automated citation screening can be very challenging. Traditional metrics used for classification tasks like precision, recall, or F-score cannot capture what we intend to measure in this task. For an automated system to be beneficial to systematic reviewers, it should save time and miss as few relevant papers as possible. Previous studies suggested that recall should not be lower than 95%, and at the same time, precision should be as high as possible [5].

Work saved over sampling at r% recall (WSS@r%) is a primary metric for evaluation of automated citation screening. It was first introduced and described by Cohen et al. [5] as "the percentage of papers that meet the original search criteria that the reviewers do not have to read (because they have been screened out by the classifier)." It estimates the human screening workload reduction by using automation tools, assuming a fixed recall level of r%. WSS@r%, given a recall of r%, is defined as follows:

$$WSS@r\% = \frac{TN + FN}{N} - (1 - r)$$

where TN is the number of true negatives, FN is the number of false negatives, and N is the total number of documents. Based on previous studies, we fix the recall at 95% and compute the WSS@95% score.

One drawback of this metric described by [5] is that it does not take into account time differences caused by varying lengths of documents and also the time needed to review a full-text article compared to only reading the title and the abstract.

A further drawback of WSS is that the maximum WSS value depends on the ratio of included/excluded samples. A perfectly balanced dataset can achieve a maximum value of WSS@95% = 0.45, whereas a highly imbalanced dataset with a 5%/95% split can obtain a maximum WSS@95% score of 0.9. Consequently, it does not make sense to compare the results nor average them across different datasets (as done in Paper A and B).

For our replicability study, we decided to use the implementations of the WSS metric provided by Papers A and B.

Cross-validation. Both papers use a stratified 10×2 cross-validation for evaluation. In this setting, data is randomly split in half: one part is used to train the classifier, and the other is left for testing. This process is then repeated ten times, and the results are accumulated from all ten runs. We also use this approach to evaluate the quality of all three models.

3.4 Code

The authors of both papers uploaded their code into public GitHub repositories:[2,3]. Both models were written in Python 3 and depend primarily on TensorFlow and Keras deep learning frameworks [1]. The whole implementation was uploaded in four commits for Paper A and one for Paper B (excluding commits containing only documentation). Except for the code, there is no information about versions of the packages used to train and evaluate the models. This missing information is crucial for replicability, as, for TensorFlow alone, in 2020, there were 27 different releases related to 6 different MINOR versions[4].

The model prepared by Paper B uses also pre-trained 100-dimensional GloVe word embeddings which we downloaded separately from the original authors' website[5] according to the instructions provided by the Paper B GitHub Readme.

Both papers did not include the original datasets they used to train and evaluate their models. Paper A provided sample data consisting of 100 documents which presents the input data format accepted by their model, making it easier to re-run the experiments. Paper B does not include sample data but describes where and how to collect and process the datasets.

4   Results and Discussion

4.1   Replicability Study

WSS@95% scores from older benchmarks and original papers, along with our replicated results, are presented in Table 2. For all datasets, both Paper A and B provide only mean WSS@95% score from cross-validation runs. Therefore, we were not able to measure statistical significance between our replicated results and the original ones. To quantify the difference, we decided to calculate the absolute delta between reported and replicated scores: $|x-y|$. Both models report a random seed for the cross-validation splits but not for the model optimisation. Usage of different seeds for model optimisation might be one of the reasons why we were not able to achieve the same results.

---

[2] https://github.com/gkontonatsios/DAE-FF.
[3] https://github.com/rvdinter/multichannel-cnn-citation-screening.
[4] https://pypi.org/project/tensorflow/#history.
[5] https://nlp.stanford.edu/data/glove.6B.zip.

Table 2. WSS@95% results for replicated models compared with original results and benchmark models. WSS@95% scores are averages across ten validation runs for each of the 23 review datasets. Underlined scores indicate the highest score within the three tested models, **bold** values indicate the highest score overall.

Dataset name	Cohen (2006)	Matwin (2010)	Cohen (2008/ 2011)	Howard (2016)	Paper A	Paper A replicated	Absolute delta	Paper B	Paper B replicated	Absolute delta	fastText classifier
ACEInhibitors	.566	.523	.733	**.801**	.787	.785	0.16%	.783	.367	41.59%	.783
ADHD	.680	.622	.526	**.793**	.665	.639	2.58%	.698	.704	0.57%	.424
Antihistamines	.000	.149	.236	.137	.310	.275	3.48%	.168	.135	3.32%	.047
Atypical Antipsychotics	.141	.206	.170	.251	.329	.190	13.92%	.212	.081	13.15%	.218
Beta Blockers	.284	.367	.465	.428	.587	.462	12.52%	.504	.399	10.51%	.419
Calcium Channel Blockers	.122	.234	.430	**.448**	.424	.347	7.66%	.159	.069	9.03%	.178
Estrogens	.183	.375	.414	**.471**	.397	.369	2.80%	.119	.083	3.56%	.306
NSAIDs	.497	.528	.672	**.730**	.723	.735	1.18%	.571	.601	2.98%	.620
Opioids	.133	.554	.364	**.826**	.533	.580	4.71%	.295	.249	4.58%	.559
Oral Hypoglycemics	.090	.085	**.136**	.117	.095	.123	2.80%	.065	.013	5.21%	.098
Proton PumpInhibitors	.277	.229	.328	.378	.400	.299	10.13%	.243	.129	11.38%	.283
Skeletal Muscle Relaxants	.000	.265	.374	**.556**	.286	.286	0.04%	.229	.300	7.14%	.090
Statins	.247	.315	.491	.435	.566	.487	7.93%	.443	.283	16.03%	.409
Triptans	.034	.274	.346	.412	.434	.412	2.24%	.266	.440	17.38%	.210
Urinary Incontinence	.261	.296	.432	**.531**	.531	.483	4.81%	.272	.180	9.21%	.439
Average Drug	.234	.335	.408	**.488**	.471	.431	5.13%	.335	.269	10.37%	.339
COPD	-	-	-	-	**.666**	.665	0.07%	-	.128	-	.312
Proton Beam	-	-	-	-	**.816**	.812	0.39%	-	.357	-	.733
Micro Nutrients	-	-	-	-	.662	**.663**	0.08%	-	.199	-	.608
Average Clinical	-	-	-	-	**.715**	.713	0.18%	-	.228	-	.551
PFOA/PFOS	-	-	-	.805	**.848**	.838	0.97%	.071	.305	23.44%	.779
Bisphenol A (BPA)	-	-	-	.752	**.793**	.780	1.34%	.792	.369	42.31%	.637
Transgenerational	-	-	-	.714	**.718**	.707	1.14%	.708	.000	70.80%	.368
Fluoride and neurotoxicity	-	-	-	.870	.799	.806	0.68%	.883	.808	7.48%	.390
Neuropathic pain	-	-	-	**.691**	.608	.598	1.03%	.620	.091	52.89%	.613
Average SWIFT	-	-	-	**.766**	.751	.748	1.03%	.615	.315	39.38%	.557
Average (all datasets)	-	-	-	-	**.564**	.537	3.59%	-	.273	17.63%	.414

For two datasets (*Bisphenol A (BPA)* and *Triptans*), Paper A reports two different results for the DAE-FF model. We suppose this was only a typing mistake, as we managed to infer the actual values based on the averaged WSS@95% score from all datasets available in the original paper.

The average delta between our replicated results and the original ones from Paper A is 3.59%. Only for three datasets is this value higher than 10%. If we consider different seeds used for training models, these results confirm the successful replication of Paper A's work.

For Paper B, the average delta is 17.63%. For 10 out of 20 datasets, this delta is more than 10%. For the two largest datasets: *Transgenerational* and *Neuropathic Pain* we were not able to successfully train the Multi-Channel CNN model. All of these results raise concerns about replicability.

Paper B also tried to replicate the DAE-FF model from Paper A. They stated that *"(...) we aimed to replicate the model (...) with open-source code via GitHub. However, we could not achieve the same scores using our dataset. After emailing the primary author, we were informed that he does not have access to his datasets anymore, which means their study cannot be fully replicated."*. Our results are contrary to findings by Paper B: we managed to replicate the results of Paper

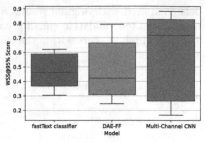

(a) ADHD review dataset.

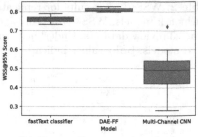

(b) Proton Beam review dataset.

Fig. 1. Example boxplots with WSS@95% scores for three models. Input features are titles and abstracts.

A successfully without having access to their original datasets. Unfortunately, Paper B does not present any quantitative results of their replicability study. Therefore, we cannot draw any conclusions regarding those results as we do not know what Paper B authors meant by *"cannot be fully replicated"*.

Figure 1 presents results for *ADHD* and *Proton Beam* datasets for all three models. The Multi-Channel CNN model has the widest range of WSS@95% scores across cross-validation runs. This is especially evident on the datasets from the Clinical group (i.e. *Proton Beam*), for which the DAE-FF and fastText models yield very steady results across every cross-validation fold. This could mean that the Multi-Channel CNN model is less stable, and its good performance is dependant on random initialisation.

Next, we compare our replicated results and the original ones from Paper A and B to previous benchmark studies. Paper A only compares their model to custom baseline methods and does not mention the previous state of the art results. None of the tested neural network-based models can improve on the results by Howard et al. [10], which uses a log-linear model with word-score and topic-weight features to classify the citations. This means that even though deep neural network models can provide significant gains in WSS@95% scores, they can still be outperformed by classic statistical methods.

4.2 Impact of Input Features

As we encountered memory problems when training the Paper B model on *Transgenerational* and *Neuropathic pain* datasets, we exclude these two datasets from our comparisons in the remaining experiments.

None of the papers provided the original input data used to train the models. We wanted to measure if the results depend on how that input data was gathered. We implemented two independent data gathering scripts using the biopython package as suggested by Paper B to obtain 18 out of 23 datasets. One implementation relied on the Medline module, where a document was represented as a dictionary of all available fields. The second implementation returned all possible fields (title, abstract, author and journal information) concatenated in a

Table 3. Influence of input document features on the WSS@95% score for three tested models. "All features" means a single string containing all possible fields. For each row, **bold** values indicate the highest score for each model, underlined scores are best overall.

Dataset name	DAE-FF				Multi-channel CNN				fastText classifier			
	All features	Title and Abstract	Abstract only	Title only	All features	Title and Abstract	Abstract only	Title only	All features	Title and Abstract	Abstract only	Title only
ACEInhibitors	.785	.709	.658	**.806**	.367	.461	.648	.525	**.783**	.776	.765	.441
ADHD	.639	.500	.404	**.651**	**.704**	.528	.692	.580	.424	**.470**	.444	.200
Antihistamines	**.275**	.168	.265	.016	.135	**.204**	.114	.105	.047	.124	.175	**.192**
Atypical Antipsychotics	.190	.221	**.230**	.046	.081	**.086**	.050	.013	**.218**	.188	.185	.095
Beta Blockers	**.462**	.451	.390	.408	**.399**	.243	.134	.211	**.419**	**.419**	.407	.262
Calcium Channel Blockers	**.347**	.337	.297	.137	.069	.083	.004	**.117**	.178	.139	.060	**.244**
Estrogens	**.369**	.358	.331	.145	.083	.076	.051	**.092**	**.306**	.199	.108	.241
NSAIDs	**.735**	.679	.690	.658	**.601**	.443	.358	.225	**.620**	.506	.512	.535
Opioids	**.580**	.513	.499	.280	.249	**.420**	.413	.287	**.559**	.558	.534	.245
Oral Hypoglycemics	.123	**.129**	.107	.019	.013	**.021**	.004	.005	**.098**	.049	.042	.016
Proton PumpInhibitors	**.299**	.291	.153	.285	**.129**	.121	.059	.118	.283	.228	.174	**.360**
Skeletal Muscle Relaxants	.286	.327	**.430**	.125	.300	**.329**	.242	.202	.090	.142	.180	**.210**
Statins	**.487**	.434	.392	.255	**.283**	.231	.120	.082	**.409**	.376	.281	.228
Triptans	**.412**	.253	.320	.199	**.440**	.404	.407	.129	.210	.205	**.211**	.075
Urinary Incontinence	.483	**.531**	.482	.372	**.180**	.161	.046	.099	**.439**	.310	.170	.434
Average Drug	**.431**	.394	.373	.293	**.269**	.254	.223	.185	**.339**	.313	.283	.252
COPD	.665	.665	.676	**.677**	.128	**.372**	.087	.093	.312	**.553**	.546	.545
Proton Beam	**.812**	.810	.790	.799	.357	.489	.408	**.559**	.733	.761	**.771**	**.771**
Micro Nutrients	.663	.648	.665	**.677**	.199	.255	.251	**.268**	**.608**	.602	.605	.601
Average Clinical	.713	.708	.670	**.718**	.228	**.372**	.249	.307	.551	.638	**.640**	.639
PFOA/PFOS	.713	.839	**.847**	.696	.305	**.405**	.391	.109	.779	**.796**	.778	.292
Bisphenol A (BPA)	**.780**	.754	.715	.631	.369	.300	**.612**	.182	**.637**	.630	.499	.079
Fluoride and neurotoxicity	.806	**.838**	.758	.726	**.808**	.688	.654	.452	**.390**	.375	.292	.250
Average SWIFT	.766	**.782**	.774	.684	.494	.464	**.552**	.247	**.602**	.600	.523	.207
Average (All datasets)	**.520**	.498	.481	.410	.295	**.301**	.274	.212	**.407**	.400	.368	.301

single string. Furthermore, we examined how robust the models are, if the input data contained only titles or abstracts of the citations. Results are presented in the Table 3.

The best average WSS@95% results are obtained for all three models when they use all available features (Fig. 2). All models achieved better results when using just the abstract data compared to the titles alone. This reaffirms our common sense reasoning that titles alone are not sufficient for citation screening. However, there are some specific datasets for which best results were obtained when the input documents contained only titles or abstracts. While this experiment does not indicate why this is the case, we can offer some potential reasons: (1) it could be that eligible citations of these datasets are more similar in terms of titles or abstract; (2) it could be that these models are not able to retrieve relevant information when there is too much noise. Intra- and inter-class dataset similarity need to be further evaluated in future studies.

As presented in Table 2, the fastText classifier model was not able to outperform the original results from Paper A and B. However, compared to our replicated results of Paper B, the fastText classifier achieves higher WSS@95% scores on 18 out of 23 datasets. It is also more robust to random initialisation compared to Multi-Channel CNN.

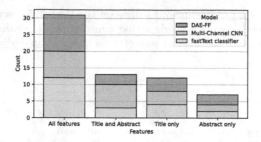

Fig. 2. A count of experiments in which a model using a specific input feature achieved the best results. Models that use all available features scored the best results 49% of times for a specific (model, dataset) combination.

4.3 Training Time

We computed the training time for each of the models. The relationship between dataset size and model training time is visualised in Fig. 3. For the DAE-FF model, we calculated both the training procedure of denoising autoencoder, feed-forward networks, and linear SVM. The DAE component is the most time-absorbing component as it consumes, on average, 93.5% of the total training time. For the fastText and Multi-Channel CNN models, we calculated the training procedure of the binary classifier.

For small datasets containing less than 1000 documents, one validation fold for fastText took on average 2 s, for Multi-Channel CNN 13 s, and DAE-FF 82 s. Training time difference increases for larger models, where the speed of fastText is even more significant. For the largest dataset, *Transgenerational*, the mean training time for fastText is 78 s, for Multi-Channel CNN 894 s and for DAE-FF, it is 18,108 s. On average, the fastText model is 72 times faster than DAE-FF and more than eight times faster than Multi-Channel CNN, although this dependency is not linear and favours fastText for larger datasets.

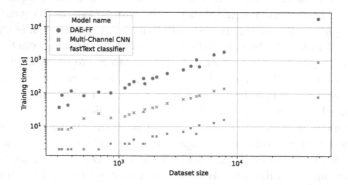

Fig. 3. The relationship between dataset size and a model training time for the three evaluated models. Both training time and dataset size are shown on a logarithmic scale.

4.4   Precision@95%recall

Finally, we measure the precision at a recall level of 95%, a metric proposed by Paper A. Table 4 shows mean scores for each model across all three review groups. Similarly to the WSS@95% metric, the best performing model is DAE-FF achieving a mean precision@95%recall on 21 datasets equal to 0.167. This method outperforms Multi-Channel CNN and fastText models by 3.2% and 4.6%, respectively. Paper A reported average precision@95%recall equal to 19% over 23 review datasets, which is comparable with our findings. Paper B does not report this score, so we cannot compare our results regarding the Multi-Channel CNN model.

Table 4. Influence of input document features on the WSS@95% score for three tested models. "All features" means a single string containing all possible fields. For each row, **bold** values indicate the highest score for each model, <u>underlined</u> scores are best overall.

	DAE-FF	Multi-channel CNN	fastText classifier
Average Drug	**.143**	.121	.112
Average Clinical	**.324**	.221	.230
Average SWIFT	**.127**	.091	.058
Average (21 datasets)	**.167**	.135	.121

5   Conclusions

This work replicates two recent papers on automated citation screening for systematic literature reviews using deep neural networks. The model proposed by Paper A consists of a denoising autoencoder combined with feed-forward and SVM layers (DAE-FF). Paper B introduces a multi-channel convolutional neural network (Multi-Channel CNN). We used the 23 publicly available datasets to measure the quality of both models. The average delta between our replicated results and the original ones from Paper A is 3.59%. Considering that we do not know the random seed used for the training of original models, we can conclude that the replication of Paper A was successful. The average delta for Paper B is 17.63%. In addition to that, this model is characterised by a significant variance, so we cannot claim successful replication of this method.

Subsequently, we evaluated the fastText classifier and compared its performance to the replicated models. This shallow neural network model based on averaging word embeddings achieved better WSS@95% results when compared to replicated scores from Paper B and, at the same time, is on average 72 and 8 times faster during training than both Paper A and B models.

None of the tested models can outperform all the others across all the datasets. DAE-FF achieves the best average results, though it is still worse when compared to a statistical method with the log-linear model. Models using all available features (title, abstract, author and journal information) perform best on the average of 21 datasets when compared to just using a title, abstract or both.

Availability of the code alone does not guarantee a replicable experimental setup. If the project was not documented for the specific software versions, it might be challenging to reconstruct these requirements based exclusively on the code, especially if the experiments were conducted some time ago. In the case of code written in Python, explicitly writing environment version with, for example, *requirements.txt* or conda's *environment.yml* files should be sufficient in most of the cases to save time for researchers trying to replicate the experiments.

Acknowledgement. This work was supported by the EU Horizon 2020 ITN/ETN on Domain Specific Systems for Information Extraction and Retrieval – DoSSIER (H2020-EU.1.3.1., ID: 860721).

References

1. Abadi, M., et al.: TensorFlow: Large-Scale Machine Learning on Heterogeneous Distributed Systems (2015). https://research.google/pubs/pub45166/
2. Bannach-Brown, A., et al.: Machine learning algorithms for systematic review: reducing workload in a preclinical review of animal studies and reducing human screening error. Syst. Rev. **8**(1), 1–12 (2019). https://doi.org/10.1186/S13643-019-0942-7, https://systematicreviewsjournal.biomedcentral.com/articles/10.1186/s13643-019-0942-7
3. Bojanowski, P., Grave, E., Joulin, A., Mikolov, T.: Enriching word vectors with Subword information. Trans. Assoc. Comput. Linguist. **5**, 135–146 (2017). https://doi.org/10.1162/tacl_a_00051, http://arxiv.org/abs/1607.04606
4. Borah, R., Brown, A.W., Capers, P.L., Kaiser, K.A.: Analysis of the time and workers needed to conduct systematic reviews of medical interventions using data from the PROSPERO registry (2017). https://doi.org/10.1136/bmjopen-2016-012545, http://bmjopen.bmj.com/
5. Cohen, A.M., Hersh, W.R., Peterson, K., Yen, P.Y.: Reducing workload in systematic review preparation using automated citation classification. J. Am. Med. Inf. Assoc. **13**(2), 206–219 (2006). https://www.ncbi.nlm.nih.gov/pmc/articles/PMC1447545/
6. Cohen, A.M.: Optimizing feature representation for automated systematic review work prioritization. In: AMIA Annual Symposium Proceedings, vol. 2008, p. 121 (2008). https://www.ncbi.nlm.nih.gov/pmc/articles/PMC2656096/
7. Devlin, J., Chang, M.W., Lee, K., Toutanova, K.: BERT: pre-training of deep bidirectional transformers for language understanding. In: NAACL HLT 2019 - 2019 Conference of the North American Chapter of the Association for Computational Linguistics: Human Language Technologies - Proceedings of the Conference, vol. 1, pp. 4171–4186, October 2018, https://arxiv.org/abs/1810.04805v2
8. van Dinter, R., Catal, C., Tekinerdogan, B.: A multi-channel convolutional neural network approach to automate the citation screening process. Appl. Soft Comput. **112**, 107765 (2021). https://doi.org/10.1016/J.ASOC.2021.107765
9. van Dinter, R., Tekinerdogan, B., Catal, C.: Automation of systematic literature reviews: a systematic literature review. Inf. Softw. Technol. **136**, 106589

(2021). https://doi.org/10.1016/j.infsof.2021.106589, https://linkinghub.elsevier.com/retrieve/pii/S0950584921000690

10. Howard, B.E., et al.: SWIFT-review: a text-mining workbench for systematic review. Syst. Rev. **5**, 1–16 (2016). https://doi.org/10.1186/s13643-016-0263-z, https://link.springer.com/article/10.1186/s13643-016-0263-z

11. Ioannidis, A.: An Analysis of a BERT Deep Learning Strategy on a Technology Assisted Review Task, April 2021. http://arxiv.org/abs/2104.08340

12. Jo, A., et al.: Systematic Reviews: CRD's guidance for undertaking reviews in health care. CRD, University of York, York, January 2009. www.york.ac.uk/inst/crd

13. Kanoulas, E., Li, D., Azzopardi, L., Spijker, R.: CLEF 2017 technologically assisted reviews in empirical medicine overview. In: CEUR Workshop Proceedings, vol. 1866, pp. 1–29, September 2017. https://pureportal.strath.ac.uk/en/publications/clef-2017-technologically-assisted-reviews-in-empirical-medicine-

14. Kanoulas, E., Li, D., Azzopardi, L., Spijker, R.: CLEF 2018 technologically assisted reviews in empirical medicine overview. In: CEUR Workshop Proceedings, vol. 2125, July 2018. https://pureportal.strath.ac.uk/en/publications/clef-2018-technologically-assisted-reviews-in-empirical-medicine-

15. Khabsa, M., Elmagarmid, A., Ilyas, I., Hammady, H., Ouzzani, M.: Learning to identify relevant studies for systematic reviews using random forest and external information. Mach. Learn. **102**(3), 465–482 (2015). https://doi.org/10.1007/S10994-015-5535-7, https://link.springer.com/article/10.1007/s10994-015-5535-7

16. Kontonatsios, G., Spencer, S., Matthew, P., Korkontzelos, I.: Using a neural network-based feature extraction method to facilitate citation screening for systematic reviews. Expert Syst. Appl. **X6**, 100030 (2020). https://doi.org/10.1016/j.eswax.2020.100030

17. Matwin, S., Kouznetsov, A., Inkpen, D., Frunza, O., O'Blenis, P.: A new algorithm for reducing the workload of experts in performing systematic reviews. J. Am. Med. Inf. Assoc. **17**(4), 446–453 (2010). https://doi.org/10.1136/JAMIA.2010.004325

18. O'Mara-Eves, A., Thomas, J., McNaught, J., Miwa, M., Ananiadou, S.: Using text mining for study identification in systematic reviews: a systematic review of current approaches. Syst. Rev. **4**(1), 5 (2015). https://doi.org/10.1186/2046-4053-4-5

19. Pennington, J., Socher, R., Manning, C.D.: GloVe: global vectors for word representation. In: EMNLP 2014 - 2014 Conference on Empirical Methods in Natural Language Processing, Proceedings of the Conference, pp. 1532–1543 (2014). https://doi.org/10.3115/V1/D14-1162, https://aclanthology.org/D14-1162

20. Sellak, H., Ouhbi, B., Frikh, B., Ben, S.M.: Using rule-based classifiers in systematic reviews: a semantic class association rules approach. In: Proceedings of the 17th International Conference on Information Integration and Web-Based Applications and Services (2015). https://doi.org/10.1145/2837185, https://doi.org/10.1145/2837185.2837279

21. Shojania, K.G., Sampson, M., Ansari, M.T., Ji, J., Doucette, S., Moher, D.: How quickly do systematic reviews go out of date? A survival analysis. Ann. Intern. Med. **147**(4), 224–233 (2007). https://doi.org/10.7326/0003-4819-147-4-200708210-00179

22. Tricco, A.C., Brehaut, J., Chen, M.H., Moher, D.: Following 411 cochrane protocols to completion: a retrospective cohort study. PLOS ONE **3**(11), e3684 (2008). https://doi.org/10.1371/JOURNAL.PONE.0003684, https://journals.plos.org/plosone/article?id=10.1371/journal.pone.0003684

23. Tsafnat, G., Glasziou, P., Choong, M.K., Dunn, A., Galgani, F., Coiera, E.: Systematic review automation technologies (2014). https://doi.org/10.1186/2046-4053-3-74

24. Tsafnat, G., Glasziou, P., Karystianis, G., Coiera, E.: Automated screening of research studies for systematic reviews using study characteristics. Syst. Rev. **7**(1), 1–9 (2018). https://doi.org/10.1186/S13643-018-0724-7, https://link.springer.com/article/10.1186/s13643-018-0724-7

25. Wallace, B.C., Trikalinos, T.A., Lau, J., Brodley, C., Schmid, C.H.: Semi-automated screening of biomedical citations for systematic reviews. BMC Bioinf. **11**(1), 1–11 (2010). https://doi.org/10.1186/1471-2105-11-55, https://bmcbioinformatics.biomedcentral.com/articles/10.1186/1471-2105-11-55

Improving Query Representations for Dense Retrieval with Pseudo Relevance Feedback: A Reproducibility Study

Hang Li[1]([⊠])[iD], Shengyao Zhuang[1][iD], Ahmed Mourad[1][iD], Xueguang Ma[2],
Jimmy Lin[2][iD], and Guido Zuccon[1][iD]

[1] The University of Queensland, St. Lucia, Australia
{hang.li,s.zhuang,a.mourad,g.zuccon}@uq.edu.au
[2] University of Waterloo, Waterloo, Canada
{x93ma,jimmylin}@uwaterloo.ca

Abstract. Pseudo-Relevance Feedback (PRF) utilises the relevance signals from the top-k passages from the first round of retrieval to perform a second round of retrieval aiming to improve search effectiveness. A recent research direction has been the study and development of PRF methods for deep language model based rankers, and in particular in the context of dense retrievers. Dense retrievers provide a trade off between effectiveness, which is often reduced compared to more complex neural rankers, and query latency, which also is reduced making the retrieval pipeline more efficient. The introduction of PRF methods for dense retrievers has been motivated as an attempt to further improve their effectiveness. In this paper, we reproduce and study a recent method for PRF with dense retrievers, called ANCE-PRF. This method concatenates the query text and that of the top-k feedback passages to form a new query input, which is then encoded into a dense representation using a newly trained query encoder based on the original dense retriever used for the first round of retrieval. While the method can potentially be applied to any of the existing dense retrievers, prior work has studied it only in the context of the ANCE dense retriever.

We study the reproducibility of ANCE-PRF in terms of both its training (encoding of the PRF signal) and inference (ranking) steps. We further extend the empirical analysis provided in the original work to investigate the effect of the hyper-parameters that govern the training process and the robustness of the method across these different settings. Finally, we contribute a study of the generalisability of the ANCE-PRF method when dense retrievers other than ANCE are used for the first round of retrieval and for encoding the PRF signal.

Keywords: Pseudo Relevance Feedback · Dense retrievers · Query representations

1 Introduction

Pseudo-Relevance Feedback (PRF) is a retrieval technique which assumes that the top-k results from the first round of retrieval are relevant. PRF,

© The Author(s), under exclusive license to Springer Nature Switzerland AG 2022
M. Hagen et al. (Eds.): ECIR 2022, LNCS 13185, pp. 599–612, 2022.
https://doi.org/10.1007/978-3-030-99736-6_40

therefore, uses this signal to improve the query representation for a second round of retrieval (or re-ranking) in a bid to obtain higher search effectiveness. PRF has been extensively studied and applied to bag-of-word retrieval models; representative techniques are Rocchio [23], KL expansion [18,32], RM3 [17] and other relevance models [9]. It is well accepted that PRF tends to improve search effectiveness, and strong bag-of-words baselines often rely on PRF (e.g., BM25+RM3 is a typical baseline combination).

Aside from its use with bag-of-words models, PRF has been recently studied in the context of Transformer [25]-based deep language models such as BERT [4] and RoBERTa [16]; examples of such Transformer-based rankers include cross-encoder architectures such as monoBERT [21]. These deep language models have been very effective for ranking although, compared to bag-of-words methods, they often require substantially more computational power and are characterised by high query latencies. Their effectiveness can be further improved by PRF – but this is at the cost of even higher query latencies, rendering the use of PRF on top of BERT-based rankers like monoBERT practically unfeasible [12].

Dense retrievers (DRs) have been proposed as alternatives to the expensive BERT-based rankers [7,8,28,33]. DRs also rely on deep language models like BERT; however instead of training a cross-encoder to encode a query and document[1] pair at the same time, it relies on a bi-encoder architecture where queries and documents are encoded separately. This separation in the encoding allows us to pre-compute document representations (which is computationally expensive for large collections) at indexing time, thus leaving only the encoding of the query and the matching between the query and document representations to be performed at query time. Dense retrievers provide a trade off between effectiveness and efficiency: while they are often less effective than the cross-encoder methods, DRs are more efficient (lower query latency). PRF with DRs then becomes suddenly more interesting than when applied to cross-encoders: PRF could provide effectiveness boosts while the additional computational cost imposed by the feedback, infeasible when considering cross-encoders, may be feasible in the context of DRs. This research direction has therefore attracted increasing interest [12,27,30].

In this paper, we consider a specific method for PRF with DRs: the ANCE-PRF method [30]. This method uses the ANCE dense retriever [28] to perform a first round of retrieval for a given query. Then, the text of the original query is concatenated with that from the top-k documents retrieved by ANCE. The output is a new text query, which is encoded using the purposely trained ANCE-PRF encoder to obtain a new dense query representation that is in turn used for computing the match with the document dense representations to determine a ranking for the query. The ANCE-PRF encoder is trained using a straightforward training procedure with negative sampling strategy based on the original ANCE

---

[1] In this paper, we use 'document' and 'passage' interchangeably. Our experiments and the methods considered are in the context of the passage retrieval task. However, the methods can generalise to deal with documents, at the cost of the development of strategies for managing the often large size of documents compared to passages [30].

model, expect that the input to the ANCE-PRF encoder is the concatenation of the query and the relevance signal (top-k documents), rather than just the query (or just the document) as in ANCE.

Given the ANCE-PRF method, we aim to replicate the initial study by Yu et al. [30] in terms of both the training of the ANCE-PRF encoder and its use for retrieval. In addition, we also aim to further extend that work by considering the factors that affect the training of the ANCE-PRF encoder, i.e., the hyper-parameters of the model, and studying their effect on the model performance and therefore its robustness across hyper-parameters settings. We also study the generalisability of the strategy underlying ANCE-PRF to other DRs. In doing so, we develop and publicly release a codebase that implements Yu et al.'s method, along with trained checkpoints of the method for both ANCE and other DRs.

2 Related Work

Pseudo-Relevance Feedback (PRF) is a classic query expansion method that aims to mitigate the mismatch between query intent and query representation [1,26], by modifying the original query with the top-k initially retrieved results. Typical PRF approaches such as Rocchio [23], query-regularized mixture model [24], KL expansion [18,32], RM3 [17], relevance models [9], and relevance-feedback matrix factorization [31] are well studied. However, most of the existing studies of PRF methods are applied on top of bag-of-words retrieval models.

With the emergence of transformer-based [25] models, many researchers have been looking into how to integrate PRF with deep language models. Zheng et al. [34] presented a BERT-based [4] PRF model, BERT-QE, which splits the PRF documents into smaller chunks and utilises the BERT model to identify the most relevant PRF document chunks and uses these chunks as PRF signals. Li et al. [11] proposed a neural PRF approach that uses a feed-forward neural network model to aggregate the query and feedback document relevance scores and provide the target document's relevance score. Yu et al. [29] utilises graph transformers to capture the PRF signals from the initial retrieved results; and Wang et al. [26] proposed a clustering method to gather the relevance signals from PRF documents. These methods show remarkable improvements, but the efficiency is significantly affected, e.g., BERT-QE inference requires $11.01\times$ more computations than BERT alone, making these models computationally infeasible for many practical applications.

Recently, *dense retrievers* [6–8,15,28] have been attracting a lot of attention from researchers. These models, which often utilise a BERT-based dual-encoder to encode queries and passages into a shared embedding space, have shown great effectiveness and efficiency in various tasks and datasets. However, most of the existing studies are focusing on different training methods, especially negative sampling techniques [5,10,28]. Most of these models encode either the query or the document to a single embedding vector [6,15,28], which fits perfectly to many vector-based PRF methods.

In recent research, because of the nature of dense retrievers that use embedding vectors to represent query and document, different methods have been studied to integrate pseudo relevance information into dense retrievers. Li et al. [12] investigated two simple approaches, Average and Rocchio, to utilise PRF information in dense retrievers (ANCE [28] and RepBERT [33]) without introducing new neural models or further training. According to the results, both models achieved superior effectiveness with these two simple approaches without hurting the efficiency significantly compared to the original models, which shows the viability of integrating PRF signals in deep language models. A more recent work attempts to utilise the pattern learning ability of transformer models to leverage the PRF signals. Yu et al. [30] replaced the query encoder in ANCE [28] model by training a new query encoder, which takes the original query text and the PRF documents text together as the new query, based on the original ANCE model as the initial training checkpoint, without changing the document encoder. However, it has several major limitations: 1) for each different PRF depths, it requires training a new query encoder; 2) the input length for the query encoder is limited, which means the PRF depth is limited; 3) the new query encoder is trained on top of the ANCE query encoder, which means for different datasets, different ANCE models need to be trained first, making this new approach hard to be generalised.

3 Improving Query Representations for Dense Retrievers with Pseudo Relevance Feedback

In this section, we briefly describe the ANCE-PRF method [30], which extends ANCE [28] to integrate the PRF signal from the top-k documents to be encoded in combination with the query to form a new query representation.

In ANCE, the score of a document d for a query q is computed by separately encoding q and d using the RoBERTa [16] pre-trained deep language model, and then calculating the inner product between the resulting dense representations:

$$f_{\text{ANCE}}(q, d) = \text{ANCE}^q(\langle s \rangle q \langle /s \rangle) \cdot \text{ANCE}^d(\langle s \rangle d \langle /s \rangle) \tag{1}$$

where ANCE^q and ANCE^d represent the query and the document encoders, respectively, and $\langle s \rangle$ and $\langle /s \rangle$ represent the [CLS] and [SEP] tokens in ANCE. Both encoders use the final layer of the $\langle s \rangle$ token embedding as the query and document dense representations. In ANCE, the document embeddings are precomputed offline and stored in an index, while the query embeddings are encoded at inference (query) time [28]. For fine-tuning Eq. 1, ANCE adopts noisy contrastive estimation loss and employs a negative sampling strategy where negative samples are dynamically retrieved from an asynchronously updated ANCE document index [28].

ANCE-PRF uses a similar schema to score documents for retrieval:

$$f_{\text{ANCE-PRF}}(q, d) = \text{ANCE}^{prf}(\langle s \rangle q \langle /s \rangle d_1 \langle /s \rangle ... d_k \langle /s \rangle) \cdot \text{ANCE}^d(\langle s \rangle d \langle /s \rangle) \tag{2}$$

where ANCE^{prf} is the newly trained PRF query encoder and $\langle s \rangle q \langle /s \rangle d_1 \langle /s \rangle$...
$d_k \langle /s \rangle$ is the text concatenation of the original query q with the feedback documents $d_1, d_2, ..., d_k$ (in addition to [CLS] and separator tokens). We denote q^{prf}
as the query embedding generated through PRF by ANCE^{prf}.

For the training of the PRF query encoder (ANCE^{prf}), ANCE-PRF uses the
standard noisy contrastive estimation loss:

$$\mathcal{L} = -log \frac{exp(q^{prf} \cdot d^+)}{exp(q^{prf} \cdot d^+) + \Sigma_{d^- \in D^-} exp(q^{prf} \cdot d^-)} \tag{3}$$

where d^+ represents a relevant document for the query, d^- represents an irrelevant document (obtained from the negative sampling technique). During the
training process, the ANCE-PRF model uses the document embeddings from the
original ANCE model. Therefore, the document embeddings remain unchanged
in the ANCE-PRF model: it is only the query embedding that changes into the
PRF query embedding q^{prf}.

Intuitively, ANCE-PRF should provide increases in search effectiveness
because the newly trained ANCE-PRF query encoder learns to extract relevant information for the query from the PRF documents using the Transformer
attention mechanism [25]. After training, the ANCE-PRF query encoder would
then pay more attention to the relevant tokens in the PRF documents, while
ignoring the irrelevant tokens from this signal. Although Yu et al. [30] do not
report the query latency of ANCE-PRF, this should be approximately twice that
of the original ANCE model.

4   Experimental Settings

4.1   Datasets

The datasets used in the original work of Yu et al. [30] are TREC DL 2019 [2],
TREC DL 2020 [3], DL Hard [19], and MS MARCO Passage Ranking V1 [20].
These datasets are based on the same corpus provided by MS MARCO Passage Ranking V1, which has ∼8.8M passages in total. Note that for TREC DL
2019/2020 queries, each query has multiple judgements on a relevance scale from
0 to 3, while MS MARCO Passage Ranking V1 only has an average of one judgement per query with binary relevance, either 0 or 1.

The original paper used the training split from MS MARCO Passage Ranking
V1 for training ANCE-PRF, which includes ∼530K queries. The trained models are evaluated on TREC DL 2019 (43 judged queries), DL 2020 (54 judged
queries), DL HARD, and MS MARCO Passage Ranking V1 Dev set (6,980
queries). For direct comparison with the ANCE-PRF model, we follow the same
process except for evaluation on TREC DL HARD (the results on this dataset
for other dense retrievers considered in this paper are not publicly available).

4.2   Models

The original work by Yu et al. [30] only considers ANCE as the initial dense retriever. To validate their hypothesis that their PRF method can be generalised to other dense retrievers, we consider two recently published dense retrievers that achieve higher performance than ANCE: TCT ColBERT V2 HN+ [15] and DistilBERT KD TASB [6]. These two dense retrievers are different from ANCE with respect to the training process, and the inference is slightly different from each other. We refer the reader to the original papers for further details. The output of these three retrievers are all embedding vectors that represent either the query or the document based on the input. The indexes for all three models are pre-computed and stored offline.

$$q_{TCT}^{prf} = \text{TCT}^{prf}([\text{CLS}]\,[\text{Q}]\,q[\text{SEP}]d_1[\text{SEP}]...d_k[\text{MASK}]*512) \qquad (4)$$

TCT ColBERT V2 HN+ uses a BERT encoder, as shown in Eq. 4, to encode queries and documents, where TCT^{prf} represents the new PRF query encoder based on the TCT ColBERT V2 HN+ query encoder. The input requires a [CLS] token, as well as a [Q] in text as prepend to the actual query text, then the PRF document texts are separated by the [SEP] token, then use the [MASK] token to pad the gap if the input is smaller than the max input size of the model, which is 512 for BERT-based models [4].

$$f_{\text{TCT-PRF}}(q,d) = q_{TCT}^{prf} \cdot \text{TCT}^d([\text{CLS}]\,[\text{D}]\,d) \qquad (5)$$

For retrieval, TCT ColBERT V2 HN+ uses a scoring function, as shown in Eq. 5, where TCT^d represents the document encoder, and the input document text is prepended with the [CLS] token and [D] in the text.

$$q_{DBERT}^{prf} = \text{DBERT}^{prf}([\text{CLS}]q[\text{SEP}]d_1[\text{SEP}]...d_k[\text{SEP}]) \qquad (6)$$

$$f_{\text{DBERT-PRF}}(q,d) = q_{DBERT}^{prf} \cdot \text{DBERT}^d([\text{CLS}]d[\text{SEP}]) \qquad (7)$$

On the other hand, DistilBERT KD TASB uses a DistilBERT encoder, as shown in Eq. 6, and a scoring functions for retrieval, as shown in Eq. 7. Similar to TCT ColBERT V2 HN+, except the input is a standard BERT input with the [CLS] token as prepend and the [SEP] token as separators to separate the PRF documents for both PRF query encoding and retrieval.

4.3   Inference and Training

Inference. To reproduce the ANCE-PRF results, the authors have provided us with a model checkpoint of PRF depth 3. Since there is no inference code available from the original authors, we utilise the open source IR toolkit Pyserini[2] [14], which has already implemented the ANCE dense retriever, by introducing a second round of ANCE retrieval with the ANCE-PRF model checkpoint. During

---

[2] https://github.com/castorini/pyserini

the inference time, the document index is the same for both the first round ANCE retrieval and the second round ANCE-PRF retrieval. The only difference in this process is that the initial retrieval uses the ANCE query encoder, while the second retrieval uses the ANCE-PRF query encoder.

Training. The authors have not released the training code. To replicate the ANCE-PRF training process, we utilise the open source dense retriever training toolkit Tevatron[3]. According to the original paper [30], all hyperparameters used in ANCE-PRF training are the same as ANCE training, and the ANCE-PRF query encoder is initialised from ANCE FirstP model[4] [28]. Although some of the parameters are still not reported in the original paper, we managed to replicate the same model as ANCE-PRF with $k = 3$ by adjusting different training settings.

We also experimented with two more effective dense retrievers, TCT Col-BERT V2 HN+ [15] and DistilBERT KD TASB [6] to investigate the generalisability of the ANCE-PRF model. Therefore, we adopted the same hyperparameters from these two models and trained with the same settings as ANCE-PRF.

All models in our experiments are trained on two Tesla V100 SMX2 32GB GPUs. In the original paper, the ANCE-PRF model is trained with per device batch size 4 and gradient accumulation step 8 for 450K steps, which is equivalent to per device batch size 32 for ~56K steps, therefore, in our training experiments, we use 10 epochs, which is roughly ~80K steps.

4.4   Evaluation Metrics

The official evaluation metric for MS MARCO Passage Ranking V1 dataset is MRR@10 [20], for TREC DL 2019 and 2020 are nDCG@10, Recall@1000 [2,3]. For the Recall@1000 evaluation metric on TREC DL 2019 and 2020, the judgements are binarized at relevance point 2 according to the official guideline. Besides the official evaluation metrics, the authors in the original work [30] also use HOLE@10 as an additional evaluation metric to measure the unjudged fraction of top 10 retrieved documents [28], to reflect the coverage of the pooled labels on these dense retrieval systems. However, in our experiments, we opt to keep the official evaluation metrics only, for the sake of comparison with other models and baselines. Statistical significance differences between models results are measured using two-tailed paired t-tests.

4.5   Research Questions

In this work, we aim to address the following research questions along with the reproducibility and replication[5] of the original method from Yu et al. [30]:

---

[3] https://github.com/texttron/tevatron.

[4] https://github.com/microsoft/ANCE.

[5] We use the terminology of reproducibility and replication in compliance with the definitions provided by ACM: https://www.acm.org/publications/policies/artifact-review-badging.

RQ1: What is the possibility of reproducing the inference results of ANCE-PRF given only a checkpoint of the trained model provided by the original authors?

RQ2: The training process is governed by a number of hyper-parameters and choices, importantly including learning rate, optimizer, and negative sampling technique settings. Given the insufficient details in the original study, it is reasonable to expect that researchers attempting to replicate the ANCE-PRF method may set these parameters to values different from those in the original study. We are then interested to study: what is the impact of ANCE-PRF training hyper-parameters on the effectiveness of the method, and in particular if this is robust to different hyper-parameter settings?

RQ3: The PRF strategy underlying ANCE-PRF can be adapted to other dense retrievers as observed by Yu et al. [30], but not empirically validated. The original ANCE-PRF model is only trained with ANCE [28] as the initial dense retriever. We are then interested to investigate: do the improvements observed for ANCE-PRF generalise to other dense retrievers, such as the two more effective models, TCT ColBERT V2 HP+ [15] and DistilBERT KD TASB [6]?

5 Results and Analysis

5.1 RQ1: Reproduce ANCE-PRF Inference

A benefit of the transformer-based neural models nowadays is its easy reproducibility; the results can be reproduced easily by using the model checkpoint. Therefore, with the PRF 3 checkpoint provided by the authors, we tried to reproduce the same results reported in the original paper; the outcomes are shown in Table 1. During the reproducibility process, we found that the ANCE-PRF model is sensitive to uppercase or lowercase letters. For the original queries used in all three datasets in this experiment, no uppercase letters existed, therefore this detail is omitted from the paper. But from our reproducibility experiments, uppercase letters exist in the corpus, and the token ids and their associated tokens embeddings are different with different cases of the same word. Therefore, for PRF queries, after concatenating the PRF documents to the original query text, the new PRF queries contain uppercase letters and leads to different tokens after tokenization, and resulting in different performance compared to what is reported in the original paper. On the other hand, if we set the tokenizer to do lowercase at inference time, then we can get the same results as the original paper. Hence, we successfully reproduced the ANCE-PRF model for inferencing by using the checkpoint provided by the authors.

To answer RQ1, we confirmed that it is possible to reproduce the same results with the model checkpoint, however one key detail that was missing in the paper is the lowercase process to the PRF query. We make our ANCE-PRF inference implementation publicly available in Pyserini toolkit[6] so that practitioners can

---

[6] https://github.com/castorini/pyserini/blob/master/docs/experiments-ance-prf.md.

Table 1. The reproduced results with the provided ANCE-PRF 3 checkpoint after inference. **unhandled** represents results with the PRF query containing both upper-case and lowercase letters. **do lowercase** indicates the results with PRF query converted to lowercase when tokenized. **ANCE-PRF 3** shows results in original paper.

Datasets	MS MARCO			TREC DL 2019		TREC DL 2020	
	MRR@10	nDCG@10	R@1000	nDCG@10	R@1000	nDCG@10	R@1000
ANCE [28]	0.330	0.388	0.959	0.648	0.755	0.646	0.776
ANCE-PRF 3 [30]	0.344	0.401	0.959	0.681	0.791	0.695	0.815
Unhandled	0.342	0.399	0.960	0.678	0.792	0.674	0.794
Do lowercase	0.344	0.402	0.960	0.681	0.791	0.695	0.815

easily reproduce the same results in the original paper with the author provided model checkpoint.

5.2 RQ2: Replicate ANCE-PRF Training

In this section, we would like to see if we can reproduce the model by following the training settings provided in the original paper. However, some details were missing and we had to consult with the authors to identify their exact settings. After clarifying the training parameters, we used the same setting to train our own ANCE-PRF model; the results are shown in Table 2.

Table 2. The replicated results with the trained ANCE-PRF 3 checkpoint after inference. **ANCE-PRF 3** shows the results from the original paper. **ANCE** represents the results from the original ANCE model. **Replicated** is the results from our replicated ANCE-PRF model.

Datasets	MS MARCO			TREC DL 2019		TREC DL 2020	
	MRR@10	nDCG@10	R@1000	nDCG@10	R@1000	nDCG@10	R@1000
ANCE [28]	0.330	0.388	0.959	0.648	0.755	0.646	0.776
ANCE-PRF 3 [30]	0.344	0.401	0.959	0.681	0.791	0.695	0.815
Replicated	0.347	0.405	0.963	0.672	0.794	0.701	0.814

From the results, once their setting was replicated, we obtained results that are close to those reported and with similar trends, at times worse, other times better, but never statistically significantly different from the results reported in the original paper. The minor differences between the two results can be potentially explained by random neuron drop out during training and the random seed while sampling the hard negatives from the initially retrieved ANCE results.

In the original study, the authors reported that they were using all hyper-parameters from ANCE [28] training, and all models are trained on two RTX

Table 3. Initial represents the results by re-initialising the linear head layer. **Inherit** represents the results by inheriting the linear head layer from ANCE. **In-Batch** represents the results by using in-batch negatives. **No In-Batch** represents the results by not using in-batch negatives. **1e-6** represents the results by using 1e-6 as the learning rate. **1e-5** represents the results by using 1e-5 as learning rate.

Datasets	MS MARCO	TREC DL 2019		TREC DL 2020	
	MRR@10	nDCG@10	R@1000	nDCG@10	R@1000
Inherit/No In-Batch/1e-5	0.347	0.672	0.794	0.701	0.814
Inherit/No In-Batch/1e-6	0.335	0.680	0.798	0.678	0.814
Inherit/In-Batch/1e-5	0.347	0.672	0.797	0.678	0.807
Initial/No In-Batch/1e-5	0.313	0.631	0.710	0.644	0.772

2080 Ti GPUs with per-GPU batch size 4 and gradient accumulation step 8 for 450K steps. However, some parameters are still unclear in ANCE training. We trained the ANCE-PRF model with two Tesla-V100 SMX2 32GB GPUs with per-GPU batch size 32, learning rate 1e-5, no in-batch negatives or cross batch negatives, and no gradient accumulation steps for 10 epochs. The reason why we chose to remove the gradient accumulation step setting is because we are using GPUs with larger memory. In the original settings, 450K steps with gradient accumulation step 8 and per-GPU batch size 4 is the same as 56,250 steps with per-GPU batch size 32. Therefore, in our training process, we used 10 training epochs, which is equivalent to 83,240 steps in total, and it is already more than the steps used in the original settings.

The optimizer in the training process for the ANCE-PRF model reported in the original study is the LAMB optimizer, which we overlooked at first, instead we used the AdamW optimizer which might lead to unsuccessful replication.

A common practice for training new models based on an existing model is to re-initialise the linear head layer and train from scratch while keeping the model body. We followed this practice at first, but it appears that the ANCE-PRF model is trained with everything inherited from the the ANCE model, including the embedding head and normalisation (linear head layer). So without keeping the linear head layer from ANCE, our trained ANCE-PRF is significantly worse than the original ANCE-PRF model, as shown in Table 3.

Recent more effective models such as RocketQA [22], uniCOIL [13] show that in-batch negatives help the model learn and achieve better performance. However, in our experiments, in-batch negatives do not help to improve the model performance, as shown in Table 3, the difference between using and not using in-batch negatives is not statistically significant though.

Learning rate also plays an important part in the training process. We have experimented with two different learning rates, 1e-5 and 1e-6; the results are shown in Table 3. Using a larger learning rate tends to improve MRR@10 for the MS MARCO dataset, while a smaller learning rate tends to improve nDCG@10 and R@1000 in TREC DL 2019. However, only MRR@10 is statistically significantly different.

Table 4. Results of training two more effective dense retrievers with the same training process as ANCE-PRF. For direct comparisons, we trained the models with PRF depth of value 3. † represents statically significant difference based on a two-tailed paired t-test.

Datasets	MS MARCO			TREC DL 2019		TREC DL 2020	
	MRR@10	nDCG@10	R@1000	nDCG@10	R@1000	nDCG@10	R@1000
ANCE [28]	0.330	0.388	0.959	0.648	0.755	0.646	0.776
ANCE-PRF 3 [30]	0.344	0.401	0.959	0.681	0.791	0.695	0.815
TCT ColBERT V2 HN+	0.359	0.420	0.970	0.720	0.826	0.688	0.843
TCT ColBERT V2 HN+ PRF 3	0.357	0.418	0.971†	0.741	0.852†	0.712†	0.840
DistilBERT KD TASB	0.344	0.407	0.977	0.721	0.841	0.685	0.873
DistilBERT KD TASB PRF 3	0.348	0.411†	0.974	0.736	0.857†	0.698†	0.866

To answer RQ2, some hyperparameters, such as learning rate, number of negatives, type of negatives, and optimizer, are crucial for reproducing the model checkpoint.

5.3 RQ3: Generalisability of ANCE-PRF Beyond ANCE

After successfully reproducing the ANCE-PRF model inference and replicating the training, we investigated if integrating this PRF strategy with other popular and more effective dense retrievers will provide some improvements in effectiveness when compared to the dense retrievers results without PRF. However, this improvement is of a smaller magnitude than that observed for ANCE, which can be observed from Table 4. This may be due to: (1) the best hyper-parameter settings for ANCE-PRF may not be adequate to generalise to other dense retrievers, and different settings may lead other dense retrievers to obtain larger improvements; this speaks to the limited robustness of ANCE-PRF's training strategy. (2) The dense retrievers we consider, TCT ColBERT V2 HN+ [15] and Distil-BERT KD TASB [6] are more effective than ANCE. The limited improvement then may be due to the fact that it is easier to improve a weaker model (ANCE) than it is to improve a more effective one.

To answer RQ3, we find that applying the same training strategy as ANCE-PRF to other more effective dense retrievers only achieves a smaller magnitude of improvement. Hence, the ANCE-PRF method may not generalize to other dense retrievers or may require specific hyper-parameter tuning.

6 Conclusion

In this paper we considered the ANCE-PRF model proposed by Yu et al. [30]. This method is the first of its kind to integrate PRF signals directly into the query encoder, without changing the document encoder or document index.

There are three research questions related to reproducing and replicating ANCE-PRF. RQ1 is aimed to address the issues when reproducing the inference results by directly adopting the model checkpoint provided by the original authors. Our experiments show that ANCE-PRF is an uncased model; it can only handle lowercase letters for queries. If the query contains uppercase letters, the ANCE-PRF model performs differently and hurts performance.

RQ2 is aimed to replicate the training process of the ANCE-PRF model by using the settings provided in the original study. However, some details are missing, which leads to unsatisfying performance of the replicated model. After consulting with the original authors and using the exact same training settings, we were able to replicate the ANCE-PRF model with insignificant differences that might be caused by the random seed in negative sampling or model initialisation. We then investigate the effects of hyper-parameters in the ANCE-PRF model training. Since some details are left out in the original study, we replicate the model by using common practice. However, in our experiments, we found that some common practice might not work in this case. For example, using the linear head layer from the ANCE model to train ANCE-PRF is significantly better than to initialise the linear head layer. In-batch negatives have been proved to be useful for training in many superior models, but in our experiment, there is no significant difference between using in-batch negatives and no in-batch negatives.

RQ3 is aimed to test the generalisability of the training method of the ANCE-PRF model. We use the same parameter settings to train the PRF model on top of TCT ColBERT V2 HN+ [15] and DistilBERT KD TASB [6], two more effective dense retrievers compared to ANCE. However, the results are mixed; the improvements with PRF are of a smaller magnitude than that observed for ANCE. This may be because the best hyper-parameter settings are not suitable for all dense retrievers; to achieve better performance one may need to adjust the parameters accordingly. Another reason may be because both newly added models are more effective than ANCE; the limited improvements may be because this training method is more suitable to improve a weaker model.

The code to reproduce the training of all the models in this work is made available at https://github.com/ielab/APR.

Acknowledgement. Hang Li is supported by a UQ Earmarked PhD Scholarship and this research is partially funded by the Grain Research and Development Corporation project AgAsk (UOQ2003-009RTX).

References

1. Clinchant, S., Gaussier, E.: A theoretical analysis of pseudo-relevance feedback models. In: Proceedings of the 2013 Conference on the Theory of Information Retrieval, pp. 6–13 (2013)
2. Craswell, N., Mitra, B., Yilmaz, E., Campos, D., Voorhees, E.M.: Overview of the TREC 2019 deep learning track. In: Text REtrieval Conference, TREC (2020)

3. Craswell, N., Mitra, B., Yilmaz, E., Campos, D., Voorhees, E.M.: Overview of the TREC 2020 deep learning Track. In: Text REtrieval Conference, TREC (2021)
4. Devlin, J., Chang, M.W., Lee, K., Toutanova, K.: BERT: pre-training of deep bidirectional transformers for language understanding. In: Proceedings of the 2019 Conference of the North American Chapter of the Association for Computational Linguistics: Human Language Technologies, pp. 4171–4186 (2019)
5. Gao, L., Dai, Z., Chen, T., Fan, Z., Van Durme, B., Callan, J.: Complement lexical retrieval model with semantic residual embeddings. In: Hiemstra, D., Moens, M.-F., Mothe, J., Perego, R., Potthast, M., Sebastiani, F. (eds.) ECIR 2021, Part I. LNCS, vol. 12656, pp. 146–160. Springer, Cham (2021). https://doi.org/10.1007/978-3-030-72113-8_10
6. Hofstätter, S., Lin, S.C., Yang, J.H., Lin, J., Hanbury, A.: Efficiently teaching an effective dense retriever with balanced topic aware sampling. In: Proceedings of the 44th International ACM SIGIR Conference on Research and Development in Information Retrieval, pp. 113–122. ACM (2021)
7. Karpukhin, V., et al.: Dense passage retrieval for open-domain question answering. In: Proceedings of the 2020 Conference on Empirical Methods in Natural Language Processing (EMNLP), pp. 6769–6781 (2020)
8. Khattab, O., Zaharia, M.: ColBERT: efficient and effective passage search via contextualized late interaction over BERT. In: Proceedings of the 43rd International ACM SIGIR conference on research and development in Information Retrieval, pp. 39–48 (2020)
9. Lavrenko, V., Croft, W.B.: Relevance based language models. In: Proceedings of the 24th annual International ACM SIGIR Conference on Research and Development in Information Retrieval, pp. 120–127 (2001)
10. Lee, K., Chang, M.W., Toutanova, K.: Latent retrieval for weakly supervised open domain question answering. In: Proceedings of the 57th Annual Meeting of the Association for Computational Linguistics, pp. 6086–6096 (2019)
11. Li, C., et al.: NPRF: a neural pseudo relevance feedback framework for Ad-hoc information retrieval. In: Proceedings of the 2018 Conference on Empirical Methods in Natural Language Processing, pp. 4482–4491 (2018)
12. Li, H., Mourad, A., Zhuang, S., Koopman, B., Zuccon, G.: Pseudo Relevance Feedback with Deep Language Models and Dense Retrievers: Successes and Pitfalls. arXiv preprint arXiv:2108.11044 (2021)
13. Lin, J., Ma, X.: A Few Brief Notes on DeepImpact, COIL, and a Conceptual Framework for Information Retrieval Techniques. arXiv preprint arXiv:2106.14807 (2021)
14. Lin, J., Ma, X., Lin, S.C., Yang, J.H., Pradeep, R., Nogueira, R.: Pyserini: An easy-to-use Python toolkit to support replicable IR research with sparse and dense representations. arXiv preprint arXiv:2102.10073 (2021)
15. Lin, S.C., Yang, J.H., Lin, J.: Distilling dense representations for ranking using tightly-coupled teachers. arXiv preprint arXiv:2010.11386 (2020)
16. Liu, Y., et al.: RoBERTa: A Robustly Optimized BERT Pretraining Approach. arXiv preprint arXiv:1907.11692 (2019)
17. Lv, Y., Zhai, C.: A comparative study of methods for estimating query language models with pseudo feedback. In: Proceedings of the 18th ACM International Conference on Information and Knowledge Management, pp. 1895–1898 (2009)
18. Lv, Y., Zhai, C.: Revisiting the divergence minimization feedback model. In: Proceedings of the 23rd ACM International Conference on Information and Knowledge Management, pp. 1863–1866 (2014)

19. Mackie, I., Dalton, J., Yates, A.: How deep is your learning: the DL-HARD annotated deep learning dataset. In: Proceedings of the 44th International ACM SIGIR Conference on Research and Development in Information Retrieval (2021)
20. Nguyen, T., et al.: MS MARCO: a human generated machine reading comprehension dataset. In: Workshop on Cognitive Computing at NIPS (2016)
21. Nogueira, R., Cho, K.: Passage Re-ranking with BERT. arXiv preprint arXiv:1901.04085 (2019)
22. Qu, Y., et al.: RocketQA: an optimized training approach to dense passage retrieval for open-domain question answering. In: Proceedings of the 2021 Conference of the North American Chapter of the Association for Computational Linguistics: Human Language Technologies, pp. 5835–5847 (2021)
23. Rocchio, J.: Relevance feedback in information retrieval. In: The SMART Retrieval System - Experiments in Automatic Document Processing, pp. 313–323 (1971)
24. Tao, T., Zhai, C.: Regularized estimation of mixture models for robust pseudo-relevance feedback. In: Proceedings of the 29th Annual International ACM SIGIR Conference on Research and Development in Information Retrieval, SIGIR 2006, pp. 162–169. Association for Computing Machinery (2006)
25. Vaswani, A., et al.: Attention is all you need. In: Proceedings of the 31st International Conference on Neural Information Processing Systems, pp. 6000–6010 (2017)
26. Wang, J., Pan, M., He, T., Huang, X., Wang, X., Tu, X.: A pseudo-relevance feedback framework combining relevance matching and semantic matching for information retrieval. Inf. Process. Manage. **57**(6), 102342 (2020)
27. Wang, X., Macdonald, C., Tonellotto, N., Ounis, I.: Pseudo-Relevance Feedback for Multiple Representation Dense Retrieval. arXiv preprint arXiv:2106.11251 (2021)
28. Xiong, L., et al.: Approximate nearest neighbor negative contrastive learning for dense text retrieval. In: International Conference on Learning Representations (2020)
29. Yu, H.C., Dai, Z., Callan, J.: PGT: pseudo relevance feedback using a graph-based transformer. In: Hiemstra, D., Moens, M.-F., Mothe, J., Perego, R., Potthast, M., Sebastiani, F. (eds.) ECIR 2021, Part II. LNCS, vol. 12657, pp. 440–447. Springer, Cham (2021). https://doi.org/10.1007/978-3-030-72240-1_46
30. Yu, H., Xiong, C., Callan, J.: Improving query representations for dense retrieval with pseudo relevance feedback. In: Proceedings of the 30th ACM International Conference on Information and Knowledge Management. ACM (2021)
31. Zamani, H., Dadashkarimi, J., Shakery, A., Croft, W.B.: Pseudo-relevance feedback based on matrix factorization. In: Proceedings of the 25th ACM International on Conference on Information and Knowledge Management, pp. 1483–1492 (2016)
32. Zhai, C., Lafferty, J.: Model-based feedback in the language modeling approach to information retrieval. In: Proceedings of the 10th ACM International Conference on Information and Knowledge Management, pp. 403–410 (2001)
33. Zhan, J., Mao, J., Liu, Y., Zhang, M., Ma, S.: RepBERT: Contextualized text embeddings for first-stage retrieval. arXiv preprint arXiv:2006.15498 (2020)
34. Zheng, Z., Hui, K., He, B., Han, X., Sun, L., Yates, A.: BERT-QE: contextualized query expansion for document re-ranking. In: Proceedings of the 2020 Conference on Empirical Methods in Natural Language Processing: Findings, pp. 4718–4728 (2020)

Another Look at DPR: Reproduction of Training and Replication of Retrieval

Xueguang Ma[✉], Kai Sun, Ronak Pradeep, Minghan Li, and Jimmy Lin

David R. Cheriton School of Computer Science, University of Waterloo,
Waterloo, Canada
{x93ma,k49sun,rpradeep,m6921i,jimmylin}@uwaterloo.ca

Abstract. Text retrieval using learned dense representations has recently emerged as a promising alternative to "traditional" text retrieval using sparse bag-of-words representations. One foundational work that has garnered much attention is the dense passage retriever (DPR) proposed by Karpukhin et al. for end-to-end open-domain question answering. This work presents a reproduction and replication study of DPR. We first verify the reproducibility of the DPR model checkpoints by training passage and query encoders from scratch using two different implementations: the original code released by the authors and another independent codebase. After that, we conduct a detailed replication study of the retrieval stage, starting with model checkpoints provided by the authors but with an independent implementation from our group's Pyserini IR toolkit and PyGaggle neural text ranking library. Although our experimental results largely verify the claims of the original DPR paper, we arrive at two important additional findings: First, it appears that the original authors under-report the effectiveness of the BM25 baseline and hence also dense–sparse hybrid retrieval results. Second, by incorporating evidence from the retriever and improved answer span scoring, we manage to improve end-to-end question answering effectiveness using the same DPR models.

Keywords: Open-domain QA · Dense retrieval

1 Introduction

Reproducibility and replicability form the foundation of the scientific enterprise. Through such studies, the community gains confidence about the veracity of previously published results. These investigations are often under-valued, especially compared to work that proposes novel models, but they nevertheless make important contributions to advancing science. To be precise, throughout this paper we use the term reproducibility and replicability in the sense articulated by the ACM,[1] characterized as "different team, same experimental setup" and "different team, different experimental setup", respectively.

---

[1] ACM Artifact Review and Badging (Version 2.0).

© The Author(s), under exclusive license to Springer Nature Switzerland AG 2022
M. Hagen et al. (Eds.): ECIR 2022, LNCS 13185, pp. 613–626, 2022.
https://doi.org/10.1007/978-3-030-99736-6_41

This paper focuses on a reproducibility and replicability study of the dense passage retriever (DPR) model proposed by Karpukhin et al. [8], as the authors have laid important cornerstones for end-to-end vector-based dense retrieval and open-domain question answering (QA). Specifically, we first conduct a reproduction of model training, verifying that we can obtain models with comparable levels of effectiveness using the released code from the authors as well as another implementation. Then, we conduct a replication of the retrieval pipeline, analyzing end-to-end retrieval effectiveness with our independent implementation. For a fair comparison and to reduce conflated factors, our replication study starts with the released checkpoints, as we have confirmed the reproducibility of model training during the first step.

DPR is worthy of detailed study because it represents an important exemplar of text retrieval using learned dense representations, which has emerged as a promising alternative to "traditional" text retrieval using sparse bag-of-words representations [5,11,16,19]. Our experiments largely verify the claims of Karpukhin et al. regarding the effectiveness of their proposed techniques. Moreover, we arrive at two important additional findings, one of which is inconsistent with the original work, the other of which presents an enhancement:

1. Focusing on retrieval, we find that the effectiveness of the sparse retrieval (BM25) baseline is higher than numbers reported in the original paper. Whereas they report that dense–sparse hybrid results do not meaningfully improve over dense retrieval alone, we arrive at the opposite conclusion, where hybrid techniques yield statistically significant gains. We are able to achieve on average a four-point improvement in top-20 accuracy over the best DPR results across five standard QA test collections.

2. Focusing on end-to-end QA effectiveness, we explore different techniques for evidence combination to extract the final answer span. Whereas the original DPR paper only uses scores from the reader to identify the final answer span, we investigate combining retriever scores and further experiment with the answer span selection technique described by Mao et al. [12]. In our best condition, we are able to achieve statistically significant improvements of around three points on exact match scores over the original DPR implementation while using the same exact DPR models.

To summarize, the main contribution of this work is the reproduction of DPR training and the replication of end-to-end retrieval experiments, where our experimental results add a number of important refinements to the original work. Code associated with our retrieval experiments is packaged in the Pyserini IR toolkit[2] [10] and code associated with our end-to-end QA experiments is part of the PyGaggle toolkit[3] for neural text ranking.

---

[2] http://pyserini.io/.

[3] http://pygaggle.ai/.

2   Methods

DPR [8] adopts a retriever–reader pipeline proposed by Chen et al. [2] for open-domain QA tasks. Both the formulation and the pipeline architecture for tackling the problem dates from at least the late 1990s [14], which means that this general approach has a long history that predates neural networks. The open-source code associated with the paper is available on GitHub (which we refer to as "the DPR repo"),[4] but it does not appear to contain code and models necessary to reproduce all results reported in the paper (more detailed discussions below).

2.1   Retriever

During retrieval, given a corpus $\mathcal{C} = \{D_1, D_2, ..., D_m\}$, the task is to return a list of the k most relevant documents (i.e., most likely to contain the answer) from $\mathcal{C}$ for each query q, where $k << |\mathcal{C}|$. In the original DPR paper and also our replication study, the corpus refers to the 2018-12-20 dump of English Wikipedia, and the "documents" are non-overlapping 100-word splits of articles.

To be clear, in most text ranking applications, the "unit of indexing" (and also retrieval) is usually referred to as a "document" D_j, although in this case it is a passage (i.e., a split) from Wikipedia. For consistency with this parlance, we use "document" and "passage" interchangeably throughout this paper. To add to the potential confusion, results of the retriever are also referred to as "contexts" that are fed to the reader.

Specifically, DPR contains a query encoder and a passage encoder, both using BERT [3] as the backbone model. Queries and passages are encoded as dense representation vectors separately as follows:

$$q^* = \mathrm{BERT}_q(q), D_j^* = \mathrm{BERT}_D(D_j)$$

where q^* and D_j^* are low dimensional vectors (768 dimensions by default). The relevance score of a passage to a query is computed by their vector dot product:

$$\mathrm{Sim}(q, D_j) = \langle q^*, D_j^* \rangle$$

Thus, the retrieval problem is carried out as nearest neighbor search in dense vector space. Operationally, this is accomplished via Facebook's Faiss library [6].

During training, given a query q, a relevant passage D^+ that contains the answer, and n non-relevant passages $D_1^-, D_2^-, ... D_n^-$, the training objective is:

$$\begin{aligned}
\mathcal{L}&(q, D^+, D_1^-, D_2^-, \cdots, D_n^-) \\
&= -\log p(D = D^+ \mid Q = q) \\
&= -\log \frac{\exp(\mathrm{Sim}(q, D_+))}{\exp(\mathrm{Sim}(q, D^+)) + \sum_{i=1}^{n} \exp(\mathrm{Sim}(q, D_i^-))},
\end{aligned}$$

---

[4] https://github.com/facebookresearch/DPR.

where $p(D = D^+ \mid Q = q)$ can be seen as a classifier given the query q evaluated at passage D^+.

Karpukhin et al. also investigated hybrid retrieval, combining results from dense retrieval (DPR) and sparse retrieval (BM25) by computing the linear combination of their respective scores to rerank the union of the two initial retrieved sets: $\lambda \cdot \mathrm{Sim}(q, D_j) + \mathrm{BM25}(q, D_j)$, where $\lambda = 1.1$, an empirical value tuned on the development set. BM25 retrieval was performed using Lucene with parameters $b = 0.4$ and $k_1 = 0.9$. However, the DPR repo does not appear to contain code for reproducing the BM25 and hybrid fusion results.

We attempt to replicate the retriever results reported in the DPR paper with Pyserini, an IR toolkit we have been developing since 2019 [10]. The toolkit supports sparse retrieval (i.e., BM25) via integration with another toolkit called Anserini [17] built on Lucene. Like in the original DPR work, Pyserini supports dense retrieval via integration with Facebook's Faiss library. Combining dense and sparse retrieval, the Pyserini toolkit supports hybrid retrieval as well.

Our efforts are divided into two distinct steps: First, we verify that the model checkpoints released by the DPR authors are reproducible by retraining the query and passage encoders from scratch. Then, for a fair comparison between our retrieval implementation and the original DPR work, we use the released checkpoints as the starting point of our replication study. Our retrieval implementation does not share any code with the DPR repo, other than evaluation scripts to ensure that results are comparable.

Similar to the original work, we calculate hybrid retrieval scores by linear combination of dense and sparse scores: $\mathrm{Sim}(q, D_j) + \alpha \cdot \mathrm{BM25}(q, D_j)$. Note that, contrary to the original work, we place the α weight on the BM25 score because this yields a more natural way to answer the pertinent research question: Given dense retrieval as a starting point, does adding BM25 as an additional relevance signal provide any value? This question is answered by comparing with a setting of $\alpha = 0$, which is equivalent to discarding BM25 results.

Finally, there are a few more details of exactly how to combine BM25 and DPR scores worth exploring. As a baseline, we use the raw scores directly in the linear combination (exactly as above). However, we notice that the range of scores from DPR and BM25 can be quite different. To potentially address this issue, we apply the following normalization technique: If a document from sparse retrieval is not in the dense retrieval results, we assign it the minimum dense retrieval score among the retrieved documents, and vice versa for the sparse retrieval score.

To arrive at a final top-k ranking, the original DPR paper generated top-k' results from DPR and top-k' results from BM25 (where $k' > k$), before considering the union of the two result sets and combining the scores to arrive at the final top-k. The original work set $k' = 2000$, but after some preliminary experimentation, we decided to fix $k' = 1000$ in our experiments since it is a more common setting in information retrieval experiments (for example, $k = 1000$ is the default in most TREC evaluations).

2.2   Reader

As is standard in a retriever–reader design, the retriever in the DPR paper returns k candidate passages (i.e., splits from Wikipedia) for each query q. The reader extracts the final answer span from the candidate contexts, where each context C_i contains the Wikipedia article title C_i^{title} and its content C_i^{text}.

The reader in DPR uses BERT-base and takes as input each candidate context C_i concatenated to the question q. Answer extraction is treated as a labeling task, and the reader identifies the answer by predicting the start and end tokens of the answer span in the contexts. To do so, the DPR reader adds a linear layer on top of BERT to predict the start logit (i.e., unnormalized probability) and end logit for each token from the final hidden layer representations. The score of an answer span is calculated by adding the start logit of the first token and the end logit of the last token. The reader returns the m highest scoring answer spans. In addition, the reader uses the learned representation of [CLS] to predict the overall relevance of the context to the question.

Mathematically, the reader operates as follows:

$$r_i, \mathcal{S} = \text{Reader}([\text{CLS}] \; q \; [\text{SEP}] \; C_i^{\text{title}} [\text{SEP}] \; C_i^{\text{text}})$$

where r_i is the overall relevance score for context C_i, and $\mathcal{S}$ comprises m potential (answer span, span score) pairs extracted from context C_i:

$$\{(S_{i,1}, s_{i,1}), (S_{i,2}, s_{i,2}), \ldots (S_{i,m}, s_{i,m})\}.$$

In the original paper, the final answer span is the candidate with the maximum span score from the context with the highest relevance score.

We attempt to replicate exactly the DPR implementation of answer extraction using our open-source PyGaggle neural reranking library, which holds the code to many of our other search-related projects. Once again, we begin with reader checkpoints released in the DPR repo, but otherwise our implementation is completely independent (other than, again, the evaluation code).

In addition to the answer extraction algorithm above, we also implement the normalized answer span scoring technique described by Mao et al. [12]. Each answer span in each candidate context C_i is re-scored according to:

$$s'_{i,j} = \text{softmax}(\vec{r})_i \cdot \text{softmax}(\vec{s_i})_j$$

where $\vec{r} = \{r_1, \cdots, r_k\}$ is the set of relevance scores of all candidate contexts and $\vec{s_i} = \{s_{i,1}, \cdots, s_{i,m}\}$ is the set of all span scores within context C_i. Duplicate answer spans across all contexts are scored by accumulating their individual scores. The answer span with the maximum score is selected as the final prediction.

In summary, we compare two answer span scoring techniques in the reader: the "original" answer span scoring technique described by Karpukhin et al. [8], and the span scoring technique described by Mao et al. [12].

2.3   Final Evidence Fusion

In the original DPR paper, the final answer span is only selected based on scores from the reader. In our replication attempt, we additionally exploit scores from the retriever to improve answer span selection. Our intuition is that predictions from both the retriever and the reader should contribute to the final answer. Concretely, instead of just using the relevance score r_i from the reader to score contexts, we fuse r_i with the retriever score R_i, calculated by: $\beta \cdot r_i + \gamma \cdot R_i$. Depending on the retrieval method, R_i can be the sparse retrieval score, the dense retrieval score, or the score after hybrid fusion. This final fused score replaces r_i as the relevance score for each context in the answer span scoring step. For example, with fusion, the answer span scoring technique from GAR [12] becomes $\mathrm{softmax}(\beta \cdot \vec{r} + \gamma \cdot \vec{R})_i \cdot \mathrm{softmax}(\vec{s_i})_j$.

Thus, to summarize, we explore four settings in our end-to-end QA replication: the original DPR span scoring technique, with and without retriever score fusion, and the answer span scoring technique of GAR [12], with and without retriever score fusion.

3   Experimental Setup

In this section, we clarify the models, datasets, metrics, and hyperparameters used in our experiments.

Reproduction of Training. We attempt to reproduce the DPR model checkpoints by training DPR from scratch, following the same settings in the original work as close as possible, with two different implementations. The first is the authors' released code in the DPR repo; experiments reported in the original paper used $8 \times$ Nvidia V100 (32 GB) GPUs, as model quality depends on a large batch size (i.e., 128). The second is code from Gao et al. [4], which is based on the original implementation but exploits gradient caching to make a large batch fit on single GPU.[5] In our reproduction, we train models on $4 \times$ V100 GPUs (the largest machine we have access to) using the authors' original code, and a single V100 GPU using the other implementation; hyperparameters are all identical to the original DPR work. The reproduced checkpoints are evaluated based on the original DPR repo's retrieval and evaluation code.

Replication of Retrieval. Our replication efforts begin with model checkpoints provided in the DPR repo. However, the authors did not release all models and datasets used in their experiments at the time of our work. Therefore, our replication experiments only use the models with released checkpoints:

- Retriever$_{\mathrm{NQ}}$: DPR encoders trained using just the NQ dataset.
- Retriever$_{\mathrm{Multi}}$: DPR encoders trained using a combination of datasets.

---

[5] https://github.com/luyug/GC-DPR.

– Reader$_{\text{NQ-Single}}$: the DPR reader trained on NQ with negative passages from retrieval results by Retriever$_{\text{NQ}}$.
– Reader$_{\text{TQA-Multi}}$: the DPR reader trained on TriviaQA with negative passages from retrieval results by Retriever$_{\text{Multi}}$.

Datasets. We evaluate retrieval effectiveness on five standard benchmark QA datasets (NQ [9], TriviaQA [7], WQ [1], CuratedTREC [14], SQuAD [13]), exactly the same as the original paper. For end-to-end QA, we evaluate on NQ and TriviaQA with the available models. More precisely, we use the Reader$_{\text{NQ-Single}}$ model to process the retrieved contexts from Retriever$_{\text{NQ}}$ for NQ and use the Reader$_{\text{TQA-Multi}}$ model to process the retrieved contexts from Retriever$_{\text{Multi}}$ for TriviaQA.

Metrics. For retrieval, we measure effectiveness in terms of top-k retrieval accuracy, defined as the fraction of questions that have a correct answer span in the top-k retrieved contexts at least once. End-to-end QA effectiveness is measured in terms of the exact match (EM) metric, defined as the fraction of questions that have an extracted answer span exactly matching the ground truth answer. Missing from the original DPR paper, we perform significance testing to assess the statistical significance of metric differences. In all cases, we apply paired t-tests at $p < 0.01$; the Bonferroni correction is applied to correct for multiple hypothesis testing as appropriate.

Hyperparameters. In the hybrid retrieval technique described in the DPR paper, the λ weight for combining dense and sparse retrieval scores is fixed to 1.1. However, our implementation replaces λ with α (see Sect. 2.1). We tune the α values on different datasets by optimizing top-20 retrieval accuracy: For datasets where we can obtain exactly same train/dev/test splits as the original DPR paper (NQ and TriviaQA), we tune the weight on the development set. For the remaining datasets, where splits are not available or the original DPR paper does not provide specific guidance, we tune the weights on a subset of the training data. We obtain the optimal weight by performing grid search in the range $[0, 2]$ with step size 0.05.

Similarly, for final evidence fusion, we tune β (i.e., the weight for the relevance score) and γ (i.e., the weight for the retriever score) on the development set of NQ and TriviaQA using grid search. For greater computational efficiency, we perform tuning in multiple passes by interweaving a coarser step size with a finer step size. For the original DPR answer span scoring technique, we fix β to one and perform a two-step grid search on γ. We start with step size 0.05 and find the optimal γ_1. Then, we use step size 0.01 in the range $[\gamma_1 - 0.04, \gamma_1 + 0.04]$ to find the optimal γ.

For the answer span scoring technique of GAR [12], we define $\delta = \frac{\gamma}{\beta}$ and perform a three-step grid search on β and δ (i.e., the weight for the retriever score becomes $\gamma = \beta \cdot \delta$). We start with step size 0.2 for both β and δ to find the optimal pair of values β_1, δ_1. We then repeat this process with step size 0.05 and 0.01 in a smaller range around the optimal β_i and δ_i from the previous pass.

Table 1. Retrieval effectiveness comparing results from the original DPR paper ("orig") and our reproduction attempt ("repro"). The symbol * on an "orig" result indicates that the corresponding checkpoint was released.

Training	NQ		TriviaQA		WQ		Curated		SQuAD	
	top20	top100	top20	top100	top20	top100	top20	top100	top20	top100
DPR-Single (orig)	78.4*	85.4*	79.4	85.0	73.2	81.4	79.8	89.1	63.2	77.2
DPR-Single (repro)	79.1	85.9	78.9	84.5	71.0	80.2	85.1	92.2	62.1	76.8
DPR-Multi (orig)	79.4*	86.0*	78.8*	84.7*	75.0*	82.9*	89.1*	93.9*	51.6*	67.6*
DPR-Multi (repro)	79.4	87.0	78.5	84.5	75.3	83.0	88.2	94.4	58.3	72.4

For final evidence fusion, we tune the weight parameters together with the number of retrieval results (k) up to 500 with a step size of 20. Optimal parameters are selected based on the highest exact match score.

4 Results

4.1 Reproduction of Training

In Table 1, we report retrieval accuracy from our reproduced model checkpoints. DPR-Single refers to the query encoder and passage encoder trained on a single dataset only and DPR-Multi refers to the model trained on the union of NQ, TriviaQA, WQ, and CuratedTREC (with WQ and CuratedTREC up-sampled by four times given their smaller sizes). To be clear, at the time of our study, the DPR repo only released training data for NQ, TriviaQA, and SQuAD. We follow the DPR paper to prepare training data for WQ and CuratedTREC, but we prepare BM25 hard negative passages by using the Pyserini toolkit because the original repo does not contain BM25 retrieval code. The DPR-Single (repro) results are from training using the authors' original code. The DPR-Multi (repro) results are from training using the code of Gao et al. [4].

The models we train from scratch arrive at a comparable level of effectiveness to the numbers reported in the original paper. Most of the differences are relatively small, within the variability commonly seen when training neural models. Interestingly, for the DPR-Multi setting, our model appears to be quite a bit better than the original model for SQuAD.

Overall, we would consider our reproduction attempt successful. In the following experiments, to reduce the number of conflated factors, we use the DPR authors' released model checkpoints.

4.2 Replication of Retrieval

Table 2 reports top-$k = \{20, 100\}$ retrieval accuracy from our replication attempt, compared to figures copied directly from the original DPR paper; here we focus on results from Retriever$_{\text{Multi}}$. The hybrid retrieval results reported in the original DPR paper is denoted Hybrid$_{\text{orig}}$, which is not directly comparable

Table 2. Comparison between the original DPR paper ("orig") and our replication attempt ("repl"). The symbol † on a BM25 result indicates effectiveness that is significantly different from DPR. The symbol ‡ indicates that the hybrid technique is significantly better than BM25 (for SQuAD) or DPR (for all remaining collections).

Condition	top20		top100	
	orig	repl	orig	repl
NQ				
DPR	79.4	79.5	86.0	86.1
BM25	59.1	62.9†	73.7	78.3†
Hybrid$_{orig}$ ($\lambda = 1.1$)	78.0	-	83.9	-
Hybrid$_{norm}$ ($\alpha = 1.30$)	-	82.6‡	-	88.6‡
Hybrid ($\alpha = 0.55$)	-	82.7‡	-	88.1‡
TriviaQA				
DPR	78.8	78.9	84.7	84.8
BM25	66.9	76.4†	76.7	83.2†
Hybrid$_{orig}$ ($\lambda = 1.1$)	79.9	-	84.4	-
Hybrid$_{norm}$ ($\alpha = 0.95$)	-	82.6‡	-	86.5‡
Hybrid ($\alpha = 0.55$)	-	82.3‡	-	86.1‡
WQ				
DPR	75.0	75.0	82.9	83.0
BM25	55.0	62.4†	71.1	75.5†
Hybrid$_{orig}$ ($\lambda = 1.1$)	74.7	-	82.3	-
Hybrid$_{norm}$ ($\alpha = 0.95$)	-	77.1‡	-	84.4‡
Hybrid ($\alpha = 0.3$)	-	77.5‡	-	84.0‡
CuratedTREC				
DPR	89.1	88.8	93.9	93.4
BM25	70.9	80.7†	84.1	89.9†
Hybrid$_{orig}$ ($\lambda = 1.1$)	88.5	-	94.1	-
Hybrid$_{norm}$ ($\alpha = 1.05$)	-	90.1	-	95.0‡
Hybrid ($\alpha = 0.7$)	-	89.6	-	94.6‡
SQuAD				
DPR	51.6	52.0	67.6	67.7
BM25	68.8	71.1†	80.0	81.8†
Hybrid$_{orig}$ ($\lambda = 1.1$)	66.2	-	78.6	-
Hybrid$_{norm}$ ($\alpha = 2.00$)	-	75.1‡	-	84.4‡
Hybrid ($\alpha = 28$)	-	75.0‡	-	84.0‡

to either of our two techniques: Hybrid$_{norm}$ (with minimum score normalization) or Hybrid (without such normalization). We make the following observations:

First, our dense retrieval results are very close to those reported in the original paper. We consider this a successful replication attempt and our efforts add veracity to the effectiveness of the DPR technique.

Second, our Pyserini BM25 implementation outperforms the BM25 results reported in the original paper across all datasets. Furthermore, the gap is larger for $k = 20$. On average, our results represent a nearly seven-point improvement in top-20 accuracy and a nearly five-point improvement in top-100 accuracy. Since the authors of DPR have not made available their code for generating the BM25 results, we are unable to further diagnose these differences.

Nevertheless, the results do support the finding that dense retrieval using DPR is (generally) more effective than sparse retrieval. We confirm that the effectiveness differences between DPR and BM25 in our replication results are statistically significant. In all datasets except for SQuAD, DPR outperforms BM25; this is consistent with the original paper. We further confirm that for SQuAD, DPR is significantly worse than BM25. As Karpukhin et al. noted, Retriever$_{Multi}$ is trained by combining training data from all datasets but excluding SQuAD; these poor results are expected, since SQuAD draws from a very small set of Wikipedia articles.

Third, the effectiveness of hybrid dense–sparse fusion appears to be understated in the original DPR paper. Karpukhin et al. found that hybrid retrieval is *less* effective than dense retrieval in most settings, which is inconsistent with our experimental results. Instead, we find that dense–sparse retrieval consistently beats sparse retrieval across all settings. The gains from both hybrid scoring techniques are statistically significant, with the exception of top-20 for CuratedTREC. Our results might be due to better BM25 effectiveness, but we are unable to further diagnose these differences because, once again, the hybrid retrieval code is not provided in the DPR repo. Further testing also finds that the differences between the two hybrid techniques are not significant. Thus, there seems to be no strong basis to prefer one hybrid technique over the other.

Table 3. The Jaccard overlap between sparse retrieval and dense retrieval results.

Condition	$k = 20$	100	500	1000
NQ	6.1	5.2	4.4	4.2
TriviaQA	9.2	6.6	5.0	4.6
WQ	5.9	5.9	5.8	5.7
CuratedTrec	6.9	7.2	6.3	5.9
SQuAD	4.5	4.1	4.0	4.0

In Table 3, we report overlap when taking different top-k results from dense retrieval and sparse retrieval. Overlap is measured in terms of Jaccard overlap, which is computed by the intersection over the union. It is apparent that the overlap between dense and sparse results is quite small, which suggests that they are effective in different ways. This provides an explanation of why hybrid retrieval is effective, i.e., it is exploiting different relevance signals. These results also justify the DPR design choice of retrieving $k' > k$ results from dense and sparse retrieval and then rescoring the union to arrive at the final top-k.

4.3   Replication of End-to-End QA

Table 4 presents results for our end-to-end question answering replication experiments on the NQ and TriviaQA datasets in terms of the exact match score. The original results are shown in the "orig" column. The "repl" column reports our attempt to replicate exactly the span scoring technique described in the original paper, whereas the "GAR" column shows results from using the technique proposed by Mao et al. [12]. The version of each technique that incorporates retriever scores (see Sect. 2.3) is denoted with a * symbol, i.e., "repl*" and "GAR*". For NQ, we used $Retriever_{NQ}$ and $Reader_{NQ\text{-}Single}$; for TriviaQA, we used $Retriever_{Multi}$ and $Reader_{TQA\text{-}Multi}$.

Table 4. End-to-end QA effectiveness in terms of the exact match score, comparing different answer span scoring techniques. The "orig" and "repl" columns are the original and replicated results; "GAR" refers to the technique by Mao et al. [12]; "*" represents fusion of retriever scores. The symbol † on a "repl*" result indicates sig. improvement over "repl"; on "GAR", over "repl"; on "GAR*", over "GAR". The symbol ‡ on "GAR*" indicates sig. improvement over "repl".

Condition	orig	repl	repl*	GAR	GAR*
NQ					
DPR	41.5	41.2	42.5†	41.5	43.5†‡
BM25	32.6	36.3	37.0	37.3†	38.4†‡
Hybrid	39.0	41.2	43.2†	41.9†	44.0†‡
TriviaQA					
DPR	56.8	57.5	58.3†	58.9†	59.5†‡
BM25	52.4	58.8	59.2	61.1†	61.6†‡
Hybrid	57.9	59.1	60.0†	61.0†	61.7†‡

With retrieval using DPR only, the "orig" and "repl" scores on both datasets are close (within a point), which suggests that we have successfully replicated the results reported in the DPR paper. With retrieval using BM25 only, our replicated results are quite a bit higher than the original DPR results; this is not a surprise given that our BM25 results are also better. When combining DPR and BM25 results at the retriever stage, the end-to-end effectiveness remains unchanged for NQ, but we observe a modest gain for TriviaQA. The gain for TriviaQA is statistically significant. So, it is *not* the case that better top-k retrieval always leads to improvement in end-to-end effectiveness.

Comparing the "repl" and "repl*" columns, we observe that combining scores from the retriever yields modest gains across all conditions. These gains are significant for four out of the six conditions, which suggests that retriever scores contribute to improving effectiveness. Comparing the "GAR" and "repl" columns, we also observe modest gains when adopting the answer span selection technique of Mao et al. [12]. These gains are significant for all except one condition. Comparing the "GAR" and "GAR*" columns, we find that in all cases, incorporating retriever scores significantly increases effectiveness.

Finally, putting everything together—using both the answer span scoring technique of Mao et al. [12] and incorporating retriever scores—we observe statistically significant gains across all retrieval conditions, as can be seen in the "GAR*" vs. "repl" columns across all rows. Compared to the best replicated results, we obtain an improvement of approximately three points in end-to-end QA effectiveness compared to the best answer extraction approach described in the original DPR paper. Note that we are able to obtain these improvements using exactly the model checkpoints provided in the DPR repo—we have simply added two relatively simple tricks to improve scoring and evidence combination.

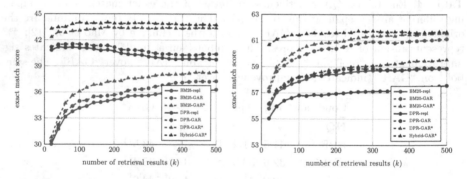

Fig. 1. End-to-end question answering effectiveness (exact match score) varying the number of retrieval results (k) for NQ (left) and TriviaQA (right).

In Fig. 1, we plot exact match scores as a function of varying k retrieval results for NQ (left) and TriviaQA (right). That is, we show how end-to-end QA effectiveness changes as the reader is provided more contexts from the retriever to consider. There are two factors here at play: On the one hand, top-k accuracy increases monotonically, i.e., as k increases, so does the likelihood that the answer appears in the contexts fed to the reader. On the other hand, the reader is asked to consider more contexts, and thus needs to discriminate the correct answer from a larger pool of candidate contexts, some of which might be low quality and thus serve as "distractors" from the correct answer. How do these factors balance out? Similar analyses in previous work with BM25 retrieval have shown that end-to-end QA effectiveness increases with increasing k [15,18]; that is, the reader does not appear to be "confused" by the non-relevant material. Indeed, in our BM25 results we also observe the same trend.

Interestingly, however, when we switch from BM25 results to DPR results, the behavior appears to change. For TriviaQA, the effectiveness curve behaves as expected, but for NQ, the exact match score trends up and then decreases after a peak. This means that while the likelihood of the reader seeing a correct answer in the candidate contexts increases with k, it is more likely to be negatively affected by increasing amounts of non-relevant contexts as well. This general behavior is also seen for the hybrid scoring techniques: as k increases, so does the exact match score, but only up to a certain point. Beyond this point, feeding the reader more candidate contexts leads to slight decreases in end-to-end effectiveness.

5   Conclusion

The breakneck pace at which NLP and IR are advancing, we argue, makes reproducibility and replicability critical to advancing science—to ensure that we are building on a firm foundation. Our study adds to the veracity of the claims made by Karpukhin et al. [8], and our work indeed confirms that DPR is an effective dense retrieval technique. Moreover, we arrive at two important findings, one of which is inconsistent with the original work, the other of which presents an enhancement. Together, they enrich our understanding of DPR.

Acknowledgment. This research was supported in part by the Canada First Research Excellence Fund and the Natural Sciences and Engineering Research Council (NSERC) of Canada. Computational resources were provided by Compute Ontario and Compute Canada.

References

1. Berant, J., Chou, A., Frostig, R., Liang, P.: Semantic parsing on Freebase from question-answer pairs. In: Proceedings of the 2013 Conference on Empirical Methods in Natural Language Processing, Seattle, Washington, pp. 1533–1544. Association for Computational Linguistics (2013)
2. Chen, D., Fisch, A., Weston, J., Bordes, A.: Reading Wikipedia to answer open-domain questions. In: Proceedings of the 55th Annual Meeting of the Association for Computational Linguistics (ACL 2017), Vancouver, British Columbia, Canada, pp. 1870–1879 (2017)
3. Devlin, J., Chang, M.W., Lee, K., Toutanova, K.: BERT: pre-training of deep bidirectional transformers for language understanding. In: Proceedings of the 2019 Conference of the North American Chapter of the Association for Computational Linguistics: Human Language Technologies, Volume 1 (Long and Short Papers), Minneapolis, Minnesota, pp. 4171–4186. Association for Computational Linguistics (2019)
4. Gao, L., Zhang, Y., Han, J., Callan, J.: Scaling deep contrastive learning batch size under memory limited setup. In: Proceedings of the 6th Workshop on Representation Learning for NLP (2021)

5. Hofstätter, S., Althammer, S., Schröder, M., Sertkan, M., Hanbury, A.: Improving efficient neural ranking models with cross-architecture knowledge distillation. arXiv:2010.02666 (2020)
6. Johnson, J., Douze, M., Jégou, H.: Billion-scale similarity search with GPUs. IEEE Trans. Big Data **7**(3), 535–547 (2021)
7. Joshi, M., Choi, E., Weld, D., Zettlemoyer, L.: TriviaQA: a large scale distantly supervised challenge dataset for reading comprehension. In: Proceedings of the 55th Annual Meeting of the Association for Computational Linguistics (Volume 1: Long Papers), Vancouver, Canada, pp. 1601–1611. Association for Computational Linguistics (2017)
8. Karpukhin, V., et al.: Dense passage retrieval for open-domain question answering. In: Proceedings of the 2020 Conference on Empirical Methods in Natural Language Processing (EMNLP), pp. 6769–6781 (2020)
9. Kwiatkowski, T., et al.: Natural questions: a benchmark for question answering research. Trans. Assoc. Comput. Linguist. **7**, 452–466 (2019)
10. Lin, J., Ma, X., Lin, S.C., Yang, J.H., Pradeep, R., Nogueira, R.: Pyserini: a Python toolkit for reproducible information retrieval research with sparse and dense representations. In: Proceedings of the 44th Annual International ACM SIGIR Conference on Research and Development in Information Retrieval (SIGIR 2021), pp. 2356–2362 (2021)
11. Lin, S.C., Yang, J.H., Lin, J.: In-batch negatives for knowledge distillation with tightly-coupled teachers for dense retrieval. In: Proceedings of the 6th Workshop on Representation Learning for NLP (RepL4NLP-2021), pp. 163–173 (2021)
12. Mao, Y., et al.: Generation-augmented retrieval for open-domain question answering. In: Proceedings of the 59th Annual Meeting of the Association for Computational Linguistics and the 11th International Joint Conference on Natural Language Processing (Volume 1: Long Papers), pp. 4089–4100. Online (2021)
13. Rajpurkar, P., Zhang, J., Lopyrev, K., Liang, P.: SQuAD: 100,000+ questions for machine comprehension of text. In: Proceedings of the 2016 Conference on Empirical Methods in Natural Language Processing, Austin, Texas, pp. 2383–2392 (2016)
14. Voorhees, E.M., Tice, D.M.: The TREC-8 question answering track evaluation. In: Proceedings of the Eighth Text REtrieval Conference (TREC-8), Gaithersburg, Maryland, pp. 83–106 (1999)
15. Xie, Y., et al.: Distant supervision for multi-stage fine-tuning in retrieval-based question answering. In: Proceedings of the Web Conference 2020 (WWW 2020), pp. 2934–2940 (2020)
16. Xiong, L., et al.: Approximate nearest neighbor negative contrastive learning for dense text retrieval. In: Proceedings of the 9th International Conference on Learning Representations (ICLR 2021) (2021)
17. Yang, P., Fang, H., Lin, J.: Anserini: enabling the use of Lucene for information retrieval research. In: Proceedings of the 40th International ACM SIGIR Conference on Research and Development in Information Retrieval (2017)
18. Yang, W., et al.: End-to-end open-domain question answering with BERTserini. In: Proceedings of the 2019 Conference of the North American Chapter of the Association for Computational Linguistics (Demonstrations), Minneapolis, Minnesota, pp. 72–77 (2019)
19. Zhan, J., Mao, J., Liu, Y., Zhang, M., Ma, S.: RepBERT: contextualized text embeddings for first-stage retrieval. arXiv:2006.15498 (2020)

Reproducing Personalised Session Search
Over the AOL Query Log

Sean MacAvaney(✉), Craig Macdonald, and Iadh Ounis

University of Glasgow, Glasgow, UK
{sean.macavaney,craig.macdonald,iadh.ounis}@glasgow.ac.uk

Abstract. Despite its troubled past, the AOL Query Log continues to be an important resource to the research community—particularly for tasks like search personalisation. When using the query log these ranking experiments, little attention is usually paid to the document corpus. Recent work typically uses a corpus containing versions of the documents collected long after the log was produced. Given that web documents are prone to change over time, we study the differences present between a version of the corpus containing documents as they appeared in 2017 (which has been used by several recent works) and a new version we construct that includes documents close to as they appeared at the time the query log was produced (2006). We demonstrate that this new version of the corpus has a far higher coverage of documents present in the original log (93%) than the 2017 version (55%). Among the overlapping documents, the content often differs substantially. Given these differences, we re-conduct session search experiments that originally used the 2017 corpus and find that when using our corpus for training or evaluation, system performance improves. We place the results in context by introducing recent adhoc ranking baselines. We also confirm the navigational nature of the queries in the AOL corpus by showing that including the URL substantially improves performance across a variety of models. Our version of the corpus can be easily reconstructed by other researchers and is included in the **ir-datasets** package.

1 Introduction

When released in 2006, the AOL Query Log [27] drew harsh criticism from the media over privacy concerns [4]. Since then, however, it has been an important resource to the research community (e.g., [25,30]). Even to this day, the AOL Query Log continues to enable studies in analysis of data leaks [13], search autocompletion [14], weak supervision for adhoc search [7,21], search result personalisation [9,16], and session-based search [1,2,6,28,33].

A key limitation of the AOL Query Log is that it does not include document contents; it only provides a user identifier, query text, query date/time, and the URL and rank of clicked documents (if any). This means that for studies that use the logs as a training and benchmark data for tasks like search result personalisation and session search, a document corpus needs to be constructed. Often the

© The Author(s), under exclusive license to Springer Nature Switzerland AG 2022
M. Hagen et al. (Eds.): ECIR 2022, LNCS 13185, pp. 627–640, 2022.
https://doi.org/10.1007/978-3-030-99736-6_42

approach used for constructing the corpus is unspecified. To the best of our knowledge, when the approach is specified, it always involves scraping *current* versions of the documents. There are two main problems with this approach. First, given that the contents of web documents are highly prone to change over time, recent versions of the documents may not reflect the contents of the documents as they appeared to the users. Second, this approach impedes reproduciblity and replicability efforts in the area, since the contents of the documents cannot be released publicly due to the potential that they contain copyrighted material.

In this paper, we study the effect that the document corpus used for AOL Query Log experiments has on reproducibility. We start by building a new document corpus that attempts to better reflect the documents present in the AOL Query Log as they appeared when the log was collected. This is accomplished by using the Internet Archive,[1] and thus we refer to our corpus as AOLIA. We find that this approach is able to cover far more of the documents that appeared in the AOL Query log (93%) when compared to a commonly-used version of the corpus that was collected in 2017 (55%, shared on request by Ahmad et al. [2]). Based on the timestamps from the Internet Archive, we are confident that the documents in AOLIA also better reflect the content of the documents as they appeared at the time, with 86% of the documents coming from during or in the three months prior to the log. We find that the content of the overlapping documents changed substantially in the 11-year period, with 28% of documents having no token overlap in the title (which is often used for session-based search [2,9,28]).

We further conduct a reproducibility and replicability study[2] of personalised session search tasks based on the AOL Query Log. We are unable to *reproduce* results using the 2017 version of the corpus, but our *replication* results (using AOLIA) are more in line with the original findings. To put the results in context, we also include a neural adhoc ranking baseline, which ultimately outperforms the methods we investigate. We also study the effect of using the document's URL as additional text and find that it improves the performance of all methods we investigate (often by a large margin), further confirming the navigational nature of the queries in the AOL Query Log. In summary, our contributions are:

1. We provide an alternative document corpus (AOLIA) for the AOL Query Log based on versions of the documents as they were likely to have appeared at the time the query log was collected.
2. We release artifacts and software such that other researchers will be able to construct AOLIA themselves, promoting reproducibility.
3. We study the reproducibility and replicability of three session-based search approaches, and find that using AOLIA alone can improve the performance of session-based search systems due to higher-quality documents, and that the training and evaluation datasets constructed from AOLIA can be considerably larger due to the increased coverage of the dataset.

---

[1] https://archive.org/.

[2] ACM version 1.1 definitions of reproducibility and replicability: https://www.acm.org/publications/policies/artifact-review-and-badging-current.

The remainder of this paper is organised as follows. In Sect. 2 we provide additional background information about the problem. Then, Sect. 3 details our process for constructing AOLIA. Section 4 provides a comparison between AOLIA and a version produced in 2017 that is used by several recent works. Section 5 then focuses on reproducing prior works using AOLIA. Finally, Sect. 6 details the limitations of our approach, and Sect. 7 draws final conclusions.

2   Background

Past works that make use of the AOL Query Log use recent versions of the log's clicked documents. Because the content of web pages can change over time, using recent versions necessitates a filtering process, which removes query-document pairs that are no longer relevant. For instance, Ahmad et al. [2] reports *"...in our preliminary experiments, we observed that many recorded clicks do not have lexical overlap concerning the queries. One possible reason is that we crawled the recorded clicks from the AOL search log in 2017 and many of the clicked documents' content updated since 2006 when the AOL log was recorded."*

Several alternatives to The Internet Archive exist as sources of data for a reproducible AOL corpus. Although the Common Crawl[3] would provide a more comprehensive corpus (i.e., it includes a more natural selection of documents, rather than only documents clicked by the user), we show in Sect. 3 that the AOL corpus at the time likely did not contain a representative sample of documents from the web, but rather focused heavily on home pages. Moreover, since the oldest version of the Common Crawl is from 2008–09, the content of the documents may already have changed since the time of the log. Finally, the size of the relevant archives (hundreds of terabytes) could add substantial difficulty in downloading and working with the data. The ClueWeb 2009[4] and 2012[5] corpora would be another option, and are appealing given that many research groups already have a copy of them. However, like the Common Crawl, they reflect the contents of documents several years after the log was constructed. Furthermore, there is low coverage of the target URLs in the ClueWeb corpora.

Other efforts investigate the stability of using mutable web resources as document corpora in IR. McCreadie et al. [23] find that naturally-occurring deletions from the Twitter corpus used by the TREC Microblog tasks do not have a substantial effect on the results of experiments that use the corpus. However, the situation for general web pages is different because the content can change over time (tweets can only be deleted, not updated). Despite these findings, Sequiera and Lin [31] investigate the use of the Internet Archive as an alternative source of data for the TREC 2013–14 Microblog corpus. Our work not only differs in terms of the document corpus targeted, but also the download mechanism; the Twitter stream they use is conveniently bundled by month by the Internet Archive, whereas there is no such bundle available for the documents present in

---

[3] https://commoncrawl.org/.
[4] https://lemurproject.org/clueweb09/.
[5] https://lemurproject.org/clueweb12/.

the AOL Query Log. Consequently, the steps involved to build our version of the AOL corpus are necessarily more complicated.

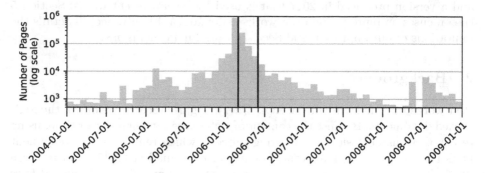

Fig. 1. Distribution of the archive dates of the web pages in the scraped collection from 2004–2008. The vast majority of pages are recovered from the time period during or right before the AOL Query Log (marked by vertical lines).

3   Reconstructing the AOL Document Corpus

In this section, we reconstruct a document corpus that better reflects the documents as they appeared to the users at the time. Through this process, we create artifacts and software that others can use to construct this dataset themselves, further promoting reproducibility in this area.

We start by building a set of all unique URLs that appear in the AOL Query Log.[6] Importantly, we acknowledge that this only represents documents that users clicked; the full list of documents indexed by AOL at the time is not available. This process results in 1,632,620 unique URLs. Nearly half (48.7%) of the URLs were only clicked a single time. All but 15 URLs specify either `http` or `https` URI schemes (14 specify `ftp` and one specifies `about`). 98.4% of the URLs refer to the home page of a website (i.e., have no path), which suggests that the search engine primarily functioned as a navigational tool at the time.

We then query archive.org's WayBack Machine's availability API[7] to request a version of the page as it appeared as close as possible to 1 March 2006 (the beginning of the AOL Query Log). Remarkably, we find that 93% of URLs were archived. Figure 1 shows the distribution of the dates of the archived pages. The vast majority of the found URLs (84%) are from the period during or in the three months prior to the log (January to May 2006). A further 8% are from before 2006, and a total of 96% of pages have an archived copy before 2007. Based on these dates, we feel that the corpus represents a reasonable approximation of the documents present in the query log at the time it was collected.

---

[6] http://www.cim.mcgill.ca/~dudek/206/Logs/AOL-user-ct-collection/aol-data.tar.gz.

[7] API Endpoint: https://archive.org/wayback/available.

We then fetch the archived versions of the documents and parse the resulting HTML using a libxml2-based parser. The title and body text are extracted (discarding content that appears in non-content tags, such as `<script>`). A small number of documents (0.3%) encountered either parsing errors or persistent download errors. We discard these documents. The median title length is 5 tokens (interquartile: (3, 9)), while the median body length is 198 tokens (interquartile: (47, 454)). As is the case for web content, some documents are substantially longer (up to 1.7M tokens). When compressed, the corpus is 3.4G in size. Using a FastText [12] language classifier[8] over the document title and body, we find that the vast majority of documents (92.5%) likely contain English text, as expected. Table 1 presents a further breakdown of the top languages in AOLIA. The breakdown is similar to that of the queries that appear in the log, when considering that language identification is more prone to errors for short texts like keyword queries.

Table 1. Top languages present in the AOLIA corpus, compared to the prevalence of the language of queries in the log.

Language	Corpus	Queries	Language	Corpus	Queries
English	92.5%	79.6%	Japanese	0.4%	0.1%
French	1.6%	3.2%	Portuguese	0.3%	0.8%
Spanish	1.4%	2.1%	Dutch	0.3%	0.9%
German	1.1%	2.6%	Russian	0.2%	0.4%
Italian	0.5%	1.8%	All others	1.5%	8.5%

Though we cannot distribute the contents of this corpus directly due to potentially copyrighted content, we take the following steps to facilitate reproducibility using this dataset:

1. We publicly release a mapping of the Internet Archive URLs so that others can fetch the same versions of the original documents.[9]
2. We provide software to download and extract the contents of these documents.[10]
3. We include a new aol-ia dataset in the ir-datasets [20] package, which provides easy access to this document corpus and the AOL log records. The package automatically downloads the log records from a public source, the Internet Archive mapping (from 1), and provides instructions to the user on how to run the extraction software (from 2). Once built, the dataset can easily be used by tools like PyTerrier [22] and OpenNIR [17].

---

[8] https://fasttext.cc/docs/en/language-identification.html.
[9] https://macavaney.us/aol.id2wb.tsv.gz.
[10] https://github.com/terrierteam/aolia-tools.

4   Comparing AOLIA with AOL17

In this section, we compare AOLIA with a version of the corpus that used more recent versions of the documents present in the log. Specifically, we use the version first used by Ahmad et al. [1], which uses documents from the AOL Query Log as they appeared in 2017 (so we call this corpus AOL17). This corpus has been used by other works (e.g., [2,9,28]). We treat AOL17 as a representative example of a contemporary version of the AOL corpus, noting that more recent versions of the corpus are likely to diverge even further from the original documents.

Table 2. Comparisons of URLs present in the AOLIA and AOL17 datasets. The Total column indicates the percentage of all URLs present in the AOL corpus.

	Count	Total
|AOLIA|	1,525,524	93.4%
|AOL17|	897,984	55.0%
|AOL17 \ AOLIA|	12,207	0.7%
|AOLIA \ AOL17|	639,747	39.2%
|AOLIA $\cup$ AOL17|	1,537,731	94.2%
|AOLIA $\cap$ AOL17|	885,777	54.3%

Table 2 provides a comparison between the URLs present in the two datasets. In terms of absolute coverage, AOLIA provides a high (albeit still incomplete) coverage of 93%. Meanwhile, AOL17 contains only 55% of the URLs found in the log. There are roughly 12k URLs found in AOL17 but not in AOLIA. *Content pages* (i.e., non-homepages) are over-represented among these documents, constituting 13% (1,637) of pages (compared to 2% of the overall corpus). AOLIA compensates for this disparity simply by virtue of size, filling in 639,747 documents missing from AOL17 (9,803 of which are content pages). Even though adding missing documents to AOLIA from AOL17 would increase the total coverage from 93.4% to 94.2%, doing so would reduce reproducibility, since those documents may contain copyrighted material and therefore cannot be distributed publicly.

We now dig into the characteristics of the 885,777 documents that overlap between the corpora. Figure 2 presents the Jaccard similarity between the set of title tokens[11] present in each version of the document. Only 17% of the titles have a perfect token overlap. Among these, 87% are exact case-insensitive sequence matches, with typical differences being the replacement or addition of punctuation in the titles, but sometimes involves the repetition of words. Table 3 shows such examples in rows 1–3.

---

[11] Tokens considered are case-folded, alphanumeric strings separated by whitespace or punctuation.

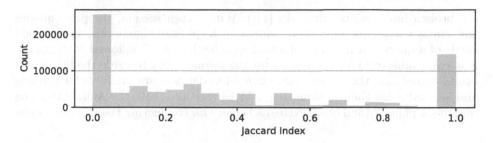

Fig. 2. Distribution of Jaccard similarities over title tokens from overlapping documents in `AOLIA` and `AOL17`.

Table 3. Examples of different titles found on pages between the `AOLIA` and `AOL17` corpora and their corresponding Jaccard index.

#	AOLIA	AOL17	Jac.
1	Welcome To Atlanta Music Group !	Welcome to Atlanta Music Group	1.0
2	Vinopolis Wine Shop - Portland , Oregon	Vinopolis Wine Shop | Portland , Oregon	1.0
3	Mechanics Savings Bank	Mechanics Savings Bank - Mechanics Savings Bank	1.0
4	Indalo Productions	InMotion Hosting	0.0
5	Kennebec Valley Organization – Home Page	*(empty)*	0.0
6	UK TV Guide	Homepage | UKTV	0.0
7	nutone // welcome	Nutone Records - Home | Facebook	0.2
8	Vedanta Press and Catalog	Books on Vedanta Philosophy	0.1
9	Venning Graphic Utilities for blending Images	Venning Graphic Utilities	0.5
10	Steinway Musical Instruments , Inc .	Steinway Musical Instruments - Steinway & Sons	0.6

The majority of documents (53%) have at most a Jaccard index of 0.25, indicating low overlap, with 28% having no overlap at all. In the cases when no overlap is present, semantically dissimilar content is often present, such as a placeholder document or a replacement with information about the web hosting provider. In semantically dissimilar cases, the queries that resulted in clicks for these documents are usually no longer relevant in `AOL17`. For instance, the query "indalo" resulted in a click of the documented represented by #4 in Table 3, which is a reasonable document in `AOLIA` but not in `AOL17`. In some cases, the content is semantically similar, such as in example #6, and still likely a reasonable document for associated queries. Based on a manually-annotated random sample of 100 documents with Jaccard similarities of 0, 23 appeared to be semantically-related (i.e., would likely satisfy similar information needs), while the remainder were not.

5 Reproduction and Replication

The AOL Query Log has been used for training and evaluating numerous search tasks. In this section, we explore the effect of `AOLIA` on one such task: session-based personalisation. In this setting, a user's sequence of searches and clicks are broken into sessions (or tasks [11]). Within each session, the prior queries and clicks can act as additional context to help disambiguate the information needs of a query. For instance, if a user searches for "cars" followed by "jaguar", it is reasonable to tailor the results for the second query towards the luxury car brand rather than the animal. Although a few datasets are available for training and/or evaluating these systems (e.g., TREC Sessions [15]), the AOL Query Log remains a popular (and often exclusive) choice for conducting these experiments.

5.1   Methods

We focus on three neural session-based personalisation techniques: M-NSRF [1], M-MatchTensor [1], and CARS [2]. We select these methods because numerous recent works in the area use them as baselines (e.g., [6,28,32,33]). All three models function as multi-task models, jointly learning to predict both document relevance and to predict the next query in the sequence. The models differ in the neural network architecture used to accomplish this. M-MatchTensor adapts the multi-task approach to the MatchTensor model [10], where MatchTensor builds a query-document similarity matrix between LSTM-encoded query and documet text and aggregates the results using CNN filters and max-pooling. M-NSRF encodes the query and document in separate bi-directional LSTM networks and combines them using a feed-forward layer to produce ranking scores. CARS builds upon M-NSRF by modeling the session interactions hierarchically using attention-based LSTM networks. For these three approaches, we use the authors' released code[12] with default parameters. In line with the code, the number of training iterations is tuned on dev data.

In addition to the above task-specific methods, we include three additional adhoc ranking baselines to put the results in context. First, we use the Terrier [26] BM25 implementation to re-rank the candidate documents with default BM25 parameters. This corresponds to the (unspecified) BM25 baseline conducted in [2]. Furthermore, in light of recent findings in *adhoc* retrieval, we also include two neural re-ranking baselines based on the T5 model [24,29]. Given that transferring relevance signals from one dataset/task to another using contextualised language models has generally been shown to be an effective technique (e.g., [18]), we include a T5 "transfer" baseline. This version is tuned on the MS MARCO dataset [3]. We also use a "tuned" baseline, which continues model tuning on the AOL session data from the MS MARCO checkpoint (batch size 8, learning rate 5×10^{-5}). In line with Nogueira et al. [24], we simply train for a fixed number of batches without tuning this or other settings (here, 10,000 batches). For all three adhoc ranking baselines, we use the PyTerrier [22] implementation.

5.2   Experimental Settings

We test the above six systems on three settings. (1) Using both sessions and documents from the AOL17 dataset. Here, we use the sessions constructed and

---

[12] https://github.com/wasiahmad/context_attentive_ir.

provided by **(author?)** [2]. This reflects the original experimental setup and is therefore a study of reproducibility. (2) Using the sessions from AOL17 (provided by [2]), but replacing the document titles with those from AOLIA. In the case where there is not a corresponding document in AOLIA, we leave the title blank (blanks present in 1.8% of documents across training, dev, and test sets). Since the same sessions are used as in (1), these results are directly comparable and isolate the impact of the document text itself. (3) Using sessions from AOLIA. In this setting, we re-create sessions as described in [2] (using same session elimination strategy and date ranges in each split), but using the full AOLIA corpus. Due to the higher coverage of documents, this results in considerably more data and longer sessions across all three splits. Table 4 presents the characteristics of the sessions built by each dataset. Both (2) and (3) are replicability studies because the experimental setting differs from the original paper (a different corpus is used than the original papers).

Table 4. Session search dataset characteristics provided by AOL17 compared to AOLIA.

	Train			Dev			Test		
	AOL17	AOLIA	%	AOL17	AOLIA	%	AOL17	AOLIA	%
# Sessions	219,748	311,877	+42%	34,090	49,522	+45%	29,369	50,944	+73%
# Queries	566,967	1,099,568	+94%	88,021	170,095	+93%	76,159	167,497	+120%
Avg. queries per session	2.58	3.53	+37%	2.58	3.43	+33%	2.59	3.29	+27%

As is commonplace for the task, we use the document title for the document content. Given that many of the queries are navigational in nature, we also test a variant of each of the 3 above settings that also appends the tokenised URL to the title, which can allow models to distinguish between pages that have the same title content and to match queries that ask for a specific URL.

In all three settings, each of the supervised methods are each trained, tuned, and tested using the data from the corresponding setting, while the unsupervised (BM25) and transfer (T5) baselines are simply run on the test set without tuning. We evaluate the results using MAP, MRR[13], which are measures commonly-used for evaluation of this task. We calculate the measures using the trec_eval implementation provided by ir-measures [19]. Note that this evaluation tool differs from the original work, which used their own implementation of the measures. Qu et al. [28] notes that this can result in differences in measures due to tie-breaking behaviour. We conduct significance tests between all pairs of systems within each setting (paired t-test, $p < 0.05$, with Bonferroni correction). We do not use a tool like repro_eval [5] to compare our results with those from the original papers because the rankings provided from the original papers are not available.

---

[13] Though this measure has been criticised [8], we report it to compare with past work.

5.3   Results

Table 5 presents the results for the three settings when using the document title as its contents. In Setting (1), where we use the same data and code as [2], we are unable to reproduce the performance reported by Ahmad et al. [2]. However, we note that the performance for CARS is not far from the results reported by Qu et al. [28], who report that the discrepancies with the original work are due to using the `trec_eval` measure implementation. Setting (2), where we use `AOLIA` document titles, yields performances closer to those reported by Ahmad et al. [2], with the CARS model outperforming M-MatchTensor and M-NSRF. Finally,

Table 5. Comparison of personalised session search baselines using various versions of the datasets. Note that the performances of systems using `AOL17` and `AOLIA` sessions cannot be directly compared. The top results we measure for each setting are bold (i.e., not including results reported by others). Non-significant differences between pairs of runs within a setting are indicated with superscript letters (paired t-test, $p < 0.05$, Bonferroni correction).

	Model	MAP	MRR	P@1
	(1) Sessions: AOL17, Documents: AOL17			
a	BM25 (unsupervised)	0.2457	0.2554	0.1454
	- from [2]	0.230	0.206	0.206
b	T5 (transfer)	0.3553	0.3649	0.2242
c	T5 (tuned)	**0.4538**	**0.4640**	**0.3001**
d	CARS	^{e}0.4280	^{e}0.4390	^{e}0.2787
	- from [2]	0.531	0.542	0.391
	- from [28]	–	0.4538	0.2940
e	M-MatchTensor	^{d}0.4335	^{df}0.4444	^{df}0.2830
	- from [2]	0.505	0.518	0.368
f	M-NSRF	0.4410	^{e}0.4521	^{e}0.2904
	- from [2]	0.491	0.502	0.391
	(2) Sessions: AOL17, Documents: AOLIA			
a	BM25 (unsupervised)	0.2942	0.3044	0.1914
b	T5 (transfer)	0.4228	0.4337	0.3021
c	T5 (tuned)	**0.5115**	**0.5223**	**0.3745**
d	CARS	0.4998	0.5107	0.3630
e	M-MatchTensor	^{f}0.4848	^{f}0.4961	^{f}0.3493
f	M-NSRF	^{e}0.4911	^{e}0.5023	^{e}0.3495
	(3) Sessions: AOLIA, Documents: AOLIA			
a	BM25 (unsupervised)	0.2413	0.2413	0.1462
b	T5 (transfer)	0.3620	0.3620	0.2260
c	T5 (tuned)	**0.4171**	**0.4171**	**0.2650**
d	CARS	0.3784	0.3784	0.2294
e	M-MatchTensor	0.3572	0.3572	0.2133
f	M-NSRF	0.4009	0.4009	0.2534

when using both sessions and documents from AOLIA (3), M-NSRF outperforms the other two session search methods. In short, we validate the findings of Ahmad et al. [2] that CARS significantly outperforms M-MatchTensor and M-NSRF, but only when using AOL17 sessions and AOLIA documents; in the other two settings, we draw the conclusion that the (simpler) M-NSRF model significantly outperforms the other two approaches.

Across all three settings, however, the adhoc (i.e., session unaware) tuned T5 model outperforms all other methods. We acknowledge that comparing T5 with CARS, M-MatchTensor, and M-NSRF is not a completely fair comparison; the T5 model benefits from a much larger model and extensive pre-training, while CARS, M-MatchTensor, and M-NSRF benefit from access to past queries and clicks within the session. Techniques for adapting contextualised language models like T5 for session search have been explored in [28], though in pilot studies we had difficulty training effective models using the released code. We also note that although T5 benefits from tuning on the target domain, it can still perform reasonably well—especially in Setting (3), where the AOLIA documents and sessions are used.

Table 6. Results when including the tokenised URL in addition to the title. The Δ column indicates the improvements compared to the results without the URL (Table 5). Non-significant differences between pairs of runs within a setting are indicated with superscript letters (paired t-test, $p < 0.05$, Bonferroni correction).

	Model	MAP	Δ	MRR	Δ	P@1	Δ
	(1) Sessions: AOL17, Documents: AOL17						
a	BM25 (unsupervised)	0.3204	+0.0747	0.3314	+0.0760	0.1991	+0.0537
b	T5 (transfer)	0.5023	+0.1470	0.5135	+0.1486	0.3572	+0.1330
c	T5 (tuned)	**0.7074**	**+0.2536**	**0.7190**	**+0.2550**	**0.6201**	**+0.3200**
d	CARS	^{f}0.6530	+0.2250	^{f}0.6643	+0.2253	^{f}0.5493	+0.2706
e	M-MatchTensor	0.6756	+0.2421	0.6871	+0.2427	0.5784	+0.2954
f	M-NSRF	^{d}0.6634	+0.2224	^{d}0.6745	+0.2224	^{d}0.5602	+0.2698
	(2) Sessions: AOL17, Documents: AOLIA						
a	BM25 (unsupervised)	0.3484	+0.0542	0.3591	+0.0547	0.2360	+0.0446
b	T5 (transfer)	0.5400	+0.1172	0.5514	+0.1177	0.3959	+0.0938
c	T5 (tuned)	**0.7071**	**+0.1956**	**0.7183**	**+0.1960**	**0.6153**	**+0.2408**
d	CARS	0.6665	+0.1667	0.6774	+0.1667	^{e}0.5660	+0.2030
e	M-MatchTensor	^{f}0.6538	+0.1690	^{f}0.6654	+0.1693	^{df}0.5569	+0.2076
f	M-NSRF	^{e}0.6520	+0.1609	^{e}0.6632	+0.1609	^{e}0.5501	+0.2006
	(3) Sessions: AOLIA, Documents: AOLIA						
a	BM25 (unsupervised)	0.2997	+0.0584	0.2997	+0.0584	0.1790	+0.0328
b	T5 (transfer)	0.4260	+0.0640	0.4260	+0.0640	0.2693	+0.0433
c	T5 (tuned)	**0.5679**	+0.1508	**0.5679**	+0.1508	**0.4418**	+0.1768
d	CARS	0.5360	+0.1576	0.5360	+0.1576	0.4082	+0.1788
e	M-MatchTensor	0.5458	**+0.1886**	0.5458	**+0.1886**	^{f}0.4297	**+0.2164**
f	M-NSRF	0.5575	+0.1566	0.5575	+0.1566	^{e}0.4336	+0.1802

Table 6 presents the results in each setting when appending the URL to the document text. We observe that in every case, this additional feature improves ranking effectiveness, sometimes by a considerable margin (up to +0.25 in MAP, +0.26 in MRR, and +0.32 in P@1). These findings underscore the importance of including this signal when queries are often navigational in nature.

Overall, we find that these experiments provide further evidence that AOLIA is well-constructed and useful. Between Settings (1) and (2), we see a consistent boost in ranking effectiveness across several models. Since the only thing we change between these settings is the document text, it suggests that the texts in AOLIA are more in line with the preferences of the users. Our experiments using T5 and URL features suggest that more care should be taken in future session-based search studies to construct evaluation data that focus on information needs that are less navigational in nature, as these can be addressed simply using established adhoc approaches and navigational signals.

6   Limitations

One limitation of using The Internet Archive are actions that the organisation takes in response to copyright claims made against archived content, which can effectively remove documents from the archive.[14] Over an approximately one-month window, 51 documents originally present in a prior version of AOLIA were no longer available on the Internet Archive, presumably due to this policy. Though this has the potential for knock-on effects downstream, prior work [23] indicates that it will likely have little effect. Specifically, McCreadie et al. [23] find that the effect of a far greater proportion of documents being deleted from the TREC Microblog 2011 corpus had a minimal effect on system evaluation, hence we expect the same to be true for AOLIA. Furthermore, when compared with the vast proportion of documents missing in more recent versions of the AOL corpus (e.g., AOL17 is missing 746,843 documents), the potential for several hundred removed documents per year seems preferable. Nevertheless, future work studying these effects may be warranted.

Although AOLIA improves the coverage and contents of documents that appear in the log, it does not attempt to fill in other documents that may have appeared in the corpus. In this way, it does not reflect a realistic sample of documents that likely existed in the entire AOL corpus at the time; it is more akin to the MS MARCO v1 passage corpus [3] (which only includes passages that were presented to annotators) than to corpora like ClueWeb (which includes documents scraped using typical web crawling techniques). As noted in Sect. 2, available web crawls would provide a less accurate picture of the documents as they appeared to the AOL users. Therefore, despite this limitation, we believe AOLIA is still valuable in many practical experimental settings.

---

[14] See their official copyright policy here: https://archive.org/about/terms.php.

7 Conclusions

In this work, we studied problems surrounding reproducibility of the AOL Query Log's corpus. We started by carefully constructing a new version of the corpus that better reflects the documents as they appeared to the users at the time of the log. We found that our approach increases the coverage of documents in the log considerably, when compared to a version that scraped documents eleven years after the log. We further found that the contents of documents are prone to considerable change over time, with the majority of document titles having very low token overlap between versions. When reproducing prior results for session search, we find that our new corpus improves the effectiveness across a variety of models (likely attributable to more realistic documents), and brings benchmarks based off the AOL Query Log more in line with adhoc ranking methods. We made access to our new version of the AOL corpus easily available to assist in future reproducibility efforts.

Acknowledgments. We acknowledge EPSRC grant EP/R018634/1: Closed-Loop Data Science for Complex, Computationally- & Data-Intensive Analytics. We thank Ahmad et al. for sharing the data and code that facilitated this study.

References

1. Ahmad, W.U., Chang, K.W., Wang, H.: Multi-task learning for document ranking and query suggestion. In: ICLR (2018)
2. Ahmad, W.U., Chang, K.W., Wang, H.: Context attentive document ranking and query suggestion. In: SIGIR (2019)
3. Bajaj, P., et al.: MS MARCO: a human generated machine reading comprehension dataset. In: CoCo@NIPS (2016)
4. Barbaro, M., Zeller, T.: A face is exposed for AOL searcher no. 4417749. The New York Times (2006). https://www.nytimes.com/2006/08/09/technology/09aol.html
5. Breuer, T., Ferro, N., Maistro, M., Schaer, P.: Repro_eval: a python interface to reproducibility measures of system-oriented IR experiments. In: ECIR (2021)
6. Cheng, Q., et al.: Long short-term session search: joint personalized reranking and next query prediction. In: The Web Conference (2021)
7. Dehghani, M., Zamani, H., Severyn, A., Kamps, J., Croft, W.B.: Neural ranking models with weak supervision. In: SIGIR (2017)
8. Fuhr, N.: Some common mistakes in IR evaluation, and how they can be avoided. In: SIGIR Forum (2018)
9. Huang, M., Peng, W., Wang, D.: TPRM: a topic-based personalized ranking model for web search. ArXiv arxiv:2108.06014 (2021)
10. Jaech, A., Kamisetty, H., Ringger, E.K., Clarke, C.: Match-tensor: a deep relevance model for search. ArXiv arxiv:1701.07795 (2017)
11. Jones, R., Klinkner, K.L.: Beyond the session timeout: automatic hierarchical segmentation of search topics in query logs. In: CIKM (2008)
12. Joulin, A., Grave, E., Bojanowski, P., Mikolov, T.: Bag of tricks for efficient text classification. ArXiv arxiv:1607.01759 (2016)
13. Kamara, S., Kati, A., Moataz, T., Schneider, T., Treiber, A., Yonli, M.: Cryptanalysis of encrypted search with LEAKER - a framework for leakage attack evaluation on real-world data. In: IACR Cryptol. ePrint Arch (2021)

14. Kang, Y.M., Liu, W., Zhou, Y.: QueryBlazer: efficient query autocompletion framework. In: WSDM (2021)
15. Kanoulas, E., Hall, M.M., Clough, P.D., Carterette, B., Sanderson, M.: Overview of the TREC 2011 session track. In: TREC (2011)
16. Ma, Z., Dou, Z., Bian, G., Wen, J.R.: PSTIE: time information enhanced personalized search. In: CIKM (2020)
17. MacAvaney, S.: OpenNIR: a complete neural ad-hoc ranking pipeline. In: WSDM (2020)
18. MacAvaney, S., Cohan, A., Goharian, N.: SLEDGE-Z: a zero-shot baseline for covid-19 literature search. In: EMNLP (2020)
19. MacAvaney, S., Macdonald, C., Ounis, I.: Streamlining evaluation with IR-measures. In: ECIR (2022)
20. MacAvaney, S., Yates, A., Feldman, S., Downey, D., Cohan, A., Goharian, N.: Simplified data wrangling with ir_datasets. In: SIGIR (2021)
21. MacAvaney, S., Yates, A., Hui, K., Frieder, O.: Content-based weak supervision for ad-hoc re-ranking. In: SIGIR (2019)
22. Macdonald, C., Tonellotto, N., MacAvaney, S., Ounis, I.: Pyterrier: declarative experimentation in python from bm25 to dense retrieval. In: CIKM (2021)
23. McCreadie, R., Soboroff, I., Lin, J.J., MacDonald, C., Ounis, I., McCullough, D.: On building a reusable twitter corpus. In: SIGIR (2012)
24. Nogueira, R., Jiang, Z., Pradeep, R., Lin, J.: Document ranking with a pretrained sequence-to-sequence model. In: Findings of EMNLP (2020)
25. Nunes, S., Ribeiro, C., David, G.: Use of temporal expressions in web search. In: ECIR (2008)
26. Ounis, I., Amati, G., Plachouras, V., He, B., Macdonald, C., Johnson, D.: Terrier information retrieval platform. In: ECIR (2005)
27. Pass, G., Chowdhury, A., Torgeson, C.: A picture of search. In: InfoScale (2006)
28. Qu, C., Xiong, C., Zhang, Y., Rosset, C., Croft, W.B., Bennett, P.N.: Contextual re-ranking with behavior aware transformers. In: SIGIR (2020)
29. Raffel, C., et al.: Exploring the limits of transfer learning with a unified text-to-text transformer. ArXiv arxiv:1910.10683 (2020)
30. Rafiei, D., Bharat, K., Shukla, A.: Diversifying web search results. In: WWW (2010)
31. Sequiera, R.D., Lin, J.: Finally, a downloadable test collection of tweets. In: SIGIR (2017)
32. Zhou, Y., Dou, Z., Wen, J.R.: Encoding history with context-aware representation learning for personalized search. In: SIGIR (2020)
33. Zhu, Y., et al.: Contrastive learning of user behavior sequence for context-aware document ranking. ArXiv arxiv:2108.10510 (2021)

Revisiting Popularity and Demographic Biases in Recommender Evaluation and Effectiveness

Nicola Neophytou[1]($^{(\boxtimes)}$), Bhaskar Mitra[2], and Catherine Stinson[3,4]

[1] The University of Manchester, Oxford Rd, Manchester M13 9PL, UK
neophytounicola@gmail.com

[2] Microsoft, 6795 Rue Marconi, Montréal, QC H2S 3J9, Canada
bmitra@microsoft.com

[3] School of Computing, Queen's University, 557 Goodwin Hall,
Kingston, ON K7L 2N8, Canada
c.stinson@queensu.ca

[4] Philosophy Department, Queen's University, John Watson Hall,
Kingston, ON K7L 3N6, Canada

Abstract. Recommendation algorithms are susceptible to popularity bias: a tendency to recommend popular items even when they fail to meet user needs. A related issue is that the recommendation quality can vary by demographic groups. Marginalized groups or groups that are under-represented in the training data may receive less relevant recommendations from these algorithms compared to others. In a recent study, Ekstrand et al. [15] investigate how recommender performance varies according to popularity and demographics, and find statistically significant differences in recommendation utility between binary genders on two datasets, and significant effects based on age on one dataset. Here we reproduce those results and extend them with additional analyses. We find statistically significant differences in recommender performance by both age and gender. We observe that recommendation utility steadily degrades for older users, and is lower for women than men. We also find that the utility is higher for users from countries with more representation in the dataset. In addition, we find that total usage and the popularity of consumed content are strong predictors of recommender performance and also vary significantly across demographic groups.

Keywords: Algorithmic fairness · Recommender Systems · Reproducibility study

1 Introduction

Recommendation systems and search tools increasingly mediate our access to information online, including news, entertainment, academic resources, and social connections. When evaluating the quality of theses results, it is common to report the mean performance over all users. Majority groups therefore tend to dominate

M. Hagen et al. (Eds.): ECIR 2022, LNCS 13185, pp. 641–654, 2022.
https://doi.org/10.1007/978-3-030-99736-6_43

overall statistics when measuring the utility of search and recommendation tools, but utility may also vary across individuals and demographic groups. Smaller demographic groups, whose needs differ from those of the largest groups, may not be well served by these algorithms that are optimized for mean performance across all users. If search and recommendation are unfair, in that the utility of search results and recommendations are systematically lower for some demographic groups, members of those groups may be hindered in their decision-making abilities, access to relevant information, and access to opportunities.

While typical methods of evaluating the effectiveness of search tools and recommender systems do not consider the disparate impact across demographic groups, several recent papers support the concern that these differences in utility do exist. Mehrotra et al. [32] investigate how the needs of different subgroups of the population are satisfied in the context of search. In particular, they study the impact on search quality by gender and age and find that both query distribution and result quality vary across these groups. Ekstrand et al. [15] perform a similar study in the context of recommender systems, which they investigate through offline top-n evaluation. They investigate whether different demographic groups experience varying utility from recommender systems, and find statistically significant differences in utility across age and gender groups.

In our work, we reproduce the findings by Ekstrand et al., and extend the analysis to incorporate additional user attributes, such as the user's country, usage, and the popularity of the content they consume. Like them, we find statistically significant differences in recommender utility by age and gender. We further investigate this effect by employing different binning strategies and metrics, and find that, on one dataset, when users are binned by age to achieve roughly equal numbers of users per bin, performance steadily degrades for older users. We also observe recommendation utility on average is higher for men than for women. In addition, we find the utility is higher for users from countries with more representation in the dataset. To understand how different demographic attributes impact recommendation quality relative to each other, we train an Explainable Boosting Machine (EBM) with user statistics and demographics as features, and recommender performance as the target variable. Our results indicate usage and popularity of consumed content are strong predictors of recommender performance. Both usage and content popularity vary significantly across groups and may provide a partial explanation for the observed differences in recommender utility, though low utility could also partially explain low usage. In summary, this work studies the following research questions in context of recommender systems:

RQ1 Does utility vary by demographic group?
RQ2 Does utility vary by usage and content popularity?
RQ3 Can usage and popularity explain demographic differences?

2   Related Work

Recommender systems predict future user-item interactions based on past user-item interactions [36]. Past interactions are often subject to biases—such as selection bias [31], conformity bias [26,30], exposure bias [28], and position bias [10,22,24]—and the collected data may reflect societal biases towards historically marginalized groups [25,40]. Recommendation algorithms trained on these datasets may further amplify these biases [39,43] resulting in homogeneity of recommendations and reduced utility to the user [8,20]. Recommender systems often demonstrate popularity bias [2,3] where popular items are recommended more frequently than warranted by their popularity, and give lower quality recommendations to users with atypical tastes [4,17,18]. These biases in recommendation raise fairness concerns for all stake-holders [1,5,35]. For content producers, unfairness may involve disparate exposure over items of comparable relevance [12,38]. For consumers of these systems, unfairness may manifest in the form of different recommendation quality across demographic groups [15]. In this work, our focus is on consumer-side fairness, building on prior work by Ekstrand et al. [15].

The fairness concerns in recommendation tasks are not just theoretical questions; they often result in real-world harms. For example, women may see fewer recommendations for high-paying jobs and career coaching services compared to men [11,27]. In the context of social networks, previous work [25,40] finds that friend recommender systems can reinforce historical biases by under-recommending minorities. Unfairness observed on microlending platforms can contribute to certain groups receiving systemically smaller loans, or higher interest rates [29]. In ride-hailing platforms, bias can lead to producer-side starvation and loss of income for drivers [41,42]. Similarly, Ekstrand and Kluver [14] find that recommender systems for books disproportionately favor male authors. The cost to publishers due to under-exposure of their content can be further aggravated by *superstar economics*, common in music and other recommendation scenarios [7,16,33,37]. For an overview of fairness and bias in recommender systems, we point the reader to a recent survey by Chen et al. [9], Ekstrand et al. [13].

3 Demographics and Popularity

As in the original work, we focus on the age and gender attributes of users in the data set, but also introduce new important variables for this study. Like Ekstrand et al., we begin our analysis with age and binary gender. For age, in addition to their bucketing scheme, which had unequal age ranges and numbers of users per bucket, we use two additional schemes, such that each age bucket: (i) is equal in age range, and (ii) includes a roughly equal number of users. This analysis with the age attribute is only possible with Last.FM (LFM360K) [6] data, since MovieLens (ML1M) [19] users can only select the age bracket they belong to, as opposed to specifying their exact age in years. This prevents the ability to manipulate age buckets for ML1M. We also look at how performance varies by country. We bucket countries by the number of users in the dataset, and by the country's gross domestic product (GDP)[1], a proxy for socioeconomic status and cultural hegemony.

---

[1] https://data.worldbank.org/indicator/NY.GDP.PCAP.CD.

Recommender systems learn from a user's past interactions with items in the collection. Users who have interacted more with the recommender system are likely to receive more relevant recommendations. To analyze how usage influences recommender utility, we bucket users by their number of interactions with items in the collection. We are also interested in the impact of popularity bias. The system may do a better job of recommending items to users who typically interact with items that are popular, compared to users with more niche interests. To investigate how item popularity affects utility, we introduce a novel *pop-index* attribute, defined as the largest value of p such that $p\%$ of items the user has interacted with have also received interactions from $p\%$ of other users. We take inspiration from the h-index [21], used to measure scholarly impact. We compare recommender utility for groups of users bucketed by pop-index.

Our experiment reproduces the same findings of statistically significant differences in recommender performance between demographics on two datasets, LFM360K and ML1M. The original paper finds differences in recommender utility between gender groups on ML1M, and between age groups on LFM360K. In our experiment, we observe these differences for both age and gender attributes on both data sets.

4   Method

4.1   Datasets

Similar to Ekstrand et al., we conduct our experiments on Last.FM and MovieLens data. LFM360K[2] represents a music recommendation task, and contains $358,868$ users and $292,385$ artists. For each user-artist pair, the dataset provides the total number of plays. There are $17,535,605$ user-artist pairs with at least one play in the dataset, which implies that the full user-artist matrix is 99.98% sparse. Entries in the user-artist matrix were collected using "user.getTopArtists()" in the Lastfm API, so include only the top artists for each user, representing a "playlist" of their favourite artists. The number of artists listened to by each user varies across users, with values between one and 166, and a mean of 50. The dataset also contains user attributes, such as binary[3] gender (67% male, 24% female, 9% missing), age (20% missing), and country (none missing).

Our second dataset ML1M[4] represents a movie recommendation task. ML1M contains $3,952$ movies and $6,040$ users who joined MovieLens in 2000. Each user-movie pair has an associated 5-point rating assigned by the user. The dataset contains $1,000,209$ ratings, corresponding to a 95.81% sparse user-movie matrix. Each user has rated at least 20 movies. The dataset also includes a binary gender, age, and occupation for each user. For the ML1M data set, users can only specify that they belong to a pre-set age bracket, as opposed to specifying exactly how

---

[2] http://ocelma.net/MusicRecommendationDataset/lastfm-360K.html.

[3] We treat gender as a binary class due to the available attributes in the dataset. We do not intend to suggest that gender identities are binary.

[4] https://grouplens.org/datasets/movielens/1m/.

old they are in years. The choice of age brackets they can choose from are displayed on the x-axis of Fig. 1g.

4.2 Model

Our study differs from the original work by use of a different collaborative filtering model; the original paper uses a range of models from the Lenskit recommender toolkit. In our experiment, we utilize the Alternating Least Squares (ALS) algorithm for Implicit feedback datasets, to investigate whether the same findings are observed on the same data sets but with another popular collaborative filtering model. As a result, what we reproduce is the statistically significant differences in recommender utility, rather than the exact scores. We therefore consider this to be a reproducibility paper, as we enact a different experimental setup to the original work. We use an ALS model for implicit feedback data [23], as implemented in the Implicit[5] code repository. We use the default hyperparameters as used by Implicit, by setting factors to 50 and the regularization constant to 0.01. We train the model for 30 iterations in all experiments. The Implicit code performs some data cleanup - as described here[6] - to deal with malformed entries in the data files. All statistics reported in Sect. 5 are computed after this cleanup.

4.3 Experiment Protocol

We conduct our experiments under a five-fold cross-validation setting. For LFM360K, each test partition contains 5, 000 randomly sampled users. For ML1M we partition the whole set of 6, 040 users into five splits containing 1, 208 users, for each iteration of cross-validation. For both datasets, we hold out 20% of the items each user has interacted with to use as test data. All other users and the rest of the test users' items are used for model training in each iteration. To avoid the cold-start problem, we remove users who listened to 40 or fewer artists in the LFM360K dataset–roughly 10% of users. The ML1M dataset only includes users who have rated over 20 or more movies, so none are removed. For evaluation, we generate 1, 000 recommendations per user, and measure the results using NDCG (normalized discounted cumulative gain), MRR (mean reciprocal rank), and RBP (rank-biased precision) metrics. To verify if differences in utility are significant across demographics, we perform Kruskall-Wallis significance tests on mean NDCG values between the demographic groups. For attributes which contain an N/A group, where the information on this attribute is not provided by the user, the N/A group is omitted from Kruskall-Wallis testing. This ensures we are only comparing groups of users who provided information on this attribute. We also run Bonferroni correction for multiple testing.

---

[5] https://github.com/benfred/implicit.

[6] https://github.com/benfred/bens-blog-code/blob/master/distance-metrics/musicdata.py#L39.

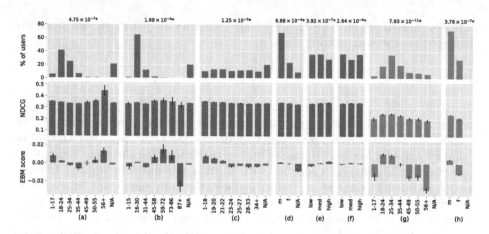

Fig. 1. Comparison of binning strategies, metrics, and datasets on recommender utility by demographic variables. Red plots represent the LFM360K dataset and grey represent ML1M. For age, we consider the original bucketing scheme from Ekstrand et al. (a and g), and buckets by equal range (b) and equal number of users (c). (d) and (h) represent gender for LFM360K and ML1M, respectively. (e) and (f) represent country ordered by number of users and by GDP for LFM360K. P-values from Kruskal-Wallis significance tests on NDCG are reported above each column. (Color figure online)

To understand the relative impact of user attributes on system performance, we train an EBM model, as implemented in the InterpretML framework [34], to predict the mean NDCG for each user as a dependent variable. We represent each user by a combination of the following features: (i) Age, (ii) Gender, (iii) Country, ordered by prevalence in the dataset and bucketed (LFM360K only), (iv) Country, ordered by GDP and bucketed (LFM360K only), (v) Usage (*i.e.*, total number of listens for LFM360K and total number of movies rated for ML1M), (vi) Pop-index, and finally (vii) The last digit of the user ID. The last digit of the user ID serves as a control feature which should have no effect on performance on either dataset. We run the EBM model once individually for each feature group, and once with all features included for cross feature-group comparison.

5   Results

Using the datasets and methods described above, we reproduce the main results from Ekstrand et al., and inquire in more detail how the quality of recommendation varies by age, gender, and country, using varied binning strategies and metrics. In addition, we study the impact of usage and item popularity on utility, and how they interplay with the other demographic variables.

RQ1 Does utility vary by demographic group?

Figure 1 shows the distribution of users, recommender utility (mean NDCG), and the EBM scores corresponding to different demographic variables. Figure 1a–1f corresponds to the LFM360K dataset (in red). Column (a) divides users into age groups according to the age range buckets Ekstrand et al. used, replicating their results. Column (b) divides users into age buckets of uniform range (15 years). Column (c) organizes users into age buckets such that the number of users in each bucket is comparable. Figure 1g and 1h presents the results for the ML1M dataset (in grey), where the age buckets again correspond to those used in Ekstrand et al., replicating their results. For each column, we run the Kruskall-Wallis significance test and on all metrics. P-values for mean NDCG are reported above each column.

5.1 Impact on Age

Ekstrand et al. find significant differences in recommender utility across different user age brackets according to the Kruskal-Wallis significance test. Our analysis confirms these findings on both datasets, as we also report significant differences based on Kruskal-Wallis significance test ($p < 0.01$) across the same age brackets (Fig. 1a and 1g). We also find significant differences when we try alternative binning strategies on LFM360K, corresponding to bins with equal age range (Fig. 1b) and bins with equal number of users (Fig. 1c). While we only report p-values corresponding to the NDCG metric for recommendation utility, we have verified the differences are also statistically significant for MRR and RBP, except for MRR for ML1M.

The first row shows on both datasets that the age distribution is skewed towards young adults, more so for LFM360K than ML1M. Because the age buckets were irregular, we show the results with buckets of uniform range (Fig. 1b). We also posit that a skewed distribution of users across age buckets may make it difficult to detect differences in utility across ages, because some age buckets contain very few users. Therefore, we additionally try buckets containing approximately equal numbers of users (Fig. 1c). When the number of users in each bucket are comparable, we find a gradual downward trend in recommender utility, as age increases. This effect was not visible in Ekstrand et al. We also observe a similar downward trend on ML1M as seen in Fig. 1g. This trend is further confirmed by the EBM scores in Figs. 1c and 1g where younger ages correspond to higher EBM scores when the number of users in each bucket are approximately equal.

5.2 Impact on Gender

Both LFM360K (Fig. 1d) and ML1M (Fig. 1h) datasets contain many more male than female users. As in Ekstrand et al., we observe statistically significant differences in utility by gender based on Kruskal-Wallis significance test ($p < 0.01$), with better recommendation utility for male than female users. This is observed in both datasets, except for MRR and RBP for LFM360K, and MRR for ML1M. Given the unbalanced user distribution across genders in these datasets, this can either be the result of a popularity bias, or a demographic bias. We revisit this question later in this section in the context of RQ3.

5.3   Impact on Country

An additional demographic variable available in the LFM360K dataset, but not in ML1M, is users' country of residence. Ekstrand et al. did not analyze whether there is evidence of recommender utility differences by country, but we perform this analysis here. We group the countries in two ways. First, according to its representation in the dataset—i.e., based on the number of users from that country, into low, medium, and high buckets—and second, by GDP, again into low, medium, and high buckets. Figures 1e and 1f show the results corresponding to the two analyses. Low GDP is used here as a proxy for social marginalization.

We find statistically significant differences by country on both measures, except for MRR and RBP for GDP. The model has higher recommender utility for users from countries with more representation in the dataset. The same trend is not observed, however, when countries are ordered by GDP.

As expected, there are no statistically significant differences found on any metric between users grouped by the last digit of their user ID, the control feature, across both data sets.

RQ2 Does utility vary by usage and content popularity?

It is not obvious when to attribute utility differences across groups of users to popularity bias, rather than bias specifically affecting demographic groups, because marginalized groups are often also less represented in training datasets. To explore this issue, we first investigate how recommender utility is affected by two measures of popularity: usage and pop-index. For a given user, high usage implies more representation in the data, while a higher pop-index corresponds to affinity towards items that are popular with other users in the dataset. In Fig. 2 we compare both these measures on the LFM360K and ML1M datasets. For both datasets there is a trend toward greater NDCG as usage increases. The EBM analysis shows the same trend, where low usage corresponds to a negative effect on the EBM score, and high usage corresponds to a positive effect. We also investigate popularity in the sense of how popular items preferred by a user are among the user population as a whole. Our hypothesis is that users whose playlists contain more popular items will likely have greater recommendation utility. On ML1M (Fig. 2d), we observe a trend which supports our hypothesis. However, on LFM360K (Fig. 2b), we observe a U-shaped trend, with higher utility associated with both groups of users with maintstream and unique tastes.

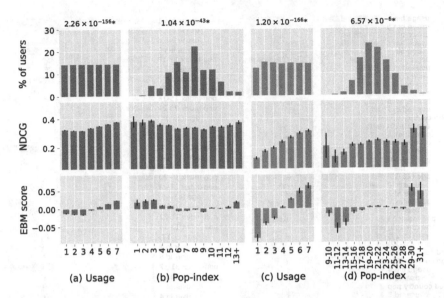

Fig. 2. Recommendation utility by usage and content popularity. Red plots represent the LFM360K dataset, grey plots represent ML1M. p-values from Kruskal-Wallis significance tests on NDCG are reported above each column. (Color figure online)

We suspect differences in observations on the two datasets may be partially explained by the semantics of user interactions in the two cases. In LFM360K, the user interacts with an artist by listening to them, and they can listen to the same artist multiple times. So, for users with more distinctive tastes, the recommender algorithm may still achieve reasonable performance by recommending items the user interacted with before. In contrast, in ML1M the user interacts with the item by providing a rating and therefore the recommender must suggest new items the user has not interacted with before, which is a more difficult challenge, specifically when the user has a distinctive taste.

RQ3 Can usage and popularity explain demographic differences?

One of our goals is to better understand the relative importance of different demographic and popularity features to explain the differences in mean recommender utility amongst users. Towards that goal, we train an EBM model to predict mean recommender utility based on these user attributes. Figure 3 shows that on both datasets (LFM360K and ML1M) the usage features emerge as the most predictive, followed by pop-index. Among the demographic attributes, some of the age-related features are ranked highest on both datasets. On LFM360K, age is followed by country (ordered by number of users) and gender as the next

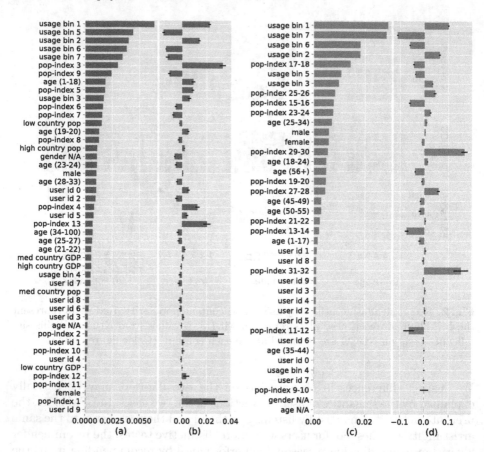

Fig. 3. Ranked features and their scores from the EBM analysis. In (a) and (c) equal numbers of users are sampled for each factor. In (b) and (d) the full database is used.

most predictive user attributes. In the absence of country information, on the ML1M dataset we observe gender to be high in the feature ranking after age. The high feature importance for usage and pop-index provides evidence than some of the demographic differences may be explained by representation in the data. This is not to argue that the recommender system under study is fair to different demographics of users. Disparity of utility across demographics may directly influence user retention [15] and usage. This creates a vicious cycle where a small difference in utility across user groups may be further amplified by subsequent disparity in system adoption and usage across demographics, leading to even bigger disparities in utility. Table 1 shows how usage and pop-index are distributed across demographic groups, further demonstrating how they may correlate with historical marginalization.

Table 1. Percentage of users in different usage and pop-index buckets for each demographic groups in LFM360K. For younger users and men a higher proportion of the population correspond to higher usage buckets. The trend for pop-index is less clear.

	Age (bucketed by equal number of users)								Gender		
	1–18	19–20	21–22	23–24	25–27	28–33	34+	N/A	m	f	N/A
Usage											
1	11%	8%	9%	11%	12%	15%	24%	21%	13%	16%	20%
2	13%	11%	12%	13%	13%	17%	17%	17%	14%	15%	15%
3	14%	15%	13%	13%	15%	15%	15%	14%	14%	16%	15%
4	16%	15%	15%	14%	14%	14%	13%	14%	14%	15%	13%
5	15%	16%	15%	16%	16%	14%	11%	13%	14%	14%	13%
6	15%	17%	18%	17%	14%	13%	11%	11%	15%	13%	12%
7	17%	18%	18%	16%	15%	12%	9%	10%	16%	11%	11%
Pop-index											
1	0%	0%	0%	0%	0%	0%	0%	0%	0%	0%	0%
2	0%	0%	0%	0%	0%	0%	1%	1%	0%	0%	0%
3	3%	3%	3%	4%	4%	5%	8%	6%	5%	4%	5%
4	2%	2%	3%	3%	3%	3%	5%	4%	4%	2%	4%
5	7%	9%	9%	10%	11%	12%	15%	11%	11%	8%	11%
6	15%	15%	14%	16%	14%	15%	15%	17%	15%	14%	17%
7	10%	10%	10%	10%	11%	10%	12%	11%	11%	10%	11%
8	23%	23%	23%	23%	24%	22%	19%	20%	21%	25%	21%
9	14%	13%	12%	12%	11%	10%	9%	11%	11%	13%	12%
10	14%	12%	13%	12%	12%	12%	10%	10%	12%	13%	8%
11	7%	8%	7%	6%	6%	6%	4%	5%	6%	7%	6%
12	2%	3%	2%	2%	2%	2%	1%	2%	2%	2%	2%
13+	2%	2%	2%	2%	1%	2%	1%	1%	2%	2%	1%

6 Discussion and Conclusion

We confirmed that recommender systems are prone to unfairness across the demographic attributes available in the datasets used here. To explore this question more thoroughly, one would need access to more detailed demographic data, and the ability to observe temporal dynamics of how recommendations affect usage and usage affects recommendations. In order to answer questions like what caused the U-shaped pattern we found in recommender utility by usage, we would need the ability to intervene on recommendations in real time.

Mehrotra et al. [32] point out that users for whom a search engine is least satisfactory can paradoxically end up having the highest measured utility. They found when utility is bad enough to make a user stop using the service for everyday needs, they still use the search engine for very easy queries that they assume even a poor search engine could get right. Such searches end up being successful, resulting in artificially high utility scores. User attrition is an issue we cannot track given the datasets used here. It may be that users who have the highest usage are a self-selecting group for whom recommenders happen to work well.

For both datasets there is a trend toward greater utility as usage increases. This is unsurprising, given that users with higher usage will provide more labels, with which the recommender can build a more accurate model of user preferences. One anomalous effect we observed is in the LastFM dataset; users with least usage have higher utility recommendations than users with slightly more usage. This could be evidence of the same effect as observed by Mehrotra et al. [32]. If LastFM gives poor recommendations for a given user, that user might stop using it for everyday music streaming, but still use it when they are looking for something very mainstream. Another possibility is since LastFM users input a few artists they like when setting up their accounts, early listens will be dominated by artists which the user identified as being among their favourites, rather than recommendations provided by the model. Utility may therefore be artificially high during early use.

The social harms that can result from unfair recommendation go well beyond some people choosing not to use a tool that others find fun and convenient. Recommendation algorithms are increasingly being used to make major life decisions, like mortgage lending, job searching, connecting with community, and basic access to information. The body of work we are adding to here demonstrates that fair recommendation is a problem requiring serious attention.

References

1. Abdollahpouri, H., Burke, R.: Multi-stakeholder recommendation and its connection to multi-sided fairness. arXiv preprint arXiv:1907.13158 (2019)
2. Abdollahpouri, H., Mansoury, M.: Multi-sided exposure bias in recommendation. arXiv preprint arXiv:2006.15772 (2020)
3. Abdollahpouri, H., Mansoury, M., Burke, R., Mobasher, B.: The connection between popularity bias, calibration, and fairness in recommendation. In: Proceedings of the RecSys, pp. 726–731 (2020)
4. Burke, R.: Hybrid recommender systems: survey and experiments. User Model. User Adap. Inter. 12(4), 331–370 (2002)
5. Burke, R.: Multisided fairness for recommendation. CoRR abs/1707.00093 (2017)
6. Celma, Ó.: Music Recommendation and Discovery: The Long Tail, Long Fail, and Long Play in the Digital Music Space. Springer, Heidelberg (2010). https://doi.org/10.1007/978-3-642-13287-2
7. Celma Herrada, Ò., et al.: Music recommendation and discovery in the long tail. Universitat Pompeu Fabra (2009)
8. Chaney, A.J.B., Stewart, B.M., Engelhardt, B.E.: How algorithmic confounding in recommendation systems increases homogeneity and decreases utility. In: Proceedings of the RecSys, pp. 224–232, RecSys 2018. Association for Computing Machinery, New York (2018)
9. Chen, J., Dong, H., Wang, X., Feng, F., Wang, M., He, X.: Bias and debias in recommender system: a survey and future directions. arXiv preprint arXiv:2010.03240 (2020)

10. Collins, A., Tkaczyk, D., Aizawa, A., Beel, J.: A study of position bias in digital library recommender systems. arXiv preprint arXiv:1802.06565 (2018)
11. Datta, A., Tschantz, M.C., Datta, A.: Automated experiments on ad privacy settings. Proc. Priv. Enh. Technol. **2015**(1), 92–112 (2015)
12. Diaz, F., Mitra, B., Ekstrand, M.D., Biega, A.J., Carterette, B.: Evaluating stochastic rankings with expected exposure. In: Proceedings of the CIKM, pp. 275–284 (2020)
13. Ekstrand, M.D., Das, A., Burke, R., Diaz, F.: Fairness and discrimination in information access systems. arXiv preprint arXiv:2105.05779 (2021)
14. Ekstrand, M.D., Kluver, D.: Exploring author gender in book rating and recommendation. User Model. User Adap. Inter. **31**, 1–44 (2021)
15. Ekstrand, M.D., et al.: All the cool kids, how do they fit in?: Popularity and demographic biases in recommender evaluation and effectiveness. In: Conference on Fairness, Accountability and Transparency, pp. 172–186 (2018)
16. Ferraro, A.: Music cold-start and long-tail recommendation: bias in deep representations. In: Proceedings of the RecSys, pp. 586–590 (2019)
17. Ghazanfar, M., Prugel-Bennett, A.: Fulfilling the needs of gray-sheep users in recommender systems, a clustering solution (2011)
18. Gras, B., Brun, A., Boyer, A.: When users with preferences different from others get inaccurate recommendations. In: Monfort, V., Krempels, K.-H., Majchrzak, T.A., Turk, Ž (eds.) WEBIST 2015. LNBIP, vol. 246, pp. 191–210. Springer, Cham (2016). https://doi.org/10.1007/978-3-319-30996-5_10
19. Harper, F.M., Konstan, J.A.: The movielens datasets: history and context. ACM Trans. Interact. Intell. Syst. **5**(4), 19:1–19:19 (2016)
20. Hashimoto, T., Srivastava, M., Namkoong, H., Liang, P.: Fairness without demographics in repeated loss minimization. In: Proceedings of the ICML, pp. 1929–1938. PMLR (2018)
21. Hirsch, J.E.: An index to quantify an individual's scientific research output. Proc. Natl. Acad. Sci. **102**(46), 16569–16572 (2005)
22. Hofmann, K., Mitra, B., Radlinski, F., Shokouhi, M.: An eye-tracking study of user interactions with query auto completion. In: Proceedings of the CIKM, pp. 549–558. ACM (2014)
23. Hu, Y., Koren, Y., Volinsky, C.: Collaborative filtering for implicit feedback datasets. In: 2008 Eighth IEEE International Conference on Data Mining, pp. 263–272. IEEE (2008)
24. Joachims, T., Granka, L., Pan, B., Hembrooke, H., Radlinski, F., Gay, G.: Evaluating the accuracy of implicit feedback from clicks and query reformulations in web search. ACM TOIS **25**(2), 7-es (2007)
25. Karimi, F., Génois, M., Wagner, C., Singer, P., Strohmaier, M.: Homophily influences ranking of minorities in social networks. Sci. Rep. **8**(1), 1–12 (2018)
26. Krishnan, S., Patel, J., Franklin, M.J., Goldberg, K.: A methodology for learning, analyzing, and mitigating social influence bias in recommender systems. In: Proceedings of the RecSys, pp. 137–144 (2014)
27. Lambrecht, A., Tucker, C.: Algorithmic bias? An empirical study of apparent gender-based discrimination in the display of stem career ads. Manage. Sci. **65**(7), 2966–2981 (2019)
28. Liu, D., Cheng, P., Dong, Z., He, X., Pan, W., Ming, Z.: A general knowledge distillation framework for counterfactual recommendation via uniform data. In: Proceedings of the SIGIR, pp. 831–840 (2020)
29. Liu, W., Guo, J., Sonboli, N., Burke, R., Zhang, S.: Personalized fairness-aware re-ranking for microlending. In: Proceedings of the RecSys, pp. 467–471 (2019)

30. Liu, Y., Cao, X., Yu, Y.: Are you influenced by others when rating? Improve rating prediction by conformity modeling. In: Proceedings of the RecSys, pp. 269–272 (2016)
31. Marlin, B.M., Zemel, R.S., Roweis, S., Slaney, M.: Collaborative filtering and the missing at random assumption. In: Proceedings of the Twenty-Third Conference on Uncertainty in Artificial Intelligence, pp. 267–275 (2007)
32. Mehrotra, R., Anderson, A., Diaz, F., Sharma, A., Wallach, H., Yilmaz, E.: Auditing search engines for differential satisfaction across demographics. In: Proceedings of the WWW, pp. 626–633 (2017)
33. Mehrotra, R., McInerney, J., Bouchard, H., Lalmas, M., Diaz, F.: Towards a fair marketplace: Counterfactual evaluation of the trade-off between relevance, fairness & satisfaction in recommendation systems. In: Proceedings of the CIKM, pp. 2243–2251 (2018)
34. Nori, H., Jenkins, S., Koch, P., Caruana, R.: InterpretML: a unified framework for machine learning interpretability. arXiv preprint arXiv:1909.09223 (2019)
35. Patro, G.K., Biswas, A., Ganguly, N., Gummadi, K.P., Chakraborty, A.: FairRec: two-sided fairness for personalized recommendations in two-sided platforms. In: Proceedings of the Web Conference, pp. 1194–1204 (2020)
36. Ricci, F., Rokach, L., Shapira, B.: Introduction to recommender systems handbook. In: Ricci, F., Rokach, L., Shapira, B., Kantor, P.B. (eds.) Recommender Systems Handbook, pp. 1–35. Springer, Boston (2011). https://doi.org/10.1007/978-0-387-85820-3_1
37. Rosen, S.: The economics of superstars. Am. Econ. Rev. **71**(5), 845–858 (1981)
38. Singh, A., Joachims, T.: Fairness of exposure in rankings. In: Proceedings of the SIGKDD, pp. 2219–2228 (2018)
39. Stinson, C.: Algorithms are not neutral: bias in collaborative filtering. arXiv preprint arXiv:2105.01031 (2021)
40. Stoica, A.A., Riederer, C., Chaintreau, A.: Algorithmic glass ceiling in social networks: the effects of social recommendations on network diversity. In: Proceedings of the WWW, pp. 923–932 (2018)
41. Sühr, T., Biega, A.J., Zehlike, M., Gummadi, K.P., Chakraborty, A.: Two-sided fairness for repeated matchings in two-sided markets: a case study of a ride-hailing platform. In: Proceedings of the SIGKDD, pp. 3082–3092 (2019)
42. Wang, G., Zhang, Y., Fang, Z., Wang, S., Zhang, F., Zhang, D.: FairCharge: a data-driven fairness-aware charging recommendation system for large-scale electric taxi fleets. Proc. ACM Interact. Mob. Wearable Ubiquitous Technol. **4**(1), 1–25 (2020)
43. Zhao, J., Wang, T., Yatskar, M., Ordonez, V., Chang, K.W.: Men also like shopping: reducing gender bias amplification using corpus-level constraints. arXiv preprint arXiv:1707.09457 (2017)

Squeezing Water from a Stone: A Bag of Tricks for Further Improving Cross-Encoder Effectiveness for Reranking

Ronak Pradeep[1](✉), Yuqi Liu[1], Xinyu Zhang[1], Yilin Li[1], Andrew Yates[2,3], and Jimmy Lin[1]

[1] University of Waterloo, Waterloo, Canada
rpradeep@uwaterloo.ca
[2] University of Amsterdam, Amsterdam, The Netherlands
[3] Max Planck Institute for Informatics, Saarbrücken, Germany

Abstract. While much recent work has demonstrated that hard negative mining can be used to train better bi-encoder models, few have considered it in the context of cross-encoders, which are key ingredients in modern retrieval pipelines due to their high effectiveness. One noteworthy exception comes from Gao et al. [13], who propose to train cross-encoders by adapting the well-known NCE loss and augmenting it with a "localized" selection of hard negative examples from the first-stage retriever, which they call the Localized Contrastive Estimation (LCE) loss. In this work, we present a replication study of LCE on a different task and combine it with several other "tricks" (e.g., replacing $BERT_{Base}$ with $ELECTRA_{Base}$ and replacing BM25 with TCT-ColBERTv2) to substantially improve ranking effectiveness. We attempt to more systematically explore certain parts of the hyperparameter space, including the choice of losses and the group size in the LCE loss. While our findings, for the most part, align with those from the original paper, we observe that for MS MARCO passage, orienting the retriever used for hard negative mining with the first-stage retriever used for inference is not as critical for improving effectiveness across all settings. Our code and documentation can be found in: https://github.com/castorini/replicate-lce.

1 Introduction

After the introduction of BERT [6] in October 2018, a simple retrieve-then-rerank approach quickly emerged in January 2019 as an effective method for applying pretrained transformers to passage retrieval [34]. This model, called monoBERT, represents the first instance of what has later become known as cross-encoders for retrieval, a class of reranking models that includes MaxP [4], CEDR [33], Birch [1], PARADE [25], and many others.

R. Pradeep and Y. Liu—Equal contribution.

M. Hagen et al. (Eds.): ECIR 2022, LNCS 13185, pp. 655–670, 2022.
https://doi.org/10.1007/978-3-030-99736-6_44

Innovations in cross-encoder models have of late stagnated in comparison to rapid developments in retrieval models based on learned dense representations [21,46] as well as learned sparse representations [5,8,12]. Part of this excitement stems from the ability of these models to directly perform ranking, as opposed to reranking based on some first-stage retrieval method to generate a list of candidates. However, reranking remains important because the output of even the best dense, sparse, or hybrid retrieval models can be further improved via reranking—and state-of-the-art effectiveness on popular benchmark datasets is achieved only by combining effective first-stage retrieval and reranking in a multi-stage ranking architecture.

Thus, although the attention of most researchers today lies beyond cross-encoders, there remain opportunities for further innovation with this class of models. In this paper, we start with the basic monoBERT model, dating back to January 2019 (which might as well be from the stone age in "neural network time"), and through a series of replication and generalization experiments, are able to improve its effectiveness by nearly 7 points absolute (20% relative) on the popular MS MARCO passage ranking task. We are, in fact, quite surprised that there is still this much effectiveness that could be squeezed out of such a mature model. How did we accomplish this? We describe below:

1. Building on the observations of Zhang et al. [53], we switched the backbone of the cross-encoder to ELECTRA$_{Base}$.
2. We replicated and then generalized the findings of Gao et al. [13], confirming the effectiveness of the LCE loss compared to hinge and cross entropy (CE) loss on MS MARCO passage ranking [2], a task not evaluated in the original paper.
3. Leveraging advances in first-stage dense retrieval methods, we used TCT-ColBERTv2 [29] to generate both the first-stage base retrieval runs for reranking and hard negatives for training our cross-encoders.
4. While Gao et al. [13] evaluated various LCE settings with up to 7 negative passages for each positive example in the batch, we extended this to 31 negatives and continued to see improvements in effectiveness.
5. Further generalizing, we noted a surprising result in our replication on MS MARCO passage ranking: it does not seem as critical as described in the original paper to train with negatives that are drawn from the first-stage retriever used for inference. That is, training with BM25 negatives or TCT-ColBERTv2 negatives both result in rerankers that perform comparably when a fixed first-stage retriever is used for reranking, for certain LCE settings. However, for inference, switching a BM25 first-stage retriever out for a TCT-ColBERTv2 first-stage retriever still brings about a significant effectiveness boost.

With the bag of tricks described above, we show that monoELECTRA$_{Base}$ can achieve an MRR@10 of 0.414 on the development set of the MS MARCO passage ranking task and an MRR@10 of 0.404 on the (blind) evaluation set. Note that this is accomplished with a standard "base" model size and without the use of any ensembles. While admittedly, none of these "tricks" in isolation are particularly noteworthy, taken together, they show that there is still room for significant improvements in a basic cross-encoder design that dates from January 2019.

2 Related Work

2.1 Cross-Encoders

As discussed, the first cross-encoder for reranking, monoBERT [34], quickly emerged after the introduction of BERT [6] itself. It followed the approach recommended by the BERT authors to handle (*query, passage*) input pairs, and demonstrated a huge leap in terms of effectiveness on the MS MARCO passage ranking [2] and TREC CAR [7] datasets. While vanilla monoBERT showed great improvement on the passage retrieval task, it was not designed to handle long input sequences as required for document retrieval. A lot of the follow-up BERT-based cross-encoder work [1,4,25,33] attempted to address this issue by either performing multiple inferences on different segments of the document or making additional architectural changes on top of BERT to better handle the longer document text.

In addition to cross-encoders relying on BERT-based pretrained Language Models (pLMs), another genre of cross-encoders takes advantage of the sequence-to-sequence pLM paradigm. Examples of these are monoT5 [35] and duoT5 [40], which use T5 [42], an extensively pretrained encoder-decoder language model. As we mostly focus on BERT-based cross-encoders in this work, we will skip the details and refer interested readers to the original papers.

There exists a strong need for better cross-encoders, which demonstrate state-of-the-art effectiveness in information retrieval tasks in various domains, even in a zero-shot setting [38,39,43,52]. They also form a vital backbone in a wide range of natural language processing tasks, including fact verification [20,37] and question answering [48].

2.2 Bi-Encoders

The success of DPR [21] and ANCE [46] revitalized bi-encoders in the new era of BERT. The goal of a bi-encoder is to learn a transformer-based mapping from queries and documents into dense fixed-width vectors wherein the inner product between the query vector and the relevant document vector is maximized. A lot of work has gone into understanding and better learning such a mapping [9,10,17,29]. A more thorough survey can be found in Lin et al. [27].

Lin et al. [28] train a bi-encoder by using on-the-fly knowledge distillation from a ColBERT [22] teacher model that computes soft-labels for in-batch negatives. This is captured by using the KL-divergence between the score distributions of the student and teacher models for all examples in the batch. They show that using this loss in addition to the standard cross entropy loss over relevance labels results in better scores.

In their follow-up work [29], a "HN+" hard-negative mining strategy is incorporated to further improve their bi-encoder, dubbed "TCT-ColBERTv2". Here, a trained TCT-ColBERT is used to first mine hard negatives to replace the BM25-based negatives. Then, the ColBERT teacher is fine-tuned using these hard negatives and the improved teacher is distilled into the bi-encoder to give TCT-ColBERTv2.

2.3   Hard Negatives

Prior work shows that the selection of negative examples is critical in training bi-encoders. Karpukhin et al. [21] compare the training effectiveness of random negatives, BM25 negatives, and in-batch negatives, and find that a mixture of BM25 and in-batch negatives yields optimal results. Xiong et al. [46] prove theoretically that local negatives are sub-optimal for dense retrieval learning. Then, they propose to prepare global negatives using the current dense retrieval model in parallel with training, which requires periodically re-indexing the corpus and retrieval. Qu et al. [41] also propose to prepare the hard negatives using the current dense retrieval model, but after the training is finished instead of on the fly. However, the paper reports that the hard negatives prepared in this way alone could degrade training and are only effective after being filtered according to an independently trained cross-encoder model. Zhan et al. [51] find that such instability caused by hard negatives could be alleviated by adding random negatives. Additionally, they periodically re-prepare the hard negatives in the ANCE [46] manner, but only update the query encoder to save the re-indexing time. All the above works confirm the importance of hard negatives and show various degrees of effectiveness.

In addition to the work described above, which focuses on hard negative training strategies for DPR-like bi-encoder fine-tuning, other works show similar observations in different methods that aid bi-encoders. Gao et al. [9] find that hard negatives are still crucial when the model is further pretrained in a way to enrich the representation of the [CLS] token, which they named Condenser. Its successor, the coCondenser [10] behaves the same after the model is additionally pre–fine-tuned on another corpus-aware unsupervised task. Hard negative mining has also been shown to be important when knowledge distillation is applied on the bi-encoders [17,29].

In contrast to the plenteous studies of hard negatives aiding bi-encoders, we only find Gao et al. [13] successfully incorporating hard negatives in cross-encoder training. To demonstrate the effectiveness of the proposed Localized Contrastive Estimation (LCE) loss, they show that training cross-encoders incorporated with the loss and harder negatives[1] can significantly improve reranking effectiveness, especially when training instances follow the same distribution as the results returned by the first-stage retrievers. Details will be introduced in Sect. 5.2.

---

[1] Here the "easy" negatives refer to the negatives sampled from BM25 results and the "hard" negatives refer to the ones sampled from HDCT [5] results.

2.4 Pretrained Transformers for Cross-Encoders

Various pretrained models have been proposed after BERT [6]. Most of the works aim at improving the general language representation ability or lowering the pretraining cost [3,24,30]. A few pretrained language models under this line have been compared in Zhang et al. [53] in *ad hoc* retrieval tasks.

Another line of work focuses on improving pLMs with IR-specific pretraining objectives. PROP [32] and B-PROP [31] propose to add a representative words prediction (ROP) task along with MLM in the pretraining stage. To prepare the training data for ROP, a document language model is used to sample a list of "pseudo queries" and their likelihoods. Then the queries are paired as (q^+, q^-) such that q^+ has a higher likelihood than q^-, and the BERT model is further pretrained to score the q^+ higher than the q^-. PROP uses a unigram language model as the document language model while its successor B-PROP uses BERT. Both are tested on a downstream retrieval task by fine-tuning a cross-encoder initialized with the ROP-pretrained BERT instead of BERT with standard pretraining. While Gao et al. [9] also propose an IR-specific pretraining task, it focuses on enriching LMs for the bi-encoder setting.

3 Loss Functions

In this section, we review common loss functions (cross entropy and hinge loss) used in cross-encoder fine-tuning and then describe the Localized Contrastive Estimation (LCE) loss function proposed by Gao et al. [13].

3.1 Cross Entropy and Hinge Loss

We begin with a quick review of how cross-encoders typically compute the relevance score given a query q and a document d, borrowing the formulation from Lin et al. [27]:

$$z_{q,d} = T_{[CLS]}W + b \tag{1}$$

where $z_{q,d}$ is the relevance score of the (query, document) pair, $T_{[CLS]}$ stands for the representation of the [CLS] token in the final layer, and W and b are the weight and the bias in the final classification layer. The dimensions of W and b might change according to the loss function—when the model is fine-tuned with the cross entropy loss, $W \in \mathbb{R}^{D \times 2}$ and $b \in \mathbb{R}^2$, whereas when it is fine-tuned with the hinge or the LCE loss, $W \in \mathbb{R}^D$ and $b \in \mathbb{R}$. That is, the output has two dimensions with cross entropy loss, one each for the relevant and non-relevant classes, while with the other two losses, the output has only one dimension, for the relevant class only.

Early cross-encoders fine-tune BERT under a classification task, using the cross entropy loss, as recommended in BERT:

$$s_{q,d} = \text{softmax}(z_{q,d})_1 \tag{2}$$

$$L_{CE} = -\sum \log(s_{q,d^+}) - \sum \log(1 - s_{q,d^-}) \tag{3}$$

where softmax$(\cdot)_1$ corresponds to the softmax score of the relevant label which by convention is indexed by 1, d^+ indicates a relevant document and d^- indicates a non-relevant document. We will use this notation from now on.

Later, MacAvaney et al. [33] fine-tune cross-encoders with the hinge loss (sometimes called max margin loss), which is more commonly used in pre-BERT neural reranker training [14,19,45]:

$$L_{\text{hinge}} = \max(0, 1 - z_{q,d^+} + z_{q,d^-}) \tag{4}$$

In the literature, cross entropy loss and hinge loss represent the two "basic" ways of training cross-encoders.

3.2   Localized Contrastive Estimation

Gao et al. [13] note that the above cross entropy loss computation considers only one document per batch per query. While not discussed in the original paper, hinge loss is similarly limited by not being able to use multiple negatives per positive unless done in a pairwise independent fashion. They also note it is important that negative examples be true negatives, especially on datasets like MS MARCO passage where many relevant passages remain unlabelled for each query. Gao et al. [13] propose the Localized Contrastive Estimation (LCE) loss to address these issues:

$$L_{LCE_q} := -\log \frac{exp(z_{q,d^+})}{\sum_{d \in G_q} exp(z_{q,d})} \tag{5}$$

$$L_{LCE} := \frac{1}{|Q|} \sum_{q \in Q, G_q \sim R_q^m} L_{LCE_q} \tag{6}$$

where R_q^m is the collection of documents top-ranked by a first-stage retriever for query q, and G_q refers to a group of documents for query q, which consists of a relevant document d^+ and $n-1$ non-relevant documents d^- sampled from R_q^m, where n is the group size.

The LCE loss combines the Noise Contrastive Estimation (NCE) [15] loss (used in, for example, Karpukhin et al. [21]) with "localized" selection of negative examples. The NCE loss scores the positive instance and multiple negative instances, normalizes all of them into probabilities, passing them through the softmax function, and encourages the model to score the positive higher than the negatives. LCE "localizes" this loss by sampling negative training examples from the top-ranked documents produced by the first-stage retriever. In combination, this loss should produce a reranker that succeeds at handling the top-ranked documents specific to a first-stage retriever while also not collapsing to match based on the confounding characteristics in the retriever's hard negative samples.

4 Experimental Setup

In this section, we describe the data and experimental configurations used in our replication. Note that *replication* indicates using a different experimental setup (e.g., implementation, framework, dataset, etc.) to generalize findings from the original paper, whereas *reproduction* indicates verifying the original paper's findings using the same experimental setup.[2]

4.1 Data

We use the MS MARCO *passage* ranking dataset [2] (MS MARCO for later reference), a large-scale *ad hoc* retrieval dataset constructed from the Bing search log. It contains 8.8 million passages and around 800K queries for training, where most of the queries have a single passage labelled relevant. These do not necessarily represent *all* true relevant passages, as it is likely that many queries in the dataset have more than one relevant passage. This setting is often called "sparse labelling". On the evaluation side, there is a small development set with 6980 queries and a blind test set with 6837 queries, both of which are similarly sparsely labelled.

We report MRR@10, the official metric, and Recall@1K (R@1K) on the small development set for all our experiments. Evaluating on the blind test set requires the submission of runs to the organizers' official leaderboard. To avoid probing the test set across various settings, we chose to submit only the test set run produced by the most effective system based on development set scores.

Note that the original work [13] uses the MS MARCO *document* ranking dataset. Thus, our experiments generalize their findings to cover the MS MARCO *passage* ranking dataset and additionally thoroughly explore certain parts in the hyperparameter space.

First-stage rankings (runs) are generated for MS MARCO's training, development, and test query sets with two retrievers: BM25 and TCT-ColBERTv2 [29]. We use the Anserini IR toolkit [47], which is built on Lucene, to generate the BM25 runs. The parameters k_1 and b are found using grid search over the range [0.6, 1.2] and [0.5, 0.9], respectively, both with step size 0.1. The tuning is based on 5 different randomly prepared query subsets, optimizing Recall@1K, following the reproduction documentation in Anserini.[3] We use the Pyserini IR toolkit [26] to generate the TCT-ColBERTv2 runs following the reproduction documentation in Pyserini.[4] We leverage the model trained with the HN+ setting as it optimizes the effectiveness of the primary metrics.

---

[2] The terms replication and reproduction are used in the sense articulated by the ACM Artifact Review and Badging (v1.1) policy; note that the definitions of the two terms are swapped in Artifact Review and Badging (v1.0).

[3] https://github.com/castorini/anserini/blob/master/docs/experiments-msmarco-passage.md.

[4] https://github.com/castorini/pyserini/blob/master/docs/experiments-tct_colbert-v2.md.

We first retrieve the top-200 passages for each query in the training query set, from which we randomly sample negative examples without replacement following Gao et al. [13]. The method of creating the training set differs from the general approach of cross-encoders which instead just relies on the official small triples training file provided by the organizers.[5] However, such an approach is common in both bi-encoders and cross-encoders when they rely on hard negative sampling.

We retrieve 1K passages for each query in the development set and test set. These form the base first-stage retriever runs which are later reranked by the cross-encoders.

4.2   Training and Inference

Our cross-encoder training and inference experiments are run on Capreolus [49,50], a toolkit for end-to-end neural ad hoc retrieval. We take advantage of its support for the MS MARCO passage ranking task, the monoBERT cross-encoder, and the training, reranking, and inference pipeline. We use the provided hinge and cross entropy loss functions, and incorporate the LCE loss into the toolkit.

The maximum numbers of tokens for the query and the entire input sequence ("[CLS] *query* [SEP] *passage* [SEP]") are set to 50 and 256, respectively. For all experiments, we initialize monoBERT with $ELECTRA_{Base}$, using the checkpoint released on HuggingFace [44].[6] We choose $ELECTRA_{Base}$ as the starting point for fine-tuning as it appears to be the most stable and effective pLM overall among those considered by Zhang et al. [53].

In all our experiments, we train monoBERT for 300K steps with a batch size of 16. We use the Adam optimizer [23] with a learning rate of $1e-5$, apply linear warm-up for the first 30K steps, and apply linear decay following warm-up. All experiments are run on Quadro RTX 8000 GPUs with TensorFlow 2.3.0. We use mixed-precision training in all the experiments.

5   Results and Discussion

In this section, we first compare the results of the three loss functions (hinge, cross entropy, and LCE) when using BM25 and TCT-ColBERTv2 first-stage retrievers. Here, we aim to show that the cross-encoders trained with the LCE loss outperform those with the other two losses on the MS MARCO passage ranking task. Then we compare their effectiveness as we vary both the source of negatives during training and the first-stage retriever during inference. Finally, we show the effect of the group size to confirm the finding that the effectiveness of cross-encoders trained with the LCE loss increases with group size, which also means it increases with more negative samples.

---

[5] https://msmarco.blob.core.windows.net/msmarcoranking/triples.train.small.tar.gz.

[6] google/electra-base-discriminator.

5.1 Loss Functions

Table 1 reports results with different loss functions and first-stage retrievers used during training and inference. In the first block, we report the scores of BM25 and TCT-ColBERTv2, which form the two baseline first-stage retrieval runs we consider for all the Capreolus rerankers.

Table 1. MRR@10 and Recall@1K with different loss functions when using BM25/TCT-ColBERTv2 as the source of hard negative and first-stage runfile. The n in the table indicates the group size. For hinge and LCE, each group always contains a positive example and $n - 1$ negative examples. For CE, each group only contains one data point, which could be either a positive or negative example. "–" indicates not applicable or the score was not reported in the original papers. Superscripts indicate significantly higher results ($p < 0.01$ with paired t-tests) after Bonferroni correction, e.g., a indicates the entry is significantly higher than the results in row (a).

	HN+ First Stage	Loss	n	MRR@10	R@1K
Baselines					
(a) BM25		–	–	0.187	0.857
(b) TCT-ColBERTv2		–	–	0.359	0.969
Prior cross-encoder work					
(c) monoBERT$_{Base}$ [34]	BM25	CE	1	0.347	–
(d) monoBERT$_{Base}$ [36]	BM25	CE	1	0.348	–
(e) monoBERT$_{Base}$ [11]	BM25	CE	1	0.353	–
(f) monoBERT$_{Base}$ [18]	BM25	CE	1	0.376	–
(g) monoBERT$_{Base}$ [34]	BM25	CE	1	0.365	–
(h) monoBERT$_{Large}$ [18]	BM25	CE	1	0.366	–
(i) monoBERT$_{Large}$ [35]	BM25	CE	1	0.372	–
(j) monoT5$_{Base}$ [35]	BM25	CE	1	0.381	–
Capreolus cross-encoders					
(1) monoELECTRA$_{Base}$	BM25	CE	1	0.378^{ab}	0.857
(2) monoELECTRA$_{Base}$		Hinge	2	0.379^{ab5}	
(3) monoELECTRA$_{Base}$		LCE	2	0.378^{ab5}	
(4) monoELECTRA$_{Base}$		LCE	8	$0.391^{ab12356}$	
(5) monoELECTRA$_{Base}$	TCT-ColBERTv2	CE	1	0.365^{a}	0.969
(6) monoELECTRA$_{Base}$		Hinge	2	0.375^{ab}	
(7) monoELECTRA$_{Base}$		LCE	2	$0.393^{ab12356}$	
(8) monoELECTRA$_{Base}$		LCE	8	$0.401^{ab123456}$	

The second block reports the scores of various comparable cross-encoders from various groups reported in the literature. We copy over the monoBERT and monoT5 scores from their original papers, rows (c), (g), and (j), respectively. We additionally include other monoBERT results reported by different groups because we observe a large variance of reported scores. This could be due to one of many reasons: different BM25 implementations, number of passages reranked, and monoBERT training hyperparameters, to name a few.

The third block of the table, rows (1–4), shows the scores when our cross-encoder is trained on BM25-sourced hard negatives and reranks the BM25 runfile. The fourth block, rows (5–8), shows the scores when the cross-encoder is trained on TCT-ColBERTv2-sourced hard negatives and reranks the TCT-ColBERTv2 runfile.

Gao et al. [13] only compare the cross entropy loss to LCE loss with a group size of 8. We generalize these results by additionally considering a group size of 2 with LCE loss and including hinge loss, which can also be viewed as having a group size of 2 but has a different loss formulation. As negative examples are sampled from the same groups of top-ranked passages, both losses benefit from the "localized" effect and the formulation is the only difference.

The cross entropy loss performs on par with the hinge loss when BM25 is used as the retriever, row (1) vs. (2). However, when using TCT-ColBERTv2 as the retriever, the hinge loss demonstrates improved effectiveness over the cross entropy loss by a slight margin, row (5) vs. (6). We suspect this is due to the pairwise loss making better use of the harder negative examples provided by TCT-ColBERTv2.

Another interesting observation is that monoBERT using LCE significantly outperforms monoBERT using the other two losses when TCT-ColBERTv2 forms the first-stage retriever, even when the group size is 2, which hinge loss uses too, row (5–7). However, LCE and hinge losses perform comparably when using BM25 as the retriever, row (2) vs. (3), and fixing the group size at 2. This indicates that the contrastive loss may itself serve as a better approach to distinguish the relevant passage from the negative ones in the ranking task, compared to the hinge loss. It additionally gains from increasing the group size, rows (3) and (7) vs. rows (4) and (8); this is more carefully examined in Sect. 5.3.

Table 2. MRR@10 and Recall@1K of all combinations of training hard negatives retriever and inference first-stage retriever on the development set of the MS MARCO passage dataset. HN refers to the source of Hard Negatives, i.e., the training retriever. All table entries use LCE with group size 8 (one positive sample with seven negative samples). Superscripts indicate significantly higher results ($p < 0.01$ with paired t-tests) after Bonferroni correction.

	HN	First-stage	MRR@10	R@1K
(a)	BM25	BM25	0.391	0.8573
(b)	TCT-ColBERTv2		0.389	
(c)	BM25	TCT-ColBERTv2	0.402^{ab}	0.9690
(d)	TCT-ColBERTv2		0.401^{ab}	

5.2 In-distributional Training Example and Hard Negative

Table 2 presents the effectiveness of the reranker when we vary the retriever for preparing training negatives and generating the development runfile. Rows (a) and (d) here correspond to rows (4) and (8) in Table 1, respectively. To obtain row (b), we use the checkpoint of row (d) to rerank the BM25 runfile. Similarly, we use the checkpoint of row (a) to rerank the TCT-ColBERTv2 runfile for row (c).

It is clear that swapping out the BM25 first-stage retriever with the dense retriever, TCT-ColBERTv2, results in significant improvement irrespective of the retriever used to mine hard negatives, rows (a–b) vs. (c–d). This is reasonable as there is a gap of around 11% in Recall@1K, meaning reranking the runfile produced by TCT-ColBERTv2 would more likely pull up the relevant passages.

We, however, observe no improvement in aligning the retriever used for hard-negative mining with that used for first-stage retrieval in evaluation. In our experiments, changing only the retriever for generating the training data does not yield significant differences in the score when we preserve the first-stage retriever to be the same, row (b) vs. (a) and row (c) vs. (d). This does not agree with the original finding of Gao et al. [13], where they find this alignment critical to the best effectiveness in the MS MARCO *document* ranking task. There are several differences in the experiments that could be responsible for this disagreement. The first is the dataset itself. Although both MS MARCO passage and MS MARCO document are from the same *ad hoc* domain, the document length may impact training data quality. Other possible causes include the range from where we sample the hard negatives, the choice of the first-stage retriever, etc. Based on these results, for the rest of the paper, we use TCT-ColBERTv2 as the first-stage retriever during inference.

Table 3. MRR@10 on the development set of the MS MARCO passage dataset across the choice of group size and retriever used to mine hard negatives. All entries use TCT-ColBERTv2 as the first stage, which has a Recall@1K of 0.9690, as seen in Table 1. Superscripts and subscripts indicate significantly higher results ($p < 0.01$ with paired t-tests) after Bonferroni correction. (e.g., $(\cdot)_{b_2}^{a2,4,8}$ indicates the entry is significantly higher than the results in row (a) with group sizes 2, 4, and 8, and the result in row (b) with group size 2.)

	HN	Group Size				
		2	4	8	16	32
(a)	TCT-ColBERTv2	0.393_{b_2}	0.400_{b_2}	0.401_{b_2}	$0.408_{b_2,4}^{a2,4,8}$	$0.414_{b_2,4,8,16}^{a2,4,8}$
(b)	BM25	0.381	0.397_{b_2}	0.402_{b_2}	0.403_{b_2}	$0.407_{b_2,4}$

5.3 LCE Group Size

We now examine the effect of the group size in the LCE loss, denoted by n, on model effectiveness. This has been studied in the original paper [13] with

$n \in \{2, 4, 6, 8\}$, where additional improvement can always be observed when the group size is increased. We explore the effect of group size in the same manner but increase the range to $\{2, 4, 8, 16, 32\}$. Additionally, we vary the retriever used for hard negative mining. We examine the MRR@10 scores across these settings in Table 3.

As noted by Gao et al. [13], we observe that the primary metric improves as the group size increases. We do so for both choices of the retriever used for hard negative mining. We surprisingly find that the metric does not seem to plateau even when the group size increases to 32 (i.e., with 31 negative samples). We did not experiment on larger group sizes due to hardware limitations,[7] but this suggests that there could be further improvements with improved hardware.

Table 2 does not note any improvements aligning the hard negative mining retriever with that used for first-stage retrieval during inference in the case with group size 8. However, we find that there do exist improvements, especially in group sizes of 2 and 32. We leave further investigation of this unusual observation as future work and use the best setting reported for the rest of the paper.

We submitted the test set run, produced from our most effective configuration, to the MS MARCO passage leaderboard.[8] Table 4 reports our scores and the systems with higher scores on the test set (at the time of our work).[9] The table shows that our best results are quite competitive to the current top results, which use ensembles of multiple cross-encoders, rows (a–c, f), or a multi-stage reranking pipeline, row (e).[10]

Table 4. MRR@10 on the official MS MARCO passage leaderboard.

Method	Dev MRR@10	Eval MRR@10
(a) coCondenser [10]	0.443	0.428
(b) C-COIL + RoBERTa [12]	0.443	0.427
(c) RocketQA + ERNIE [41]	0.439	0.426
(d) DR-BERT	0.420	0.419
(e) expando-mono-duo-T5 [40]	0.420	0.408
(f) DeepCT + TF-Ranking Ensemble [16]	0.420	0.408
(g) monoELECTRA	0.414	0.404

---

[7] The experiment involving a group size of 32 requires 4 Quadro RTX 8000 GPUs (48G memory each) to train with a batch size of 16.

[8] https://microsoft.github.io/MSMARCO-Passage-Ranking-Submissions/leaderboard.

[9] We copy the best results from each group and discard anonymous results.

[10] We cannot compare with the DR-BERT system, as we do not find its resources publicly available online.

6   Conclusion

In this paper, we replicate the LCE loss proposed by Gao et al. [13] on a different codebase and generalize their findings to the MS MARCO passage dataset. We confirm the superiority of LCE loss to the cross entropy and hinge loss on the passage ranking task, with improved effectiveness when using a better first-stage retrieval method like TCT-ColBERTv2 during inference. However, we argue that more exploration is necessary to conclude if the alignment between the training and inference first-stage retriever is essential across group sizes. Finally, we confirm that the effectiveness can be further strengthened by increasing the number of hard negatives in each group.

Acknowledgments. This research was supported in part by the Canada First Research Excellence Fund and the Natural Sciences and Engineering Research Council (NSERC) of Canada. Computational resources were provided by Compute Ontario and Compute Canada.

References

1. Akkalyoncu Yilmaz, Z., Yang, W., Zhang, H., Lin, J.: Cross-domain modeling of sentence-level evidence for document retrieval. In: Proceedings of the 2019 Conference on Empirical Methods in Natural Language Processing and the 9th International Joint Conference on Natural Language Processing (EMNLP-IJCNLP), pp. 3490–3496, November 2019
2. Bajaj, P., et al.: MS MARCO: a human generated machine reading comprehension dataset. arXiv preprint arXiv:1611.09268v3 (2018)
3. Clark, K., Luong, M.T., Le, Q.V., Manning, C.D.: ELECTRA: pre-training text encoders as discriminators rather than generators. arXiv preprint arXiv:2003.10555 (2020)
4. Dai, Z., Callan, J.: Deeper text understanding for IR with contextual neural language modeling. In: Proceedings of the 42nd Annual International ACM SIGIR Conference on Research and Development in Information Retrieval (SIGIR 2019), pp. 985–988 (2019)
5. Dai, Z., Callan, J.: Context-aware document term weighting for ad-hoc search. In: Proceedings of The Web Conference 2020, p. 1897–1907 (2020)
6. Devlin, J., Chang, M.W., Lee, K., Toutanova, K.: BERT: pre-training of deep bidirectional transformers for language understanding. In: Proceedings of the 2019 Conference of the North American Chapter of the Association for Computational Linguistics: Human Language Technologies, Volume 1 (Long and Short Papers), pp. 4171–4186 (2019)
7. Dietz, L., Verma, M., Radlinski, F., Craswell, N.: TREC complex answer retrieval overview. In: Proceedings of the Twenty-Seventh Text REtrieval Conference (TREC 2018) (2018)
8. Formal, T., Piwowarski, B., Clinchant, S.: SPLADE: sparse lexical and expansion model for first stage ranking. In: Proceedings of the 44th International ACM SIGIR Conference on Research and Development in Information Retrieval (SIGIR 2021), pp. 2288–2292 (2021)

9. Gao, L., Callan, J.: Condenser: a pre-training architecture for dense retrieval. In: Proceedings of the 2021 Conference on Empirical Methods in Natural Language Processing, pp. 981–993, November 2021

10. Gao, L., Callan, J.: Unsupervised corpus aware language model pre-training for dense passage retrieval. arXiv preprint arXiv:2108.05540 (2021)

11. Gao, L., Dai, Z., Callan, J.: Understanding BERT rankers under distillation. In: Proceedings of the 2020 ACM SIGIR on International Conference on Theory of Information Retrieval. ICTIR 2020, pp. 149–152 (2020)

12. Gao, L., Dai, Z., Callan, J.: COIL: Revisit exact lexical match in information retrieval with contextualized inverted list. In: Proceedings of the 2021 Conference of the North American Chapter of the Association for Computational Linguistics: Human Language Technologies, pp. 3030–3042, June 2021

13. Gao, L., Dai, Z., Callan, J.: Rethink training of BERT rerankers in multi-stage retrieval pipeline. In: Hiemstra, D., Moens, M.-F., Mothe, J., Perego, R., Potthast, M., Sebastiani, F. (eds.) ECIR 2021. LNCS, vol. 12657, pp. 280–286. Springer, Cham (2021). https://doi.org/10.1007/978-3-030-72240-1_26

14. Guo, J., Fan, Y., Ai, Q., Croft, W.B.: A deep relevance matching model for ad-hoc retrieval. In: Proceedings of the 25th ACM International on Conference on Information and Knowledge Management, CIKM 2016, pp. 55–64 (2016)

15. Gutmann, M., Hyvärinen, A.: Noise-contrastive estimation: A new estimation principle for unnormalized statistical models. In: Proceedings of the Thirteenth International Conference on Artificial Intelligence and Statistics. Proceedings of Machine Learning Research, vol. 9, pp. 297–304 (2010)

16. Han, S., Wang, X., Bendersky, M., Najork, M.: Learning-to-rank with BERT in TF-ranking. arXiv preprint arXiv:2004.08476 (2020)

17. Hofstätter, S., Lin, S.C., Yang, J.H., Lin, J., Hanbury, A.: Efficiently teaching an effective dense retriever with balanced topic aware sampling. In: Proceedings of the 44th International ACM SIGIR Conference on Research and Development in Information Retrieval (SIGIR 2021), SIGIR 2021, pp. 113–122 (2021)

18. Hofstätter, S., Zlabinger, M., Hanbury, A.: Interpretable & time-budget-constrained contextualization for re-ranking. In: Proceedings of the 24th European Conference on Artificial Intelligence (ECAI 2020), Santiago de Compostela, Spain, pp. 513–520 (2020)

19. Hui, K., Yates, A., Berberich, K., de Melo, G.: PACRR: a position-aware neural IR model for relevance matching. In: Proceedings of the 2017 Conference on Empirical Methods in Natural Language Processing, pp. 1049–1058 (2017)

20. Jiang, K., Pradeep, R., Lin, J.: Exploring listwise evidence reasoning with T5 for fact verification. In: Proceedings of the 59th Annual Meeting of the Association for Computational Linguistics and the 11th International Joint Conference on Natural Language Processing (Volume 2: Short Papers), pp. 402–410 (2021)

21. Karpukhin, V., et al.: Dense passage retrieval for open-domain question answering. In: Proceedings of the 2020 Conference on Empirical Methods in Natural Language Processing (EMNLP), pp. 6769–6781 (2020)

22. Khattab, O., Zaharia, M.: ColBERT: efficient and effective passage search via contextualized late interaction over BERT. In: Proceedings of the 43rd International ACM SIGIR Conference on Research and Development in Information Retrieval, pp. 39–48 (2020)

23. Kingma, D.P., Ba, J.: Adam: a method for stochastic optimization. arXiv preprint arXiv:1412.6980 (2014)

24. Lan, Z., Chen, M., Goodman, S., Gimpel, K., Sharma, P., Soricut, R.: ALBERT: a lite BERT for self-supervised learning of language representations. arXiv preprint arXiv:1909.11942 (2019)
25. Li, C., Yates, A., MacAvaney, S., He, B., Sun, Y.: PARADE: passage representation aggregation for document reranking. arXiv preprint arXiv:2008.09093 (2020)
26. Lin, J., Ma, X., Lin, S.C., Yang, J.H., Pradeep, R., Nogueira, R.: Pyserini: a Python toolkit for reproducible information retrieval research with sparse and dense representations. In: Proceedings of the 44th Annual International ACM SIGIR Conference on Research and Development in Information Retrieval (SIGIR 2021), pp. 2356–2362 (2021)
27. Lin, J., Nogueira, R., Yates, A.: Pretrained transformers for text ranking: BERT and beyond. arXiv preprint arXiv:2010.06467 (2020)
28. Lin, S.C., Yang, J.H., Lin, J.: Distilling dense representations for ranking using tightly-coupled teachers. arXiv preprint arXiv:2010.11386 (2020)
29. Lin, S.C., Yang, J.H., Lin, J.: In-batch negatives for knowledge distillation with tightly-coupled teachers for dense retrieval. In: Proceedings of the 6th Workshop on Representation Learning for NLP (RepL4NLP-2021), pp. 163–173 (2021)
30. Liu, Y., et al.: RoBERTa: a robustly optimized BERT pretraining approach. arXiv preprint arXiv:1907.11692 (2019)
31. Ma, X., Guo, J., Zhang, R., Fan, Y., Ji, X., Cheng, X.: B-PROP: bootstrapped pretraining with representative words prediction for ad-hoc retrieval. In: Proceedings of the 44th International ACM SIGIR Conference on Research and Development in Information Retrieval (SIGIR 2021) (2021)
32. Ma, X., Guo, J., Zhang, R., Fan, Y., Ji, X., Cheng, X.: PROP: pre-training with representative words prediction for ad-hoc retrieval. In: Proceedings of the 14th ACM International Conference on Web Search and Data Mining (2021)
33. MacAvaney, S., Yates, A., Cohan, A., Goharian, N.: CEDR: contextualized embeddings for document ranking. In: Proceedings of the 42nd International ACM SIGIR Conference on Research and Development in Information Retrieval, pp. 1101–1104 (2019)
34. Nogueira, R., Cho, K.: Passage re-ranking with BERT. arXiv preprint arXiv:1901.04085 (2019)
35. Nogueira, R., Jiang, Z., Pradeep, R., Lin, J.: Document ranking with a pretrained sequence-to-sequence model. In: Findings of the Association for Computational Linguistics: EMNLP 2020, pp. 708–718 (2020)
36. Nogueira, R., Yang, W., Cho, K., Lin, J.: Multi-stage document ranking with BERT. arXiv preprint arXiv:1910.14424 (2019)
37. Pradeep, R., Ma, X., Nogueira, R., Lin, J.: Scientific claim verification with VerT5erini. In: Proceedings of the 12th International Workshop on Health Text Mining and Information Analysis, pp. 94–103 (2021)
38. Pradeep, R., Ma, X., Nogueira, R., Lin, J.: Vera: Prediction techniques for reducing harmful misinformation in consumer health search. In: Proceedings of the 44th Annual International ACM SIGIR Conference on Research and Development in Information Retrieval (SIGIR 2021) (2021)
39. Pradeep, R., Ma, X., Zhang, X., Cui, H., Xu, R., Nogueira, R., Lin, J.: H2oloo at TREC 2020: when all you got is a hammer... deep learning, health misinformation, and precision medicine. In: Proceedings of the Twenty-Ninth Text REtrieval Conference (TREC 2020) (2020)
40. Pradeep, R., Nogueira, R., Lin, J.: The expando-mono-duo design pattern for text ranking with pretrained sequence-to-sequence models. arXiv preprint arXiv:2101.05667 (2021)

41. Qu, Y., et al.: RocketQA: an optimized training approach to dense passage retrieval for open-domain question answering. In: Proceedings of the 2021 Conference of the North American Chapter of the Association for Computational Linguistics: Human Language Technologies, pp. 5835–5847 (2021)

42. Raffel, C., et al.: Exploring the limits of transfer learning with a unified text-to-text transformer. J. Mach. Learn. Res. **21**(140), 1–67 (2020)

43. Thakur, N., Reimers, N., Rücklé, A., Srivastava, A., Gurevych, I.: BEIR: a heterogenous benchmark for zero-shot evaluation of information retrieval models. arXiv preprint arXiv:2104.08663, April 2021

44. Wolf, T., et al.: HuggingFace's transformers: state-of-the-art natural language processing. arXiv preprint arXiv:1910.03771 (2019)

45. Xiong, C., Dai, Z., Callan, J., Liu, Z., Power, R.: End-to-end neural ad-hoc ranking with kernel pooling. In: Proceedings of the 40th International ACM SIGIR Conference on Research and Development in Information Retrieval, SIGIR 2017, pp. 55–64 (2017)

46. Xiong, L., et al.: Approximate nearest neighbor negative contrastive learning for dense text retrieval. In: Proceedings of the 9th International Conference on Learning Representations (ICLR 2021) (2021)

47. Yang, P., Fang, H., Lin, J.: Anserini: enabling the use of Lucene for information retrieval research. In: Proceedings of the 40th Annual International ACM SIGIR Conference on Research and Development in Information Retrieval (SIGIR 2017), pp. 1253–1256 (2017)

48. Yang, W., et al.: End-to-end open-domain question answering with BERTserini. In: Proceedings of the 2019 Conference of the North American Chapter of the Association for Computational Linguistics (Demonstrations), pp. 72–77 (2019)

49. Yates, A., Arora, S., Zhang, X., Yang, W., Jose, K.M., Lin, J.: Capreolus: a toolkit for end-to-end neural ad hoc retrieval. In: Proceedings of the 13th International Conference on Web Search and Data Mining, pp. 861–864 (2020)

50. Yates, A., Jose, K.M., Zhang, X., Lin, J.: Flexible IR pipelines with Capreolus. In: Proceedings of the 29th International Conference on Information and Knowledge Management (CIKM 2020) (2020)

51. Zhan, J., Mao, J., Liu, Y., Guo, J., Zhang, M., Ma, S.: Optimizing dense retrieval model training with hard negatives. In: Proceedings of the 44th International ACM SIGIR Conference on Research and Development in Information Retrieval (SIGIR 2021), pp. 1503–1512 (2021)

52. Zhang, E., et al.: Covidex: neural ranking models and keyword search infrastructure for the COVID-19 open research dataset. In: Proceedings of the First Workshop on Scholarly Document Processing, pp. 31–41 (2020)

53. Zhang, X., Yates, A., Lin, J.: Comparing score aggregation approaches for document retrieval with pretrained transformers. In: Hiemstra, D., Moens, M.-F., Mothe, J., Perego, R., Potthast, M., Sebastiani, F. (eds.) ECIR 2021, Part II. LNCS, vol. 12657, pp. 150–163. Springer, Cham (2021). https://doi.org/10.1007/978-3-030-72240-1_11

An Evaluation Study of Generative Adversarial Networks for Collaborative Filtering

Fernando Benjamín Pérez Maurera[1,2]($\boxtimes$) (ID), Maurizio Ferrari Dacrema[1] (ID), and Paolo Cremonesi[1] (ID)

[1] Politecnico di Milano, Milan, Italy
{fernandobenjamin.perez,maurizio.ferrari,paolo.cremonesi}@polimi.it
[2] ContentWise, Milan, Italy
fernando.perez@contentwise.com

Abstract. This work explores the reproducibility of CFGAN. CFGAN and its family of models (TagRec, MTPR, and CRGAN) learn to generate personalized and fake-but-realistic rankings of preferences for top-N recommendations by using previous interactions. This work successfully replicates the results published in the original paper and discusses the impact of certain differences between the CFGAN framework and the model used in the original evaluation. The absence of random noise and the use of real user profiles as condition vectors leaves the generator prone to learn a degenerate solution in which the output vector is identical to the input vector, therefore, behaving essentially as a simple autoencoder. The work further expands the experimental analysis comparing CFGAN against a selection of simple and well-known properly optimized baselines, observing that CFGAN is not consistently competitive against them despite its high computational cost. To ensure the reproducibility of these analyses, this work describes the experimental methodology and publishes all datasets and source code.

Keywords: Generative Adversarial Networks · Recommender Systems · Collaborative Filtering · Reproducibility

1 Introduction

In recent years, Generative Adversarial Networks (GANs) have become the state-of-the-art technique inside the group of generative methods, i.e., methods that learn how to generate fake data from the real one. Their primary use has been in the computer vision domain [13, 16–18]. They have also been used in Information Retrieval [34] and Recommender Systems, the most notable example being Collaborative Filtering GAN (CFGAN) [4], and the family of models based on it, such as TagRec [5], CRGAN [35], MTPR [36, 37].

This work contributes to the trend of evaluation studies in Machine Learning, Information Retrieval, and Recommender Systems domains [10, 11, 21, 22, 38].

© The Author(s), under exclusive license to Springer Nature Switzerland AG 2022
M. Hagen et al. (Eds.): ECIR 2022, LNCS 13185, pp. 671–685, 2022.
https://doi.org/10.1007/978-3-030-99736-6_45

This work discusses the implications of certain differences between the CFGAN framework and the model that was used in the experimental evaluation, which would adversely affect its learning ability, providing a reference for future works. In particular, the generator is left prone to reach a degenerate solution and behave as a simple autoencoder, therefore, belonging to the same family of previous recommendation models such as [28,32]. This discussion is based on the findings of [23], which highlights the importance of describing not only *how* a model works, but also *what* works and *why* it works, as well as how experimental inquiries that aim to deepen our understanding are valuable research contributions even when no new algorithm is proposed. Furthermore, this work analyzes the replicability, reproducibility, and recommendation quality of CFGAN [4] as well as its numerical stability which is known to be a challenge for GANs [9,25]. The main research questions of this work are:

RQ1: Is CFGAN replicable and numerically stable? i.e., does CFGAN achieve the claimed results using the same experimental setup as in [4]?

RQ2: What is the impact of the differences between the CFGAN framework and the model used for the evaluation in [4], and why do they raise theoretical and methodological concerns regarding the learning ability of the model?

RQ3: Is CFGAN reproducible, achieving the claimed recommendation quality when compared to properly-tuned baselines? How does CFGAN compare along other dimensions such as beyond-accuracy and scalability metrics?

2　Collaborative Filtering Generative Adversarial Networks

GANs have been successfully applied to numerous prediction and classification tasks. This work addresses a family of generative models originated from GANs used in Recommender Systems. Briefly, a GAN[1] consists of two neural networks that are trained together in an adversarial setting until they reach convergence. The first neural network is called the *generator*, denoted as G, while the second network is called the *discriminator*, denoted as D [3,8,13,14]. CFGAN[2] is the most notable GAN recommendation algorithm [5,37]. Its main attribute is that it generates personalized user or item profiles, mainly by solely using previous interactions, but is able to learn from sources of information as well [4].

CFGAN Training Process. Figure 1 shows an illustration of the training process of CFGAN. Every epoch starts by feeding the generator G with random noise z and a condition vector c. The generator creates preferences of users towards items (or vice versa) which are then masked (see Masking). The discriminator D then receives the real profiles, the masked profiles, and the condition. The discriminator tells the probability that each masked and real profiles come

---

[1] The supplemental material [29] contains the formal formulation of GANs.

[2] For a detailed explanation of CFGAN we refer the reader to the reference article [4].

from the real data. The discriminator is updated based on how well it is able to correctly distinguish fake data from real data. The generator is updated based on how much it could generate *fake but realistic* data.

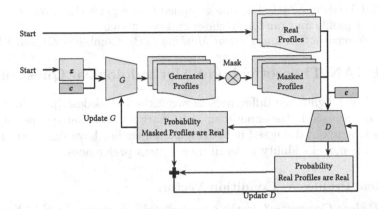

Fig. 1. Training process of CFGAN. G, D, z and c are the generator network, discriminator network, random noise, and condition vectors, respectively. Real profiles are not masked.

Modes. CFGAN has two modes: *user-based* (u) or *item-based* (i). The first learns to generate user profiles, while the second learns to generate item profiles.

Masking. CFGAN applies a mask to the generated profiles by performing an element-wise product between these and the real profiles. If the variant is *Partial Masking*, then the mask changes (see Variants).

Architecture. Both the generator and discriminator of CFGAN are fully connected feed-forward neural networks independent from each other where each has its own hyper-parameters, e.g., number of hidden layers, learning rate, regularization, and others. If the *mode* is *user-based*, then the number of input neurons is the number of *items* in the dataset. Conversely, the number of input neurons for an *item-based* CFGAN is the number of *users* in the dataset.

Recommendations. In a top-N item recommendation scenario, the trained generator creates user profiles containing the preference scores of users toward items. Recommendations are built by ranking the items from the highest to lowest score and selecting the top-N.

Variants. CFGAN has three variants:

- *Zero Reconstruction (ZR)*: Changes the loss function of the generator. It ensures that a sample of non-interacted items are given zero-weights in the generated profiles.
- *Partial Masking (PM)*: The mask applied to the generated profiles combines the user profile and randomly-chosen unseen items.
- *Zero Reconstruction and Partial Masking (ZP)*: Combines ZR and PM.

3   CFGAN Theoretical and Methodological Questions

This work highlights key differences between the initial description of CFGAN and the model used in the experimental evaluation of that same paper [4]. These differences were not discussed in the original paper but have significant implications on the model's ability to learn user or item preferences.

3.1   Real Profiles as Condition Vectors

What Raises Concerns? In the experimental evaluation of CFGAN, the condition vector provided to both the generator and the discriminator is the real user/item profile, i.e., the interactions that CFGAN is learning to generate.

Why Is It a Concern? As a consequence, CFGAN is prone to generate a trivial solution. The generator could learn the identity function between the condition vector and the output, therefore easily deceiving the discriminator without learning to generate new profiles. On the other hand, the discriminator could learn that the generated user profile should be identical to the condition vector to be real, again learning a trivial function. In practice, this will push the generator to behave as an *autoencoder* [20], which reconstructs as output the same input (condition) it was provided with.

How to Avoid This Concern? Since the condition vector can contain any information, a simple strategy would be to use other feature data related to the items or users or other contextual information. In a pure collaborative recommendations scenario, where no features or context is available, a possible strategy is to change the condition vector to be the user/item classes (i.e., unique identifiers) depending on the CFGAN mode. This decision is aligned with previous works on GANs [26]. In Recommender Systems, using the user/item classes provides a mechanism to generate *personalized recommendations* to every user. In contrast to the original CFGAN reference, using the user/item classes excludes the possibility that the generator and discriminator learn a trivial solution.

3.2   No Random Noise

What Raises Concerns? The reference article states that the random noise is not provided as input to the generator in its experiments because the goal is to generate the single best recommendation list rather than multiple ones.

Why Is It a Concern? This violates the framework defined in the same article and the design principles of GANs. In practice, discarding noise is problematic because it drastically reduces the input space and the generator will be trained on a very sparse set of user profiles. This assumes that the user profiles will not change, which will make CFGAN non-robust in a real application where the data change rapidly. This is known as the *dataset shift* problem. Since the data change over time as new interactions are collected, and models are not continuously retrained, models should be robust to and be able to use the *new* data that was not present during training [27, 30].

How to Avoid This Concern? Feed the generator with a random noise vector z and the condition vector c. z is drawn from a normal distribution with zero mean and unit variance, i.e., $z \sim \mathcal{N}(\mu, \sigma^2)$ where $\mu = 0$ and $\sigma^2 = 1$ as suggested by other works [13, 26]. The size of z is a key element while training GANs. However, previous works do not have consensus concerning the size of z [8]. We use a heuristic to set the size of the random vector and try different values depending on the number of input neurons: 50%, 100%, or 200% of them. In practice, the condition c and the random vector z are concatenated, and this new vector becomes the input to the first layer of the generator network.

3.3 Methodological Questions

What Raises Concerns? The CFGAN description does not state how to choose the number of training epochs nor the stopping criterion for the training phase.

Why Is It a Concern? The number of training epochs and the stopping criterion are two key methodological aspects for most machine learning models. With the current GAN formulation, these two are defined by hand instead of automatically chosen by the continuous evaluation of GAN, which might lead to a non-optimal model, misuse of computational resources, and negatively affect the published results' replicability. There are well-known objective ways to measure the recommendation quality in offline scenarios without human intervention in the Recommender Systems domain, e.g., with accuracy metrics.

How to Avoid This Concern? Use an early-stopping mechanism based on the one used in previous works for other machine learning recommenders, such as matrix factorization or linear regression [10, 11]. An early-stopping mechanism periodically evaluates CFGAN on validation data while CFGAN is being trained on train data. The training stops when the CFGAN quality does not improve over the best evaluation for a fixed number of evaluations.

4 Experimental Methodology

The experiments, results, and discussion are based on one of the following two experiments: (i) execution of the source code provided in the CFGAN reference article as-is to assess the result replicability; (ii) hyper-parameter tuning of different recommenders using a well-known evaluation framework to study the reproducibility of the results and evaluate along different dimensions (see [10, 11]) The source code of the experiments is available online[3].

Datasets. The experiments use the same datasets[4] (a sub-sampled version of Ciao[5] [4, 33], ML100K [15], and ML1M [15]) and splits (*train* and *test*) provided with the CFGAN reference article [4]. For scenarios that required a *validation* split, we created one by applying the same strategy as the reference: random holdout at 80% of the *train* split. Given the modest size of these datasets, all experiments are done on the CPU.

Technologies. The implementation of all experiments, is based on the evaluation framework published in [10], which includes the implementation of some simple yet competitive state-of-the-art baselines for Recommender Systems. For the replication study, the original implementation has been used as provided. For the reproducibility study and the other experiments, the original CFGAN source code has been adapted to the framework with no changes to the core algorithm.

4.1 Methodology for the Replicability of CFGAN

The original CFGAN source code includes the implementation of CFGAN and its training loop using a fixed set of hyper-parameters that are dataset-dependent. The training procedure is the following: it fits a CFGAN recommender using the *train* split of the selected dataset and evaluates the recommender using the *test* split. With respect to the evaluation metrics, this source code evaluates CFGAN on *accuracy* metrics: precision (PREC), recall (REC), Mean Reciprocal Rank (MRR), and Normalized Discounted Cumulative Gain (NDCG) at recommendation list length 5 and 20. The limitations of this source code are the lack of the implementation of the baselines and the hyper-parameter tuning of all recommenders, e.g., baselines and CFGAN. Due to this, the replication study is only possible for CFGAN.

---

[3] https://github.com/recsyspolimi/ecir-2022-an-evaluation-of-GAN-for-CF and [29].
[4] The Watcha [4] dataset was not provided with the reference article.
[5] The reference article does not provide instructions to reproduce this version of the dataset. We contacted the authors for clarifications but did not receive a reply.

4.2   Methodology for the Reproducibility of CFGAN

The reproducibility study expands the original CFGAN evaluation by including: (i) new baselines that were shown to provide high recommendation quality; (ii) a well-defined hyper-parameter optimization strategy; (iii) a well-defined early-stopping strategy; and (iv) a comparison against accuracy, beyond-accuracy, and scalability metrics.

In particular, the goal of (i) and (ii) is to assess the recommendation quality of CFGAN against a wider set of recommendation models which are properly tuned under the same conditions. The models we report range from non-personalized, neighborhood-based, and non-neural machine learning approaches. This decision is aligned with results obtained by previous evaluation studies in the domain [10,11]. Regarding the hyper-parameter optimization of CFGAN, it should be noted that the search-space described in the reference article, considering that it is done using a grid-search, contains more than $3 \cdot 10^8$ cases, which cannot be reproduced in a reasonable time. Due to this, this work adopts a different optimization strategy: Bayesian Search as used in [10]. The hyper-parameter ranges and distributions of CFGAN are reported in Table 1. The Bayesian Search starts with 16 initial random searches and performs a total of 50 cases for each algorithm. Each model in this search is fit with the *train* split and evaluated against the *validation* one. The best hyper-parameters are chosen as those with the highest NDCG at 10. Once the optimal hyper-parameters set is chosen, it trains the final models using this set and the union of the train and validation splits, evaluating the final models against the test set.

Evaluation Metrics. Recommenders are evaluated using the original accuracy metrics (PREC, REC, MRR, and NDCG) and against the following *beyond-accuracy* metrics: novelty [39], item coverage (Cov. Item, quota of recommended items), and distributional diversity (Div. MIL [39] and Div. Gini [1]). Using these new metrics provides a broader picture of the quality of all recommenders.

Baselines. Due to space limitations, this work provides only a list of baseline recommenders. A thorough description of all baselines, and the list, range and distribution of their hyper-parameters are in [10]. The baselines list is the following: **Top Popular** [10] as a non-personalized approach. **UserKNN CF** and **ItemKNN CF** [10] as neighborhood-based CF (similarities: cosine, dice, jaccard, asymmetric cosine, and tversky) and shrinkage term. **RP3beta** [6] as a graph-based approach. **PureSVD** [7] and **MF BPR** [31] as matrix factorization models. **SLIM ElasticNet** [10,28] as a machine learning approach. Lastly, **EASE R** as a fast linear autoencoder [32].

CFGAN Recommenders. The hyper-parameter tuning is done on a total of 18 different CFGAN models: three datasets (Ciao, ML100K, and ML1M), two modes (item-based i and user-based u), and three variants (ZR, PM, and ZP).

To ensure a clear stopping criteria and a fair training for CFGAN, it is trained using the early-stopping criteria defined in [10] and presented in Sect. 3. The number of minimum and maximum epochs is in Table 1. The early-stopping selects the best number of epochs by using the validation data. The optimal number of epochs is used to train the final model. We recall that the original description of CFGAN *does not provide* an early-stopping mechanism.

Table 1 lists all hyper-parameters of CFGAN, where hyper-parameters like *optimizer, activation* are left unchanged with respect to the reference article. Apart from the number of training epochs, the optimizer, and activation, the rest of the hyper-parameters are set by the Bayesian Search.

Table 1. Hyper-parameters for CFGAN. These are divided in two groups. The first group contains specific hyper-parameters of CFGAN. The second group are hyper-parameters of the generator and discriminator neural networks, values between networks can be different.

Hyper-Parameter	Type	Range	Distribution
# of Epochs	Integer	$200-400^a$	early-stopping
ZR Coefficient	Real	$0-1$	uniform
ZR Ratio	Integer	$10-90$	uniform
PM Ratio	Integer	$10-90$	uniform
# of Hidden Layers	Integer	$1-4$	uniform
# of Hidden Features	Integer	$50-300$	uniform
# of Steps	Integer	$1-4$	uniform
$l2$ Regularization	Real	$1 \cdot 10^{-4} - 1 \cdot 10^{-1}$	log-uniform
Learning Rate	Real	$1 \cdot 10^{-4} - 5 \cdot 10^{-3}$	log-uniform
Batch Size	Integer	$32-256$	uniform
Optimizer	Categorical	ADAM [19]	–
Activation	Categorical	sigmoid	–

[a]Due to how the training is performed, this range is close to the 1.000 and 1.500 epochs used in the reference article

5 Experiments Results and Discussion

5.1 RQ1: CFGAN Replicability and Numerical Stability

To address **RQ1**, we report the results of the replication study, as described in Sect. 4.1, by using the original source code and data. This experiment has two goals: (i) verify that published results are replicable; and (ii) measure the numerical stability of CFGAN given the stochastic nature of its architecture [9,25].

Table 2 shows the results of the experiment, we only report two metrics due to space limitations[6]. The results reported in the reference article are denoted as

---

[6] A table with all metrics is available in the supplemental materials of this work.

Table 2. Comparison between the accuracy metrics in the reference article [4] and those obtained in the replicability experiment (see Section 5.1) at recommendation list length of 20. Statistics calculated over 30 executions, evaluating on the last epoch using recommendation lists of length 20. We consistently obtain *lower* results across the three datasets on average. For the Ciao dataset, the original source code trains a different variant (in bold) than the reported in the reference article.

Dataset	Variant	Stats	PREC	NDCG
Ciao	**iZR**	Mean ± Std	0.0402 ± 0.0014	0.1135 ± 0.0038
	iZP	Reference [4]	0.0450	0.1240
ML100K	iZP	Mean ± Std	0.2851 ± 0.0025	0.4207 ± 0.0048
	iZP	Reference [4]	0.2940	0.4330
ML1M	iZP	Mean ± Std	0.3079 ± 0.0011	0.4035 ± 0.0016
	iZP	Reference [4]	0.3090	0.4060

Reference. Due to the stochastic nature of CFGAN models, we do not expect to achieve *exact* numerical replicability. For all datasets, we see that the replicated results are *lower* than those reported in the reference article. For the ML1M dataset, the difference between the average and reported NDCG is -0.62%. On the smaller ML100K, the results are more varied: -2.84% between the average and reported NDCG. For the Ciao dataset, the results could not be replicated due to two factors: (i) the original source code trained a different variant (iZR) than the reported in the reference article (iZP); and (ii) lack of reproducible hyperparameters sets for this dataset in the reference article. Lastly, with respect to the numerical stability, under 30 executions of this replication, the results indicate that the reference implementation of CFGAN is numerically stable.

5.2 RQ2: Impact of Theoretical and Methodological Concerns

This section reports the results of the experiments related to **RQ2**, those used to measure the impact of the theoretical and methodological concerns raised in Sect. 3. Table 3 compares the results of the reference CFGAN (denoted as Reference), the models tuned in Sect. 5.3 (presented in Table 4), and the variants of this experiment.

Impact of Random Noise. As seen in Sect. 2, CFGAN receives random noise as part of its input. However, in the experiments of the reference article, the random noise is removed. This experiment included three different sizes of random noise. The results indicate that the recommendation quality improves slightly by removing the random noise, however, as stated in Sect. 3, it comes at the cost of risking lower generalizability and lower robustness of the generator in a practical use case. We argue the random noise should always be present. However, we recall that doing an exhaustive analysis of the impact of random noise in GAN and CFGAN is beyond the scope of this paper.

Table 3. Accuracy and beyond-accuracy values for different CFGAN models for the ML1M dataset at recommendation list length of 20. The suffix Reference is the model in the reference article (where − denotes non published values). The suffix ES indicates that the model uses early-stopping (see Table 4), NO-ES indicates it does not. The suffix CC indicates that the model uses the user/item class as the condition vector. The suffix RN-X means that the model uses random noise of size X. Hyper-parameter sets of variants are chosen as described in Section 4.2 except for those with the Reference suffix.

Variant	PREC	NDCG	Cov. Item	Variant	PREC	NDCG	Cov. Item
iZP Reference [4]	0.3090	0.4060	−	uZP Reference [4]	−	−	−
iZP ES	0.2407	0.2972	0.4894	uZP ES	0.2764	0.3620	0.1833
iZP NO-ES	0.2494	0.3111	0.4041	uZP NO-ES	0.2797	0.3639	0.1882
iZP CC	0.0384	0.0507	0.0296	uZP CC	0.0916	0.1106	0.0231
iZP RN-3020	0.2059	0.2475	0.3995	uZP RN-1841	0.2737	0.3591	0.1841
iZP RN-6040	0.1683	0.2000	0.4663	uZP RN-3682	0.2781	0.3651	0.1839
iZP RN-12080	0.1304	0.1471	0.5076	uZP RN-7364	0.2759	0.3626	0.1955

Impact of Condition Vector. Similarly as before, in the experiments of the reference article, the condition vector is set to be the user/item profiles, which increases the risk of reaching a trivial solution. This experiment changed the condition vector to be the user/item classes. The results show that changing the condition vector with the current CFGAN architecture dramatically lowers the model's ability to learn to generate accurate profiles. This constitutes a *negative result*, as that the current architecture does not appear to be suitable to handle the user/item classes as the condition vector. Identifying an appropriate architecture to do so and an appropriate condition vector to use in scenarios where only past user interactions are available is an open research question that goes beyond the scope of this paper.

Impact of Early-Stopping. The reference article does not provide an early-stopping mechanism for CFGAN, although models in Recommender Systems typically benefit from one, as discussed in Sect. 3. This experiment removed the early-stopping and set the maximum number of epochs as 400 (this is the maximum number of epochs set for the early-stopping as seen in Table 1). Results show that using early-stopping slightly decreases the recommendation quality of CFGAN, however, we argue that the benefits of using it outweigh the downsides of it, especially if scalability is taken into account. For instance, the iZP variant trains on 645 and 1200 epochs with and without early-stopping, respectively, i.e., a decrease of 46.25% in training time and 4.47% in NDCG.

Table 4. Accuracy and beyond-accuracy metrics for tuned baselines and CFGAN on the ML1M dataset at recommendation list length of 20. Higher accuracy values than CFGAN models reached by baselines in bold. ItemKNN and UserKNN use asymmetric cosine. CFGAN results are different than Table 2 due to the hyper-parameter tuning. CFGAN models use early-stopping.

	PREC	REC	MRR	NDCG	Novelty	Cov. Item	Div. MIL	Div. Gini
Random	0.0099	0.0056	0.0326	0.0108	0.0732	1.0000	0.9946	0.8977
TopPop	0.1552	0.1146	0.3852	0.1938	0.0473	0.0299	0.4529	0.0095
UserKNN CF	0.2891	**0.2570**	**0.6595**	**0.3888**	0.0513	0.3286	0.8921	0.0655
ItemKNN CF	0.2600	0.2196	0.6254	0.3490	0.0497	0.2097	0.8148	0.0362
RP3beta	0.2758	0.2385	**0.6425**	0.3700	0.0506	0.3427	0.8565	0.0528
PureSVD	0.2913	0.2421	**0.6333**	0.3783	0.0516	0.2439	0.9142	0.0712
SLIM ElasticNet	**0.3119**	**0.2695**	**0.6724**	**0.4123**	0.0514	0.3153	0.8984	0.0696
MF BPR	0.2485	0.2103	0.5753	0.3242	0.0512	0.3126	0.8855	0.0631
EASE R	**0.3171**	**0.2763**	**0.6795**	**0.4192**	0.0518	0.3338	0.9146	0.0803
CFGAN iZR	0.2862	0.2547	0.6312	0.3770	0.0542	0.4123	0.9583	0.1459
CFGAN iPM	0.2505	0.1950	0.5454	0.3138	0.0523	0.3669	0.9218	0.0901
CFGAN iZP	0.2407	0.1742	0.5230	0.2972	0.0530	0.4894	0.9256	0.0901
CFGAN uZR	0.2955	0.2473	0.6222	0.3799	0.0523	0.2167	0.9205	0.0837
CFGAN uPM	0.2367	0.1928	0.5513	0.3054	0.0516	0.1782	0.8962	0.0550
CFGAN uZP	0.2764	0.2342	0.6208	0.3620	0.0513	0.1833	0.9062	0.0617

5.3 RQ3: Reproducibility Evaluation Against Properly Tuned Baselines

To address **RQ3**, we report the recommendation quality of CFGAN and baseline recommenders using a Bayesian hyper-parameter tuning approach, as described in Sect. 4.2. The goal is to evaluate [4] on the same top-N recommendation scenario of the reference paper against a set of properly tuned baselines on accuracy and beyond-accuracy metrics and study if published results are reproducible.

Table 4 shows the results of accuracy and beyond-accuracy metrics of properly tuned recommenders. Due to space constraints, the focus of this discussion is on the dataset with the highest number of interactions studied in the reference article [4], i.e., ML1M. Results with other datasets are comparable[7].

The results indicate that CFGAN is outperformed by three simple baselines in NDCG, sometimes by almost 10%, in particular by other autoencoder-based recommendation models like EASE R and SLIM Elastic Net. These findings are consistent to those reported in several other evaluation studies [2,10–12,24]. The accuracy across CFGAN models varies depending on the CFGAN mode and variant. For instance, the most and least accurate variants are uZR and iZP, respectively, with approximately 21.76% difference in their NDCG metrics. Under the current methodology, we cannot confirm the claim that item-based models or ZP variants outperform other variants, as indicated in the reference

---

[7] The full results are in the supplemental material [29].

article [4]. In fact, our most accurate variant is uZR. When looking at beyond-accuracy metrics, item-based CFGAN models have equal or higher diversity than baselines. In particular, iZR has the highest novelty, item coverage, and distributional diversity, while also being the second-most accurate variant with respect to NDCG. User-based CFGAN models have less coverage than all baselines.

It can be seen that the results of the replicability study using hyper-parameter optimization and early-stopping reported in Table 4 are lower than those reported in the replication study in Table 2. This indicates that the non-reproducible hyper-parameter search and early-stopping criteria have an important impact on the recommendation quality. As a last observation, using the results reported in the reference article CFGAN would not be competitive against the baselines.

Scalability. Concerning the recommendation time, all algorithms are able to create recommendations lists to all users in a total time between 7 and 20 s. Differently from other neural models [10], CFGAN models provide fast recommendations. Due to the lack of random noise, they generate static recommendation lists.

Concerning the training time, CFGAN models take more time to train than any baseline. We categorize models into three groups: (i) ItemKNN, UserKNN, PureSVD, RP3beta, and EASE R take between 2 and 25 s on average; (ii) machine learning approaches, i.e., SLIM and MF BPR take between 3 and 9 min to train on average; and (iii) all CFGAN models take between 25 and 40 min to train on average. Even on a comparatively small dataset as ML1M, the difference in training time between the first and the last group is two orders of magnitude. Using more performing hardware, i.e., GPU could reduce this gap.

Under this offline evaluation, which is the same as in the original article [4], CFGAN does not generate more accurate recommendations than simple baselines. As CFGAN is a neural approach, bigger datasets with more complex relations between users, items, and their interactions might increase the accuracy of CFGAN. However, this is unpractical due to the higher computational cost of CFGAN models, therefore, we do not report experiments with bigger datasets.

6   Conclusions

This work presents an evaluation study of the family of models of CFGAN, addressing three research questions under the same top-N recommendation scenario as the reference article [4]. Are previously published results of CFGAN replicable? What is the impact of the differences between the CFGAN framework and the model evaluated in the reference article? Are previously published results of CFGAN reproducible? Regarding the model's architecture, using as condition vector the user profile and removing the random noise leaves the model prone to a trivial and not useful solution in which the generator behaves as a simple autoencoder, negatively affecting the model's ability to generalize. Due to this, we argue a different approach should be used, which is still an open research question. The experimental results indicate that CFGAN is replicable and numerically stable, but not reproducible as it can be outperformed by simple but properly tuned baselines. This result adds to the recent evidence that properly tuned baselines can outperform complex methods and suggest CFGAN is not yet a mature recommendation algorithm.

References

1. Adomavicius, G., Kwon, Y.: Improving aggregate recommendation diversity using ranking-based techniques. IEEE Trans. Knowl. Data Eng. **24**(5), 896–911 (2012). https://doi.org/10.1109/TKDE.2011.15
2. Armstrong, T.G., Moffat, A., Webber, W., Zobel, J.: Improvements that don't add up: ad-hoc retrieval results since 1998. In: Proceedings of the 18th ACM Conference on Information and Knowledge Management, CIKM 2009, Hong Kong, China, 2–6 November 2009, pp. 601–610. ACM (2009). https://doi.org/10.1145/1645953.1646031
3. Borji, A.: Pros and cons of GAN evaluation measures. Comput. Vis. Image Underst. **179**, 41–65 (2019). https://doi.org/10.1016/j.cviu.2018.10.009
4. Chae, D., Kang, J., Kim, S., Lee, J.: CFGAN: a generic collaborative filtering framework based on generative adversarial networks. In: Proceedings of the 27th ACM International Conference on Information and Knowledge Management, CIKM 2018, Torino, Italy, 22–26 October 2018, pp. 137–146. ACM (2018). https://doi.org/10.1145/3269206.3271743
5. Chen, H., Wang, S., Jiang, N., Li, Z., Yan, N., Shi, L.: Trust-aware generative adversarial network with recurrent neural network for recommender systems. Int. J. Intell. Syst. **36**(2), 778–795 (2021). https://doi.org/10.1002/int.22320
6. Christoffel, F., Paudel, B., Newell, C., Bernstein, A.: Blockbusters and wallflowers: accurate, diverse, and scalable recommendations with random walks. In: Proceedings of the 9th ACM Conference on Recommender Systems, RecSys 2015, Vienna, Austria, 16–20 September 2015, pp. 163–170. ACM (2015). https://doi.org/10.1145/2792838.2800180
7. Cremonesi, P., Koren, Y., Turrin, R.: Performance of recommender algorithms on top-N recommendation tasks. In: Proceedings of the 2010 ACM Conference on Recommender Systems, RecSys 2010, Barcelona, Spain, 26–30 September 2010, pp. 39–46. ACM (2010). https://doi.org/10.1145/1864708.1864721
8. Creswell, A., White, T., Dumoulin, V., Arulkumaran, K., Sengupta, B., Bharath, A.A.: Generative adversarial networks: an overview. IEEE Signal Process. Mag. **35**(1), 53–65 (2018). https://doi.org/10.1109/MSP.2017.2765202
9. Fellicious, C., Weissgerber, T., Granitzer, M.: Effects of random seeds on the accuracy of convolutional neural networks. In: Nicosia, G., et al. (eds.) LOD 2020, Part II. LNCS, vol. 12566, pp. 93–102. Springer, Cham (2020). https://doi.org/10.1007/978-3-030-64580-9_8
10. Ferrari Dacrema, M., Boglio, S., Cremonesi, P., Jannach, D.: A troubling analysis of reproducibility and progress in recommender systems research. ACM Trans. Inf. Syst. **39**(2), 20:1–20:49 (2021). https://doi.org/10.1145/3434185
11. Ferrari Dacrema, M., Cremonesi, P., Jannach, D.: Are we really making much progress? A worrying analysis of recent neural recommendation approaches. In: Proceedings of the 13th ACM Conference on Recommender Systems, RecSys 2019, Copenhagen, Denmark, 16–20 September 2019, pp. 101–109. ACM (2019). https://doi.org/10.1145/3298689.3347058

12. Ferrari Dacrema, M., Parroni, F., Cremonesi, P., Jannach, D.: Critically examining the claimed value of convolutions over user-item embedding maps for recommender systems. In: CIKM 2020: The 29th ACM International Conference on Information and Knowledge Management, Virtual Event, Ireland, 19–23 October 2020, pp. 355–363. ACM (2020). https://doi.org/10.1145/3340531.3411901

13. Goodfellow, I.J., et al.: Generative adversarial nets. In: Advances in Neural Information Processing Systems 27: Annual Conference on Neural Information Processing Systems 2014, Montreal, Quebec, Canada, 8–13 December 2014, pp. 2672–2680 (2014). https://proceedings.neurips.cc/paper/2014/hash/5ca3e9b122f61f8f06494c97b1afccf3-Abstract.html

14. Goodfellow, I.J., et al.: Generative adversarial networks. Commun. ACM 63(11), 139–144 (2020). https://doi.org/10.1145/3422622

15. Harper, F.M., Konstan, J.A.: The MovieLens datasets: history and context. ACM Trans. Interact. Intell. Syst. 5(4), 19:1–19:19 (2016). https://doi.org/10.1145/2827872

16. Isola, P., Zhu, J., Zhou, T., Efros, A.A.: Image-to-image translation with conditional adversarial networks. In: 2017 IEEE Conference on Computer Vision and Pattern Recognition, CVPR 2017, Honolulu, HI, USA, 21–26 July 2017, pp. 5967–5976. IEEE Computer Society (2017). https://doi.org/10.1109/CVPR.2017.632

17. Karras, T., Laine, S., Aila, T.: A style-based generator architecture for generative adversarial networks. In: IEEE Conference on Computer Vision and Pattern Recognition, CVPR 2019, Long Beach, CA, USA, 16–20 June 2019, pp. 4401–4410. Computer Vision Foundation/IEEE (2019). https://doi.org/10.1109/CVPR.2019.00453

18. Karras, T., Laine, S., Aittala, M., Hellsten, J., Lehtinen, J., Aila, T.: Analyzing and improving the image quality of StyleGAN. In: 2020 IEEE/CVF Conference on Computer Vision and Pattern Recognition, CVPR 2020, Seattle, WA, USA, 13–19 June 2020, pp. 8107–8116. Computer Vision Foundation/IEEE (2020). https://doi.org/10.1109/CVPR42600.2020.00813

19. Kingma, D.P., Ba, J.: Adam: a method for stochastic optimization. In: 3rd International Conference on Learning Representations, ICLR 2015, San Diego, CA, USA, 7–9 May 2015, Conference Track Proceedings (2015)

20. Kingma, D.P., Welling, M.: An introduction to variational autoencoders. Found. Trends Mach. Learn. 12(4), 307–392 (2019). https://doi.org/10.1561/2200000056

21. Lin, J.: The neural hype and comparisons against weak baselines. SIGIR Forum 52(2), 40–51 (2019). https://doi.org/10.1145/3308774.3308781

22. Lin, J.: The neural hype, justified! A recantation. SIGIR Forum 53(2), 88–93 (2021). https://doi.org/10.1145/3458553.3458563

23. Lipton, Z.C., Steinhardt, J.: Troubling trends in machine learning scholarship. ACM Queue 17(1), 80 (2019). https://doi.org/10.1145/3317287.3328534

24. Ludewig, M., Jannach, D.: Evaluation of session-based recommendation algorithms. User Model. User Adapt. Interact. 28(4-5), 331–390 (2018). https://doi.org/10.1007/s11257-018-9209-6

25. Madhyastha, P., Jain, R.: On model stability as a function of random seed. In: Proceedings of the 23rd Conference on Computational Natural Language Learning, CoNLL 2019, Hong Kong, China, 3–4 November 2019, pp. 929–939. Association for Computational Linguistics (2019). https://doi.org/10.18653/v1/K19-1087

26. Mirza, M., Osindero, S.: Conditional generative adversarial nets. CoRR abs/1411.1784 (2014). http://arxiv.org/abs/1411.1784

27. Moreno-Torres, J.G., Raeder, T., Alaíz-Rodríguez, R., Chawla, N.V., Herrera, F.: A unifying view on dataset shift in classification. Pattern Recognit. **45**(1), 521–530 (2012). https://doi.org/10.1016/j.patcog.2011.06.019

28. Ning, X., Karypis, G.: SLIM: sparse linear methods for top-N recommender systems. In: 11th IEEE International Conference on Data Mining, ICDM 2011, Vancouver, BC, Canada, 11–14 December 2011, pp. 497–506. IEEE Computer Society (2011). https://doi.org/10.1109/ICDM.2011.134

29. Pérez Maurera, F.B., Ferrari Dacrema, M., Cremonesi, P.: An Evaluation of Generative Adversarial Networks for Collaborative Filtering - Supplemental Material (2022). https://doi.org/10.5281/zenodo.5879345

30. Quionero-Candela, J., Sugiyama, M., Schwaighofer, A., Lawrence, N.D.: Dataset Shift in Machine Learning. The MIT Press, New York (2009)

31. Rendle, S., Freudenthaler, C., Gantner, Z., Schmidt-Thieme, L.: BPR: Bayesian personalized ranking from implicit feedback. In: UAI 2009, Proceedings of the Twenty-Fifth Conference on Uncertainty in Artificial Intelligence, Montreal, QC, Canada, 18–21 June 2009, pp. 452–461. AUAI Press (2009)

32. Steck, H.: Embarrassingly shallow autoencoders for sparse data. In: The World Wide Web Conference, WWW 2019, San Francisco, CA, USA, 13–17 May 2019, pp. 3251–3257. ACM (2019). https://doi.org/10.1145/3308558.3313710

33. Tang, J., Gao, H., Liu, H.: mTrust: discerning multi-faceted trust in a connected world. In: Proceedings of the Fifth International Conference on Web Search and Web Data Mining, WSDM 2012, Seattle, WA, USA, 8–12 February 2012, pp. 93–102. ACM (2012). https://doi.org/10.1145/2124295.2124309

34. Wang, J., et al.: IRGAN: a minimax game for unifying generative and discriminative information retrieval models. In: Proceedings of the 40th International ACM SIGIR Conference on Research and Development in Information Retrieval, Shinjuku, Tokyo, Japan, 7–11 August 2017, pp. 515–524. ACM (2017). https://doi.org/10.1145/3077136.3080786

35. Wang, Z., Xu, Q., Ma, K., Jiang, Y., Cao, X., Huang, Q.: Adversarial preference learning with pairwise comparisons. In: Proceedings of the 27th ACM International Conference on Multimedia, MM 2019, Nice, France, 21–25 October 2019, pp. 656–664. ACM (2019). https://doi.org/10.1145/3343031.3350919

36. Xia, B., Bai, Y., Yin, J., Li, Q., Xu, L.: MTPR: a multi-task learning based poi recommendation considering temporal check-ins and geographical locations. Appl. Sci. **10**(19) (2020). https://doi.org/10.3390/app10196664

37. Xie, F., Li, S., Chen, L., Xu, Y., Zheng, Z.: Generative adversarial network based service recommendation in heterogeneous information networks. In: 2019 IEEE International Conference on Web Services, ICWS 2019, Milan, Italy, 8–13 July 2019, pp. 265–272. IEEE (2019). https://doi.org/10.1109/ICWS.2019.00053

38. Yang, W., Lu, K., Yang, P., Lin, J.: Critically examining the "neural hype": weak baselines and the additivity of effectiveness gains from neural ranking models. In: Proceedings of the 42nd International ACM SIGIR Conference on Research and Development in Information Retrieval, SIGIR 2019, Paris, France, 21–25 July 2019, pp. 1129–1132. ACM (2019). https://doi.org/10.1145/3331184.3331340

39. Zhou, T., Kuscsik, Z., Liu, J.G., Medo, M., Wakeling, J.R., Zhang, Y.C.: Solving the apparent diversity-accuracy dilemma of recommender systems. Proc. Natl. Acad. Sci. **107**(10), 4511–4515 (2010). https://doi.org/10.1073/pnas.1000488107

Seed-Driven Document Ranking for Systematic Reviews: A Reproducibility Study

Shuai Wang(✉)🆔, Harrisen Scells🆔, Ahmed Mourad🆔, and Guido Zuccon🆔

The University of Queensland, St. Lucia, Australia
{shuai.wang2,h.scells,a.mourad,g.zuccon}@uq.edu.au

Abstract. Screening or assessing studies is critical to the quality and outcomes of a systematic review. Typically, a Boolean query retrieves the set of studies to screen. As the set of studies retrieved is unordered, screening all retrieved studies is usually required for high-quality systematic reviews. Screening prioritisation, or in other words, ranking the set of studies, enables downstream activities of a systematic review to begin in parallel. We investigate a method that exploits seed studies – potentially relevant studies used to seed the query formulation process – for screening prioritisation. Our investigation aims to reproduce this method to determine if it is generalisable on recently published datasets and determine the impact of using multiple seed studies on effectiveness. We show that while we could reproduce the original methods, we could not replicate their results exactly. However, we believe this is due to minor differences in document pre-processing, not deficiencies with the original methodology. Our results also indicate that our reproduced screening prioritisation method, (1) is generalisable across datasets of similar and different topicality compared to the original implementation, (2) that when using multiple seed studies, the effectiveness of the method increases using our techniques to enable this, (3) and that the use of multiple seed studies produces more stable rankings compared to single seed studies. Finally, we make our implementation and results publicly available at the following URL: https://github.com/ielab/sdr.

Keywords: Systematic reviews · Document ranking · Re-ranking

1 Introduction

A systematic review is a focused literature review that synthesises all relevant literature for a specific research topic. Identifying relevant publications for medical systematic reviews is a highly tedious and costly exercise, often involving multiple reviewers to screen (i.e., assess) upwards of tens of thousands of studies. It is a standard practice to screen each study retrieved for a systematic review by a Boolean query. However, in recent years, there has been a dramatic rise in Information Retrieval methods that attempt to re-rank this set of studies for a variety

M. Hagen et al. (Eds.): ECIR 2022, LNCS 13185, pp. 686–700, 2022.
https://doi.org/10.1007/978-3-030-99736-6_46

of reasons, such as stopping the screening early (once a sufficient number of studies have been found) or beginning downstream phases of the systematic review process earlier (such as the acquisition of the full-text of studies). However, a known problem with many of these methods is that they use a different query from the Boolean query used to perform the initial literature search. Instead, most methods typically resort to less informationally representative sources for queries that can be used for ranking, such as the title of the systematic review, e.g., [18] (containing narrow information about the retrieval topic), or concatenating the clauses of the Boolean query together, e.g., [3] (negating the structural information in Boolean clauses). We instead turn our attention to methods that use more informative sources of information to perform re-ranking.

Indeed, we focus this reproducibility study on one such method: seed-driven document ranking (SDR) from Lee and Sun [16]. SDR exploits studies that are known a priori to develop the research focus and search strategy for the systematic review. These studies are often referred to as 'seed studies' and are commonplace in the initial phases of the systematic review creation process. This method and others such as CLF [21] (which directly uses the Boolean query for ranking) have been shown to significantly outperform other methods that use a naïve query representation. Despite this, the SDR method was published when there was little data for those seeking to research this topic, and there have been methods published since that did not include SDR as a comparison. To this end, we devise the following research questions (RQs) to guide our investigation into why we are interested in reproducing the SDR method:

RQ1 *Does the effectiveness of SDR generalise beyond the CLEF TAR 2017 dataset?* The original study was only able to be investigated on a single dataset of systematic review topics. In this study, we plan to use our replicated implementation of SDR to examine the effectiveness of this method across more recent datasets, and datasets that are more topically varied (CLEF TAR 2017 only contains systematic reviews about diagnostic test accuracy).

RQ2 *What is the impact of using multiple seed studies collectively on the effectiveness of SDR?* The original study focused on two aspects of their method: an initial ranking using a single seed study and an iterative ranking which further uses the remaining seed studies one at a time. We focus on investigating the first aspect concerning the impact of multiple seed studies (multi-SDR) used collectively for input to produce an initial ranking.

RQ3 *To what extent do seed studies impact the ranking stability of single- and multi-SDR?* In a recent study by Scells et al. [23] to generate Boolean queries from seed studies, it was found that seed studies can have a considerable and significant effect on the effectiveness of resulting queries. We perform a similar study that aims to measure the variance in effectiveness of SDR in single- and multi- seed study settings.

With the investigation into the above research questions, we will (1) demonstrate the **novelty** of the method by performing experiments on more datasets (**RQ1**), and experiments that reveal more about the effectiveness of the method (**RQ2**,

RQ3), (2) assess the **impact** of SDR towards the Information Retrieval community and the wider systematic review community, (3) investigate the **reliability** of SDR by comparing it to several baselines on publicly available datasets, and (4) make our complete reproduced implementation of SDR publicly **available** for others to use as a baseline in future work on re-ranking for systematic reviews.

2   Replicating SDR

In the original paper of Lee and Sun, they devise two experimental settings for SDR: an initial ranking of retrieved studies using a seed study and iteratively re-ranking by updating the query used for SDR with one seed study at a time to simulate the manual screening process. We focus on the initial ranking stage for two reasons: (1) screening prioritisation is an accepted practice in the systematic review creation process as all studies must still be screened [4]; and (2) an effective initial ranking will naturally result in a more effective and efficient re-ranking of studies, as more studies that are relevant will be identified faster. The intuition for SDR is that relevant studies are similar to each other. The original paper makes two important observations about seed studies to support this intuition: (1) that relevant studies are more similar to each other than they are to non-relevant studies; and (2) that relevant studies share many *clinical terms*. These two observations are used to inform the representation and scoring of studies, given a seed study. We attempt to replicate these observations below to verify both that our implementation follows the same steps to make similar observations and whether the assumptions derived from them hold.

Observation 1. For a given systematic review, its relevant documents share higher pair-wise similarity than that of irrelevant documents.

We find that this observation is valid in our reproduction, as demonstrated by Fig. 1. In order to produce this plot, irrelevant studies were randomly undersampled ten times. The number of non-relevant studies is always the same as the number of relevant studies for each topic. This means it is unlikely that we will produce the exact result initially found for this observation by Lee and Sun. Furthermore, one reason that the average pairwise similarity for the relevant studies may not match the original results is that the textual content of studies on PubMed may have changed or been updated. Rather than using a dump of PubMed from 2017, we used the latest version of studies on PubMed, as it is unknown the exact date that studies were extracted from PubMed in the original paper, and the CLEF TAR dataset does not give an exact date.

Observation 2. Relevant documents for a given systematic review share high commonality in terms of clinical terms.

We found that this observation is also valid in our reproduction, as demonstrated in Fig. 2. It can be seen that the commonality of terms for the bag of words (BOW) and bag of clinical words (BOC) representations closely match those

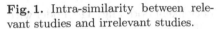

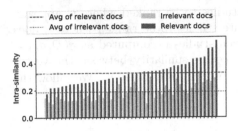

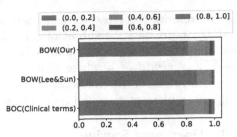

Fig. 1. Intra-similarity between relevant studies and irrelevant studies.

Fig. 2. Distribution of terms in relevant studies.

reported by Lee and Sun. However, we also found that with some minor modifications to the pre-processing of studies, we achieved a similar (yet still lower) commonality for terms using the BOW representation. We believe that the BOC representation shares a higher commonality of terms because the vocabulary size is smaller than the BOW representation. Naturally, with a smaller vocabulary, it is more likely for studies to share common terms. When pre-processing studies using the method described in original paper, we find that BOC terms count for 4.6% of the vocabulary, while they account for 31.2% using our pre-processing. In fact, our BOW vocabulary is only 14.8% their BOW vocabulary. Note that BOC is a distinct subset of BOW.

2.1 Document Representation

Given Observation 1 about relevant studies for this task, Lee and Sun chose to represent studies as a 'bag of clinical words' (BOC). They chose to use the Unified Medical Language System (UMLS) as their ontology of clinical terms. UMLS is an umbrella ontology that combines many common medical ontologies such as SNOMED-CT and MeSH. In order to identify UMLS concepts (and therefore the clinical terms) within the studies, Lee and Sun combine the outputs of the NCBO Bioportal [20] API[1] and QuickUMLS [24]. We follow their process as described, however we are not aware if it is not possible to set a specific version for the NCBO API. We use QuickUMLS version 1.4.0 with UMLS 2016AB.

2.2 Term Weighting

SDR weights terms based on the intuition that terms in relevant studies are more similar to each other (or occur with each other more frequently) than non-relevant studies. The weight of an individual term in a seed study is estimated by measuring to what extent it separates similar (pseudo-relevant) and dissimilar (pseudo-non-relevant) studies. Formally, each term t_i in a seed document d_s $(t_i \in d_s)$ is weighted using the function $\varphi(t_i, d_s) = \ln\left(1 + \frac{\gamma(D_{t_i}, d_s)}{\gamma(D_{\bar{t}_i}, d_s)}\right)$, where

---

[1] http://data.bioontology.org/documentation.

D_{t_i} represents the subset of candidate studies to be ranked where t_i appears, and $D_{\bar{t}_i}$ represents the subset of candidate studies to be ranked where t_i does not appear. The average similarity between studies is computed as $\gamma(D, d_s) = \frac{1}{|D|} \sum_{d_j \in D} sim(d_j, d_s)$, where sim is the cosine similarity between the vector representations of the candidate study d_j and the seed study d_i. We follow the original implementation and represent studies as tf-idf vectors.

2.3 Document Scoring

The original SDR implementation uses the query likelihood language model with Jelenik-Mercer smoothing for scoring studies. Typically, this ranking function is derived as indicated by QLM shown in Eq. 1, where $c(t_i, d_s)$ represents the count of a term in a seed study, $c(t_i, d)$ represents the count of a term in a candidate study, L_d represents the number of terms in a study, $p(t_i|\mathbb{C})$ represents the probability of a term in a background collection, and λ is the Jelenik-Mercer smoothing parameter. To incorporate the term weights as described in Subsect. 2.2, the original paper includes φ function into the document scoring function as shown in Eq. 1:

$$
score(d, d_s) = \sum_{t_i \in d, d_s} \overbrace{\varphi(t_i, d_s)}^{\text{Term Weight}} \cdot c(t_i, d_s) \cdot \log\left(1 + \overbrace{\frac{1 - \lambda}{\lambda} \cdot \frac{c(t_i, d)}{L_d \cdot p(t_i|\mathbb{C})}}^{\text{QLM}}\right) \quad (1)
$$

where $p(t_i|\mathbb{C})$ is estimated using maximum likelihood estimation over the entire candidate set of studies C. In the original paper, when additional seed studies were ranked in the top-k set of candidate seed studies (denoted as $d_{s'}$), a re-ranking was initiated by expanding each t_i in d_s with the new terms from $d_{s'}$. For our replication study, we only consider the initial ranking of candidate studies, as an abundance of baseline methods can be used as a comparison for this task. It is also arguably the most important step as a poor initial ranking will naturally result in a less effective and less efficient re-ranking.

2.4 Multi-SDR

One assumption in the original paper is that only a single seed study can be used at a time for ranking candidate studies. We propose a modification by studying the impact of using multiple seed studies collectively. In practice, it is common for Boolean queries (i.e., the search strategies used to retrieve the set of candidate studies we use for ranking) to be developed with a handful of seed studies, not just a single seed study. We hypothesise that the effectiveness of SDR will increase when multiple seed studies are used. Each relevant study must be used as a seed study for ranking, as the seed studies are not known in any of the collections we used. Therefore the average performance across topics was recorded (i.e., leave-one-out cross-validation). This study follows the methodology for the single-SDR method described in the subsections above. How we adapt single-SDR for a multi-SDR setting, and how we make this comparable to single-SDR is described as follows.

Grouping Seed Studies. To study multi-SDR, we adopt a similar approach to the original paper; however, we instead randomly group multiple seed studies together and perform leave-one-out cross-validation over these groups. To account for any topic differences that may impact performance, we use a sliding window across the list of seed studies so that a seed study can appear in multiple groups. The number of seed studies to fill each group was chosen to be 20% of the total seed studies. Rather than use a fixed number of seed studies, choosing different proportions simulates the use of seed studies in practice, i.e., different amounts of seed studies may be known before conducting a review.

Combining Seed Studies for Multi-SDR. The way we exploit multiple seed studies for SDR is, we believe, similar to how Lee and Sun used multiple seed studies in their relevance feedback approach to SDR. We concatenate seed studies together such that the resulting representation can be used directly with the existing single-SDR framework. We acknowledge that there may be more sophisticated approaches to exploit multi-SDR. However, we leave this as future work as it is out of the scope for this reproducibility study.

When computing term weights for multi-SDR, we also encountered computational infeasibility for large groups of seed studies. To this end, we randomly under-sampled the number of irrelevant studies to 50 each time we compute φ.

Comparing Single-SDR to Multi-SDR. Directly comparing the results of multi-SDR to single-SDR is not possible due to the leave-one-out cross-validation style of evaluation used for single-SDR. To address this, we apply an oracle to identify the most effective single-SDR run out of all the seed studies used for a given multi-SDR run in terms of MAP. We then remove the other seed studies used in the multi-SDR run from the oracle-selected single-SDR run so that both runs share the same number of candidate studies for ranking.

3   Experimental Setup

3.1   Datasets

When the original SDR paper was published, only a single collection with results of baseline method implementations was available. We intend to assess the generalisability of their SDR method on several new collections which have been released since. The collections we consider are:

CLEF TAR 2017 [9] This is the original dataset that was used to study SDR. We include this dataset to confirm that we achieve the same or similar results as the original paper. This collection includes 50 systematic review topics on diagnostic test accuracy – a type of systematic review that is challenging to create. The 50 topics are split into 20 training topics and 30 testing topics. In our evaluation, we removed topics CD010653, CD010771, CD010386, CD012019, CD011549 as they contained only a single or no relevant studies to use as seed studies. For our experiments using multiple seed studies, we further removed topics CD010860, CD010775, CD010896, CD008643, CD011548, CD010438, CD010633, CD008686 due to low numbers of relevant studies.

CLEF TAR 2018 [11] This collection adds 30 diagnostic test accuracy systematic reviews as topics to the existing 2017 collection; however, it also removes eight because they are not 'reliable for training or testing purposes. In total, this collection contains 72 topics. Our evaluation only used 30 additional reviews of the 2018 dataset and removed topics CD012216, CD009263, CD011515, CD011602, and CD010680 as they contained only a single or no relevant studies to use as seed studies. We also removed topic CD009263 because we ran into memory issues when running experiments on this topic due to many candidate documents (approx. 80,000). For our experiments using multiple seed studies, we removed topics CD012083, CD012009, CD010864, CD011686, CD011420 due to low numbers of relevant studies.

CLEF TAR 2019 [10] This collection further develops on the previous years' by also including systematic reviews of different types. From this collection, we use the 38 systematic reviews of interventions (i.e., a different type of diagnostic test accuracy).[2] We use this collection to study the generalisability of SDR on other kinds of systematic reviews. In our evaluation, we removed topics CD010019, CD012342, CD011140, CD012120, CD012521 as they contained only a single or no relevant studies to use as seed studies. For our experiments using multiple seed studies, we further removed topics CD011380, CD012521, CD009069, CD012164, CD007868, CD005253, CD012455 due to low numbers of relevant studies.

3.2   Baselines

The baselines in the original paper included the best performing method from the CLEF TAR 2017 participants, several seed-study-based methods, and variations of the scoring function used by SDR. For our experiments, we compare our reproduction of SDR to all of the original baselines that we have also reproduced from the original paper. The baselines in the original paper include: BM25-{BOW, BOC}, QLM-{BOW, BOC}, SDR-{BOW, BOC}, and AES-{BOW, BOC}. The last method, AES, is an embedding-based method that averages the embeddings for all terms in the seed studies. The AES method uses pre-trained word2vec embeddings using PubMed and Wikipedia (as specified in the original paper). We also include a variation that uses only PubMed embeddings (AES-P). Finally, we also include the linear interpolation between SDR and AES, using the same parameter as the original paper ($\alpha = 0.3$). We use the same versions of the pre-trained embeddings as the original paper.

---

[2] Although the overview paper claims there are 40 interventions topics, there are two topics that appear in both training and testing splits. However, like the previous datasets, we ignore these splits and combine the training and testing splits.

3.3 Evaluation Measures

For comparison to the original paper, we use the same evaluation measures. These include MAP, precision@k, recall@k, LastRel%, and Work Saved over Sampling (WSS). LastRel is a measure introduced at CLEF TAR'17 [9]. It is calculated as the rank position of the last relevant document. LastRel% is the normalised percentage of studies that must be screened in order to obtain all relevant studies. Work Saved Over Sampling; a measure initially proposed to measure classification effectiveness [7], is calculated instead here, by computing the fraction of studies that can be removed from screening to obtain all relevant documents; i.e., $WSS = \frac{|C| - LastRel}{|C|}$. Where C is the number of studies originally retrieved (i.e., the candidate set for re-ranking). For precision@k and recall@k, we report much deeper levels of k: the original paper reported $k = \{10, 20, 30\}$; where we report $k = \{10, 100, 1000\}$. Furthermore, we also report nDCG at these k-values, as it provides additional information about relevant study rank positions. We compute LastRel% and WSS using the scripts used in CLEF TAR 2017. For all other evaluation measures we use trec_eval (version 9.0.7).

3.4 Document Pre-processing

It is widely known that document pre-processing (e.g., tokenisation, stopwords, or stemming) can have a profound effect on ranking performance [8]. Although the original paper provides information about the versions of the libraries it uses for ranking, there were fewer details, such as how documents were tokenised or which stopword list was used. We reached out to the original authors to confirm the exact experimental settings. From the original paper, documents were split using space, then stopwords were removed using nltk.

The modifications we made to the document pre-processing pipeline were that documents were first pre-processed to remove punctuation marks and then tokenised using gensim version 3.2.0 tokeniser. For stopwords, as the original authors have not specified the nltk version, we used the latest version at the time of publishing, version 3.6.3. Then terms used are lowercased for in all methods except for AES. No stemming has been applied in either pre-processing pipeline.

4 Results

Before we investigate the three research questions of our reproducibility study, we first examine the extent to which we were able to replicate the results of Lee and Sun. In this study, we were unable to exactly replicate the results due to what we believe to be minor differences in document pre-processing and evaluation setup. Despite these difference, the results in Table 1 show a similar performance across the baselines and evaluation measures compared to what Lee and Sun originally reported in their paper for our pre-processing pipeline.

The results observed comparing the document pre-processing pipeline for the BOW representation as described by Lee and Sun (*-LEE) to our document pre-processing pipeline show that the BOW baselines may not have been as strong as if the original authors had performed a similar pipeline as us. We find that although the results comparing their baseline is statistically significant with our best performing method, our baseline is not significantly different. Finally, we find that the SDR-BOW-AES-LEE method, which corresponds to their most effective method, is significantly worse than our most effective method for 2017, SDR-BOW-AES-P.

In terms of the BOC representation we were unable to identify a more effective pipeline for extracting clinical terms. Here, we applied the clinical term extraction tools over *individual terms* in the document (following the pre-processing of Lee and Sun), and not *the entire document*. Although we find this to be counter-intuitive, as tools like QuickUMLS and the NCBO API use text semantics to match n-grams, the result of applying the tools to individual terms has the effect of reducing the vocabulary of a seed study to the key concepts.

Finally, comparing our evaluation setup to Lee and Sun, we find that there were a number of topics in the CLEF TAR 2017 dataset that were incompatible with SDR. Rather than attempting to replicate their results, we simply do not compare their original results with ours, since we do not have access to their run files or precise evaluation setup. Furthermore, when we compare the results we report from to the best performing participant at CLEF TAR 2017 that did not use relevance feedback [3], we remove the same topics from the run file of this participant for fairness. Although this method cannot be directly compared to, we can see that even relatively unsophisticated methods that use seed studies such as BM25-BOW are able to outperform the method by this participant.

4.1 Generalisability of SDR

We next investigate the first research question: *Does the effectiveness of SDR generalise beyond the CLEF TAR 2017 dataset?* In Table 2, we can see that the term weighting of SDR almost always increases effectiveness compared to using only QLM, and that interpolation with AES can have further benefits to effectiveness. However, we note that few of these results are statistically significant.

While we are unable to include all of the results for space reasons, we find that SDR-BOC-AES-P was not always the most effective SDR method. Indeed on the 2019 dataset, SDR-BOW was the most effective. The reason for this may be due to the difference in topicality of the 2019 dataset. This suggests that not only is the method of identifying clinical terms not suitable for these intervention systematic review topics, but that the interpolation between SDR and AES may require dataset-specific tuning.

Table 1. Reproduction results of baselines and SDR methods on the CLEF TAR 2017 dataset. For BOW methods, the pre-processing pipeline used by Lee and Sun is denoted by '-LEE'. BOW methods that do not have this demarcation correspond to our pipeline. For AES methods, word2vec PubMed embeddings are denoted by '-P'. AES methods that do not have this demarcation correspond to word2vec embeddings that include PubMed and Wikipedia. Statistical significance (Student's two-tailed paired t-test with Bonferonni correction, $p < 0.05$) between the most effective method (SDR-BOC-AES-P) and all other methods is indicated by †.

Method	MAP	Prec. 10	Prec. 100	Prec. 1000	Recall 10	Recall 100	Recall 1000	nDCG 10	nDCG 100	nDCG 1000	LR%	WSS
Sheffield-run-2 [3]	0.1706	0.1367	0.0703	0.0156	0.1759	0.5133	0.8353	0.2089	0.3342	0.4465	0.4660	0.5340
BM25-BOW-LEE	0.1710†	0.2027†	0.0867†	0.0195†	0.1543	0.5118†	0.8798†	0.2439†	0.3419†	0.4770†	0.4902†	0.5098†
BM25-BOW	0.1810	0.2128†	0.0898†	0.0200	0.1646	0.5232†	0.8928	0.2560	0.3534†	0.4899†	0.4427†	0.5573†
BM25-BOC	0.1764†	0.2145†	0.0895†	0.0200	0.1562	0.5215†	0.8944	0.2539	0.3496†	0.4871†	0.4401†	0.5599†
QLM-BOW-LEE	0.1539†	0.1846†	0.0778†	0.0184†	0.1367†	0.4664†	0.8508†	0.2198†	0.3091†	0.4454†	0.4662†	0.5338†
QLM-BOW	0.1973	0.2360	0.0964	0.0203	**0.1855**	0.5464	0.9081	**0.2827**	0.3772	0.5100	0.3851	0.6149
QLM-BOC	0.1894	0.2330	0.0951	0.0202	0.1809	0.5376	0.9032	0.2771	0.3684	0.5018	0.3936	0.6064
SDR-BOW-LEE	0.1533†	0.1777†	0.0780†	0.0185†	0.1304†	0.4710†	0.8576†	0.2142†	0.3088†	0.4460†	0.4660†	0.5340†
SDR-BOW	0.1972	0.2264	0.0952	0.0204	0.1718	0.5398	0.9083	0.2739	0.3728	0.5081	0.3742	0.6258
SDR-BOC	0.1953	0.2329	**0.0974**	**0.0206**	0.1751	0.5530	0.9151	0.2756	0.3751	0.5086	0.3689	0.6311
AES-BOW	0.1516†	0.1768†	0.0785†	0.0190†	0.1369†	0.4611†	0.8794†	0.2163†	0.3106†	0.4552†	0.4549†	0.5451†
AES-BOW-P	0.1604†	0.1872†	0.0809†	0.0193†	0.1480†	0.4954†	0.8895†	0.2274†	0.3255†	0.4669†	0.4088†	0.5912†
SDR-BOW-LEE-AES	0.1716†	0.2008†	0.0870†	0.0197	0.1484†	0.5250†	0.8988†	0.2389†	0.3429†	0.4792†	0.4148†	0.5852†
SDR-BOW-AES	0.1958	0.2309	0.0957	0.0203	0.1750	0.5568	0.9163	0.2756	0.3764	0.5090	0.3880†	0.6120†
SDR-BOC-AES	0.1964	0.2364	0.0972	0.0204	0.1770	0.5699	0.9195	0.2800	0.3813	0.5117	0.3830†	0.6170†
SDR-BOW-LEE-AES-P	0.1764†	0.2058†	0.0883†	0.0199	0.1570	0.5349†	0.9081†	0.2448†	0.3500†	0.4865†	0.3796†	0.6204†
SDR-BOW-AES-P	0.1983	0.2322	0.0961	0.0204	0.1740	0.5673	0.9206	0.2768	0.3812	0.5128	0.3608	0.6392
SDR-BOC-AES-P	**0.1984**	**0.2369**	0.0970	0.0205	0.1788	**0.5737**	**0.9241**	0.2807	**0.3837**	**0.5147**	**0.3566**	**0.6434**

4.2 Effect of Multiple Seed Studies

Next, we investigate the second research question: *What is the impact of using multiple seed studies collectively on the effectiveness of SDR?* Firstly, several topics were further removed for these experiments. Therefore, the results of single-SDR in Table 3 are not directly comparable to the results in Tables 1 and 2. In order to measure the effect multiple studies has on SDR compared to single seed studies, we also remove the same topics for single-SDR.

We find that across all three datasets, compared to single-SDR, multi-SDR can significantly increase the effectiveness. We also find that the largest increases in effectiveness are seen on shallow metrics across all three CLEF TAR datasets. This has implications for the use of SDR in practice, as typically, multiple seed studies are available before conducting the screening process. Therefore, when multiple seed studies are used for the initial ranking process, active learning methods that iteratively rank unjudged studies will naturally be more effective (as more relevant studies are retrieved in the early rankings). However, we argue that the assumption that relevant studies are a good surrogate for seed studies made by Lee and Sun [16] and by others in other work such as Scells et al. [23] may be weak and that methods that utilise relevant studies for this purpose overestimate effectiveness. In reality, seed studies may not be relevant studies.

Table 2. Generalisability of results on the CLEF TAR 2017, 2018 and 2019 datasets. Representations used in this table are all BOC. Statistical significance (Student's two-tailed paired t-test with Bonferonni correction, $p < 0.05$) between the most effective method (SDR-AES-P) and other methods is indicated by †.

	Method	MAP 10	Prec. 100	Prec. 1000	Prec. 10	Recall 100	Recall 1000	Recall 10	nDCG 100	nDCG 1000	nDCG	LR%	WSS
2017	QLM	0.1894	0.2330	0.0951	0.0202	**0.1809**	0.5376	0.9032	0.2771	0.3684	0.5018	0.3936	0.6064
	SDR	0.1953	0.2329	**0.0974**	**0.0206**	0.1751	0.5530	0.9151	0.2756	0.3751	0.5086	0.3689	0.6311
	SDR-AES-P	**0.1984**	**0.2369**	0.0970	0.0205	0.1788	**0.5737**	**0.9241**	**0.2807**	**0.3837**	**0.5147**	**0.3566**	**0.6434**
2018	QLM-BOC	0.2344	0.2594	0.1130	0.0219	0.1821	**0.6214**	0.9104	0.3141	0.4156	0.5312	0.3317†	0.6683†
	SDR	0.2374	0.2549	0.1136	0.0221	0.1798	0.6176	0.9174	0.3117	0.4163	0.5351	0.3024	0.6976
	SDR-AES-P	**0.2503**	**0.2688**	**0.1161**	**0.0222**	**0.1957**	0.6036	**0.9234**	**0.3259**	**0.4243**	**0.5445**	**0.2695**	**0.7305**
2019	QLM	0.2614	0.2599	0.0881	0.0169	0.2748	0.7032	0.9297	0.3458	0.4700	0.5482	0.4085	0.5915
	SDR	0.2790	0.2663	**0.0899**	**0.0169**	**0.3048**	0.7151	0.9337	0.3594	0.4846	0.5602	0.3819	**0.6181**
	SDR-AES-P	**0.2827**	**0.2667**	0.0898	0.0168	0.2973	**0.7174**	**0.9378**	**0.3649**	**0.4913**	**0.5672**	**0.3876**	0.6124

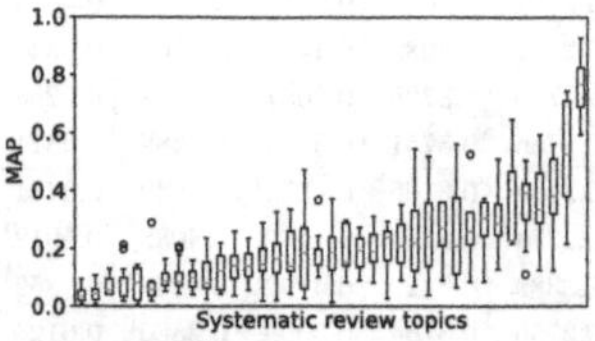

(a) Single-SDR; 2017 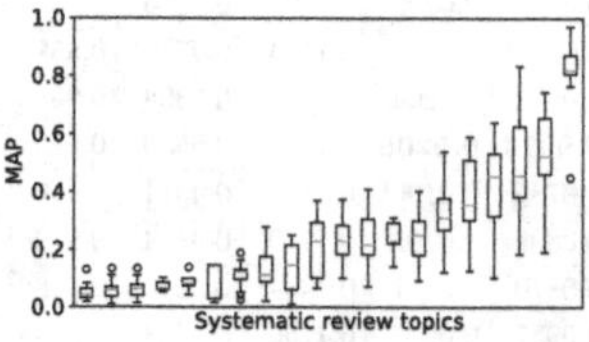   (b) Single-SDR; 2018 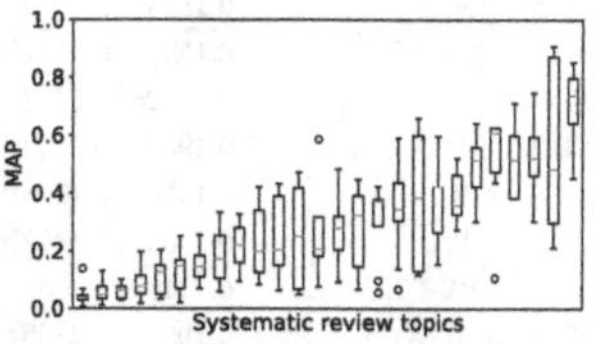   (c) Single-SDR; 2019

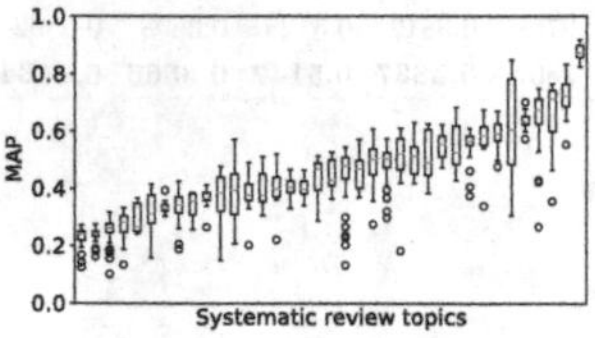

(d) Multi-SDR; 2017 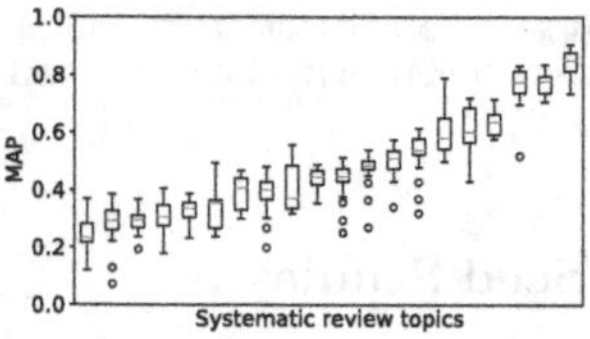   (e) Multi-SDR; 2018 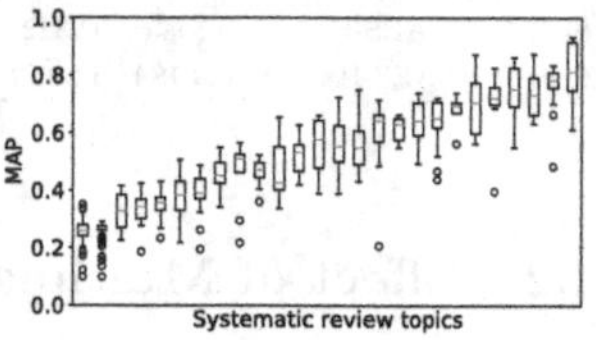   (f) Multi-SDR; 2019

Fig. 3. Topic-by-topic distribution of effectiveness (MAP) for the oracle-selected single-SDR-BOC-AES-P method (top figures) versus multi-SDR-BOC-AES-P.

They may be discarded once a Boolean query has been formulated (e.g., they may not be randomised controlled trials or unsuitable for inclusion in the review).

4.3   Variability of Seed Studies on Effectiveness

Finally, we investigate the last research question: *To what extent do seed studies impact the ranking stability of single- and multi-SDR?* We investigate this research question by comparing the topic-by-topic distribution of performance for the same results present in Table 3. These results are visualised in Fig. 3. That is, we compare the multi-SDR results to the oracle single-SDR results, described in Sect. 2.4 so that we can fairly compare the variance of one to the other. We find that the variance obtained by multi-SDR is generally higher than that of single-SDR using DTA systematic review topics (Fig. 3a vs. Fig. 3d – and Fig. 3b

Table 3. Results comparing single-SDR and multi-SDR on the CLEF TAR 2017, 2018, and 2019 datasets. Note that the results for single-SDR are not directly comparable to the above tables as explained in Sect. 2.4. Statistical differences (Student's paired two-tailed t-test, $p < 0.05$) are indicated pairwise between the single- and multi- SDR BOC and BOW methods for each year (e.g., single-SDR-BOC-AES-P vs. multi-SDR-BOC-AES-P for 2017). % Change indicates the average difference between single- and multi-{BOW+BOC}.

	Method	MAP	Prec. 10	Prec. 100	Prec. 1000	Recall 10	Recall 100	Recall 1000	nDCG 10	nDCG 100	nDCG 1000	LR%	WSS
2017	Single-BOC	0.3116	0.4235	0.1463	0.0255	0.2219	0.6344	0.9469	0.4830	0.5330	0.6595	0.3699	0.6301
	Single-BOW	0.3098	0.4076	0.1465	0.0255	0.2158	0.6366	0.9472	0.4679	0.5312	0.6566	0.3687	0.6313
	Multi-BOC	0.4554†	0.5804†	0.1752†	0.0272†	0.2917†	0.7151†	0.9661†	0.6817†	0.6765†	0.7835†	0.3427	0.6573
	Multi-BOW	0.4610†	0.5910†	0.1762†	0.0272†	0.2951†	0.7155†	0.9659†	0.6924†	0.6805†	0.7866†	0.3450	0.6550
	% Change	47.4801	41.0234	20.0132	6.6705	34.1131	12.5557	2.0029	44.5398	27.5202	19.3035	-6.8792	4.0283
2018	Single-BOC	0.3345	0.4443	0.1671	0.0285	0.2041	0.6181	0.9280	0.5011	0.5296	0.6551	0.2641	0.7359
	Single-BOW	0.3384	0.4433	0.1678	0.0286	0.2062	0.6197	0.9383	0.4955	0.5301	0.6579	0.2577	0.7423
	Multi-BOC	0.4779†	0.6130†	0.1979†	0.0307†	0.2821†	0.6997†	0.9592†	0.7199†	0.6823†	0.7908†	0.2394†	0.7606†
	Multi-BOW	0.4809†	0.6109†	0.1978†	0.0306†	0.2813†	0.6968†	0.9585†	0.7218†	0.6835†	0.7924†	0.2396	0.7604
	% Change	42.5011	37.8814	18.1509	7.2657	37.3377	12.8217	2.7561	44.6754	28.8870	20.5797	-8.1919	2.8990
2019	Single-BOC	0.3900	0.4249	0.1285	0.0221	0.3196	0.7261	0.9368	0.5365	0.6164	0.6897	0.4304	0.5696
	Single-BOW	0.3925†	0.4418	0.1272	0.0222	0.3366	0.7243	0.9386	0.5516	0.6164	0.6916	0.4285	0.5715
	Multi-BOC	0.5341†	0.5746†	0.1533†	0.0243†	0.3962†	0.7896†	0.9622†	0.7105†	0.7458†	0.8091†	0.3852†	0.6148†
	Multi-BOW	0.5374†	0.5864†	0.1521†	0.0244†	0.4031†	0.7853†	0.9616†	0.7223†	0.7466†	0.8114†	0.3877†	0.6123†
	% Change	36.9305	33.9958	19.3948	9.9327	21.8599	8.5825	2.5819	31.6927	21.0510	17.3213	-10.0189	7.5424

vs. Fig. 3e). We compute the mean variance across all topics, and find that the variance of multi-SDR (4.49e−2) is 10.89% higher than single-SDR (4.44e−2) result for the 2017 dataset, and 11.76 % for the 2018 dataset (single: 3.43e−2; multi: 4.17e−2). For the 2019 dataset, we find that the variance of multi-SDR (7.93e−2) is 6.51% lower than single-SDR (8.48e−2).

However, when we randomly sample seed studies from each group for single-SDR, we find that the variance of multi-SDR is significantly lower: 53.2% average decrease across 2017, 2018, and 2019. For space reasons, we do not include the full results. This suggests that the choice of seed study is considerably more important for single-SDR than for multi-SDR and that multi-SDR produces much more stable rankings, regardless of the seed studies chosen for re-ranking.

5   Related Work

Currently, it is a requirement for most high-quality systematic reviews to retrieve literature using a Boolean query [4,6]. Given that a Boolean query retrieves studies in an unordered set, it is also a requirement that all of the studies must be screened (assessed) for inclusion in the systematic review [4]. It is currently becoming more common for a ranking to be induced over this set of studies in order to begin downstream processes of the systematic review earlier [19], e.g., acquiring the full-text of studies or results extraction. This ranking of studies has come to be known as 'screening prioritisation', as popularised by the

CLEF TAR tasks which aimed to automate these early stages of the systematic review creation pipeline [9–11]. As a result, in recent years there has been an uptake in Information Retrieval approaches to enable screening prioritisation [1–3,5,15,16,18,21,22,25,27]. The vast majority of screening prioritisation use a different representation than the original Boolean query for ranking. Often, a separate query must be used to perform ranking, which may not represent the same information need as the Boolean query. Instead, the SDR method by Lee and Sun [16] forgoes the query all together and uses studies that have a high likelihood of relevance, seed studies [6], to rank the remaining studies. These are studies that are known a priori to the query formulation step. The use of documents for ranking is similar to the task of query-by-document [17,26] which has also been used extensively in domain-specific applications [12–14]. However, as Lee and Sun note, the majority of these methods try to extract key phrases or concepts from these documents to use for searching. SDR differentiates itself from these as the intuition is that the entire document is a relevance signal, rather than certain meaningful sections. Given the relatively short length of documents here (i.e., abstracts of studies), this intuition is more meaningful than other settings where the length of a document may be much longer.

6 Conclusions

We reproduced the SDR for systematic reviews method by Lee and Sun [16] on all the available CLEF TAR datasets [9–11]. Across all three of these datasets, we found that the 2017 and 2018 datasets share a similar trend in results than to the 2019 dataset. We believe that this is due to topical differences between the datasets and that proper tuning of SDR would result in results that better align with those seen in 2017 and 2018. We also performed several pre-processing steps that revealed that the BOW representation of relevant studies could also share a relatively high commonality of terms compared to the BOC representation. Furthermore, we found that the BOC representation for SDR is generally beneficial and that term weighting generally improves the effectiveness of SDR. We also found that multi-SDR was able to outperform single-SDR consistently. Our results also used an oracle to select the most effective seed studies to compare multi-SDR to single-SDR. This means that the actual gap in effectiveness between single-SDR and multi-SDR may be considerably larger. Finally, in terms of the impact of seed studies on ranking stability, we found that although multi-SDR was able to achieve higher performance than single-SDR, multi-SDR generally had a higher variance in effectiveness.

For future work, we believe that deep learning approaches such as BERT and other transformer-based architectures will provide richer document representations that may better discriminate relevant from non-relevant studies. Finally, we believe that the technique used to sample seed studies in the original paper and this reproduction paper may overestimate the actual effectiveness. This is because a seed study is not necessarily a relevant study, and that seed studies may be discarded after the query has been formulated. For this, we suggest that a new collection is required that includes the seed studies that were originally used to formulate the Boolean query, in addition to the studies included in the analysis portion of the systematic review.

Further investigation into SDR will continue to accelerate systematic review creation, thus increasing and improving evidence-based medicine as a whole.

Acknowledgment. Shuai Wang is supported by a UQ Earmarked PhD Scholarship and this research is funded by the Australian Research Council Discovery Projects programme ARC DP DP210104043.

References

1. Abualsaud, M., Ghelani, N., Zhang, H., Smucker, M.D., Cormack, G.V., Grossman, M.R.: A system for efficient high-recall retrieval. In: Proceedings of the 41st Annual International ACM SIGIR Conference on Research and Development in Information Retrieval, pp. 1317–1320 (2018)
2. Alharbi, A., Briggs, W., Stevenson, M.: Retrieving and ranking studies for systematic reviews: university of Sheffield's approach to CLEF eHealth 2018 task 2. In: CEUR Workshop Proceedings: Working Notes of CLEF 2018: Conference and Labs of the Evaluation Forum, vol. 2125. CEUR Workshop Proceedings (2018)
3. Alharbi, A., Stevenson, M.: Ranking abstracts to identify relevant evidence for systematic reviews: the university of sheffield's approach to CLEF eHealth 2017 task 2. In: CEUR Workshop Proceedings: Working Notes of CLEF 2017: Conference and Labs of the Evaluation Forum (2017)
4. Chandler, J., Cumpston, M., Li, T., Page, M.J., Welch, V.A.: Cochrane Handbook for Systematic Reviews of Interventions. Wiley, Hoboken (2019)
5. Chen, J., et al.: ECNU at 2017 eHealth task 2: technologically assisted reviews in empirical medicine. In: CEUR Workshop Proceedings: Working Notes of CLEF 2017: Conference and Labs of the Evaluation Forum (2017)
6. Clark, J.: Systematic reviewing. In: Suhail, A.R. Doi, G.M.W. (ed.) Methods of Clinical Epidemiology (2013)
7. Cohen, A., Hersh, W., Peterson, K., Yen, P.: Reducing workload in systematic review preparation using automated citation classification. J. Am. Med. Inform. Assoc. **13**(2), 206–219 (2006)
8. Croft, W.B.: Combining approaches to information retrieval. In: Croft, W.B. (eds.) Advances in Information Retrieval. The Information Retrieval Series, vol. 7, pp. 1–36. Springer, Boston (2002). https://doi.org/10.1007/0-306-47019-5_1
9. Kanoulas, E., Li, D., Azzopardi, L., Spijker, R.: CLEF 2017 technologically assisted reviews in empirical medicine overview. In: CEUR Workshop Proceedings: Working Notes of CLEF 2017: Conference and Labs of the Evaluation Forum (2017)
10. Kanoulas, E., Li, D., Azzopardi, L., Spijker, R.: CLEF 2019 technology assisted reviews in empirical medicine overview. In: CEUR Workshop Proceedings: Working Notes of CLEF 2018: Conference and Labs of the Evaluation Forum, vol. 2380 (2019)
11. Kanoulas, E., Spijker, R., Li, D., Azzopardi, L.: CLEF 2018 technology assisted reviews in empirical medicine overview. In: CEUR Workshop Proceedings: Working Notes of CLEF 2018: Conference and Labs of the Evaluation Forum (2018)

12. Kim, Y., Croft, W.B.: Diversifying query suggestions based on query documents. In: Proceedings of the 37th International ACM SIGIR Conference on Research & Development In Information Retrieval, pp. 891–894 (2014)

13. Kim, Y., Croft, W.B.: Improving patent search by search result diversification. In: Proceedings of the 2015 International Conference on The Theory of Information Retrieval, pp. 201–210 (2015)

14. Kim, Y., Seo, J., Croft, W.B., Smith, D.A.: Automatic suggestion of phrasal-concept queries for literature search. Inform. Process. Manag. **50**(4), 568–583 (2014)

15. Lagopoulos, A., Anagnostou, A., Minas, A., Tsoumakas, G.: Learning-to-rank and relevance feedback for literature appraisal in empirical medicine. In: Bellot, P., et al. (eds.) CLEF 2018. LNCS, vol. 11018, pp. 52–63. Springer, Cham (2018). https://doi.org/10.1007/978-3-319-98932-7_5

16. Lee, G.E., Sun, A.: Seed-driven document ranking for systematic reviews in evidence-based medicine. In: Proceedings of the 41st Annual International ACM SIGIR Conference on Research and Development in Information Retrieval, pp. 455–464 (2018)

17. Lv, Y., Moon, T., Kolari, P., Zheng, Z., Wang, X., Chang, Y.: Learning to model relatedness for news recommendation. In: Proceedings of the 20th International Conference on World Wide Web, pp. 57–66 (2011)

18. Miwa, M., Thomas, J., O'Mara-Eves, A., Ananiadou, S.: Reducing systematic review workload through certainty-based screening. J. Biomed. Inform. **51**, 242–253 (2014)

19. Norman, C.R., Leeflang, M.M., Porcher, R., Névéol, A.: Measuring the impact of screening automation on meta-analyses of diagnostic test accuracy. Syst. Rev. **8**(1), 243 (2019)

20. Noy, N.F., et al.: Bioportal: ontologies and integrated data resources at the click of a mouse. Nucleic Acids Res. **37**(Suppl._2), W170–W173 (2009)

21. Scells, H., Zuccon, G., Koopman, B.: You *Can* teach an old dog new tricks: rank fusion applied to coordination level matching for ranking in systematic reviews. In: Jose, J.M., et al. (eds.) ECIR 2020. LNCS, vol. 12035, pp. 399–414. Springer, Cham (2020). https://doi.org/10.1007/978-3-030-45439-5_27

22. Scells, H., Zuccon, G., Deacon, A., Koopman, B.: QUT ielab at CLEF eHealth 2017 technology assisted reviews track: initial experiments with learning to rank. In: CEUR Workshop Proceedings: Working Notes of CLEF 2017: Conference and Labs of the Evaluation Forum (2017)

23. Scells, H., Zuccon, G., Koopman, B.: A comparison of automatic Boolean query formulation for systematic reviews. Inform. Retr. J., 1–26 (2020)

24. Soldaini, L., Goharian, N.: QuickUMLS: a fast, unsupervised approach for medical concept extraction. In: Medical Information Retrieval Workshop (2016)

25. Wu, H., Wang, T., Chen, J., Chen, S., Hu, Q., He, L.: ECNU at 2018 eHealth task 2: technologically assisted reviews in empirical medicine. Methods Companion Methods Enzymol. **4**(5), 7 (2018)

26. Yang, Y., Bansal, N., Dakka, W., Ipeirotis, P., Koudas, N., Papadias, D.: Query by document. In: Proceedings of the Second ACM International Conference on Web Search and Data Mining, pp. 34–43 (2009)

27. Zou, J., Li, D., Kanoulas, E.: Technology assisted reviews: finding the last few relevant documents by asking Yes/No questions to reviewers. In: Proceedings of the 41st Annual International ACM SIGIR Conference on Research and Development in Information Retrieval, pp. 949–952 (2018)

Author Index

Printed in the United States
by Baker & Taylor Publisher Services